The Development Reader

Edited by

Sharad Chari

and

Stuart Corbridge

Routledge
Taylor & Francis Group

LONDON AND NEW YORK

First published 2008
by Routledge
2 Park Square, Milton Park, Abingdon, Oxon OX14 4RN

Simultaneously published in the USA and Canada
by Routledge
270 Madison Avenue, New York, NY 10016

Routledge is an imprint of the Taylor & Francis Group, an informa business

© 2008 selection and editorial matter Sharad Chari and Stuart Corbridge

Typeset in Amasis MT and Akzidenz Grotesque by
RefineCatch Limited, Bungay, Suffolk
Printed and bound in Great Britain
by MPG Books Ltd, Bodmin

British Library Cataloguing in Publication Data
A catalogue record for this book is available from the British Library

Library of Congress Cataloging in Publication Data
A catalog record for this book has been requested

ISBN13: 978–0–415–41505–7 (pbk)
ISBN13: 978–0–415–41504–0 (hbk)
ISBN13: 978–0–203–92653–6 (ebk)

ISBN10: 0–415–41505–5 (pbk)
ISBN10: 0–415–41504–7 (hbk)
ISBN10: 0–203–92653–6 (ebk)

THE DEVELOPMENT READER

Two hundred and fifty years ago, the gap between the richest and poorest regions of the world was rather narrow. Then came the Great Divergence, spurred on by the industrial revolution in north-west Europe and parts of North America. Some people in these parts of the world now began to get a sense that time was not cyclical. The future might be radically different from the past. Progress was a word that was much discussed. Later on, talk turned to 'development'. Over the past 50 or so years, the possibility of development has been extended worldwide and a new industry has grown up to promote Development in the 'Third World'.

The Development Reader brings together 54 key readings on development history, theory and policy: Adam Smith and Frantz Fanon meet, among others, Jean Oi, Amartya Sen and Jeffrey Sachs. It shows how debates around development have been structured by different readings of the roles played by markets, empire, nature and difference in the organisation of world affairs. For example, present-day concerns about economic liberalisation echo long-standing debates around free trade, extended divisions of labour and national economic policy. Likewise, old debates about empire are reappearing in critical perspectives on US policy in the Middle East. And while there is little room today for old-fashioned environmental or cultural determinism, the attention now being given to climate change and a clash of civilisations shows that questions of nature and difference remain at the centre of development politics. Part and individual extract introductions guide students through the material and bind the readings into a coherent whole. Organised chronologically as well as thematically, it offers an intellectual history of the debates and political struggles that swirl around development.

By bringing together intellectual history and contemporary development issues in this way, *The Development Reader* breaks fresh ground. It will have broad appeal across the humanities and social sciences, and is essential reading for students of contemporary development issues, practitioners and campaigners.

Sharad Chari is Lecturer in Human Geography at the London School of Economics and Honorary Research Fellow in the School of Development Studies at the University of KwaZulu-Natal, South Africa. He works on the historical ethnography of labour, work, activism, gender, state-sanctioned racism, and development in India and South Africa. He is the author of *Fraternal Capital: Peasant-Workers, Self-Made Men, and Globalization in Provincial India* (Stanford University Press, 2004), and is working on a monograph on space, race and activism in twentieth-century South Africa.

Stuart Corbridge is Professor of Development Studies at the London School of Economics. He has written widely on economic and political change in India and the history of development thought. His most recent book (with Williams, Srivastava and Véron) is *Seeing the State: Governance and Governmentality in India* (Cambridge University Press, 2005).

Contents

Contributors

Jeffrey Sachs is Director of the Earth Institute at Columbia University. He has written widely on poverty and development, and is Special Advisor to the Secretary-General of the United Nations.

Andrew Mellinger is an economist and geospatial analyst who has worked on issues linking geography to poverty and patterns of differential development.

John Gallup is an Adjunct Professor of Economics at the University of California, Berkeley, who writes on geography, poverty and economic development.

Mike Davis is a Professor of History at the University of California, Davis, who combines social activism with critical writing on the state of the world's cities, among other things.

Anne McClintock is Simone de Beauvoir Professor of English at the University of Wisconsin, Madison, and a leading authority on race, gender and sexuality in colonial contexts.

Heinz Arndt (1915–2002) was an historian of economic ideas who wrote widely on development issues, particularly with regard to Indonesia.

Adam Smith (1723–1790) was a key figure in the Scottish Enlightenment and a leading authority in the linked fields of political economy and moral philosophy.

Karl Marx (1818–1883) was a radical political economist and revolutionary thinker who provided both a critique of capitalist development and a manifesto for communism (with Friedrich Engels).

Herbert Spencer (1820–1903) was an evolutionary social theorist and liberal political thinker who popularised the phrase 'the survival of the fittest'.

Mohandas Gandhi (1869–1948) was a radical political thinker who pioneered the use of civil disobedience and non-violence as strategies to secure the recovery of India from British rule.

John Maynard Keynes (1883–1946) challenged the foundations of neoclassical economics and contributed significantly to new employment and monetary policies in the 1930s and 1940s.

Karl Polanyi (1886–1964) was a central European intellectual who wrote widely on the cultural, social and political underpinnings of apparently 'pure' economic and market relationships.

John Sydenham Furnivall (1878–1960) was a colonial administrator in Burma, as well as being a scholar who reflected on British and Dutch colonialism and their consequences in South-East Asia.

Alexander Gerschenkron (1904–1978) was born in Russia but moved to the USA where he taught history at Harvard University and was a pioneering theorist of state-led development.

Frantz Fanon (1925–1961) was born and grew up in Martinique and became a leading critic of racism, the psychology of colonisation, and the politics of decolonisation.

Arturo Escobar is a Professor of Anthropology at the University of North Carolina, Chapel Hill, and a leading theorist of 'post-developmentalism' and Latin American social activism.

Walt Whitman Rostow (1916–2003) popularised ideas about the stages of economic growth and served as a Special Assistant for National Security Affairs to President Lyndon B. Johnson.

Kingsley Davis (1908–1997) was an American demographer and sociologist who did pioneering work in the field of demographic transition studies, not least in South Asian contexts.

William Arthur Lewis (1915–1991) grew up in St Lucia and won the Nobel Prize in Economics for work in the field of economic development, particularly his writings on dual economies.

Hans Singer (1910–2006) was a pioneering figure in post-war development economics who remains best known for his critique of standard trade theories.

Alec Nove (1915–1994) taught at Glasgow University and wrote on the economics of socialism, particularly with respect to the Soviet Union and its satellites.

Rhoads Murphey is an Emeritus Professor of History and Asian Studies at the University of Michigan who has lived and worked in many regions of East Asia.

Paul Baran (1910–1964) was a Marxist associated with the *Monthly Review Press* in New York, itself known for its critiques of imperialism and Third World 'underdevelopment'.

Harold Wolpe (1926–1996) was an anti-Apartheid activist and radical political economist known for his work on the relationship between South African capitalism and state-sanctioned racism.

Rachel Carson (1907–1964) was a zoologist, marine biologist, civil servant, and foundational thinker of the environmental movement in the US and beyond.

Ester Boserup (1910–1999) was an economist who set the terms of feminist scholarship in the areas of agrarian livelihoods, fertility, and gender divisions of labour.

Michael Lipton is a Research Professor at the University of Sussex, UK, and a specialist in agriculture, land reform, poverty, nutrition, and demography in developing countries.

William Mangin is an Emeritus Professor of Anthropology at Syracuse University who is best known for his work on urbanisation, ethnicity, education, health, politics and alcohol use in Peru.

Peter Bauer (1915–2002) was an economist associated with Friedrich von Hayek who helped to pioneer the counter-revolution in development studies in the late twentieth century.

Deepak Lal is an economist at the University of California, Los Angeles, who is a trenchant critic of statist development and defender of classical economic liberalism.

John Williamson is a Senior Fellow of the Institute for International Economics, Washington DC, known for his work on the dollar, the US in the world economy, and 'the Washington Consensus'.

James Scott is Sterling Professor of Political Science at Yale University and a pioneering political theorist of peasant politics, domination and resistance, and 'high-modernist' development.

Michael Edwards, a geographer and Director of the Ford Foundation's Governance and Civil Society Unit in New York City, is a leading authority on non-governmental organisations (NGOs).

Diane Elson is a Professor of Sociology at the University of Essex and groundbreaking feminist critic of development processes and institutions.

James Ferguson is a Professor of Anthropology at Stanford University and an innovative ethnographer of development, the state, and 'globalisation' as seen from African perspectives.

Dani Rodrik, a Professor of International Political Economy at Harvard University, is an institutional economist associated with the contemporary remaking of development economics.

James Mahon is a Professor of Political Science at Williams College where he has worked on comparative politics, US–Latin America foreign relations, and Latin American politics.

Jean Oi, the William Haas Professor of Chinese Politics at Stanford University, has produced pioneering work on post-Maoist institutions in China's transition from socialism to capitalism.

Patrick Heller is Associate Professor of Sociology at Brown University and an important political theorist of class, state, and democratic development in India, South Africa, and Brazil.

Mahmood Mamdani is Herbert Lehman Professor at Columbia University and a noted commentator on African political economy, postcolonial politics, and Islamophobia.

David Mosse is Professor of Social Anthropology at the School of Oriental and African Studies, London, and an expert on irrigation, caste, South India, and the anthropology of development.

Martin Wolf is chief economist at the *Financial Times* and a leading expert on international trade policy.

Robert Wade is Professor of Political Economy and Development at the London School of Economics and a key thinker on East Asian development and international political economy.

Partha Dasgupta, the Frank Ramsey Professor of Economics at Cambridge, is best known for his work at the intersection of environmental and development economics.

Amartya K. Sen, the Lamont University Professor at Harvard University, won the Nobel Prize for his work on poverty, famine, gender, public action and political theory.

Bina Agarwal is a Professor of Economics at the Institute of Economic Growth, Delhi University, and a leading student of gendered access to resources, particularly land.

Brooke Schoepf is an economic and medical anthropologist at Harvard University Medical School working on the dynamics of infectious disease and gendered vulnerability in Africa.

Charles Hirschkind is Associate Professor of Anthropology at the University of California, Berkeley, and an ethnographer of Islam, media technologies, and aural/phonic traditions in Egypt.

Saba Mahmood is Associate Professor of Anthropology at the University of California, Berkeley, and an important critic of liberalism, feminism, Islamism, and embodiment in Egypt.

Stephen Radelet, a Senior Fellow at the Centre for Global Development in the US, is a development professional who has linked scholarship with policy work for various governments.

Paul Collier is a Professor of Economics at Oxford University who has written powerfully on civil war, aid, globalisation and poverty, mainly with reference to sub-Saharan Africa.

David Harvey is Distinguished Professor of Anthropology at the City University of New York and a foundational thinker on the Marxist theory of space, uneven development, and US imperialism.

Ching Kwan Lee is Associate Professor of Sociology at the University of Michigan and an important ethnographer of gender, labour, and subaltern politics in contemporary China.

Frederick Cooper is Professor of History at New York University, and a key scholar working on labour, development, capitalism, empire, decolonisation, and emancipation in Africa.

Fernando Coronil is Associate Professor of Anthropology and History at the University of Michigan and a key postcolonial ethnographer of capital, nature, oil, modernity, and Venezuela.

Arjun Appadurai is the John Dewey Professor in the Social Sciences at the New School University and an anthropologist of religion, commodities, and post-industrial Mumbai.

Tim Dyson is a Professor at the London School of Economics who has made important contributions to the demography of food, war, infectious disease, and global warming.

PART ONE

The object of development

INTRODUCTION TO PART ONE

Most of us have given some thought to the matter of development. It is one of the most heavily used words of our time and one of the most contentious. For many people in the global North, development is something that happens in the South or the Third World. It is something that happens to other people. But even a few moments' thought shows this can't be right. Development in the sense of social, economic and political change is happening all the time, everywhere. None of us stands still. 'No man is an island, entire of itself' (to quote the poet, John Donne). In any case, the rich countries of today weren't always so rich. It is widely agreed that the economic gap between Europe, Asia and Africa was far less around 1750 than it was in 1950, a point we'll come back to shortly. Something happened during this two-hundred-year period that led to enormous geographical differences in the global map of poverty and plenty.

The so-called 'great divergence' led some people to argue that the richer countries of today are developed, by which they mean their peoples enjoy high levels of material wealth. Economists first indexed this version of development in terms of per capita incomes. We can estimate this figure by dividing a country's gross domestic product (GDP) by the number of people living there. GDP refers to the market value of a country's final goods and services in a given time period. In 2006, for example, data from the International Monetary Fund (IMF) suggested that Luxemburg was the richest country in the world in these terms. Luxemburg's GDP per capita was $87,955, placing the country's inhabitants comfortably ahead of the Norwegians ($72,306), in second place, and the people of Qatar ($62,914) in third. At the bottom of this league table were countries such as Liberia, Ethiopia and Malawi, where average annual incomes were under $200. People living in these countries generally have to survive on less than one dollar a day. Along with many more people throughout sub-Saharan Africa and South Asia they make up the population of what Paul Collier (2007, and see Part 9) calls 'the bottom billion'. These are the people who have been left behind over the last fifty or sixty years, in the period when Development was capitalised and turned into something more intentional than change alone.

When most of us think of development today it is precisely Development in this sense. Big-D Development is something that governments and private companies direct, perhaps in combination with leading global agencies like the World Bank. And yet another moment's thought tells us that even big-D Development is difficult to pin down and define, never mind to make happen in practice – there is no end point to the escalator. Measuring a country's level of development in terms of GDP per capita is a first step, but it tells us nothing about how incomes are distributed across a country. Where levels of income inequality are very high – as in the United States and Brazil, for example – the median income will be considerably higher than the modal income. It also tells us nothing about how far one dollar will stretch in a country. For that we need GDP per capita figures adjusted for purchasing power parities (PPPs): a figure that takes into account the fact that we can buy far more food for $10 in Ethiopia than in Luxemburg. Going back to those two countries, the IMF's data suggest that GDP per capita in PPP terms in Luxemburg in 2006 was $80,471, still very high, while that for Ethiopia was $1,044 in 2005. The PPP figure for Qatar meanwhile was just $33,049, a big change from the unadjusted figure. And this is just to stick with incomes. Robert Kennedy once noted that the national product of the United States includes spending on air pollution, napalm and cigarette advertising, but 'does not include the beauty of our poetry or the strength of our marriages, or the intelligence of our public debate or the integrity of our public officials. It measures everything, in short, except that which makes life worthwhile.'

Measuring what makes life worthwhile, however, is notoriously difficult, as we all have different ideas about the good life. Still, the development economist and Nobel Prize winner, Amartya Sen, is surely not wrong to insist that meaningful definitions of development must stretch to include the basic capabilities and freedoms that people need to lead satisfying lives. In his view, we should not consider countries developed where large numbers of women and men are without voting rights, basic health care or the ability to read and write. Acting on advice from Sen and the Pakistani economist, Mahbub ul Haq, the United Nations Development Programme (UNDP) has devised a Human Development Index (HDI) that ranks countries according to a composite measure of GDP per capita (in PPP terms), life expectancy and adult literacy/school enrolment rates. The 2006 Report had Norway in first place, Luxemburg in twelfth, and Qatar not listed in the leading forty countries. Mali, Sierra Leone and Niger were at the bottom of this particular league table, suggesting that while countries can buck general trends, HDI scores tend on average to be highly correlated with GDP scores. Some income-rich countries will exclude women from the public sphere and allocate them fewer voting rights and de facto medical care than men (see Sen in Part 8). They will score relatively low in HDI terms. Some poor countries or regions might invest heavily in core public goods, as China did in the 1950s and 1960s, along with the south Indian state of Kerala. They will have quite high HDI scores. For the most part, though, much as we might wish it otherwise, it costs money to provide decent health care and education systems – money that comes in the main from concerted economic growth.

(A quick note here. The Beatles declared in 1964 that 'money can't buy you love', and that might be right. By the same token, some reputable economists, including Richard Layard (2006) of the London School of Economics, maintain that what we should be trying to maximise globally is 'happiness'. The King of Bhutan agrees. He has promoted the idea of measuring 'gross national happiness'. You might think that *The Development Reader* has little to say on this issue, and that's true at one level. We are mainly concerned with debates around more established conceptions of development. Nevertheless, one of the core themes of this reader, which we'll come on to shortly, is bound up with ideas of 'difference'. In Part 2, we'll see Mohandas Gandhi speaking back to Western conceptions of civilisation and progress just as sharply as the King of Bhutan might be speaking back to them today, one hundred years later.)

Measurements of development, then, are clearly wrapped up with conceptions of development. Something has to be identified before it can be measured. But where do these conceptions come from? And how new are they? How do they feed into public policy, and with what effects? These are the main questions that animate *The Development Reader*. There is plenty in the readings that follow which will excite readers who want to get to grips with important debates on contemporary development policy. We have tried to touch base with such key issues as global climate change and its consequences, the importance of institutions in determining development strategies, and the gendering of development policies and trajectories. We have also tried to select readings that speak to different regions of the world, and to step outside our own areas of expertise (Southern Africa and South Asia).

Most especially, however, we have also tried to select papers that give a sense of the history of debates around development. We certainly do not pretend that the way that contemporary students think about the role of markets or nature, say, in relation to development is the same as it was one hundred or two hundred years ago. Not at all. But we do strongly agree with Jeffrey Sachs and his colleagues when they note that social theorists have been captivated by the question of 'why some countries [are] stupendously rich [while] others [are] horrendously poor ... since the late eighteenth century, when Scottish economist Adam Smith addressed the issue in his magisterial work *The Wealth of Nations*'. Development Studies as an academic subject has existed for only fifty years, but serious studies of development have been around for more than two hundred years. One criticism that is commonly voiced of Development Studies is that it is depthless and ahistorical. We hope in this volume of readings to show how much present-day debates on development can be linked to prior engagements with what we shall call 'the object of development'. By this, we mean to refer to the way that issues around development are conceptualised, contested and placed in the domain of public policy making.

For the most part, we explore these connections by coming forward in time. *The Development Reader* is organised chronologically. But we can also hope to explain our purpose by moving backward in time, which is how we have organised this first set of three readings. We start with the reading by Jeffrey D. Sachs, Andrew D. Mellinger and John L. Gallup. As we have seen, the starting point for these three authors is the present-day map of the world: its stupendous riches and horrendous poverty. How should we interpret and explain this map? What can be done to diminish or remove that grinding poverty?

Jeffrey Sachs (as we will see in Part 9) is a leading figure in the contemporary development business. In addition to holding a prestigious academic post at Columbia University in New York City, Sachs worked closely with Kofi Annan at the United Nations on the Millennium Development Goals: a project to reduce different measures of poverty by half from 1990 to 2015. His article with Mellinger and Gallup was written for a general audience. Its core points are as follows: (a) the map of world wealth and poverty is determined mainly by geography: rich countries and regions are to be found in temperate locations and/or near coastal regions and navigable waterways; poor countries are burdened by harsh tropical climates and poor inland locations; (b) poverty, or the curse of bad geography, can be lifted by large-scale programmes of foreign aid that are targeted against diseases like malaria and HIV-AIDS and in favour of agricultural growth, biotechnology and infrastructural provision. Sachs *et al.* end by calling for a modest increase of donor financing. An extra $25 billion per year over twenty years (say), or just $28 per person per year from people living in the wealthy nations, could make all the difference, and promote people in the bottom billion into the ranks of the four billion or so people above them who have made progress in the post-war Age of Development.

These are fascinating arguments and, if true, quite encouraging. It suggests that a big push is in order to solve the problems caused by bad geography. But are they credible arguments? Are they true? Some other economists and political scientists have rushed to take issue with the arguments of Sachs *et al.* William Easterly (2006), for example, who we will meet again in the Introduction to Part 6, pours scorn on the idea that governments and outside experts can deliver the world's poorest from their poverty by means of planning, money and technology. How will this be done? Who will solve the coordination problems that must arise? Who will guard against the corrupt misuse of taxpayer monies? Easterly and Levine (2003), too, along with Rodrik *et al.* (2004), have also challenged the intellectual rigour of the bad geography thesis. They accept there is a correlation between geography and development. But they dispute the idea that apparently observable patterns are in any way *caused* by tropicality or location. The causation, rather, comes from institutions, by which they mean the presence or absence of strong property rights regimes and the rule of law. Geography, in short, is something like a screen, or even a false variable. Poor countries might well have bad geographies, but these geographies are largely the results of malfunctioning institutions and regimes of government. Good governments work with private agents to build efficient roads and irrigation systems, bad governments don't. Good governments enforce legal contracts and guard against the abuse of public office for private gain, bad governments don't. It follows that big aid might not be the answer to bad geography. What matters far more, these critics suggest, is the far more difficult task of changing long-standing path dependencies and forms of rule – what they would call endogenous institutions.

Still other economists and political scientists suggest that present-day patterns of development were forged in the Americas, Africa and Asia long ago. Local institutions are largely the result of the particular ways that colonised countries were settled (or not) by Europeans. Take North America, for example. Daron Acemoglu *et al.* (2001, 2002) have suggested that low levels of endemic disease, and low existing levels of population density, encouraged the British to settle in present-day Canada and the United States. The north-eastern seaboard, especially, lacked natural resources which could be exploited for the benefit of London. In consequence, the settlers got down to the business of building new agrarian and industrial economies. They also built societies that were informed by and yet deepened British systems of governance and property rights regimes. In short, institutions were put in place that would later support regional systems of capital accumulation and economic growth. Contrast this with regions blighted by malaria, or by easy-to-mine supplies of precious minerals or (later) oil. In these regions, Acemoglu *et al.* argue, the incentives for settlers both to put down roots and to be technologically innovative were limited. It was too easy to live off the fat of the land and the labour of others. Systems of rule tended to grow up in these colonies – including in the southern United States – that protected the rights of small white minorities and which failed to support non-extractive forms of capitalist development. Mines and plantations were worked by indentured and slave labour recruited from abroad, or from among those within the native populations who survived the imposition of colonial rule.

We will come back to these debates at several points in *The Development Reader*. The first point we wish to make here is that maps of the world need to be read with care. Present a child of 10 with a map of the world showing that poor countries are located mainly in the tropics, and s/he might reasonably deduce that poverty is caused by climate. But s/he might also be wrong. The true object of development policy might be something quite different to climate or geography. Now consider the Third World. Many people will know that this term is quite recent, and was invented in the late 1940s, in French, as *Les Tiers Monde*. We can't have a Third World until we have First and Second worlds, and these blocs came more sharply into being during the Cold War between the United States and the USSR (see Part 4). But in another sense, as Mike Davis shows us in the second reading, the Third World of today was mainly produced in the nineteenth century. This is when the legacies of European colonialism really made themselves felt. Significantly, too, in light of present-day concerns about global climate change, Davis shows that scholars and policy makers were concerned more than one hundred years ago about the connections between El Niño events (later, the El Niño–Southern Oscillation (ENSO)) and the production of famines.

Geography again. And politics. For, as Davis shows, while we live in a world that is manifestly physical and buffeted by non-human forces, the vulnerabilities that can be caused by climate changes are always shaped by geographies of power. People die in famines when they can't access food from the land directly or when they can't exchange for food by first selling their labour or traded goods. Yet, as Davis argues here, 'The looms of India and China were defeated not so much by market competition as they were forcibly dismantled by war, invasion, opium and a Lancashire-imposed system of one-way tariffs.' History, he suggests, following Marx, is a very bloody business – far more bloody than most Whigs allow, or indeed scholars like Acemoglu, Johnson and Robinson (AJR). AJR focus on patterns of settler mortality and natural resource endowments to understand the production of present-day geographies. For his part, Davis draws attention to the deliberate violence with which European states and capitalists were able to produce famine and deindustrialisation in the colonised world. If Jeffrey Sachs installs nature as a key theme for development studies, and AJR that of markets and institutions, Davis links markets to empire and a tradition of writing that connects development in one region to underdevelopment in another. Same maps, different explanations.

Markets, empire and nature: here are three recurrent themes in our reader. We come back to them time and again, picking up continuities and discontinuities in the way they are theorised, modelled and brought into combination. We also explore how difficult it can be to link intellectual perspectives on development to questions of development policy or politics. Suppose for a moment that Mike Davis is right about the colonial development of underdevelopment. Would this change our attitudes to development policy? Possibly not. Unless we believe that exactly the same processes are at work today, or that the ex-colonised countries are owed a huge debt of restitution by the colonial powers, it is not clear what Davis might propose as an antidote to the sins of imperial mal-development. Particular framings of the object of development – as an absence or excess of capitalism, as bad geography, as poor institutions – may or may not lead to credible development policies. That said, the work of scholars like Marx and Davis, or AJR, does suggest that we should be sceptical of claims which suggest that the weight of past history can be swiftly set aside by planned development in the present. There is a politics to these claims – and vested interests, too, if we consider the funding of leading development institutions – which needs to be rendered visible, thought through and challenged.

Lastly, there is the fourth strand of our book, that of culture or difference. Here too we are centrally concerned with the framing of development issues, or with how development becomes an object for analysis or public policy. But this time we want to move away from economics and politics. We want to open up a space for those dissenting voices, like Gandhi, who call into question assumptions about the rightness of development as material progress. Rather more so, however, we want to take seriously certain arguments from postcolonial studies, and also from feminism, about the importance of standing back from taken-for-granted terms and practices, or what are often called dominant discourses. Colonial settlement policies and imperialism, for example, were not simply about market relationships, geopolitics or (un)equal trade, any more than US-led wars in West Asia today are simply about oil. What assumptions does the development business make about those peoples around the world who are to be 'developed', whether by an aid-led big push or by free trade? What absences are supposed to define their lives? Are African countries simply lacking in money and technology, or are they still being understood in terms of discourses which present African people (as if there could be any such singularity) as the victims of unreason, or possibly even of 'race' (no matter that this is not voiced so openly these days)? Are they not 'modern' enough, and if they are not, precisely how are modern development subjects produced? What words and images are we using to structure today's development debates? What practices are being promoted to further what Tania Murray Li (2007) calls 'the will to improve'?

Few people today speak of the white man's burden in Africa, although Bill Easterly has been keen to write of Jeffrey Sachs's endeavours in just these terms. Likewise, few people are inclined to join with the great sociologist Max Weber, writing just one hundred years ago, in describing Hindus as a class of people who 'lived in dread of the magical evil of innovation'. Tell that to Indian software engineers. In short, few people would admit to seeing the world today in terms of the strongly racialised and sexualised discourses that Anne McClintock sets out in her reading here, when she describes the 'porno-tropics' of imperial discovery and penetration. Most will be scandalised by the Western discourses of cleanliness and commodity racism that McClintock describes.

But what's always far harder to see is what is scandalous about our own perspectives. How will our ancestors judge us? How do people elsewhere judge us now? (Consider how devout high-caste Hindus might look with disgust on beef-eating Westerners, or how some Westerners look on Islamic women wearing the *burkha*). Max Weber also said that the true function of social science was to render problematic that which is conventionally self-evident, and that's advice we have tried to take on board here. Perhaps the greatest danger in work on development is that the idea itself is taken for granted. *The Development Reader* is designed throughout to make this more difficult.

REFERENCES AND FURTHER READING

Acemoglu, D., Johnson, S. and Robinson, J. (2001) 'The colonial origins of comparative development: an empirical investigation', *American Economic Review* 91: 1369–1401.

Acemoglu, D., Johnson, S. and Robinson, J. (2002) 'Reversal of fortune, geography and institutions in the making of modern world income distribution', *Quarterly Journal of Economics* 117: 1231–94.

Collier, P. (2007) *The Bottom Billion: Why the Poorest Countries are Failing and What Can Be Done About It*, Oxford: Oxford University Press.

Diamond, J. (2005) *Collapse: How Societies Choose to Fail or Succeed*, New York: Viking.

Easterly, W. (2006) *The White Man's Burden: Why the West's Efforts to Aid the Rest Do So Much Ill and So Little Good*, Oxford: Oxford University Press.

Easterly, W. and Levine, R. (2003) 'Tropics, germs and crops: how endowments influence economic development', *Journal of Monetary Economics* 50 (1): 3–39.

Jenkins, R. (2006) 'Where development meets history', *Commonwealth and Comparative Politics* 44: 2–15.

Lange, M., Mahoney, J. and Vom Hau, M. (2006) 'Colonialism and development: a comparative analysis of Spanish and British colonies', *American Journal of Sociology* 111 (5): 1412–62.

Layard, R. (2006) *Happiness: Lessons from a New Science*, London: Penguin.

Li, T. M. (2007) *The Will to Improve: Governmentality, Development and the Practice of Politics*, Durham, NC: Duke University Press.

Pomeranz, K. (2000) *The Great Divergence: China, Europe and the Making of the Modern World*, Princeton, NJ: Princeton University Press.

Rodrik, D., Subramanian, A. and Trebbi, F. (2004) 'Institutions rule: the primacy of institutions over integration and geography in economic development', *Journal of Economic Growth* 9 (2): 131–65.

Said, E. (1979) *Orientalism*, New York: Vintage.

Sen, A. K. (2000) *Development as Freedom*, New York: Anchor Books.

'The Geography of Poverty and Wealth'

Scientific American (2001)

Jeffrey D. Sachs, Andrew D. Mellinger and John L. Gallup

Editors' Introduction

Jeffrey Sachs, Andrew Mellinger and John Gallup first began to write extensively on geography and development in the late 1990s when they were colleagues at Harvard University's Center for International Development. Early work on the density of coastal populations and the economic costs of malaria led to a more ambitious linking of bad geography to slow economic growth and low human development scores. Not dissimilar ideas had been put forward in the mid-1970s by Andrew Kamarck (1976) at the World Bank. What was distinctive about the work of Sachs, Mellinger and Gallup, however, whether writing together or with other colleagues, was its use of large and sometimes new (remotely sensed) data sets.

Some of that work is on display in the short article reprinted here, although these same ideas have been explored at greater length and in more depth in a large number of scholarly articles. As the reader will quickly see, geography, whether measured in terms of climate or proximity, becomes the independent variable in accounts of differential development at the global and continental scales. But this isn't to say that geography is restored to a place in the sciences which it claimed to enjoy in the days of environmental determinism. The perspective here is not that 'man' is the prisoner of nature, as was widely believed one hundred years ago; it is rather that human societies need to take large-scale actions now to ensure that the curse of bad geography is lessened and in time overcome. This can best be done, Sachs and his colleagues conclude, by what some have called a big push for Africa. Roads and railways need to be built, along with more efficient port complexes and electronic communications networks. At the same time, a large amount of money – most of which must come from foreign donors – needs to be poured into a war on debilitating diseases, of which malaria remains the most important.

As we shall see later in this book, the bad geography thesis has been challenged by a wide range of critics. Perhaps the most substantive challenge has come from Dani Rodrik *et al.* (2004), who have argued that once we control for the quality of institutions (notably property rights and the rule of law), the apparently independent effects of geography on economic development disappear. This being so, they conclude, public policy should begin with a good governance agenda, rather than with an aid-led big push for development.

We will pick up this and related debates at several points through *The Development Reader*. What is at stake in these debates are issues of evidence (how we collect meaningful data sets), theory (how we make sense of this data) and public policy. All of them relate to what we are calling 'the object of development'.

The reading is printed here without the Map and Table referred to at page 11.

Key references

J. Sachs, A. Mellinger and J. Gallup (2001) 'The geography of poverty and wealth', *Scientific American*, March: 70–6.

A. Mellinger, J. Sachs and J. Gallup (2002) 'Climate, coastal proximity and development', in G. Clark, M. Feldman and M. Gertler *et al.* (eds) *The Oxford Handbook of Economic Geography*, Oxford: Oxford University Press.

J. Sachs (2001) 'Tropical underdevelopment', *NBER Working Paper* 8119.

A. Kamarck (1976) *The Tropics and Economic Development: A Provocative Inquiry Into the Poverty of Nations*, Baltimore, MD: Johns Hopkins University Press and the World Bank.

D. Rodrik, A. Subramanian and F. Trebbi (2004) 'Institutions rule: the primacy of institutions over integration and geography in economic development', *Journal of Economic Growth* 9 (2): 131–65.

Why are some countries stupendously rich and others horrendously poor? Social theorists have been captivated by this question since the late 18th century, when Scottish economist Adam Smith addressed the issue in his magisterial work *The Wealth of Nations*. Smith argued that the best prescription for prosperity is a free-market economy in which the government allows businesses substantial freedom to pursue profits. Over the past two centuries, Smith's hypothesis has been vindicated by the striking success of capitalist economies in North America, western Europe and East Asia and by the dismal failure of socialist planning in eastern Europe and the former Soviet Union.

Smith, however, made a second notable hypothesis: that the physical geography of a region can influence its economic performance. He contended that the economies of coastal regions, with their easy access to sea trade, usually outperform the economies of inland areas. Although most economists today follow Smith in linking prosperity with free markets, they have tended to neglect the role of geography. They implicitly assume that all parts of the world have the same prospects for economic growth and long-term development and that differences in performance are the result of differences in institutions. Our findings, based on newly available data and research methods, suggest otherwise. We have found strong evidence that geography plays an important role in shaping the distribution of world income and economic growth.

Coastal regions and those near navigable waterways are indeed far richer and more densely settled than interior regions, just as Smith predicted. Moreover, an area's climate can also affect its economic development. Nations in tropical climate zones generally face higher rates of infectious disease and lower agricultural productivity (especially for staple foods) than do nations in temperate zones. Similar burdens apply to the desert zones. The very poorest regions in the world are those saddled with both handicaps: distance from sea trade and a tropical or desert ecology.

A skeptical reader with a basic understanding of geography might comment at this point, "Fine, but isn't all of this familiar?" We have three responses. First, we go far beyond the basics by systematically quantifying the contributions of geography, economic policy and other factors in determining a nation's performance. We have combined the research tools used by geographers – including new software that can create detailed maps of global population density – with the techniques and equations of macroeconomics. Second, the basic lessons of geography are worth repeating, because most economists have ignored them. In the past decade the vast majority of papers on economic development have neglected even the most obvious geographical realities.

Third, if our findings are true, the policy implications are significant. Aid programs for developing countries will have to be revamped to specifically address the problems imposed by geography. In particular, we have tried to formulate new strategies that would help nations in tropical zones raise their agricultural productivity and reduce the prevalence of diseases such as malaria.

THE GEOGRAPHICAL DIVIDE

The best single indicator of prosperity is gross national product (GNP) per capita – the total value of a country's economic output, divided by its population. A map showing the world distribution of GNP per capita immediately reveals the vast gap between rich and poor nations. Notice that the great majority of the poorest countries lie in the geographical tropics – the area between the tropic of Cancer and the tropic of Capricorn. In contrast, most of the richest countries lie in the temperate zones.

A more precise picture of this geographical divide can be obtained by defining tropical regions by climate rather than by latitude. The map divides the world into five broad climate zones based on a classification scheme developed by German climatologists Wladimir P. Köppen and Rudolph Geiger. The five zones are tropical-subtropical (hereafter referred to as tropical), desert-steppe (desert), temperate-snow (temperate), highland and polar. The zones are defined by measurements of temperature and precipitation. We excluded the polar zone from our analysis because it is largely uninhabited.

Among the 28 economies categorized as high income by the World Bank (with populations of at least one million), only Hong Kong, Singapore and part of Taiwan are in the tropical zone, representing a mere 2 percent of the combined population of the high-income regions. Almost all the temperate-zone countries have either high-income economies (as in the cases of North America, western Europe, Korea and Japan) or middle-income economies burdened by socialist policies in the past (as in the cases of eastern Europe, the former Soviet Union and China). In addition, there is a strong temperate–tropical divide within countries that straddle both types of climates. Most of Brazil, for example, lies within the tropical zone, but the richest part of the nation – the southernmost states – is in the temperate zone.

The importance of access to sea trade is also evident in the world map of GNP per capita. Regions far from the sea, such as the landlocked countries of South America, Africa and Asia, tend to be considerably poorer than their coastal counterparts. The differences between coastal and interior areas show up even more strongly in a world map delineating GNP density – that is, the amount of economic output per square kilometer. This map is based on a detailed survey of global population densities in 1994. Geographic information system software is used to divide the world's land area into five-minute-by-five-minute sections (about 100 square kilometers at the equator). One can estimate the GNP density for each section by multiplying its population density and

its GNP per capita. Researchers must use national averages of GNP per capita when regional estimates are not available. To make sense of the data, we have classified the world's regions in broad categories defined by climate and proximity to the sea. We call a region "near" if it lies within 100 kilometers of a seacoast or a sea-navigable waterway (a river, lake or canal in which oceangoing vessels can operate) and "far" otherwise. Regions in each of the four climate zones we analyzed can be either near or far, resulting in a total of eight categories. The table shows how the world's population, income and land area are divided among these regions.

The breakdown reveals some striking patterns. Global production is highly concentrated in the coastal regions of temperate climate zones. Regions in the "temperate-near" category constitute a mere 8.4 percent of the world's inhabited land area, but they hold 22.8 percent of the world's population and produce 52.9 percent of the world's GNP. Per capita income in these regions is 2.3 times greater than the global average, and population density is 2.7 times greater. In contrast, the "tropical-far" category is the poorest, with a per capita GNP only about one third of the world average.

INTERPRETING THE PATTERNS

In our research we have examined three major ways in which geography affects economic development. First, as Adam Smith noted, economies differ in their ease of transporting goods, people and ideas. Because sea trade is less costly than land- or air-based trade, economies near coastlines have a great advantage over hinterland economies. The per-kilometer costs of overland trade within Africa, for example, are often an order of magnitude greater than the costs of sea trade to an African port. Here are some figures we found recently: The cost of shipping a six-meter-long container from Rotterdam, the Netherlands, to Dar-es-Salaam, Tanzania – an air distance of 7,300 kilometers – was about $1,400. But transporting the same container overland from Dar-es-Salaam to Kigali, Rwanda – a distance of 1,280 kilometers by road – cost about $2,500, or nearly twice as much.

Second, geography affects the prevalence of disease. Many kinds of infectious diseases are endemic to the tropical and subtropical zones. This tends to be true of diseases in which the pathogen spends part of its life cycle outside the human host: for instance, malaria (carried by mosquitoes) and helminthic infections (caused by parasitic worms). Although epidemics of malaria have occurred sporadically as far north as Boston in the past century, the disease has never gained a lasting foothold in the temperate zones, because the cold winters naturally control the mosquito-based transmission of the disease. (Winter could be considered the world's most effective public health intervention.) It is much more difficult to control malaria in tropical regions, where transmission takes place year-round and affects a large part of the population.

According to the World Health Organization, 300 million to 500 million new cases of malaria occur every year, almost entirely concentrated in the tropics. The disease is so common in these areas that no one really knows how many people it kills annually – at least one million and perhaps as many as 2.3 million. Widespread illness and early deaths obviously hold back a nation's economic performance by significantly reducing worker productivity. But there are also long-term effects that may be amplified over time through various social feedbacks.

For example, a high incidence of disease can alter the age structure of a country's population. Societies with high levels of child mortality tend to have high levels of fertility: mothers bear many children to guarantee that at least some will survive to adulthood. Young children will therefore constitute a large proportion of that country's population. With so many children, poor families cannot invest much in each child's education. High fertility also constrains the role of women in society, because child rearing takes up so much of their adult lives.

Third, geography affects agricultural productivity. Of the major food grains – wheat, maize and rice – wheat grows only in temperate climates, and maize and rice crops are generally more productive in temperate and subtropical climates than in tropical zones. On average, a hectare of land in the tropics yields 2.3 metric tons of maize, whereas a hectare in the temperate zone yields 6.4 tons. Farming in tropical rain-forest environments is hampered by the fragility of the soil: high temperatures mineralize the organic materials, and the intense rainfall leaches them out of the soil. In tropical environments that have wet and dry seasons – such as the African savanna – farmers must contend with the rapid loss of soil moisture

resulting from high temperatures, the great variability of precipitation, and the ever present risk of drought. Moreover, tropical environments are plagued with diverse infestations of pests and parasites that can devastate both crops and livestock.

Many of the efforts to improve food output in tropical regions – attempted first by the colonial powers and then in recent decades by donor agencies – have ended in failure. Typically the agricultural experts blithely tried to transfer temperate-zone farming practices to the tropics, only to watch livestock and crops succumb to pests, disease and climate barriers. What makes the problem even more complex is that food productivity in tropical regions is also influenced by geologic and topographic conditions that vary greatly from place to place. The island of Java, for example, can support highly productive farms because the volcanic soil there suffers less nutrient depletion than the non-volcanic soil of the neighboring islands of Indonesia.

Moderate advantages or disadvantages in geography can lead to big differences in long-term economic performance. For example, favorable agricultural or health conditions may boost per capita income in temperate-zone nations and hence increase the size of their economies. This growth encourages inventors in those nations to create products and services to sell into the larger and richer markets. The resulting inventions further raise economic output, spurring yet more inventive activity. The moderate geographical advantage is thus amplified through innovation.

In contrast, the low food output per farm worker in tropical regions tends to diminish the size of cities, which depend on the agricultural hinterland for their sustenance. With a smaller proportion of the population in urban areas, the rate of technological advance is usually slower. The tropical regions therefore remain more rural than the temperate regions, with most of their economic activity concentrated in low-technology agriculture rather than in high-technology manufacturing and services.

We must stress, however, that geographical factors are only part of the story. Social and economic institutions are critical to long-term economic performance. It is particularly instructive to compare the post-World War II performance of free-market and socialist economies in neighboring countries that share the same geographical characteristics: North and South Korea, East and West Germany, the Czech

Republic and Austria, and Estonia and Finland. In each case we find that free-market institutions vastly outperformed socialist ones.

The main implication of our findings is that policy-makers should pay more attention to the developmental barriers associated with geography – specifically, poor health, low agricultural productivity and high transportation costs. For example, tropical economies should strive to diversify production into manufacturing and service sectors that are not hindered by climate conditions. The successful countries of tropical Southeast Asia, most notably Malaysia, have achieved stunning advances in the past 30 years, in part by addressing public health problems and in part by moving their economies away from climate-dependent commodity exports (rubber, palm oil and so on) to electronics, semiconductors and other industrial sectors. They were helped by the high concentration of their populations in coastal areas near international sea lanes and by the relatively tractable conditions for the control of malaria and other tropical diseases. Sub-Saharan Africa is not so fortunate: most of its population is located far from the coasts, and its ecological conditions are harsher on human health and agriculture.

The World Bank and the International Monetary Fund, the two international agencies that are most influential in advising developing countries, currently place more emphasis on institutional reforms – for instance, overhauling a nation's civil service or its tax administration – than on the technologies needed to fight tropical diseases and low agricultural productivity. One formidable obstacle is that pharmaceutical companies have no market incentive to address the health problems of the world's poor. Therefore, wealthier nations should adopt policies to increase the companies' motivation to work on vaccines for tropical diseases. In one of our own initiatives, we called on the governments of wealthy nations to foster greater research and development by pledging to buy vaccines for malaria, HIV/AIDS and tuberculosis from the pharmaceutical companies at a reasonable price. Similarly, biotechnology and agricultural research companies need more incentive to study how to improve farm output in tropical regions.

The poorest countries in the world surely lack the resources to relieve their geographical burdens on their own. Sub-Saharan African countries have per capita income levels of around $1 a day. Even when such countries invest as much as 3 or 4 percent of their GNP in public health – a large proportion of national income for a very poor country – the result is only about $10 to $15 per year per person. This is certainly not enough to control endemic malaria, much less to fight other rampant diseases such as HIV/AIDS, tuberculosis and helminthic infections.

A serious effort at global development will require not just better economic policies in the poor countries but far more financial support from the rich countries to help overcome the special problems imposed by geography. A preliminary estimate suggests that even a modest increase in donor financing of about $25 billion per year – only 0.1 percent of the total GNP of the wealthy nations, or about $28 per person – could make a tremendous difference in reducing disease and increasing food productivity in the world's poorest countries.

'The Origins of the Third World'

from *Late Victorian Holocausts* (2001)

Mike Davis

Editors' Introduction

Mike Davis was born in 1946 in Fontana, California, after his family migrated west following the Great Depression. He grew up in El Cajon, in the back country of Southern California, a region that has remained an abiding focus in his writing and political activism. After his father died, Davis started work at 16 as a meat cutter. He also became involved in the militant non-violent activism of the Congress for Racial Equality (CORE) in the early 1960s. After a short period of study at Reed College, Davis set off to work for Students for a Democratic Society (SDS), the emerging voice of the New Left. By the late 1960s, Davis was a member of the most militant and anti-Stalinist branch of the Communist Party in the United States, the Southern California chapter, but even here he proved to be too much of a freethinker. Davis worked as a truck driver and tour bus driver, and learnt the underside of his native Los Angeles driving around and organising bus workers. A butchers' union scholarship allowed Davis to pursue studies in economics and history at the University of California, Los Angeles, and then, three years later, to study Irish labour history at Edinburgh University, Scotland. Davis's education criss-crossed between Los Angeles, Belfast and London, driven by academic as much as by activist considerations.

In 1981, Davis joined the editorial board of *New Left Review*, an important independent Marxist journal linked to the publishing house Verso in London. He started the Haymarket Series of books for Verso to critique North American society. Part of this series, his own *Prisoners of the American Dream*, was a searing indictment of working-class lives in America under President Ronald Reagan. When he returned to the United States in 1987, Davis was not allowed to submit the draft of his now classic book on Los Angeles, *City of Quartz*, for his PhD, without attending required courses. Instead, he spent the next decade as a peripatetic temporary lecturer, teaching on, and across, the length and breadth of the sprawling city of Los Angeles. Davis's writing from the 1990s reflects a growing knowledge of neighbourhoods, gangs, labour unions, city politics, urban myths and fantasies, and periodic catastrophes. His writings on Los Angeles have become a staple for urban and labour studies alike. In contrast to the article by Sachs *et al.*, Davis has tried to stress the way in which 'bad geography' is a product of race and class struggle. Davis has not just written about LA's urban ills, he has participated in securing peace between gangs, and justice for immigrant workers.

In 1998, Davis received a MacArthur 'genius' Award. He has also been a fellow of the Getty Institute in Los Angeles, a professor at the University of California, Irvine, and an editor at Verso and *New Left Review*. Davis has recently shifted to writing on wider themes. *Planet of Slums* explores the global spread of urban informal settlements. It is a sweeping argument that has been subject to criticism, particularly for its dire portrayal of African cities (see, for instance, the special issue of *Mute* 2/3, online at http://www.metamute.org/en/Naked-Cities-Struggle-in-the-Global-Slums, accessed 4 October 2007).

In *Late Victorian Holocausts*, his book from which the reading here comes, Davis argues that a series of droughts across North Africa, India and China in the late nineteenth century were in fact linked. Today, they would be called El Niño conditions. In the 1870s and 1890s, response to drought was catastrophic for social and political reasons. 'Bad geography' is a product of power and violence. More than 50 million people starved to death as a consequence of political, economic and environmental processes: processes of development. In this short, rich excerpt, Davis asks how it was that famine and deindustrialisation came to China and Bengal and marked the 'beginning of the Third World'. Davis shows how China and India were crucial to keeping British imperial hegemony afloat in the late nineteenth century. He also shows that the price of this subsidy was paid with the blood of rural populations who died because of the dramatic increase in their vulnerability to food system collapse – or famine. Like Marx in Part 2, Davis argues that markets, particularly in a time of pervasive famine, are made by force. It is important to unravel precise social histories in order to understand who has borne the costs of development, and who profited in 'the making of the Third World' as the British Empire neared its end.

Key references

Mike Davis (2001) *Late Victorian Holocausts: El Niño Famines and the Making of the Third World*, London: Verso.
— (1986) *Prisoners of the American Dream: Politics and Economy in the History of the U.S. Working Class*, London: Verso.
— (1990/2006) *City of Quartz: Excavating the Future in Los Angeles*, London: Verso.
— (2006) *Planet of Slums*, London: Verso.
Michael J. Watts (1982) *Silent Violence: Food, Famine and Peasantry in Northern Nigeria*, Berkeley: University of California Press.
Giovanni Arrighi (1994) *The Long Twentieth Century: Money, Power and the Origins of Our Times*, London: Verso.

Emaciated people, disease, ribs showing, shriveled bellies, corpses, children with fly-encircled eyes, with swollen stomachs, children dying in the streets, rivers choked with bodies, people; living, sleeping, lying, dying on the streets in misery, beggary, squalor, wretchedness, a mass of aboriginal humanity . . .

Harold Isaacs

What historians, then, have so often dismissed as "climatic accidents" turn out to be not so accidental after all.[1] Although its syncopations are complex and quasi-periodic, ENSO has a coherent spatial and temporal logic. And, contrary to Emmanuel Le Roy Ladurie's famous (Eurocentric?) conclusion in *Times of Feast, Times of Famine* that climate change is a "slight, perhaps negligible" shaper of human affairs, ENSO is an episodically potent force in the history of tropical humanity.[2] If, as Raymond Williams once observed, "Nature contains, though often unnoticed, an extraordinary amount of human history," we are now learning that the inverse is equally true: there is an extraordinary amount of hitherto unnoticed environmental instability in modern history.[3] The

power of ENSO events indeed seems so overwhelming in some instances that it is tempting to assert that great famines, like those of the 1870s and 1890s (or, more recently, the Sahelian disaster of the 1970s), were "caused" by El Niño, or by El Niño acting upon traditional agrarian misery. This interpretion, of course, inadvertently echoes the official line of the British in Victorian India as recapitulated in every famine commission report and viceregal allocution: millions were killed by extreme weather, not imperialism.[4] Was this true?

'BAD CLIMATE' VERSUS 'BAD SYSTEM'

At this point it would be immensely useful to have some strategy for sorting out what the Chinese pithily contrast as "bad climate" versus "bad system." Y. Kueh, as we have seen, has attempted to parameterize the respective influences of drought and policy upon agricultural output during the Great Leap Forward famine of 1958–61. The derivation of his "weather index," however, involved fifteen years of arduous research and the resolution of "a series of complicated methodological and technical problems" including a necessary comparative regression to the 1930s. Although his work is methodologically rich, his crucial indices depend upon comprehensive meteorological and econometric data that are simply not available for the nineteenth century. A direct statistical assault on the tangled causal web of the 1876–77 and 1896–1902 famines thus seems precluded.[5]

An alternative is to construct a "natural experiment." As Jared Diamond has advocated in a recent sermon to historians, such an experiment should compare systems "differing in the presence or absence (or in the strong or weak effect) of some putative causative factor."[6] We ideally need, in other words, an analogue for the late Victorian famines in which the natural parameters are constant but the social variables significantly differ. An excellent candidate for which we possess unusually detailed documentation is the El Niño event of 1743–44 (described as "exceptional" by Whetton and Ruther-furd) in its impact on the north China plain.[7] Although not as geographically far-reaching as the great ENSO droughts of 1876–78 or 1899–1900, it otherwise pre-figured their intensities. The spring monsoon failed two years in a row, devastating winter wheat in Hebei (Zhili) and northern Shandong. Scorching winds withered crops and farmers dropped dead in their fields from sunstroke. Provincial grain supplies were utterly inadequate to the scale of need. Yet unlike the late nineteenth century, there was no mass mortality from either starvation or disease. Why not?

Pierre-Etienne Will has carefully reconstructed the fascinating history of the 1743–44 relief campaign from contemporary records. Under the skilled Confucian administration of Fang Guancheng, the agricultural and hydraulic expert who directed relief operations in Zhili, the renowned "ever-normal gran-aries" in each county immediately began to issue rations (without any labor test) to peasants in the officially designated disaster counties.[8] (Local gentry had already organized soup kitchens to ensure the survival of the poorest residents until state distribu-tions began.) When local supplies proved insufficient, Guancheng shifted millet and rice from the great store of tribute grain at Tongcang at the terminus of the Grand Canal, then used the Canal to move vast quantities of rice from the south. Two million peasants were maintained for eight months, until the return of the monsoon made agriculture again possible. Ultimately 85 percent of the relief grain was borrowed from tribute depots or granaries outside the radius of the drought.[9]

As Will emphasizes, this was famine defense in depth, the "last word in technology at the time." No contemporary European society guaranteed subsist-ence as a human right to its peasantry (ming-sheng is the Chinese term), nor, as the Physiocrats later marveled, could any emulate "the perfect timing of [Guancheng's] operations: the action taken always kept up with developments and even anticipated them."[10] Indeed, while the Qing were honoring their social contract with the peasantry, contemporary Europeans were dying in the millions from famine and hunger-related diseases following arctic winters and summer droughts in 1740–43. "The mortality peak of the early 1740s," emphasizes an authority, "is an outstanding fact of European demographic history."[11] In Europe's Age of Reason, in other words, the "starving masses" were French, Irish and Calabrian, not Chinese.

Moreover "the intervention carried out in Zhili in 1743 and 1744 was not the only one of its kind in the eighteenth century, nor even the most extensive."[12] Indeed, as Table 1 indicates, the Yellow River flooding of the previous year (1742/43) involved much larger expenditures over a much broader region. (In addition to the ENSO-correlated droughts and floods shown in the table, Will has also documented seven other flood disasters that involved massive relief mobilization.) Although comparable figures are unavailable, Beijing also acted aggressively to aid Shandong officials in preventing famine during the series of El Niño droughts that afflicted that province (and much of the tropics) between 1778 and 1787.[13] The contrast with the chaotic late-Qing relief efforts in 1877 and 1899 (or, for that matter, Mao's monstrous mishandling of the 1958–61 drought) could not be more striking. State capacity in eighteenth-century China, as Will and his collabor-ators emphasize, was deeply impressive: a cadre of skilled administrators and trouble-shooters, a unique national system of grain price stabilization, large crop surpluses, well-managed granaries storing more than a million bushels of grain in each of twelve provinces, and incomparable hydraulic infra-structures.[14]

The capstone of Golden Age food security was the invigilation of grain prices and supply trends by the emperor himself. Although ever-normal granaries

	Quinn Intensity	Provinces	Amount of Relief
1720/21	Very strong	Shaanxi	Unknown
1742/43	(Flooding)	Jiangsu/Anhui	17 million taels; 2.3 million shi
1743/44	Moderate+	Hebei	0.87 million taels; 1 million shi
1778	Strong	Henan	1.6 million taels, 0.3 million shi
1779/80	La Niña	Henan	same
1785	?	Henan	2.8 million taels

Table 1 ENSO disasters relieved by the Qing
Source: Constructed from Table VII, Whetton and Rutherfurd, p. 244; Table 20, Will, *Bureaucracy and Famine*, pp. 298–9.

were an ancient tradition, price monitoring was a chief innovation of the Qing. "Great care was exercised by the eighteenth-century Emperors in looking over the memorials and price lists in search of inconsistencies." On the fifth of every month *hsien* magistrates forwarded detailed price reports to the prefectures, who summarized them for the provincial governors who, in turn, reported their content in memorials to the central government.[15] Carefully studied and annotated by the emperors, these "vermillion rescripts" testify to an extraordinary engagement with the administration of food security and rural well-being. "In the 1720s and 1730s," R. Bin Wong writes, "the Yongzheng emperor personally scrutinized granary operations, as he did all other bureaucratic behavior; his intense interest in official efforts and his readiness to berate officials for what he considered failures partially explain the development of granary operations beyond the levels achieved in the late Kangxi period."[16] Yongzheng also severely sanctioned speculation by the "rich households [who] in their quest for profit habitually remove grain by the full thousand or full myriad bushels."[17]

His successor, Qianlong, ordered the prefects to send the county-level price reports directly to the Bureau of Revenue in Beijing so he could study them firsthand. The emperors' intense personal involvement ensured a high standard of accuracy in price reporting and, as Endymion Wilkinson demonstrates, frequently led to significant reform.[18] This was another *differentia specifica* of Qing absolutism. It is hard to imagine a Louis XVI spending his evenings scrupulously poring over the minutiae of grain prices from Limoges or the Auvergne, although the effort might have ultimately saved his head from the guillotine.

Nor can we easily picture a European monarch intimately involved in the esoteria of public works to the same degree that the Qing routinely immersed themselves in the details of the Grand Canal grain transport system. "The Manchu emperors," Jane Leonard points out, "had since the early reigns involved themselves deeply in Canal management, not just in broad questions of policy, but in the control and supervision of lower-level administrative tasks." When, for example, flooding in 1824 destroyed sections of the Grand Canal at the critical Huai–Yellow River junction, the Tao-kuang emperor personally assumed command of reconstruction efforts.[19]

In contrast, moreover, to later Western stereotypes of a passive Chinese state, government during the high Qing era was proactively involved in famine prevention through a broad program of investment in agricultural improvement, irrigation and waterborne transportation. As in other things, Joseph Needham points out, the eighteenth century was a golden age for theoretical and historical work on flood control and canal construction. Civil engineers were canonized and had temples erected in their honor.[20] Confucian activists like Guancheng, with a deep commitment to agricultural intensification, "tended to give top priority to investments in infrastructure and to consider the organization of food relief merely a makeshift." Guancheng also wrote a famous manual (the source of much of Will's account) that codified historically tested principles of disaster planning and relief management: something else that has little precedent in backward European tradition.[21]

Finally, there is plentiful evidence that the northern China peasantry during the high Qing was more nutritionally self-reliant and less vulnerable to climate stress than their descendants a century later. In the eighteenth century, after the Kangxi emperor permanently froze land revenue at the 1712 level, China experienced "the mildest agrarian taxation it had ever known in the whole of its history."[22] Dwight Perkins estimates that the formal land tax was a mere 5 to 6 percent of the harvest and that a large portion was expended locally by *hsien* and provincial governments.[23] Likewise, the exchange ratio between silver and copper coinage, which turned so disastrously against the poor peasantry in the nineteenth century, was stabilized by the booming output of the Yunnan copper mines (replacing Japanese imports) and the great inflow of Mexican bullion earned by China's huge trade surplus.[24] Unlike their contemporary French counterparts, the farmers of the Yellow River plain (the vast majority of whom owned their land) were neither crushed by exorbitant taxes nor ground down by feudal rents. North China, in particular, was unprecedentedly prosperous by historical standards, and Will estimates that the percentage of the rural population ordinarily living near the edge of starvation – depending, for example, on husks and wild vegetables for a substantial part of their diet – was less than 2 percent.[25] As a result, epidemic disease, unlike in Europe, was held in check for most of the "Golden Age."[26]

Still, could even Fang Guancheng have coped with drought disasters engulfing the larger part of north China on the scale of 1876 or even 1899? It is important to weigh this question carefully, since drought-famines were more localized in the eighteenth century, and because the 1876 drought, as we have seen, may have been a 200-year or even 500-year frequency event. Moreover, the late Victorian droughts reached particular intensity in the loess highlands of Shanxi and Shaanxi, where transport costs were highest and bottlenecks unavoidable. It is reasonable, therefore, to concede that a drought of 1876 magnitude in 1743 would inevitably have involved tens, perhaps even hundreds, of thousands of deaths in more remote villages.

Such a drought, however, would have been unlikely, as in the late nineteenth century, to grow into a veritable holocaust that consumed the greater part of the populations of whole prefectures and counties. In contrast to the situation in 1876–77, when granaries were depleted or looted and prices soared out of control, eighteenth-century administrators could count on a large imperial budget surplus and well-stocked local granaries backed up by a huge surplus of rice in the south. Large stockpiles of tribute grain

at strategic transportation nodes in Henan and along the Shanxi–Shaanxi border were specially designated for the relief of the loess provinces, and an abundance of water sources guaranteed the Grand Canal's navigability year-round.[27] Whereas in 1876 the Chinese state – enfeebled and demoralized after the failure of the Tongzhi Restoration's domestic reforms – was reduced to desultory cash relief augmented by private donations and humiliating foreign charity, in the eighteenth century it had both the technology and political will to shift grain massively between regions and, thus, relieve hunger on a larger scale than any previous polity in world history.[28]

'LAWS OF LEATHER' VERSUS 'LAWS OF IRON'

What about famine in pre-British India? Again, there is little evidence that rural India had ever experienced subsistence crises on the scale of the Bengal catastrophe of 1770 under East India Company rule or the long siege by disease and hunger between 1875 and 1920 that slowed population growth almost to a standstill. The Moguls, to be sure, did not dispose of anything like the resources of the centralized Qing state at its eighteenth-century zenith, nor was their administrative history as well documented. As Sanjay Sharma has pointed out, "The problems of intervening in the complex network of caste-based local markets and transport bottlenecks rendered an effective state intervention quite difficult."[29]

On the other hand, benefiting perhaps from a milder ENSO cycle, Mogul India was generally free of famine until the 1770s. There is considerable evidence, moreover, that in pre-British India before the creation of a railroad-girded national market in grain, village-level food reserves were larger, patrimonial welfare more widespread, and grain prices in surplus areas better insulated against speculation.[30] (As we have seen, the perverse consequence of a unitary market was to export famine, via price inflation, to the rural poor in grain-surplus districts.) The British, of course, had a vested interest in claiming that they had liberated the populace from a dark age of Mogul despotism: "One of the foundations of Crown Rule was the belief that . . . India's past was full of depravity."[31] But, as Bose and Jalal point out, "The picture of an emaciated and oppressed peasantry, mercilessly exploited by the emperor and his nobility, is being seriously altered in the light of new interpretations of the evidence."[32] Recent research by Ashok Desai indicates that "the mean standard of food consumption in Akbar's empire was appreciably higher than in the India of the early 1960s."[33]

The Mogul state, moreover, "regarded the protection of the peasant as an essential obligation," and there are numerous examples of humane if sporadic relief operations.[34] Like their Chinese contemporaries, the Mogul rulers Akbar, Shahjahan and Aurangzeb relied on a quartet of fundamental policies – embargoes on food exports, antispeculative price regulation, tax relief and distribution of free food without a forced-labor counterpart – that were an anathema to later British Utilitarians.[35] They also zealously policed the grain trade in the public interest. As one horrified British writer discovered, these "oriental despots" punished traders who shortchanged peasants during famines by amputating an equivalent weight of merchant flesh.[36]

In contrast to the Raj's punitive taxation of irrigation and its neglect of traditional wells and reservoirs, the Moguls used tax subsidies to promote water conservation. As David Hardiman explains in the case of Gujarat: "Local officials had considerable discretion over tax assessment, and it seems to have been their practice to encourage well-construction by granting tax concessions. In the Ahmedabad region, for example, it was common to waive the tax on a 'rabi' crop raised through irrigation from a recently constructed well. The concession continued until the tax exemptions were held to have equalled the cost of construction."[37]

Occasionally, the British paid appropriate tribute to the policies of their "despotic" predecessors. The first Famine Commission Report in 1880, for example, cited Aurangzeb's extraordinary relief campaign during the (El Niño?) drought-famine of 1661: "The Emperor opened his treasury and granted money without stint. He gave every encouragement to the importation of corn and either sold it at reduced prices, or distributed it gratuitously amongst those who were too poor to pay. He also promptly acknowledged the necessity of remitting the rents of the cultivators and relieved them for the time being of other taxes. The vernacular chronicles of the period attribute the salvation of millions of lives and the preservation of many provinces to his strenuous exertions."[38]

Food security was also probably better in the Deccan during the period of Maratha rule. As Mountstuart Elphinstone admitted retrospectively after the British conquest, "The Mahratta country flourished, and the people seem to have been exempt from some of the evils which exist under our more perfect Government."[39] His contemporary, Sir John Malcolm, "claimed that between 1770 and 1820 there had been only three very bad seasons in the Maratha lands and, though some years had been 'indifferent,' none had been as 'bad as to occasion any particular distress.' "[40] D.E.U. Baker cites a later British administrative report from the Central Provinces that contrasted the desultory relief efforts of the East India Company during the droughts of the 1820s and 1830s ("a few thousand rupees") with the earlier and highly effective Maratha policy of forcing local elites to feed the poor ("enforced charity of hundreds of rich men").[41] Indeed the resilient Maratha social order was founded on a militarized free peasantry and "very few landless laborers existed." In contrast to the British-imposed *raiyatwari* system, occupancy rights in the Maratha Deccan were not tied to revenue payment, taxes varied according to the actual harvest, common lands and resources were accessible to the poor, and the rulers subsidized local irrigation improvements with cheap *taqavi* (or *tagai*) loans.[42] In addition, Elphinstone observed, the "sober, frugal, industrious" Maratha farmers lived in generally tolerant coexistence with the Bhils and other tribal peoples. Ecological and economic synergies balanced the diverse claims of plains agriculture, pastoralism and foothill swidden.[43]

In contrast to the rigidity and dogmatism of British land-and-revenue settlements, both the Moguls and Marathas flexibly tailored their rule to take account of the crucial ecological relationships and unpredictable climate fluctuations of the subcontinent's drought-prone regions. The Moguls had "laws of leather," wrote journalist Vaughan Nash during the famine of 1899, in contrast to the British "laws of iron."[44] Moreover, traditional Indian elites, like the great Bengali *zamindars*, seldom shared Utilitarian obsessions with welfare cheating and labor discipline. "Requiring the poor to work for relief, a practice begun in 1866 in Bengal under the influence of the Victorian Poor Law, was in flat contradiction to the Bengali premise that food should be given ungrudgingly, as a father gives food to his children."[45] Although the British insisted that they had rescued

India from "timeless hunger," more than one official was jolted when Indian nationalists quoted from an 1878 study published in the prestigious *Journal of the Statistical Society* that contrasted thirty-one serious famines in 120 years of British rule against only seventeen recorded famines in the entire previous two millennia.[46]

India and China, in other words, did not enter modern history as the helpless "lands of famine" so universally enshrined in the Western imagination. Certainly the intensity of the ENSO cycle in the late nineteenth century, perhaps only equaled on three or four other occasions in the last millennium, must loom large in any explanation of the catastrophes of the 1870s and 1890s. But it is scarcely the only independent variable. Equal causal weight, or more, must be accorded to the growing social vulnerability to climate variability that became so evident in south Asia, north China, northeast Brazil and southern Africa in late Victorian times. As Michael Watts has eloquently argued in his history of the "silent violence" of drought-famine in colonial Nigeria: "Climate risk . . . is not given by nature but . . . by 'negotiated settlement' since each society has institutional, social, and technical means for coping with risk. . . . Famines [thus] are social crises that represent the failures of particular economic and political systems."[47]

PERSPECTIVES ON VULNERABILITY

Over the last generation, scholars have produced a bumper-crop of revealing social and economic histories of the regions teleconnected to ENSO's episodic disturbances. The thrust of this research has been to further demolish orientalist stereotypes of immutable poverty and overpopulation as the natural preconditions of the major nineteenth-century famines. There is persuasive evidence that peasants and farm laborers became dramatically more pregnable to natural disaster after 1850 as their local economies were violently incorporated into the world market. What colonial administrators and missionaries – even sometimes creole elites, as in Brazil – perceived as the persistence of ancient cycles of backwardness were typically modern structures of formal or informal imperialism.

From the perspective of political ecology, the vulnerability of tropical agriculturalists to extreme climate events after 1870 was magnified by simultaneous restructurings of household and village linkages to regional production systems, world commodity markets and the colonial (or dependent) state. "It is, of course, the constellation of these social relations," writes Watts, "which binds households together and project them into the marketplace, that determines the precise form of the household vulnerability. It is also these same social relations that have failed to stimulate or have actually prevented the development of the productive forces that might have lessened this vulnerability." Indeed, new social relations of production, in tandem with the New Imperialism, "not only altered the extent of hunger in a statistical sense but changed its very etiology."[48] Three points of articulation with larger socioeconomic structures were especially decisive for rural subsistence in the late Victorian "proto-third world."

First, the forcible incorporation of smallholder production into commodity and financial circuits controlled from overseas tended to undermine traditional food security. Recent scholarship confirms that it was *subsistence adversity* (high taxes, chronic indebtedness, inadequate acreage, loss of subsidiary employment opportunities, enclosure of common resources, dissolution of patrimonial obligations, and so on), not entrepreneurial opportunity, that typically promoted the turn to cash-crop cultivation. Rural capital, in turn, tended to be parasitic rather than productivist as rich landowners redeployed fortunes that they built during export booms into usury, rack-renting and crop brokerage. "Marginal subsistence producers," Hans Medick points out, ". . . did not benefit from the market under these circumstances; they were devoured by it."[49] Medick, writing about the analogous predicament of marginal smallholders in "proto-industrial" Europe, provides an exemplary description of the dilemma of millions of Indian and Chinese poor peasants in the late nineteenth century:

> For them [even] rising agrarian prices did not necessarily mean increasing incomes. Since their marginal productivity was low and production fluctuated, rising agrarian prices tended to be a source of indebtedness rather than affording them the opportunity to accumulate surpluses. The "anomaly of the agrarian markets" forced the marginal subsistence producers into an unequal exchange relationship through the market. . . . Instead of profiting from exchange, they were

forced by the market into the progressive deterioration of their conditions of production, i.e. the loss of their property titles. Especially in years of bad harvests, and high prices, the petty producers were compelled to buy additional grain, and, worse, to go into debt. Then, in good harvest years when cereal prices were low, they found it hard to extricate themselves from the previously accumulated debts; owing to the low productivity of their holdings they could not produce sufficient quantities for sale.[50]

As a result, the position of small rural producers in the international economic hierarchy equated with downward mobility, or, at best, stagnation. There is consistent evidence from north China as well as India and northeast Brazil of falling household wealth and increased fragmentation or alienation of land. Whether farmers were directly engaged by foreign capital, like the Berari *khatedars* and Cearan *parceiros* who fed the mills of Lancashire during the Cotton Famine, or were simply producing for domestic markets subject to international competition like the cotton-spinning peasants of the Boxer hsiens in western Shandong, commercialization went hand in hand with pauperization without any silver lining of technical change or agrarian capitalism.

Second, the integration of millions of tropical cultivators into the world market during the late nineteenth century was accompanied by a dramatic deterioration in their terms of trade. Peasants' lack of market power vis-à-vis crop merchants and creditors was redoubled by their commodities' falling international purchasing power. The famous Kondratief downswing of 1873–1897 made dramatic geographical discriminations. As W. Arthur Lewis suggests, comparative productivity or transport costs alone cannot explain an emergent structure of global unequal exchange that valued the products of tropical agriculture so differently from those of temperate farming. "With the exception of sugar, all the commodities whose price was lower in 1913 than in 1883 were commodities produced almost wholly in the tropics. All the commodities whose prices rose over this thirty-year period were commodities in which the temperate countries produced a substantial part of total supplies. The fall in ocean freight rates affected tropical more than temperate prices, but this should not make a difference of more than five percentage points."[51]

Third, formal and informal Victorian imperialism, backed up by the supernational automatism of the Gold Standard, confiscated local fiscal autonomy and impeded state-level developmental responses – especially investments in water conservancy and irrigation – that might have reduced vulnerability to climate shocks. As Curzon once famously complained to the House of Lords, tariffs "were decided in London, not in India; in England's interests, not in India's."[52] Moreover, as we shall see in the next chapter, any grassroots benefit from British railroad and canal construction was largely canceled by official neglect of local irrigation and the brutal enclosures of forest and pasture resources. Export earnings, in other words, not only failed to return to smallholders as increments in household income, but also as usable social capital or state investment.

In China, the "normalization" of grain prices and the ecological stabilization of agriculture in the Yellow River plain were undermined by an interaction of endogenous crises and the loss of sovereignty over foreign trade in the aftermath of the two Opium Wars. As disconnected from world market perturbations as the starving loess provinces might have seemed in 1877, the catastrophic fate of their populations was indirectly determined by Western intervention and the consequent decline in state capacity to ensure traditional welfare. Similarly the depletion of "ever-normal" granaries may have resulted from a vicious circle of multiple interacting causes over a fifty-year span, but the coup de grace was certainly the structural recession and permanent fiscal crisis engineered by Palmerston's aggressions against China in the 1850s. As foreign pressure intensified in later decades, the embattled Qing, as Kenneth Pomeranz has shown, were forced to abandon both their traditional mandates: abandoning both hydraulic control and grain stockpiling in the Yellow River provinces in order to concentrate on defending their endangered commercial littoral.[53]

British control over Brazil's foreign debt and thus its fiscal capacity likewise helps explain the failure of either the empire or its successor republic to launch any antidrought developmental effort in the sertão. The zero-sum economic conflicts between Brazil's rising and declining regions took place in a structural context where London banks, above all the Rothschilds, ultimately owned the money-supply. In common with India and China, the inability to politically regulate interaction with the world market at the very time when mass subsistence increasingly depended upon food entitlements acquired in international trade became a sinister syllogism for famine. Moreover in the three cases of the Deccan, the Yellow River basin and the Nordeste, former "core" regions of eighteenth-century subcontinental power systems were successively transformed into famished peripheries of a London-centered world economy.

The elaboration of these theses, as always in geo-historical explanation, invites closer analysis at different magnifications. Before considering case-studies of rural immiseration in key regions devastated by the 1870s and 1890s El Niño events or looking at the relationships among imperialism, state capacity and ecological crisis at the village level, it is necessary to briefly discuss how the structural positions of Indians and Chinese (the big battalions of the future Third World) in the world economy changed over the course of the nineteenth century. Understanding how tropical humanity lost so much economic ground to western Europeans after 1850 goes a long way toward explaining why famine was able to reap such hecatombs in El Niño years. As a baseline for understanding the origins of modern global inequality (and that is the key question), the herculean statistical labors of Paul Bairoch and Angus Maddison over the last thirty years have been complemented by recent comparative case-studies of European and Asian standards of living.

THE DEFEAT OF ASIA

Bairoch's famous claim, corroborated by Maddison, is that differences in income and wealth between the great civilizations of the eighteenth century were relatively slight: "It is very likely that, in the middle of the eighteenth century, the average standard of living in Europe was a little bit lower than that of the rest of the world."[54] When the *sans culottes* stormed the Bastille, the largest manufacturing districts in the world were still the Yangzi Delta and Bengal, with Lingan (modern Guangdong and Guangxi) and coastal Madras not far behind.[55] India alone produced one-quarter of world manufactures, and while its "pre-capitalist agrarian labour productivity was probably less than the Japanese-Chinese level, its commercial capital surpassed that of the Chinese."[56]

	1700	1820	1890	1952
China	23.1	32.4	13.2	5.2
India	22.6	15.7	11.0	3.8
Europe	23.3	26.6	40.3	29.7

Table 2 Shares of world GDP (percent)
Source: Angus Maddison, *Chinese Economic Performance in the Long Run*, Paris 1998, p. 40.

As Prasannan Parthasarathi has recently shown, the stereotype of the Indian laborer as a half-starved wretch in a loincloth collapses in the face of new data about comparative standards of living. "Indeed, there is compelling evidence that South Indian labourers had higher earnings than their British counterparts in the eighteenth century and lived lives of greater financial security." Because the productivity of land was higher in South India, weavers and other artisans enjoyed better diets than average Europeans. More importantly, their unemployment rates tended to be lower because they possessed superior rights of contract and exercised more economic power. But even outcaste agricultural labourers in Madras earned more in real terms than English farm laborers.[57] (By 1900, in contrast, Romesh Chunder Dutt estimated that the average British household income was 21 times higher.)[58]

New research by Chinese historians also challenges traditional conceptions of comparative economic growth. Referring to the pathbreaking work of Li Bozhong, Philip Huang notes that "the outstanding representative of this new academic tendency has even argued the overall economic development of the Yangzi Delta in the Qing exceeded that of 'early modern' England."[59] Similarly, Bin Wong has recently emphasized that the "specific conditions associated with European proto-industrialization – expansion of seasonal crafts, shrinking farm size, and good marketing systems – may have been even more widespread in China [and India] than in Europe."[60] "Basic functional literacy," adds F. Mote, "was more widespread than in Western countries at that time, including among women at all social levels."[61]

Moreover, in the recent forum "Re-thinking 18th Century China," Kenneth Pomeranz points to evidence that ordinary Chinese enjoyed a higher standard of consumption than eighteenth-century Europeans:

Chinese life expectancy (and thus nutrition) was at roughly English levels (and so above Continental ones) even in the late 1700s. (Chinese fertility was actually lower than Europe's between 1550 and 1850, while its population grew faster; thus mortality must have been low.) Moreover, my estimates of "non-essential" consumption come out surprisingly high. Sugar consumption works out to between 4.3 and 5.0 pounds per capita ca. 1750 – and much higher in some regions – compared with barely 2 pounds per capita for Europe. China circa 1750 seems to have produced 6–8 lbs. of cotton cloth per capita; its richest area, the Yangzi Delta (population roughly 31 million), probably produced between 12 and 15 lbs. per capita. The UK, even in 1800, produced roughly 13 lbs. of cotton, linen and wool cloth combined per resident, and Continental output was probably below China's.[62]

Pomeranz has also calculated that "the Lower Yangzi appears to have produced roughly as much cotton cloth per capita in 1750 as the UK did cotton, wool, linen and silk cloth combined in 1800 – plus an enormous quantity of silk."[63] In addition, as Maddison demonstrates, the Chinese GDP in absolute terms grew faster than that of Europe throughout the eighteenth century, dramatically enlarging its share of world income by 1820.

The usual stereotype of nineteenth-century economic history is that Asia stood still while the Industrial Revolution propelled Britain, followed by the United States and eventually the rest of Western Europe, down the path of high-speed GNP growth. In a superficial sense, of course, this is true, although the data gathered by Bairoch and Maddison show that Asia lost its preeminence in the world economy later than most of us perhaps imagine. The future Third World, dominated by the highly developed commercial and handicraft economies of India and China, surrendered ground very grudgingly until 1850 (when it still generated 65 percent of global GNP), but then declined with increasing rapidity through the rest of the nineteenth century (only 38 percent of world GNP in 1900 and 22 percent in 1960).[64]

The deindustrialization of Asia via the substitution of Lancashire cotton imports for locally manufactured textiles reached its climax only in the decades after the construction of the Crystal Palace. "Until 1831," Albert Feuerwerker points out, "Britain purchased more 'nankeens' (cloth manufactured in Nanking and other places in the lower Yangzi region) each year than she sold British-manufactured cloth to China."[65] Britain exported 51 million yards of cloth to Asia in 1831; 995 million in 1871; 1413 million in 1879; and 2000 million in 1887.[66]

But why did Asia stand in place? The rote answer is because it was weighted down with the chains of tradition and Malthusian demography, although this did not prevent Qing China, whose rate of population increase was about the same as Europe's, from experiencing extraordinary economic growth throughout the eighteenth century. As Jack Goldstone recently argued, China's "stasis" is an "anachronistic illusion that come[s] from reading history backwards."[67] The relevant question is not so much why the Industrial Revolution occurred first in England, Scotland and Belgium, but why other advanced regions of the eighteenth-century world economy failed to adapt their handicraft manufactures to the new conditions of production and competition in the nineteenth century.

As Marx liked to point out, the Whig view of history deletes a great deal of very bloody business. The looms of India and China were defeated not so much by market competition as they were forcibly

	1750	1800	1830	1860	1880	1900
Europe	23.1	28.0	34.1	53.6	62.0	63.0
UK	1.9	4.3	9.5	19.9	22.9	18.5
Tropics	76.8	71.2	63.3	39.2	23.3	13.4
China	32.8	33.3	29.8	19.7	12.5	6.2
India	24.5	19.7	17.6	8.6	2.8	1.7

Table 3 Shares of world manufacturing output, 1750–1900 (percent)
Source: Derived from B. R. Tomlinson, "Economics: The Periphery," in Andrew Porter (ed.), *The Oxford History of the British Empire: The Nineteenth Century*, Oxford 1990, p. 69 (Table 3.8).

	Western Europe		China	
1400	430	(43)	500	(74)
1820	1034	(122)	500	(342)
1950	4902	(412)	454	(547)

Table 4 Standing in place: China vs. Europe (dollars per capita GDP/(population in millions))
Source: Lu Aiguo, *China and the Global Economy Since 1840*, Helsinki 2000, p. 56 (Table 4.1 as derived from Maddison).

dismantled by war, invasion, opium and a Lancashire-imposed system of one-way tariffs. (Already by 1850, imposed Indian opium imports had siphoned 11 percent of China's money-supply and 13 percent of its silver stock out of the country.)[68] Whatever the internal brakes on rapid economic growth in Asia, Latin America or Africa, it is indisputable that from about 1780 or 1800 onward, every serious attempt by a non-Western society to move over into a fast lane of development or to regulate its terms of trade was met by a military as well as an economic response from London or a competing imperial capital. Japan, prodded by Perry's black ships, is the exception that proves the rule.

The use of force to configure a "liberal" world economy (as Marx and later Rosa Luxemburg argued) is what Pax Britannica was really about. Palmerston paved the way for Cobden. The Victorians, according to Brian Bond's calculations, resorted to gunboats on at least seventy-five different occasions.[69] The simultaneous British triumphs in the Mutiny and the "Arrow" War in 1858, along with Japan's yielding to Perry in the same year, were the epochal victories over Asian economic autonomy that made a Cobdenite world of free trade possible in the second half of the nineteenth century (Thailand had already conceded a 3 percent tariff in 1855).[70] The Taiping Revolution – "more revolutionary in its aims than the Meiji Restoration, insisting on gender equality and democratizing literacy" – was a gigantic attempt to revise that verdict, and was, of course, defeated only thanks to the resources and mercenaries that Britain supplied to the embattled Qing.[71]

This is not to claim that the Industrial Revolution necessarily depended upon the colonial conquest or economic subjugation of Asia; on the contrary, the slave trade and the plantations of the New World were much more strategic streams of liquid capital and natural resources in boosting the industrial take-off in Britain, France and the United States. Although Ralph Davis has argued that the spoils of Plessy contributed decisively to the stability of the Georgian order in an age of revolution, the East India Company's turnover was small change compared to the great trans-Atlantic flow of goods and capital.[72] Only the Netherlands, it would appear, depended crucially upon Asian tribute – the profits of its brutal *culturrstelsel* – in financing its economic recovery and incipient industrialization between 1830 and 1850.

Paradoxically, monsoon Asia's most important "moment" in the Victorian world economy was not at the beginning of the epoch, but towards its end. "The full value of British rule, the return on political investments first made in the eighteenth century," write Cain and Hopkins in their influential history of British imperialism, "was not realised until the second half of the nineteenth century, when India became a vital market for Lancashire's cotton goods and when other specialised interests, such as jute manufacturers in Dundee and steel producers in Sheffield, also greatly increased their stake in the sub-continent."[73] The coerced levies of wealth from India and China were not essential to the rise of British hegemony, but they were absolutely crucial in postponing its decline.

THE LATE VICTORIAN WORLD ECONOMY

During the protracted period of stop-and-go growth from 1873 to 1896 (what economic historians mis-leadingly used to call the "Great Depression"), the rate of capital formation and the growth of productivity of both labor and capital in Britain began a dramatic slowdown.[74] She remained tied to old products and technologies while behind their tariff barriers Germany and the United States forged leadership in cutting-edge oil, chemical and electrical industries. Since British imports and overseas investment still dynamized local growth from Australia to Denmark, the potential "scissors" between UK productivity and consumption threatened the entire structure of world trade. It was in this conjuncture that the starving Indian and Chinese peasantries were wheeled in as unlikely saviors. For a generation they braced the entire system of international settlements, allowing England's continued financial supremacy to temporarily coexist with its relative industrial decline. As Giovanni Arrighi emphasizes, "The large surplus in

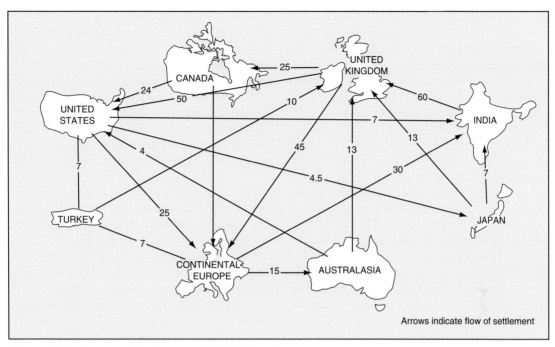

Figure 1 World system of settlements, 1910 (£ millions)
Source: S. Saul, *Studies in British Overseas Trade, 1870–1914*, Liverpool 1960, p. 58.

the Indian balance of payments became the pivot of the enlarged reproduction of Britain's world-scale processes of capital accumulation and of the City's mastery of world finance."[75]

The operation of this crucial circuit was simple and ingenious. Britain earned huge annual surpluses in her transactions with India and China that allowed her to sustain equally large deficits with the United States, Germany and the white Dominions. True, Britain also enjoyed invisible earnings from shipping, insurance, banking and foreign investment, but without Asia, which generated 73 percent of British trade credit in 1910, Anthony Latham argues, Britain "presumably would have been forced to abandon free trade," while her trading partners would have been forced to slow their own rates of industrialization. The liberal world economy might otherwise have fragmented into autarkic trading blocs, as it did later during the 1930s:

> The United States and industrial Europe, in particular Germany, were able to continue their policy of tariff protection only because of Britain's surplus with Asia. Without that Asian surplus, Britain would no longer have been able to subsidise their growth. So what emerges is that Asia

in general, but India and China in particular, far from being peripheral to the evolution of the international economy at this time, were in fact crucial. Without the surpluses which Britain was able to earn there, the whole pattern of international economic development would have been severely constrained.[76]

India, of course, was the greatest captive market in world history, rising from third to first place among consumers of British exports in the quarter century after 1870.[77] "British rules," writes Marcello de Cecco in his study of the Victorian gold standard system, "deliberately prevented Indians from becoming skilled mechanics, refused contracts to Indian firms which produced materials that could be got from England, and generally hindered the formation of an autonomous industrial structure in India."[78] Thanks to a "government stores policy that reserved most government purchases to British products and by the monopoly of British agency houses in organizing the import-export trade," India was forced to absorb Britain's surplus of increasingly obsolescent and noncompetitive industrial exports.[79] By 1910 this included two-fifths of the UK's finished cotton goods and three-fifths of its exports of electrical products, railway equipment, books and pharmaceuticals. As a result, observes de Cecco, Britain avoided "having to restructure her industry and was able to invest her capital in the countries where it gave the highest return." Thanks to India, "British financiers were not compelled to 'tie' their loans to British exports because the Imperial outlet was always available for British products."[80]

The subcontinent was equally important to the rentier strata. The climate-detonated crisis of English agriculture in the late 1870s and the subsequent decline of farm output produced a sharp fall in agricultural rents in England and Wales from £53 million in 1876 to only £37 million in 1910.[81] Indian army and civil service sinecures were accordingly famous for rescuing the fortunes of Britain's landed aristocracy. But, as Cain and Hopkins have argued in making their case for a hegemonic "gentlemanly capitalism," even bigger spoils were returned to the middle classes of London and the Home Counties as government-guaranteed interest on railroad debentures and Indian bonds. "This constituency of southern investors, and its institutional representatives in banking and shipping, fell in readily behind the flag of empire and gave full support to policies of free trade and sound money. If British rule in India was helpful to British industry, it was vital to British investment."[82] As Hobsbawm points out, "not even the free-traders wished to see this goldmine escape from British control."[83]

But how, in an age of famine, could the subcontinent afford to subsidize its conquerer's suddenly precarious commercial supremacy?[84] In a word, it couldn't, and India was forced-marched into the world market, as we shall see, by revenue and irrigation policies that compelled farmers to produce for foreign consumption at the price of their own food security. This export drive was the hallmark of the new public finance strategy introduced by James Wilson – founder of *The Economist* and finance member of the Council of India – in the first years of direct rule. The opening of the Suez Canal and the growth of steam shipping drastically reduced the transport costs of bulk commodity export from the subcontinent. As a result India's seaborne foreign trade increased more than eightfold between 1840 and 1886.[85] In addition to opium cultivation in Bengal, new export monocultures of indigo, cotton, wheat and rice supplanted millions of acres of subsistence crops. Part of this production, of course, was

designed to assure low grain prices in the metropolis after the debacle of English agriculture in the 1870s. Between 1875 and 1900, years that included the worst famines in Indian history, annual grain exports increased from 3 million to 10 million tons: a quantity that, as Romesh Dutt pointed out, was equivalent to the annual nutrition of 25 million people. By the turn of the century, India was supplying nearly a fifth of Britain's wheat consumption as well as allowing London grain merchants to speculate during shortages on the Continent.[86]

But Indian agriculture's even more decisive contribution to the imperial system, from the East India Company's first illegal shipment of opium to Canton, was the income it earned in the rest of the Eastern Hemisphere. Especially in the 1880s and 1890s, the subcontinent's permanent trade and current account imbalances with Britain were financed by its trade surpluses of opium, rice and cotton thread vis-à-vis the rest of Asia. Indeed England's systematic exploitation of India depended in large part upon India's commercial exploitation of China. This triangular trade between India, China and Britain had a strategic economic importance in the Victorian world system that transcended other far larger flows of commerce. If China generated only a tiny 1.3 percent of the total volume of world trade in the late nineteenth century, it was nonetheless immensely valuable to the British Empire, which monopolized fully 80 percent of China's foreign trade in the 1860s and 60 percent as late as 1899. (British firms, which controlled two-thirds of coastal shipping, also took an important slice of China's domestic commerce.)[87]

From the beginning of the nineteenth century, the East India Company had relied on opium exports from Bengal to Canton (which in 1832 earned a net profit "at least fourteen times the prime cost") to finance the growing deficits generated by its expensive military operations on the subcontinent. By forcibly enlarging the Chinese demand for the narcotic and, thus, the taxes collected on its export, the two Opium Wars (1839–42 and 1856–58) and the punitive Treaty of Tianjin (1858) revolutionized the revenue base of British India. "Opium," says John Wong, "serviced the cost of imperial expansion in India."[88] Opium shipments from India reached a peak of 87,000 chests in 1879, the biggest drug transaction in world history.[89]

This extraordinarily one-sided trade – in 1868 India supplied over 35 percent of China's imports but bought less than 1 percent of its exports – also subsidized the imports of US cotton that fueled the industrial revolution in Lancashire.[90] "The sale of Bengal opium to China," Latham explains, "was a great link in the chain of commerce with which Britain had surrounded the world. The chain worked like this: The United Kingdom paid the United States for cotton by bills upon the Bank of England. The Americans took some of those bills to Canton and swapped them for tea. The Chinese exchanged the bills for Indian opium. Some of the bills were remitted to England as profit; others were taken to India to buy additional commodities, as well as to furnish the money remittance of private fortunes in India and the funds for carrying on the Indian government at home."[91]

When, after 1880, the Chinese unofficially resorted to domestic cultivation of opium (an early example of "import-substitution") to reduce their trade deficit, British India found a lucrative new advantage in the export of factory-spun cotton yarn, which, as we shall see, had a devastating impact on Chinese folk textiles. Moreover, in the later nineteenth century Britain herself started earning a substantial surplus in the China trade for the first time. The Second Opium War – or "Arrow" War – which increased British exports to China tenfold in a single decade was the turning point.[92] Britain's dominant role in Chinese foreign trade, built by Victorian *narcotraficantes* with gunboats, thus leveraged the whole free-trade imperium. "China," summarizes Latham, "directly through Britain and indirectly through India, enabled Britain to sustain her deficits with the United States and Europe on which those countries depended for export stimulus and, in the case of the United States, capital inflow to some degree."[93]

Moreover, China was forced at bayonet point to cede control over tariffs to the British inspector-general of the Imperial Maritime Customs Administration, a de facto imperial proconsul who "came to enjoy more influence with the Foreign Office than did the British Minister in Peking."[94] China's growing trade deficit became intractable by 1884. "Not a single year [in the rest of the nineteenth century] showed a surplus; the average annual deficit rose to 26.6 million taels – roughly about 10 percent of the yearly total trade, but over 20 percent of the annual imports or just under 30 percent of the annual exports."[95] Among its traditional monopolies, tea was undercut in the world market by Indian production while Japanese silk competed with the famous brands of southern China. Unlike India, China was unable to finance any of its "consistent and growing overall deficit" via trade surpluses with a third party, nor could it siphon compensatory incomes, like Britain, from its overseas colonies. As a result, the Qing became increasingly dependent upon foreign exchange remittances from 5 million Chinese emigrants in southeast Asia, Oceania, Peru, the Caribbean and the United States.[96] Although the government publicly expressed its disgust with the coolie trade, it had little alternative but to collaborate in its expansion. The so-called "yellow peril" that English writers would help to popularize was thus a direct consequence of Asia's increasing subsidization of faltering British hegemony. Emigrant Chinese plantation workers and railroad laborers, like Indian ryots, balanced England's accounts on their bent backs.

[. . .]

NOTES

The epigraph is from Isaacs, *Scratches on Our Minds: American Images of China and India*, New York 1958, p. 273.

1 For a typically cavalier view, see Roland Lardinols, "Famine, Epidemics and Mortality in South India: A Reappraisal of the Demographic Crisis of 1876–1878," *Economic and Political Weekly* 20:111 (16 March 1985), p. 454.
2 Emmanuel Le Roy Ladurie, *Tmes of Feast, Times of Famine: A History of Climate Since the Year 1000*, Garden City, N.Y. 1971, p. 119.
3 Raymond Williams, *Problems in Materialism and Culture*, London 1980, p. 67.
4 When it served their interests, of course, the British could switch epistemologies. In the case of late-nineteenth-century China, for example, the British and their allies primarily blamed Qing corruption, not drought, for the millions of famine deaths.
5 Kueh, pp. 4–5.
6 Jared Diamond, *Guns, Germs, and Steel: The Fates of Human Societies*, New York 1997, pp. 424–5.
7 Re 1743–44: "another exceptional period in the eastern hemisphere, which corresponds with QN El Niño of 1744, although conditions were more markedly dry in the east in 1743" (Whetton and Rutherfurd, pp. 243–6).
8 "The first Qing emperor envisioned ever-normal granaries in county seats, charity granaries in major towns, and community granaries in the countryside.

Ever-normal granaries were to be managed by members of the magistrate's staff, who were directed to sell, lend, or give away grain in the spring and to make purchases, collect loans, and solicit contributions in the autumn" (Pierre-Etienne Will and R. Bin Wong [with James Lee, Jean Oi and Peter Perdue], *Nourish the People: The State Civilian Granary System in China, 1650–1850*, Ann Arbor, Mich. 1981, p. 19).

9 Will, *Bureaucracy and Famine*, Chapters 7 and 8.

10 Ibid., pp. 86 and 189.

11 John Post, *Food Shortage, Climatic Variability, and Epidemic Disease in Preindustrial Europe: The Mortality Peak in the Early 1740s*, Ithaca, N.Y. 1985, p. 30.

12 Will, p. 270.

13 Jean Oi and Pierre-Etienne Will, "North China: Shandong During the Qianlong Period," in Will and Wong, pp. 369–70. ENSO correlations based on Quinn chronology.

14 "Introduction," in Will and Wong, p. 21. China's roads, on the other hand, remained miserable, and were a major obstacle to market integration as well as famine relief.

15 Wilkinson, pp. 122–9.

16 R. Bin Wong, "Decline and Its Opposition, 1781–1850," in Will and Wong, p. 76.

17 Helen Dunstan, *Conflicting Counsels to Confuse the Age: A Documentary Study of Political Economy in Qing China, 1644–1840*, Ann Arbor, Mich. 1996, p. 251.

18 Wilkinson, pp. 122–9. See also Will, "The Control Structure," in Will and Wong, pp. 220–21.

19 Jane Leonard, " 'Controlling from Afar': Open Communications and the Tao-Kuang Emperor's Control of Grand Canal–Grain Transport Management, 1824–26," *Modern Asian Studies* 22:4 (1988), p. 666.

20 Joseph Needham, *Science and Civilization in China*, vol. 4, Cambridge 1971, p. 326.

21 Will, p. 257.

22 Jacques Gernet, *A History of Chinese Civilization*, 2nd edn., Cambridge 1996, p. 468.

23 Dwight Perkins, *Agricultural Development in China, 1368–1968*, Chicago 1969, p. 176.

24 Endymion Wilkinson, "Studies in Chinese Price History," Ph.D. diss., Princeton University 1970, p. 31.

25 Will, p. 32.

26 J. A. G. Roberts, *A Concise History of China*, Cambridge, Mass. 1990, p. 173.

27 On the special tribute granaries at Luoyang and Shanzhou organized during the Kangxi reign, see Will and Wong, pp. 32 and 301.

28 Food security in the mid-eighteenth century may have consumed 10 percent of annual Qing revenue. As Wong emphasizes, "For a state to spend such sums for this purpose on a regular basis for well over a century is likely unique in the early modern world" ("Qing

Granaries and Late Imperial History," in Will and Wong, p. 477).

29 Sanjay Sharma, "The 1837–38 Famine in U.P.: Some Dimensions of Popular Action," *IESHR* 30:3 (1993), p. 359.

30 Bhatia, p. 9.

31 Darren Zook, "Developing India: The History of an Idea in the Southern Countryside, 1860–1990," Ph.D. diss., University of California, Berkeley 1998, p. 158. The Raj was built upon mythology and hallucination. As Zook points out, the British universally attributed the ruins scattered through the Indian countryside to the decadence of native civilizations, when, in fact, many were direct memorials to the violence of British conquest (p. 157).

32 Sugata Bose and Ayesha Jalal, *Modern South Asia*, Delhi 1999, p. 43.

33 Ashok Desai, "Population and Standards of Living in Akbar's Time," *IESHR* 9:1 (1972), p. 61.

34 Chetan Singh, "Forests, Pastoralists and Agrarian Society in Mughal India," in David Arnold and Raachandra Guha (eds.), *Nature, Culture, Imperialism: Essay on the Environmental History of South Asia*, Delhi 1996, p. 22.

35 Habibul Kondker, "Famine Policies in Pre-British India and the Question of Moral Economy," *South Asia* 9:1 (June 1986), pp. 25–40; and Kuldeep Mahtur and Niraja Jayal, *Drought, Policy and Politics*, New Delhi 1993, p. 27. Unfortunately, contemporary discussion of famine history before 1763 has been contaminated by Hindu-versus-Muslim bickering. See, for example, the apparent anti-Muslim bias in Mushtag Kaw, "Famines in Kashmir, 1586–1819: The Policy of the Mughal and Afghan Rulers," *IESHR* 33:1 (1996), pp. 59–70.

36 C. Blair, *Indian Famines*, London 1874, pp. 8–10.

37 David Hardiman, "Well Irrigation in Gujarat: Systems of Use, Hierarchies of Control," *Economic and Political Weekly*, 20 June 1998, p. 1537.

38 Commission quoted in W. R. Aykroyd, *The Conquest of Famine*, London 1974, p. 51. See also John Richards, *The Mughal Empire* (*The New Cambridge History of India, 1:5*), Cambridge 1993, p. 163.

39 Bagchi, pp. 11–12 and 27.

40 J. Malcolm, *A Memoir of Central India*, vol. 1, London 1931, p. 7, quoted in D. E. U. Baker, *Colonialism in an Indian Hinterland: The Central Provinces, 1820–1920*, Delhi 1993, p. 28.

41 Baker, p. 52.

42 J. Richards and Michelle McAlpin, "Cotton Cultivating and Land Clearing in the Bombay Deccan and Karnatak: 1818–1920," in Richard Tucker and J. Richards (eds.), *Global Deforestation and the Nineteenth-Century World Economy*, Durham 1983, pp. 71 and 74.

43 Ibid.

44 Nash, p. 92.

45 Greenough, *Prosperity and Misery*, p. 59.

46 C. Walford, "The Famines of the World: Past and Present," *Journal of the Statistical Society* 41:13 (1878), pp. 434–42. I cite Walford elsewhere from the expanded 1879 book version of this article.

47 Michael Watts, *Silent Violence: Food, Famine and Peasantry in Northern Nigeria*, Berkeley 1983, pp. 462–3. This "negotiation," of course, is two-sided and must include climate shock as an independent variable.

48 Watts, pp. 267 and 464.

49 Hans Medick, "The Proto-Industrial Family Economy and the Structures and Functions of Population Development under the Proto-Industrial System," in P. Kriedte et al. (eds.), *Industrialization Before Industrialization*, Cambridge 1981, p. 45.

50 Ibid., pp. 44–5.

51 Lewis, *Growth and Fluctuations*, p. 189.

52 Cited in Clive Dewey, "The End of the Imperialism of Free Trade," p. 35.

53 Kenneth Pomeranz, *The Making of a Hinterland: State, Society, and Economy in Inland North China, 1853–1937*, Berkeley 1993.

54 Paul Bairoch, "The Main Trends in National Economic Disparities Since the Industrial Revolution," in Paul Bairoch and Maurice Levy-Leboyer (eds.), *Disparities in Economic Development Since the Industrial Revolution*, London 1981, p. 7.

55 Paul Bairoch, "International Industrialization Levels from 1750–1980," *Journal of European Economic History* 11 (1982), p. 107.

56 Fritjof Tichelman, *The Social Evolution of Indonesia*, The Hague 1980, p. 30.

57 Prasannan Parthasarathi, "Rethinking Wages and Competitiveness in Eighteenth Century Britain and South India," *Past and Present*, 158 (Feb. 1998), pp. 82–7 and 105–6.

58 Dutt, cited in Eddy, p. 21.

59 Philip Huang, *The Peasant Family and Rural Development in the Yangzi Delta, 1350–1988*, Stanford, Calif. 1990.

60 Wong, p. 38.

61 F. W. Mote, *Imperial China, 900–1800*, Cambridge, Mass. 1999, p. 941.

62 Kenneth Pomeranz, "A High Standard of Living and Its Implications," contribution to "E.H.R. Forum: Rethinking 18th Century China," Internet, 19 Nov. 1997.

63 Pomeranz, "Two Worlds of Trade, Two Worlds of Empire: European State-Making and Industrialization in a Chinese Mirror," in David Smith et al., *States and Sovereignty in the Global Economy*, London 1999, p. 78 (my emphasis).

64 See S. Patel, "The Economic Distance Between Nations: Its Origin, Measurement and Outlook, *Economic Journal*, March 1964. (There is some discrepancy between his figures for the aggregate non-European world and the later estimates of Bairoch and Maddison.)

65 Albert Feuerwerker, *The Chinese Economy, 1870–1949*, Ann Arbor, Mich. 1995, pp. 32–3.

66 Paul Bairoch, "Geographical Structure and Trade Balance of European Foreign Trade, from 1800–1970," *Journal of European Economic History* 3:3 (Winter 1978), p. 565. Ch'en cites 1866 as the beginning of the serious penetration of imported textiles into China (p. 64).

67 Jack Goldstone, "Review of David Landes, *The Wealth and Poverty of Nations*," *Journal of World History* 2:1 (Spring 2000), p. 109.

68 Carl Trocki, *Opium, Empire and the Global Political Economy*, London 1999, p. 98.

69 Brian Bond, *Victorian Military Campaigns*, London 1967, pp. 309–11.

70 See O'Rourke and Williamson, pp. 53–4.

71 Historians traditionally contrast the Meiji and Tonzhang restorations, but as Goldstone suggests, the more significant comparison is between the Taipings and Japan. "What if China's old imperial regime, like Japan's, had collapsed in the mid nineteenth century, and not fifty years later, what then? What if the equivalent of Chiang Kai-shek's new model army had begun formation in the 1860s and not the 1920s? Would Japan still have been able to colonize Korea and Taiwan? What would have been the Asian superpower?" (Goldstone, ibid.).

72 "India wealth supplied the funds that bought the national debt back from the Dutch and others, first temporarily in the interval of peace between 1763 and 1774, and finally after 1783, leaving Britain nearly free from overseas indebtedness when it came to face the great French wars from 1793" (Ralph Davis, *The Industrial Revolution and British Overseas Trade*, Leicester 1979, pp. 55–6).

73 P. Cain and A. Hopkins, *British Imperialism: Innovation and Expansion, 1688–1914*, London 1993, p. 334.

74 For a recent review, see Young Goo-Park, "Depression and Capital Formation: The UK and Germany, 1873–96," *Journal of European Economic History* 26:3 (Winter 1997), especially pp. 511 and 516.

75 Giovanni Arrighi, *The Long Twentieth Century: Money, Power and the Origins of Our Times*, London 1994, p. 263.

76 A. Latham, *The International Economy and the Undeveloped World, 1865–1914*, London 1978, p. 70. Latham, it should be noted, is notoriously apologistic for British colonialism in India, arguing that the subcontinent's "relatively low growth overall is due largely to climatic factors, not to any deleterious effect of British colonial policy" (see A. Latham, "Asian Stagnation: Real or Relative?", in Derek Aldcroft and Ross Catterall (eds.), *Rich Nations – Poor Nations: The Long-Run Perspective*, Cheltenham 1996, p. 109).

77 Robin Moore, "Imperial India, 1858–1914," in Andrew Porter (ed.), *The Oxford History of the British Empire: The Nineteenth Century*, Oxford 1999, p.441.

78 Marcello de Cecco, *The International Gold Standard: Money and Empire*, New York 1984, p. 30.

79 Ravi Palat, et al., "Incorporation of South Asia," p. 185. According to these authors, the apparent exceptions to Indian deindustrialization in fact proved the rule: cotton spinning was integral to the production of an export surplus from the China trade while jute manufacture was an "island of British capital . . . initiated, organized, and controlled by British civil servants and merchants" (p. 186).

80 Ibid., pp. 37–8.

81 J. Stamp, *British Incomes and Property*, London 1916, p. 36.

82 Cain and Hopkins, pp. 338–9.

83 Eric Hobsbawm, *Industry and Empire: An Economic History of Britain Since 1750*, London 1968, p. 123.

84 The same question, of course, could be asked of Indonesia, which in the late nineteenth century generated almost 9 percent of the Dutch national domestic product. See Angus Maddison, "Dutch Income in and from Indonesia, 1700–1938," *Modern Asian Studies* 23:4 (1989), p. 647.

85 Eric Stokes, "The First Century of British Colonial Rule in India: Social Revolution or Social Stagnation?" *Past and Present* 58 (Feb. 1873), p. 151.

86 Dietmar Rothermund, *An Economic History of India*, New York 1988, p. 36; Dutt, *Open Letters*, p. 48.

87 Lu Aiguo, *China and the Global Economy Since 1840*, Helsinki 2000, pp. 34, 37 and 39 (Table 2.4).

88 J.W. Wong, *Deadly Dreams: Opium and the Arrow War (1856–1860) in China*, Cambridge 1998, pp. 390 and 396. The British tea imports from China, which opium also financed, were the source of the lucrative tea duty that by mid-century almost compensated for the cost of the Royal Navy (pp. 350–55).

89 Lu Aiguo, p. 36.

90 Latham, *The International Economy*, p. 90. India (including Burma) also earned important income from rice exports to the Dutch East Indies.

91 Ibid., pp. 409–10. See also M. Greenberg, *British Trade and the Opening of China*, Cambridge 1951, p. 15.

92 Latham, pp. 453–4.

93 Ibid., pp. 81–90. After Japan's victory in 1895, however, its textile exports began to crowd India and Britain out of the Chinese market (p. 90).

94 Cain and Hopkins, p. 425.

95 Jerome Ch'en, *State Economic Polices of the Ch'ing Government, 1840–1895*, New York 1980, p. 116.

96 Latham, ibid.

'The Lay of the Land'

from *Imperial Leather: Race, Gender and Sexuality in the Colonial Contest* (1995)

Anne McClintock

Editors' Introduction

Anne McClintock is Simone de Beauvoir Professor of English at the University of Wisconsin at Madison, USA. McClintock completed her PhD from Columbia University in 1989, after which she published biographies of iconic feminist thinker Simone de Beauvoir (who famously said 'one is not born a woman, one becomes one'), and the early twentieth-century South African writer, Olive Schreiner. Following her widely acclaimed *Imperial Leather*, McClintock has written on postcoloniality, gender, sexuality and literature, and has co-edited a collection on these themes, *Dangerous Liaisons*, through the journal *Social Text*. McClintock is also an artist and a writer of creative non-fiction.

The selection from *Imperial Leather* that follows bears sharply on the object of development. McClintock draws from the important ideas of the philosopher Michel Foucault on forms of modern power and control. Foucault saw modern power as working through discourses that worked across institutions, regulating the conduct of individuals and, importantly, making them regulate their own conduct. Important to Foucault's framework is the notion that norms are supported by means of the control of various deviants on society's margins. Ideas of the bourgeois family, the disciplined worker, the good citizen and the healthy body, for instance, are upheld by defining and controlling the criminal, the unemployed, the insane, the homosexual, the pervert, and so on. Ideas of 'degeneracy' and 'contagion' were employed in late nineteenth-century modernising projects to control the 'social body'. Foucault argues that Victorian society was obsessed with sex and yet that discourses of sexuality were often about other things: about creating waged labourers, disciplined families and orderly social classes. Foucault calls sexuality a 'transfer point' of intersecting power relations.

The historian and anthropologist Ann Laura Stoler extends Foucault's work on social control and state racism in *Race and the Education of Desire*, showing that colonies acted as laboratories of imperial racism, in class and gender-specific ways. Stoler also draws out Foucault's notion that sexuality acts as a dense transfer point in relations of power. Moral panic about sexuality – whether in the control of marriage and concubinage, or in child-rearing and domestic relations – is often about the instabilities of imperial control. As we show in this reader, gender and sexuality have never just been about women as a constituency, even if women, gender and sexuality are often the grounds on which social crisis plays out. We will see this in Part 8 in Schoepf's work on AIDS in Africa, as well as in Hirschkind and Mahmood's argument about US militarism in Afghanistan.

In this reading, McClintock turns to ideas of degeneracy as the other side of the pursuit of 'progress', as linked objects of development. The concept of 'race' in the second half of the nineteenth century allowed people to think of many races of different origins, some of which were innately degenerate. Ideas of Africa, McClintock argues, were inevitably tied to ideas of sex, and specifically of sexual excess and profligacy that needed to be civilised and controlled. But ideas of race, gender and class were always intertwined, both in colonies and in the imperial heartland. Degenerates were always used to distinguish the norm, the bourgeoisie or properly domesticated middle classes.

In order to link race, gender, sexuality, imperialism and capitalism, McClintock turns to the idea of the 'fetish'. By fetishism, McClintock refers to the use of an object (the fetish) as a means to displace social contradictions. Fetishism is a kind of repetitive or compulsive displacement of contradictions onto the fetish object. McClintock ends the extract by arguing that the commodities of the British Empire were exhibited for the British public as fetish objects; they helped to displace the gender, class and race tensions that were pulsing through the violent late Victorian capitalist world. In other words, commodities acted as fetish objects to keep imperial citizens in Britain from seeing their complicity with a violent world system that was forcibly starving people in the name of the market. Commodity fetishism only presents the final products as magical objects, as if they were not created by violent world-historic processes. McClintock provides different tools from Davis to analyse these objects of development and the processes through which they come to be seen. Take a look at the packaging of any supermarket product and see how it does, and does not, reflect the actual labours that have made it what it is today. What Marx calls 'commodity fetishism' – the concealment of living labour and social struggle behind the objects we think we need and desire – is also a crucially important object of development.

Key references

Anne McClintock, (1995) *Imperial Leather: Race Gender and Sexuality in the Colonial Contest*, London: Routledge.
Anne McClintock, Aamir Mufti and Ella Shohat (eds) (1997) *Dangerous Liaisons: Gender, Nation, and Postcolonial Perspectives*, Minneapolis: University of Minnesota Press.
Ann Laura Stoler (1995) *Race and the Education of Desire: Foucault's* History of Sexuality *and the Colonial Order of Things*, Raleigh, NC: Duke University Press.
Michel Foucault (1990) *The History of Sexuality – Volume 1*, New York: Vintage.
Mary Poovey (1995) *Making a Social Body – British Cultural Formation, 1830–1864*, Chicago, IL: University of Chicago Press.

[. . .]

DEGENERATION

A triangulated discourse

From the outset, the idea of progress that illuminated the nineteenth century was shadowed by its somber side. Imagining the degeneration into which humanity could fall was a necessary part of imagining the exaltation to which it could aspire. The degenerate classes, defined as departures from the normal human type, were as necessary to the self-definition of the middle class as the idea of degeneration was to the idea of progress, for the distance along the path of progress traveled by some portions of humanity could be measured only by the distance others lagged behind.[1] Normality thus emerged as a product of

deviance, and the baroque invention of clusters of degenerate types highlighted the boundaries of the normal.

The poetics of degeneration was a poetics of social crisis. In the last decades of the century, Victorian social planners drew deeply on social Darwinism and the idea of degeneration to figure the social crises erupting relentlessly in the cities and colonies. By the end of the 1870s, Britain was foundering in severe depression, and throughout the 1880s class insurgency, feminist upheavals, the socialist revival, swelling poverty and the dearth of housing and jobs fed deepening middle class fears. The crises in the cities were compounded by crises in the colonies as Britain began to feel the pinch of the imperial rivalry of Germany and the United States. The atmosphere of impending catastrophe gave rise to major changes in social theory, which drew on the poetics of degeneration for legitimation. Suffused as it was with Lamarckian thinking, the eugenic discourse of degeneration was deployed both as a regime of discipline imposed on a deeply distressed populace, as well as a reactive response to very real popular resistance.

Biological images of disease and contagion served what Sander Gilman has called "the institutionalization of fear," reaching into almost every nook and cranny of Victorian social life, and providing the Victorian elite with the justification it needed to discipline and contain the "dangerous classes."[2] As the century drew to a close, biological images of disease and pestilence formed a complex hierarchy of social metaphors that carried considerable social authority. In *Outcast London* Gareth Stedman Jones shows how London became the focus of wealthy Victorians' growing anxieties about the unregenerate poor, variously described as the "dangerous" or "ragged" classes, the "casual poor," or the "residuum."[3] The slums and rookeries were figured as the hotbeds and breeding haunts of "cholera, crime and chartism."[4] "Festering" in dark and filthy dens, the scavenging and vagrant poor were described by images of putrefaction and organic debility. Thomas Plint described the "criminal class" as a "moral poison" and "pestiferous canker," a "non-indigenous" and predatory body preying on the healthy.[5] Carlyle saw the whole of London as an infected wen, a malignant ulcer on the national body politic.

The image of bad blood was drawn from biology but degeneration was less a biological fact than it was a social figure. Central to the idea of degeneration was the idea of *contagion* (the communication of disease, by touching, from body to body), and central to the idea of contagion was the peculiarly Victorian paranoia about boundary order. Panic about blood contiguity, ambiguity and *metissage* expressed intense anxieties about the fallibility of white male and imperial potency. The poetics of contagion justified a politics of exclusion and gave social sanction to the middle class fixation with boundary sanitation, in particular the sanitation of sexual boundaries. Body boundaries were felt to be dangerously permeable and demanding continual purification, so that sexuality, in particular women's sexuality, was cordoned off as the central transmitter of racial and hence cultural contagion. Increasingly vigilant efforts to control women's bodies, especially in the face of feminist resistance, were suffused with acute anxiety about the desecration of sexual boundaries and the consequences that racial contamination had for white male control of progeny, property and power. Certainly the sanitation syndromes were in part genuine attempts to combat the "diseases of poverty," but they also served more deeply to rationalize and ritualize the policing of boundaries between the Victorian ruling elite and the "contagious" classes, both in the imperial metropoles and in the colonies.

Controlling women's sexuality, exalting maternity and breeding a virile race of empire-builders were widely perceived as the paramount means for controlling the health and wealth of the male imperial body politic, so that, by the turn of the century, sexual purity emerged as a controlling metaphor for racial, economic and political power.[6] In the metropolis, as Anna Davin shows, population was power and societies for the promotion of public hygiene burgeoned, while childrearing and improving the racial stock became a national and imperial duty. State intervention in domestic life increased apace. Fears for the military prowess of the imperial army were exacerbated by the Anglo-Boer war, with the attendant discovery of the puny physiques, bad teeth and general ill health of the working class recruits. Motherhood became rationalized by the weighing and measuring of babies, the regimentation of domestic schedules and the bureaucratic administration of domestic education. Special opprobrium fell on "nonproductive" women (prostitutes, unmarried mothers, spinsters) and on "nonproductive men" (gays, the unemployed, the impoverished). In the eyes of policymakers and administrators, the bounds of empire could be secured and upheld only by proper domestic discipline and decorum, sexual probity and moral sanitation.

If, in the metropolis, as Ann Stoler writes, "racial deterioration was conceived to be a result of the moral turpitude and the ignorance of working class mothers, in the colonies the dangers were more pervasive, the possibilities of contamination worse."[7] Towards the end of the century, increasingly vigilant administrative measures were taken against open or ambiguous domestic relations, against concubinage, against mestizo customs. "*Metissage* (interracial unions) generally and concubinage in particular, represented the paramount danger to racial purity and cultural identity in all its forms. Through sexual contact with women of color European men 'contracted' not only disease but debased sentiments, immoral proclivities and extreme susceptibility to decivilized states.[8] In the chapters that follow, I explore how women who were ambiguously placed on the imperial divide (nurses, nannies, governesses, prostitutes and servants) served as boundary markers and mediators. Tasked with the purification and maintenance of boundaries, they were especially fetishized as dangerously ambiguous and contaminating.

The social power of the image of degeneration was twofold. First, social classes or groups were described with telling frequency as "races," "foreign groups," or "nonindigenous bodies," and could thus be cordoned off as biological and "contagious," rather than as social groups. The "residuum" were seen as irredeemable outcasts who had turned their backs on progress, not through any social failure to cope with the upheavals of industrial capitalism, but because of an organic degeneration of mind and body. Poverty and social distress were figured as biological flaws, an organic pathology in the body politic that posed a chronic threat to the riches, health and power of the "imperial race."

Second, the image fostered a sense of the legitimacy and urgency of state intervention, not only in public life but also in the most intimate domestic arrangements of metropolis and colony. After the 1860s, there was a faltering of faith in the concepts of individual progress and perfectibility.[9] If Enlightenment philosophy attempted to rewrite history in terms of the individual subject, the nineteenth century posed a number of serious challenges to history as the heroics of individual progress. Laissez-faire policies alone could not be trusted to deal with the problems of poverty or to allay fears of working class insurgence. "In such circumstances, the problem of degeneration and its concomitant, chronic poverty, would ultimately have to be resolved by the state."[10] The usefulness of the quasi-biological metaphors of "type," "species," "genus" and "race" was that they gave full expression to anxieties about class and gender insurgence without betraying the social and political nature of these distinctions. As Condorcet put it, such metaphors made "nature herself an accomplice in the crime of political inequality."[11]

DEGENERATION AND THE FAMILY TREE

The day when, misunderstanding the inferior occupations which nature has given her, women leave the home and take part in our battles; on this day a social revolution will begin and everything that maintains the sacred ties of the family will disappear.

Le Bon

In the poetics of degeneracy we find two anxious figures of historical time, both elaborated within the metaphor of the family. One narrative tells the story of the familial progress of humanity from degenerate native child to adult white man. The other narrative presents the converse: the possibility of racial decline from white fatherhood to a primordial black degeneracy incarnated in the black mother. The scientists, medical men and biologists of the day tirelessly pondered the evidence for both, marshaling the scientific "facts" and elaborating the multifarious taxonomies of racial and sexual difference, baroque in their intricacy and flourish of detail.

Before the 1850s two narratives of the origins of the races were in play. The first and more popular account, monogenesis, described the genesis of all races from the single creative source in Adam. Drawing on the Plotinian notion of corruption as distance from the originary source, scientists saw different races as having fallen unevenly from the perfect Edenic form incarnated in Adam. Simply by dwelling in different climates, races had degenerated unequally, creating an intricately shaded hierarchy of decline. By midcentury, however, a second, competing narrative had begun to gain ground – polygenesis, according to which theory different races had sprung up in different places, in different "centers of creation."[12] In this view, certain races in certain places were seen to be originally, naturally and inevitably degenerate.[13] Freedom itself came to be defined as an unnatural zone for Africans. Woe betide the race that migrated from its place.

After 1859, however, evolutionary theory swept away the creationist rug that had supported the intense debate between monogenists and polygenists, but it satisfied both sides by presenting an even better rationale for their shared racism. The monogenists continued to construct linear hierarchies of races according to mental and moral worth; the polygenists now admitted a common ancestry in the prehistoric mists but affirmed that the races had been separate long enough to evolve major inherited differences in talent and intelligence.[14]

At this time, evolutionary theory entered an "unholy alliance" with the allure of numbers, the amassing of measurements and the science of statistics.[15] This alliance gave birth to "scientific" racism, the most authoritative attempt to place social ranking and social disability on a biological and "scientific" footing. Scientists became enthralled by the magic of measurement. Anatomical criteria were sought for determining the relative position of races in the human series.[16] Francis Galton (1822–1911), pioneer statistician and founder of the eugenics movement, and Paul Broca, clinical surgeon and founder of the Anthropological Society of Paris (1859) inspired other scientists who followed them in the vocation of measuring racial worth off the geometry of the human body. To the earlier criterion of cranial capacity as the primary measure of racial and sexual ranking was now added a welter of new "scientific" criteria: the length and shape of the head, protrusion of the jaw, the distance between the peak of the head and brow, flatheadedness, a "snouty" profile, a long forearm (the characteristic of apes), underdeveloped calves (also apelike), a simplified and lobeless ear (considered a stigma of sexual excess notable in prostitutes), the placing of the hole at the base of the skull, the straightness of the hair, the length of the nasal cartilage, the flatness of the nose, prehensile feet, low foreheads, excessive wrinkles and facial hair. The features of the face spelled out the character of the race.

Increasingly, these stigmata were drawn on to identify and discipline atavistic "races" within the European race: prostitutes, the Irish, Jews, the unemployed, criminals and the insane. In the work of men such as Galton, Broca and the Italian physician, Cesare Lombroso, the geometry of the body mapped the psyche of the race.

What is of immediate importance here is that the welter of invented criteria for distinguishing degeneracy was finally gathered up into a dynamic, historical narrative by one dominant metaphor: the Family of Man. What had been a disorganized and inconsistent inventory of racial attributes was now drawn together into a genesis narrative that offered, above all, a figure of historical change.

Ernst Haeckel, the German zoologist, provided the most influential idea for the development of this metaphor.[17] His famous catchphrase, "ontogeny recapitulates phylogeny," captured the idea that the ancestral lineage of the human species could be read off the stages of a child's growth. Every child rehearses in organic miniature the ancestral progress of the race. The theory of recapitulation thus depicted the child as a type of social bonsai, a miniature family tree. As Gould put it, every individual as it grows to maturity "climbs its own family tree."[18] The irresistible value of the idea of recapitulation was that it offered an apparently absolute biological criterion not only for racial but also for sexual and class ranking. If the white male child was an atavistic throwback to a more primitive adult ancestor, he could be scientifically compared with other living races and groups to rank their level of evolutionary inferiority. A vital analogy had thus appeared:

> The adults of inferior groups must be like the children of superior groups, for the child represents a primitive adult ancestor. If adult blacks and women are like white male children, then they are living representatives of an ancestral stage in the evolution of white males. An anatomical theory for ranking races – based on entire bodies – had been found.[19]

Haggard summed up the analogy: "In all essentials the savage and the child of civilization are identical." Mayhew, likewise, described the London street-seller as an atavistic regression, a racial "child," who would "without training, go back to its parent stock – the vagabond savage."[20] G. A. Henty, like Haggard a popular and influential author of boy's stories, argued similarly: "The intelligence of an average negro is about equal to that of a European child of ten years old."[21] Thus the family metaphor and the idea of recapitulation entered popular culture, children's literature, travel writing and racial "science" with pervasive force.

The scope of the discourse was enormous. A host of "inferior" groups could now be mapped, measured and ranked against the "universal standard" of the white male child – within the organic embrace of the family metaphor and the Enlightenment regime of "rational" measurement as an optics of truth. In sum, a three-dimensional map of social difference had emerged, in which minute shadings of racial, class and gender hierarchy could be putatively measured across space: the measurable space of the empirical body [Fig. 1].

Figure 1 Racial measurement as an optics of truth. Nast's cartoon in *Harper's Weekly* (9 December 1876) stages an analogy between the racial and political weight of a freed slave and an Irishman.

"WHITE NEGROES" AND "CELTIC CALIBANS"

Antinomies of race

> He was a young Irishman . . . he had the silent enduring beauty of a carved ivory mask . . . that momentary but revealed immobility . . . a time-lessness . . . which negroes express sometimes without ever aiming at it; something old, old, old and acquiescent in the race!
>
> *D. H. Lawrence*

In the last decades of the nineteenth century, the term "race" was used in shifting and unstable ways, sometimes as synonymous with "species," some-times with "culture," sometimes with "nation," some-times to denote biological ethnicity or sub-groups within national groupings: the English "race" com-pared, say, with the "Irish" race. A small but dedicated group of doctors, antiquarians, clergymen, historians and geologists set out to uncover the minute shadings of difference that distinguished the "races" of Britain. Dr John Beddoe, founding member of the Ethnological Society, devoted thirty years of his life to measuring what he called the "Index of Nigrescence" (the amount of residual melanin in the skin, hair and

eyes) in the peoples of Britain and Ireland and con-cluded that the index rose sharply from east to west and south to north.[22]

In 1880, Gustave de Molinari (1819–1912) wrote that England's largest newspapers "allow no occasion to escape them of treating the Irish as an inferior race – as a kind of white negroes [sic]."[23] Molinari's phrase "white negroes" appeared in translation in a leader in *The Times* and was consistent with the popular assumption after the 1860s that certain physical and cultural features of the Irish marked them as a race of "Celtic Calibans" quite distinct from the Anglo-Saxons. As a visitor to Ireland commented: "Shoes and stockings are seldom worn by these beings who seem to form a different race from the rest of mankind."[24]

But Ireland presented a telling dilemma for pseudo-Darwinian imperial discourse. As Britain's first and oldest colony, Ireland's geographic proximity to Britain, as David Lloyd points out, resulted in its "undergoing the transition to hegemonic colonialism far earlier than any other colony."[25] But, as Claire Wills notes, the difficulty of placing the pale-skinned Irish in the hierarchy of empire was "compounded by the absence of the visual marker of skin colour difference which was used to legitimate domination in other colonized societies."[26] The English stereo-type of the Irish as a simianized and degenerate race also complicates postcolonial theories that skin color (what Gayatri Spivak usefully calls "chromatism") is the crucial sign of otherness. Chromatism, Wills notes, is a difference "which naturally does not apply to the relationship between the Irish and their English colonizers."[27] Certainly, great efforts were made to liken the Irish physiognomy to those of apes but, Wills argues, English racism concentrated primarily on the "barbarism" of the Irish accent.[28]

I suggest, however, that English racism also drew deeply on the notion of the *domestic* barbarism of the Irish as a marker of racial difference. In an exemplary image, an Irishman is depicted lazing in front of his hovel – the very picture of domestic disarray [Fig. 2]. The house is out of kilter, the shutter is askew. He lounges cheerily on an upturned wash-basin, visible proof of a slovenly lack of dedication to domestic order. What appears to be a cooking pot perches on his head. In the doorway, the boundary between private and public, his wife displays an equally cheerful slothfulness. In both husband and wife, the absence of skin color as a marker of degeneration is

Figure 2 "Celtic Calibans." *Puck*, Vol. 10, #258 (15 February 1882, p. 378). The title of Frederick B. Opper's cartoon "The King of A Shantee" suggests an analogy between the Irish and Africans.

compensated for by the simianizing of their physiognomies: exaggerated lips, receding foreheads, unkempt hair and so on. In the chapters that follow, I suggest that the iconography of *domestic degeneracy* was widely used to mediate the manifold contradictions in imperial hierarchy – not only with respect to the Irish but also to the other "white negroes": Jews, prostitutes, the working class, domestic workers, and so on, where skin color as a marker of power was imprecise and inadequate.

Racial stigmata were systematically, if often contradictorily, drawn on to elaborate minute shadings of difference in which social hierarchies of race, class and gender overlapped each other in a three-dimensional graph of comparison. The rhetoric of race was used to invent distinctions between what we would now call *classes*.[29] T. H. Huxley compared the East London poor with the Polynesian savage, William Booth chose the African pygmy, and William Barry thought that the slums resembled nothing so much as a slave ship.[30]

White women were seen as an inherently degenerate "race," akin in physiognomy to black people and apes. Gustave le Bon, author of the influential study of crowd behavior *La Psychologie des Foules*, compared female brain size with that of the gorilla and evoked this comparison as signaling a lapse in development:

All psychologists who have studied the intelligence of women, as well as poets and novelists, recognize today that they represent the most inferior forms of human evolution and that they are closer to children and savages than to an adult, civilized man.[31]

At the same time, the rhetoric of *gender* was used to make increasingly refined distinctions among the different *races*. The white race was figured as the male of the species and the black race as the female.[32] Similarly, the rhetoric of *class* was used to inscribe minute and subtle distinctions between other *races*. The Zulu male was regarded as the "gentleman" of the black race, but was seen to display features typical of females of the white race [Fig. 3]. Carl Vogt, for example, the preeminent German analyst of race in the midcentury, saw similarities between the skulls of white male infants and those of the white female working class, while noticing that a mature black male shared his "pendulous belly" with a white woman who had had many children.[33] On occasion, Australian aborigines, or alternatively Ethiopians, were regarded as the most debased "lower class" of the African races, but more often than not the female Khoisan (derogatorily known as "Hottentots" or "Bushmen") were located at the very nadir of human degeneration, just before the species left off its human form and turned bestial [Fig. 4].[34]

In cameo, then, the English middle-class male was placed at the pinnacle of evolutionary hierarchy (generally, the middle- or upper-middle-class male was regarded as racially superior to the degenerate aristocrat who had lapsed from supremacy). White English middle-class women followed. Irish or Jewish men were represented as the most inherently degenerate "female races" within the white male gender, approaching the state of apes.[35] Irish working-class women were depicted as lagging even farther behind in the lower depths of the white race.

Domestic workers, female miners and working-class prostitutes (women who worked publicly and visibly for money) were stationed on the threshold between the white and black races, figured as having fallen farthest from the perfect type of the white male and sharing many atavistic features with "advanced" black men [Fig. 5]. Prostitutes – as the metropolitan analogue of African promiscuity – were marked as especially atavistic and regressive. Inhabiting, as they did, the threshold of marriage and market, private

Figure 3 Feminizing African men

Figure 4 Militant woman as degenerate

Figure 5 Working woman as degenerate

and public, prostitutes flagrantly demanded money for services middle-class men expected for free.[36] Prostitutes visibly transgressed the middle-class boundary between private and public, paid work and unpaid work, and in consequence were figured as "white Negroes" inhabiting anachronistic space, their 'racial' atavism anatomically marked by regressive signs: "Darwin's ear," exaggerated posteriors, unruly hair and other sundry "primitive" stigmata.[37]

At this time, the idea of the Family of Man was itself confirmed through ubiquitous metaphoric analogies with science and biology. Bolstered by

pseudo-scientific racism after the 1850s and commodity racism after the 1880s, the monogamous patriarchal family, headed by a single, white father, was vaunted as a biological fact, natural, inevitable and right, its lineage imprinted immemorially in the blood of the species – during the same era, one might add, when the social functions of the family household were being replaced by the bureaucratic state.

A triangulated, switchboard analogy thus emerged between racial, class and gender deviance as a critical element in the formation of the modern, imperial imagination. In the symbolic triangle of

Figure 6 Global progress consumed at a glance

deviant money, deviant sexuality and deviant race, the so-called degenerate classes were metaphorically bound in a regime of surveillance, collectively figured by images of sexual pathology and racial aberration as atavistic throwbacks to a primitive moment in human prehistory, surviving ominously in the heart of the modern, imperial metropolis. Depicted as transgressing the natural distributions of money, sexual power and property and as thereby fatally threatening the fiscal and libidinal economy of the imperial state, these groups became subject to increasingly vigilant and violent state control.

IMPERIALISM AS COMMODITY SPECTACLE

In 1851, the topoi of progress and the Family of Man, panoptical time and anachronistic space found their architectural embodiment in the World Exhibition at the Crystal Palace in London's Hyde Park. At the Exhibition, the progress narrative began to be consumed as mass spectacle. The Exhibition gathered under one vaulting glass roof a monumental display of "the Industry of All Nations." Covering fourteen acres of park, it featured exhibitions and artifacts from thirty-two invited members of the "family of Nations." Crammed with industrial commodities, decorative merchandise, ornamental gardens,

machinery, musical instruments and industrial ore and thronged by thousands of marveling spectators, the Great Exhibition became a monument not only to a new form of mass consumption but also to a new form of commodity spectacle.

The Crystal Palace housed the first consumer dreams of a unified world time. As a monument to industrial progress, the Great Exhibition embodied the hope that all the world's cultures could be gathered under one roof – the global progress of history represented as the commodity progress of the Family of Man. At the same time, the Exhibition heralded a new mode of marketing history: the mass consumption of time as a commodity spectacle. Walking about the Exhibition, the spectator (admitted into the museum of modernity through the payment of cash) consumed history as a commodity. The dioramas and panoramas (popular, naturalistic replicas of scenes from empire and natural history) offered the illusion of marshaling all the globe's cultures into a single, visual pedigree of world time. In an exemplary image, the Great Exhibition literally drew the world's people toward the monumental display of the commodity: global progress consumed visually in a single image [Fig. 6]. Time became global, a progressive accumulation of panoramas and scenes arranged, ordered and catalogued according to the logic of imperial capital. At the same time, it was clearly implicit, only the west had the technical skill and innovative spirit to render the historical pedigree of the Family of Man in such perfect, technical form.

The Exhibition had its political equivalent in the Panopticon, or Inspection House. In 1787, Jeremy Bentham proposed the Panopticon as the model for an architectural solution to social discipline. The organizing principle of the Panopticon was simple. Factories, prisons, workhouses and schools would be constructed with an observation tower as the center. Unable to see inside the inspection tower, the inhabitants would presume they were under perpetual surveillance. Daily routine would be conducted in a state of permanent visibility. The elegance of the idea was the principle of self-surveillance; its economy lay, supposedly, in its elimination of the need for violence. The inmates, thinking they were under constant observation, would police themselves. The Panopticon thus embodied the bureaucratic principle of dispersed, hegemonic power. In the Inspection House, the regime of the spectacle (inspection, observation, sight) merged with the regime of power.

As Foucault observed, the crucial point of the Panopticon is that anyone, in theory, can operate the Inspection House. The inspectors are infinitely interchangeable and any member of the public may visit the Inspection House to inspect how affairs are run. As Foucault notes: "This Panopticon, subtly arranged so that an observer may observe, at a glance, so many individuals, also enables everyone to come and observe any of the observers. The seeing machine . . . has become a transparent building in which the exercise of power may be supervised by society as a whole."

The innovation of the Crystal Palace, that exemplary glass inspection house, lay in its ability to merge the pleasure principle with the discipline of the spectacle. In the glass seeing-machine, thousands of civic inspectors could observe the observers: a voyeuristic discipline perfectly embodied in the popular feature of the panorama. Seated about the circular observation-tower of the panorama, spectators consumed the moving views that swept before them, indulging the illusion of traveling at speed through the world. The panorama inverted the panoptical principle and put it at the disposal of consumer pleasure, converting panoptical surveillance into commodity spectacle – the consumption of the globe by voyeurs. Yet, all the while caught in the enchantment of surveillance, these imperial

monarchs-of-all-they-survey offered their immobile backs to the observation of others.[38]

The Crystal Palace converted panoptical surveillance into consumer pleasure. As Susan Buck-Morss points out: "The message of the world exhibitions was the promise of social progress for the masses without revolution."[39] The Great Exhibition was a museum without history, a market without labor, a factory without workers. In the industrial booths, technology was staged as if giving birth effortlessly, ready-made, to the vast emporium of the world's merchandise.

At the same time, in the social laboratory of the Exhibition, a crucial political principle took shape: the idea of democracy as the voyeuristic consumption of commodity spectacle. Most crucially, an emerging national narrative began to include the working class into the Progress narrative as consumers of national spectacle. Implicit in the Exhibition was the new experience of *imperial* progress consumed as a *national* spectacle [Fig. 7]. At the Exhibition, white British workers could feel included in the imperial nation, the voyeuristic spectacle of racial "superiority" compensating them for their class subordination [Fig. 8].[40]

During what Luke Gibbons calls "the twilight of colonialism," a child's toy was manufactured for the "Big Houses" of the Irish ascendancy, which promised to give the "British Empire at a Glance."[41] Gibbons describes the toy thus: "It took the form of a map of the world, mounted on a wheel complete with small apertures which revealed all that was worth knowing about the most distant corners of the Empire. One of the apertures gave a breakdown of each colony in terms of its 'white' and 'native' population, as if both categories were mutually exclusive."[42] This toy-world perfectly embodies the scopic megalomania that animates the panoptical desire to consume the world whole. It also embodies its failure, for, as Gibbons adds: "When it came to Ireland, the wheel ground to a halt for here was a colony whose subject population was both 'native' and 'white' at the same time. This was a corner of the Empire, apparently, that could not be taken in at a glance."[43] The toy-world marks a transition – from the imperial science of the surface to commodity racism and imperial kitsch. Imperial kitsch and commodity spectacle made possible what the colonial map could only promise: the mass marketing of imperialism as a global system of signs.

Figure 7 Commodity fetishism goes global

Figure 8 Sugar-coating imperialism

[. . .]

NOTES

1 The degenerate classes were not perceived as synonymous with the 'respectable' working classes, who had availed themselves of the benefits of sober and diligent toil during the comparative boom of the late 1860s and early 1870s. As Henry Mayhew neatly put it: "I shall consider the whole of the metropolitan poor under three separate phases, according as they will work, they can't work, and they won't work." Henry Mayhew, "Labour and the Poor," *Chronicle*, October 19, 1849.

2 See Sander Gilman, ed., *Degeneration: The Dark Side of Progress* (New York: Columbia University Press, 1985), p. xiv. See also Gilman, *Difference and Pathology: Stereotypes of Sexuality, Race and Madness* (Ithaca: Cornell University Press, 1985); Nancy Stepan, "Race and Gender: The Role of Analogy in Science," *Isis 77* (June 1986): 261–277; and Richard D. Walter, "What Became of the Degenerate? A Brief History of a Concept," *Journal of the History of Medicine and the Allied Sciences* 11 (1956): 42–49.

3 Gareth Stedman Jones, *Outcast London* (New York: Pantheon, 1971), p. 11. See also Henry Mayhew, *London Labour and the London Poor*, III, John Rosenberg, ed. (New York: Dover, 1968), pp. 376–377; Gertrude Himmelfarb, *The Idea of Poverty* (New York: Vintage Books, 1985), p. 361.

4 Mayhew, *London Labour*, p. 167.

5 Thomas Plint, *Crime in England: Its Relation, Character and Extent, as Developed from 1801 to 1848* (New York: Arno, [1851] 1974), pp. 148–149.

6 See Anna Davin, "Imperialism and Motherhood," *History Workshop* 5 (Spring 1978): 9–65.

7 Ann Laura Stoler, "Carnal Knowledge and Imperial Power: Gender, Race, and Morality in Colonial Asia," in Micaela di Leonardo, ed., *Gender and the Cross-roads of Knowledge: Feminist Anthropology in the Postmodern Era* (Berkeley: University of California Press, 1991), p. 74.

8 Stoler, "Carnal Knowledge and Imperial Power: Gender, Race, and Morality in Colonial Asia," p. 78.

9 It is no accident that Darwin entitled his work *On the Origin of Species* rather than, say, the origin of man.

10 Jones, *Outcast London*, p. 313.

11 Quoted in Stephen Jay Gould, *The Mismeasure of Man* (New York: Norton, 1981), p. 21.

12 See Samuel G. Morton, "Value and the Word Species in Zoology," *American Journal of Science and Arts* 11 (May 1851): 275; and Gould, ibid., p. 73.

13 Prompted by fears of miscegenation and the free movement of black people after the abolition of slavery in America and the colonies, and arguing from the evidence of the Egyptian mummies, polygenesists held that different races had always been fixed and separate creations properly at home in different zones and climates of the world. Freed slaves, for example, were seen as "doomed to degenerate as they moved northward into white, temperate territory, and as they moved socially and politically into freedom." Stepan, "Race and Gender," p. 100.

14 Gould, *The Mismeasure of Man*, p. 73.

15 Gould, *The Mismeasure of Man*, p. 74.

16 In the 1820s, Samual G. Morton had began to gather together his vast collection of human skulls from around the world, blending an untiring measurement of their cranial capacities with his own special flair for interpretive invention and ingenuity, elaborating on this basis his famous treatise on the character of race, *Crania Americana* (Philadelphia: John Pennington, 1839).

17 See the selections from Haeckel in Theodore D. McCown and Kenneth A. R. Kennedy, eds., *Climbing Man's Family Tree: A Collection of Writings on Human Phylogeny, 1699 to 1971* (Engelwood Cliffs, N.J.: Prentice, 1972), pp. 133–148. For a detailed discussion, see Gould, *Ontogeny and Phylogeny* (Cambridge, Mass,: Harvard University Press, 1977), esp. pp. 126–135.

18 Gould, *The Mismeasure of Man*, p. 114. Gould points out that recapitulation became the enabling idea for the late-nineteenth-century obsession with retracing the evolution of ancestral lineages and played a vital role not only in the professions of embryology, comparative morphology and paleontology but also in the articulation of psychoanalytic theory.

19 *The Mismeasure of Man*, p. 326.

20 *The Mismeasure of Man*, p. 320.

21 G. A. Henty, *By Sheer Pluck: A Tale of the Asbanti War* (London: Blackie and Son, 1884), p. 118.

22 John Beddoe, *The Races of Britain: A Contribution to the Anthropology of Western Europe* (Bristol: J. W. Arrowsmith, 1885). On the racial stereotyping of the Irish, see L. Perry Curtis, Jr., *Apes and Angels: The Irishman in Victorian Caricature* (Newton Abbot: David and Charles, 1971); Richard Ned Lebow, *White Britain and Black Ireland: The Influence of Stereotypes on Colonial Policy* (Philadelphia: Institute for the Study of Human Issues, 1976); and Thomas William Hodgson Crosland, *The Wild Irishman* (London: T. Werner Laurie, 1905).

23 Molinari's phrase "une variété de négres blancs" appeared in translation in a leader in *The Times* of London on September 18, 1880. See Curtis, *Apes and Angels*, p. 1.

24 Philip Luckombe, *A Tour Through Ireland: Wherein the Present State of That Kingdom is Considered* (London: T. Lowndes, 1783), p. 19.

25 David Lloyd, *Nationalism and Minor Literature* (Berkeley: University of California Press, 1988), p. 3.

26 Claire Wills, "Language Politics, Narrative, Political Violence," in "Neocolonialism," ed. Robert Young. *The Oxford Literary Review* 13 (1991): 21.

27 Wills, "Language Politics," p. 56.

28 See also Richard Kearney, ed., *The Irish Mind* (Dublin: Wolfhound Press, 1985); L. P. Curtis, Jr., *Anglo-Saxons and Celts: A Study of Anti-Irish Prejudice in Victorian England* (Bridgeport: Conference on British Studies of University of Bridgeport, 1968); Seamus Deane, "Civilians and Barbarians" *Ireland's Field Day* (London: Hutchinson, 1985), pp. 33–42.

29 Seth Luther, for example, was confident that "the wives and daughters of the rich manufacturers would no more associate with a factory girl than they would with a negro slave," *Address to the Working Men of New*

England, pamphlet reprinted in Philip Taft and Leo Sten, eds., *Religion, Reform and Revolution. Labor Panaceas in the Nineteenth Century* (New York: Arno, 1970), p. 1.

30 William Booth, *In Darkest England and the Way Out* (London: International Headquarters of the Salvation Army, 1890); William Barry, *The New Antigone* (London: Barry, 1887).

31 Gustave le Bon, *La Psychologie des Foules* (1879), pp. 60–61. Quoted in Gould (1981), p. 105; English trans. from Robert K. Merton, *The Crowd: A Study of the Popular Mind* (New York, Viking, 1960).

32 See Stepan, "Race and Gender."

33 Carl Vogt, *Lectures on Man: His Place in Creation and in the History of the Earth* ed. James Hunt (London: Longman, Green and Roberts, 1864), p. 81. For the analogy of the "pathological" sexuality of "lower races" and women, see Eugene S. Talbot, *Degeneracy: Its Causes, Signs and Results* (London: W. Scott, 1898), p. 319–323. See also Havelock Ellis, *Man and Woman: A Study of Secondary Sexual Characteristics* (London: Black, 1926), pp. 106–107. For the working of the analogy in scientific discourse, see Stepan, "Race and Gender," pp. 261–277. For the relation between female sexuality and degeneration, see Jill Conway, "Stereotypes of Femininity in a Theory of Sexual Evolution," *Victorian Studies* 14 (1970): 47–62; and Fraser Harrison, *The Dark Angel: Aspects of Victorian Sexuality* (London: Sheldon, 1977).

34 Philip Thickness thought that black people in Britain, "their legs without any inner calf, and their broad flat foot, and long toes . . . have much the resemblance of the Orang Outang, or Jocko . . . and other quadrupeds of their own climate." *A Year's Journey through France and Part of Spain*, second edition (1778): 102–105. Quoted in Fryer, p. 162.

35 Charles Kingsley, author of *Westward Ho* and *The Water Babies* wrote after a trip to Sligo in 1860: "I am haunted by the human chimpanzees I saw along that hundred miles of horrible country. . . . To see white chimpanzees is dreadful; if they were black, one would not feel it so much." Letter to his wife, July 4, 1860, in *Charles Kingsley: His Letters and Memories of His Life*, ed. Francis E. Kingsley (London: Henry S. King and Co, 1877), p. 107.

36 I explore the relation between prostitution, race and the law in "Screwing the System: Sexwork, Race and the Law," *Boundary 11* 19, 2 (Summer 1992): 70–95.

37 See Gilman's analysis of the racializing of prostitutes in *Difference and Pathology*.

38 Mary Louise Pratt uses the term "monarch-of-all-I-survey" to describe the imperial stance of converting panoramic spectacle, especially at the moment of "discovery," into a position of authority and power.

39 Susan Buck-Morss, *The Dialectics of Seeing: Walter Benjamin and the Arcades Project* (Cambridge, Mass.: The MIT Press, 1990), p. 128.

40 If the World's Fairs were largely festivities for the paying middle class, vigorous efforts were made to encourage workers to the mass consumption of commodities as spectacle. Assembled under one roof, the workers of the world could admire and gawk at the marvels they had produced but could not themselves own. In 1867, 400,000 French workers were given free tickets to the Paris Fair; foreign workers were housed at government expense. Susan Buck-Morss, *The Dialectics of Seeing*, p. 86.

41 I am grateful to Luke Gibbons, who writes about this toy in "Race Against Time: Racial Discourse and Irish History," in "Neocolonialism," ed. Robert Young, *Oxford Literary Review* 13 (1991): 95.

42 Gibbons, "Race Against Time," p. 95.

43 Gibbons, "Race Against Time," p. 95.

PART TWO

Markets, empire, nature, and difference

INTRODUCTION TO PART TWO

One of the most important arguments made by Cowen and Shenton in their book, *Doctrines of Development*, is that 'modern development doctrine is based upon a reversal in the order of the positive and negative dimensions of development as a process. It was the apprehension of the destructive dimension of a process of development which . . . was the starting point for the modern intention to develop' (1996: ix).

This is a challenging observation and it bears restating. It also needs to be qualified in two important respects. Cowen and Shenton are suggesting that processes immanent to the production of the modern condition – those associated with capitalism, most obviously – were considered so threatening in the nineteenth century that they had to be tempered by immanent development's supposedly benign twin, intentional development. Thus it was that nineteenth-century doctrines of development first took shape around a notion of trusteeship. This was true, for example, of the Saint-Simonians and positivists in France. It was also true of Cardinal Newman's 1845 *Essay on the Development of Christian Doctrine* in the UK. These and other nineteenth-century thinkers worried about the social consequences of the industrial growth they saw all around them. They worried about the production of unproductive populations. They worried about the growth of urban crime and destitution. They worried about a loss of moral character, and even about civilisation itself, in a world in which men and women seemed not to be in charge of their own lives. They worried, with Marx and Engels, and with the great European novelists and painters, that 'all that was solid would melt into air, all that was holy would be profaned' (to paraphrase *The Communist Manifesto* of 1848).

How the beast was to be tamed, however, prompted huge debate, then as now, and we want to give a feeling for this here. In particular, we want to convey our sense that discussions of development, for all that they have changed markedly over the past two centuries, and for all that they have moved well beyond Europe and India (our main subject regions here), have long been concerned with the four linked issues we set out in Part 1. That is, with the growth and regulation of market economies, with empire and geopolitics, with transformations of nature, and with questions of cultural difference. This is also why we feel obliged to present two qualifications to Cowen and Shenton's thesis. For while it is largely true that US-style modernisation theory in the 1950s and 1960s did accentuate the positive dimensions of development to an almost absurd degree, there were dissident voices even then that worried about the costs of rapid economic growth, both in regard to social solidarities (e.g. Samuel Huntington) and in respect of its ecological consequences (see Parts 4 and 5). Likewise, while it is true that many nineteenth-century thinkers were alarmed by the negative dimensions of immanent development, this was less true of the great eighteenth-century thinker, Adam Smith, or many of his disciples. Smith essayed a more positive view of the possibilities that could be unleashed by extended divisions of labour and free trade, or by the invisible hand of the market.

In this Part, then, we want to establish some beginning points for important discontinuities *and* continuities in the ways that development issues, as we would now call them, have been discussed over a period of two hundred years. We reprint some classic voices on markets, empire, nature and culture from the time of the Industrial Revolution in the United Kingdom until the beginnings of the First World War. We start, however, with a short essay by Heinz Arndt that will help readers place these contributions in broader context.

Arndt provides a semantic history of economic development that reaches well beyond Western conceptions of development, or what Smith called material progress. The work of the Chinese nationalist leader, Sun-Yat sen, is mentioned here, as Arndt moves into the 1920s (a period picked up in Part 3). But Arndt begins with Adam Smith and so do we. Writing in the late eighteenth century, Smith foresaw a world of capitalist production and exchange that could work for the benefit of all. Smith believed that productivity gains accrue from specialisation – for example, in the workplace or through extended divisions of labour – and that these gains are best distributed through systems of free trade. But Smith was not naive enough to expect that what *could* work for the benefit of all *would* always work that way. To the contrary, Smith's work is replete with references to the undermining of competitive capitalism by systems of governance that reward private greed at the expense of ordinary people, and where agents are unconstrained by principals. The reading reproduced here, from Book IV of *An Inquiry into the Nature and Causes of the Wealth of Nations*, is remarkably incisive, and contemporary, for this very reason.

Smith begins by noting the extraordinary economic potential contained in 'the two most important events recorded in the history of mankind' – namely, the 'discovery of America and that of a passage to the East Indies by the Cape of Good Hope'. These early moments in the history of globalisation linked entire continents. New international divisions of labour were created which brought with them the potential to enrich people across hitherto distant and largely separate lands. Unhappily, however, Smith continues, the main vehicles for the production of these new geographies were giant trading companies, like the Dutch and English East India Companies. These companies established systems of preferential trading that guaranteed them monopoly rents. These rents in turn were underpinned by violence and the force of arms. As a consequence, an empowering network of free and competitive exchange was debased to serve a limited private interest. Worse, Smith concludes, the inability of company bosses to police their self-serving 'servants' led to locally oppressive forms of rule which cheated native peoples of the benefits of progress. The success of these companies in monopolising the East Indies trade also led to huge price mark-ups for Europeans.

For Smith, the most effective antidotes to these depredations were trade liberalisation and better regimes of governance. As he makes clear at the end of the excerpt reproduced here, Smith did not mean to 'throw any odious imputation upon the general character of the servants of the East India Company, and much less upon that of any particular persons'. Smith blamed the 'system of government in which they [were] placed'. Corruption was a function of circumstance. The overriding need was to set capitalism free from the restrictions imposed upon it by monopolistic trading houses and their mercantilist backers among European governments.

Writing more than sixty years later, the German political economist, Friedrich List, took a very different view. List doubted the sincerity of English arguments in favour of free trade, believing that England had raised 'her manufacturing power ... by means of protective duties and restrictions on navigation' (1841: chapter 33). Instead, he argued the case for limited protectionism in order that latecomer countries could build up their own systems of industrial capitalism. In doing so he set in train a notion that remains at the heart of development studies (see also the Introduction to Part 9).

Karl Marx went much further, as is evident from the reading reproduced here. Far more than Comte or Newman, Marx believed that processes of capitalist development were inherently violent and dislocating. Capitalist development was a bargain made with the Devil. It was a Faustian pact that brought all manner of tortures to the men and women at the sharp end of processes of primitive accumulation. But while capitalism tore apart taken-for-granted institutions like the family, it also promised new freedoms and pleasures. Marx ends the passage here with a brief quotation from Goethe to this effect.

For Marx, capitalism was not a system of economic production and exchange that could be tamed or made to serve broader social goals. In this specific respect, Marx was less romantic than many of his contemporaries. But he was also a man of his time. One of Marx's great heroes was Charles Darwin, and there is more than a hint of evolutionary theory in Marx's commentary on 'The British Rule in India'. Marx provides chapter and verse on the miseries imposed upon Hindostan by the British. He writes of the destruction of public works and of India's handicraft industries. But Marx ends his article with a discussion of the evils of 'pre-British rule in India' – the caste system and the slavery it induced, petty village tyrannies and Oriental despotism, and Barbarian egotism and superstition. For Marx, British rule, though 'actuated only by the vilest interests', was 'the unconscious tool of history' in bringing about 'a fundamental revolution in the social state of Asia'.

In his wider corpus of work Marx argued that capitalism would give way in due course to socialism or communism. Marx was a rationalist, much like Comte and Saint-Simon in this respect. He looked forward to a day when secular and scientific world views would replace the superstitious outlooks that he associated with organised religion. He also predicted that the contradictions of capitalism – the inequalities that it produced, its necessary tendencies to uneven development and boom and bust – would spur organised rebellions. The rule of capital would come to be replaced by regimes of accumulation that were focused on people's needs and abilities. Lenin later qualified and expanded this analysis. He suggested that competitive capitalism would give way to an era of monopoly capitalism. This would generate pressures for imperialism and inter-imperialist wars, and it would be these pressures and dangers that finally would push working people towards the less barbaric systems of socialism and communism – to the systems of collective rule that Marx and Lenin looked forward to in their own versions of the end of history.

Marx's faith in the capacity of men and women to remake themselves within different modes of pro-duction – selfish under capitalism, sharing under socialism – was more radical than Smith's account of the production of social conventions under different systems of governance. And it was certainly at odds with another body of work that was being codified in the second half of the nineteenth century, this time under the heading of Social Darwinism. As always, different contributions to this body of work varied in their levels of sophistication. In Western Europe, the organic analogy that Herbert Spencer refers to here was bluntly on display in the geographer, Friedrich Ratzel's, attempts to explain the sources of German strength and future prosperity. The analogy that Ratzel proposed was between the strength of the individual, predominantly male, body and the strength of the body of the nation. Countries grew, and could compete successfully with other countries for space and resources, to the extent that their populations grew in size and in terms of 'quality' (a link backwards through eugenics), *and* to the extent that they could command strategic choke-points in a rapidly 'closing' (we would now say globalising) world system.

On that same global stage, and its crudest forms, the brute biologism of Social Darwinism also gave rise to the proposition that Black African populations were not suited for development, or at any rate not for auto-development. Development would have to be brought from outside to population groups marked down by inferior cranial development (the pseudo-science of phrenology and cephalic indices), and/or by the disadvantages of geography and 'Tropicality'. Spencer's work is more sophisticated than this. But here too we see a fascination for societal typologies and a rejection of the possibility of Mankind. What is proposed instead are 'essential natures', or the demarcation of social groups on the basis of internal nature (race, genetics) or external nature (the environment). Biology and physical geography direct socio-logy. The imagined worlds conjured into being by static regimes of societal classification describe real geographies that have been torn apart. What bridges these worlds is a form of trusteeship in which the civilised Occident is required to take on responsibility for slowly improving the social welfare of the 'native'. The White Man's burden is emphatically not the form of trusteeship proposed by Smith or Comte: a form of trusteeship in which untrammelled capitalist development would be saved from itself by well-meaning governments and social reformers.

In contemporary parlance, what is proposed by Social Darwinism is a sharply inscribed regime of Othering. White is to Civilised as Black is to Barbaric. White shows the potential for Development where Black, when left alone, is condemned to Underdevelopment. As Edward Said so often reminded us, however, the West's attempts to create an Other for itself were never homogeneous. Some in the West objected to that region's apparent fall from a state of grace. A minor key within Western thought was sternly opposed to the development of Development and all or most of what went with it: environmental despoliation, loss of community, rampant egoism and materialism, and so on. Ruskin, Tolstoy and Gauguin are three well-known cases in point from the late nineteenth century. No one, however, spoke back to the West's accounts of its own civilisational supremacy as directly as did Mohandas Karamchand Gandhi in India. We close this part with a reading from Gandhi's dialogue with himself, *Hind Swaraj*, a monograph written at furious pace in 1910 on a ship from England to South Africa. As Gandhi so memorably puts it, 'Civilisation is not an incurable disease, but it should never be forgotten that the English people are at present afflicted by it.'

The same claim is now being made in some quarters of anti-developmentalism, a point that should come as no surprise. According to the activist, Gustava Esteva (1992), 'If you live in Rio or Mexico City, you need to be very rich or very stupid not to notice that development stinks.' As we have tried to suggest here, many of the key debates in contemporary development studies – on globalisation, industrial policy, good governance, resource conflicts, the meanings and purposes of development itself – were already being debated more than one hundred years ago. In important respects, indeed, the parameters of these debates continue to be underpinned by the four recurrent themes that structure this and other sections of the reader: markets (their constitution and regulation), empire (geopolitics and imperialism), nature (political ecology and the environment) and difference (culture and the challenges of postcolonial theory).

REFERENCES AND FURTHER READING

Arndt, H. (1987) *Economic Development: The History of an Idea*, Chicago, IL: University of Chicago Press.

Berman, M. (1983) *All That Is Solid Melts Into Air: The Experience of Modernity*, London: Verso.

Chakrabarty, D. (2000) *Provincialising Europe: Postcolonial Thought and Historical Difference*, Princeton, NJ: Princeton University Press.

Comte, A. (1875) *System of Positive Polity, Volume 1*, London: Longmans, Green.

Cowen, M. and Shenton, R. (1996) *Doctrines of Development*, London: Routledge.

Dutt, R. (1970 [1904]) *The Economic History of India*, New York: Burt Franklin.

Esteva, G. (1992) 'The right to stop development', NGONET UNCED Feature, 13 June.

Friedman, B. (2005) *The Moral Consequences of Economic Growth*, New York: Vintage Books, especially Chapter 2, 'Perspectives from the Enlightenment and Its Roots'.

Haller, J. (1975) *Outcasts from Evolution: Scientific Attitudes of Racial Inferiority, 1859–1900*, New York: McGraw-Hill.

Hirschman, A. (1982) 'Rival interpretations of market society: civilizing, destructive, or feeble?', *Journal of Economic Literature* 20: 1463–84.

Lenin, V. I. (1970) *Imperialism: The Highest Stage of Capitalism*, Peking: Foreign Languages Press.

List, F. (1856 [1841]) *The National System of Political Economy*, Philadelphia, PA: J.B. Lippincott and Co.

Mackinder, H. (1904) 'The geographical pivot of history', *Geographical Journal* 23: 421–42.

Marx, K. and Engels, K. (1998 [1848]) *The Communist Manifesto*, Harmondsworth: Penguin.

Newman, J. (1845) *An Essay on the Development of Christian Doctrine*, London: James Toovey.

Ratzel, F. (1897) *Politische Geographie*, Munich.

Said, E. (1979) *Orientalism*, New York: Pantheon.

'Economic Development: A Semantic History'

from *Economic Development and Cultural Change* (1981)

Heinz Arndt

Editors' Introduction

Heinz Arndt was born in Breslau (then in Germany) in 1915 and moved to Britain when Hitler came to power. Arndt studied Modern Greats at Oxford University, specialising in politics and economics, and then pursued doctoral work in the history of political thought. Thereafter, Arndt moved through the London School of Economics, Chatham House and Manchester University to a position at Sydney University in 1946. In 1950 he was appointed to a Chair in Economics at Canberra University College – later, the Australian National University (ANU) – where he remained until his death in 2002.

Heinz Arndt read very widely and this shows in the article reprinted here. His first area of expertise at ANU was in the field of monetary economics and the banking system. In 1958–59, however, he spent three months in India, where he worked with P. C. Mahalanobis at the Indian Statistical Institute. It seems probable that his time in India took Arndt back to an earlier body of work at Chatham House, where he had interacted with Paul Rosenstein-Rodan, one of the founding fathers of development economics. Arndt now also began to work on the Indonesian economy, which remained the focal point for much of his later research.

Having lived through the birth and expansion of development studies, Arndt was well equipped to write about the subject, and to put the contributions of newer writers in proper historical context. This is what we find Arndt doing here. Arndt's short article remains of interest for the way that it reaches beyond the conventional heartlands of early development studies. He also reminds us that the idea of development is hardly new, however much it has changed and been contested since the time of Adam Smith. The job at hand, Arndt suggests, is to provide a semantic history of Economic Development. This is what Raymond Williams would call a keyword for our time. The meanings of Development need to be carefully excavated and understood, a task that Arndt begins here and which is central to the aims of *The Development Reader*.

Key references

Heinz Arndt (1981) 'Economic development: a semantic history', *Economic Development and Cultural Change* (April): 457–66.

— (1944) *The Economic Lessons of the Nineteen-Thirties*, London: Oxford University Press.

— (1957) *The Australian Trading Banks*, Melbourne: Cheshire.

— (1978) *The Rise and Fall of Economic Growth: A Study in Contemporary Thought*, Melbourne: Longman Cheshire.
— (1993) *Fifty Years of Development Studies*, Canberra: ANU Press.
R. Williams (1985) *Keywords: A Vocabulary of Culture and Society*, Oxford: Oxford University Press.

When I began work on my book on *Economic Development: The History of an Idea*, I asked myself how and when the term 'economic development' entered the English language. The resulting detective work uncovered a story which surprised me and others.

So commonplace has the concept of 'economic development' become to this generation that it comes as a surprise to find the Oxford English Dictionary still unaware of 'development' as a technical term in economics, as contrasted with its use in mathematics, biology, music, or photography. Nor, incidentally, is there an entry on 'economic development' in the Encyclopedia of the Social Sciences. The story of how the term 'economic development' entered the English language and came, for a time at least, to be identified with growth in per capita income is both curious and illuminating.

Adam Smith spoke, not of economic development, but of 'the progress of England towards opulence and improvement' (Smith 1776 Vol. 1:367). 'Material progress' was the expression almost invariably used by mainstream economists, from Adam Smith until World War II, when they referred to what we would now call the economic development of the West during those two centuries.[1] When Colin Clark in 1940 published his monumental comparative study of economic development, he still called it *The Conditions of Economic Progress* (the title Marshall had had in mind for the fourth volume of his *Principles*, which he had planned but never wrote) (Clark 1925:vii).

Economists and economic historians wrote about the rise of capitalism, the industrial revolution, the evolution of trade, or 'The Growth of Free Industry and Enterprise' (Marshall 1920 Vol. 1 Appendix A:723). But this historical process appears rarely if ever to have been described as economic development. As a policy objective, economic development became increasingly prominent during the nineteenth century, first in Germany and Russia and other countries in Europe, later in Japan and China and elsewhere, in what we now call the 'Third World'. But it was generally referred to as 'modernization' or 'westernization' or, not infrequently, 'industrialization'.

When Alfred Marshall used the word 'development', it was in a literal sense, denoting merely emergence over time, as in 'the development of speculation in every form', or 'the development of social institutions' (Marshall 1920 ibid.: 752). This remained generally true, at least in the British and American literature, until the 1930s.

However, there were a few exceptions. One is J.A. Schumpeter's *Theory of Economic Development*, but this, though published in German in 1911 as *Theorie der wirtschaftlichen Entwicklung*, was not translated into English until 1934 (Schumpeter 1934). A second exception is the use of the term 'economic development' by economic historians in the 1920s. Lilian Knowles, reader in economic history at the London School of Economics (LSE), in 1924 published her book, *The Economic Development of the British Overseas Empire*, and mentioned in the preface that a unit with the same title had recently been made a compulsory subject for the Bachelor of Commerce degree of London University (Knowles 1924:ix). A few years later, Vera Anstey, also at the LSE, followed Knowles with *The Economic Development of India* (Anstey 1929). Another LSE economic historian, R.H. Tawney, in his book on China written in 1931 spoke of the 'long process of development' that had occurred in the West and of the 'forces which have caused the economic development of China' and referred to the analogy between China's twentieth century economic condition and that of Europe in the Middle Ages as implying 'a comparison of stages of economic development' (Tawney 1932:18).

These intriguing exceptions provide the clue to the two quite distinct channels through which the term 'economic development' entered English usage. Tawney, like Schumpeter, knew his Marx. Lilian Knowles and Vera Anstey were historians of Empire.

MARXIST ORIGINS

In one sense, the birthplace of 'economic development' in English would seem to be the first English translation of Marx's Capital, and the date 1887.

The preface to the first German edition contains the famous statement that 'it is not a question of the higher or lower degree of development of the social antagonisms that result from the natural laws of capitalist production, it is a question of these laws themselves, of these tendencies working with iron necessity towards inevitable results. The country that is more developed industrially only shows, to the less developed, the image of its own future' (Marx and Engels 1969 (2):87). Here, as in the subsequent passage when he referred to 'the historical circumstances that prevented, in Germany, the development of the capitalist mode of production, and consequently the development, in that country, of modern bourgeois society' (Marx and Engels 1969(2):92), Marx used the word 'development' in the sense in which it forms the key concept of his economic interpretation of history.

As Schumpeter put it, in Marx's schema of thought, 'Development was ... the central theme. And he concentrated his analytical powers on the task of showing how the economic process, changing itself by virtue of its own inherent logic, incessantly changes the social framework – the whole of society in fact' (Schumpeter 1954:573).

As has often been pointed out, Marx derived his concept of development, including the notion of phases or stages of development which unfold in a dialectic process according to an inexorable law, from Hegel. Hegel, in turn, stood in a long tradition – from Aristotle, with his concept of development as the realization of 'potential' matter in 'actual' form, to Fichte, who was the first to argue that 'history proceeds dialectically' (cf. Passmore 1970 Ch.11). Some of Hegel's formulations strike notes strangely familiar to students of recent development literature. 'The principle of Development involves ... the existence of a latent germ of being – a capacity or potentiality striving to realize itself ... The history of the world ... is the process of development ... [This] development, therefore, does not present the harmless tranquility of mere growth' (Hegel 1956:54–5).[2] But it was Marx who gave development a specifically economic connotation.

Marx's notion of stages of economic development is a constant theme in later Marxist literature, but it is difficult to find references in this literature to more or less 'developed' countries or nations. When the Second Congress of the Communist International of 1920 reached the important conclusion

that, *pace* Marx, the capitalist stage of economic development is not one through which all countries must pass, the distinction drawn is between 'oppressing' and 'oppressed' or between 'advanced' and 'backward' nations: 'In all colonies and backward countries ... with the aid of the proletariat of the most advanced countries, the backward countries may pass to the Soviet system and, after passing through a definite stage of development, to communism, without passing through the capitalist stage of development' (quoted in d'Encausse and Schram 1969:159).[3]

COLONIAL DEVELOPMENT

'Economic development' as used by the British historians of Empire of the 1920s is a concept quite different from the Marxist one, with a considerably longer history. What Lilian Knowles set out to write about in her history of the economic development of the British Overseas Empire was 'the remarkable economic achievements within the Empire during the past centuries ... the hacking down of the forest or the sheep rearing or the gold mining which made Canada, Australia and South Africa into world factors ... or the struggle with the overwhelming forces of nature which took shape in the unromantic guise of "Public Works" in India' (Knowles 1924:vii). A few years earlier, Lord Milner had warned, in an official memorandum, that 'it is more than ever necessary that the economic resources of the Empire should be developed to the utmost' (quoted in Lugard 1921:489) and in 1929 the British Parliament passed a Colonial Development Act.

Whereas for Marx and Schumpeter, economic development was a historical process that happened without being consciously willed by anyone, economic development for Milner and others concerned with colonial policy was an activity, especially, though not exclusively, of government. In Marx's sense, it is a society or an economic system that 'develops'; in Milner's sense, it is natural resources that are 'developed'. Economic development in Marx's sense derives from the intransitive verb, in Milner's sense from the transitive verb.[4]

The origins of the transitive concept of economic development which, by the 1920s, was in fairly common use in the specialist British literature of colonial history and policy are to be found, not in nineteenth century British (or American) writings about economics and economic history but in Australian (and to a lesser extent, Canadian) writings, and they go a long way back. The directors of the Van Dieman's Land Company, which held large tracts of land on the Australian mainland, expressed in their thirteenth annual report of 1838 the dominant local opinion about the needs of the young colonies: 'Population is the only thing wanted to develop the Company's locations' (quoted in Roberts 1924:68). A colonial politician made the same point a few years later, in 1845, in the Legislative Council of New South Wales: 'The resources of a new country can only be developed by constant additions to its population' (quoted in Goodwin 1966:423). The case for construction of railways was put in similar terms in 1854 – 'The best and most economical means of developing the vast resources of the interior' (Goodwin 1966:269) – and in 1861 one Charles Mayes in Melbourne published a pamphlet, entitled *Essay on the Manufactures More Immediately Required for the Economic Development of the Resources of the Colony* (Goodwin 1966:318) which the Oxford English Dictionary in its next edition might well list as the earliest (so far) known use of the term 'economic development'.

In Canada, too, as early as 1846, the *Canadian Economist* argued that 'Canada is now thrown upon her own resources, and if she wishes to prosper, these resources must be developed' (quoted in Innis and Lower 1933:303). But whereas in Australia the transitive use of 'development' was continuous and common from the middle of the nineteenth century onward – side by side with synonyms such as 'opening up our natural resources' or 'the steady occupancy and proper advancement of the Colony' (La Trobe, lieutenant-governor of Victoria 1851, quoted in Roberts 1924:287)[5] – the Canadian example is the only one so far discovered before the 1880s, and in the United States it does not seem to have been used at all in the nineteenth century.[6]

That 'economic development' in the transitive sense entered the language and became common in Australia, while being used much less in Canada and not at all in the United States, is no historical accident. In the United States, and for much of the time also in Canada, economic development happened, as immigrants from Europe streamed in; settlers went west to take up fertile land; communities established towns and cities; private companies constructed railways; and mining, logging, manufac-turing, banking, and other enterprises grew, within (and sometimes without) legal rules made by government. In Australia's hostile environment, where settlers from the earliest convict days had to contend with drought, flood, pests, distance, and more drought, economic development did not happen. It was always seen to need government initiative, action to 'develop' the continent's resources by bringing people and capital from overseas, by constructing railways, and by making settlement possible through irrigation and other 'developmental' public works (Eggleston 1931). So well established did this notion become in Australia that by the 1920s it was referred to as 'the doctrine of development before settlement' (Eggleston 1931:197).

DEVELOPMENT AND WELFARE

Development of natural resources was not always viewed as a task of government. The British authority on colonial policy, J.S. Furnivall, referred to 'the development of the material resources of Burma through trade and economic enterprise' (Furnivall 1931:i), and it was probably also in this sense that the term was used in an International Labour Office study of Brazil which identified 'continuous occupation and development of the country, in space as in time', as 'the primary condition for the economic exploitation [of its resources]' (Mauretta 1937:9). But whether the agent was government or private enterprise, 'development' in this transitive sense was for long kept quite distinct from the process of economic development, usually still referred to as the 'progress' of society or of the nation's wealth. Nor did it, in itself, connote a rise in living standards. It was development of resources, not of people. The three distinct concepts are nicely found together in the preface to the first official statistical yearbook of Australia in 1890, which described its purpose as being 'to afford information by which the progress of these Colonies may be gauged . . . So much has been accomplished in the development of the material resources of the new land, and the social well-being of its people' (Coghlin 1890).

How little economic development and welfare were synonymous until quite recently is most clearly demonstrated by the doctrine which, in British colonial theory, came to be called the 'dual mandate'. One of the features of imperialism in its late nineteenth century hey-day was the emergence of notions of 'trusteeship' for the welfare of the native peoples. Colonial government, it came to be thought, had two distinct functions, development and welfare. As a historian of French colonial policy has expressed it, the new colonial theorists demanded 'a policy by which the conqueror would be most able to develop the conquered region economically, but also one in which the conqueror realized his responsibility to the native's . . . mental and physical well-being' (Betts 1961:120). It was very much in this spirit that the British government in 1939 replaced the earlier Colonial Development Act by a Colonial Development and Welfare Act. W.K. Hancock commented on the latter in 1942: ' "Development and welfare" will probably be the cry of the generation which follows the present one . . . In the nineteenth century development occurred as a by-product of profit.' The new concept is quite different: 'It gives a positive economic and social content to the philosophy of colonial trusteeship by affirming the need for minimum standards of nutrition, health and education' (Hancock 1942:267).

An interesting exception, though one that may prove the rule, is the Chinese nationalist leader, Sun Yat-sen. In 1922, he published (in English) a remarkable book, *The International Development of China*, in which he proposed a massive program for the economic development of China with the aid of foreign capital. In breadth of imagination, it anticipates by a generation much of the post 1945 literature on economic development. 'China must not only regulate private capital, but she must also develop state capital and promote industry . . . build means of production, railroads, and waterways, on a large scale. Open new mines . . . hasten to foster manufacturing.' The reason for questioning whether Sun Yat-sen should be regarded as an exception is partly that his thinking was probably influenced by the October Revolution in Russia and thus indirectly by the Marxist tradition and partly that his use of 'economic development' is, after all, closer to that of Milner than of Marx: 'The natural resources of China are great and their proper development could create an unlimited market for the whole world'.[7]

Another exception, outside the mainstream of economic writing in the English language, was the use of 'economic development' in Australia (and probably the other Dominions) where the distinction between the transitive and intransitive meanings became blurred to the point of obliteration. When, in 1931, D.B. Copland edited a special issue on Australia for *The Annals of the American Academy of Political and Social Science*, he referred to it as 'a survey of recent trends in Australian economic development', and he wrote in the last chapter that 'by the end of 1929 Australia had reached the close of a period of rapid development and high prosperity' and that the growth of Australian manufacturing production during the years 1913–26 had represented 'a natural development in a country that had first pursued primary production', though 'somewhat forced' (Copland 1931:260). In such passages, neither he nor his readers, one suspects, were any longer conscious of the transitive, as contrasted with the intransitive, meaning; during the 1930s, 'economic development' was constantly used in Australia and increasingly elsewhere, in this ambivalent sense.

In 1939, Eugene Staley, starting where Sun Yat-sen had left off, proposed a 'world development plan' (Staley 1939:68). After the outbreak of war, the idea was taken up by many others in the spate of plans for the post-war world. Another former member of the International Labour Office secretariat, Wilfred Benson, was probably the first to speak in 1942 of 'underdeveloped areas' in the postwar sense, and in 1944 Rosenstein-Rodan expounded his ideas for 'The International Development of Economically Backward Areas' (Benson 1942:10; Rosenstein-Rodan 1944:157–65).

In the immediate postwar years, 'economic development' became virtually synonymous with growth in per capita income in the less developed countries. Arthur Lewis in 1944 declared the object of a program of rapid economic development to be to 'narrow the gap' in per capita income between rich and poor countries (Lewis 1944:156). One of the first United Nations documents on development plans stated in 1947 that 'the governments' ultimate aim in economic development is to raise the national welfare of the entire population' (United Nations 1947:xv). Lewis's great book on economic development appeared in 1955 under the title *The Theory of Economic Growth*, and in Rostow's hands Marx's stages of economic development became *The Stages of Economic Growth*. Gunnar Myrdal was merely reflecting established usage when, in 1957, he referred to 'the definition of economic development as a rise in the levels of living of the common people' (Lewis 1955; Rostow 1960; Myrdal 1957).

A few years earlier, Hla Myint had made an attempt to reverse the trend. Protesting against the practice of crystallizing 'low income per head' into the definition of backward countries, he proposed a return to the earlier distinction between 'under-developed' natural resources and 'backward' peoples. He thought it 'more illuminating . . . to give these terms different connotations by using the former to mean underdeveloped resources, and the latter to refer to the backward people in a given area', funda-mentally because he agreed with Furnivall that efficient development of natural resources does not necessarily reduce the backwardness of people (Myint 1954:132–63). But it was too late. Before long, standard textbooks defined economic development as 'a sustained, secular improvement in material well-being . . . reflected in an increasing flow of goods and services' (Okun and Richardson 1961:230), or even announced right at the start that 'the terms "economic development" and "economic growth" will be used to refer to a sustained increase in per capita income' (Baldwin 1966:1).

What many development economists have tried to do in the last twenty years is to get away from this identification of 'economic development' and 'eco-nomic growth'. One form this endeavour has taken is to breathe into 'development' some of the Hegelian connotations that had got lost on the way.

NOTES

This paper was originally published as 'Economic develop-ment: a semantic history', *Economic Development and Cultural Change*, 1981, April:457–66.

1   For quotations from J.S. Mill, A. Marshall, K. Wicksell, L. Robbins, A.G.B. Fisher, and others, see Arndt 1978:Ch.2.
2   In the last sentence of the quotation, Hegel's *Entwick-lung* was unaccountably translated in the English version as 'expansion'.
3   The original text was presumably in Russian, but there is no reason to doubt that these words were accurately translated.
4   'Economic development' in the intransitive sense had another potential source besides the line of thought that led from Hegel to Marx and Schumpeter, but its con-tribution was virtually stillborn and deserves only a footnote: This was the biological theory of evolution. Some years before Darwin's *Origin of Species*, Herbert

Spencer had begun to give the concept of biological evolution a social application (see especially his 'Pro-gress: Its Law and Course' and 'The Social Organism', reprinted in Spencer (1891)). He argued that social progress was more than 'simple growth'; rather, like the evolution of a biological organism, it was an 'evolution of the simple into the complex'. At one point he mentioned 'it has not been by command of any ruler that some men have become manufacturers, while others have remained cultivators . . . it has arisen under the pressure of human wants'. However, he did not pur-sue the idea any further. The only economist to take it up appears to have been the Australian, W.E. Hearn, whose *Plutology* (1863) received honourable mentions by Jevons and Marshall. In a chapter on 'The Industrial Evolution of Society', he expounded the Spencerian analogy. The evolution of both an individual and a society 'consists in, or at least is invariably attended by, an increase of bulk, a greater complexity of structure'. He used the idea to support a case for balanced growth between agriculture and industry: 'In societies as in organisms, growth and development, increased bulk and increased complexity of structure, ought always to proceed with equal pace.' He also, apparently unconscious of any inconsistency, introduced it in support of his *laissez-faire* principles: 'Every attempt to interfere with the ordinary development of a country', e.g. by protecting some industries or restraining others, 'tends to produce uniformity in the occupation of that country and so to arrest its development and retard its progress' (Hearn:384, 393, 437). But Hearn made nothing more of the idea, nor was it taken up by any of the later social Darwinists in the United States and elsewhere who were more interested in the survival of the fittest.
5   The fact that these synonyms began to be displaced by 'development' in the 1830s and 1840s may be explained by the vogue which ideas of evolution and development were enjoying about that time in natural sciences, such as biology and geology; the Oxford Eng-lish Dictionary cites uses in more general literature in the same period.
6   Thus the word 'development' does not occur once in two works about aspects of nineteenth century eco-nomic history in the United States, railway policy and public lands policy, the Australian counterparts of which use it constantly (Miller 1939).
7   Sun Yat-sen had expounded his grandiose ideas for railway development in China before World War I, and although the book was not published until 1922, two years after a visit to the Soviet Union, it was based on lectures he gave in 1918. But even at that time, what was happening in Russia made as great an impression on him as news of the Great Leap Forward in China was to make on Indian opinion forty years later.

REFERENCES

Anstey, V., 1929. *The Economic Development of India*, Longmans, Green, London.

Arndt, H.W., 1978. *The Rise and Fall of Economic Growth*, Longman Cheshire, Melbourne.

Baldwin, D.A., 1966. *Economic Development and American Foreign Policy, 1943–62*, University of Chicago Press, Chicago.

Benson, W., 1942. 'The economic advancement of under-developed areas', in *The Economic Basis of Peace*, National Peace Council, London: 10.

Betts, R.E. 1961. *Assimilation and Association in French Colonial Theory*, Columbia University Press, New York.

Clark, Colin, 1925. *The Conditions of Economic Progress*, Macmillan Publishing Co., London.

Coghlin, T.A., 1890. *A Statistical Account of the Seven Colonies of Australasia*, Sydney, 1890.

Copland, D.B., 1931. 'The national income and economic prosperity', in *An Economic Survey of Australia*, Annals of the American Academy of Political and Social Science, 158, November:260.

d'Encausse, H.C. and Schram, S.R., 1969. *Marxism and Asia*, Penguin Press, London.

Eggleston, F.W., 1931. 'Australian Loan and Developmental Policy' in *An Economic Survey of Australia*, Annals of the American Academy of Political and Social Science 158, November: 193–201.

Furnivall, J.S., 1931. *An Introduction to the Political Economy of Burma*, 3rd ed.; reprint ed., Rangoon, People's Literature Committee and House, 1957.

Goodwin, C.D.W., 1966. *Economic Enquiry in Australia*, Duke University Press, Durham, N.C.

Hancock, W.K., 1942. *Survey of British Commonwealth Affairs, vol. 2, Problems of Economic Policy 1918–1939*, 2 parts, Oxford University Press, London.

Hearn, W.E., 1963. *Plutology*, George Robertson, Melbourne.

Hegel, G.W.F., 1956. *The Philosophy of History*, Dover Publications, New York.

Innis, H.A. and Lower, A.R.M., 1933. *Select Documents in Canadian Economic History 1783–1885*, University of Toronto Press, Toronto.

Knowles, L.C.A., 1924. *The Economic Development of the British Overseas Empire*, George Routledge and Sons, London.

Lewis, W.A., 1944. 'An economic plan for Jamaica', *Agenda*, November(4):156.

——, 1955. *The Theory of Economic Growth*, Allen and Unwin, London, 1955.

Lugard, F.D., 1921. *The Dual Mandate in British Tropical Africa*, William Blackwood and Sons, Edinburgh.

Marshall, A., 1920. *Principles of Economics*, 2 vols., 9th ed., Macmillan Publishing Co, 1961.

Marx, K. and Engels, F., 1969. *Selected Works in Three Volumes*, Progress Publishers, Moscow.

Mauretta, F., 1937. *Some Social Aspects of Present and Future Economic Development of Brazil, Studies and Reports*, ser. B, no.25, International Labour Office, Geneva.

Miller, S.L., 1939. *Inland Transportation*, Peter Smith, New York.

Myint, Hla, 1954. 'An interpretation of economic backwardness', *Oxford Economic Papers*, 6:132–63.

Myrdal, Gunnar, 1957. *Economic Theory and Under-developed Regions*, Gerald Duckworth and Co., London.

Okun B. and Richardson, R.W. (eds), 1961. *Studies in Economic Development*, Holt, Rinehart and Winston, New York.

Passmore, J., 1970. *The Perfectibility of Man*, Gerald Duckworth and Co., London.

Roberts, S.H., 1924. *History of Land Settlement in Australia*, Melbourne University Press, Melbourne.

Rosenstein-Rodan, P.N., 1944. 'The international development of economically backward areas', *International Affairs*, April, 20:157–65.

Rostow, W.W., 1960. *The Stages of Economic Growth*, Cambridge University Press, Cambridge.

Schumpeter, J.A., 1934. *The Theory of Economic Development*, trans. R. Opie, Harvard Economic Studies, vol. 46, Harvard University Press, Cambridge, Mass.

——, 1954. *History of Economic Analysis*, Oxford University Press, Oxford.

Smith, A., 1776. *The Wealth of Nations*, E. Cannan (ed.), 2 vols; reprint ed., London, University Paperbacks, 1961.

Spencer, H., 1981. *Essay, Scientific, Political and Speculative*, 3 vols, Williams and Norgate, London, 1891, vol. 1.

Staley, Eugene, 1939. *World Economy in Transition*, Council on Foreign Relations, New York.

Sun Yat-sen, 1922. *The International Development of China*, G.P. Putnam's Sons, New York.

Tawney, R.H., 1932. *Land and Labour in China*, Allen and Unwin, London.

United Nations, 1947. *Economic Development in Selected Countries: Plans, Programmes and Agencies*, United Nations, New York, October.

'Of the Advantages which Europe has derived from the Discovery of America, and from that of a Passage to the East Indies by the Cape of Good Hope'

from *An Inquiry into the Nature and Causes of the Wealth of Nations* (1776)

Adam Smith

Editors' Introduction

Adam Smith was born in the Scottish town of Kirkcaldy in 1723 and died in Edinburgh in 1790. Although he spent some time at Oxford University he lived most of his life in Scotland and was a major figure in the Scottish Enlightenment. Smith was a friend of the philosopher, David Hume, and while he is often hailed as the father of modern economics he is best seen as a broad-ranging intellectual with particular interests in political economy and moral philosophy. His most famous works are *The Theory of Moral Sentiments*, first published in 1759, and *An Inquiry into the Nature and Causes of the Wealth of Nations* (published first in 1776 and generally known as *The Wealth of Nations*).

The Wealth of Nations is best known for its suggestion that the rough and tumble of 'free markets' is guided (or should be guided) by an 'invisible hand' that ensures that resources are properly allocated to produce the goods and services that consumers demand. If a particular good is in short supply its price will rise, causing more people to produce it. Likewise, if there is no demand for a given product, a company will go out of business. But there is more to Smith's work than this simple if elegant idea. Smith critiqued Physiocrat ideas about the importance of land as a source of value. Smith placed his emphasis on labour and on the productivity of labour. He proposed that economic efficiency would be increased by specialisation and extended divisions of labour. Smith also wrote *The Wealth of Nations* as an attack upon mercantilist doctrines. Smith became an advocate of free trade at a time when most governments still clung to the idea that countries should use foreign trade to build up large stores of bullion.

Smith is well known for his advocacy of self-interest on the part of economic agents. 'It is not from the benevolence of the butcher, the brewer, or the baker that we expect our dinner, but from their regard to their own interest.' It is perhaps less well known, however, and certainly less well publicised today, that Smith was wary of the tendency for private monopolies to be formed by the successful exercise of self-interest and 'greed'. In the reading that we reproduce here, from Book IV of *The Wealth of Nations*, we see Smith drawing a stark distinction between the possibilities for economic accumulation opened up after the discovery of a sea passage to the East Indies, and the ways in which these advantages were monopolised by officers of the English East India Company. The dominant voice is one of disgust, or at least of disappointment. Smith ends, however, true to form, by calling for a change in the 'system of government' which allowed the East India Company to behave as it did. The reading is thus contemporary in all sorts of ways. It raises issues related to what we would now call globalisation (and trade), as well as questions to do with the management of large firms or corporations and regimes of good (or bad) governance. The reader will want to ask questions about what has changed since the time that Smith was writing. Are there any modern-day equivalents of the East India Company? If so, which corporations would best fit the bill, and why?

Key references

Adam Smith (1993 [1776]) *An Inquiry into the Nature and Causes of the Wealth of Nations*, Oxford: Clarendon Press.
— (2002[1759]) *The Theory of Moral Sentiments*, Cambridge: Cambridge University Press.
— (2004) *Selected Philosophical Writings*, edited by James Otteson, London: Imprint.
J. Buchan (2007) *The Authentic Adam Smith: His Life and Ideas*, New York: W. W. Norton.
E. Rothschild (2002) *Economic Sentiments: Adam Smith, Condorcet and the Enlightenment*, Cambridge, MA: Harvard University Press.

[. . .]

The discovery of America, and that of a passage to the East Indies by the Cape of Good Hope, are the two greatest and most important events recorded in the history of mankind. Their consequences have already been very great: but, in the short period of between two and three centuries which has elapsed since these discoveries were made, it is impossible that the whole extent of their consequences can have been seen. What benefits, or what misfortunes to mankind may hereafter result from those great events no human wisdom can foresee. By uniting, in some measure, the most distant parts of the world, by enabling them to relieve one another's wants, to increase one another's enjoyments, and to encourage one another's industry, their general tendency would seem to be beneficial. To the natives, however, both of the East and West Indies, all the commercial benefits which can have resulted from those events have been sunk and lost in the dreadful misfortunes which they have occasioned. These misfortunes, however, seem to have arisen rather from accident than from any thing in the nature of those events themselves. At the particular time when these discoveries were made, the superiority of force happened to be so great on the side of the Europeans, that they were enabled to commit with impunity every sort of injustice in those remote countries. Hereafter, perhaps, the natives of those countries may grow stronger, or those of

Europe may grow weaker, and the inhabitants of all the different quarters of the world may arrive at that equality of courage and force which, by inspiring mutual fear, can alone overawe the injustice of independent nations into some sort of respect for the rights of one another. But nothing seems more likely to establish this equality of force than that mutual communication of knowledge and of all sorts of improvements which an extensive commerce from all countries to all countries naturally, or rather necessarily, carries along with it.

In the mean time one of the principal effects of those discoveries has been to raise the mercantile system to a degree of splendor and glory which it could never otherwise have attained to. It is the object of that system to enrich a great nation rather by trade and manufactures than by the improvement and cultivation of land, rather by the industry of the towns than by that of the country. But, in consequence of those discoveries, the commercial towns of Europe, instead of being the manufacturers and carriers for but a very small part of the world (that part of Europe which is washed by the Atlantic ocean, and the countries which lie round the Baltick and Mediterranean seas), have now become the manufacturers for the numerous and thriving cultivators of America, and the carriers, and in some respects the manufacturers too, for almost all the different nations of Asia, Africa, and America. Two new worlds have been opened to their industry, each of them much greater and more extensive than the old one, and the market of one of them growing still greater and greater every day. . . .

In the trade to America every nation endeavours to engross as much as possible the whole market of its own colonies, by fairly excluding all other nations from any direct trade to them. During the greater part of the sixteenth century, the Portugueze endeavoured to manage the trade to the East Indies in the same manner, by claiming the sole right of sailing in the Indian seas, on account of the merit of having first found out the road to them. The Dutch still continue to exclude all other European nations from any direct trade to their spice islands. Monopolies of this kind are evidently established against all other European nations, who are thereby not only excluded from a trade to which it might be convenient for them to turn some part of their stock, but are obliged to buy the goods which that trade deals in somewhat dearer, than if they could import them themselves directly from the countries which produce them.

But since the fall of the power of Portugal, no European nation has claimed the exclusive right of sailing in the Indian seas, of which the principal ports are now open to the ships of all European nations. Except in Portugal, however, and within these few years in France, the trade to the East Indies has in every European country been subjected to an exclusive company. Monopolies of this kind are properly established against the very nation which erects them. The greater part of that nation are thereby not only excluded from a trade to which it might be convenient for them to turn some part of their stock, but are obliged to buy the goods which that trade deals in, somewhat dearer than if it was open and free to all their countrymen. Since the establishment of the English East India company, for example, the other inhabitants of England, over and above being excluded from the trade, must have paid in the price of the East India goods which they have consumed, not only for all the extraordinary profits which the company may have made upon those goods in consequence of their monopoly, but for all the extraordinary waste which the fraud and abuse, inseparable from the management of the affairs of so great a company, must necessarily have occasioned. The absurdity of this second kind of monopoly, therefore, is much more manifest than that of the first.

Both these kinds of monopolies derange more or less the natural distribution of the stock of the society: but they do not always derange it in the same way.

Monopolies of the first kind always attract to the particular trade in which they are established, a greater proportion of the stock of the society than what would go to that trade of its own accord.

Monopolies of the second kind, may sometimes attract stock towards the particular trade in which they are established, and sometimes repel it from that trade according to different circumstances. In poor countries they naturally attract towards that trade more stock than would otherwise go to it. In rich countries they naturally repel from it a good deal of stock which would otherwise go to it. . . .

The English and Dutch companies, though they have established no considerable colonies, . . . have both made considerable conquests in the East Indies. But in the manner in which they both govern their new subjects, the natural genius of an exclusive company has shown itself most distinctly. In the spice islands the Dutch are said to burn all the spiceries which a fertile season produces beyond what they expect to dispose of in Europe with such a profit as they think sufficient. In the islands where they have no settlements, they give a premium to those who collect the young blossoms and green leaves of the clove and nutmeg trees which naturally grow there, but which this savage policy has now, it is said, almost compleatly extirpated. Even in the islands where they have settlements they have very much reduced, it is said, the number of those trees. If the produce even of their own islands was much greater than what suited their market, the natives, they suspect, might find means to convey some part of it to other nations; and the best way, they imagine, to secure their own monopoly, is to take care that no more shall grow than what they themselves carry to market. By different arts of oppression they have reduced the population of several of the Moluccas nearly to the number which is sufficient to supply with fresh provisions and other necessaries of life their own insignificant garrisons, and such of their ships as occasionally come there for a cargo of spices. Under the government even of the Portugueze, however, those islands are said to have been tolerably well inhabited. The English company have not yet had time to establish in Bengal so perfectly destructive a system. The plan of their government, however, has had exactly the same tendency. It has not been uncommon, I am well assured, for the chief, that is, the first clerk of a factory, to order a peasant to plough up a rich field of poppies, and sow it with rice or some other grain. The pretence was, to prevent a scarcity of provisions; but the real reason, to give the chief an opportunity of selling at a better price a large quantity of opium, which he happened then to have upon hand. Upon other occasions the order has been reversed; and a rich field of rice or other grain has been ploughed up, in order to make room for a plantation of poppies; when the chief foresaw that extraordinary profit was likely to be made by opium. The servants of the company have upon several occasions attempted to establish in their own favour the monopoly of some of the most important branches, not only of the foreign, but of the inland trade of the country. Had they been allowed to go on, it is impossible that they should not at some time or another have attempted to restrain the production of the particular articles of which they had thus usurped the monopoly, not only to the quantity which they themselves could purchase, but to that which they could expect to sell with such a profit as they might think sufficient. In the course of a century or two, the policy of the English company would in this manner have probably proved as compleatly destructive as that of the Dutch.

Nothing, however, can be more directly contrary to the real interest of those companies, considered as the sovereigns of the countries which they have conquered, than this destructive plan. In almost all countries the revenue of the sovereign is drawn from that of the people. The greater the revenue of the people, therefore, the greater the annual produce of their land and labour, the more they can afford to the sovereign. It is his interest, therefore, to increase as much as possible that annual produce. But if this is the interest of every sovereign, it is peculiarly so of one whose revenue, like that of the sovereign of Bengal, arises chiefly from a land-rent. That rent must necessarily be in proportion to the quantity and value of the produce, and both the one and the other must depend upon the extent of the market. The quantity will always be suited with more or less exactness to the consumption of those who can afford to pay for it, and the price which they will pay will always be in proportion to the eagerness of their competition. It is the interest of such a sovereign, therefore, to open the most extensive market for the produce of his country,

to allow the most perfect freedom of commerce, in order to increase as much as possible the number and the competition of buyers; and upon this account to abolish, not only all monopolies, but all restraints upon the transportation of the home produce from one part of the country to another, upon its exportation to foreign countries, or upon the importation of goods of any kind for which it can be exchanged. He is in this manner most likely to increase both the quantity and value of that produce, and consequently of his own share of it, or of his own revenue.

But a company of merchants are, it seems, incapable of considering themselves as sovereigns, even after they have become such. Trade, or buying in order to sell again, they still consider as their principal business, and by a strange absurdity, regard the character of the sovereign as but an appendix to that of the merchant, as something which ought to be made subservient to it, or by means of which they may be enabled to buy cheaper in India, and thereby to sell with a better profit in Europe. They endeavour for this purpose to keep out as much as possible all competitors from the market of the countries which are subject to their government, and consequently to reduce, at least, some part of the surplus produce of those countries to what is barely sufficient for supplying their own demand, or to what they can expect to sell in Europe with such a profit as they may think reasonable. Their mercantile habits draw them in this manner, almost necessarily, though perhaps insensibly, to prefer upon all ordinary occasions the little and transitory profit of the monopolist to the great and permanent revenue of the sovereign, and would gradually lead them to treat the countries subject to their government nearly as the Dutch treat the Moluccas. It is the interest of the East India company, considered as sovereigns, that the European goods which are carried to their Indian dominions, should be sold there as cheap as possible; and that the Indian goods which are brought from thence should bring there as good a price, or should be sold there as dear as possible. But the reverse of this is their interest as merchants. As sovereigns, their interest is exactly the same with that of the country which they govern. As merchants their interest is directly opposite to that interest.

But if the genius of such a government, even as to what concerns its direction in Europe, is in this manner essentially and perhaps incurably faulty, that of its administration in India is still more so. That administration is necessarily composed of a council of merchants, a profession no doubt extremely respectable, but which in no country in the world carries along with it that sort of authority which naturally over-awes the people, and without force commands their willing obedience. Such a council can command obedience only by the military force with which they are accompanied, and their government is therefore necessarily military and despotical. Their proper business, however, is that of merchants. It is to sell, upon their masters account, the European goods consigned to them, and to buy in return Indian goods for the European market. It is to sell the one as dear and to buy the other as cheap as possible, and consequently to exclude as much as possible all rivals from the particular market where they keep their shop. The genius of the administration, therefore, so far as concerns the trade of the company, is the same as that of the direction. It tends to make government subservient to the interest of monopoly, and consequently to stunt the natural growth of some parts at least of the surplus produce of the country to what is barely sufficient for answering the demand of the company.

All the members of the administration, besides, trade more or less upon their own account, and it is in vain to prohibit them from doing so. Nothing can be more compleatly foolish than to expect that the clerks of a great counting-house at ten thousand miles distance, and consequently almost quite out of sight, should, upon a simple order from their masters, give up at once doing any sort of business upon their own account, abandon for ever all hopes of making a fortune, of which they have the means in their hands, and content themselves with the moderate salaries which those masters allow them, and which, moderate as they are, can seldom be augmented, being commonly as large as the real profits of the company trade can afford. In such circumstances, to prohibit the servants of the company from trading upon their own account, can have scarce any other effect than to enable the superior servants, under pretence of executing their masters order, to oppress such of the inferior ones as have had the misfortune to fall under their displeasure. The servants naturally endeavour to establish the same monopoly in favour of their own private trade as of the publick trade of the company. If they are suffered to act as they could wish, they will establish this monopoly openly and directly, by fairly prohibiting all other people from trading in the articles in which they chuse to deal; and this, perhaps, is the best and least oppressive way of establishing it. But if by an order from Europe they are prohibited from doing this, they will, notwithstanding, endeavour to establish a monopoly of the same kind, secretly and indirectly, in a way that is much more destructive to the country. They will employ the whole authority of government, and pervert the administration of justice, in order to harass and ruin those who interfere with them in any branch of commerce which, by means of agents, either concealed, or at least not publickly avowed, they may chuse to carry on. But the private trade of the servants will naturally extend to a much greater variety of articles than the publick trade of the company. The publick trade of the company extends no further than the trade with Europe, and comprehends a part only of the foreign trade of the country. But the private trade of the servants may extend to all the different branches both of its inland and foreign trade. The monopoly of the company can tend only to stunt the natural growth of that part of the surplus produce which, in the case of a free trade, would be exported to Europe. That of the servants tends to stunt the natural growth of every part of the produce in which they chuse to deal, of what is destined for home consumption, as well as of what is destined for exportation; and consequently to degrade the cultivation of the whole country, and to reduce the number of its inhabitants. It tends to reduce the quantity of every sort of produce, even that of the necessaries of life, whenever the servants of the company chuse to deal in them, to what those servants can both afford to buy and expect to sell with such a profit as pleases them.

From the nature of their situation too the servants must be more disposed to support with rigorous severity their own interest against that of the country which they govern, than their masters can be to support theirs. The country belongs to their masters, who cannot avoid having some regard for the interest of what belongs to them. But it does not belong to the servants. The real interest of their masters, if they were capable of understanding it, is the same with that of the country,[1] and it is from ignorance chiefly, and the meanness of mercantile prejudice, that they ever oppress it. But the real interest of the servants is by no means the same with that of the country, and the most perfect information would not necessarily put an end to their oppressions. The regulations accordingly which have been sent out from Europe, though they have been frequently weak, have upon most occasions been well-meaning. More intelligence and perhaps less good-meaning has sometimes appeared in those established by the servants in India. It is a very singular government in which every member of the administration wishes to get out of the country and consequently to have done with the government, as soon as he can, and to whose interest, the day after he has left it and carried his whole fortune with him, it is perfectly indifferent though the whole country was swallowed up by an earthquake.

I mean not, however, by any thing which I have here said, to throw any odious imputation upon the general character of the servants of the East India company, and much less upon that of any particular persons. It is the system of government, the situation in which they are placed, that I mean to censure, not the character of those who have acted in it. They acted as their situation naturally directed, and they who have clamoured the loudest against them would, probably, not have acted better themselves. In war and negociation, the councils of Madras and Calcutta have upon several occasions conducted themselves with a resolution and decisive wisdom which would have done honour to the senate of Rome in the best days of that republick. The members of those councils, however, had been bred to professions very different from war and politicks. But their situation alone, without education, experience, or even example, seems to have formed in them all at once the great qualities which it required, and to have inspired them both with abilities and virtues which they themselves could not well know that they possessed. If upon some occasions, therefore, it has animated them to actions of magnanimity which could not well have been expected from them, we should not wonder if upon others it has prompted them to exploits of somewhat a different nature.

Such exclusive companies, therefore, are nuisances in every respect; always more or less inconvenient to the countries in which they are established, and destructive to those which have the misfortune to fall under their government.

NOTE

1 The interest of every proprietor of India Stock, however, is by no means the same with that of the country in the government of which his vote gives him some influence. See Book V. Chap. i. Part 3d.

'The British Rule in India'

New York Daily Tribune (25 June 1853)

Karl Marx

Editors' Introduction

Karl Marx was born in Trier, Germany in 1818 and died in London in 1883. He was a giant of the nineteenth century, both in intellectual and in political circles. Arguably, he is best known today for his work from 1848 with Friedrich Engels, *The Communist Manifesto* – a call to arms made at a time of widespread social upheaval in Europe, and a text which prefigures Marx's extraordinary accounts of the processes of creative destruction unleashed by capitalism. 'All that is solid melts into air', wrote Marx and Engels, giving a far better sense than any contemporary critic of the ceaseless energy of an economic system in which 'to stand still is to perish'. Marx's greatest work, however, was *Das Kapital* (or Capital), only the first volume of which was published (in 1867) before his death; Volumes II and III of a projected six volumes were published posthumously by Engels. In that work, Marx undertook to provide both a theoretical and an empirical account of what might be called the deep structures, or laws of motion, of capitalism. In the view of Marx, and in that of many of his followers, this account of capitalism – the sense it gives of economic dynamism amid the production of huge inequalities and instability – was 'scientific': it was not to be conflated with the moralising critiques of capitalism and its effects that were being penned at about the same time by Pierre-Joseph Proudhon or Charles Dickens. Marx also intended his work to have political purchase, as is well known. Some have suggested that there are echoes in his work of the evolutionism that he would have found in the work of Charles Darwin, one of Marx's heroes. Certainly, there is a hint of this in the Marxian suggestion that societies would follow a standard course of history, moving from primitive communist and slave modes of production, through feudalism and capitalism and on to socialism and communism.

As ever, though, with a thinker as gifted and as hard-working as Marx (as also Adam Smith and Maynard Keynes), there is much that cannot be conveyed in a simple biopic. The work of these writers changed over time, and in response to events and new intellectual ideas. The subtlety of their thought has sometimes been corroded by the less than subtle thoughts of their followers (Smithians, Marxists or Keynesians). In addition, in Marx's case, funds were often short, and he was compelled to distil his thoughts in paid articles for newspapers, including – writing from London – the *New York Daily Tribune*. The reading that we have selected here comes from this journal in June 1853, and offers one of Marx's comparatively rare writings on non-European settings. Marx never visited India, or Hindostan, so what we get is a sense of what he had read about the subcontinent, filtered and refracted through his emerging studies of political economy. And what we get is classic Marx: an essay that highlights the damaging effects of British rule and colonial capitalism in India, but which also celebrates England's role – even if 'actuated only by the vilest interests' – in pushing India away from caste-based oppression towards its 'destiny' as a modern country. The suggestion here is a classic of developmentalism, and it would be strongly echoed one hundred years later in the 1950s. Development, change and social revolution will be tough, they will cause damage and upheaval, but they are all processes that societies have to go through before men and women can live in a better world. For Marx, this meant a post-capitalist and post-imperialist world where, ultimately, the fruits of economic growth and technological change would be used to address the real needs of human beings.

The reader might want to compare Marx in the 1850s with the article by David Harvey in Part 9 which describes the 'new imperialism' of the American era. It is sometimes said that Marxism is dead in the water, or that it has no intellectual purchase since the end of the Cold War. But how compelling is this suggestion? Even if one discounts Marxism as a guide to practical politics – and not everyone does – might it not still be argued that Marxists like Harvey have more to tell us about the shaping of the modern world than academics writing from intellectual traditions less attuned to asymmetries of power or the contradictions of capitalism? And if not, why not? What assumptions does Marx make, does Harvey make, that we might want to challenge – and why?

Key references

Karl Marx (1853) 'The British rule in India', *New York Daily Tribune*, in *Karl Marx, Friedrich Engels Collected Works*, vol. 12, London: Lawrence and Wishart; Moscow: International Publishers, 1979.

Karl Marx and Friedrich Engels (1998 [1848]) *The Communist Manifesto*, London: Penguin.

Karl Marx (1973) *Grundrisse*, Harmondsworth: Penguin.

— (1973–76) *Capital* (3 volumes), Harmondsworth: Penguin.

D. Harvey (1982) *The Limits to Capital*, Oxford: Blackwell.

D. McClellan (2006) *Karl Marx: A Biography*, fourth edition, London: Palgrave-Macmillan.

LONDON, FRIDAY, JUNE 10, 1853

Telegraphic dispatches from Vienna announce that the pacific solution of the Turkish, Sardinian and Swiss questions, is regarded there as a certainty.

Last night the debate on India was continued in the House of Commons, in the usual dull manner. Mr. Blackett charged the statements of Sir Charles Wood and Sir J. Hogg with bearing the stamp of optimist falsehood. A lot of Ministerial and Directorial advocates rebuked the charge as well as they could, and the inevitable Mr. Hume summed up by calling on Ministers to withdraw their bill. Debate adjourned.

Hindostan is an Italy of Asiatic dimensions, the Himalayas for the Alps, the Plains of Bengal for the Plains of Lombardy, the Deccan for the Apennines, and the Isle of Ceylon for the Island of Sicily. The same rich variety in the products of the soil, and the same dismemberment in the political configuration. Just as Italy has, from time to time, been compressed by the conqueror's sword into different national masses, so do we find Hindostan, when not under the pressure of the Mohammedan, or the Mogul, or the Briton, dissolved into as many independent and con-

flicting States as it numbered towns, or even villages. Yet, in a social point of view, Hindostan is not the Italy, but the Ireland of the East. And this strange combination of Italy and of Ireland, of a world of voluptuousness and of a world of woes, is anticipated in the ancient traditions of the religion of Hindostan. That religion is at once a religion of sensualist exuberance, and a religion of self-torturing asceticism; a religion of the Lingam and of the juggernaut; the religion of the Monk, and of the Bayadere.

I share not the opinion of those who believe in a golden age of Hindostan, without recurring, however, like Sir Charles Wood, for the confirmation of my view, to the authority of Khuli-Khan. But take, for example, the times of Aurangzeb; or the epoch, when the Mogul appeared in the North, and the Portuguese in the South; or the age of Mohammedan invasion, and of the Heptarchy in Southern India; or, if you will, go still more back to antiquity, take the mythological chronology of the Brahman himself, who places the commencement of Indian misery in an epoch even more remote than the Christian creation of the world.

There cannot, however, remain any doubt but that the misery inflicted by the British on Hindostan is of an essentially different and infinitely more intensive kind than all Hindostan had to suffer before. I do not allude to European despotism, planted upon Asiatic despotism, by the British East India Company, forming a more monstrous combination than any of the divine monsters startling us in the Temple of Salsette. This is no distinctive feature of British Colonial rule, but only an imitation of the Dutch, and so much so that in order to characterise the working of the British East India Company, it is sufficient to literally repeat what Sir Stamford Raffles, the *English* Governor of Java, said of the old Dutch East India Company:

"The Dutch Company, actuated solely by the spirit of gain, and viewing their [Javan] subjects, with less regard or consideration than a West India planter formerly viewed a gang upon his estate, because the latter had paid the purchase money of human property, which the other had not, employed all the existing machinery of despotism to squeeze from the people their utmost mite of contribution, the last dregs of their labor, and thus aggravated the evils of a capricious and semi-barbarous Government, by working it with all the practised ingenuity of politicians, and all the monopolizing selfishness of traders."

All the civil wars, invasions, revolutions, conquests, famines, strangely complex, rapid, and destructive as the successive action in Hindostan may appear, did not go deeper than its surface. England has broken down the entire framework of Indian society, without any symptoms of reconstitution yet appearing. This loss of his old world, with no gain of a new one, imparts a particular kind of melancholy to the present misery of the Hindoo, and separates Hindostan, ruled by Britain, from all its ancient traditions, and from the whole of its past history.

There have been in Asia, generally, from immemorial times, but three departments of Government; that of Finance, or the plunder of the interior; that of War, or the plunder of the exterior; and, finally, the department of Public Works. Climate and territorial conditions, especially the vast tracts of desert, extending from the Sahara, through Arabia, Persia, India, and Tartary, to the most elevated Asiatic highlands, constituted artificial irrigation by canals and water-works the basis of Oriental agriculture. As in Egypt and India, inundations are used for fertilizing the soil in Mesopotamia, Persia, &c.; advantage is taken of a high level for feeding irrigative canals. This prime necessity of an economical and common use of water, which, in the Occident, drove private enterprise to voluntary association, as in Flanders and Italy, necessitated, in the Orient where civilization was too low and the territorial extent too vast to call into life voluntary association, the interference of the centralizing power of Government. Hence an economical function devolved upon all Asiatic Governments, the function of providing public works. This artificial fertilization of the soil, dependent on a Central Government, and immediately decaying with the neglect of irrigation and drainage, explains the otherwise strange fact that we now find whole territories barren and desert that were once brilliantly cultivated, as Palmyra, Petra, the ruins in Yemen, and large provinces of Egypt, Persia, and Hindostan; it also explains how a single war of devastation has been able to depopulate a country for centuries, and to strip it of all its civilization.

Now, the British in East India accepted from their predecessors the department of finance and of war, but they have neglected entirely that of public works. Hence the deterioration of an agriculture which is not capable of being conducted on the British principle of free competition, of *laissez-faire* and *laissez-aller*. But in Asiatic empires we are quite accustomed to see agriculture deteriorating under one government and reviving again under some other government. There the harvests correspond to good or bad government, as they change in Europe with good or bad seasons. Thus the oppression and neglect of agriculture, bad as it is, could not be looked upon as the final blow dealt to Indian society by the British intruder, had it not been attended by a circumstance of quite different importance, a novelty in the annals of the whole Asiatic world. However changing the political aspect of India's past must appear, its social condition has remained unaltered since its remotest antiquity, until the first decennium of the 19th century. The hand-loom and the spinning-wheel, producing their regular myriads of spinners and weavers, were the pivots of the structure of that society. From immemorial times, Europe received the admirable textures of Indian labor, sending in return for them her precious metals, and furnishing thereby his material to the goldsmith, that indispensable member of Indian society, whose love of finery is so great that even the lowest class, those who go about nearly naked, have commonly a pair of golden earrings and a gold ornament of some kind hung round their necks. Rings on the fingers and toes have also been common. Women as well as children frequently wore massive bracelets and anklets of gold or silver, and statuettes of divinities in gold and silver were met with in the households. It was the British intruder who broke up the Indian handloom and destroyed the spinning-wheel. England began with driving the Indian cottons from the European market; it then introduced twist into Hindostan, and in the end inundated the very mother country of cotton with cottons. From 1818 to 1836 the export of twist from Great Britain to India rose in the proportion of 1 to 5,200. In 1824 the export of British muslins to India hardly amounted to 1,000,000 yards, while in 1837 it surpassed 64,000,000 yards. But at the same time the population of Dacca decreased from 150,000 inhabitants to 20,000. This decline of Indian towns celebrated for their fabrics was by no means the worst consequence. British steam and science uprooted, over the whole surface of Hindostan, the union between agriculture and manufacturing industry.

These two circumstances – the Hindoo, on the one hand, leaving, like all Oriental peoples, to the Central Government the care of the great public works, the prime condition of his agriculture and commerce, dispersed, on the other hand, over the surface of the country, and agglomerated in small centers by the domestic union of agricultural and manufacturing pursuits – these two circumstances had brought about, since the remotest times, a social system of particular features – the so-called *village system*, which gave to each of these small unions their independent organization and distinct life. The peculiar character of this system may be judged from the following description, contained in an old official report of the British House of Commons on Indian affairs:

"A village, geographically considered, is a tract of country comprising some hundred or thousand acres of arable and waste lands; politically viewed it resembles a corporation or township. Its proper establishment of officers and servants consists of the following descriptions: The *potail*, or head inhabitant, who has generally the superintendence of the affairs of the village, settles the disputes of the inhabitants attends to the police, and performs the duty of collecting the revenue within his village, a duty which his personal influence and minute acquaintance with the situation and concerns of the people render him the best qualified for this charge. The *kurnum* keeps the accounts of cultivation, and registers everything connected with it. The *tallier* and the *totie*, the duty of the former of which consists [...] in gaining information of crimes and offenses, and in escorting and protecting persons travelling from one village to another; the province of the latter appearing to be more immediately confined to the village, consisting, among other duties, in guarding the crops and assisting in measuring them. The boundary-man, who preserves the limits of the village, or gives evidence respecting them in cases of dispute. The Superintendent of Tanks and Watercourses distributes the water [...] for the purposes of agriculture. The Brahmin, who performs the village worship. The schoolmaster, who is seen teaching the children in a village to read and write in the sand. The calendar-brahmin, or astrologer, etc. These officers and servants generally constitute the establishment of a village; but in some parts of the country it is of less extent, some of the duties and functions above described being united in the same person; in others it exceeds the above-named number of individuals. [...] Under this simple form of municipal government, the inhabitants of the country have lived from time immemorial. The boundaries of the villages have been but seldom altered; and though the villages themselves have been sometimes injured, and even desolated by war, famine or disease, the same name, the same limits, the same interests, and even the same families have continued for ages. The inhabitants gave themselves no trouble about the breaking up and divisions of kingdoms; while the village remains entire, they care not to what power it is transferred, or to what sovereign it devolves; its internal economy remains unchanged. The *potail* is still the head inhabitant, and still acts as the petty judge or magistrate, and collector or renter of the village."

These small stereotype forms of social organism have been to the greater part dissolved, and are disappearing, not so much through the brutal interference of the British tax-gatherer and the British soldier, as to the working of English steam and English free trade. Those family-communities were based on domestic industry, in that peculiar combination of hand-weaving, hand-spinning and hand-tilling agriculture which gave them self-supporting power. English interference having placed the spinner in Lancashire and the weaver in Bengal, or sweeping away both Hindoo spinner and weaver, dissolved these small semi-barbarian, semi-civilized communities, by blowing up their economical basis, and thus produced the greatest, and to speak the truth, the only social revolution ever heard of in Asia.

Now, sickening as it must be to human feeling to witness those myriads of industrious patriarchal and inoffensive social organizations disorganized and dissolved into their units, thrown into a sea of woes, and their individual members losing at the same time their ancient form of civilization, and their hereditary means of subsistence, we must not forget that these idyllic village-communities, inoffensive though they may appear, had always been the solid foundation of Oriental despotism, that they restrained the human mind within the smallest possible compass, making it the unresisting tool of superstition, enslaving it beneath traditional rules, depriving it of all grandeur and historical energies. We must not forget the barbarian egotism which, concentrating on some miserable patch of land, had quietly witnessed the ruin of empires, the perpetration of unspeakable cruelties, the massacre of the population of large towns, with no other consideration bestowed upon them than on natural events, itself the helpless prey of any aggressor who deigned to notice it at all. We must not forget that this undignified, stagnatory, and vegetative life, that this passive sort of existence evoked on the other part, in contradistinction, wild, aimless, unbounded forces of destruction and rendered murder itself a religious rite in Hindostan. We must not forget that these little communities were contaminated by distinctions of caste and by slavery, that they subjugated man to external circumstances instead of elevating man the sovereign of circumstances, that they transformed a self-developing social state into never changing natural destiny, and thus brought about a brutalizing worship of nature, exhibiting its degradation in the fact that man, the sovereign of nature, fell down on his knees in adoration of Kanuman, the monkey, and Sabbala, the cow.

England, it is true, in causing a social revolution in Hindostan, was actuated only by the vilest interests, and was stupid in her manner of enforcing them. But that is not the question. The question is, can mankind fulfil its destiny without a fundamental revolution in the social state of Asia? If not, whatever may have been the crimes of England she was the unconscious tool of history in bringing about that revolution.

Then, whatever bitterness the spectacle of the crumbling of an ancient world may have for our personal feelings, we have the right, in point of history, to exclaim with Goethe:

"Sollte diese Qual uns quälen
Da sie unsre Lust vermehrt,
Hat nicht myriaden Seelen
Timur's Herrschaft aufgezehrt?"

["Should this torture then torment us
Since it brings us greater pleasure?
Were not through the rule of Timur
Souls devoured without measure?"]
[From Goethe's "An Suleika", *Westöstlicher Diwan*]

Karl Marx

'The Organic Analogy Reconsidered' and 'Societal Typologies'

from *On Social Evolution: Selected Writings* (1972)

Herbert Spencer

Editors' Introduction

Herbert Spencer was born in 1820 in Derby. He died in 1903 as a controversial proponent of 'sociology' as a universal system of thought combining biology, metaphysics, ethics and psychology. Living within a provincial Midlands town during the Industrial Revolution, Spencer's family were active as religious Dissenters and Methodists. Spencer himself first worked as a railway engineer. At this time, between 1837 and 1848, he participated on and off in radical politics: in the Anti-Corn Law League, in the Complete Suffrage Union and in the Anti-State Church Association. Like other Midlands radicals, Spencer was critical of the distant alliance of aristocracy, Church and State. Like others of his time, too, Spencer experienced ideas of the European Enlightenment in practice. But he also held on to some of the ideals of English Evangelism: of an emerging better society, of the perfectibility of human nature (and hence variation), and of 'society' as a balancing force against an overbearing 'state'. This would be the bedrock of Spencer's lifelong scholarly interest in 'sociology', an interest that deepened when he moved to London in 1848 to work for *The Economist*.

In London, biology and evolution came to occupy the centre of Spencer's revision of Enlightenment notions of human nature and progress, as is evident in his first book, from 1851, on *Social Statics: or, the Conditions Essential to Human Happiness Specified, and the First of Them Developed*. Human nature is incredibly varied, argued Spencer, and it would be a mistake to infer an ideal from this variation. Rather, varied populations are compelled to 'adapt' in the long term to an ideal of perfection. There are Christian traces here, but this is also a new form of reasoning derived from the scientific thought of Lyell, Lamarck and Darwin. Spencer attended scientific lectures in London and read on advances in biological thought, from animal morphology to embryology. The 'organic analogy' suggested several things to Spencer. First, population growth could spur the race to perfection, contra Malthus. Second, all aspects of progress were actually in parallel processes of adaptation from chaotic homogeneity to ordered heterogeneity; he planned to show this in a multi-volume series on Synthetic Philosophy, beginning with *The Principles of Psychology*, and in a volume on metaphysics called *First Principles*. Third, state intervention in society was an obstacle to evolution. Like the French Positivists, Spencer saw society as a holistic, coordinated system, which favoured political liberalism. However, Spencer took this system to be a natural machine made up of specialised parts which would ultimately not be controllable by human, let alone state, intervention. Fourth, given the declining significance of religion, ethics could be grounded in nature. This principle of 'nature' as the primary force of immanent change and variation is an enduring legacy in the stream of thought that Spencer represents.

In the first reading that follows here, 'The Organic Analogy Reconsidered', we see Spencer mapping this evolutionary schema on to a notion of transition from primitive to nineteenth-century societies, one which also involves a transition from 'militancy' to 'industrialism'. These ideas may have been rooted in his own Midlands-eye view of industrialisation and Evangelical critique, including stereotypes of savage industrial labourers. Be that as it may, Spencer expresses a universal logic of immanent development in which ordered heterogeneity – of deepening divisions of labour, for instance – also brings altruism and cooperation. In the second reading, 'Societal Typologies', we see Spencer linking these ideas to the typological urge of biological classification applied to human variation. At this time, Spencer was also directing a massive multi-volume set of studies called *Descriptive Sociology*, with volumes from 'English' to 'Lowest races, Negrito races and Malayo-Polynesian races'. What is important in this reading is that Spencer seeks to deduce from this cross-sectional data how simple societies aggregate and form more consolidated political entities *because of* the immanent tendency towards complex societies. Conflict is necessary to this process, not unlike Marx's view of revolution, but to produce a more efficient rather than a socially just future. Curiously, Spencer's last works were critical of welfare and 'collectivism', but also of imperialism and what he called the 'cannibalism' of strong nations 'devouring' weaker ones. Spencer's thought fell out of favour, but his legacies persist in 'functionalism' in sociology and anthropology, in the idea of stages of growth (Rostow, Part 4), and in notions of human hierarchies tending towards Western modernity. Spencer's work is at the more sophisticated end of a spectrum of work on hierarchised human difference. We caution the reader to be aware that ideas of 'nature' and 'human nature' can be harnessed to projects of racism in different ways, sometimes at odds with the intentions of evolutionist thinkers like Spencer. Nonetheless, Spencer demonstrates one way in which ideas of nature, human hierarchy, Christianity and secularism become part of an organic understanding of society. For the basis of these comments, and more on Spencer's life and work, see the brilliant introduction by J. D. Y. Peel in *On Social Evolution,* from which the reading here has been excerpted.

Key references

Herbert Spencer (1851) *Social Statics: or, the Conditions Essential to Human Happiness Specified, and the First of Them Developed*, London: Chapman.
— (1874) *Descriptive Sociology, or Groups of Social Facts: No. 3 Types of Lowest Races, Negrito Races, and Malayo-Polynesian Races*, London: Williams & Norgat.
J. D. Y. Peel (ed.) (1972) *Herbert Spencer On Social Evolution: Selected Writings*, Chicago, IL: University of Chicago Press.
— (1971) *Herbert Spencer: The Evolution of a Sociologist*, New York: Basic Books.
George Stocking, (1987) *Victorian Anthropology*, New York: Free Press.

THE ORGANIC ANALOGY RECONSIDERED[1]

What is a society?

This question has to be asked and answered at the outset. Until we have decided whether or not to regard a society as an entity; and until we have decided whether, if regarded as an entity, a society is to be classed as absolutely unlike all other entities or as like some others; our conception of the subject-matter before us remains vague.

It may be said that a society is but a collective name for a number of individuals. Carrying the controversy between nominalism and realism into another sphere, a nominalist might affirm that just as there exist only the members of a species, while the species considered apart from them has no existence; so the units of a society alone exist, while the existence of the society is but verbal. Instancing a lecturer's audience as an aggregate which by disappearing at the close of the lecture, proves itself to be not a thing but only a certain arrangement of persons, he might argue that the like holds of the citizens forming a nation.

But without disputing the other steps of his argument, the last step may be denied. The arrangement, temporary in the one case, is lasting in the other; and it is the permanence of the relations among component parts which constitutes the individuality of a whole as distinguished from the individualities of its parts. A coherent mass broken into fragments ceases to be a thing; while, conversely, the stones, bricks, and wood, previously separate, become the thing called a house if connected in fixed ways.

Thus we consistently regard a society as an entity, because, though formed of discrete units, a certain concreteness in the aggregate of them is implied by the maintenance, for generations and centuries, of a general likeness of arrangement throughout the area occupied. And it is this trait which yields our idea of a society. For, withholding the name from an ever-changing cluster such as primitive men form, we apply it only where some constancy in the distribution of parts has resulted from settled life.

But now, regarding a society as a thing, what kind of thing must we call it? It seems totally unlike every object with which our senses acquaint us. Any likeness it may possibly have to other objects, cannot be manifest to perception, but can be discerned only by reason. If the constant relations among its parts make it an entity; the question arises whether these constant relations among its parts are akin to the constant relations among the parts of other entities. Between a society and anything else, the only conceivable resemblance must be one due *to parallelism of principle in the arrangement of components*.

There are two great classes of aggregates with which the social aggregate may be compared – the inorganic and the organic. Are the attributes of a society, considered apart from its living units, in any way like those of a not-living body? or are they in any way like those of a living body? or are they entirely unlike those of both?

The first of these questions needs only to be asked to be answered in the negative. A whole of which the parts are alive, cannot, in its general characters, be like lifeless wholes. The second question, not to be thus promptly answered, is to be answered in the affirmative. The reasons for asserting that the permanent relations among the parts of a society, are analogous to the permanent relations among the parts of a living body, we have now to consider.

A society is an organism

When we say that growth is common to social aggregates and organic aggregates, we do not thus entirely exclude community with inorganic aggregates: some of these, as crystals, grow in a visible manner; and all of them, on the hypothesis of evolution, are concluded to have arisen by integration at some time or other. Nevertheless, compared with things we call inanimate, living bodies and societies so conspicuously exhibit augmentation of mass, that we may fairly regard this as characteristic of them both. Many organisms grow throughout their lives; and the rest grow throughout considerable parts of their lives. Social growth usually continues either up to times when the societies divide, or up to times when they are overwhelmed.

Here, then, is the first trait by which societies ally themselves with the organic world and substantially distinguish themselves from the inorganic world.

It is also a character of social bodies, as of living bodies, that while they increase in size they increase in structure. A low animal, or the embryo of a high one, has few distinguishable parts; but along with its acquirement of greater mass, its parts multiply and simultaneously differentiate. It is thus with a society. At first the unlikenesses among its groups of units are inconspicuous in number and degree; but as it becomes more populous, divisions and sub-divisions become more numerous and more decided. Further, in the social organism as in the individual organism, differentiations cease only with that completion of the type which marks maturity and precedes decay.

Though in inorganic aggregates also, as in the entire solar system and in each of its members, structural differentiations accompany the integrations; yet these are so relatively slow, and so relatively simple, that they may be disregarded. The multiplication of contrasted parts in bodies politic and in living bodies, is so great that it substantially constitutes another common character which marks them off from inorganic bodies.

This community will be more fully appreciated on

observing that progressive differentiation of structures is accompanied by progressive differentiation of functions.

The multiplying divisions, primary, secondary, and tertiary, which arise in a developing animal, do not assume their major and minor unlikenesses to no purpose. Along with diversities in their shapes and compositions there go diversities in the actions they perform: they grow into unlike organs having unlike duties. Assuming the entire function of absorbing nutriment at the same time that it takes on its structural characters, the alimentary system becomes gradually marked off into contrasted portions; each of which has a special function forming part of the general function. A limb, instrumental to locomotion or prehension, acquires divisions and sub-divisions which perform their leading and their subsidiary shares in this office. So is it with the parts into which a society divides. A dominant class arising does not simply become unlike the rest, but assumes control over the rest; and when this class separates into the more and the less dominant, these, again, begin to discharge distinct parts of the entire control. With the classes whose actions are controlled it is the same. The various groups into which they fall have various occupations: each of such groups also, within itself, acquiring minor contrasts of parts along with minor contrasts of duties.

And here we see more clearly how the two classes of things we are comparing distinguish themselves from things of other classes; for such differences of structure as slowly arise in inorganic aggregates, are not accompanied by what we can fairly call differences of function.

Why in a body politic and in a living body, these unlike actions of unlike parts are properly regarded by us as functions, while we cannot so regard the unlike actions of unlike parts in an inorganic body, we shall perceive on turning to the next and the most distinctive common trait.

Evolution establishes in them both, not differences simply, but definitely-connected differences – differences such that each makes the others possible. The parts of an inorganic aggregate are so related that one may change greatly without appreciably affecting the rest. It is otherwise with the parts of an organic aggregate or of a social aggregate. In either of these the changes in the parts are mutually determined, and the changed actions of the parts are mutually dependent. In both, too, this mutuality increases as the evolution advances. The lowest type of animal is all stomach, all respiratory surface, all limb. Development of a type having appendages by which to move about or lay hold of food, can take place only if these appendages, losing power to absorb nutriment directly from surrounding bodies, are supplied with nutriment by parts which retain the power of absorption. A respiratory surface to which the circulating fluids are brought to be aerated, can be formed only on condition that the concomitant loss of ability to supply itself with materials for repair and growth, is made good by the development of a structure bringing these materials. So is it in a society. What we call with perfect propriety its organization, has a necessary implication of the same kind. . . .

Here let it once more be pointed out that there exist no analogies between the body politic and a living body, save those necessitated by that mutual dependence of parts which they display in common. Though, in foregoing chapters, comparisons of social structures and functions to structures and functions in the human body, have in many cases been made, they have been made only because structures and functions in the human body furnish the most familiar illustrations of structures and functions in general. The social organism, discrete instead of concrete, asymmetrical instead of symmetricial, sensitive in all its units instead of having a single sensitive centre, is not comparable to any particular type of individual organism, animal or vegetal. All kinds of creatures are alike in so far as each shows us co-operation among its components for the benefit of the whole; and this trait, common to them, is a trait common also to communities. Further, among the many types of individual organisms, the degree of this co-operation measures the degree of evolution; and this general truth, too, is exhibited among social organisms. Once more, to effect increasing co-operation, creatures of every order show us increasingly-complex appliances for transfer and mutual influence; and to this general characteristic, societies of every order furnish a corresponding characteristic. Community in the fundamental principles of organization is thus the only community asserted.

But now let us drop this alleged parallelism between individual organizations and social organizations. I have used the analogies elaborated, but as a scaffolding to help in building up a coherent body of sociological inductions. Let us take away the scaffolding: the inductions will stand by themselves.

We saw that societies are aggregates which grow; that in various types of them there are great varieties in the degrees of growth reached; that types of successively larger sizes result from the aggregation and re-aggregation of those of smaller sizes; and that this increase by coalescence, joined with interstitial increase, is the process through which have been formed the vast civilized nations.

Along with increase of size in societies goes increase of structure. Primitive wandering hordes are without established unlikenesses of parts. With growth of them into tribes habitually come some differences; both in the powers and occupations of their members. Unions of tribes are followed by more differences, governmental and industrial – social grades running through the whole mass, and contrasts between the differently-occupied parts in different localities. Such differentiations multiply as the compounding progresses. They proceed from the general to the special: first the broad division between ruling and ruled; then within the ruling part divisions into political, religious, military, and within the ruled part divisions into food-producing classes and handi-craftsmen; then within each of these divisions minor ones, and so on.

Passing from the structural aspect to the functional aspect, we note that while all parts of a society have like natures and activities there is hardly any mutual dependence, and the aggregate scarcely forms a vital whole. As its parts assume different functions they become dependent on one another, so that injury to one hurts others; until in highly-evolved societies, general perturbation is caused by derangement of any portion. This contrast between undeveloped and developed societies, is due to the fact that, with increasing specialization of functions comes increasing inability in each part to perform the functions of other parts.

The organization of every society begins with a contrast between the division which carries on relations, habitually hostile, with environing societies, and the division which is devoted to procuring necessaries of life; and during the earlier stages of development these two divisions constitute the whole. Eventually there arises an intermediate division serving to transfer products and influences from part to part. And in all subsequent stages, evolution to the two earlier systems of structures depends on evolution of this additional system.

While the society as a whole has the character of its sustaining system determined by the general character of its environment, inorganic and organic, the respective parts of this system differentiate in adaptation to the circumstances of the localities; and, after primary industries have been thus localized and specialized, secondary industries dependent upon them arise in conformity with the same principle. Further, as fast as societies become compounded and recompounded and the distributing system develops, the parts devoted to each kind of industry, originally scattered, aggregate in the most favourable localities; and the localized industrial structures, unlike the governmental structures, grow regardless of the original lines of division.

Increase of size, resulting from the massing of groups, necessitates means of communication; both for achieving combined offensive and defensive actions, and for exchange of products. Scarcely traceable tracks, paths, rude roads, finished roads, successively arise; and as fast as intercourse is thus facilitated, there is a transition from direct barter to trading carried on by a separate class; out of which evolves, in course of time, a complex mercantile agency of wholesale and retail distributors. The movement of commodities effected by this agency, beginning as a slow flux to and reflux from certain places at long intervals, passes into rhythmical, regular, rapid currents; and materials for sustentation distributed hither and thither, from being few and crude become numerous and elaborated. Growing efficiency of transfer with greater variety of transferred products, increases the mutual dependence of parts at the same time that it enables each part to fulfil its function better.

Unlike the sustaining system, evolved by converse with the organic and inorganic environments, the regulating system is evolved by converse, offensive and defensive, with environing societies. In primitive headless groups temporary chieftainship results from temporary war; chronic hostilities generate permanent chieftainship; and gradually from the military control results the civil control. Habitual war, requiring prompt combination in the actions of parts, necessitates subordination. Societies in which there is little subordination disappear, and leave outstanding those in which subordination is great; and so there are established societies in which the habit fostered by war and surviving in peace, brings about permanent submission to a government. The centralized regulating system thus evolved is in early stages the sole regulating system. But in large societies that become predominantly industrial, there is added a decentralized regulating system for the industrial structures; and this, at first subject in every way to the original system, acquires at length substantial independence. Finally there arises for the distributing structures also, an independent controlling agency.

SOCIETAL TYPOLOGIES[2]

A glance at the respective antecedents of individual organisms and social organisms, shows why the last admit of no such definite classification as the first. Through a thousand generations a species of plant or animal leads substantially the same kind of life; and its successive members inherit the acquired adaptations. When changed conditions cause divergences of forms once alike, the accumulating differences arising in descendants only superficially disguise the original identity – do not prevent the grouping of the several species into a genus; nor do wider divergences that began earlier, prevent the grouping of genera into orders and orders into classes. It is otherwise with societies. Hordes of primitive men, dividing and subdividing, do, indeed, show us successions of small social aggregates leading like lives, inheriting such low structures as had resulted, and repeating those structures. But higher social aggregates propagate their respective types in much less decided ways. Though colonies tend to grow like their parents, yet the parent societies are so comparatively plastic, and the influences of new habitats on the derived societies are so great, that divergences of structure are inevitable. In the absence of definite organizations established during the similar lives of many societies descending one from another, there cannot be the precise distinctions implied by complete classification.

Two cardinal kinds of differences there are, however, of which we may avail ourselves for grouping societies in a natural manner. Primarily we may arrange them according to their degrees of composition, as simple, compound, doubly-compound, trebly-compound; and secondarily, though in a less specific way, we may divide them into the pre-

dominantly militant and the predominantly industrial – those in which the organization for offence and defence is most largely developed, and those in which the sustaining organization is most largely developed.

We have seen that social evolution begins with small simple aggregates; that it progresses by the clustering of these into larger aggregates; and that after consolidating, such clusters are united with others like themselves into still larger aggregates. Our classification, then, must begin with societies of the first or simplest order.

We cannot in all cases say with precision what constitutes a simple society; for, in common with products of evolution generally, societies present transitional stages which negative sharp divisions. As the multiplying members of a group spread and diverge gradually, it is not always easy to decide when the groups into which they fall become distinct. Here the descendants of common ancestors inhabiting a barren region, have to divide while yet the constituent families are near akin; and there, in a more fertile region, the group may hold together until clusters of families remotely akin are formed: clusters which, diffusing slowly, are held by a common bond that slowly weakens. By and by comes the complication arising from the presence of slaves not of the same ancestry, or of an ancestry but distantly allied; and these, though they may not be political units, must be recognized as units sociologically considered. Then there is the kindred complication arising where an invading tribe becomes a dominant class. Our only course is to regard as a simple society, one which forms a single working whole unsubjected to any other, and of which the parts cooperate, with or without a regulating centre, for certain public ends. Here is a table, presenting with as much definiteness as may be, the chief divisions and sub-divisions of such simple societies.

On contemplating these uncivilized societies which, though alike as being uncompounded, differ in their sizes and structures, certain generally-associated traits may be noted. Of the groups without political organization, or with but the vaguest traces of it, the lowest are those small wandering ones which live on the wild food sparsely distributed in forests, over barren tracts, or along seashores. Where small simple societies remain without chiefs though settled, it is where circumstances allow them to be habitually peaceful. Glancing down the table we find

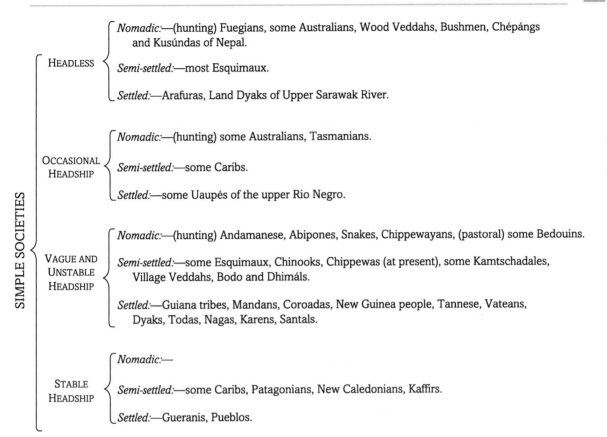

SIMPLE SOCIETIES

HEADLESS
- *Nomadic:*—(hunting) Fuegians, some Australians, Wood Veddahs, Bushmen, Chépángs and Kusúndas of Nepal.
- *Semi-settled:*—most Esquimaux.
- *Settled:*—Arafuras, Land Dyaks of Upper Sarawak River.

OCCASIONAL HEADSHIP
- *Nomadic:*—(hunting) some Australians, Tasmanians.
- *Semi-settled:*—some Caribs.
- *Settled:*—some Uaupés of the upper Rio Negro.

VAGUE AND UNSTABLE HEADSHIP
- *Nomadic:*—(hunting) Andamanese, Abipones, Snakes, Chippewayans, (pastoral) some Bedouins.
- *Semi-settled:*—some Esquimaux, Chinooks, Chippewas (at present), some Kamtschadales, Village Veddahs, Bodo and Dhimáls.
- *Settled:*—Guiana tribes, Mandans, Coroadas, New Guinea people, Tannese, Vateans, Dyaks, Todas, Nagas, Karens, Santals.

STABLE HEADSHIP
- *Nomadic:*—
- *Semi-settled:*—some Caribs, Patagonians, New Caledonians, Kaffirs.
- *Settled:*—Gueranis, Pueblos.

reason for inferring that the changes from the hunting life to the pastoral, and from the pastoral to the agricultural, favour increase of population, the development of political organizations, of industrial organization, and of the arts; though these causes do not of themselves produce these results.

The second table contains societies which have passed to a slight extent, or considerably, or wholly, into a state in which the simple groups have their several governing heads subordinated to a general head. The stability or instability alleged of the headship in these cases, refers to the headship of the composite group, and not to the headships of the simple groups. As might be expected, stability of this compound headship becomes more marked as the original unsettled state passes into the completely settled state: the nomadic life obviously making it difficult to keep the heads of groups subordinate to a general head. Though not in all cases accompanied by considerable organization, this coalescence evidently conduces to organization. The completely-settled compound societies are mostly characterized by division into ranks, four, five, or six, clearly marked

off; by established ecclesiastical arrangements; by industrial structures that show advancing division of labour, general and local; by buildings of some permanence clustered into places of some size; and by improved appliances of life generally.

In the succeeding table are placed societies formed by the recompounding of these compound groups, or in which many governments of the types tabulated above have become subject to a still higher government. The first notable fact is that these doubly-compound societies are all completely settled. Along with their greater integration we see in many cases, though not uniformly, a more elaborate and stringent political organization. Where complete stability of political headship over these doubly-compound societies has been established, there is mostly, too, a developed ecclesiastical hierarchy. While becoming more complex by division of labour, the industrial organization has in many cases assumed a caste structure. To a greater or less extent, custom has passed into positive law; and religious observances have grown definite, rigid, and complex. Towns and roads have become general; and

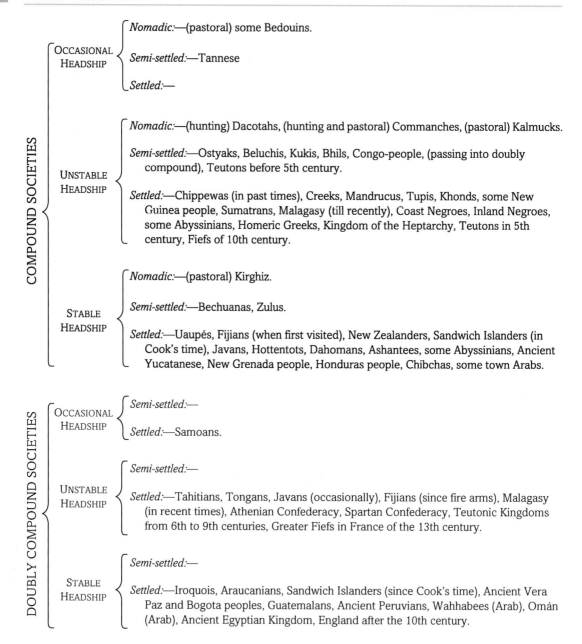

considerable progress in knowledge and the arts has taken place.

There remain to be added the great civilized nations which need no tabular form, since they mostly fall under one head – trebly compound. Ancient Mexico, the Assyrian Empire, the Egyptian Empire, The Roman Empire, Great Britain, France, Germany, Italy, Russia, may severally be regarded as having reached this stage of composition, or perhaps, in some cases, a still higher stage. Only in respect of

the stabilities of their governments may they possibly require classing apart – not their political stabilities in the ordinary sense, but their stabilities in the sense of continuing to be the supreme centres of these great aggregates. So defining this trait, the ancient trebly-compound societies have mostly to be classed as unstable; and of the modern, the Kingdom of Italy and the German Empire have to be tested by time.

As already indicated, this classification must not be taken as more than a rough approximation to the truth. In some cases the data furnished by travellers and others are inadequate; in some cases their accounts are conflicting; in some cases the composition is so far transitional that it is difficult to say under which of two heads it should come. Here the gens or the phratry may be distinguished as a local community; and here these groups of near or remote kinsmen are so mingled with other such groups as practically to form parts of one community. Evidently the like combination of several such small communities, passing through stages of increasing cohesion, leaves it sometimes doubtful whether they are to be regarded as many or as one. And when, as with the larger social aggregates, there have been successive conquests, resulting unions, subsequent dissolutions, and re-unions otherwise composed, the original lines of structure become so confused or lost that it is difficult to class the ultimate product.

But there emerge certain generalizations which we may safely accept. The stages of compounding and re-compounding have to be passed through in succession. No tribe becomes a nation by simple growth; and no great society is formed by the direct union of the smallest societies. Above the simple group the first stage is a compound group inconsiderable in size. The mutual dependence of parts which constitutes it a working whole, cannot exist without some development of lines of intercourse and appliances for combined action; and this must be achieved over a narrow area before it can be achieved over a wide one. When a compound society has been consolidated by the co-operation of its component groups in war under a single head – when it has simultaneously differentiated somewhat its social ranks and industries, and proportionately developed its arts, which all of them conduce in some way to better co-operation, the compound society becomes practically a single one. Other societies of the same order, each having similarly reached a stage of organization alike required and made possible by this co-ordination of actions throughout a larger mass, now form bodies from which, by conquest or by federation in war, may be formed societies of the doubly-compound type. The consolidation of these has again an accompanying advance of organization distinctive of it – an organization for which it affords the scope and which makes it practicable – an organization having a higher complexity in its regulative, distributive, and industrial systems. And at later stages, by kindred steps, arise the still larger aggregates having still more complex structures. In this order has social evolution gone on, and only in this order does it appear to be possible. Whatever imperfections and incongruities the above classification has, do not hide these general facts – that there are societies of these different grades of composition; that those of the same grade have general resemblances in their structures; and that they arise in the order shown.

NOTES

1   From *The Principles of Sociology* (London: Williams and Norgate, 1876), vol. 1, part 2, pp. 465–69, 613–16.
2   *Ibid.*, pp. 569–76.

'Civilisation' and 'What is True Civilisation?'

from *Hind Swaraj* (1910)

M. K. Gandhi

Editors' Introduction

M. K. Gandhi was born in 1869 in the town of Porbandar in Kathiawar, a princely state in British India. After an early marriage, Gandhi attended University College London in 1888 to train as a barrister. In the imperial capital, Gandhi engaged with vegetarians and birth control advocates about bodily restraint, and with Theosophists about universal religion. After returning to India, a frustrated career in Bombay and Rajkot led Gandhi to accept a job with an Indian firm in Natal, South Africa, where he lived from 1893 to 1914. It was in Natal that Gandhi first became a social and political activist. He and his family initiated attempts in communal living, not without internal conflict. In Durban and Pietermaritzburg, and famously in the train between them, Gandhi experienced a blunt form of racism that led him to a life of politics.

Gandhi's first campaign was a failed attempt at protecting Indians' rights to vote. It would be fair to say that Gandhi's political stance was limited to Indian incorporation into the mainstream, rather than black enfranchisement more generally, and debates continue on the extent to which political exigencies forced inclusion of the majority of Indian indentured workers. In South Africa, Gandhi developed his particular form of non-violent ideology – *satyagraha*, or 'truth-force' – and its related practice of passive resistance. *Satyagraha* combined philosophical and Orientalist ideas acquired in London with experiments in communal living in Natal, in what he saw as a distinctly non-Western form of protest backed by metaphysical certainty in the universal truth. There was also a certain political savvy built into this protest form, as Gandhi became the arbiter of moral authority rather than ceding power to rank-and-file protesters who may have expressed their militancy differently. Indeed, this executive power allowed Gandhi to make important exceptions to his rule of non-violence, in the South African War and First World War, but not the Second World War, by which point Gandhi refused Indian cooperation for a war fought for 'freedom' denied to colonised populations. One of Gandhi's lasting contributions to South Africa was the formation of the Natal Indian Congress. Another key contribution was the Gujarati writings first published together in English translation as *Hind Swaraj* in Johannesberg, in 1910, from which the reading is excerpted.

On returning to India in 1914, Gandhi became involved in a series of campaigns, sometimes, as in Champaran and Kheda, linking specific disputes to the broader machinery of anti-colonialism, and at other times using the long-distance protest march as a means of galvanising the movement across the country, as in the Salt *Satyagraha*. Increasingly, the Indian National Congress found his populism useful, as Jawaharlal Nehru would later note in his autobiography. Gandhi was given the title 'Mahatma', or great soul, and 'Bapu', or father. Gandhi's work also continued on social issues, most provocatively on the question of 'untouchability' and its relationship to caste and Hinduism. Gandhi was predisposed to argue against a hard-and-fast connection, preferring to think of the reform of caste, and a kind of tutelage of people outside caste who he referred to as 'Harijans', or children of the Hindu god Hari. This would be strongly rejected by B. R. Ambedkar, the leader of India's 'Depressed Classes'. After the Quit India Movement during the Second World War, the British agreed to begin a process of decolonisation, but this would be on the basis of the Partition of British India into India and Pakistan. Gandhi was deeply against the Partition, he mourned the riots that ensued, and he fasted until the riots subsided.

In the reading that follows from *Hind Swaraj*, Gandhi asks, and answers through a mock interview, why India has become a subject nation. His answer diverges from the commonly held view that Gandhi was a latecomer to nineteenth-century romanticism. Gandhi's argument hinges on a morality intrinsic to India's 'true civilisation', compromised by falling for the allure of 'modern' civilisation. What is important is that Gandhi concedes that there is an immanent developmental process at work in the West, involving industrial revolution and societal transformation. He only sees it as part of an affliction from which *both* the English and the colonised Indian can be rescued. Wary that he may be accused of romanticism, Gandhi concludes this excerpt by arguing that Indian civilisation is not without several internal faults, but that colonial rule may have made Indians more ready to combat them. The reader should be attentive to Gandhi's rhetoric and logic, as he attempts to maintain an argument about India's civilisational difference while building a defence of anti-colonial nationalism through 'non-Western' tools of bodily regulation, non-violence and *Satyagraha*.

Key references

M. K. Gandhi (1997 [1910]) *Hind Swaraj and Other Writings*, ed. A. J. Parel, Cambridge: Cambridge University Press.
— (1957) *An Autobiography: The Story of My Experiments with Truth*, Boston, MA: Beacon Press.
— (1994) *The Collected Works of Mahatma Gandhi*, New Delhi: Publications Division, Ministry of Information and Broadcasting, Govt. of India.
David Arnold (2001) *Gandhi (Profiles in Power)*, New York: Longman.
Partha Chatterjee (1986) *Nationalist Thought and the Colonial World: A Derivative Discourse*, Minneapolis: University of Minnesota Press.

CIVILISATION

READER: Now you will have to explain what you mean by civilisation.

EDITOR: It is not a question of what I mean. Several English writers refuse to call that civilisation which passes under that name. Many books have been written upon that subject. Societies have been formed to cure the nation of the evils of civilisation. A great English writer has written a work called 'Civilization: its Cause and Cure.' Therein he has called it a disease.

READER: Why do we not know this generally?

EDITOR: The answer is very simple. We rarely find people arguing against themselves. Those who are intoxicated by modern civilisation are not likely to write against it. Their care will be to find out facts and arguments in support of it, and this they do unconsciously, believing it to be true. A man, whilst he is dreaming, believes in his dream; he is undeceived only when he is awakened from his sleep. A man labouring under the bane of civilisation is like a dreaming man. What we usually read are the works of defenders of modern civilisation, which undoubtedly claims among its votaries very brilliant and even some very good men. Their writings hypnotise us. And so, one by one, we are drawn into the vortex.

READER: This seems to be very plausible. Now will you tell me something of what you have read and thought of this civilisation?

EDITOR: Let us first consider what state of things is described by the word 'civilisation'. Its true test lies in the fact that people living in it make bodily welfare the object of life. We will take some examples. The people of Europe today live in better built houses than they did a hundred years ago. This is considered an emblem of civilisation, and this is also a matter to promote bodily happiness. Formerly, they wore skins, and used as their weapons spears. Now, they wear long trousers, and, for embellishing their bodies, they wear a variety of clothing, and, instead of spears, they carry with them revolvers containing five or more chambers. If people of a certain country, who have hitherto not been in the habit of wearing much clothing, boots, etc., adopt European clothing, they are supposed to have become civilised out of savagery. Formerly, in Europe, people ploughed their lands mainly by manual labour. Now, one man can plough a vast tract by means of steam-engines, and can thus amass great wealth. This is called a sign of civilisation. Formerly, the fewest men wrote books that were most valuable. Now, anybody writes and prints anything he likes and poisons people's mind. Formerly, men travelled in wagons; now they fly through the air in trains at the rate of four hundred and more miles per day. This is considered the height of civilisation. It has been stated that, as men progress, they shall be able to travel in airships and reach any part of the world in a few hours. Men will not need the use of their hands and feet. They will press a button and they will have their clothing by their side. They will press another button and they will have their newspaper. A third, and a motorcar will be in waiting for them. They will have a variety of delicately dished-up food. Everything will be done by machinery. Formerly, when people wanted to fight with one another, they measured between them their bodily strength; now it is possible to take away thousands of lives by one man working behind a gun from a hill. This is civilisation. Formerly, men worked in the open air only so much as they liked. Now, thousands of workmen meet together and for the sake of maintenance work in factories or mines. Their condition is worse than that of beasts. They are obliged to work, at the risk of their lives, at most dangerous occupations, for the sake of millionaires. Formerly, men were made slaves under physical compulsion, now they are enslaved by temptation of money and of the luxuries that money can buy. There are now diseases of which people never dreamt before, and an army of doctors is engaged in finding out their cures, and so hospitals have increased. This is a test of civilisation. Formerly, special messengers were required and much expense was incurred in order to send letters; today, anyone can abuse his fellow by means of a letter for one penny. True, at the same cost, one can send one's thanks also. Formerly, people had two or three meals consisting of home-made bread and vegetables; now, they require something to eat every two hours, so that they have hardly leisure for anything else. What more need I say? All this you can ascertain from several authoritative books. These are all true tests of civilisation. And, if anyone speaks to the contrary, know that he is ignorant. This civilisation takes note neither of morality nor of religion. Its votaries calmly state that their business is not to teach religion. Some even consider it to be a superstitious growth. Others put on the cloak of religion, and prate about morality. But, after twenty years' experience, I have come to the conclusion that immorality is often taught in the name of morality. Even a child can understand that in all I have described above there can be no inducement to morality. Civilisation seeks to increase bodily comforts, and it fails miserably even in doing so.

This civilisation is irreligion, and it has taken such a hold on the people in Europe that those who are in it appear to be half mad. They lack real physical strength or courage. They keep up their energy by intoxication. They can hardly be happy in solitude. Women, who should be the queens of households, wander in the streets, or they slave away in factories. For the sake of a pittance, half a million women in England alone are labouring under trying circumstances in factories or similar institutions. This awful fact is one of the causes of the daily growing suffragette movement.

This civilisation is such that one has only to be patient and it will be self-destroyed. According to the teaching of Mahomed this would be considered a Satanic civilisation. Hinduism calls it the Black Age. I cannot give you an adequate conception of it. It is eating into the vitals of the English nation. It must be shunned. Parliaments are really emblems of slavery. If you will sufficiently think over this, you will entertain the same opinion, and cease to blame the English. They rather deserve our sympathy. They are a shrewd nation and I, therefore, believe that they will cast off the evil. They are enterprising and industrious, and their mode of thought is not inherently immoral. Neither are they bad at heart. I, therefore, respect them. Civilisation is not an incurable disease, but it should never be forgotten that the English people are at present afflicted by it.

[...]

WHAT IS TRUE CIVILISATION?

READER: You have denounced railways, lawyers and doctors. I can see that you will discard all machinery. What, then, is civilisation?

EDITOR: The answer to that question is not difficult. I believe that the civilisation India has evolved is not to be beaten in the world. Nothing can equal the seeds sown by our ancestors. Rome went, Greece shared the same fate, the might of the Pharaohs was broken, Japan has become westernised, of China nothing can be said, but India is still, somehow or other, sound at the foundation. The people of Europe learn their lessons from the writings of the men of Greece or Rome, which exist no longer in their former glory. In trying to learn from them, the Europeans imagine that they will avoid the mistakes of Greece and Rome. Such is their pitiable condition. In the midst of all this, India remains immovable, and that is her glory. It is a charge against India that her people are so uncivilised, ignorant and stolid, that it is not possible to induce them to adopt any changes. It is a charge really against our merit. What we have tested and found true on the anvil of experience, we dare not change. Many thrust their advice upon India, and she remains steady. This is her beauty; it is the sheet-anchor of our hope.

Civilisation is that mode of conduct which points out to man the path of duty. Performance of duty and observance of morality are convertible terms. To observe morality is to attain mastery over our mind and our passions. So doing, we know ourselves. The Gujarati equivalent for civilisation means 'good conduct'.

If this definition be correct, then India, as so many writers have shown, has nothing to learn from anybody else, and this is as it should be. We notice that mind is a restless bird; the more it gets the more it wants, and still remains unsatisfied. The more we indulge our passions, the more unbridled they become. Our ancestors, therefore, set a limit to our indulgences. They saw that happiness was largely a mental condition. A man is not necessarily happy because he is rich, or unhappy because he is poor. The rich are often seen to be unhappy, the poor to be happy. Millions will always remain poor. Observing all this, our ancestors dissuaded us from luxuries and pleasures. We have managed with the same kind of plough as it existed thousands of years ago. We have retained the same kind of cottages that we had in former times, and our indigenous education remains the same as before. We have had no system of life-corroding competition. Each followed his own occupation or trade, and charged a regulation wage. It was not that we did not know how to invent machinery, but our forefathers knew that, if we set our hearts after such things, we would become slaves and lose our moral fibre. They therefore, after due deliberation, decided that we should only do what we could with our hands and feet. They saw that our real happiness and health consisted in a proper use of our hands and feet. They further reasoned that large cities were a snare and a useless encumbrance, and that people would not be happy in them, that there would be gangs of thieves and robbers, prostitution and vice flourishing in them, and that poor men would be robbed by rich men. They were, therefore, satisfied with small villages. They saw that kings and their swords were inferior to the sword of ethics, and they, therefore, held the sovereigns of the earth to be inferior to the Rishis and the Fakirs. A nation with a constitution like this is fitter to teach others than to learn from others. This nation had courts, lawyers and doctors, but they were all within bounds. Everybody knew that these professions were not particularly superior; moreover, these vakils and vaids did not rob people; they were considered people's dependants, not their masters. Justice was tolerably fair. The ordinary rule was to avoid courts. There were no touts to lure people into them. This evil, too, was noticeable only in and around capitals. The common people lived independently, and followed their agricultural occupation. They enjoyed true Home Rule.

And where this cursed modern civilisation has not reached, India remains as it was before. The inhabitants of that part of India will very properly laugh at your new-fangled notions. The English do not rule over them, nor will you ever rule over them. Those in whose name we speak we do not know, nor do they know us. I would certainly advise you and those like you who love the motherland to go into the interior that has yet not been polluted by the railways, and to live there for six months; you might then be patriotic and speak of Home Rule.

Now you see what I consider to be real civilisation. Those who want to change conditions such as I have described are enemies of the country and are sinners.

READER: It would be all right if India were exactly as you have described it, but it is also India where there are hundreds of child widows, where two-year-old babies are married, where twelve-year-old girls are mothers and housewives, where women practise polyandry, where the practice of Niyog obtains, where, in the name of religion, girls dedicate themselves to prostitution, and where, in the name of religion, sheep and goats are killed. Do you consider these also symbols of the civilisation that you have described?

EDITOR: You make a mistake. The defects that you have shown are defects. Nobody mistakes them for ancient civilisation. They remain in spite of it. Attempts have always been made, and will be made, to remove them. We may utilise the new spirit that is born in us for purging ourselves of these evils. But what I have described to you as emblems of modern civilisation are accepted as such by its votaries. The Indian civilisation as described by me has been so described by its votaries. In no part of the world, and under no civilisation, have all men attained perfection. The tendency of Indian civilisation is to elevate the moral being, that of the Western civilisation is to propagate immorality. The latter is godless, the former is based on a belief in God. So understanding and so believing, it behoves every lover of India to cling to the old Indian civilisation even as a child clings to its mother's breast.

PART THREE

Reform, revolution, resistance

The largely nineteenth-century contexts within which Marx, Spencer and Gandhi formed their views were changed markedly by economic and political developments between 1910 and 1950. The First World War of 1914–18 led in Russia to the Bolshevik revolution of 1917 and soon thereafter to Soviet support for the politics of anti-imperialism. A number of important leaders from the colonised world visited the Soviet Union in the 1920s and 1930s. Most were impressed by the ways in which planning and technocratic zeal were put in the service of a development process that seemed to secure high rates of economic growth for the benefit of all citizens. Jawaharlal Nehru, the future prime minister of India (1947–64) was one such figure, although he remained sceptical about the social costs of enforced growth. The First World War also led to the radicalisation of many nationalist organisations, and most obviously the one that Nehru would later help to lead, the Indian National Congress. Nationalist politicians noted that tens of thousands of Indians had died fighting for the British Empire. The expected rewards for this sacrifice, however, had not been forthcoming. The massacre of civilians at Jallianwala Bagh, Amritsar, by General Dyer and his troops in 1919, seemed to confirm that Empire was growing more and not less repressive. Resistance was mandatory, just as it would be for some people in the Age of Development that succeeded formal Empire.

Meanwhile, in Latin America, the period that stretched from 1914 to 1945 was seen as a time of relative detachment from the United States and Western Europe. It was a period when Latin Americans could experiment with their own versions of industrial development. These were now extensively debated and these debates would later move into the mainstream of development studies (see Part 4). The Depression years of the late 1920s and 1930s also encouraged intellectuals in Western Europe to reconsider the economic and moral case for market capitalism. Some amongst them, including John Maynard Keynes and Karl Polanyi, began to write about the political consequences of an economic system that was prone to continuous and possibly worsening cycles of boom and bust.

Keynes is mainly known in academic circles for his *General Theory of Employment, Interest and Money*. This famous book from 1936 challenged the prevailing view that labour markets would move to a full employment equilibrium of their own accord. In political circles Keynes is best known as one of the architects of the Bretton Woods system. Keynes was the UK's main representative at the July 1944 conference held at Bretton Woods, New Hampshire, to discuss post-war international monetary arrangements. He was also the chief proponent of the International Bank for Reconstruction and Development, now part of the World Bank Group. However, Keynes's proposal for a truly global currency, Bancor, was squashed by the American delegation. Instead, the dollar was made to serve as the world's reserve currency. For just about twenty-five years the price of the US dollar was fixed against gold. The prices of all other leading currencies were then fixed against the US dollar. (As many critics have remarked, the system worked only so long as the United States was prepared to subordinate its national interests in favour of the greater global good of a stable international economy. This would change in the 1960s when the United States began to print dollars to pay for its war in Vietnam. This weakened the once stable link between the greenback and gold. The United States ended the gold window in 1971. By 1973, the exchange values of most leading currencies were floating against each other.) In the reading reproduced here we see Keynes looking forward from 1928–30 to the end of the century and beyond. We also see his tremendous optimism, so much on display at Bretton Woods. Writing in the midst of the Depression, Keynes predicted that 'mankind is solving its economic problem . . . the standard of life in progressive countries one hundred years hence will be between four and eight times as high as it is today'.

This extraordinary prediction was largely borne out, at least in what became known as the countries of the Organization for Economic Cooperation and Development (OECD). Whether Keynes foresaw the even more impressive growth of East Asia is hard to say. It is also clear that Keynes's optimism was underpinned by his faith in the capacity of an educated class of managers to promote technical change for the greater good. This idea became central to the Promethean Age of Development that began after the Second World War (see Part 4). At the same time, however, we also see in Keynes's essay the classic Bloomsbury or Fabian view that love of money is degrading. 'The love of money as a possession – as distinguished from the love of money as a means to the enjoyments and realities of life – will be recognised for what it is, a somewhat disgusting morbidity, one of those semi-criminal, semi-pathological propensities which one hands over with a shudder to the specialists in mental disease.' Keynes, in short, a man who once lampooned bankers as the least realistic and most romantic of men, held the deeply romantic view that capitalism could be tamed, managed and detached from the grubby money motive. There was a measure of arrogance here, as well. Capitalism was to be run not by bankers but by an educated and urbane elite, by the chosen few who could see a future Paradise for the common man (another perspective which informed Developmentalism in the 1960s and 1970s). Everything is possible, Keynes concludes, so long as Reason takes precedence over Commerce, and 'assuming no important wars and no important increase in population'. (Keynes was attracted to eugenics, as were many Fabian socialists: here is a link back to social Darwinism.)

Optimism is also apparent in the Polanyi reading which follows. Indeed, the similarities between the two essays are quite striking. Polanyi joins with Keynes in suggesting that the inter-war period marked another 'great transformation' in the social affairs of Europe. This time it was not from feudalism to capitalism, but away from capitalism as a self-regulating market system. '[W]e are witnessing a development under which the economic system ceases to lay down the law to society and the primacy of society over that system is secured.' In Polanyi's view, labour, land and money are in the process of being removed from the market system. The coming era would see widespread fixing of staple food prices and the de facto nationalisation of money and land markets. In addition, wage rates would be determined by forms of corporatism in which the power of trade unions would be key.

The major political question for Polanyi, as for Keynes, concerned the character of the corporate settlements that would pertain, and within which freedoms could be exercised. Late in his life Keynes was taken to task by Friedrich von Hayek, the great Austrian economist and political philosopher. Keynes had proposed a body of economic theory and practice that offered a road to serfdom, Hayek argued in 1944. Keynes, for his part, felt the need to declare that he was not a Keynesian, and that he remained at heart a social and economic Liberal. He even called Hayek's *The Road to Serfdom* a 'grand book' with which he was 'in deeply moved agreement'. Polanyi, in contrast, associates Hayekian liberalism with free market dogmatism, and even libertarianism, and comes closer to an avowedly 'Keynesian' position. In his view, mankind has to be saved from Liberalism, just as much as it has to be saved from Fascism and Communism. None of these systems guarantees real freedoms, he suggests: negative freedoms, of course, but also the positive freedoms to work, eat, gain an education, and so on. (Here, Polanyi anticipates some of the arguments that would be made more than fifty years later by Amartya Sen in his celebrated book *Development as Freedom*, albeit from a more Owenite perspective.)

Polanyi's sensitivity to the different ways in which capitalism is organised, and managed politically, prefigures some recent work on governance (see Part 7). Indeed, it is interesting that Polanyi has been rediscovered of late, and that his work can be cited against views of Development which propose a singular route to modernity or some other state of bliss. We see a similar concern for particularities (of Empire, of the agrarian question) in the next two readings, those by J. S. Furnivall and Alexander Gerschenkron. The Furnivall excerpt sets the scene for a book-length analysis of colonial practices in

Burma and Netherlands India. Furnivall's point of entry is the suggestion that, 'In Burma the British have from the first relied on western principles of rule, on the principles of law and economic freedom; in Netherlands India the Dutch have tried to conserve and adapt to modern use the tropical principles of custom and authority.' Although not uncontroversial, versions of this suggestion are now at the heart of one set of debates on the meanings of good governance and the path-dependence of institutional settlements. We pick them up in Part 7. For his part, Gerschenkron explores the connections between bread and democracy in pre-1914 Germany. Gerschenkron is best known in development studies for his work on the problems and possibilities of late industrialisation. We have selected this reading, however, because it perfectly anticipates work in the 1960s by Barrington Moore, Jr. on the social origins of dictatorship and democracy. It also heralds more recent work by Daron Acemoglu and James Robinson on the economic origins of the same. This too is institutionalism *avant la lettre*.

Finally in this Part, and we might have placed it elsewhere in the reader, we include a brief essay by Frantz Fanon. Once again, we have not excerpted from the author's best-known work (*The Wretched of the Earth*). But classic Fanonian themes are still on display in this account of the politics of radio in Algeria. Fanon was not a follower of Gandhi, although like Gandhi he experienced the racism of the West first-hand in the capital city of a metropolitan power – Paris for Fanon, London for Gandhi. Fanon was also forced to think through the process of colonial subjection in a very different way than Gandhi. British and French colonial ideology differed in their methods of harnessing race, culture and imperial citizenship to projects of accumulation and rule. But if Fanon rejected Gandhi's tactic of non-violence, he did share with Gandhi a propensity to work with and against the grain of colonial modes of power. We have thus chosen an excerpt from *A Dying Colonialism*. Here, Fanon shows us how the Algerian people turned to their advantage technologies that were first advanced as forms of domination. The radio is a classic case in point. Radio broadcasting in Algeria was funded by the French government for much the same purposes that the United States was then beginning to fund *Voice of America*. As the revolution in Algeria wore on, however, the *Voice of Algeria* became a key tool for anti-colonial organising in the qasbahs and villages, in those dispersed and necessarily secret sites that knitted together the sensibility of being part of a national liberation movement. Like Gandhi, Fanon reminds us that the modern fruits of development – languages, technologies, practices and sentiments – can prove useful in the arduous work of sustaining anti-colonial revolution.

REFERENCES AND FURTHER READING

Acemoglu, D. and Robinson, J. (2006) *Economic Origins of Dictatorship and Democracy*, Cambridge: Cambridge University Press.

Adas, M. (1989) *Machines as the Measures of Men: Science, Technology and Ideologies of Western Dominance*, Ithaca, NY: Cornell University Press.

Dorfman, R. (1991) 'Economic development from the beginning to Rostow', *Journal of Economic Literature* 29: 573–91.

Dubois, L. (2004) *Avengers of the New World: The Story of the Haitian Revolution*, Cambridge, MA: Belknap Press.

Fanon, F. (1992 [1965]) *The Wretched of the Earth*, Harmondsworth: Penguin.

Gerschenkron, A. (1962) *Economic Backwardness in Historical Perspective*, Cambridge, MA: Harvard University Press.

Hayek, F. von (2001 [1944]) *The Road to Serfdom*, London: Routledge.

James, C. L. R. (1989) *The Black Jacobins*, New York: Vintage Books.

Keynes, J. M. (1973 [1936]) *The General Theory of Employment, Interest and Money*, Basingstoke: Macmillan.

Lenin, V. I. (1974 [1905]) *The Development of Capitalism in Russia*, Moscow: Progress.

Mannoni, O. (1950) *Psychologie de la Colonisation*, Paris: Editions du Seuil.

Moore, B. (1966) *The Social Origins of Dictatorship and Democracy: Lord and Peasant in the Making of the Modern World*, Boston, MA: Beacon Press.

Nehru, J. (1962 [1936]) *An Autobiography*, Bombay: Allied.

Nurkse, R. (1953) *Problems of Capital Formation in Underdeveloped Countries*, New York: Oxford University Press.

Rosenstein-Radan, P. (1943) 'Problems of industrialization in eastern and south-eastern Europe', *Economic Journal* 53: 202–11.

Sen, A. K. (2000) *Development as Freedom*, New York: Anchor Books.

Skidelsky, R. (2001) *John Maynard Keynes: Volume III – Fighting for Britain, 1937–1946*, London: Papermac.

'Economic Possibilities for our Grandchildren'

from *Essays in Persuasion* (1931)

John Maynard Keynes

Editors' Introduction

John Maynard Keynes was born in Cambridge, England, in 1883 and died in 1946. By the time of his death, Keynes was widely recognised as the most important and influential economist of the first half of the twentieth century. Trained in mathematics and economics at Cambridge University, Keynes later worked both for the university as an academic and for the British Treasury as an adviser to the Chancellor of the Exchequer (Finance Minister). Keynes observed the 1919 Peace Conference as a Treasury representative. His book on *The Economic Consequences of the Peace* appeared in the same year and established Keynes as a leading critic of the Allies' decision to impose crippling war reparations on the defeated Germans. Partly as a result of this work, and of his subsequent writings on monetary policy and reform, Keynes came to the view that variations in the overall level of economic activity were determined principally by fluctuations in aggregate demand. In his classic work, *The General Theory of Employment, Interest and Money* (1936), Keynes argued that governments should spend heavily on public works in times of high unemployment, squeezing credit and money when the economy is heated up. Labour markets could not be expected to clear – or reach a full employment equilibrium – on the basis of working people choosing lower wages for themselves.

Throughout much of the Second World War Keynes worked with colleagues in the UK and the United States – notably the American, Harry Dexter White – on plans for a post-war international monetary system. The system would need to be less tied to gold than hitherto, and less prone to boom and bust and damaging competitive devaluations. This work led to the Bretton Woods conference of 1944 and ultimately to the founding of the International Bank for Reconstruction and Development (IBRD, or World Bank) and the International Monetary Fund (IMF). Keynes's suggestion that the IBRD should be able to issue a truly global currency, Bancor, was rejected by the United States, which wanted the powers of seignorage that attach to the capacity to issue the main international unit of account (the US dollar). American hegemony was also demonstrated in the decision to house the headquarters of both Bretton Woods institutions in Washington, DC; Keynes argued the merits of London.

The article that is reproduced here comes from a collection of Keynes's essays published under the title *Essays in Persuasion*. Keynes often gave talks on radio in the UK and wrote for newspapers and magazines. He was a well-known public intellectual, married to a Russian ballerina from 1925, and was a leading light in the so-called Bloomsbury group of Cambridge and London-based intellectuals (including Virginia Woolf, Duncan Grant and Lytton Strachey). In this reading, Keynes tries to reassure his broader public that, while times are tough just now (the early 1930s), in the long run all will come good. Barring new wars or excessive population growth, Keynes suggests, the chances are high that ordinary men and women in the UK will be four to eight (and most likely eight) times better off in 2030 than in 1930. The main problem for future generations, he concludes, will be how to live wisely. How will people use their greater leisure time? How will they lead cultivated lives, freeing themselves from the disease of money worship? How will people grasp the opportunities for freedom that constant accumulation must provide? These remain key issues for many people in more affluent societies, although some readers will want to challenge Keynes's view that economic growth and prosperity will lead inevitably to more leisure time. What evidence is there that people in richer countries are working less hours, or less productively, than their grandparents? Where too is the question of female participation in the organised labour force in Keynes's work? Are there any signs in this reading that Keynes foresaw the massive growth of paid female employment that has been apparent in many countries since the 1960s? If not, why not? What (implicit) assumptions was Keynes making about the domains of 'work' and the 'home' (see also the essays by Boserup and Elson in Parts 5 and 6).

Key references

John Maynard Keynes (1985) *The Collected Writings of John Maynard Keynes, Vol. IX Essays in Persuasion*, London: Macmillan, Cambridge University Press for the Royal Economic Society.
— (1960) *The General Theory of Employment, Interest and Money*, London: Macmillan.
— (1971) *The Economic Consequences of the Peace*, London: Macmillan.
R. Skidelsky (2004) *John Maynard Keynes 1883–1946: Economist, Philosopher, Statesman*, London: Pan.
F. von Hayek (2001) *The Road to Serfdom*, London: Routledge.

This essay was first presented in 1928 as a talk to several small societies, including the Essay Society of Winchester College and the Political Economy Club at Cambridge. In June 1930 Keynes expanded his notes into a lecture on 'Economic Possibilities for our Grandchildren' which he gave at Madrid. It appeared in literary form in two instalments in the *Nation and Athenaeum*, 11 and 18 October 1930, in the midst of the slump.

I

We are suffering just now from a bad attack of economic pessimism. It is common to hear people say that the epoch of enormous economic progress which characterised the nineteenth century is over; that the rapid improvement in the standard of life is now going to slow down – at any rate in Great Britain; that a decline in prosperity is more likely than an improvement in the decade which lies ahead of us.

I believe that this is a wildly mistaken interpretation of what is happening to us. We are suffering, not from the rheumatics of old age, but from the growing-pains of over-rapid changes, from the painfulness of

readjustment between one economic period and another. The increase of technical efficiency has been taking place faster than we can deal with the problem of labour absorption; the improvement in the standard of life has been a little too quick; the banking and monetary system of the world has been preventing the rate of interest from falling as fast as equilibrium requires. And even so, the waste and confusion which ensue relate to not more than 7½ per cent of the national income; we are muddling away one and sixpence in the £, and have only 18s 6d, when we might, if we were more sensible, have £1; yet, nevertheless, the 18s 6d mounts up to as much as the £1 would have been five or six years ago. We forget that in 1929 the physical output of the industry of Great Britain was greater than ever before, and that the net surplus of our foreign balance available for new foreign investment, after paying for all our imports, was greater last year than that of any other country, being indeed 50 per cent greater than the corresponding surplus of the United States. Or again – if it is to be a matter of comparisons – suppose that we were to reduce our wages by a half, repudiate four-fifths of the national debt, and hoard our surplus wealth in barren gold instead of lending it at 6 per cent or more, we should resemble the now much-envied France. But would it be an improvement?

The prevailing world depression, the enormous anomaly of unemployment in a world full of wants, the disastrous mistakes we have made, blind us to what is going on under the surface – to the true interpretation of the trend of things. For I predict that both of the two opposed errors of pessimism which now make so much noise in the world will be proved wrong in our own time – the pessimism of the revolutionaries who think that things are so bad that nothing can save us but violent change, and the pessimism of the reactionaries who consider the balance of our economic and social life so precarious that we must risk no experiments.

My purpose in this essay, however, is not to examine the present or the near future, but to disembarrass myself of short views and take wings into the future. What can we reasonably expect the level of our economic life to be a hundred years hence? What are the economic possibilities for our grandchildren?

From the earliest times of which we have record – back, say to two thousand years before Christ – down to the beginning of the eighteenth century, there was no very great change in the standard of life of the average man living in the civilised centres of the earth. Ups and downs certainly. Visitations of plague, famine, and war. Golden intervals. But no progressive, violent change. Some periods perhaps 50 per cent better than others – at the utmost 100 per cent better – in the four thousand years which ended (say) in A.D. 1700.

This slow rate of progress, or lack of progress, was due to two reasons – to the remarkable absence of important technical improvements and to the failure of capital to accumulate.

The absence of important technical inventions between the prehistoric age and comparatively modern times is truly remarkable. Almost everything which really matters and which the world possessed at the commencement of the modern age was already known to man at the dawn of history. Language, fire, the same domestic animals which we have today, wheat, barley, the vine and the olive, the plough, the wheel, the oar, the sail, leather, linen and cloth, bricks and pots, gold and silver, copper, tin, and lead – and iron was added to the list before 1000 B.C. – banking, statecraft, mathematics, astronomy, and religion. There is no record of when we first possessed these things.

At some epoch before the dawn of history – perhaps even in one of the comfortable intervals before the last ice age – there must have been an era of progress and invention comparable to that in which we live today. But through the greater part of recorded history there was nothing of the kind.

The modern age opened, I think, with the accumulation of capital which began in the sixteenth century. I believe – for reasons with which I must not encumber the present argument – that this was initially due to the rise of prices, and the profits to which that led, which resulted from the treasure of gold and silver which Spain brought from the New World into the Old. From that time until today the power of accumulation by compound interest, which seems to have been sleeping for many generations, was reborn and renewed its strength. And the power of compound interest over two hundred years is such as to stagger the imagination.

Let me give in illustration of this a sum which I have worked out. The value of Great Britain's foreign investments today is estimated at about £4,000 million. This yields us an income at the rate of about 6½ per cent. Half of this we bring home and enjoy; the other half, namely, 3¼ per cent, we leave to accumulate abroad at compound interest. Something of this sort has now been going on for about 250 years.

For I trace the beginnings of British foreign investment to the treasure which Drake stole from Spain in 1580. In that year he returned to England bringing with him the prodigious spoils of the *Golden Hind*. Queen Elizabeth was a considerable shareholder in the syndicate which had financed the expedition. Out of her share she paid off the whole of England's foreign debt, balanced her budget, and found herself with about £40,000 in hand. This she invested in the Levant Company – which prospered. Out of the profits of the Levant Company, the East India Company was founded; and the profits of this great enterprise were the foundation of England's subsequent foreign investment. Now it happens that £40,000 accumulating at 3¼ per cent compound interest approximately corresponds to the actual volume of England's foreign investments at various dates, and would actually amount today to the total of £4,000 million which I have already quoted as being what our foreign investments now are. Thus, every £1 which Drake brought home in 1580 has now become £100,000. Such is the power of compound interest!

From the sixteenth century, with a cumulative crescendo after the eighteenth, the great age of science and technical inventions began, which since the beginning of the nineteenth century has been in full flood – coal, steam, electricity, petrol, steel, rubber, cotton, the chemical industries, automatic machinery and the methods of mass production, wireless, printing, Newton, Darwin, and Einstein, and thousands of other things and men too famous and familiar to catalogue.

What is the result? In spite of an enormous growth in the population of the world, which it has been necessary to equip with houses and machines, the average standard of life in Europe and the United States has been raised, I think, about fourfold. The growth of capital has been on a scale which is far beyond a hundred-fold of what any previous age had known. And from now on we need not expect so great an increase of population.

If capital increases, say, 2 per cent per annum, the capital equipment of the world will have increased by a half in twenty years, and seven and a half times in a hundred years. Think of this in terms of material things – houses, transport, and the like.

At the same time technical improvements in manufacture and transport have been proceeding at a greater rate in the last ten years than ever before in history. In the United States factory output per head was 40 per cent greater in 1925 than in 1919. In Europe we are held back by temporary obstacles, but even so it is safe to say that technical efficiency is increasing by more than 1 per cent per annum compound. There is evidence that the revolutionary technical changes, which have so far chiefly affected industry, may soon be attacking agriculture. We may be on the eve of improvements in the efficiency of food production as great as those which have already taken place in mining, manufacture, and transport. In quite a few years – in our own lifetimes I mean – we may be able to perform all the operations of agriculture, mining, and manufacture with a quarter of the human effort to which we have been accustomed.

For the moment the very rapidity of these changes is hurting us and bringing difficult problems to solve. Those countries are suffering relatively which are not in the vanguard of progress. We are being afflicted with a new disease of which some readers may not yet have heard the name, but of which they will hear a great deal in the years to come – namely, *technological unemployment*. This means

unemployment due to our discovery of means of economising the use of labour outrunning the pace at which we can find new uses for labour.

But this is only a temporary phase of maladjustment. All this means in the long run *that mankind is solving its economic problem*. I would predict that the standard of life in progressive countries one hundred years hence will be between four and eight times as high as it is today. There would be nothing surprising in this even in the light of our present knowledge. It would not be foolish to contemplate the possibility of a far greater progress still.

II

Let us, for the sake of argument, suppose that a hundred years hence we are all of us, on the average, eight times better off in the economic sense than we are today. Assuredly there need be nothing here to surprise us.

Now it is true that the needs of human beings may seem to be insatiable. But they fall into two classes – those needs which are absolute in the sense that we feel them whatever the situation of our fellow human beings may be, and those which are relative in the sense that we feel them only if their satisfaction lifts us above, makes us feel superior to, our fellows. Needs of the second class, those which satisfy the desire for superiority, may indeed be insatiable; for the higher the general level, the higher still are they. But this is not so true of the absolute needs – a point may soon be reached, much sooner perhaps than we all of us are aware of, when these needs are satisfied in the sense that we prefer to devote our further energies to non-economic purposes.

Now for my conclusion, which you will find, I think, to become more and more startling to the imagination the longer you think about it.

I draw the conclusion that, assuming no important wars and no important increase in population, the *economic problem* may be solved, or be at least within sight of solution, within a hundred years. This means that the economic problem is not – if we look into the future – *the permanent problem of the human race.*

Why, you may ask, is this so startling? It is startling because – if, instead of looking into the future, we look into the past – we find that the economic problem, the struggle for subsistence, always has been hitherto the primary, most pressing problem of the human race – not only of the human race, but of the whole of the biological kingdom from the beginnings of life in its most primitive forms.

Thus we have been expressly evolved by nature – with all our impulses and deepest instincts – for the purpose of solving the economic problem. If the economic problem is solved, mankind will be deprived of its traditional purpose.

Will this be a benefit? If one believes at all in the real values of life, the prospect at least opens up the possibility of benefit. Yet I think with dread of the readjustment of the habits and instincts of the ordinary man, bred into him for countless generations, which he may be asked to discard within a few decades.

To use the language of today – must we not expect a general 'nervous breakdown'? We already have a little experience of what I mean – a nervous breakdown of the sort which is already common enough in England and the United States amongst the wives of the well-to-do classes, unfortunate women, many of them, who have been deprived by their wealth of their traditional tasks and occupations – who cannot find it sufficiently amusing, when deprived of the spur of economic necessity, to cook and clean and mend, yet are quite unable to find anything more amusing.

To those who sweat for their daily bread leisure is a longed-for sweet – until they get it.

There is the traditional epitaph written for herself by the old charwoman:

Don't mourn for me, friends, don't weep for
    me never,
For I'm going to do nothing for ever and ever.

This was her heaven. Like others who look forward to leisure, she conceived how nice it would be to spend her time listening-in – for there was another couplet which occurred in her poem:

With psalms and sweet music the
    heavens'll be ringing,
But I shall have nothing to do with the singing.

Yet it will only be for those who have to do with the singing that life will be tolerable – and how few of us can sing!

Thus for the first time since his creation man will be faced with his real, his permanent problem – how to use his freedom from pressing economic cares, how to occupy the leisure, which science and compound interest will have won for him, to live wisely and agreeably and well.

The strenuous purposeful money-makers may carry all of us along with them into the lap of economic abundance. But it will be those peoples, who can keep alive, and cultivate into a fuller perfection, the art of life itself and do not sell themselves for the means of life, who will be able to enjoy the abundance when it comes.

Yet there is no country and no people, I think, who can look forward to the age of leisure and of abundance without a dread. For we have been trained too long to strive and not to enjoy. It is a fearful problem for the ordinary person, with no special talents, to occupy himself, especially if he no longer has roots in the soil or in custom or in the beloved conventions of a traditional society. To judge from the behaviour and the achievements of the wealthy classes today in any quarter of the world, the outlook is very depressing! For these are, so to speak, our advance guard – those who are spying out the promised land for the rest of us and pitching their camp there. For they have most of them failed disastrously, so it seems to me – those who have an independent income but no associations or duties or ties – to solve the problem which has been set them.

I feel sure that with a little more experience we shall use the new-found bounty of nature quite differently from the way in which the rich use it today, and will map out for ourselves a plan of life quite otherwise than theirs.

For many ages to come the old Adam will be so strong in us that everybody will need to do *some* work if he is to be contented. We shall do more things for ourselves than is usual with the rich today, only too glad to have small duties and tasks and routines. But beyond this, we shall endeavour to spread the bread thin on the butter – to make what work there is still to be done to be as widely shared as possible. Three-hour shifts or a fifteen-hour week may put off the problem for a great while. For three hours a day is quite enough to satisfy the old Adam in most of us!

There are changes in other spheres too which we must expect to come. When the accumulation of wealth is no longer of high social importance, there will be great changes in the code of morals. We shall be able to rid ourselves of many of the pseudo-moral principles which have hag-ridden us for two hundred years, by which we have exalted some of the most distasteful of human qualities into the position of the highest virtues. We shall be able to afford to dare to assess the money-motive at its true value. The love of money as a possession – as distinguished from the love of money as a means to the enjoyments and realities of life – will be recognised for what it is, a somewhat disgusting morbidity, one of those semi-criminal, semi-pathological propensities which one hands over with a shudder to the specialists in mental disease. All kinds of social customs and economic practices, affecting the distribution of wealth and of economic rewards and penalties, which we now maintain at all costs, however distasteful and unjust they may be in themselves, because they are tremendously useful in promoting the accumulation of capital, we shall then be free, at last, to discard.

'Freedom in a Complex Society'

from *The Great Transformation: The Political and Economic Origins of Our Time* (1944)

Karl Polanyi

Editors' Introduction

Karl Polanyi was born in 1886 in Budapest, Hungary, to a prosperous family engaged in distinguished intellectual, artistic and radical circles. Polanyi studied Philosophy at the University of Budapest, and earned a PhD in 1908, followed by a law degree in 1914. While finishing his doctorate, Polanyi formed the Club Galilei in 1908, which called for a society shorn of feudal baggage, open to liberal ideas about politics and education. The club also tried to promote the idea of a national culture that could more effectively represent workers and peasants. Polanyi's interactions with intellectuals in his milieu, including Georg Lukács and Karl Mannheim, would be key to his growth. An interest in developments in Russia, and the debates around populism, led Polanyi to join the National Bourgeois Radical Party in 1914. When the Austro-Hungarian empire collapsed after the First World War, Polanyi witnessed the quick succession of a liberal regime in Hungary followed by the Republic of Soviets, which also collapsed within two months of Polanyi's departure to Vienna. Polanyi left with a compressed and engaged sense of multiple regimes and political possibilities. He was by now critical of both liberal democracy and Soviet socialism, and more drawn to democratic socialist alternatives in Vienna's municipal socialism. Polanyi remained committed to workers' education in Vienna and London, where he fled from Austrian fascism in 1933.

In Vienna, Polanyi worked as a journalist for the prestigious *Der Oesterreichische Volkswirt*, at which time he became a strong critic of the abstract economics of the Austrian School, with its strict adherence to methodological individualism, or the notion that society is reducible to atomised, utility-maximising individuals. Polanyi could not have thought more differently. With Lukács, Polanyi thought capitalist markets turned relationships into things – a process Lukács calls 'reification'. Polanyi began to think through Lukács's philosophical critique of commodification in comparative sociological and historical cases. These thoughts, and the historical research he undertook in London, culminated in his writing *The Great Transformation*, published in 1944 and excerpted in the reading that follows. Polanyi wrote this after his move to the United States, where he taught at Bennington College in Vermont. *The Great Transformation* is written at the historical crossroads that Polanyi witnessed in the contexts he had moved through. One of its key arguments is that the idea of the 'self-regulating market', taken for granted by the Austrian School, is what Marx would call a concrete abstraction, a real concept that emerges with the generalisation of commodification. This fiction spurs commodification in ever greater realms, but it also runs into various obstacles, particularly in treating those entities as commodities which are not in fact products of the capitalist labour process. What are these things? Land, labour and money are not produced in factories as their capacities are spent in the process of commodification. Land, or environments, persist through ecological processes; labour through sex, childbirth and care-giving, through gender and kinship systems; and money through state intervention. Polanyi's insight is that land and labour are fundamentally 'outside' the commodity circuit. What is more, their commodification through the myth of the 'self-regulating market', leads to various kinds of exhaustion – soil depletion, pollution, global climate change, exacerbated gender inequalities, racialised exposure – prompting various kinds of defence of their reproduction. This 'double movement' between commodification and 'social protection' (of fictitious commodities) sees development as an interplay between an immanent process and an intentional enterprise. As Michael Burawoy (2003) puts it in his incisive essay, Polanyi provides a theory of a new form of society, 'active society', which emerges in defence of land and labour, against the excesses of treating the market as if it were 'self-regulated'.

The reading comes from the powerful concluding chapter to *The Great Transformation*, drawing together Polanyi's sociological and historical explanation of market society and its contradictions. Writing in a moment of danger not unlike our own, Polanyi calls attention to the many meanings of the keyword 'freedom', as it is used to point to quite different developmental futures. Freedom can be viewed at two levels: institutionally as a balancing act, or as a more fundamental moral question about what it means to inhabit industrial society. At the first level, Polanyi's general argument is that the institutional separation of 'economy' and 'politics' endangers society itself, prompting the suspension of various liberties in the interests of the few. But this need not be the case. Freedom can be 'a prescriptive right extending far beyond the narrow confines of the political sphere into the intimate organization of society itself' (p. 265). Planning and control might enable a broadening of freedom, including the protection of nonconformity and dissidence. When land, labour and society itself is endangered under 'liberal economy', or what we would today call the 'free market system', 'society' becomes apparent as something in crisis. Polanyi sees two roads ahead: fascism or socialism. Both uphold 'society', one by emptying freedom of its moral content, and the other by upholding the intrinsic value of expanded freedom. We write as the 'War on Terror' continues to be prosecuted in the name of 'freedom', at a time when Polanyi could not be more relevant.

Between 1947 and 1953, Polanyi taught at Columbia University in New York City, commuting from Canada as his wife was prevented from entering the United States on the basis that she was a former communist. Polanyi's work in the 1950s shifted from modern to ancient empires, and to various types of pre-modern exchange systems. There appears to have been a break in his work in the mid-1940s as he turned to the anthropology of non-market societies. Like the socialist anthropologist Marcel Mauss, Polanyi's interest was driven by an attempt to find the logic of socialism outside the West, and outside Leninist orthodoxy, in the everyday practices of exchange in non-market societies. This work culminated in his edited *Trade and Market in the Early Empires*, and in the posthumously published *Dahomey and the Slave Trade* and *Primitive, Archaic and Modern Economics*. Karl Polanyi died in 1964.

Key references

Karl Polanyi (2001 [1944]) *The Great Transformation: The Political and Economic Origins of Our Time*, Boston, MA: Beacon Press.
— (1957) *Trade and Market in the Early Empires*, New York: Free Press.
— (1966) *Dahomey and the Slave Trade*, Seattle, WA: University of Washington Press.
— (1968) *Primitive, Archaic and Modern Economics: Essays of Karl Polanyi*, New York: Anchor Books.
Michael Burawoy (2003) 'For a sociological Marxism: the complementary convergence of Antonio Gramsci and Karl Polanyi', *Politics and Society* 31(2): 193–261.

Nineteenth-century civilization was not destroyed by the external or internal attack of barbarians; its vitality was not sapped by the devastations of World War I nor by the revolt of a socialist proletariat or a fascist lower middle class. Its failure was not the outcome of some alleged laws of economics such as that of the falling rate of profit or of underconsumption or over-production. It disintegrated as the result of an entirely different set of causes: the measures which society adopted in order not to be, in its turn, annihilated by the action of the self-regulating market. Apart from exceptional circumstances such as existed in North America in the age of the open frontier, the conflict between the market and the elementary requirements of an organized social life provided the century with its dynamics and produced the typical strains and stresses which ultimately destroyed that society. External wars merely hastened its destruction.

After a century of blind 'improvement' man is restor-ing his 'habitation.' If industrialism is not to extinguish the race, it must be subordinated to the requirements of man's nature. The true criticism of market society is not that it was based on economics – in a sense, every and any society must be based on it – but that its economy was based on self-interest. Such an organization of economic life is entirely unnatural, in the strictly empirical sense of *exceptional*. Nineteenth-century thinkers assumed that in his economic activity man strove for profit, that his materialistic propensities would induce him to choose the lesser instead of the greater effort and to expect payment for his labor; in short, that in his economic activity he would tend to abide by what they described as economic rationality, and that all contrary behavior was the result of outside interference. It followed that markets were natural institutions, that they would spontaneously arise if only men were let alone. Thus, nothing could be more normal than an economic sys-tem consisting of markets and under the sole control of market prices, and a human society based on such markets appeared, therefore, as the goal of all pro-gress. Whatever the desirability or undesirability of such a society on moral grounds, its practicability – this was axiomatic – was grounded in the immutable characteristics of the race.

Actually, as we now know, the behavior of man both in his primitive state and right through the course of history has been almost the opposite from that implied in this view. Frank H. Knight's "no specifically human motive is economic" applies not only to social life in general, but even to economic life itself. The tendency to barter, on which Adam Smith so confidently relied for his picture of primitive man, is not a common tendency of the human being in his economic activities, but a most infrequent one. Not

only does the evidence of modern anthropology give the lie to these rationalistic constructs, but the history of trade and markets also has been completely dif-ferent from that assumed in the harmonistic teachings of nineteenth-century sociologists. Economic history reveals that the emergence of national markets was in no way the result of the gradual and spontaneous emancipation of the economic sphere from govern-mental control. On the contrary, the market has been the outcome of a conscious and often violent inter-vention on the part of government which imposed the market organization on society for noneconomic ends. And the self-regulating market of the nine-teenth century turns out on closer inspection to be radically different from even its immediate predeces-sor in that it relied for its regulation on economic self-interest. *The congenital weakness of nineteenth-century society was not that it was industrial but that it was a market society.* Industrial civilization will continue to exist when the utopian experiment of a self-regulating market will be no more than a memory.

Yet the shifting of industrial civilization onto a new nonmarketing basis seems to many a task too desperate to contemplate. They fear an institutional vacuum or, even worse, the loss of freedom. Need these perils prevail?

Much of the massive suffering inseparable from a period of transition is already behind us. In the social and economic dislocation of our age, in the tragic vicissitudes of the depression, fluctuations of currency, mass unemployment, shiftings of social status, spectacular destruction of historical states, we have experienced the worst. Unwittingly we have been paying the price of the change. Far as mankind still is from having adapted itself to the use of machines, and great as the pending changes are, the restoration of the past is as impossible as the trans-ferring of our troubles to another planet. Instead of eliminating the demonic forces of aggression and conquest, such a futile attempt would actually ensure the survival of those forces, even after their utter military defeat. The cause of evil would become endowed with the advantage, decisive in politics, of representing the possible, in opposition to that which is impossible of achievement however good it may be of intention.

Nor does the collapse of the traditional system leave us in the void. Not for the first time in history may makeshifts contain the germs of great and per-manent institutions.

Within the nations we are witnessing a development under which the economic system ceases to lay down the law to society and the primacy of society over that system is secured. This may happen in a great variety of ways, democratic and aristocratic, constitutionalist and authoritarian, perhaps even in a fashion yet utterly unforeseen. The future in some countries may be already the present in others, while some may still embody the past of the rest. But the outcome is common with them all: the market system will no longer be self-regulating, even in principle, since it will not comprise labor, land, and money.

To take labor out of the market means a transformation as radical as was the establishment of a competitive labor market. The wage contract ceases to be a private contract except on subordinate and accessory points. Not only conditions in the factory, hours of work, and modalities of contract, but the basic wage itself, are determined outside the market; what role accrues thereby to trade unions, state, and other public bodies depends not only on the character of these institutions but also on the actual organization of the management of production. Though in the nature of things wage differentials must (and should) continue to play an essential part in the economic system, other motives than those directly involved in money incomes may outweigh by far the financial aspect of labor.

To remove land from the market is synonymous with the incorporation of land with definite institutions such as the homestead, the cooperative, the factory, the township, the school, the church, parks, wild life preserves, and so on. However widespread individual ownership of farms will continue to be, contracts in respect to land tenure need deal with accessories only, since the essentials are removed from the jurisdiction of the market. The same applies to staple foods and organic raw materials, since the fixing of prices in respect to them is not left to the market. That for an infinite variety of products competitive markets continue to function need not interfere with the constitution of society any more than the fixing of prices outside the market for labor, land, and money interferes with the costing-function of prices in respect to the various products. The nature of property, of course, undergoes a deep change in consequence of such measures since there is no longer any need to allow incomes from the title of property to grow without bounds, merely in order to ensure employment, production, and the use of resources in society.

The removal of the control of money from the market is being accomplished in all countries in our day. Unconsciously, the creation of deposits effected this to a large extent, but the crisis of the gold standard in the 1920s proved that the link between commodity money and token money had by no means been severed. Since the introduction of 'functional finance' in all-important states, the directing of investments and the regulation of the rate of saving have become government tasks.

To remove the elements of production – land, labor, and money – from the market is thus a uniform act only from the viewpoint of the market, which was dealing with them as if they were commodities. From the viewpoint of human reality that which is restored by the disestablishment of the commodity fiction lies in all directions of the social compass. In effect, the disintegration of a uniform market economy is already giving rise to a variety of new societies. Also, the end of market society means in no way the absence of markets. These continue, in various fashions, to ensure the freedom of the consumer, to indicate the shifting of demand, to influence producers' income, and to serve as an instrument of accountancy, while ceasing altogether to be an organ of economic self-regulation.

In its international methods, as in these internal methods, nineteenth-century society was constricted by economics. The realm of fixed foreign exchanges was coincident with civilization. As long as the gold standard and – what became almost its corollary – constitutional regimes were in operation, the balance of power was a vehicle of peace. The system worked through the instrumentality of those Great Powers, first and foremost Great Britain, who were the center of world finance, and pressed for the establishment of representative government in less-advanced countries. This was required as a check on the finances and currencies of debtor countries with the consequent need for controlled budgets, such as only responsible bodies can provide. Though, as a rule, such considerations were not consciously present in the minds of statesmen, this was the case only because the requirements of the gold standard ranked as axiomatic. The uniform world pattern of monetary and representative institutions was the result of the rigid economy of the period.

Two principles of nineteenth-century international

life derived their relevance from this situation: anarchistic sovereignty and 'justified' intervention in the affairs of other countries. Though apparently contradictory, the two were interrelated. Sovereignty, of course, was a purely political term, for under unregulated foreign trade and the gold standard governments possessed no powers in respect to international economics. They neither could nor would bind their countries in respect to monetary matters – this was the legal position. Actually, only countries which possessed a monetary system controlled by central banks were reckoned sovereign states. With the powerful Western countries this unlimited and unrestricted national monetary sovereignty was combined with its complete opposite, an unrelenting pressure to spread the fabric of market economy and market society elsewhere. Consequently, by the end of the nineteenth century the peoples of the world were institutionally standardized to a degree unknown before.

This system was hampering both on account of its elaborateness *and* its universality. Anarchistic sovereignty was a hindrance to all effective forms of international cooperation, as the history of the League of Nations strikingly proved; and enforced uniformity of domestic systems hovered as a permanent threat over the freedom of national development, especially in backward countries and sometimes even in advanced, but financially weak countries. Economic cooperation was limited to private institutions as rambling and ineffective as free trade, while actual collaboration between peoples, that is, between governments, could never even be envisaged.

The situation may well make two apparently incompatible demands on foreign policy: it will require closer cooperation between friendly countries than could even be contemplated under nineteenth-century sovereignty, while at the same time the existence of regulated markets will make national governments more jealous of outside interference than ever before. However, with the disappearance of the automatic mechanism of the gold standard, governments will find it possible to drop the most obstructive feature of absolute sovereignty, the refusal to collaborate in international economics. At the same time it will become possible to tolerate willingly that other nations shape their domestic institutions according to their inclinations, thus transcending the pernicious nineteenth-century dogma of the

necessary uniformity of domestic regimes within the orbit of world economy. Out of the ruins of the Old World, cornerstones of the New can be seen to emerge: economic collaboration of governments *and* the liberty to organize national life at will. Under the constrictive system of free trade neither of these possibilities could have been conceived of, thus excluding a variety of methods of cooperation between nations. While under market economy and the gold standard the idea of federation was justly deemed a nightmare of centralization and uniformity, the end of market economy may well mean effective cooperation with domestic freedom.

The problem of freedom arises on two different levels: the institutional and the moral or religious. On the institutional level it is a matter of balancing increased against diminished freedoms; no radically new questions are encountered. On the more fundamental level the very possibility of freedom is in doubt. It appears that the means of maintaining freedom are themselves adulterating and destroying it. The key to the problem of freedom in our age must be sought on this latter plane. Institutions are embodiments of human meaning and purpose. We cannot achieve the freedom we seek, unless we comprehend the true significance of freedom in a complex society.

On the institutional level, regulation both extends and restricts freedom; only the balance of the freedoms lost and won is significant. This is true of juridical and actual freedoms alike. The comfortable classes enjoy the freedom provided by leisure in security; they are naturally less anxious to extend freedom in society than those who for lack of income must rest content with a minimum of it. This becomes apparent as soon as compulsion is suggested in order to more justly spread out income, leisure and security. Though restriction applies to all, the privileged tend to resent it, as if it were directed solely against themselves. They talk of slavery, while in effect only an extension to the others of the vested freedom they themselves enjoy is intended. Initially, there may have to be reduction in their own leisure and security, and, consequently, their freedom so that the level of freedom throughout the land shall be raised. But such a shifting, reshaping and enlarging of freedoms should offer no ground whatsoever for the assertion that the new condition must necessarily be less free than was the old.

Yet there are freedoms the maintenance of which is of paramount importance. They were, like peace, a by-product of nineteenth-century economy, and we have come to cherish them for their own sake. The institutional separation of politics and economics, which proved a deadly danger to the substance of society, almost automatically produced freedom at the cost of justice and security. Civic liberties, private enterprise and wage-system fused into a pattern of life which favored moral freedom and independence of mind. Here again, juridical and actual freedoms merged into a common fund, the elements of which cannot be neatly separated. Some were the corollary of evils like unemployment and speculator's profits; some belonged to the most precious traditions of Renaissance and Reformation. We must try to maintain by all means in our power these high values inherited from the market-economy which collapsed. This, assuredly, is a great task. Neither freedom nor peace could be institutionalized under that economy, since its purpose was to create profits and welfare, not peace and freedom. We will have consciously to strive for them in the future if we are to possess them at all; they must become chosen aims of the societies toward which we are moving. This may well be the true purport of the present world effort to make peace and freedom secure. How far the will to peace can assert itself once the interest in peace which sprang from nineteenth-century economy has ceased to operate will depend upon our success in establishing an international order. As to personal liberty; it will exist to the degree in which we will deliberately create new safeguards for its maintenance and, indeed, extension. In an established society the right to nonconformity must be institutionally protected. The individual must be free to follow his conscience without fear of the powers that happen to be entrusted with administrative tasks in some of the fields of social life. Science and the arts should always be under the guardianship of the republic of letters. Compulsion should never be absolute; the "objector" should be offered a niche to which he can retire, the choice of a "second-best" that leaves him a life to live. Thus will be secured the right to nonconformity as the hallmark of a free society.

Every move toward integration in society should thus be accompanied by an increase of freedom; moves toward planning should comprise the strengthening of the rights of the individual in society. His indefeasible rights must be enforceable under the law even against the supreme powers, whether they be personal or anonymous. The true answer to the threat of bureaucracy as a source of abuse of power is to create spheres of arbitrary freedom protected by unbreakable rules. For however generously devolution of power is practiced, there will be strengthening of power at the center, and, therefore, danger to individual freedom. This is true even in respect to the organs of democratic communities themselves, as well as the professional and trade unions whose function it is to protect the rights of each individual member. Their very size might make him feel helpless, even though he had no reason to suspect ill-will on their part. The more so, if his views or actions were such as to offend the susceptibilities of those who wield power. No mere declaration of rights can suffice: institutions are required to make the rights effective. Habeas corpus need not be the last constitutional device by which personal freedom was anchored in law. Rights of the citizen hitherto unacknowledged must be added to the Bill of Rights. They must be made to prevail against all authorities, whether state, municipal, or professional. The list should be headed by the right of the individual to a job under approved conditions, irrespective of his or her political or religious views, or of color and race. This implies guarantees against victimization however subtle it be. Industrial tribunals have been known to protect the individual member of the public even from such agglomerations of arbitrary power as were represented by the early railway companies. Another instance of possible abuse of power squarely met by tribunals was the Essential Works Order in England, or the "freezing of labor" in the United States, during the emergency, with their almost unlimited opportunities for discrimination. Wherever public opinion was solid in upholding civic liberties, tribunals or courts have always been found capable of vindicating personal freedom. It should be upheld at all cost – even that of efficiency in production, economy in consumption or rationality in administration. An industrial society can afford to be free.

The passing of market-economy can become the beginning of an era of unprecedented freedom. Juridical and actual freedom can be made wider and more general than ever before; regulation and control can achieve freedom not only for the few, but for all. Freedom not as an appurtenance of privilege, tainted at the source, but as a prescriptive right extending far beyond the narrow confines of the political sphere into the intimate organization of society itself. Thus will old freedoms and civic rights be added to the fund of new freedom generated by the leisure and security that industrial society offers to all. Such a society can afford to be both just and free.

Yet we find the path blocked by a moral obstacle. Planning and control are being attacked as a denial of freedom. Free enterprise and private ownership are declared to be essentials of freedom. No society built on other foundations is said to deserve to be called free. The freedom that regulation creates is denounced as unfreedom; the justice, liberty and welfare it offers are decried as a camouflage of slavery. In vain did socialists promise a realm of freedom, for means determine ends: the U.S.S.R., which used planning, regulation and control as its instruments, has not yet put the liberties promised in her Constitution into practice, and, probably, the critics add, never will. . . . But to turn against regulation means to turn against reform. With the liberal the idea of freedom thus degenerates into a mere advocacy of free enterprise – which is today reduced to a fiction by the hard reality of giant trusts and princely monopolies. This means the fullness of freedom for those whose income, leisure, and security need no enhancing, and a mere pittance of liberty for the people, who may in vain attempt to make use of their democratic rights to gain shelter from the power of the owners of property. Nor is that all. Nowhere did the liberals in fact succeed in reestablishing free enterprise, which was doomed to fail for intrinsic reasons. It was as a result of their efforts that big business was installed in several European countries and, incidentally, also various brands of fascism, as in Austria. Planning, regulation, and control, which they wanted to see banned as dangers to freedom, were then employed by the confessed enemies of freedom to abolish it altogether. Yet the victory of fascism was made practically unavoidable by the liberals' obstruction of any reform involving planning, regulation, or control.

Freedom's utter frustration in fascism is, indeed, the inevitable result of the liberal philosophy, which claims that power and compulsion are evil, that freedom demands their absence from a human community. No such thing is possible; in a complex society this becomes apparent. This leaves no alternative but either to remain faithful to an illusionary idea of freedom and deny the reality of society, or to accept that reality and reject the idea of freedom.

The first is the liberal's conclusion; the latter the fascist's. No other seems possible.

Inescapably we reach the conclusion that the very possibility of freedom is in question. If regulation is the only means of spreading and strengthening freedom in a complex society, and yet to make use of this means is contrary to freedom per se, then such a society cannot be free.

Clearly, at the root of the dilemma there is the meaning of freedom itself. Liberal economy gave a false direction to our ideals. It seemed to approximate the fulfillment of intrinsically utopian expectations. No society is possible in which power and compulsion are absent, nor a world in which force has no function. It was an illusion to assume a society shaped by man's will and wish alone. Yet this was the result of a market view of society which equated economics with contractual relationships, and contractual relations with freedom. The radical illusion was fostered that there is a nothing in human society that is not derived from the volition of individuals and that could not, therefore, be removed again by their volition. Vision was limited by the market which "fragmentated" life into the producers' sector that ended when his product reached the market, and the sector of the consumer for whom all goods sprang from the market. The one derived his income "freely" from the market, the other spent it "freely" there. Society as a whole remained invisible. The power of the state was of no account, since the less its power, the smoother the market mechanism would function. Neither voters, nor owners, neither producers, nor consumers could be held responsible for such brutal restrictions of freedom as were involved in the occurrence of unemployment and destitution. Any decent individual could imagine himself free from all responsibility for acts of compulsion on the part of a state which he, personally, rejected; or for economic suffering in society from which he, personally, had not benefited. He was "paying his way," was "in nobody's debt," and was unentangled in the evil of power and economic value. His lack of responsibility for them seemed so evident that he denied their reality in the name of his freedom.

But power and economic value are a paradigm of social reality. They do not spring from human volition; noncooperation is impossible in regard to them. The function of power is to ensure that measure of conformity which is needed for the survival of the group; its ultimate source is opinion – and who could help holding opinions of some sort or other? Economic value ensures the usefulness of the goods produced; it must exist prior to the decision to produce them; it is a seal set on the division of labor. Its source is human wants and scarcity – and how could we be expected not to desire one thing more than another? Any opinion or desire will make us participants in the creation of power and in the constituting of economic value. No freedom to do otherwise is conceivable.

We have reached the final stage of our argument.

The discarding of the market utopia brings us face to face with the reality of society. It is the dividing line between liberalism on the one hand, fascism and socialism on the other. The difference between these two is not primarily economic. It is moral and religious. Even where they profess identical economics, they are not only different but are, indeed, embodiments of opposite principles. And the ultimate on which they separate is again freedom. By fascists and socialists alike the reality of society is accepted with the finality with which the knowledge of death has molded human consciousness. Power and compulsion are a part of that reality; an ideal that would ban them from society must be invalid. The issue on which they divide is whether in the light of this knowledge the idea of freedom can be upheld or not; is freedom an empty word, a temptation, designed to ruin man and his works, or can man reassert his freedom in the face of that knowledge and strive for its fulfillment in society without lapsing into moral illusionism?

This anxious question sums up the condition of man. The spirit and content of this study should indicate an answer.

We invoked what we believed to be the three constitutive facts in the consciousness of Western man: knowledge of death, knowledge of freedom, knowledge of society. The first, according to Jewish legend, was revealed in the Old Testament story. The second was revealed through the discovery of the uniqueness of the person in the teachings of Jesus as recorded in the New Testament. The third revelation came to us through living in an industrial society. No one great name attaches to it; perhaps Robert Owen came nearest to becoming its vehicle. It is the constitutive element in modern man's consciousness.

The fascist answer to the recognition of the reality of society is the rejection of the postulate of freedom. The Christian discovery of the uniqueness of the individual and of the oneness of mankind is negated by fascism. Here lies the root of its degenerative bent.

Robert Owen was the first to recognize that the Gospels ignored the reality of society. He called this the "individualization" of man on the part of Christianity and appeared to believe that only in a cooperative commonwealth could "all that is truly valuable in Christianity" cease to be separated from man. Owen recognized that the freedom we gained through the teachings of Jesus was inapplicable to a complex society. His socialism was the upholding of man's claim to freedom *in such a society*. The post-Christian era of Western civilization had begun, in which the Gospels did not any more suffice, and yet remained the basis of our civilization.

The discovery of society is thus either the end or the rebirth of freedom. While the fascist resigns himself to relinquishing freedom and glorifies power which is the reality of society, the socialist resigns himself to that reality and upholds the claim to freedom, in spite of it. Man becomes mature and able to exist as a human being in a complex society. To quote once more Robert Owen's inspired words: "Should any cause of evil be irremovable by the new powers which men are about to acquire, they will know that they are necessary and unavoidable evils; and childish, unavailing complaints will cease to be made."

Resignation was ever the fount of man's strength and new hope. Man accepted the reality of death and built the meaning of his bodily life upon it. He resigned himself to the truth that he had a soul to lose and that there was worse than death, and founded his freedom upon it. He resigns himself, in our time, to the reality of society which means the end of that freedom. But, again, life springs from ultimate resignation. Uncomplaining acceptance of the reality of society gives man indomitable courage and strength to remove all removable injustice and unfreedom. As long as he is true to his task of creating more abundant freedom for all, he need not fear that either power or planning will turn against him and destroy the freedom he is building by their instrumentality. This is the meaning of freedom in a complex society; it gives us all the certainty that we need.

'The Background of Colonial Policy and Practice'

from *Colonial Policy and Practice: A Comparative Study of Burma and Netherlands India* (1948)

J. S. Furnivall

Editors' Introduction

John Sydenham Furnivall was born in Essex, England, in 1878 and graduated from Cambridge University in 1900. In 1902 he joined the Indian Civil Service (ICS) and was posted in various government jobs in Burma. He founded the Burma Research Society in 1910 and married a Burmese woman. In later years he forged friendships with, and sometimes worked directly for, leading nationalist politicians, particularly so after he left the ICS in 1923. Furnivall's wide writings on Burma led to a lectureship at Cambridge University from 1935 to 1941. Thereafter he lived until his death in 1961 both in the UK and in independent Burma.

Furnivall is best known for his work on the 'plural society' in Burma, and indeed elsewhere in South-East Asia. This idea certainly had recourse to notions of race and ethnicity, and what Furnivall called the problems of immigration, but in his major works the concept was also rooted in economic concerns. Furnivall was mainly concerned with the social effects in Burma and Asia of Western political and economic rule. He was particularly concerned that 'traditional' societies and economies could not cope with the onslaught of capitalism, and that, even where capital accumulation did lead to more rapid economic growth, it would not lead to greater 'happiness' for ordinary people. In this specific respect, his work echoes that of Keynes. Furnivall mistrusted the money motive and what he saw as the deadening – or levelling – cultural effects of utilitarianism. While deeply critical of British rule in Burma, Furnivall nonetheless looked to cultured and enlightened administrators as one line of defence of 'the native' from brute capitalism. He also argued the claims of welfare and autonomy (independence) as preconditions for democracy and development, rather than the other way round. That is to say, he challenged prevailing claims that the imperial powers were developing their charges for democracy, and that no such mission could be achieved outside a colonial framework.

In the reading which follows, Furnivall contrasts the experience of British rule in Burma with that of Dutch rule in Netherlands India (Java). We reproduce it here because Furnivall disrupts key assumptions about the uniformity of the colonial experience. He establishes important similarities between the two imperial regimes, but also highlights key areas of difference – most notably in respect of timing (of occupation) and the use of what he calls 'tropical principles of custom and authority'. Modern scholars would pay far more attention to local (and especially) nationalist voices than Furnivall does, although these do emerge later in this, his most important book. Nevertheless, Furnivall does give a reasonable sense of the different colonial legacies and path-dependencies that British and Dutch rule set up in South and South-East Asia. This is surely where the legacy of his own work lies.

Key references

J. S. Furnivall (1948) *Colonial Policy and Practice: A Comparative Study of Burma and Netherlands India*, Cambridge: Cambridge University Press.
— (1931) *An Introduction to the Political Economy of Burma*, Rangoon: Burma Book Club.
— (1941) *Progress and Welfare in Southeast Asia: A Comparison of Colonial Policy and Practice*, New York: Secretariat, Institute of Pacific Relations.
— (1945) *The Tropical Far East*, London: Oxford University Press.
N. Tarling (ed.) (2000) *The Cambridge History of South-East Asia*, Cambridge: Cambridge University Press.

1. COLONIES AND DOMINIONS

One outstanding feature of world history during the last few centuries is the expansion of Europe. In some regions, chiefly in North and South America, settlements of European origin have become independent powers. Other settlements, in what are now the British dominions, are independent in all but name and sentiment, though owing a voluntary allegiance to the Crown. But wide areas, mostly in the tropics, still remain under the rule of a few western powers, reckoning as western powers the countries and dominions of western origin. Some, at the outset of the late war, were ruled by Japan, which had recently established itself in the position of a western power. In law these subordinate regions exhibit a wide variety of status, but they may all conveniently be termed colonies, and the powers that have assumed responsibility for their welfare may be termed colonial powers. Colonization originally implied settlement, but the tropics have been colonized with capital rather than with men, and most tropical countries under foreign rule are dependencies rather than colonies, though in practice both terms are used indifferently. Some tropical countries, such as Siam, though in law politically independent, resemble colonies because, in their economic relations, they are largely dependent on the West.

Up to the end of the last century, or even later, tropical peoples seemed, after the initial troubles of pacification, to accept, and even to welcome, foreign rule. Of British dependencies it has been said that 'a sentiment of alienation or estrangement has seldom shown itself in the earlier days of British administration. . . . Where the sentiment occurs, it has tended to manifest itself most clearly as the people of the dependencies advance in education and standards of living.'[1] That is generally true of tropical dependencies in their relations with colonial powers. Towards the end of the nineteenth century impatience of foreign control appeared both in India and the Philippines, and during the present century a reaction against western rule, and even against western civilization, has spread rapidly and widely. This raised new problems for those concerned directly in the administration or development of tropical countries, but it hardly touched the ordinary man except to suggest that his investments might be safer elsewhere. Recent events, however, and especially the rapid collapse of western rule throughout the Tropical Far East, have thrown down a rude challenge to western dominion in the tropics, and even those who know little of the outer world, beyond what they gather from the newspaper, have come to feel that colonial policy and practice need reconsideration.

The general growth of discontent and unrest, and the general sudden collapse in the Pacific, suggest a common cause in some defect inherent in colonial relations. The situation is all the more remarkable by the striking contrast between the dependencies and the dominions. Formerly it was held that colonies resembled fruit which falls from the parent tree when it is ripe. Yet the links of interest and affection between Britain and the dominions seem to grow stronger, while dependencies clamour for independence even when to outside view they seem unripe for it, and we have come to regard it as natural and right that they should claim independence. It would seem possible for Burma to gain much more than Canada from an association with Britain. But, as a former Chief Secretary to the Government of Burma said recently, 'it may be candidly admitted that our rule was not beloved by the Burmans'.[2] Why should the ties with Canada grow closer while Burma finds them irksome? What common factor is there making for estrangement with the dependencies? All human relations have a particular character arising out of the personality of those concerned, and this is no less true of colonial relations. The relations between Englishmen and Hindus cannot be identical with the relations between French and Annamese, or between Dutch and Javanese. Yet Europeans have much in common that contrasts with features common to tropical peoples in general, and colonial relations in themselves present many common problems that are little if at all affected by the particular nationality or race of those concerned. Let us then glance briefly at some of the common factors in colonial policy and practice.

2. THE BACKGROUND OF COLONIAL POLICY

It would appear superfluous to argue that the western powers which have come to exercise dominion in the tropics have a fundamental unity. No one would suggest that Portuguese and Spaniard, Dutch, English, French and American are all alike; all have their distinctive national character which is an invisible export on every ship from Europe. The towns which they have built in the tropics are as different as those who built them. Similarly, each colonial power has stamped its own imprint on its system of administration. But the West has a common civilization derived from Greece and Rome under the vitalizing impulse of Christianity. When this civilization first made contact with the tropics the rebirth of reason had already begun to liberate fresh sources of material power by the rationalization of economic life; from that time onwards there has been a growing tendency in the West to base economic and social relations on reason, impersonal law and individual rights.

Similarly, throughout the tropics one finds everywhere a basic resemblance in the social structure. There is a wide range of variation, from primitive tribal communities to settled agricultural societies with a complex social and political organization, yet in all the variations, from the most primitive to the most advanced, one can trace a common pattern. Under native rule everywhere, even in settled agricultural communities, social and political relations were customary, not legal; authority was personal,

based on Will and not on Law; and both custom and authority were closely bound up with religion. The social order rested not on impersonal law and individual rights, but on personal authority and customary obligations, and authority and custom derived their sanction not from reason but from religion.

Wherever a western power has gained dominion in the tropics, or even, as in Siam, a large measure of economic control without actual political sovereignty, an intimate contact has been established between these two contrary principles of social life: between the eastern system, resting on religion, personal authority and customary obligations, and the western system, resting on reason, impersonal law and individual rights. That is a common feature of all colonial relations. And it is common to them also that the contact has been established for the advantage of the colonial power, and that the first condition of establishing dominion is the maintenance of order.

Ordinarily, the motive of colonial expansion has been economic advantage. Considerations of prestige and military strategy have played their part, but in the main economic considerations have prevailed. It is, indeed, generally true that colonization has arisen out of commerce, and not commerce out of colonization: the doctrine that trade follows the flag is quite modern, and in history the flag has followed trade. During the earlier stages of its expansion Europe produced little that tropical peoples required, and the main object of colonial policy was to get tropical produce at the lowest cost. That was the economic end of empire in the days of Portugal and Spain. The same end was attained more effectively by the great chartered companies, concentrating on economic activities, enjoying, as regards their nationals, a monopoly over eastern produce, and exercising sovereign powers which enabled them to obtain part of their supplies free of cost as tribute. But, with the growing rationalization of life in the western world, man has achieved a greater mastery over material resources, and there has been a continual readjustment of colonial policy to meet new conditions arising through economic progress. The first great change was the Industrial Revolution, which enabled Europe to sell its products in the tropics. This required a new colonial policy, no longer directed towards obtaining tropical produce cheaply through chartered companies, but towards opening up markets for the sale of European manufactures by the encouragement of native enterprise. Further economic development in Europe built up large-scale enterprise, needing a secure command over vast quantities of raw materials which natives working by their own methods were unable to supply; at the same time it created a surplus of capital available for overseas investment. Thus, from about 1870, with the opening of the Suez Canal, the tropical market ceased to be the main object of colonial policy, which now aimed at obtaining the necessary raw materials of industry by the rapid development of the tropics through western enterprise. Meanwhile, in some of the more closely populated regions, the abundant supply of cheap unorganized labour was attracting capital to oriental industry and manufacture, with the result that during recent years imports from India and Japan have reduced the value of the tropics as a market for cheap European manufactures, though increasing rather than diminishing their demand for capital goods. This has encouraged in the colonial powers a more sympathetic attitude towards local manufactures in their dependencies, and towards economic nationalism; but the full effect on colonial policy of this new trend, which was interrupted by the recent war, is not yet clearly visible. In this gradual transformation of world economy during the building of the modern world, all colonial powers and all dependencies have been involved, and it has been necessary, as a condition of survival, that colonial powers should revise their policy to meet the challenge of new circumstances.

The course of economic progress has had a further reaction on colonial policy by its effect on domestic politics within the colonial powers. In Britain the transition from a policy of endowing a chartered company with a monopoly over tropical products to a policy of opening a free market for British goods signalized, and was partly the result of, the transfer of power from the aristocracy and London merchants to the manufacturers of Manchester. When, in due course, policy turned in the direction of developing the natural resources of the tropics by western enterprise, Manchester had to meet the growing rivalry of heavy industry in Birmingham, and of financial interests in the City. But at this stage a new factor intervened. In order to secure its interests against the increasing power of capital, labour entered politics. For some time this

had no obvious reaction on colonial policy. Manchester and Birmingham alike furnished employment, and imports of tropical produce encouraged production and reduced prices; colonial rule was profitable, more or less, to all classes. But the clash between labour and capital in Parliament encouraged a tendency towards colonial autonomy under capitalist control, answered by labour with demands for colonial autonomy on a democratic basis. Then, with the growth of tropical industry and manufacture, the conditions of labour in the tropics became a matter of economic interest to various classes in Britain, and welfare measures began to assume a new importance in colonial policy. In other lands, also, colonial policy has been moulded by the interplay of conflicting domestic interests, and in general on much the same pattern, though in some countries the rivalry of industry and agriculture has had a greater influence on colonial policy than in Britain.

Naturally, colonial policy is framed with reference to the interests, real or imagined, of the colonial power. But modern colonization is an affair of capital and not of men, and capital knows no country. The capitalist, the owner of capital, may be swayed by national or ethical considerations, but reason impels him to disregard them and lay out his capital where, on a long view, it will obtain the best return, and within recent times capitalist activities, reaching out rapidly beyond national boundaries, have acquired an extra-national character. Labour interests have had to counter this by uniting in an international association with a common policy. Out of the adjustment of the relations between capital and labour on an international plane has arisen a common recognition that the economic development of the tropics is a matter of world welfare.

Not infrequently, however, there is a conflict between world welfare and particular national interests, and colonial powers, pursuing their own interests, tend to disregard world welfare and the interests of other powers. The danger of rivalry in the colonial sphere first found public recognition in the Berlin Conference of 1884–5, which marked a new stage in the evolution of a common colonial policy. Before that one can trace a common resemblance in the colonial policy of different powers because they all had *similar* interests; from the closing years of the nineteenth century the common features of colonial policy are accentuated because colonial powers recognize, at least in name, a *common* interest in the development of backward areas. Colonial policy, though still directed with primary reference to the interests, real or imagined, of the colonial power, must now be justified to world opinion with reference to world welfare.

But colonial policy is only one aspect of national policy; each generation evolves its own social philosophy with no direct reference to colonial affairs, and inevitably colonial policy is framed in terms of the broad general conceptions which dominate the national outlook, and reflects the emergence of new general ideas. Moreover, although colonial relations arise out of the search for material advantage, men like to justify their activities on moral grounds and colour them with the warm glow of humanitarianism. 'I do not pretend', says the historian of the Portuguese Empire, 'that our sole aim was to preach, if others will allow that it was not only to trade.' When men came to accept the principle of freedom of person, property and trade, consistency required that it should be extended to colonial relations. Similarly, the idea of social justice that took shape in Europe during the nineteenth century permeated colonial policy, and so likewise did the ideas of democracy and self-determination or nationalism. It was self-determination in Europe for which Asquith drew the sword in 1914, but before the war was ended the right of colonies to self-determination had been recognized in principle. Again, popular Darwinism of the last century held that progress was achieved through the survival of the fittest, of the more efficient; accordingly, in the name of progress and efficiency, men set themselves to run the colonies on more business-like, efficient, lines. Thus colonial policy tends to follow, even if at a distance, domestic policy. Liberty, Social Justice, Democracy, if approved as sauce for the domestic goose, are served up a little later with the colonial gander; Free Trade is good for Britain, and good, therefore, for India; social legislation protects British labour and should, therefore, promote welfare in the tropics; democracy strengthens the political fibre of Europe and should, therefore, help dependencies towards autonomy. As one aspect of national policy, colonial policy reflects ideas transcending economic aims. It is true that, in retrospect, ideals claiming a moral basis have often worked in practice to the advantage of the colonial power rather than of the dependency, and men are apt to dismiss the humanitarianism of a former age as

humbug and its zealous advocates as hypocrites. That is partly because on the whole morality is a paying proposition; yet, though it pays to be honest, men are not honest merely because it pays. For ideas have an independent vitality; men accept them even to their apparent prejudice. And the ideas may outlast the empire which they help to build, just as Roman law survived the fall of Rome and provides a steel framework to the fabric of the modern world, and as in the Spanish colonies the Faith survived the fall of Spain. So in Asia and Africa the ideas of Law and Liberty and Social Justice may have a longer life than western rule. But the point immediately relevant is that the succession of these general ideas is common to western civilization and therefore to all western colonial powers, and forms a common feature in the evolution of colonial policy.

If, then, colonial relations have so much in common, and colonial policy reflects the application to colonial relations of ideas which are common to all colonial powers, it might seem that a comparative survey of pronouncements on colonial policy should lead to the enunciation of certain principles of colonial rule. But the matter is not so simple. One obstacle to the progress of the social sciences is the difficulty that students find in maintaining an objective attitude. The rules for statesmen which the economist prescribes in his academic cloister reflect his political opinions; educationists vehemently debate what education should and might be, and what it could or must do with little heed to what education actually is and does; 'political philosophy has chiefly concerned itself with how men *ought* to live, and what form of government they *ought* to have, rather than with what *are* their political habits and institutions'.[3] Similarly, projects of colonial policy lay down the goal at which it ought to aim, but the measures advocated to attain the goal, even if free from unconscious bias and likely to succeed in Europe, often lead in tropical dependencies in a different or even a contrary direction. In policy, as in law, men must be held to intend the natural consequences of their acts, and it is from the results of colonial policy rather than from statements of its objects that its true character may be ascertained. In the study of colonial affairs statements of policy need scrutiny in the light of practice. That is far more difficult. For if a survey of colonial policy reveals a fundamental identity in colonial relations, the study of colonial practice gives a first impression of diversity.

3. THE BACKGROUND OF COLONIAL PRACTICE

In respect of colonial practice, however, one can distinguish two factors characteristic of colonial relations in general. Since colonial relations are predominantly economic, colonial practice is conditioned by economic laws, and we shall see that in the tropics the working of these laws has certain features common to all dependencies. The second factor common to all colonial practice is that the responsibility for maintaining order is assumed by the colonial power; an organic autonomous society maintains order with more or less success in virtue of its inherent vitality, but a dependency is kept alive, as it were, by artificial respiration, by pressure exercised mechanically from outside and above. These two factors are common to all colonial relations, though it is in connection with them that divergences in colonial practice first arise.

The prime care of any colonial power must be to maintain order, for order is essential to such advantages as it anticipates from imposing its rule on the dependency. In maintaining order the colonial power must choose between the western principle of law and the tropical system of relying on personal authority, between direct and indirect rule. Secondly, to attain the prospective advantages the colonial power must choose between the western principle of freedom and the tropical system of compulsion. These are the two main lines along which colonial practice divides, and the line actually followed is a resultant of complex factors. General ideas carry some weight: compulsion will be less uncongenial in an age which values welfare more than freedom. So also do national traditions: reliance on authority will be less uncongenial to a western power with a strongly centralized administration. Of greater importance is the relative strength of the ruler and the ruled: a chartered company reinforced once a year by a small fleet of sailing vessels must show more regard to native sentiment than a Governor who at need can summon by radio the assistance of modern naval and military forces. These things affect colonial practice, but the dominating factor is the nature of the advantage envisaged by the colonial power from its relations with the dependency; whether it is chiefly concerned to sell its own goods or to obtain tropical produce. All these conditions vary from one colonial power to another, from one dependency to another, in different dependencies of the same colonial power, and in the same dependency at different times; and colonial practice varies accordingly. Naturally then in any comparison of colonial relations it is the diversity of practice which first impresses the observer, and only by tracing their evolution can the first impression be corrected.

Yet even a historical study may be misleading. For colonial relations arise out of the impact of western civilization on the tropics, and the violence of this impact varies with the width of the gap between the particular western and tropical cultures, and the rate at which the process of bringing them together is effected. In their mastery over the material world, India and Europe in the sixteenth century were far closer together than modern Europe and the modern tropics; the connection dates from before the Industrial Revolution, and both have grown up in the modern world together. Again, Indian society stands alone in resting on the institution of caste, which has fortified it to some extent against the impact of the West. For both these reasons care is needed in drawing parallels between conditions in India and in the tropics generally. Where tropical conditions elsewhere resemble those of India, one may presume that they are inherent in colonial relations. For example, one feature of the modern tropics that is not confined to India is the growth of agricultural indebtedness; another feature is the increasing impatience of foreign rule. On the other hand, if India is in some measure exempt from the 'atomization' of society that is found elsewhere in the tropics, the explanation may lie in the fact that the impact of the West has been less violent, and that caste has afforded a considerable measure of protection.

Still it remains true that, if one would understand colonial practice, one must study it both comparatively and historically. Such a survey could hardly be compressed within the limits of a single volume, even of the dimensions of Lord Hailey's *African Survey*, and would need abilities to which the present writer can make no pretence. It is, however, as a contribution to such a study that the following comparison of British rule in Burma with Dutch rule in Netherlands India is intended. It is a comparison with many points of interest. Both countries came under the effective rule of the home Government shortly after the Industrial Revolution had transformed the economic, and therefore the social and political, relations between Europe and the tropics, and when the social philosophy of Europe was dominated by liberal ideas; in both, therefore, it is possible to trace the reaction on colonial practice of subsequent economic progress and of the procession of general ideas. Both countries, like the tropics in general, lacked the consolidating bond of caste to protect them against the solvent influence of western thought and economic forces. In these things both countries are alike. But in respect of colonial practice they show a striking contrast. In Burma the British have from the first relied on western principles of rule, on the principles of law and economic freedom; in Netherlands India the Dutch have tried to conserve and adapt to modern use the tropical principles of custom and authority. In each case the choice was dictated in the first instance by the interests of the colonial power. When Britain first established contact with Burma it could already sell its produce in the East and looked to Burma as a market. The Dutch at that time had neither manufactures nor capital; they wanted tropical produce at the lowest cost and could best obtain it through the native chieftains. For an understanding of subsequent developments it is necessary to have some knowledge of physical conditions and past history in the two regions.

[. . .]

NOTES

1 Hailey, Lord, *Britain and Her Dependencies* (1943), p. 36.
2 Leach, *Quarterly Review*, Jan. 1944, p. 50.
3 Fortes, M. and Evans-Pritchard, E. E., *African Political Systems* (1940), p. 4.

'The Impasse'

from *Bread and Democracy in Germany* (1943)

Alexander Gerschenkron

Editors' Introduction

Alexander Gerschenkron was born in Odessa, Russia, in 1904 and died in 1978, four years after retiring from a long academic career based mainly at Harvard University. Gerschenkron is best known in development studies for his 1951 essay, 'Economic backwardness in historical perspective'. One of the key arguments of that essay was that 'the development of a backward country may, by very virtue of its backwardness, tend to differ fundamentally from that of an advanced country'. In this manner, Gerschenkron sought to imply that there was no singular path to development, no stages of growth as (later) Rostow and (earlier) Marx perhaps liked to imply. To the contrary, underdeveloped countries could turn backwardness into an advantage. They could import sophisticated technology and industrial machinery from more advanced countries and avoid a lengthy process of learning-by-doing. In this way, too, well-organised governments in backward countries could meet the aspirations of ordinary people for 'development' in a world where the gap between rich and poor countries was more evident than hitherto (because of new communications technologies) and yet also less acceptable (following the end of colonial rule). Basing his theory on the experiences of Central and Eastern Europe, Gerschenkron advised backward countries to specialise in capital-intensive industrialisation if they wanted to catch up with early industrialisers like the UK and Germany.

It is not clear that many developing countries acted directly on Gerschenkron's advice. Import-substitution in much of Latin America was geared more to consumption goods than to producer goods, and there is no evidence that we are aware of that planners in India, which did follow a capital goods-based strategy from 1956 to 1966, took inspiration from Gerschenkron's paper. In any case, as we mean to imply by the reading which follows, there is far more to Gerschenkron than this one essay. Like many of the first generation of 'development economists' – Nurkse, Rosenstein-Radan, Singer, Prebisch and others – Gerschenkron forged his views on government and the economy in the inter-war period, and with reference to the Depression, the revolution in Russia, and the rise of Fascism in Italy and Germany. His family left Russia after the 1917 revolution and settled in Vienna, Austria, where Alexander earned a PhD. Gerschenkron in turn left Austria in 1938 as the Nazis seized power there, moving to Berkeley before joining the faculty at Harvard. Arguably his most important work, therefore, is his analysis of *Bread and Democracy in Germany* – an outstanding book which traces the 'delayed' (and partial) democratisation of Germany back to policies of agricultural protectionism before the Second World War. Gerschenkron also understood, and documents, the instabilities that were induced in German politics by a system of 'rye protection' that was intended to serve the class interests of Prussia's Junkers. *Bread and Democracy in Germany* is a classic analysis of the 'agrarian question' and of the politics of food production and urban – rural relations more broadly. It is mainly here, we believe, that Gerschenkron's work has relevance today. (Readers interested in Gerschenkron might like to read the extraordinary account of his life and times written by his grandson, Nicholas Dawidoff (2002.))

Key references

Alexander Gerschenkron (1989 [1943]) *Bread and Democracy in Germany*, Ithaca, NY: Cornell University Press.
— (1962) *Economic Backwardness in Historical Perspective*, Cambridge, MA: Harvard University Press.
— (1968) *Continuity in History and Other Essays*, Cambridge, MA: Harvard University Press.
N. Dawidoff (2002) *The Fly Swatter: How My Grandfather Made His Way in the World*, New York: Pantheon Books.
R. Gwynne (2006) 'Alexander Gerschenkron', in D. Simon (ed.) *Fifty Key Thinkers on Development*, London: Routledge.

Heavy as was the cost of the gift which the German nation was making to the Junkers, the system was also, as a long-term policy, far from being sufficient. On the contrary, just because German protection was first and foremost a system of rye protection, it carried within itself the germs of instability.

However much the interests of the consumer were neglected by those who were shaping the policy of rye protection, the long-run demand schedule for rye could not be forever ignored. It is a historical fact that, as far as bread consumption is concerned, rye has been continually pushed back ever since the effects of the Industrial Revolution began to find expression in rising standards of living and in changes of social and political outlook. Rye had been for centuries the bread of the common man in the whole of Europe. Even in England, until the revolutionary upheavals of the seventeenth century, rye consumption was very considerable. In France, up to the end of the eighteenth century, wheat was the food of a privileged minority. Yet this condition was impossible to maintain. Industrialization and urbanization brought about changed working and living conditions – the replacement of work in the open air by work of more sedentary character in closed rooms – and provided the physiological background for dietary changes which were most unfavorable to the poorly digestible rye bread. Moreover, the progress of democracy, as well as the democratization of the general philosophy of men, worked most strongly in the same direction.

Bread had been for thousands of years the reward of human labor, the symbol of wealth, home life, happiness, and hospitality. All religions used and many still use bread in their holy rites. It symbolized the mystic relation between man and God, between man and the friendliness of nature. It is no wonder that the French Revolution gave a deadly blow to the consumption of rye bread in France. When the idea of natural law imposed itself on the minds of men, when people began to believe that nature had created all men equal, then inequality as between men with regard to this chief gift of nature became intolerable. Therefore, the progress of democracy in Europe was long associated with the progress of wheat and the decline of rye in human consumption. Only in the harsh climates of the largely agricultural Scandinavian democracies, the speed at which wheat gained ascendancy over rye was considerably slowed down. It is, under these circumstances, one of the grim humors of history that the Junkers, the most reactionary group in Germany, were vitally interested in the production of the "reactionary" grain, rye.

Democratic development in Germany had never received the stimulus of a great revolution. An acute sense of group distinctions remained a characteristic of German society and was carefully preserved by those who had a vested interest in this caste system. It was a natural result of this social structure that equality of bread consumption never became an accomplished fact in Germany. The armies of the French Revolution consumed wheat bread. Later, for a number of years, a small admixture of rye was in use, but since 1822 the French soldiers have always received pure wheat bread. In Germany, heavy rye bread remained an attribute of army life, and rye bread continued to play an outstanding role in the food consumption of the country as the bread of the lower classes.

Nevertheless, the beginning of this century witnessed new developments of rye consumption in Germany. Per capita wheat consumption increased between 1895 and 1905, whereas rye consumption in the same period experienced on the whole a slight decline. From then on, per capita consumption of both grains fell till 1914. In other words, the specific conditions in Germany retarded the development so much that rye consumption did not start its decline before the next stage of evolution was reached, when the share of both grains began to be reduced by new changes in the diet of the population, marked by an increased consumption of high-grade foodstuffs.

As far as the proportion of rye in the total consumption of grains in Germany was concerned, a comparison of the years 1899–1900 and 1913–1914 is very interesting. The computations made by Jasny show that in this period the proportion of barley and maize increased, whereas that of rye, wheat, and oats decreased. Specifically, the proportion of rye fell from 34.5 per cent to 32.6 per cent; wheat fell from 18.8 per cent to 18.3 per cent, and oats from 27.3 per cent to 26.4 per cent. The proportion of barley and maize increased from 19.4 to 22.6 per cent. In the words of Jasny, whose lucid exposition has been summarized in the preceding pages, "Consumption of rye, which of all grains had experienced the greatest expansion of production, contracted more than the consumption of any other grain."[1] Conversely, the extent of the area sown to barley decreased between 1900 and 1913, and barley production was reduced in the same period.

Under these conditions it is understandable that rye was being exported from Germany at an ever-increasing rate (see above). On foreign markets the reduction of demand for German rye expressed itself in that period of rising grain prices in a continual fall of price for this commodity in relation to other grains, and in sharp fluctuations from year to year even in the period of rising prices. The time was rapidly approaching when the trend would be reversed, and an absolute decline of the price for German rye would set in.[2] The way out of this situation would have been either to do away with the compromise between the Junkers and the peasants and to raise the tariffs on fodder grains, primarily barley and maize, in order to increase the use of rye as a fodder, or to increase the duty on rye even more, thus increasing the value of the import certificate.

The first alternative was extremely difficult politically. The tariff compromise between the grain-growing big estates and the smaller farms was heavily in favor of the former. The low tariffs on fodder grains within the aggregate system of agrarian protection satisfied no more than the minimum of the needs of peasant economies.

An attempt to abolish or to diminish the differential between the duty on rye and the fodder grains might easily have led to a peasant rebellion against the leadership of the Junkers. Not only would the maintenance of grain protection have been endangered, but in all probability such action might have provoked changes in the political constitution of the country. In particular, if the peasants had withdrawn their support or cooperation from the Conservative Party, a continuation of the Junker regime in the Prussian Diet would have been greatly jeopardized. Therefore, the only practical possibility would have been an increase of the export premium on rye at the expense of the whole community which, of course, would necessitate further increases in the tariff on wheat lest rapid shifts of consumption to the latter grain take place. In this event, at least some increase of duties on the principal products of peasant economy would likewise be unavoidable in order to strengthen the cohesion of the agrarian bloc.

There was every indication that preparations for a new campaign for increased tariffs on the principal bread grains was under way in Germany on the eve of the war. The trade agreement with Russia was due to expire in 1917, and this would have provided an excellent opportunity for a revision of the tariff. The chances of success were not unfavorable. The strides Germany had made toward a vertical monopolistic concentration in industry, together with the system of industrial export premiums, had diminished friction between the individual industries. In contrast to the early 'nineties, German industry was sharing fully in a period of prosperity and Germany was enjoying a period of rapid industrialization, which largely checked the stream of emigrants from the country. The fear of losing the Russian market by a new increase in tariffs was certainly a deterrent circumstance. Russia had grown stronger politically and economically since the conclusion of the Bülow trade treaty in 1904. Yet the international tension had grown too. Europe had witnessed the Annexation Crisis, the Libyan and Balkan wars, and the second Morocco Crisis. Since grain production was an essential element of the *Wehrwirtschaft* ("war preparedness") ideology of Germany, the international situation would most assuredly have been used by the Junkers and by the government as a strong argument in favor of increased protection. Under these circumstances, there was no small probability that a coalition of the Conservative and Center parties would find sufficient support from various quarters of industry so that increases of agricultural tariffs could be carried out by the familiar construction of a solidarity bloc. A general depression which hit both industry and agriculture was the classical setting for the policy of solidarity blocs in Germany. It was the transition from depression to recovery which was bound to place great strain on the bloc and threaten its disruption. But once industrial prosperity was well on the way, industrial circles were not disinclined to support the demands of agrarian protectionists and to bear part of the cost of additional protection. For, apart from the idea that this protection was a necessary prerequisite to the political might of Germany, the maintenance of close ties between industry and agriculture was instrumental in achieving the political isolation of labor.

The last prewar years were marked by popular movements against the rising cost of living. Disorders and street demonstrations were taking place in the cities. A tariff campaign would have provoked even more bitter conflicts than those witnessed at the time of Bülow's revision. It would have widened the rift between labor and the farmers and by the same token have delayed the process of democratization in the country. Also, in this sense the specific technique of grain protectionism in Germany, with the structural changes and the recurrent upward revisions it entailed, was an antidemocratic institution. The particular situation which arose on the eve of the war in 1914 with regard to rye makes it very difficult to accept the statement that at that time Germany was safely en route to complete democracy and parliamentarism.

Such statements are frequently made. Stolper, for instance, insists that in 1914 Germany had been "on the road of democratization" under the chancellorship of Bethmann-Hollweg, the "modern-minded" successor of Bülow after 1909.[3] Doubts of the validity of these statements are permissible, unless they are meant to refer primarily to conditions in South German states, where electoral reforms at the beginning of the century greatly widened the scope of franchise and where, as in Baden, democratic majorities of Socialists and Liberals existed in the Diet. Bethmann-Hollweg was a convinced opponent of parliamentary government and in his speeches in the Reichstag went so far as to predict that such a government would never become reality in Germany. To be sure the new chancellor tried to reform the Prussian Diet with its three-class electoral system, which, assigning to 82 per cent of the voters the role

of a *quantité négligeable*, should be considered the most reactionary electoral system in Europe at the time.[4] But it is significant that Bethmann-Hollweg's bill on the subject was so much of a half- or quarter-measure as to provoke objections even from people who could not be accused of being fundamentally biased against the East Elbian aristocracy and Prussian traditions. Only a few years earlier Bethmann-Hollweg publicly denied that the Prussian Diet was in need of reform. The speech in which he introduced the reform bill in the Diet was an apologia for the Prussian system, rather than a defense of his own bill, and was full of invectives against general franchise, secret ballot, parliamentarism, and democracy in general. The bill was duly defeated by a majority of Conservatives and the Center – the Catholic peasants again towed in the wake of the Protestant Junkers, after a temporary break in the years 1907–1909.[5]

It was only years later, in 1917, when the coming collapse began to cast its shadow before, that Bethmann-Hollweg issued an Imperial message promising the introduction of the general franchise in the State of Prussia after the war. Even this action was, as Bülow rightly insists, actuated by fear rather than by political principles.[6] In December, 1917, the Junkers in the Prussian Diet refused to redeem the Imperial promise. As late as May, 1918, the Conservative Party with help from the National Liberals and the Center still voted down a proposal to place the Diet on the basis of equal franchise.

The democratization of the country appeared effectively blocked on the eve of the war. But this was not all. Plans to proceed in the opposite direction were considered. The Prussian courts and Prussian police were harassing labor organizations. Various far-reaching measures against the labor movement were under consideration, including, as it seems, disbandment of the trade unions and the Social-Democratic Party, numerically the strongest political group in the country, which had polled four and a half million votes at the last election.[7] This ill agrees with the picture of a Germany moving irresistibly toward democracy.

The outbreak of the war closed the period of intense agricultural protection which had been initiated by Bülow in Germany. It obviated the necessity of dealing with the problem of increase of grain protection. On the contrary, one of the first wartime measures of the German government was the abolition of all duties and other restrictions on the import of agricultural products. It was not only consideration of supply problems which prompted this decision. At a time when national unity was the paramount need and labor votes for war credits a most important manifestation of this unity, it became impossible to continue the policy of agrarian protection, which had been responsible for a wide rift between two large sections of the people.

Since 1914 it has became customary in the writings of adherents of the principle of grain protection in Germany to take the outbreak of the war as a supreme proof of the wisdom of the artificial maintenance of grain protection in the country.[8] Whereas free trade, it was argued, had led to the competition with England and thus had made the war between the two powers inevitable, the maintained and increased amount of grain production within the country rendered it possible for Germany to provide for years the bread supply of the people in face of blockade by the Entente.

A few critical remarks must suffice here. First, agricultural protectionism in Germany, and particularly the agreement of Norderney in 1904 which Bülow had extorted at a moment of Russian defeats and humiliations, had left a great deal of resentment in Russia, in government circles as well as with the public. Discussions and computations of the Russian economic losses caused by grain protection in Germany were prominent in newspapers, journals, and books. This resentment, greatly fomented by the Russian nobility, played its part in the ensuing rapprochement between England and Russia.[9] Most notably, the fall of rye prices after 1909 caused severe hardships to Russian rye producers and to the Russian grain traders, whereas the Germans with the help of export premiums even found it profitable to sell their rye on Russian markets.

This led to the imposition by Russia of countervailing duties on grains, which in turn led to belligerent criticisms in the agrarian press in Germany. On the Russian side the suspicion was voiced in the Duma and endorsed by the government (February, 1914) that Germany might attempt to involve Russia in external political entanglements in order to improve Germany's bargaining position in the coming negotiations for a new trade agreement, thus creating a situation akin to that of 1904.[10] Exaggerated as some of these interpretations may have been, they nevertheless demonstrate the close relation which

existed between the problem of grain protection in Germany and the sphere of "high politics." The German rye policy certainly had played its part in creating the envenomed international atmosphere of 1914.

Second, it was precisely the import-certificate system which rendered it possible to export grains immediately after the harvest, whereas the imports in the West of Germany were distributed over the whole year. Had not the war started just before the harvest, grain protection would have been of very little avail; in the summer of 1914 there were no stocks of grains accumulated in the country.[11]

It is certainly striking to see the extent to which the economic interests of the Junkers took precedence over the task of war preparedness, which had been so much emphasized for long years by the propaganda of the Union of Agriculturists as one of the greatest benefits of grain protection. There is little doubt that this propaganda had materially enhanced the bellicose mood of the German public and was largely responsible for its acceptance of the war in complete ignorance as to how little the domestic grain production would be able to mitigate the penury which was to be caused by the Entente blockade. Finally, it was agricultural protectionism which had hopelessly delayed the democratization of the country, had preserved the feudal groups as a powerful element in its political structure, and had left an irresponsible maniac on the throne endowed with supreme authority over the government and the army. It is indeed difficult to ignore the large part which the system of German grain protection played in involving the country and the world in the first World War.[12]

NOTES

1 Naum Jasny, "Die Zukunft des Roggens," *Vierteljahrshefte zur Konjunkturforschung.* Sonderheft 20 (Berlin, 1930), pp. 62–63.

2 *Ibid.*, p. 82.

3 Gustav Stolper, *German Economy, 1870–1940* (New York, 1940), p. 13.

4 Perhaps, as P. N. Miliukov insists, with the exception of the Russian suffrage to the Duma, as created by Stolypin's reactionary coup d'état in 1906. See P. N. Miliukov, *Rossiia na perelome* (Paris, 1927), 1:7.

5 Gustav Schmoller, "Die preussische Wahlrechtsreform

von 1910," *Jahrbuch für Gesetzgebung, Verwaltung und Volkswirtschaft im Deutschen Reich*, XXXIV: 1263, 1273–1278; *Stenographische Berichte über die Verhandlungen des Reichstages*, XII Legislaturperiode, II Session, 258:1670, 259:1410.

6 Bernhard von Bülow, *Denkwürdigkeiten*, III: 259.

7 Albert C. Grzesinski, *Inside Germany* (New York, 1939), p. 34; Hermann Wendel, "Scharfer Wind," *Die Neue Zeit*, Vol. 32:2 (June 12, 1914), pp. 457–460; *Stenographische Berichte über die Verhandlungen des Reichstages*, XIII Legislatur-periode, I Session, 295:8914–8917.

8 Cf., for example, Michael Hainisch, *Das Getreide-monopol*, Schriften des Vereines für Sozialpolitik (Munich-Leipzig, 1916), Vol. 155:1, p. 356; and *idem, Die Landflucht* (Jena, 1924), p. 304.

9 Gregor Alexinski, *Russia and the Great War* (London, 1915), p. 65.

10 *Mezhdunarodnyie otnosheniia v epochu imperialisma, Documenty iz arkhivov tsarskogo i vremennogo pravitel'stv, 1818–1917*, Series III, I (Moscow-Leningrad, 1931): 265; II (Moscow-Leningrad, 1933): 269, 389.

11 Friedrich Aereboe, *Der Einfluss des Krieges auf die land-wirtschaftliche Produktion in Deutschland*, Veröffent-lichungen der Carnegie Stiftung für internationalen Frieden, Wirtschafts- und Sozialgeschichte des Weltkrieges (Berlin, 1927), p. 30. Also, Theodor Plaut, *Deutsche Handelspolitik* (Leipzig-Berlin, 1929), p. 106.

12 Cf. Eckart Kehr, "Englandhass und Weltpolitik," pp. 500–526; also: Lujo Brentano, *Ist das System Brentano zusammengebrochen?* (Berlin, 1918), p. 67.

THREE

'This is the Voice of Algeria'

from *A Dying Colonialism* (1959)

Frantz Fanon

Editors' Introduction

Frantz Fanon was born in 1925 in Martinique, then as now a department of France. As a 'mixed race' young man growing up just as Martinique came under Vichy fascism in 1940, young Frantz witnessed the everyday brutalities of colonial racism. He decided at the age of 18 to join the Free French Forces in Dominica, and then the French Army in France. Wounded in the Battle of Alsace, Fanon was both awarded the *Croix de Guerre* and relieved of his service along with other black soldiers as their 'bleached' regiment crossed the Rhine to celebrate the liberation of France. Returning to Martinique in 1945, Fanon participated in the parliamentary campaign of the great poet Aimé Césaire. He then returned to France to study medicine and psychiatry in Lyon, qualifying as a psychiatrist in 1951. While in France, Fanon wrote his first book, later translated as *Black Skin, White Masks* (1967 [1952]). Here, he reflected on the psychological effects of colonialism upon a black person who had come through French higher education but who would always be subject to racism.

In 1953, Fanon moved to Algeria to continue his practice at the Blida-Joinville Psychiatric Hospital, until his resignation in 1956. He instituted socio-therapy in the hospital, connecting treatment with patients' backgrounds. In 1954, Fanon joined the Algerian Revolution and the Front de Libération Nationale (FLN). The reading that follows comes from *Year 5 of the Algerian Revolution*, republished as *A Dying Colonialism* (1965 [1959]), an important study of the process of decolonisation through the many fronts in the fight for liberation. *A Dying Colonialism* argues that various aspects of 'traditional' culture, propped up and derided by the coloniser, have been transformed through the practice of revolution into tools for revolution itself. So it is with the veil, an instrument of gender domination first used by the coloniser to mark the backward Algerian Muslim, then used by the revolutionary to spirit explosives into the white quarters (brilliantly portrayed in Gillo Pontecorvo's 1966 film *The Battle of Algiers*). So it is with colonial medicine: first used to humiliate, supervise and possibly injure the native; later, through the revolution, modern medicine becomes an instrument of basic health education. The reading tells a similar story about the transformation of the radio from a tool of colonial propaganda into an essential device for an underground and dispersed revolutionary movement that must broadcast the new nation into being. The reader might want to read this text against Gandhi (Part 2) and other defences of 'tradition' as opposed to 'modernity'.

Fanon was expelled from Algeria in 1957 after his public resignation and exposure of his political activities. He left France in secret for Tunis, where he wrote as part of the editorial collective of *El Moudjahid*. He became the Ambassador of the Provisional Algerian Government to Ghana, bringing together the disparate ends of his life on the colonial periphery of France. Returning to Tunis, Fanon worked on the attempt to create an external front for Algeria's decolonisation. Diagnosed with leukaemia at this time, Fanon was first sent to the Soviet Union for treatment. On his return, he dictated *The Wretched of the Earth* (1963 [1961]), and delivered lectures to the Armée de Libération Nationale on the Algerian border. Fanon met Jean-Paul Sartre in Rome before being assisted by the CIA to receive treatment in the United States, where he died as 'Ibrahim Fanon' in Bethesda, Maryland, in December 1961.

Fanon's classic, *The Wretched of the Earth*, with a powerful introduction by Sartre, became an iconic text for national liberation movements of the twentieth century. This book works through the question of violence in struggles for decolonisation. The important chapter called 'The pitfalls of national consciousness' shows how the regime that succeeds decolonisation might easily be hijacked by a narrow elite that does not represent the nation's developmental interests as a whole. This chapter sows the seeds for the emergence of 'postcolonial' thought at the end of the twentieth century, after the developmental failures of many societies appear to confirm Fanon's warning.

Key references

Frantz Fanon (1965 [1959]) *A Dying Colonialism*, trans. H. Chevalier, New York: Grove Press (originally published in French in 1959 under the title *L'An V de la Révolution Algérienne*).
— (1967 [1952]) *Black Skin, White Masks*, trans. C. L. Markmann, New York: Grove Press.
— (1965) *The Wretched of the Earth*, trans. C. Farrington, New York: Grove Press.
Nigel Gibson (ed.) (1999) *Rethinking Fanon: The Continuing Legacy*, Amherst, NY: Humanity Books.
David Macey (2000) *Frantz Fanon. A Biography*, New York: Picador.

[. . .]

We have already noted the accelerated speed with which the radio was adopted by the European society. The introduction of the radio in the colonizing society proceeded at a rate comparable to that of the most developed Western regions. We must always remember that in the colonial situation, in which, as we have seen, the social dichotomy reaches an incomparable intensity, there is a frenzied and almost laughable growth of middle-class gentility on the part of the nationals from the metropolis. For a European to own a radio is of course to participate in the eternal round of Western petty-bourgeois ownership, which extends from the radio to the villa, including the car and the refrigerator. It also gives him the feeling that colonial society is a living and palpitating reality, with its festivities, its traditions eager to establish themselves, its progress, its taking root. But especially, in the hinterland, in the so-called colonization centers, it is the only link with the cities, with Algiers, with the metropolis, with the world of the civilized. It is one of the means of escaping the inert, passive, and sterilizing pressure of the "native" environment. It is, according to the settler's expression, "the only way to still feel like a civilized man."

On the farms, the radio reminds the settler of the reality of colonial power and, by its very existence, dispenses safety, serenity. Radio-Alger is a con-firmation of the settler's right and strengthens his certainty in the historic continuity of the conquest, hence of his farm. The Paris music, extracts from the metropolitan press, the French government crises, constitute a coherent background from which colonial society draws its density and its justification. Radio-Alger sustains the occupant's culture, marks it off from the non-culture, from the nature of the occupied. Radio-Alger, the voice of France in Algeria, constitutes the sole center of reference at the level of news. Radio-Alger, for the settler, is a daily invitation not to "go native," not to forget the rightfulness of his culture. The settlers in the remote outposts, the pioneering adventurers, are well aware of this when they say that "without wine and the radio, we should already have become Arabized."[1]

In Algeria, before 1945, the radio as a technical news instrument became widely distributed in the dominant society. It then, as we have seen, became both a means of resistance in the case of isolated Europeans and a means of cultural pressure on the dominated society. Among European farmers, the radio was broadly regarded as a link with the civilized world, as an effective instrument of resistance to the corrosive influence of an inert native society, of a society without a future, backward and devoid of value.

For the Algerian, however, the situation was totally different. We have seen that the more well-to-do family hesitated to buy a radio set. Yet no explicit, organized, and motivated resistance was to be observed, but rather a dull absence of interest in that piece of French presence. In rural areas and in regions remote from the colonization centers, the situation was clearer. There no one was faced with the problem, or rather, the problem was so remote from the everyday concerns of the native that it was quite clear to an inquirer that it would be outrageous to ask an Algerian why he did not own a radio.

A man conducting a survey during this period who might be looking for satisfactory answers would find himself unable to obtain the information he needed. All the pretexts put forth had of course to be carefully weighed. At the level of actual experience, one cannot expect to obtain a rationalization of attitudes and choices.

Two levels of explanation can be suggested here. As an instrumental technique in the limited sense, the radio receiving set develops the sensorial, intellectual, and muscular powers of man in a given society. *The radio in occupied Algeria is a technique in the hands of the occupier which, within the framework of colonial domination, corresponds to no vital need insofar as the "native" is concerned.* The radio, as a symbol of French presence, as a material representation of the colonial configuration, is characterized by an extremely important negative valence. The possible intensification and extension of sensorial or intellectual powers by the French radio are implicitly rejected or denied by the native. The technical instrument, the new scientific acquisitions, when they contain a sufficient charge to threaten a given feature of the native society, are never perceived in themselves, in calm objectivity. The technical instrument is rooted in the colonial situation where, as we know, the negative or positive coefficients always exist in a very accentuated way.

At another level, as a system of information, as a bearer of language, hence of message, the radio may be apprehended within the colonial situation in a special way. Radiophonic technique, the press, and in a general way the systems, messages, sign transmitters, exist in colonial society in accordance with a well-defined statute. Algerian society, the dominated society, never participates in this world of signs. The messages broadcast by Radio-Alger are listened to solely by the representatives of power in Algeria, solely by the members of the dominant authority and seem magically to be avoided by the members of the "native" society. The non-acquisition of receiver sets by this society has precisely the effect of strengthening this impression of a closed and privileged world that characterizes colonialist news. In the matter of daily programs, before 1954, eulogies addressed to the occupation troops were certainly largely absent. From time to time, to be sure, there might be an evocation over the radio of the outstanding dates of the conquest of Algeria, in the course of which, with an almost unconscious obscenity, the occupier would belittle and humiliate the Algerian resistant of 1830. There were also the commemorative celebrations in which the "Moslem" veterans would be invited to place a wreath at the foot of the statue of General Bugeaud or of Sergeant Blandan, both heroes of the conquest and liquidators of thousands of Algerian patriots. But on the whole it could not be said that the clearly racialist or anti-Algerian content accounted for the indifference and the resistance of the native. *The explanation seems rather to be that Radio-Alger is regarded by the Algerian as the spokesman of the colonial world. Before the war the Algerian, with his own brand of humor, had defined Radio-Alger as "Frenchmen speaking to Frenchmen."*

[. . .]

The acquisition of a radio set in Algeria, in 1955, represented the sole means of obtaining news of the Revolution from non-French sources. This necessity assumed a compelling character when the people learned that there were Algerians in Cairo who daily drew up the balance-sheet of the liberation struggle. From Cairo, from Syria, from nearly all the Arab countries, the great pages written in the *djebels* by brothers, relatives, friends flowed back to Algeria.

Meanwhile, despite these new occurrences, the introduction of radio sets into houses and the most remote *douars* proceeded only gradually. There was no enormous rush to buy receivers.

It was at the end of 1956 that the real shift occurred. At this time tracts were distributed announcing the existence of a Voice of Free Algeria. The broadcasting schedules and the wavelengths were given. This voice "that speaks from the *djebels*," not geographically limited, but bringing to all Algeria the great message of the Revolution, at once acquired an essential value. In less than twenty days the entire stock of radio sets was bought up. In the *souks*[2] trade in used receiver sets began. Algerians who had served their apprenticeship with European radio-electricians opened small shops. Moreover, the dealers had to meet new needs. The absence of electrification in immense regions in Algeria naturally created special problems for the consumer. For this reason battery-operated receivers, from 1956 on, were in great demand on Algerian territory. In a few weeks several thousand sets were sold to Algerians, who bought them as individuals, families, groups of houses, *douars, mechtas.*

Since 1956 the purchase of a radio in Algeria has meant, not the adoption of a modern technique for getting news, but the obtaining of access to the only means of entering into communication with the Revolution, of living with it. In the special case of the portable battery set, an improved form of the standard receiver operating on current, the specialist in technical changes in underdeveloped countries might see a sign of a radical mutation. The Algerian, in fact, gives the impression of finding short cuts and of achieving the most modern forms of news-communication without passing through the inter-mediary stages.[3] In reality, we have seen that this "progress" is to be explained by the absence of electric current in the Algerian *douars.*

The French authorities did not immediately realize the exceptional importance of this change in attitude of the Algerian people with regard to the radio. Traditional resistances broke down and one could see in a *douar* groups of families in which fathers, mothers, daughters, elbow to elbow, would scrutinize the radio dial waiting for the *Voice of Algeria.* Suddenly indifferent to the sterile, archaic modesty and antique social arrangements devoid of brother-hood, the Algerian family discovered itself to be immune to the off-color jokes and the libidinous references that the announcer occasionally let drop.

Almost magically – but we have seen the rapid and dialectical progression of the new national requirements – the technical instrument of the radio receiver lost its identity as an enemy object. The radio set was no longer a part of the occupier's arsenal of cultural oppression. In making of the radio a primary means of resisting the increasingly over-whelming psychological and military pressures of the occupant, Algerian society made an autonomous decision to embrace the new technique and thus tune itself in on the new signaling systems brought into being by the Revolution.

The Voice of Fighting Algeria was to be of capital importance in consolidating and unifying the people. We shall see that the use of the Arab, Kabyle and French languages which, as colonialism was obliged to recognize, was the expression of a non-racial con-ception, had the advantage of developing and of strengthening the unity of the people, of making the fighting Djurdjura area real for the Algerian patriots of Batna or of Nemours. The fragments and splinters of acts gleaned by the correspondent of a newspaper more or less attached to the colonial domination, or communicated by the opposing military authorities, lost their anarchic character and became organized into a national and Algerian political idea, assuming their place in an overall strategy of the reconquest of the people's sovereignty. The scattered acts fitted into a vast epic, and the Kabyles were no longer "the men of the mountains," but the brothers who with Ouamrane and Krim made things difficult for the enemy troops.

Having a radio meant paying one's taxes to the nation, buying the right of entry into the struggle of an assembled people.

The French authorities, however, began to realize the importance of this progress of the people in the technique of news dissemination. After a few months of hesitancy legal measures appeared. The sale of radios was now prohibited, except on presentation of a voucher issued by the military security or police services. The sale of battery sets was absolutely pro-hibited, and spare batteries were practically with-drawn from the market. The Algerian dealers now had the opportunity to put their patriotism to the test, and they were able to supply the people with spare batteries with exemplary regularity by resorting to various subterfuges.[4]

The Algerian who wanted to live up to the Revolution, had at last the possibility of hearing an official voice, the voice of the combatants, explain the combat to him, tell him the story of the Liberation on the march, and incorporate it into the nation's new life.

Here we come upon a phenomenon that is sufficiently unusual to retain our attention. The highly trained French services, rich with experience acquired in modern wars, past masters in the practice of "sound-wave warfare," were quick to detect the wave-lengths of the broadcasting stations. The pro-grams were then systematically jammed, and the *Voice of Fighting Algeria* soon became inaudible. A new form of struggle had come into being. Tracts were distributed telling the Algerians to keep tuned in for a period of two or three hours. In the course of a single broadcast a second station, broadcasting over a different wave-length, would relay the first jammed station. The listener, enrolled in the battle of the waves, had to figure out the tactics of the enemy, and in an almost physical way circumvent the strategy of the adversary. Very often only the operator, his ear glued to the receiver, had the unhoped-for opportunity of hearing the *Voice*. The other Algerians present in the room would receive the echo of this voice through the privileged interpreter who, at the end of the broadcast, was literally besieged. Specific questions would then be asked of this incarnated voice. Those present wanted to know about a particular battle mentioned by the French press in the last twenty-four hours, and the interpreter, embarrassed, feeling guilty, would sometimes have to admit that the *Voice* had not mentioned it.

But by common consent, after an exchange of views, it would be decided that the *Voice* had in fact spoken of these events, but that the interpreter had not caught the transmitted information. A real task of reconstruction would then begin. Everyone would participate, and the battles of yesterday and the day before would be re-fought in accordance with the deep aspirations and the unshakable faith of the group. The listener would compensate for the fragmentary nature of the news by an autonomous creation of information.

Listening to the *Voice of Fighting Algeria* was motivated not just by eagerness to hear the news, but more particularly by the inner need to be at one with the nation in its struggle, to recapture and to assume the new national formulation, to listen to and to repeat the grandeur of the epic being accomplished up there among the rocks and on the *djebels*. Every morning the Algerian would communicate the result of his hours of listening in. Every morning he would complete for the benefit of his neighbor or his comrade the things not said by the *Voice* and reply to the insidious questions asked by the enemy press. He would counter the official affirmations of the occupier, the resounding bulletins of the adversary, with official statements issued by the Revolutionary Command.

Sometimes it was the militant who would circulate the assumed point of view of the political directorate. Because of a silence on this or that fact which, if prolonged, might prove upsetting and dangerous for the people's unity, the whole nation would snatch fragments of sentences in the course of a broadcast and attach to them a decisive meaning. Imperfectly heard, obscured by an incessant jamming, forced to change wave-lengths two or three times in the course of a broadcast, the *Voice of Fighting Algeria* could hardly ever be heard from beginning to end. It was a choppy, broken voice. From one village to the next, from one shack to the next, the *Voice of Algeria* would recount new things, tell of more and more glorious battles, picture vividly the collapse of the occupying power. The enemy lost its density, and at the level of the consciousness of the occupied, experienced a series of essential setbacks. Thus the *Voice of Algeria*, which for months led the life of a fugitive, which was tracked by the adversary's powerful jamming networks, and whose "word" was often inaudible, nourished the citizen's faith in the Revolution.

This *Voice* whose presence was felt, whose reality was sensed, assumed more and more weight in proportion to the number of jamming wave-lengths broadcast by the specialized enemy stations. It was the power of the enemy sabotage that emphasized the reality and the intensity of the national expression. By its phantom-like character, the radio of the *Moudjahidines*, speaking in the name of Fighting Algeria, recognized as the spokesman for every Algerian, gave to the combat its maximum of reality.

Under these conditions, claiming to have heard the *Voice of Algeria* was, in a certain sense, distorting the truth, but it was above all the occasion to proclaim one's clandestine participation in the essence of the Revolution. It meant making a deliberate choice, though it was not explicit during the first months, between the enemy's congenital lie and the people's own lie, which suddenly acquired a dimension of truth.

[. . .]

Accepting the radio technique, buying a receiver set, and participating in the life of the fighting nation, all these coincided. The frenzy with which the people exhausted the stock of radio sets gives a rather accurate idea of its desire to be involved in the dialogue that began in 1955 between the combatant and the *nation*.

In the colonial society, Radio-Alger was not just one among a number of voices. *It was the voice of the occupier*. Tuning in Radio-Alger amounted to accepting domination; it amounted to exhibiting one's desire to live on good terms with oppression. It meant giving in to the enemy. Switching on the radio meant validating the formula, "This is Algiers, the French Radio Broadcast." The acquiring of a radio handed the colonized over to the enemy's system and prepared for the banishing of hope from his heart.

The existence of the *Voice of Fighting Algeria*, on the other hand, profoundly changed the problem. Every Algerian felt himself to be called upon and wanted to become a reverberating element of the vast network of meanings born of the liberating combat. The war, daily events of military or political character, were broadly commented on in the news programs of the foreign radios. In the foreground the voice of the *djebels* stood out. We have seen that the phantom-like and quickly inaudible character of this voice in no way affected *its felt reality and its power*. Radio-Alger, Algerian Radio-Broadcasting, lost their sovereignty.

Gone were the days when mechanically switching on the radio amounted to an invitation to the enemy. For the Algerian the radio, as a technique, became transformed. The radio set was no longer directly and solely tuned in on the occupier. To the right and to the left of Radio-Alger's broadcasting band, on different and numerous wave-lengths, innumerable stations could be tuned in to, among which it was possible to distinguish the friends, the enemies' accomplices, and the neutrals. Under these conditions, having a receiver was neither making oneself available to the enemy, nor giving him a voice, nor yielding on a point of principle. On the contrary, on the strict level of news, it was showing the desire to keep one's distance, to hear other voices, to take in other prospects. It was in the course of the struggle for liberation and thanks to the creation of a *Voice of Fighting Algeria* that the Algerian experienced and concretely discovered the existence of voices other than the voice of the dominator which formerly had been immeasurably amplified because of his own silence.

The old monologue of the colonial situation, already shaken by the existence of the struggle, disappeared completely by 1956. *The Voice of Fighting Algeria* and all the voices picked up by the receiver now revealed to the Algerian the tenuous, very relative character, in short, the imposture of the French voice presented until now as the only one. The occupier's voice was stripped of its authority.

The nation's *speech*, the nation's spoken *words* shape the world while at the same time renewing it.

Before 1954, native society as a whole rejected the radio, turned a deaf ear to the technical development of methods of news dissemination. There was a non-receptive attitude before this import brought in by the occupier. In the colonial situation, the radio did not satisfy any need of the Algerian people.[5] On the contrary, the radio was considered, as we have seen, a means used by the enemy to quietly carry on his work of depersonalization of the native.

The national struggle and the creation of *Free Radio Algeria* have produced a fundamental change in the people. The radio has appeared in a massive way at once and not in progressive stages. What we have witnessed is a radical transformation of the means of perception, of the very world of perception. Of Algeria it is true to say that there never was, with respect to the radio, a pattern of listening habits, of audience reaction. Insofar as mental processes are concerned, the technique had virtually to be invented. *The Voice of Algeria*, created out of nothing, brought the nation to life and endowed every citizen with a new status, *telling him so explicitly*.

After 1957, the French troops in operation formed the habit of confiscating all the radios in the course of a raid. At the same time listening in on a certain number of broadcasts was prohibited. Today things have progressed. *The Voice of Fighting Algeria* has multiplied. From Tunis, from Damascus, from Cairo, from Rabat, programs are broadcast to the people. The programs are organized by Algerians. The French services no longer try to jam these powerful and numerous broadcasts. The Algerian has the opportunity every day of listening to five or six different broadcasts in Arabic or in French, by means of which he can follow the victorious development of the Revolution step by step. As far as news is concerned, the word of the occupier has been seen to suffer a progressive devaluation. After having imposed the national voice upon that of the dominator, the radio welcomes broadcasts from all the corners of the world. The "Week of Solidarity with Algeria," organized by the Chinese people, or the resolutions of the Congress of African Peoples on the Algerian war, link the *fellah* to an immense tyranny-destroying wave.

Incorporated under these conditions into the life of the nation, the radio will have an exceptional importance in the country's building phase. After the war a disparity between the people and what is intended to speak for them will no longer be possible. The revolutionary instruction on the struggle for liberation must normally be replaced by a revolutionary instruction on the building of the nation. The fruitful use that can be made of the radio can well be imagined. Algeria has enjoyed a unique experience. For several years, the radio will have been for many, one of the means of saying "no" to the occupation and of believing in the liberation. The identification of the voice of the Revolution with the fundamental truth of the nation has opened limitless horizons.

NOTES

1   Radio-Alger is in fact one of the mooring-lines maintained by the dominant society. Radio-Monte-Carlo, Radio-Paris, Radio-Andorre likewise play a protective role against "Arabization."
2   *souk* – market or shop. (Translator's note)
3   In the realm of military communications, the same phenomenon is to be noted. In less than fifteen months the National Army of Liberation's "liaison and tele-communications system" became equal to the best that is to be found in a modern army.
4   The arrival in Algeria by normal channels of new sets and new batteries obviously became increasingly difficult. After 1957 it was from Tunisia and Morocco, via the underground, that new supplies came. The regular introduction of these means of establishing contact with the official voice of the Revolution became as important for the people as acquiring weapons or munitions for the National Army.
5   In this connection may be mentioned the attitude of the French authorities in present-day Algeria. As we know, television was introduced into Algeria several years ago. Until recently, a simultaneous bilingual commentary accompanied the broadcasts. Some time ago, the Arabic commentary ceased. This fact once again confirms the aptness of the formula applied to Radio-Alger: "Frenchmen speaking to Frenchmen."

PART FOUR

Promethean visions

We have seen that ideas about the importance of material progress, and even about economic growth and development, were widespread by the inter-war years, and not just in North America, Japan and Western Europe. Nationalist movements in India and China talked about development, and so too did political leaders in Latin America. The Soviet Union was by then committed to a strategy of rapid industrialisation that had devastating consequences for many peasant farmers, or *kulaks*. Their surpluses were squeezed to make resources available for the urban-industrial sector. Even in Africa, there were signs by the 1940s that the colonial powers were beginning to move on from ideas of stewardship and the White Man's burden. There was new talk about development corporations and the governmental duty of welfare.

Against this background, it might seem perverse for Arturo Escobar to claim that Development was invented as a set of social and economic practices in the 1950s. Or that, more specifically, the Age of Development first took shape in the wake of a speech by President Truman on 20 January 1949: a speech that committed the United States to large-scale disbursements of foreign aid under the Point 4 Program. But in a most important respect Escobar is right. What marked the 1950s and 1960s was the invention of forms of development discourse, or Developmentalism, which celebrated the possibility of rapid economic growth and cultural change in those parts of the globe that henceforth would be known as the Third World. One corollary of this, as Escobar shows in the reading reproduced here, was the 'discovery of mass poverty' in Asia, Africa and Latin America. In his view, the Third World was first invented as a residual category (not the First World, not the Second World). It then became the object of political struggle between the United States and the USSR (for which reason Developmentalism must be seen as a Cold War ideology). It was also pathologised as a site of mass poverty and other absences (absence of education, absence of industry, absence of a strong work ethic, etc.). Finally, the Third World was offered development as a cure for its supposed maladies. Development would turn the Third World into the First World (or the Second World). It became a recipe for the abolition of most of the world.

As several critics have pointed out, Escobar's willingness to think of Developmentalism as something singular, or with a capital D, causes problems. It makes it hard for him to acknowledge the very different sets of developmental practices that have been produced since 1950. In our terms, he is not sufficiently attuned to the complex ways in which technologies of rule for the transformation of the economy, or the polity, or of nature or difference, can be combined within a more general commitment to 'development'. (This is so whether we define development as economic growth, enhanced freedom or improved human development scores.) For example, in the 1950s and 1960s there were important debates about the roles of planning and trade in the process of economic development. We shall see this later in Part Four. There were also spirited debates about the relative weights that countries should give to economic growth and social equity. Michael Cowen and Robert Shenton are right to hold that 'modern development doctrine is based [upon a fierce accentuation of] the positive dimensions of development as a process' (1996: ix). Nevertheless, ruling elites were always aware of the political, social and economic costs of 'development', particularly in the form of rapid industrialisation. They also understood that development meant rather more than a desire to become the next United States (or Soviet Union). This is one reason why we should be cautious in assuming that Developmentalism in the 1950s was fully dominated by political and academic ideas coming out of the Unites States, or 'the West' more broadly. Third World countries were perfectly capable of supplying their own visions of a better future.

And yet it does make sense to start this section with the United States. This is true notwithstanding that Development Studies has struggled to gain ground in the United States, where accounts of the Third World more often emanate from within comparative politics or international relations. The US and US-based intellectuals were in at the birth of post-war developmentalism in two supremely important ways: as the main architects of the post-war global political economy, and as the progenitors of what is generally called 'modernisation theory'. (If space allowed, this should more properly be called 'US post-war modernisation theory'.)

We shall pick up the architecture of the post-war world system more extensively in later Parts. Suffice it to say here that what we must call modernisation theory was born in a world where territorial empires were rapidly being dismantled. It also emerged in a world where the reinvigoration of First World economies was taking shape under the new rules of the Bretton Woods system. The re-industrialisation of West Germany and Japan seemed to herald an era that few people had imagined could take shape within just fifteen years of the Second World War. (In 1944 there were proposals for the pastoralisation of Germany.) So too did a rapid increase in trade between richer countries. Particularly in the United States, this was an age of consumerism. Working Americans were now able to purchase television sets and refrigerators. They also observed a new phenomenon amongst their children. The teenager emerged in the 1950s, an adolescent with money in his or her pocket. It was a self-confident era, a time when, as British Prime Minister Harold Macmillan told his compatriots, 'people had never had it so good'.

But what about people in the Third World? How could they be made to share in the affluent society? How could they be made to share its values? How could they be persuaded to work harder, gain a scientific outlook, and pursue clearly specified personal goals? How could their governments build up the strong industrial economies that development seemed both to imply and to demand?

In an important book, Nils Gilman (2003) has explored how a diverse but linked body of work on modernisation theory was put forward in the United States to answer this broad raft of questions. (If the questions were posed most sharply in the West, the assumptions they betrayed – about pre-scientific world views and the lack of a Protestant work ethic – were widely shared by modernising elites in Asia, Africa and Latin America.) Most of this work came from a small number of academics based at Harvard and Princeton universities, the universities of Chicago and Pennsylvania, and the Massachusetts Institute of Technology. Their work drew on Talcott Parsons's monumental studies of social organisation and dislocation. Based at Harvard, Parsons was inspired by the earlier work of Emile Durkheim on mechanically and organically integrated societies. He tried to specify the deeper structures and transformational rules that governed the workings of pre-modern and modern societies. What most united a first generation of modernisation studies, however, was Parsons's faith that modernity (or First World status) was an end-point for all societies, and that the contemporary United States best represented this exalted state. As Gilman points out, the basic creed of post-war modernisation theory held that modern societies were to be recognised by certain key traits or indicators. These included industrialisation, state-provided welfare, a consensual model of social organisation based around limited class differences, and a high regard for science and expertise as guides to good government. In short, a modern society was what American intellectuals wanted, and to some degree did claim to see in the United States of the 1950s. And now this same society was to be built abroad.

But how, precisely? Notwithstanding that modernisation theory promoted a linear conception of development from traditional to modern societies, there was some disagreement about how this evolution might best be achieved. What might be called the more liberal strand within US modernisation theory – Walt Rostow, for example, and possibly Edward Shils – placed its faith in dynamic, modernising elites that would propel ex-colonial countries to modernity (for Rostow, an age of high mass consumption). They would exploit the fact that their countries were economic, political and social latecomers. Crudely stated, this version of modernisation theory held that the United States was in a race with the Soviet Union for the hearts and minds of people in the global periphery. It was important to demonstrate that socialism and communism were 'deviant' ideologies which had directed a wrong turn from the true road to development. It also needed to be shown that capitalist development in the Third World could be kick-started by democratic states minded to organise a big push for, or take-off into, high levels of sustained economic growth. The full title of Rostow's best-known academic book was *The Stages of Economic Growth: A Non-Communist Manifesto* (1960). There, Rostow maintained that governments in the developing world could best mimic the successes of the already developed world by mobilising high levels of savings for industrial development within a broadly capitalist framework. Development involved a measure of deferred gratification, or belt-tightening, and here Soviet-style ideologies seemed at first to hold an advantage. For Rostow, however, the inevitable pain of paying for industrial-led growth could be ameliorated by flows of foreign capital and aid from abroad, notably from the United States and its allies. In any case, Soviet-style planning invited political repression and something close to the death of the soul.

Rostow was reasonably liberal in his views about the domestic United States and rampantly optimistic about the possibility of mimicry abroad. His longer-term involvements with 'development', however, proved to be remarkably dark. Rostow first came to prominence politically as a speech-writer for John F. Kennedy in the run-up to the 1960 election. Not long after Kennedy's assassination in 1963 Rostow became the US National Security Advisor to President Lyndon B. Johnson. In this capacity he was involved in key decisions to escalate the war in Vietnam. Much like Samuel Huntington, a conservative critic of rushed modernisation, Rostow was driven in practice to support violent and authoritarian means to promote order and modernisation in South-East Asia. As Gilman puts it, perhaps with one eye on the neo-conservatives who would lead the United States to another brutal war in Iraq, 'Tension existed between the modernization theorists' sincere desire to imagine better lives for the global masses and their increasingly authoritarian approach to achieving this vision' (2003: 20).

There is another point to be made here: American intellectuals like Rostow and Huntington courted a public life. For such men, as perhaps also for Gabriel Almond, theorising about the production of modern societies was not enough. Their aim was to put ideas into practice, an unconscious echo perhaps of Keynes's dictum from the 1930s, that 'it is ideas, not vested interests that are important for good or ill'. In Huntington's case this commitment was always of a more 'realist' bent than was to be found in Rostow. He was ever more willing to prioritise order over democracy. Less romantic than Rostow, Huntington maintained that state-building and economic development in the 'free world' would often have to be pursued by authoritarian or military elites. Democracy was more likely to be a product of economic development than a precondition for it, as Rostow perhaps liked to imagine. Democracy presupposed the growth of a middle class, and/or political agitations around state taxation and expenditure policies ('no taxation without representation': see also the 'good governance' debates in Part 7).

Huntington also maintained the conservative position that culture matters, and that cultures, or what more recently he has called 'civilisations', are both durable and reasonably distinct from one another. In a widely read and widely criticised book from the mid-1990s he prophesied not the end of history but a coming clash of civilisations, notably between the 'West' and 'Islam'. In this respect, too, Huntington, so often talked about as a modernisation theorist, stands outside the mainstream. Modernisation theorists in the 1950s and 1960s rather advocated the possibility, indeed the likelihood, of social homogenisation, of the Other becoming more like Us (the American). Processes of industrialisation and secularisation would see to this, particularly in the medium term. Religiosity and ethnic ties would whither away, a view that Huntington, to his credit, strongly disputed. In the short run, however – and for the most part modernisation theorists assumed very compressed time horizons, as witness the declaration of the 1960s as *the* United Nations' Development Decade – the production of modern men and women in Third World countries could be speeded up by the introduction of modern educational systems and modern birth control technologies.

It was the minds and bodies of men and women in the Third World that were, finally, the targets of the technologies of rule that grew out of modernisation theory. Kingsley Davis becomes a key figure in this respect. In the late 1940s Davis became one of the first authors of demographic transition theory. This was a theory that held that development itself was the best form of contraception. Households would adjust their fertility decisions after local death rates had come down. This would be in response to rising incomes and new developments in health care and sanitation. At just this time, however, as the historian Simon Sretzer has noted, that is, in late 1948 and 1949, 'those in the United States still dreaming of a globe emerging from colonial servitude into a regime of liberal democratic free trade were awakening to a nightmare, experiencing a strong sense of loss of control in a dangerous and alien world' (1993: 676).

The US State Department now began to worry about feeding growing numbers of poor people through the 1950s and 1960s. In due course, the Green Revolution was proposed as a peaceful alternative to Red Revolution. It offered a suitably technocratic way of putting food into the bellies of potentially rebellious peasants and landless workers. Less obviously, perhaps, while post-war modernisation theory largely dispensed with race as an organising category (its universalism proposed that we could all become developed, or even American), in practice key sponsors of modernisation were concerned about the rapid growth of non-white populations in the Third World. Sretzer suggests that Davis himself may have come under pressure from the US State Department to argue for a speeded-up version of the demographic transition, whether by voluntary or coercive means. Such pressure would have been brought to bear through his employer, the Office of Population Research (OPR) at Princeton University. The OPR was closely associated with and partly funded by the State Department.

In the reading reproduced here, which comes from Kingsley Davis's classic account of *The Population of India and Pakistan*, we see Davis cleaving to a middle line. He discusses a range of options for dealing with the 'population problems' of the two countries. Davis warns that a failure to adopt planned parenthood policies in India and Pakistan will lead 'to perpetual poverty for their citizens or . . . absolute catastrophes' (1951: 231). Nevertheless, he remains at some distance from the forms of authoritarian Malthusianism that would be trumpeted twenty years later by scholars including Garrett Hardin and Paul Ehrlich.

Davis's work is both of American modernisation theory and a partial corrective to it. In this last respect it has something in common with that of W. Arthur Lewis and Hans Singer, as is revealed in the next two readings. Lewis is certainly committed to many of the core propositions of modernisation theory, and in particular to the absolute priority that should be attached to urban-led industrial growth. But his reasoning is more conventionally economic: the structural transformation of economies involves the transfer of labour from less to more productive jobs. It is also inflected by the technocratic faith both of Keynesianism and of mid-twentieth-century thought more generally. Like Rostow, too, who also draws upon Keynes, Lewis is adamant that, 'The central fact of economic development is that the distribution of incomes is altered in favour of the saving class'. His famous two-sector model of a closed economy with unlimited supplies of labour is structured by and subordinate to this basic observation. Savings converted into investment in the urban-industrial economy drive rural to urban migration. This continues up to the point where unproductive labour can no longer be decanted, at no real cost, from the countryside to the city.

Lewis's work in the 1950s, along with work on input–output models, was picked up on, mirrored and in key respects developed by versions of planned modernisation emanating from within the Third World. The Second and Third Five-Year plans in India (1956–66) are cases in point. Designed largely by a Planning Commission led by Prime Minister Jawaharlal Nehru and chief economist-statistician P. C. Mahalanobis, these two plan periods enunciated a tremendous faith in wise government, urban-industrial development, and the educating effects of concrete. Nehru famously urged villagers close to the new dams built along the Damodar river to look upon India's 'modern temples'. He also insisted that economic development must be broad-based, socially equitable and regionally dispersed. Like most mid-century intellectuals and policy makers, Nehru rejected the idea that economic development could be prosecuted by private capitalists alone. Poor countries faced foreign exchange constraints, and governments had to ration how scarce dollars were allocated and spent. Governments also had to ensure that future hard currencies could be earned from manufactured goods produced domestically. If India was not to spend its future earnings on foreign cars, it had first to build steel mills at home. Local industries had to substitute for imported goods, particularly in what was then called the 'commanding heights of the economy' (steel, chemicals, basic infrastructure). Ordinary people also had to be provided with schools and clinics.

In all of this the state had to take a lead. Nehru was never a fan of Americanisation, either as an economic model or as a way of life. Nor was he unduly impressed with the Soviet alternative, although both he and his Planning Commission noted Soviet achievements in respect of planned economic growth. Nehru preferred to think of the Third World – or of Third Worldism and non-alignment – as a middle way between capitalism and socialism, and for a while India played both sides against the middle. Many of its major industrial projects in the 1950s and 1960s were supported with foreign aid from the First World and from the Second World. Sadly, by the time of Nehru's death in 1964, this vision of international amity was being undone. India was defeated in a short war by its Chinese 'brother' in 1962. Shortly afterwards, in 1966–67, when India was being led by Nehru's daughter, Indira Gandhi, and with the country approaching famine conditions, Washington used its food aid programme as a weapon to secure important changes in India's economic and strategic policies. Indira Gandhi, much like W. Arthur Lewis, as it transpired, would later become a sharp critic of the American-dominated international economic system. They both noted the difficulties of prosecuting ostensibly national models of economic development within a global political economy dominated by superpower conflict and increasingly open international capital flows.

By the mid-1960s the United States was already embroiled in Vietnam and challenges to Rostovian optimism about 'development as Americanisation' were blooming in many quarters. Important challenges came from the *Monthly Review* school of American Marxism, as we shall see in Part 5. We should also reiterate here that US policies towards the Third World in the 1950s and 1960s never conformed to just one version of modernisation theory. Nor were they informed only by Cold War struggles or by a general-ised commitment to free trade. In any case, as we have seen, the broad goals of economic development and cultural change – modernisation – were compatible with economic policies that allocated very differ-ent roles to government. This was clearly the case in respect of industrial policy, but it was also true in relation to trade and international economic integration.

Critical here was a body of work produced ten years before Rostow's *The Stages of Economic Growth*. In the vanguard were the Argentine economist Raoul Prebisch, with colleagues from the UN's Economic Commission for Latin America, and the UK-based economist Hans Singer. Joined together by history, the Singer–Prebisch critique of static comparative advantage theory argued that underdeveloped countries had been converted into food and raw material exporters by previous rounds of investment by the indus-trialised (colonial) world. It further maintained that there was a long-run tendency for the income terms of trade to move against primary commodities. In other words, the price of primary commodities tended to increase more slowly than the price of manufactured goods and services. Import-substitution industrialisation was thus a rational and perhaps even a necessary policy instrument in latecomer societies bent on catching up with the West. 'Absorption of the fruits of technical progress in primary production is not enough,' Singer warned, 'what is wanted is absorption for reinvestment.'

Hans Singer was undoubtedly committed to the modernisation of the Third World, as indeed was W. Arthur Lewis. But both in their early work, and also later in their careers, they steered clear of the Americanisation thesis that is at the heart of what more conventionally is known as modernisation theory. Their intellectual inspiration came more from Keynes and List than from neoclassical economics or Parsonian sociology. The work of Singer and Lewis, and indeed of Baran and the Marxists (Part 5), all of whom believed in a broad version of modernisation (whether capitalist, welfarist or socialist), further illustrates the limits of talking about modernisation theory except in a limited and well-defined manner.

It is also important to recall that a very powerful commitment to state-directed development was made in the Soviet Union in the 1920s and 1930s, and was wrapped around with important debates between the likes of Preobrazhensky and Bukharin, as well as Lenin and Stalin. Alec Nove briefly rehearses these contributions in our next reading. He also reflects on the failings of modernisation, in this case in its socialist guise. In the end, he suggests, the Promethean visions of the mid-twentieth century were undone by the simplifications they promoted, a point that is made with still greater force in the last reading in this Part, that by Rhoads Murphey. Murphey looks at the quite extraordinary campaigns that were being waged against 'nature' in the 1950s and 1960s by Chairman Mao and the Chinese Communist Party. In these campaigns, 'Nature is explicitly seen as an enemy, against which [communist] man must fight an unending war, with more conviction and fervour and with a brighter vision of the ultimate results than even the Darwinian–Spencerian West held.' Slogans including 'The Desert Surrenders', 'We Will Bend Nature To Our Will', and 'Chairman's Thoughts Are Our Guide To Scoring Victories in the Struggle Against Nature' only begin to suggest the violence of the uber-modernising project that was unleashed in the name of a political ideology which denied any 'environmental limits on what man can do'.

As we have tried to suggest already in this reader, all forms of developmentalism propose transformations in human relationships with (external) nature, and indeed in most cases in 'human nature' itself (as we learn to become modern). In China, however, this headlong rush to modernity – cf. the grotesquely misnamed Great Leap Forward of 1959–62 – exceeded anything imagined by Walt Rostow or his colleagues in the United States, and was set in motion by a political regime that brooked little dissent. Little wonder, then, that Murphey felt moved to warn that 'All of this [was happening] in a totalitarian system, [which] multiplies the consequences of any planning errors, [and] which may reach mammoth proportions before they can, at enormous expense, be corrected.'

We will pick up this warning again in the next two Parts, when we meet the work of Rachel Carson and James Scott. As they also point out, it would be the job of future work on development to re-complicate the worlds that modernisation theory proposed so radically to simplify. As this work took shape, moreover, we begin to see the invention of development studies as an academic subject that escapes the clutches of Parsonian sociology (not to mention hardline Maoism), and which differentiates itself from comparative politics and international relations theory. We also see a shift in the centre of gravity of academic work away from the United States and back towards Europe and the Third World. The catastrophe of America's war in Vietnam hastened this double movement along.

REFERENCES AND FURTHER READING

Adas, M. (2003) 'Modernization theory and the American revival of the scientific and technological standards of social achievement and human worth', in D. Engerman, N. Gilman, M. Haefele and M. Latham (eds) *Staging Growth: Modernization, Development and the Global Cold War*, Amherst: University of Massachusetts Press.

Apter, D. (1965) *The Politics of Modernization*, Princeton, NJ: Princeton University Press.

Bell, D. (1960) *The End of Ideology: On the Exhaustion of Political Ideas in the 1950s*, New York: Free Press.

Cowen, M. and Shenton, R. (1996) *Doctrines of Development*, London: Routledge.

Fukuyama, F. (1992) *The End of History and the Last Man*, New York: Free Press.

Gerschenkron, A. (1965) *Economic Backwardness in Historical Perspective: A Book of Essays*, New York: Praeger.

Gilman, N. (2003) *Mandarins of the Future*, Baltimore, MD: Johns Hopkins University Press.

Huntington, S. (1968) *Political Order in Changing Societies*, New Haven, CT: Yale University Press.

Huntington, S. (1996) *The Clash of Civilizations and the Remaking of World Order*, New York: Simon and Schuster.

Klingensmith, D. (2003) 'Building India's modern temples: Indians and Americans in the Damodar Valley Corporation, 1945–60', in K. Sivaramakrishnan and A. Agrawal (eds) *Regional Modernities*, Stanford, CA: Stanford University Press.

Latham, M. (2000) *Modernization as Ideology: American Social Science and 'Nation Building' in the Kennedy Era*, Chapel Hill: University of North Carolina Press.

Nehru, J. (1946) *The Discovery of India*, Calcutta.

Pletsch, C. (1981) 'The three worlds, or the division of social scientific labor, circa 1950–1975', *Comparative Studies in Society and History* 23: 565–90.

Prebisch, R. (1950) *The Economic Development of Latin America and Its Principal Problems*, New York: United Nations Economic Commission on Latin America.

Sretzer, S. (1993) 'The idea of demographic transition and the study of fertility change: a critical intellectual history', *Population and Development Review* 19: 659–701.

'The Problematization of Poverty: The Tale of Three Worlds and Development'

from *Encountering Development: The Making and Unmaking of the Third World* (1995).

Arturo Escobar

Editors' Introduction

Arturo Escobar was born in Manizales, Colombia, in 1952. He was trained first as a chemical engineer in Cali before he moved to Cornell University in the United States to study food science and nutrition, graduating in 1978. Escobar then enrolled in an interdisciplinary PhD programme at Berkeley and began to publish several papers which sought to use the insights of Foucault on discourse theory and the archaeology of knowledge to fashion a critical account of development and Developmentalism. Not surprisingly, his account of the 'making and unmaking of the Third World' was widely anticipated when it was published in 1995. His work raised important questions about the production of the development business/industry – why from the late 1940s in the United States? How did this relate to Cold War geopolitics? What new assumptions were being made, and why, about the prospects for economic development and cultural change in a formally ex-colonial world? – and about the effects upon different countries, social groups and individuals of technologies of rule prosecuted in the name of Development (five-year plans, foreign aid, anti-poverty campaigns, and so on). *Encountering Development* sought to argue that the dream of development had turned into a nightmare for those on the wrong end of it. Forty years of development had ended in pathetic failure.

Whether or not this is true is something we explore throughout *The Development Reader*. What Escobar seems mainly to be concerned about, in any case, is the effects upon the less economically accomplished countries of doctrines of modernisation that at once invent the 'Third World' as that which is not modern, and which propose various domestic and foreign-led interventions for its de-traditionalisation. These interventions include broad programmes of education, secularisation and sanitation, as well as programmes to support industrialisation and urbanisation. (The remaining readings in this Part discuss some of these programmes in more detail.) In the reading that we excerpt here, Escobar suggests that, 'One of the many changes that occurred in early post-World War II period was the "discovery" of mass poverty in Asia, Africa and Latin America.' This was how the Third World became known and was later defined by donor agencies: as a more or less uniform site of poverty (and not, for example, of different cultural achievements or political problems). The Third World was pathologised, and a War on Poverty was declared by the United States (as by the government of India in the 1970s) as one important means by which the reorganisation (or Development) of the 'Third World' could be taken out of the hands of ordinary people (the poor and illiterate) and entrusted to external experts. Thus was colonialism continued under another name, Escobar suggests, and thus must it be resisted. (For a spirited critique of post-development theory, check out the essay by Ray Kiely (1999).)

Key references

Arturo Escobar (1995) *Encountering Development: The Making and Unmaking of the Third World*, Princeton, NJ: Princeton University Press.

— (2001) 'Culture sits in places: reflections on globalism and subaltern strategies of localization', *Political Geography* 20: 139–74.

— (2005) 'Economics and the space of modernity', *Cultural Studies* 19: 139–75.

S. Alverez, E. Dagnino and A. Escobar (eds) (1998) *Culture of Politics/Politics of Culture: Revisioning Latin American Social Movements*, Boulder, CO: Westview.

R. Kiely (1999) 'The last refuge of the noble savage? A critical assessment of post-development theory', *European Journal of Development Research* 11: 30–55.

The word "poverty" is, no doubt, a key word of our times, extensively used and abused by everyone. Huge amounts of money are spent in the name of the poor. Thousands of books and expert advice continue to offer solutions to their problems. Strangely enough, however, nobody, including the proposed "beneficiaries" of these activities, seems to have a clear, and commonly shared, view of poverty. For one reason, almost all the definitions given to the word are woven around the concept of "lack" or "deficiency." This notion reflects only the basic relativity of the concept.

What is necessary and to whom? And who is qualified to define all that?"

Majid Rahnema, *Global Poverty: A Pauperizing Myth*, 1991

One of the many changes that occurred in the early post-World War II period was the "discovery" of mass poverty in Asia, Africa, and Latin Americá. Relatively inconspicuous and seemingly logical, this discovery was to provide the anchor for an important restructuring of global culture and political economy. The discourse of war was displaced onto the social domain and to a new geographical terrain: the Third World. Left behind was the struggle against fascism. In the rapid globalization of U.S. domination as a world power, the "war on poverty" in the Third World began to occupy a prominent place. Eloquent facts were adduced to justify this new war: "Over 1,500,000 million people, something like two-thirds of the world population, are living in conditions of acute hunger, defined in terms of identifiable nutritional disease. This hunger is at the same time the cause and effect of the poverty, squalor, and misery in which they live" (Wilson 1953, 11).

Statements of this nature were uttered profusely throughout the late 1940s and 1950s (Orr 1953; Shonfield 1950; United Nations 1951). The new emphasis was spurred by the recognition of the chronic conditions of poverty and social unrest existing in poor countries and the threat they posed for more developed countries. The problems of the poor areas irrupted into the international arena. The United Nations estimated that per capita income in the United States was $1,453 in 1949, whereas in Indonesia it barely reached $25. This led to the realization that something had to be done before the levels of instability in the world as a whole became intolerable. The destinies of the rich and poor parts of the world were seen to be closely linked. "Genuine world prosperity is indivisible," stated a panel of experts in 1948. "It cannot last in one part of the world if the other parts live under conditions of poverty and ill health" (Milbank Memorial Fund 1948, 7; see also Lasswell 1945).

Poverty on a global scale was a discovery of the post-World War II period. As Sachs (1990) and Rahnema (1991) have maintained, the conceptions and treatment of poverty were quite different before 1940. In colonial times the concern with poverty was conditioned by the belief that even if the "natives" could be somewhat enlightened by the presence of the colonizer, not much could be done about their poverty because their economic development was pointless. The natives' capacity for science and technology, the basis for economic progress, was seen as nil (Adas 1989). As the same authors point out, however, within Asian, African, and Latin or Native American societies – as well as throughout most of European history – vernacular societies had developed ways of defining and treating poverty that accommodated visions of community, frugality, and sufficiency. Whatever these traditional ways might have been, and without idealizing them, it is true that massive poverty in the modern sense appeared only when the spread of the market economy broke down community ties and deprived millions of people from access to land, water, and other resources. With the consolidation of capitalism, systemic pauperization became inevitable.

Without attempting to undertake an archaeology of poverty, as Rahnema (1991) proposes, it is important to emphasize the break that occurred in the conceptions and management of poverty first with the emergence of capitalism in Europe and subsequently with the advent of development in the Third World. Rahnema describes the first break in terms of the advent in the nineteenth century of systems for dealing with the poor based on assistance provided by impersonal institutions. Philanthropy occupied an important place in this transition (Donzelot 1979). The transformation of the poor into the assisted had profound consequences. This "modernization" of poverty signified not only the rupture of vernacular relations but also the setting in place of new mechanisms of control. The poor increasingly appeared as a social problem requiring new ways of intervention in society. It was, indeed, in relation to poverty that the modern ways of thinking about the meaning of life, the economy, rights, and social management came into place. "Pauperism, political economy, and the discovery of society were closely interwoven" (Polanyi 1957, 84).

The treatment of poverty allowed society to conquer new domains. More perhaps than on industrial and technological might, the nascent order of capitalism and modernity relied on a politics of poverty the aim of which was not only to create consumers but to transform society by turning the poor into objects of knowledge and management. What was involved in this operation was "a techno-

discursive instrument that made possible the conquest of pauperism and the invention of a politics of poverty" (Procacci 1991, 157). Pauperism, Procacci explains, was associated, rightly or wrongly, with features such as mobility, vagrancy, independence, frugality, promiscuity, ignorance, and the refusal to accept social duties, to work, and to submit to the logic of the expansion of "needs." Concomitantly, the management of poverty called for interventions in education, health, hygiene, morality, and employment and the instillment of good habits of association, savings, child rearing, and so on. The result was a panoply of interventions that accounted for the creation of a domain that several researchers have termed "the social" (Donzelot 1979, 1988, 1991; Burchell, Gordon, and Miller 1991).

As a domain of knowledge and intervention, the social became prominent in the nineteenth century, culminating in the twentieth century in the consolidation of the welfare state and the ensemble of techniques encompassed under the rubric of social work. Not only poverty but health, education, hygiene, employment, and the poor quality of life in towns and cities were constructed as social problems, requiring extensive knowledge about the population and appropriate modes of social planning (Escobar 1992). The "government of the social" took on a status that, as the conceptualization of the economy, was soon taken for granted. A "separate class of the 'poor' " (Williams 1973, 104) was created. Yet the most significant aspect of this phenomenon was the setting into place of apparatuses of knowledge and power that took it upon themselves to optimize life by producing it under modern, "scientific" conditions. The history of modernity, in this way, is not only the history of knowledge and the economy, it is also, more revealingly, the history of the social.[1]

As we will see, the history of development implies the continuation in other places of this history of the social. This is the second break in the archaeology of poverty proposed by Rahnema: the globalization of poverty entailed by the construction of two-thirds of the world as poor after 1945. If within market societies the poor were defined as lacking what the rich had in terms of money and material possessions, poor countries came to be similarly defined in relation to the standards of wealth of the more economically advantaged nations. This economic conception of poverty found an ideal yardstick in the annual per capita income. The perception of poverty on a global scale "was nothing more than the result of a comparative statistical operation, the first of which was carried out only in 1940" (Sachs 1990, 9). Almost by fiat, two-thirds of the world's people were transformed into poor subjects in 1948 when the World Bank defined as poor those countries with an annual per capita income below $100. And if the problem was one of insufficient income, the solution was clearly economic growth.

Thus poverty became an organizing concept and the object of a new problematization. As in the case of any problematization (Foucault 1986), that of poverty brought into existence new discourses and practices that shaped the reality to which they referred. That the essential trait of the Third World was its poverty and that the solution was economic growth and development became self-evident, necessary, and universal truths. This chapter analyzes the multiple processes that made possible this particular historical event. It accounts for the 'developmentalization' of the Third World, its progressive insertion into a regime of thought and practice in which certain interventions for the eradication of poverty became central to the world order. This chapter can also be seen as an account of the production of the tale of three worlds and the contest over the development of the third. The tale of three worlds was, and continues to be despite the demise of the second, a way of bringing about a political order "that works by the negotiation of boundaries achieved through ordering differences" (Haraway 1989a, 10). It was and is a narrative in which culture, race, gender, nation, and class are deeply and inextricably intertwined. The political and economic order coded by the tale of three worlds and development rests on a traffic of meanings that mapped new domains of being and understanding, the same domains that are increasingly being challenged and displaced by people in the Third World today.

THE INVENTION OF DEVELOPMENT

The emergence of the new strategy

From July 11 to November 5, 1949, an economic mission, organized by the International Bank for Reconstruction and Development, visited Colombia with the purpose of formulating a general development program for the country. It was the first mission of this kind sent out by the International Bank to an underdeveloped country. The mission included fourteen international advisers in the following fields: foreign exchange; transportation; industry, fuel, and power; highways and waterways; community facilities; agriculture; health and welfare; financing and banking; economics; national accounts; railroads; and petroleum refineries. Working closely with the mission was a similar group of Colombian advisers and experts.

Here is how the mission saw its task and, consequently, the character of the program proposed:

> We have interpreted our terms of reference as calling for a comprehensive and internally consistent program. . . . The relationships among various sectors of Colombian economy are very complex, and intensive analysis of these relationships has been necessary to develop a consistent picture. . . . This, then, is the reason and justification for an overall program of development. Piecemeal and sporadic efforts are apt to make little impression on the general picture. Only through a generalized attack throughout the whole economy on education, health, housing, food and productivity can the vicious circle of poverty, ignorance, ill health and low productivity be decisively broken. But once the break is made, the process of economic development can become self-generating.
>
> (International Bank 1950, xv)

The program called for a "multitude of improvements and reforms" covering all important areas of the economy. It constituted a radically new representation of, and approach to, a country's social and economic reality. One of the features most emphasized in the approach was its comprehensive and integrated character. Its comprehensive nature demanded programs in all social and economic aspects of importance, whereas careful planning, organization, and allocation of resources ensured the integrated character of the programs and their successful implementation. The report also furnished a detailed set of prescriptions, including goals and quantifiable targets, investment needs, design criteria, methodologies, and time sequences.

It is instructive to quote at length the last paragraph of the report, because it reveals several key features of the approach that was then emerging:

One cannot escape the conclusion that reliance on natural forces has not produced the most happy results. Equally inescapable is the conclusion that with knowledge of the underlying facts and economic processes, good planning in setting objectives and allocating resources, and determination in carrying out a program for improvements and reforms, a great deal can be done to improve the economic environment by shaping economic policies to meet scientifically ascertained social requirements. . . . Colombia is presented with an opportunity unique in its long history. Its rich natural resources can be made tremendously productive through the application of modern techniques and efficient practices. Its favorable international debt and trade position enables it to obtain modern equipment and techniques from abroad. International and foreign national organizations have been established to aid underdeveloped areas technically and financially. All that is needed to usher a period of rapid and widespread development is a determined effort by the Colombian people themselves. In making such an effort, Colombia would not only accomplish its own salvation but would at the same time furnish an inspiring example to all other underdeveloped areas of the world.

(International Bank 1950, 615)

The messianic feeling and the quasi-religious fervor expressed in the notion of salvation are noticeable. In this representation, "salvation" entails the conviction that there is one right way, namely, development; only through development will Colombia become an "inspiring example" for the rest of the underdeveloped world. Nevertheless, the task of salvation/development is complex. Fortunately, adequate tools (science, technology, planning, and international organizations) have already been created for such a task, the value of which has already been proved by their successful application in the West. Moreover, these tools are neutral, desirable, and universally applicable. Before development, there was nothing: only "reliance on natural forces," which did not produce "the most happy results." Development brings the light, that is, the possibility to meet "scientifically ascertained social requirements." The country must thus awaken from its lethargic past and follow the one way to salvation, which is, undoubt-

edly, "an opportunity unique in its long history" (of darkness, one might add).

This is the system of representation that the report upholds. Yet, although couched in terms of humanitarian goals and the preservation of freedom, the new strategy sought to provide a new hold on countries and their resources. A type of development was promoted which conformed to the ideas and expectations of the affluent West, to what the Western countries judged to be a normal course of evolution and progress. As we will see, by conceptualizing progress in such terms, this development strategy became a powerful instrument for normalizing the world. The 1949 World Bank mission to Colombia was one of the first concrete expressions of this new state of affairs.

Precursors and antecedents of the development discourse

As we will see in the next section, the development discourse exemplified by the 1949 World Bank mission to Colombia emerged in the context of a complex historical conjunction. Its invention signaled a significant shift in the historical relations between Europe and the United States, on the one hand, and most countries in Asia, Africa, and Latin America, on the other. It also brought into existence a new regime of representation of these latter parts of the world in Euramerican culture. But "the birth" of the discourse must be briefly qualified; there were, indeed, important precursors that presaged its appearance in full regalia after World War II.

The slow preparation for the launching of development was perhaps most clear in Africa, where, a number of recent studies suggest (Cooper 1991; Page 1991), there was an important connection between the decline of the colonial order and the rise of development. In the interwar period, the ground was prepared for the institution of development as a strategy to remake the colonial world and restructure the relations between colonies and metropoles. As Cooper (1991) has pointed out, the British Development Act of the 1940s – the first great materialization of the development idea – was a response to challenges to imperial power in the 1930s and must thus be seen as an attempt to reinvigorate the empire. This was particularly clear in the settler states in southern Africa, where preoccupations with questions of labor and food supplies led to strategies for the modernization of segments of the African population, often, as Page (1991) argues, at the expense of Afrocentric views of food and community defended by women. These early attempts were to crystallize in community development schemes in the 1950s. The role of the League of Nations in negotiating decolonization through the system of mandates was also important in many cases in Asia and Africa. After the Second World War, this system was extended to a generalized decolonization and the promotion of development by the new system of international organizations (Murphy and Augelli 1993).

Generally speaking, the period between 1920 and 1950 is still ill understood from the vantage point of the overlap of colonial and developmentalist regimes of representation. Some aspects that have received attention in the context of north and/or sub-Saharan Africa include the constitution of a labor force and a modernized class of farmers marked by class, gender, and race, including the displacement of African self-sufficient systems of food and cultural production; the role of the state as architect, for instance, in the "detribalization" of wage labor, the escalation of gender competition, and the struggle over education; the ways in which discourses and practices of agricultural experts, health professionals, urban planners, and educators were deployed in the colonial context, their relation to metropolitan discourses and interests, and the metaphors furnished by them for the reorganization of the colonies; the modification of these discourses and practices in the context of the colonial encounter, their imbrication with local forms of knowledge, and their effect on the latter; and the manifold forms of resistance to the colonial power/knowledge apparatuses (see, for instance, Cooper and Stoler 1989; Stoler 1989; Packard 1989; Page 1991; Rabinow 1989; Comaroff 1985; Comaroff and Comaroff 1991; Rau 1991).

The Latin American case is quite different from the African, although the question of precursors of development must also be investigated there. As is well known, most Latin American countries achieved political independence in the early decades of the nineteenth century, even if on many levels they continued to be under the sway, of European economies and cultures. By the beginning of the twentieth century, the ascendancy of the United States was felt in the entire region. United States – Latin American relations took on a double-edged significance early in the century. If on the one hand those in power perceived that opportunities for fair exchange existed, on the other hand the United States felt increasingly justified in intervening in Latin American affairs. From the interventionist big stick policy of the early part of the century to the good neighbor principle of the 1930s, these two tendencies coexisted in U.S. foreign policy toward Latin America, the latter having much more important repercussions than the former.

Robert Bacon, former U.S. secretary of state, exemplified the "fair exchange" position. "The day has gone," he stated in his 1916 report of a trip to South America, "when the majority of these countries, laboriously building up a governmental structure under tremendous difficulties, were unstable, tottering and likely to fall from one month to another . . . They 'have passed,' to use the words of Mr. Root, 'out of the condition of militarism, out of the condition of revolution, into the condition of industrialism, into the path of successful commerce, and are becoming great and powerful nations'" (Bacon 1916, 20). Elihu Root, whom Bacon mentioned in a positive light, actually represented the side of active interventionism. A prominent statesman and an expert in international law, Root was a major force in shaping U.S. foreign policy and took active part in the interventionist policy of the earlier part of the century, when the U.S. military occupied most Central American countries. Root, who was awarded the Nobel Peace Prize in 1912, played a very active role in the separation of Colombia from Panama. "With or without the consent of Colombia," he wrote on that occasion, "we will dig the canal, not for selfish reasons, not for greed or gain, but for the world's commerce, benefiting Colombia most of all. . . . We shall unite our Atlantic and Pacific coasts, we shall render inestimable service to mankind, and we shall grow in greatness and honor and in the strength that comes from difficult tasks accomplished and from the exercise of the power that strives in the nature of a great constructive people" (Root 1916, 190).

Root's position embodied the conception of international relations then prevailing in the United States.[2] The readiness for military intervention in the pursuit of U.S. strategic self-interest was tempered from Wilson to Hoover. With Wilson, intervention was accompanied by the goal of promoting "republican" democracies, meaning elite, aristocratic regimes. Often these attempts were fueled by ethnocentric and racist positions. Attitudes of superiority "convinced the United States it had the right and ability to intervene politically in weaker, darker, poorer countries" (Drake 1991, 7). For Wilson, the promotion of democracy was the moral duty of the U.S. and of "good men" in Latin America. "I am going to teach the South American republics to elect good men," he summed up (quoted in Drake 1991, 13). As Latin American nationalism mounted after World War I, the United States reduced open interventionism and proclaimed instead the principles of the open door and the good neighbor, especially after the mid-twenties. Attempts were made to provide some assistance, particularly regarding financial institutions, the infrastructure, and sanitation. During this period the Rockefeller Foundation became active for the first time in the region (Brown 1976). On the whole, however, the 1912–1932 period was ruled by a desire on the part of the United States to achieve "ideological as well as military and economic hegemony and conformity, without having to pay the price of permanent conquest" (Drake 1991, 34).

Although this state of relations revealed an increasing U.S. interest in Latin America, it did not constitute an explicit, overall strategy for dealing with Latin American countries. This situation was profoundly altered during the subsequent decades and especially after the Second World War. Three inter-American conferences – held at Chapultepec in Mexico (February 21–March 8, 1945), Rio de Janeiro (August 1947), and Bogotá (March 30–April 30, 1948) – were crucial in articulating new rules of the game. As the terrain for the cold war was being fertilized, however, these conferences made evident the serious divergence of interests between Latin America and the United States, marking the demise of the good neighbor policy. For while the United States insisted on its military and security objectives, Latin American countries emphasized more than ever economic and social goals (López Maya 1993).[3]

At Chapultepec, several Latin American presidents made clear the importance of industrialization in the consolidation of democracy and asked the United States to help with a program of economic transition from war production of raw materials to industrial production. The United States, however, insisted on questions of hemispheric defense, reducing economic policy to a warning to Latin American countries to abandon "economic national-

ism." These disagreements grew at the Rio Conference on Peace and Security. Like the Bogotá conference of 1948 – which marked the birth of the Organization of American States – the Rio conference was dominated by the growing anti-Communist crusade. As U.S. foreign policy became more militarized, the need for appropriate economic policies, including the protection of the nascent industries, became more and more central to the Latin American agenda. The United States to some extent finally acknowledged this agenda in Bogotá. Yet then secretary of state General Marshall also made clear that Latin America could in no way expect something similar to the Marshall Plan for Europe (López Maya 1993).

In contrast, the United States insisted on its open door policy of free access of resources to all countries and on the encouragement of private enterprise and the "fair" treatment of foreign capital. U.S. experts on the area completely misread the Latin American situation. A student of U.S. foreign policy toward Latin America during the late 1940s put it thus:

> Latin America was closest to the United States and of far greater economic importance than any other Third World region, but senior U.S. officials increasingly dismissed it as an aberrant, benighted area inhabited by helpless, essentially childish peoples. When George Kennan [head of State Department policy planning] was sent to review what he described as the "unhappy and hopeless" background there, he penned the most acerbic dispatch of his entire career. Not even the Communists seem viable "because their Latin American character inclines them to individualism [and] to undiscipline." . . . Pursuing the motif of the "childish" nature of the area, he condescendingly argued that if the United States treated the Latin Americans like adults, then perhaps they would have to behave like them.
>
> (Kolko 1988, 39, 40)[4]

Like Currie's image of "salvation," the representation of the Third World as a child in need of adult guidance was not an uncommon metaphor and lent itself perfectly to the development discourse. The infantilization of the Third World was integral to development as a "secular theory of salvation" (Nandy 1987).

It must be pointed out that the economic demands Latin American countries made were the reflection of changes that had been taking place for several decades and that also prepared the ground for development – for instance, the beginning of industrialization in some countries and the perceived need to expand domestic markets; urbanization and the rise of professional classes; the secularization of political institutions and the modernization of the state; the growth of organized labor and social movements, which disputed and shared the industrialization process; increased attention to positivist sciences; and various types of modernist movements. Some of these factors were becoming salient in the 1920s and accelerated after 1930.[5] But it was not until the World War II years that they began to coalesce into a clearer momentum for national economic models. In Colombia, talk of industrial development and, occasionally, the economic development of the country appeared in the early to mid-1940s, linked to a perceived threat by the popular classes. State interventionism became more noticeable, even if within a general model of economic liberalism, as an increase in production began to be seen as the necessary route to social progress. This awareness was accompanied by a medicalization of the political gaze, to the extent that the popular classes began to be perceived not in racial terms, as until recently, but as diseased, underfed, uneducated, and physiologically weak masses, thus calling for unprecedented social action (Pécaut 1987, 273–352).[6]

Despite the importance of these historical processes, it is possible to speak of the invention of development in the early post-World War II period. In the climate of the great postwar transformations, and in scarcely one decade, relations between rich and poor countries underwent a drastic change. The conceptualization of these relations, the form they took, the scope they acquired, the mechanisms by which they operated, all of these were subject to a substantial mutation. Within the span of a few years, an entirely new strategy for dealing with the problems of the poorer countries emerged and took definite shape. All that was important in the cultural, social, economic, and political life of these countries – their population, the cultural character of their people, their processes of capital accumulation, their agriculture and trade, and so on – entered into this new strategy.

[. . .]

NOTES

1 Foucault (1979, 1980a; 1980b, 1991) refers to this aspect of modernity – the appearance of forms of knowledge and regulatory controls centered on the production and optimization of life – as "biopower." Biopower entailed the "governmentalization" of social life, that is, the subjection of life to explicit mechanisms of production and administration by the state and other institutions. The analysis of biopower and governmentality should be an integral component of the anthropology of modernity (Urla 1993).

2 Root's words also reflect a salient feature of North American consciousness, namely, the utopian desire to bring progress and happiness to all peoples not only within the confines of their own country but beyond their shores as well. At times, within this kind of mentality the world becomes a vast surface burdened with problems to be solved, a disorganized horizon that has to be set "on the path of ordered liberty" once and for all, "with or without the consent" of those to be reformed. This attitude was also at the root of the dream of development.

3 For an in-depth treatment of U.S. foreign policy toward Latin America and the Third World, see Kolko (1988) and Bethell (1991). See also Cuevas Cancino (1989); Graebner (1977); Whitaker (1948); Yerguin (1977); Wood, B. (1985); and Haglund (1985). It must be pointed out that most scholars have missed the significance of the emergence of the development discourse in the late 1940s and early 1950s. López Maya, on whose work the account of three conferences is based, is an exception.

4 Ethnocentric remarks were at times expressed quite openly during the first half of the century. Wilson's ambassador to England, for instance, explained that the United States would intervene in Latin America to "make 'em vote and live by their decisions." If this did not work, "We'll go in again and make 'em vote again. . . . The United States will be there for two hundred years and it can continue to shoot men for that little space till they learn to vote and rule themselves" (quoted in Drake 1991, 14). The "Latin mind" was believed to "scorn democracy" and be ruled by emotion, not by reason.

5 Cardoso and Faletto (1979) discuss some of these changes for Latin America as a whole. The rise of social movements in Colombia in the 1920s is analyzed in Archila (1980).

6 The interpretation of this period of Colombia's history is highly disputed. Economic historians (see, for instance, Ocampo, ed. 1987) generally believe that the Great Depression and World War II pushed the ruling class toward industrialization as the only viable alternative for development. This view, held by many in

Latin America, has been disputed recently. Sáenz Rovner (1989, 1992) rejects the idea that growth and development were goals that the Colombian elite shared in the 1940s, adding that the government did not seriously consider the Currie report. Antonio García's (1953) paper provides important clues to assess the status of planning in Colombia with reference to the Currie mission. For García, planning activities in the 1940s were highly ineffective not only because of narrow conceptions of the planning process but because the various planning bodies had no power to implement the desired goals and programs. Although he found the Currie report unobjectionable from the economic viewpoint, he took issue with it on social grounds, advocating instead the kind of planning process that Jorge Eliécer Gaitán presented to congress in 1947.

By the late 1940s, García had a fully worked out alternative to capitalist development models, which has not been given the attention it merits by economic and social historians (see García 1948, 1950). This alternative, based on a sophisticated structural and dialectical interpretation of "backwardness" – in ways that resembled and presaged Paul Baran's (1957) work of a few years later – was based on a distinction between economic growth and the overall development of society. This was revolutionary, given the fact that a liberal model of development was becoming consolidated at this point, as Pécaut (1987) has shown in detail. More research needs to be done on this period from the perspective of the rise of development. Although nineteenth-century-style "economic essay" was the rule until the 1940s – for instance, in the works of Luis López de Mesa (1944) and Eugenio Gómez (1942) – in the 1930s several authors were calling for new styles of inquiry and decision making, based on greater objectivity, quantification, and programming. See, for instance, López (1976) and García Cadena (1956). Some of these issues are dealt with in Escobar (1989).

REFERENCES

Adas, Michael. 1989. Machines as the Measure of Men. Ithaca: Cornell University Press.

Archila, Mauricio. 1980. Los Movimientos Sociales entre 1920 y 1924: Una Aproximación Methodológica. Cuadernos de Filosofiá y Letras 3 (3): 181–230.

Bacon, Robert. 1916. For Better Relations with Our Latin American Neighbors. Washington, D.C.: Carnegie Endowment for Peace.

Baran, Paul. 1957. The Political Economy of Growth. New York: Monthly Review Press.

Bethell, Leslie. 1991. From The Second World War to the Cold War: 1944–1954. In Exporting Democracy: The United States and Latin America, edited by Abraham F. Lowenthal, 41–71. Baltimore: Johns Hopkins University Press.

Brown, Richard. 1976. Public Health in Imperialism: Early Rockefeller Programs at Home and Abroad. American Journal of Public Health 66 (9): 897–903.

Burchell, Graham, Colin Gordon, and Peter Miller, eds. 1991. The Foucault Effect. Chicago: University of Chicago Press.

Cardoso, Fernando Henrique and Enzo Faletto. 1979. Dependency and Development in Latin America. Berkeley: University of California Press.

Comaroff, Jean. 1985. Body of Power, Spirit of Resistance. Chicago: University of Chicago Press.

—— and John Comaroff. 1991. Of Revelation and Revolution. Chicago: University of Chicago Press.

Cooper, Frederick. 1991. Development and the Remaking of the Colonial World. Paper presented at SSRC meeting on Social Science and Development, Berkeley, Calif., November 15–16.

—— and Ann Stoler. 1989. Introduction: Tensions of Empire: Colonial Control and Visions of Rule. American Ethnologist 16 (4): 609–22.

Cuevas Cancino, Francisco. 1989. Roosevelt y la Buena Vecindad. México, D. E.: Fondo de Cultura Económica.

Donzelot, Jacques. 1979. The Policing of Families. New York: Pantheon Books.

—— 1988. The Promotion of the Social. Economy and Society 17 (3): 217–34.

—— 1991. Pleasure in Work. In The Foucault Effect, edited by Graham Burchell, Colin Gordon, and Peter Miller, 251–80. Chicago: University of Chicago Press.

Drake, Paul. 1991. From Good Men to Good Neighbors: 1912–1932. In Exporting Democracy: The United States and Latin America, edited by Abraham F. Lowenthal, 3–41. Baltimore: Johns Hopkins University Press.

Escobar, Arturo. 1989. The Professionalization and Institutionalization of 'Development' in Colombia in the Early Post-World War II Period. International Journal of Educational Development 9 (2): 139–54.

—— 1992. Planning. In The Development Dictionary, edited by Wolfgang Sachs, 112–45. London: Zed Books.

Foucault, Michel. 1979. Discipline and Punish. New York: Vintage Books.

—— 1980a. Power/Knowledge. New York: Pantheon Books.

—— 1980b. The History of Sexuality. Introduction. New York: Vintage Books.

—— 1986. The Use of Pleasure. New York: Pantheon Books.

—— 1991. Governmentality. In The Foucault Effect, edited by Graham Burchell, Colin Gordon, and Peter Miller, 87–104. Chicago: University of Chicago Press.

García, Antonio. 1948. Bases de la Economía Contemporánea. Elementos para una Economía de la Defensa. Bogotá: RFIOC.

—— 1950. La Democracia en la Teoría y en la Práctica. Una Posición Frente al Capitalismo y al Comunismo. Bogotá: Iqueíma.

—— 1953. La Planificación de Colombia. El Trimestre Economico 20:435–63.

García Cadena, A. 1956. Unas Ideas Elementales sobre Problemas Colombianos. Bogotá: Banco de la República.

Gómez, Eugenio. 1942. Problemas Colombianos. Sociología e Historia. Bogotá: Editorial Santa Fé.

Graebner, Norman. 1977. Cold War Diplomacy: American Foreign Policy, 1945–1975. New York: D. Van Nostrand.

Haglund, David. 1985. Latin America and the Transformation of U.S. Strategic Thought. Albuquerque: University of New Mexico Press.

International Bank for Reconstruction and Development. 1950. The Basis of a Development Program for Colombia. Baltimore: Johns Hopkins University Press.

Kolko, Gabriel, 1988. Confronting the Third World: United States Foreign Policy, 1945–1980. New York: Pantheon Books.

Lasswell, Harold. 1945. World Politics Faces Economics. New York: McGraw-Hill.

López, Alejandro. 1976. Escritos Escogidos. Bogotá: Colcultura.

López de Mesa, Luis. 1944. Posibles Rumbos de la Economía Colombiana. Bogotá: Imprenta Nacional.

López Maya, Margarita. 1993. Cambio de Discursos en la Relacion entre los Estados Unidos y América Latina de la Segunda Guerra Mundial a la Guerra Fría (1945–1948). Presented at the Thirty-fourth Annual Convention of the International Studies Association, Acapulco, March 23–27.

Milbank Memorial Fund. 1948. International Approaches to Problems of Underdeveloped Countries. New York: Milbank Memorial Fund.

Murphy, Craig, and Enrico Augelli. 1993. International Institutions, Decolonization, and Development. International Political Science Review 14 (1): 71–85.

Nandy, Ashis. 1987. Traditions, Tyranny, and Utopias. Delhi: Oxford University Press.

Ocampo, José Antonio, ed. 1987. Historia Económica de Colombia. Colombia: Siglo XXI.

Orr, John Boyd. 1953. The White Man's Dilemma. London: G. Allen and Unwin.

Packard, Randall. 1989. The "Healthy Reserve" and the "Dressed Native": Discourses on Black Health and the Language of Legitimation in South Africa. American Ethnologist 16 (4): 686–704.

Page, Helán. 1991. Historically Conditioned Aspiration and Gender/Race/Class Relations in Colonial and Post-Colonial Zimbabwe. Photocopy.

Pécaut, Daniel. 1987. Orden y Violencia: Colombia 1930–1954. Bogotá: Siglo XXI Editores.

Polanyi, Karl. 1957. The Great Transformation. Boston: Beacon Press.

Procacci, Giovanna. 1991. Social Economy and the Government of Poverty. In The Foucault Effect, edited by Graham Burchell, Colin Gordon, and Peter Miller, 151–68. Chicago: University of Chicago Press.

Rabinow, Paul. 1989. French Modern: Norms and Forms of the Social Environment. Cambridge: MIT Press.

—— and William Sullivan, eds. 1987. Interpretive Social Science: A Second Look. Berkeley: University of California Press.

Rahnema, Majid. 1991. Global Poverty: A Pauperizing Myth. Interculture 24 (2): 4–51.

Rau, Bill. 1991. From Feast to Famine. London: Zed Books.

Root, Elihu. 1916. Addresses on International Subjects. Cambridge: Harvard University Press.

Sachs, Wolfgang. 1990. The Archaeology of the Development Idea. Interculture 23 (4): 1–37.

Sáenz Rovner, Eduardo. 1989. Industriales, Proteccionismo y Política en Colombia: Intereses, Conflictos y Violencia. Universidad de los Andes. Facultad de Administración. Monografía no. 13.

—— 1992. La Ofensiva Empresarial. Industriales, Políticos y Violencia en los Años 40 en Colombia. Bogotá: Tercer Mundo.

Shonfield, Andrew. 1950. Attack on World Poverty. New York: Random House.

Stoler, Ann. 1989. Making Empire Respectable: The Politics of Race and Sexual Morality in Twentieth Century Colonial Cultures. American Ethnologist 16 (4): 634–61.

—— 1991. When the Moon Waxes Red. New York: Routledge.

United Nations, Department of Social and Economic Affairs. 1951. Measures for the Economic Development of Underdeveloped Countries. New York: United Nations.

Urla, Jacqueline. 1993. Cultural Politics in the Age of Statistics: Numbers, Nations, and the Making of Basque Identities. American Ethnologist 20 (4): 818–43.

Whitaker, Arthur. 1948. The United States and South America: The Northern Republics. Cambridge: Harvard University Press.

Williams, Raymond. 1973. The Country and the City. New York: Oxford University Press.

Wilson, Harold. 1953. The War on World Poverty. London: Gollancz.

Wood, Geof. 1985. The Politics of Development Policy Labelling. Development and Change 16 (3): 347–73.

Yerguin, Daniel. 1977. Shattered Peace: The Origins of the Cold War and the National Security State, Boston: Houghton Mifflin.

'Marxism, Communism and the Stages-of-Growth'

from *The Stages of Economic Growth: A Non-Communist Manifesto* (1960)

W. W. Rostow

Editors' Introduction

Walt Whitman Rostow was born in New York City in 1916 and died in 2003 after a long career in government and the academy. Rostow worked for a while before the Second World War at Columbia University, and in 1945/46 worked at Oxford University as the Harmsworth Professor of American History. He later lectured briefly at Cambridge University before taking up a position, which he held through the 1950s, in economic history at the Massachusetts Institute of Technology (MIT). From 1969 until his death he lived and worked in Austin, where he held an endowed Chair at the Lyndon B. Johnson School of Public Affairs at the University of Texas. In between these various academic postings, Rostow worked for the Office of Strategic Services in London during the Second World War; he also served as a scriptwriter for John F. Kennedy in the run-up to the 1960 presidential election, as President Kennedy's Deputy Special Assistant for National Security Affairs, and, from 1966, to January 1969, as President Johnson's National Security Advisor. Rostow's prospective return to MIT in 1969 was blocked by colleagues furious at his role in the escalation of the war in Vietnam.

It is absurd to separate Rostow's political and academic careers, although this has been done with some regularity in development studies and associated disciplines. Rostow was a true believer in the American Dream, including its more liberal incarnations. A high achiever in his early life, with what is often described as a restlessly upbeat personality, Rostow proposed his stages of growth model both as an endorsement of American values and as a blueprint for traditional societies to follow if they were to imitate the United States. Communism, for Rostow, was anathema, a deviant ideology that had to be guarded against in the academy, and by the US government in South-East Asia. In a curious echo of Marx, albeit with less nuance, Rostow was driven to a linear theory of societal evolution and to a philosophy where the ends justified the means (including warfare). In Rostow's case, however, while modernisation was to be shaped by the promotion of science and technology, as well as by increased injections of savings and investment (supplemented by foreign funds), it would also be driven by innovative entrepreneurs and a broader capitalist class that deserved high rates of return for taking risks.

In the reading that follows, we try to keep Rostow's academic and political selves firmly in play. Instead of reprinting the well-known second chapter of *The Stages of Economic Growth*, where Rostow sets out his five stages of growth, we present the first half of the book's last chapter, where Rostow compares his work with that of Marx. Readers will draw their own conclusions about the merits of a comparison that is weighted so strongly in favour of its author.

Key references

W. W. Rostow (1960) *The Stages of Economic Growth: A Non-Communist Manifesto*, London: Cambridge University Press.
— (1971) *Politics and the Stages of Growth*, Cambridge: Cambridge University Press.
— (1978) *The World Economy: History and Prospect*, London: Macmillan.
A. Fishlow (1965) 'Empty economic stages?' *Economic Journal* 75: 112–25.
M. Haefele (2003) 'Walt Rostow's stages of economic growth: ideas and action', in D. Engerman, N. Gilman, M. Haefele and M. Latham (eds) *Staging Growth: Modernization, Development and the Global Cold War*, Amherst: University of Massachusetts Press.

■ ■ ■ ■ ■ ■ ■

This final chapter considers how the stages-of-growth analysis compares with Marxism; for, in its essence, Marxism is also a theory of how traditional societies came to build compound interest into their structures by learning the tricks of modern industrial technology; and of the stages that will follow until they reach that ultimate stage of affluence which, in Marx's view, was not Socialism, under the dictatorship of the proletariat, but true Communism. As against our stages – the traditional society; the preconditions; take-off; maturity; and high mass-consumption – we are setting, then, Marx's feudalism; bourgeois capitalism; Socialism; and Communism.

We shall proceed by first summarizing the essence of Marx's propositions. We shall then note the similarities between his analysis and the stages-of-growth; and the differences between the two systems of thought, stage by stage. This will provide a way of defining the status and meaning of Marxism, as seen from the perspective of the stages-of-growth sequence. Finally, we shall look briefly at the evolution of Marxist thought and Communist policy, from Lenin forward; and draw some conclusions.

THE SEVEN MARXIST PROPOSITIONS

Marxist thought can be summarized in seven propositions, as follows.

First, the political, social and cultural characteristics of societies are a function of how the economic process is conducted. And, basically, the political, social and cultural behaviour of men is a function of their economic interests. All that follows in Marx derives from this proposition until the stage of Com-

munism is reached, when the burden of scarcity is to be lifted from men and their other more humane motives and aspirations come to dominance.[1]

Second, history moves forward by a series of class struggles, in which men assert their inevitably conflicting economic interests in a setting of scarcity.

Third, feudal societies – in our phrase, traditional societies[2] – were destroyed because they permitted to grow up within their framework a middle class, whose economic interests depended on the expansion of trade and modern manufactures; for this middle class successfully contended against the traditional society and succeeded in imposing a new political, social, and cultural superstructure, conducive to the pursuit of profit by those who commanded the new modern means of production.

Fourth, similarly, capitalist industrial societies would, Marx predicted, create the conditions for their destruction because of two inherent characteristics: because they created a mainly unskilled working force, to which they continued to allocate only a minimum survival real wage; and because the pursuit of profit would lead to a progressive enlargement of industrial capacity, leading to a competitive struggle for markets, since the purchasing power of labour would be an inadequate source of demand for potential output.

Fifth, this innate contradiction of capitalism – relatively stagnant real wages for labour and the build-up of pressure to find markets for expanding capacity – would produce the following specific mechanism of self-destruction: an increasingly self-conscious and assertive proletariat goaded, at last, to seize the means of production in the face of increasingly severe crises of unemployment. The seizure would be made easier because, as the competition for markets mounted, in the most mature stage of capitalism, monopolies would be formed; and the setting for transfer of ownership to the State would be created.

Sixth – this is a Leninist extension of Marxism – the mechanism of capitalism's downfall would consist not only in successive crises of increasingly severe unemployment, but also in imperialist wars, as the competition for trade and outlets for capital, induced by markets inadequate to capacity, led on not only to monopolies but also to a world-wide colonial struggle among the national monopolies of the capitalist world. The working class would thus seize power and install socialism not only in a setting of chronic, severe unemployment but also of disruption caused by imperialist wars, to which the capitalist world would be driven in order to avoid unemployment and to evade and divert the growing assertiveness of an increasingly mobilized and class-conscious proletariat, led and educated by the Communists within its ranks.

Seventh, once power is seized by the Socialist state, acting on behalf of the industrial proletariat – in the phase called 'the dictatorship of the proletariat' – production would be driven steadily forward, without crises; and real income would expand to the point where true Communism would become possible. This would happen because Socialism would remove the inner contradictions of capitalism. Let me quote Marx's vision of the end of the process: 'In a higher phase of Communist society, after the enslaving subordination of individuals under the division of labour, and therefore also the anti-thesis between mental and physical labour has vanished; after labour, from a mere means of life has of itself become the prime necessity of life; after the productive resources have also increased with the all-round development of the individual and all the springs of co-operative wealth flow more abundantly – only then can the narrow horizon of the bourgeois law be fully left behind and society inscribe on its banners: from each according to his ability, to each according to his needs.'[3]

SIMILARITIES WITH STAGES-OF-GROWTH ANALYSIS

Now let us identify the broad similarities between Marx's historical sequence and the stages-of-growth analysis.

First, they are both views of how whole societies evolve, seen from an economic perspective; both are explorations of the problems and consequences for whole societies of building compound interest into their habits and institutions.

Second, both accept the fact that economic change has social, political, and cultural consequences; although the stages-of-growth rejects the notion that the economy as a sector of society – and economic advantage as a human motive – are necessarily dominant.

Third, both would accept the reality of group and class interests in the political and social process, linked to interests of economic advantage; although the stages-of-growth would deny that they have been the unique determining force in the progression from traditional societies to the stage of high mass-consumption.

Fourth, both would accept the reality of economic interests in helping determine the setting out of which certain wars arose; although the stages-of-growth would deny the primacy of economic interests and motives as an ultimate cause of war-making; and it would relate economic factors and war in ways quite different from those of Marx and Lenin.

Fifth, both would pose, in the end, the goal or the problem of true affluence – of the time when, in Marx's good phrase – labour 'has of itself become the prime necessity of life'; although the stages-of-growth has something more to say about the nature of the choices available.

Sixth, in terms of economic technique, both are based on sectoral analyses of the growth process; although Marx confined himself to consumption goods and capital goods sectors, while the stages-of-growth are rooted in a more disaggregated analysis of leading sectors which flows from a dynamic theory of production.

CENTRAL THEMES OF STAGES-OF-GROWTH

With these two catalogues as background we can now isolate more precisely and more positively how the stages-of-growth analysis attempts, stage by stage, to deal with and to solve the problems with which Marx wrestled, and to avoid what appear to be Marx's basic errors.

The first and most fundamental difference

between the two analyses lies in the view taken of human motivation. Marx's system is, like classical economics, a set of more or less sophisticated logical deductions from the notion of profit maximization, if profit maximization is extended to cover, loosely, economic advantage. The most important analytic assertion in Marx's writings is the assertion in the Communist Manifesto that capitalism 'left no other nexus between man and man than naked self-interest, than callous "cash payment" '.

In the stages-of-growth sequence man is viewed as a more complex unit. He seeks, not merely economic advantage, but also power, leisure, adventure, continuity of experience and security; he is concerned with his family, the familiar values of his regional and national culture, and a bit of fun down at the local. And beyond these diverse homely attachments, man is also capable of being moved by a sense of connexion with human beings everywhere, who, he recognizes, share his essentially paradoxical condition. In short, net human behaviour is seen not as an act of maximization, but as an act of balancing alternative and often conflicting human objectives in the face of the range of choices men perceive to be open to them.

This notion of balance among alternatives perceived to be open is, of course, more complex and difficult than a simple maximization proposition; and it does not lead to a series of rigid, inevitable stages of history. It leads to patterns of choice made within the framework permitted by the changing setting of society: a setting itself the product both of objective real conditions and of the prior choices made by men which help determine the current setting which men confront.[4]

We shall not explore here the formal properties of such a dynamic system; but it follows directly from this view of how individuals act that the behaviour of societies is not uniquely determined by economic considerations. The sectors of a society interact: cultural, social, and political forces, reflecting different facets of human beings, have their own authentic, independent impact on the performance of societies, including their economic performance. Thus, the policy of nations and the total performance of societies – like the behaviour of individuals – represent acts of balance rather than a simple maximization procedure.

On this view it matters greatly how societies go about making their choices and balances. Specifically,

it follows that the central phenomenon of the world of post-traditional societies is not the economy – and whether it is capitalist or not – it is the total procedure by which choices are made. The stages-of-growth would reject as inaccurate Marx's powerful but over-simplified assumption that a society's decisions are simply a function of who owns property. For example, what Marx regards as capitalist societies at no stage, even in their purest form, ever made all their major decisions simply in terms of the free-market mechanism and private advantage. In Britain, for example, even at the height of the drive to maturity – in let us say the 1815–50 period, when the power of the industrial capitalist was least dilute – in these years factory legislation was set in motion; and after the vote was extended in the Second and Third Reform Bills, the policy of the society was determined by a balance between interests of profit and relative utility maximization on the one hand, and, on the other, interests of welfare as made effective on a 'one man one vote' basis through the political process. Capitalism, which is the centre of Marx's account of the post-feudal phase, is thus an inadequate analytic basis to account for the performance of Western societies. One must look directly at the full mechanism of choice among alternative policies, including the political process – and, indeed, the social and religious processes – as independent arenas for making decisions and choices.

To be more concrete, nothing in Marx's analysis can explain how and why the landed interests in the end accepted the Reform Bill of 1832, or why the capitalists accepted the progressive income tax, or the welfare state; for it is absolutely essential to Marxism that it is over property that men fight and die. In fact one must explain such phenomena with reference to a sense of commitment to the national community and to the principles of the individualist-utilitarian creed that transcended mere profit advantage. Similarly, nothing in Marx's analysis explains the patient acceptance of the framework of private capitalism by the working class, when joined to the democratic political process, despite continued disparities in income.

Marx – and Hegel – were correct in asserting that history moves forward by the clash of conflicting interests and outlooks; but the outcome of conflict in a regularly growing society is likely to be governed by ultimate considerations of communal continuity which a Boston lawyer, Charles Curtis – old in the ways of advocacy and compromise – recently put as follows:

> I suggest [he said] that things get done gradually only between opposing forces. There is no such thing as self-restraint in people. What looks like it is indecision. . . . It may be that truth is best sought in the market of free speech, but the best decisions are neither bought nor sold. They are the result of disagreement, where the last word is not 'I admit you're right', but 'I've got to live with the son of a bitch, haven't I'.[5]

This ultimate human solvent, Karl Marx – a lonely man, profoundly isolated from his fellows – never understood. He regarded it, in fact, as cowardice and betrayal, not the minimum condition for organized social life, any time, anywhere.

And, as developed in chapter 8, a simple analysis of war, in terms of economic advantage, breaks down in the face of a consideration of the different types of armed conflict and how they actually came about. Nationalism—and all that goes with it in terms of human sentiment and public policy—is a hangover from the world of traditional societies.[6]

One need look no farther than the primacy colonial peoples give to independence over economic development, or the hot emotions Arab politicians can generate in the street crowds, to know that economic advantage is an insufficient basis for explaining political behaviour; and all of modern history sustains the view that what we now see about us in Asia, the Middle East, and Africa is typical of human experience, when confronted with the choices faced in transitional societies.

Thus, the account of the break-up of traditional societies offered here is based on the convergence of motives of private profit in the modern sectors with a new sense of affronted nationhood. And other forces play their part as well, for example the simple perception that children need not die so young or live their lives in illiteracy: a sense of enlarged human horizons, independent of both profit and national dignity. And when independence or modern nationhood are at last attained, there is no simple, automatic switch to a dominance of the profit motive and economic and social progress. On the contrary there is a searching choice and problem of balance among the three directions policy might go: external

assertion; the further concentration of power in the centre as opposed to the regions; and economic growth.

Then, indeed, when these choices are at last sorted out, and progress has gripped the society, history has decreed generally a long phase when economic growth is the dominant but not exclusive activity: the take-off and the sixty years or so of extending modern techniques. It is in the drive to maturity that societies have behaved in the most Marxist way, but each in terms of its own culture, social structure and political process; for growing societies, even growing capitalist societies, have differed radically in these respects. There has been no uniform 'superstructure' in growing societies. On the contrary, the differing nature of the 'superstructures' has strongly affected the patterns which economic growth assumed. And even in the drive to maturity we must be careful not to identify what was done – the energetic extension of modern technique – with a too-simple hypothesis about human motivation. We know that during take-offs and during the drive to maturity societies did, in fact, tend substantially to set aside other objectives and clear the way for activities which would, within human and resource and other societal restraints, maximize the rate of growth. But this is not to say that the profit motive itself was dominant. It certainly played a part. But in the United States after the Civil War, for example (perhaps the most materialist phase of any capitalist society, superficially examined), men did the things necessary to industrialize a great, rich continent, not merely to make money, but because power, adventure, challenge, and social prestige were all to be found in the market-place of a society where Church and State were relatively unimportant. The game of expansion and money-making was rewarding at this stage, not merely in terms of money, but in terms of the full range of human motives and aspirations. How, otherwise, can one explain the ardent striving of men long after they made more money than they or their children could conceivably use? And similar modifications in the Marxist view of human motivation would be required in an accurate account of the German, Japanese, Swedish, French, British, and – indeed – the Russian drives to maturity.

At this stage we come, of course, to Marx's familiar technical errors: his implicit Malthusian theory of population, and his theory of stagnant real wages.

It is an old game to point out that, in fact, population did not so move as to maintain a reserve army of unemployed, and that the workings of competitive capitalism yielded not stagnant real wages but rising real wages. Robinson and Kaldor have recently, for example, emphasized these deep flaws in Marx's economics.[7] And indeed they are, in formal terms, quite technical errors in judging how the economic process would operate. But they are more. They directly reflect Marx's basic proposition about societies; for neither political power, social power, nor, even, economic power neatly followed the fact that property was privately owned. Competition did not give way to monopoly; and competition, even imperfect, permitted wages to approximate net marginal value product; and this technical aspect of the market mechanism was buttressed by an acceptance of trade unions by the society and by a growing set of political interventions allowed and encouraged by the democratic political process. Moreover, the fact of mass progress itself, ruled out in Marx's analysis, made men rethink the calculus of having children; and it yielded a non-Malthusian check on the birth-rate: a check based not on poverty and disease but on progress itself. Think here not only of the older cases of declining birth-rates in history but of the radical fall in the birth-rate in resurgent Japan and Italy of the 1950s.

And so, when compound interest took hold, progress was shared between capital and labour; the struggle between classes was softened; and when maturity was reached they did not face a cataclysmic impasse. They faced, merely, a new set of choices; that is, the balance between the welfare state; high mass-consumption; and a surge of assertiveness on the world scene.

Thus, compound interest and the choices it progressively opened up by raising the average level of real income becomes a major independent variable in the stages-of-growth; whereas, in Marx's theory, compound interest appears in the perverse form of mounting profits, capable only of being distributed in high capitalist living, unusable capacity, and war. Put another way, the income-elasticity of demand is a living force in the stages-of-growth analysis; whereas it is virtually ruled out in Marx's powerful simplifications.

Now the Leninist question: whether capitalism, having an alleged built-in tendency for profits to decline, causes monopolies to rise, and crises to become progressively more severe, and leads to a desperate competitive international struggle for markets, and to wars.

First, the question of industrial concentration. Here we would merely assert that the evidence in the United States, at least, in no way suggests that the degree of concentration has increased significantly in, say, the last fifty years. And where it has increased it has done so more on the basis of the economies of large-scale research and development than because the market environment has been too weak to sustain small firms. And I doubt that the story would be very different in other mature societies of the West. Moreover, where concentrations of economic power have persisted, they have been forced to operate increasingly on terms set by the political process rather than merely the maximization procedures of the market-place itself.

Second, the question of increasingly severe crises. Down to 1914 there is no evidence whatsoever that the amplitude of cycles in unemployment increased. On the contrary, the evidence is of a remarkable uniformity in the cycles of the nineteenth century, whether viewed in terms of such statistics of unemployment as we have, or in terms of years of increasing and decreasing economic activity. There was, of course, the unique depression of the 1930s. But, if the view developed in chapter 6 is correct, the relative interwar stagnation in Western Europe was due not to long-run diminishing returns but to the failure of Western Europe to create a setting in which its national societies moved promptly into the age of high mass-consumption, yielding new leading sectors. And this failure, in turn, was due mainly to a failure to create initial full employment in the post-1920 setting of the terms of trade. Similarly the protracted depression of the United States in the 1930s was due not to long-run diminishing returns, but to a failure to create an initial renewed setting of full employment, through public policy, which would have permitted the new leading sectors of suburban housing, the diffusion of automobiles, durable consumers' goods and services to roll forward beyond 1929.

There is every reason to believe, looking at the sensitivity of the political process to even small pockets of unemployment in modern democratic societies, that the sluggish and timid policies of the 1920s and 1930s with respect to the level of employment will no longer be tolerated in Western societies. And now the technical tricks of that trade – due to the Keynesian revolution – are widely understood. It should not be forgotten that Keynes set himself the task of defeating Marx's prognosis about the course of unemployment under capitalism; and he largely succeeded.

As for that old classical devil 'diminishing returns' – which Marx took over in the form of his assumption of the declining level of profits – we cannot be dogmatic over the very long run; but the scale and pace of scientific enterprise in the modern world (which, as a sector, is at a rapid growth-stage) make it unlikely that we will lack things to do productively if people prefer productive activity to leisure. Besides, societies have it open to them, if they wish to continue the strenuous life, to follow the American lead and reimpose a Malthusian surge of population, when they get bored with gadgets.

Finally, the question of mature capitalism's dependence on colonies. Here we need only note that, while colonialism is virtually dead, capitalism in the Western Hemisphere, Western Europe and Japan is enjoying an extraordinary surge of growth. It is perfectly evident that, whatever the economic troubles of the capitalist societies, they do not stem primarily from a dependence on imperialism. If anything, their vulnerability now derives from an unwillingness to concern themselves sufficiently with – and to allocate adequate resources to – the world of underdeveloped nations. Domestic demand is not so inadequate as to force attention outward: it is too strong to make it possible for governments to mobilize adequate resources for external affairs. The current hope of Communism lies not in the exploitation of confusion and crises brought on by a compulsive struggle to unload exports, but from an excessive absorption of the capitalist world with the attractions of domestic markets.

This brings us to a comparison between Marx's view of Communism and the stage beyond mass consumption in the stages-of-growth. On this issue Marx was a nineteenth-century romantic. He looked to men, having overcome scarcity, permitting their better natures to flower; to work for the joy of personal expression in a setting where affluence had removed the need and temptation for avarice. This is indeed a decent and legitimate hope; an aspiration; and, even, a possibility. But, it is not the only alternative. There are babies and boredom, the development of new inner human frontiers, outer space and trivial

pleasures – or, maybe, destruction, if the devil makes work for idle hands. But while this is man's ultimate economic problem, if all goes well, it is a problem that we of this generation can set aside, to a degree, given the agenda that faces us in a world of nuclear weapons and in the face of the task of making a peaceful world community that will embrace the older and newer nations which have learned the tricks of growth.

[. . .]

NOTES

1   The exact form of the function relating economic interest to non-economic behaviour varies in Marx's writings and in the subsequent Marxist literature. Much in the original texts – and virtually all the operational conclusions derived from them – depend on a simple and direct function relating economic interest to social and political behaviour. In some parts of the Marxist literature, however, the function is developed in a more sophisticated form. Non-economic behaviour is seen as related not immediately and directly to economic self-interest but to the ideology and loyalties of class. Since, however, class interests and ideologies are presented as, essentially, a function of the techniques of production and the social relationships arising from them, this indirect formulation yields much the same results as the more primitive statement of connexion. In the main stream of Marxist literature, from beginning to end, it is only in seeking, protecting and enlarging property and income that men are really serious. Finally, there are a few passages in Marx – and more in Engels – which reveal a perception that human behaviour is affected by motives which need not be related to or converge with economic self-interest. This perception, if systematically elaborated, would have altered radically the whole flow of the Marxist argument and its conclusions. Marx, Engels, and their successors have turned their backs on this perception, in ideological formulations; although, as suggested later in this chapter, Lenin and his successors in Communist politics have acted vigorously on this perception.

2   Marx's concept of feudalism is too restrictive to cover all the traditional societies, a number of which did not develop a class of nobility, linked to the Crown, owning large tracts of land. Marxist analyses of traditional China, for example, have been strained on this point.

3   Quoted from 'Critique of the Gotha Programme', in J. Eaton, *Political Economy, a Marxist Textbook* (London, 1958), p. 187.

4   In the stages-of-growth some of the characteristics which have a persistent effect on the whole sequence of growth are rooted in the traditional society and its culture. They constitute an initial condition for the growth process with consequences for a time-period which transcends the sweep from the preconditions forward. See the author's *British Economy of the Nine-teenth Century* (Oxford, 1948), chapter vi, especially pp. 128n. and 140.

5   C. Curtis, *A Commonplace Book* (New York, 1957), pp. 112–13.

6   This theme is developed by Schumpeter in his writings about Marx and in his essay on Imperialism (J. Schumpeter, *Imperialism* (ed. B. Hoselitz, Meridian Books, New York, 1955), especially pp. 64 ff.; and *Ten Great Economists* (London, 1952), especially pp. 20 and 61 ff.). Whereas Schumpeter emphasized the per-sistence of irrational and romantic nationalist attitudes, the present analysis would underline two other factors. First, the role of certain groups and attitudes derived from the traditional society, in the growth process itself. Second, the structural fact that, once national sov-ereignty was accepted as a rule of the world arena, nations found themselves gripped in an almost inescapable oligopolistic struggle for power which did have elements of rationality.

7   Joan Robinson, *Marx, Marshall, and Keynes* (Delhi, 1955); and N. Kaldor, 'A Model of Economic Growth', *Economic Journal*, December 1957, especially pp. 618–21.

'Population Policy and the Future'

from *The Population of India and Pakistan* (1951)

Kingsley Davis

Editors' Introduction

Kingsley Davis was born in Texas in 1908 and died in 1997. After a PhD in Sociology at Harvard University, Davis first worked at the Pennsylvania State University before moving as a research associate to the Office of Population Research at Princeton University, where he later became a Professor of Sociology and Anthropology. In 1948, Davis moved to Columbia University in New York City, and then again in 1955 to the University of California at Berkeley. In 1977 he moved to the University of Southern California in Los Angeles, where he later combined a position as Distinguished Professor of Sociology with a senior research fellowship at Stanford's Hoover Institution.

Some of Kingsley Davis's best-known work was produced early in his career, notably while working in the Office of Population Research (OPR) at Princeton. Davis was one of the central figures in the 1940s who worked on demographic transition theory, a body of work he helped name. While at OPR, Davis also carried out work on population change in a number of European countries. This was partly financed by the League of Nations. In 1945 the OPR received a contract from the Office of the Geographer in the US State Department to carry out demographic work on Asia. This was at a time when the United States began to concern itself about Asia's rising population. How would it be fed? What threat did it pose, if any? The 1945 contract was partly used to fund Davis's work on population issues in India (from 1947, India and Pakistan), and thus his famous book from 1951, *The Population of India and Pakistan*. This book became the definitive study of the population of India and Pakistan from about 1880 to 1940. The reading which we reproduce here is chapter 23 of the book, 'Population policy and the future'. Davis assumes that the basic aim of the governments of India and Pakistan is to improve the per capita incomes of their citizens. He also maintains that the density of population in the Indian subcontinent is already too high to permit rapid economic expansion. It follows that both countries need population control policies. The rest of Davis's chapter addresses the types of population policies – including emigration and birth control – that would best serve the needs of the two countries. He concludes by saying that, 'The demographic account will have to be balanced sometime. Over any extended period it is impossible to control deaths and not control births.' Insights such as this were later used with some abandon by the United States as it sought to fund birth control policies in Asia, raising profound questions about the impacts of state interventions on the bodies of poor women and men. More hardline views than Davis supported were later exploited by the government of India during the Emergency years (1975–77, when a suspension of democratic rule coincided with forced sterilisations in north India), and by communist China with its one-child policy (officially from 1979). The long-term effects of the last of these interventions remain to be seen.

Key references

Kingsley Davis (1951) *The Population of India and Pakistan*, Princeton, NJ: Princeton University Press.
— (1949) *Human Society*, New York: Macmillan.
— (1972) *World Urbanization, 1950–1970*, Berkeley, CA: Institute of International Studies.
— (1973) *Cities: Their Origin, Growth and Human Impact*, New York: W. H. Freeman.
— S. Sretzer (1993) 'The idea of demographic transition and the study of fertility change: a critical intellectual history', *Population and Development Review* 19: 659–701.

If a "policy" is an official program for organized action, it involves two elements: an end to be attained and the means for attaining this end. A "population policy," then, must be one that conceives a possible but currently non-existent demographic condition either as an end in itself or as a necessary means to some further end. It implies that the actual demographic situation is not satisfactory, and that a more favorable one could be attained by pursuing the policy in question.

As for India and Pakistan, few observers regard the existing situation as satisfactory, because poverty is conjoined with a high density and with a fast rate of growth based on extreme fertility and a lower but still high mortality. Whether or not one desires a population policy, however, depends both on the goal that one has in mind and the supposed connection of the demographic situation with the goal.

POLICY GOALS

Since India and Pakistan are comparatively poor, the goal that receives most attention is greater real income. Yet each of these nations, just emerged from a struggle for its independence, is imbued with another aim, the attainment of national power. Although usually regarded as mutually beneficial, these two goals may occasionally conflict. Both the Soviet Union and Nazi Germany in peacetime, and all countries in wartime, have shown that an increase in industrial and military power does not necessarily require a corresponding rise in purchasing power. Rigid consumer controls, forced labor, enforced saving, and subsidy of armaments industries may all increase national power at the expense of living standards for the masses.

"A higher standard of living is, all else being equal,

a positive factor in a country's military strength vis-a-vis another country. However, this holds true only within limits. It does not [necessarily] hold true . . . of an industrial establishment devoted principally to the satisfaction of certain strata of society, as for example, bureaucracy, skilled labor, and the army, at the expense of the living standards of the population as a whole. It does not follow as a matter of course that such a state will be militarily 'inefficient.' Such a maldistribution of national income can be perpetuated indefinitely without causing much distress to the ruling groups, and may be accepted stoically, perhaps even enthusiastically, by the masses . . ."[1]

In 1949 Pakistan was spending, according to the *New York Times*, 67 per cent of her estimated annual revenues for national defense and was planning to spend still more, in spite of the fact that her need for economic development was very great.[2] India was spending a lesser but still high proportion of her budget (about 45 per cent) on military affairs.

Another end that may conflict with both national power and higher income is individualistic democracy. Such a social order, as the case of France shows, may impede concerted action both in the military and the economic spheres.

The truth is, of course, that no public policy can pursue one end to the exclusion of all others. Some sort of balance must be struck between different goals, and much of the dispute as to policy concerns not the exclusion but the relative emphasis that will be given the various major goals. What policies are favored with respect to population depends in part on which goals are stressed.

But with respect to population another complication arises. Because the customs and institutions governing demographic behavior are deeply rooted in the mores, such behavior is seldom regarded simply as a means to an ulterior end. For instance, the reduction of mortality is universally considered an end in itself; consequently an increase in mortality is not acceptable as a means to a given goal, even though in fact it could serve in this way. Only on occasion, as in the case of a war declaration, is an increase in mortality accepted as an inevitable but undesirable consequence of a given policy. Similarly, a high fertility or the customs that inevitably lead to a high fertility may be considered as an automatic good and therefore not a means to any other aim.[3] Such goals set a narrow limit to population policy. As yet it is not clear to what extent the governments and peoples of the Indian subcontinent will be willing to treat demographic matters as means.

For the sake of our present discussion, however, let us assume that the major though by no means exclusive goal is a higher per capita income, and that the question of population policy is thus oriented toward the problem of mitigating Indian and Pakistani poverty. This allows us to analyze different possibilities in the light of conditions in the region.

IS A POPULATION POLICY NEEDED?

Assuming a willingness to regard demographic behavior from an instrumental point of view, one may still ask if it is necessary to do so. Do population density and rate of growth have anything to do with poverty? If not, a population policy is hardly necessary. Previous chapters have already dealt with the relation of population to poverty, but a few more words on it seem necessary from the policy standpoint.

Since per capita income depends upon resources, technology, and economic organization as well as upon population, the question is whether or not one needs to worry about population at all. Why not simply accept the expanding population as a basic fact and direct one's policies toward improving the economic situation? This point of view has been repeatedly maintained in India, though less often than in some other countries. Its basic difficulty, the reader will recall, is that population change and economic development are interlinked. Economic expansion cannot forever compensate for a constant increase of population, because economic potentialities are affected by population. The people of the Indian subcontinent have apparently already reached the point where density and rapid growth are impeding economic development. Therefore, it seems somewhat unrealistic to attempt to do something on the economic side and yet do nothing on the population side. To make the discussion more concrete, let us take a recent example of the exclusively economic point of view.

An Indian economist, Dr. Baljit Singh, published a book[4] showing that the Indian diet is 22 per cent below physical requirements. The shortage, he showed, is not due to an absolute lack of food potentialities, but rather to other conditions. A considerable amount of food is wasted in wedding feasts and other celebrations. A great amount is lost in the fields and in storage due to animal, bird, and insect pests. Still more is lost through poor distribution and class inequalities. Inefficiency also arises from the inflexible nature of Indian food habits, the people being divided into rice-eaters and bread-eaters, unwilling to use substitutes. Some food is exported. And Indian agriculture is so undercapitalized and badly organized that the land produces far less than it is capable of producing. For this reason Dr. Singh, although he acknowledged the massive population growth in India, rejected the idea of trying to stop this growth by direct measures. ". . . There is no call for the pessimism implied in the demand to consciously curtail and limit the country's population by a widespread adoption of the so-called neo-Malthusianism. . . . The right course would be to husband economic resources in agriculture, industry and trade with the aid of science and modern inventions to realize optimum conditions of living for the growing population."[5] He maintained that by cutting wastes, stepping up production, and changing eating habits, the food problem could be solved. To this end he would stop feasts, kill pests, ban food exports, make rationing universal, consolidate agricultural holdings, shift the Indian dietary in the direction of potatoes, and import foodstuffs.

Clearly, so far as the *existing* population is concerned, Dr. Singh is right. If his proposed measures were taken, this population could eat very well. But in the long run (and a not very extended "long run") his solution is no solution at all. As soon as the food situation improved, the population would grow even faster than it has been growing. Soon the time would be reached when the food supply would again be deficient, not because of waste or bad habits, but because of increased numbers. At that point no further remedial measures could be taken on the food side, except at greatly increased cost, because they would already have been taken. The kind of relief offered, therefore, would be at best a temporary relief, and it would result in a larger and more insoluble problem later on.[6]

Although the author admitted that some of his proposals, such as pig and poultry raising and potato-eating, are probably too radical for his countrymen, none of them is so radical as to cause the food supply to keep pace indefinitely with rapid population growth. Furthermore, he said nothing about the effect of a redundant population in preventing some of the very food reforms by which he proposed to alleviate the situation. The conclusion seems inevitable that something must be done on the population side if Indian poverty is to be reduced. As Notestein has said: "It is not the problem of doubling, or perhaps even tripling, the product of backward regions that staggers the imagination; it is the need for an indefinite continuation of such an expansion in order to keep up with an unending growth. The demographic problem is not that of putting an immediate end to growth, but of checking growth before the populations become unmanageably large – for example, before the present numbers are doubled."[7]

Dr. Singh's neglect of the population variable placed him in a queer position, a position that one might call totalitarian Puritanism. He would abolish feasts by law, require the eating of potaoes, regulate the milling of rice, make rationing universal, abolish moneylenders and restrict middlemen, encourage sexual abstinence, close the bazaars, and kill most of the wild animals. To what purpose? Not, in the long run, so that Indians could eat better, but rather so that there could be more of them. This strange view, that the purpose of life is to sweat and strain in order that the maximum number may be supported, is completely at variance with the goal of a higher standard of life.

In the end any attempt to compensate indefinitely on the economic side for population increase is bound to fail, because human beings live in a finite world. Atomic energy, use of the sun's rays, harnessing of the tides, all may enormously increase the food supply, but they cannot forever take care of an ever growing population. To see this one has only to put it in an extreme form. If the human species is 200,000 years old, and if from an initial pair it had increased at the modest rate of 1 per cent per annum, the number

of human beings would now run to approximately 10^{1720}. If each of these individuals weighed on an average 100 pounds, their total weight would be 10^{1722} pounds. Since the weight of the entire earth is only about 10^{25} pounds, it can be seen that the human species would long since have reached the point where *all the substance of the earth would be in their bodies*. It would be idle to talk about feeding them, because there would be absolutely nothing to feed them with. But long before this point were reached the growth rate would have stopped, because the piling up of human beings in layers would have smothered those lying a few feet below the human surface. In long-run terms, then, it is preposterous to talk about economic measures being able "to take care of" population growth. The latter, if continued steadily, would in only a few centuries use up all the resources that even the wildest technological enthusiast might conceive.

It may be urged, however, that the problem of population growth in the Indian region is not a matter of 200,000 years or several centuries, but of here and now. This is exactly the point. In the Indian subcontinent, given world conditions as they are and not as a utopian might wish, the existing density and growth rate leave doubt that the standard of living can be raised if the growth is continued. Under ideal economic conditions, with the whole world focused on the problem of maintaining Indians and Pakistanis, a population of more than two billion could probably live in this area. But the populations in other parts of the world are growing too. The world economic system is not and cannot be focused on India alone. Even over a short period, say thirty to forty years, it is questionable that the Indian rate of population growth can be maintained and at the same time a significant rise in the standard of living be achieved. In any case it will take a herculean effort. Since the demographic situation is thus apparently handicapping economic progress, it seems foolish to forswear any demographic policy and simply try to step up economic production. This would be as foolish as simply forswearing any economic policy and trying to do it all on the population side. The task of raising the standard of living in India and Pakistan will be hard enough without making it even harder by a blind unwillingness to deal with relevant factors. Both countries desperately need a population policy. Of course, when economic plans are undertaken with a view to, and in a way susceptible of, affecting the rate of population growth, they become a part of population policy (as shown below). But pursued without reference to population or as a perpetual compensation for population change, purely economic measures do not constitute a population policy.

ALTERNATIVE POLICIES

To the extent that the Indian region has a population problem, it is obviously a problem of over- rather than under-population. The logical approach to improving the Indian living standard would therefore be to slow down the growth rate. Demographically speaking, there are only three ways of doing this – by raising the death rate, encouraging emigration, or lowering the birth rate.

Mortality and population policy

Not only are health and longevity ends in themselves, but they are part of a high living standard. Therefore it would be self-contradictory to say that the death rate should be increased in order to improve the standard of living. It is precisely a high death rate, among other things, that a population policy is designed to avoid. If people get poorer and poorer they will inevitably begin dying off faster and faster until their number fails to grow. At that point the problem of population growth will have been solved, but not the problem of poverty. Countries like India and Pakistan face the question of (a) how to stop population growth before a rise in mortality automatically stops it, and (b) how to lower mortality still more without defeating this aim by a corresponding rise in numbers.

Yet low mortality is not the only element in a high standard of living. There are other elements having little or nothing to do with longevity. Therefore, a *temporary* rise in mortality would not necessarily represent a regression in the *total* standard of living. Its effect would depend largely on the duration, causes, and circumstances of the rise. For instance, as mentioned before, a sudden epidemic that quickly killed 50 million people in India and Pakistan would greatly increase the average real income of the remaining population, especially if its incidence were highest in the non-productive ages. Such a sudden

increase in deaths would only temporarily disrupt the economy, and it would in one stroke eliminate a huge portion of the surplus population. Conceivably it might open the way to social reforms that would otherwise be more difficult and thus help to break the vicious circle of poverty.

This is what Gandhi probably meant when, pushed into a corner by a journalist's insistence on the population problem, he said, "Then perhaps we need some good epidemics."[8] Gandhi laughed when he said this, because he recognized that a policy of promoting epidemics could not be seriously envisaged. It is important, however, to distinguish between the negative evaluation of high mortality on the one hand and its actual consequences on the other. Notestein seemingly confuses the two. "Policies designed to yield [a rising death rate] are occasionally suggested as a temporary expedient to obtain release from pressure, pending a decline in fertility. However . . . the suggestion is based on a misconception of the factors governing growth. A period of increasing mortality would in fact impede the developments essential to induce a decline in fertility. Rising mortality in the areas under consideration means in reality rising population pressure, and not a solution of that pressure."[9]

This and similar passages imply that a heightened death rate is rejected as policy because it always reduces economic efficiency. But increased mortality may not have a net adverse economic effect. Whether it does or not depends on how long it is sustained and what its causes are. If it is sustained or rises over a long period, it is a sign that conditions are getting worse. If it is sudden but temporary, as in the influenza epidemic of 1918 or the earlier famines, it may bring unusual prosperity through relief from excess numbers until the population builds up again to its former level. Even the immediately bad effects of a civil war may be compensated for by a subsequent period of lessened population pressure. This being true, the rejection of increased mortality as a policy does not rest on its economic effects. It rests on the fact that human life, except under extreme group necessity, is viewed as an end in itself and not as a means to an end. This reason explains why official domestic policy with reference to deaths is nearly always in one direction – limitation.[10]

The limitation of deaths by health measures, although not pursued as a population policy, nevertheless has demographic effects. In so far as it succeeds while other demographic variables remain fixed, it tends to increase the population. Consequently, a policy of reducing deaths but not reducing births can lead to an unbalanced situation. The United States may seem to be an exception, because here public health has advanced rapidly, there has been no *official* policy favoring birth control, and yet the rate of population growth is declining. There has, however, been an unofficial or *private* policy of limiting offspring, and it is precisely this factor that makes possible the continued success of public health work. If it were not for birth control in the United States, the great reduction in mortality would soon produce such excessive population that the death rate would be forced up. Public health success would thus defeat itself.

We have seen that India's semi-colonial position has seemingly resulted in an unequal acquisition of Western culture. Scientific techniques of preventing death have been imported and accepted with enthusiasm. By a combination of these techniques and a minimum disturbance of the Indian way of life, the spectacular causes of death (famine, epidemic, war) have been considerably controlled. But because the texture of Indian life was left relatively unchanged, and because India's industrialization and urbanization were slow under the British, the birth rate has not changed. Hence the demographic situation has got out of balance, even compared to the imbalance usually produced by the industrial revolution. Instead of going through the growth phase of the population transition rather quickly (as recent industrial countries are doing) the Indian sub-continent seems, through the force of circumstances, to have reached this phase prematurely and to be lingering in it overlong. Having neither a public nor a private policy of birth limitation to accompany the improvement in mortality, public health work in India and Pakistan is now up against greater and greater obstacles. Unless birth control begins to occur on a wide scale or a miraculous outlet in emigration occurs, the death rate must ultimately rise again and public health policy fail.

Emigration as a policy

The Union of India would undoubtedly benefit if six million of her people, grouped in families, could emigrate each year for the next thirty years. Similarly Pakistan would benefit if about two million left from there each year. Such emigration would remove the annual population increase and would reduce the present population by a little. It would thus give the two countries a 30-year breathing spell in which to industrialize. But in all probability neither government would favor that much emigration; if either one did, it would insist that the emigrants be well provided for, that they retain their cultural, religious, and national allegiance, and that they not remove essential skills or much wealth from the country. Such conditions would effectively prevent mass emigration of the size mentioned.

Although neither government is likely to favor or carry through a huge emigration policy, there is a current of thought in Indian circles (though apparently not in Pakistan) which definitely favors large-scale emigration – or at least the *right* to make such an exodus. One prominent exponent of this view is Radhakamal Mukerjee, whose ideas are worth examining.

Like many Indians and a few Westerners,[11] Mukerjee feels that the European peoples have acquired the major share of the world without settling it as fully as the Asiatic peoples could settle it. From most of these domains, particularly the comparatively "empty" ones (e.g. Australia, western North America, parts of Africa) Asiatics are excluded by the color bar. At the same time, by virtue of their economic and political ascendancy over Asiatic countries, the Western peoples have achieved a higher standard of living than the Oriental peoples. The net result is that comparatively small numbers command the greatest area and wealth of the world, while the bulk of humanity is compressed into a relatively small area and suffers from extreme poverty. The obvious solution of this inequality is to open up the empty spaces, especially the tropical areas, to mass emigration from India, China, Java, and Japan. Unless such equalization occurs fairly soon, war will be inevitable.

"The Pacific is, to a large degree, an Asiatic Ocean, and the island, large or small, including the sub-continent of Australia and New Zealand, may be said to belong to a pan-Asiatic system. In this part of the globe, which is largely uninhabited, the doctrine of Asiatic *Lebensraum* cannot be dismissed offhand nor the doctrine of the White Man's reserve taken for granted. . . . The acid test of the Atlantic Charter will be the satisfactory settlement of the racial issue of the Indians in South Africa and the revision of the White Australian Policy. The British Empire, which comprises the largest empty spaces of the globe, has now evolved into a Commonwealth of Nations. . . . The essence of the commonwealth idea is utterly incompatible with the color-bar in respect of Oriental migration to the open lands. . . .

"In one part of [the Pacific basin] millions live on 3- to 5-acre holdings and go on subnutritional and subphysiological standards; while in another, tractors, sheep, and cattle luxuriate on the open spaces and men's artificially bolstered-up standard of living is protected by government tariffs, subsidies, and bans on foreign immigration. Such an economic and social contrast is entirely incompatible with world peace."[12]

Although these quotations concern the Pacific area, Mukerjee's vision goes beyond that region to huge territories in South America, North America, and Africa. And his conception of the "carrying capacity" of these areas is by no means modest. "Rice grown in the tropical rain forests of South America may support about 2,400,000,000, and again in the forests and savannas of Africa with a population capacity of another 2,300,000,000."[13] He seemingly accepts Matsuoka's belief that Oceania, could support 600 to 800 million persons,[14] and he thinks that the Philippines could take 90 million settlers.[15] He apparently feels that hundreds of millions of Asiatics could settle in the United States and Canada. "Vast arid areas in North America which are now settled only by cattlemen can be brought under the plough and the harrow if Chinese and Indian immigration is encouraged on a reasonable scale."[16] Obviously, this view of the world's carrying capacity implies that the annual emigration of 7 million from the Indian subcontinent could be easily sustained for a very long time – a mere drop in the bucket in a world capable of supporting about ten billion people.

A pertinent question, however, is how *well* would these people be supported? What would be their level of living? On this matter Professor Mukerjee is optimistic, saying that Asiatic migrants acquire a higher standard of living in the new areas to which they go.[17] Since he cites no evidence, one may question the statement. Certainly for a time, until the usual Indian population pressure develops, the real income may rise. This is the case in Fiji. And if, as in South Africa, the Indians form a commercial and urban minority,

they may well enjoy a higher income than Indians generally; and the same will be true if, as in Burma, they manage to dominate the local population. But if they take to peasant agriculture, growing rice under subsistence conditions, and living together by the tens of millions, it seems only a question of time until they will overflow the land and sink back to the usual Indian standard. Certainly a place like Mauritius does not reveal descendants of Indian immigrants having a high standard of life.

How low a living standard our author would deem satisfactory for the Indian migrants is shown by the following passages: "In the case of Asiatic subsistence farmers who would settle in the pioneer lands, substantial capital investments for planned railway development and supply of social services will not be required as a precondition of colonization. For the Asiatics, unreclaimed land is for potential subsistence and arable land is food. With European colonists land is . . . merely one factor in the complicated problem of keeping up a standard of living on the land so as not to sink . . . to the level of peasants.

"The European colonist . . . is a farmer. The Asiatic colonist is a peasant. He belongs to the land. No title is more accurately given.

"There is no doubt that the entry of Indian or Chinese agriculturalists as small-scale subsistence farmers depending upon family labor will speed up the change from the European estate-system to subsistence agriculture."[18]

To Western ears this does not sound like a high standard of living but more like what is now prevalent in teeming Asia. Indeed, one begins to see on what basis Professor Mukerjee assumes such huge carrying capacities for the various parts of the world. If Asiatics settled there at their customary standard of living, these areas could indeed hold additional hundreds of millions. But in this case what would be the purpose of settling them there? Professor Mukerjee never clearly states the goal he has in view, other than equalizing the world's wealth between Asiatics and non-Asiatics. One is entitled to suspect that there is an implicit imperialistic aim. He seemingly wants to see Asiatics spread over the world because he prefers Asiatics to Europeans. If there is no such implicit aim as this, then one may ask the same kind of question that was previously asked in connection with Baljit Singh's views: Why use more of the earth's surface merely to support more people at subsistence?

As the *sole* relief for population pressure, emigration is a palliative rather than a solution. To be effective it must not only remove people from the region but, by hastening social change, aid in reducing fertility. Even as a palliative, however, it is hardly suitable to India and Pakistan. Their populations are so large that outlets for their total natural increase would be impossible to find. The rest of the world would become filled with Indians and Pakistanis. Minority problems and conflicts would develop. Moreover, the proud governments of the two countries would not admit the failure implied by such mass emigration, nor would they tolerate an undignified status for the emigrants in new areas. A palliative emigration large enough to carry off the natural increase of the Indian sub-continent therefore seems extremely unlikely as either a policy or a fact.

When used in conjunction with other measures, emigration may in some cases assist in alleviating the basic situation. If it can remove population in sufficient proportion, it may give an impetus to reform at home, to a rise in the standard of living, and to industrial expansion. It will have this effect still more if the emigrants are negatively selected with reference to skills, if the cost of emigration is borne by someone else than the sending country, and if, once abroad, the emigrants remit large capital funds to the people at home. However, in practice such ideal conditions of emigration are virtually never encountered. Seldom do emigrants leave in sufficient number to make a serious dent in the population.[19] They often have higher skills than the average among the population that remains.[20] And their remittances are usually too small to represent a powerful stimulant.[21] Furthermore, emigration is much more feasible as a help for overpopulation in small countries like Ireland, Puerto Rico, Norway, and Jamaica than it is in large ones like India, Pakistan, China, Java, Japan, and Italy.[22] Certainly there is little indication that the movement of Indians to overseas areas, as discussed in Chapter 14, had any effect in easing India's population problem. The volume of emigration was for many years quite substantial in absolute numbers, yet in relation to the total population of India it was insignificant. It brought very little financial return to India, nor did it stimulate much foreign trade. Conceivably future migration might have a different effect, but, for reasons given above, the volume of emigration will probably be insufficient to have much effect on the masses left at home. As either an exclusive or an ancillary solution, therefore, emigration would seem a weak reed for India and Pakistan to lean on in their population policy. This is realized by nearly all Indian students themselves.[23]

Birth control

The promotion of birth control as an official population policy has been rare in human society. Most governments have tolerated it as a private practice (in spite of restrictive laws), and some have favored it in one way or another.[24] But an official campaign to teach, encourage, and spread the use of contraception among the general public has been conspicuous by its absence. Governmental policy with reference to births has thus generally been the opposite of the policy with respect to deaths. Public funds have been liberally expended for public health, including medical research, education, treatment, and prevention. These efforts have yielded large dividends in the saving of lives. But the idea of lowering birth rates by similar expenditures has met either with hostility or with indifference in official circles.

The reason for this extreme contrast in public policy is rather clear. Prior to the Industrial Revolution all human societies had a high death rate, as judged by modern standards. In order to survive they therefore had to have a high birth rate. Their high death rate, however, was primarily involuntary. All peoples wished to enjoy good health and a long life; their failure to achieve this end was due to an inability to control deaths beyond a certain point. In general those societies that managed to reduce deaths below the world average gained in their competition with other societies for survival. At the same time a high birth rate was also an advantage in survival. It follows that throughout human history the societies that managed to control their deaths reasonably well and to have at the same time a high birth rate were the ones that continued to exist. The others fell by the wayside. As a consequence, the customs and institutions of the surviving societies express both a high evaluation of health and longevity and a high evaluation of fertility.[25]

When the Industrial Revolution made it possible for Western societies to make extraordinary gains in the control of vital processes, the application of this control was naturally guided by the traditional values. Since death control was already an established value, the new science and technology were put to work for this purpose. But on the fertility side the established value was in the direction of more rather than fewer births; consequently the new science and technology were not directed toward lowering fertility but rather, if anything, toward raising it. In other words, the movement to limit fertility in unaccustomed ways met with strong opposition as being contrary to an established value, whereas attempts to preserve life, even in unaccustomed ways, met with approval.[26] It was only after the successful preservation of life had resulted in larger families, and these larger families had proved an embarrassment to the individual in the highly urbanized and mobile structure of modern society, that he sought a way around the full practice of his high fertility mores. He left the customary evaluation intact, but tended to violate it to a certain degree in his own private behavior.[27] Thus the lag of birth control behind death control – a lag which gave a tremendous spurt to population growth – was implicit in the growing rationalism of modern life, which first attacked the negative value (death) and only later the positive value (high fertility).

Since the lag has almost invariably accompanied the Industrial Revolution, its presence now in India and Pakistan is not strange. And because it eventually disappears, it will someday disappear in India. The case of Japan indicates that it can happen in an Oriental as well as a Western country, given the industrial basis. But from a policy standpoint the crucial question is whether the change of attitude as regards fertility must wait upon the gradual unfoldment of the Industrial Revolution or whether it can be induced more quickly. Evidence in Chapter 10 gave no indication that the high fertility in India and Pakistan will soon be reduced by any automatic process. If fertility is to be substantially lowered soon, it will be only by some strong and unique policy.

Theoretically, there are two conceivable ways of reducing fertility quickly. One – the direct method – is to bring birth control immediately to the people. The other – the indirect method – is to industrialize at once and thus create overnight the conditions that will cause people voluntarily to limit their fertility. Although the two are not necessarily antithetical, we shall discuss them separately – the first method in the present section and the second method in the section that follows.

Those who zealously wish to start an all-out birth control campaign in India and Pakistan are impelled not only by humanitarian motives but also by a greater appreciation of technology than of sociology. Forgetting the cultural obstacles, they are tempted to believe that if a quick, easy, inexpensive, and semi-permanent contraceptive could be found which, like an injection or a pill, would produce harmless sterility for six months or more, it could be brought to the Indian and Pakistani population much as smallpox vaccination has been brought to them. From a purely physical point of view birth control is easier than death control. It involves the management of only one type of germ and only one kind of contagion, as against hundreds of types in health work. It involves only one period of life, as against all periods subject to disease; and only one type of medical specialist, as against dozens in fighting sickness. It involves relatively simple and easy principles that the layman can grasp, as against complicated ones that he cannot grasp in general medicine. The money it requires cannot compare to that required for other kinds of medical attention. Indeed so simple is the process of contraception, so clear the principle, that it is absurd to think that science, which has accomplished so much in so many more complex matters, cannot find suitable techniques for accomplishing this goal. In fact, we know that when there is a will to limit family size, even crude techniques will greatly reduce fertility.[28]

But the simplicity of the problem from the *physical* point of view merely underlines the fact that the obstacles are not primarily technological but socio-logical. An immediate birth control campaign in countries like India and Pakistan would leave social conditions the same. The peoples' old values and sentiments would remain intact, as would their old illiteracy and conservatism. They would therefore lack the incentive to adopt contraception even if it were handed to them. They would not have the habits and personal circumstances that go with family limitation. In other words, the very forces responsible for the current high fertility would make the adoption of such a policy difficult and unlikely, in spite of its demographic advantages.

This negative conclusion, however, should not be maintained dogmatically. It is true that as yet the technique of contraception has not been perfected. Very little medical and biological research has been expended on improving contraceptive technology, and the existing techniques, evolved in the middle class cultures of industrial countries, are not very well suited to the Indian and Pakistani populations.[29] A marked improvement in this sphere – a technological revolution in contraception – might considerably enhance the success of a birth control campaign. Furthermore, since no all-out government campaign using every available educational and propagandistic resource to bring contraception to an agricultural people has ever been tried, we do not know simply on *a priori* grounds that it will not work. We do know that in cases where birth control clinics have been set up in backward communities, the response has not been satisfactory.[30] We know that the cost of an adequate birth control service for a population of 420 million would be enormous and would strain the exchequers of both India and Pakistan. And finally – most relevant of all – we know that the people of the Indian subcontinent have already had some exposure to contraception and have thus given some evidence of what their reactions to a large-scale campaign might be.

In the cities and among the intellectual classes of the Indian subcontinent, modern contraception has been an imported culture trait for some time. Prior to the modern period, and still among the rural population, folk methods (mainly magical) were in use.[31] But the international birth control movement, epitomized by Margaret Sanger, did not fail to penetrate to India. In 1911 the President of the Calcutta Municipal Council, Babu Nilambara Mukerji, strongly advocated instruction in birth control. In 1928 the Madras Neo-Malthusian League was founded. Sometime prior to 1936 a Birth Control Information Centre was founded at Calicut, and clinics were in existence in Delhi, Nagpur, Lahore, Lucknow, Akola, Satara, Indore, Calcutta, Poona, Mysore City, and Bangalore. In 1934 the journal, *Marriage Hygiene*, was started in Bombay and fought valiantly for birth control in India. Mrs. How-Martyn, an English birth control advocate, toured India three different times, the first time in 1935. On her third tour, in 1936, she addressed the All-India Medical Conference, the Institute of Population Research at Lucknow, the Marriage Welfare and Child Guidance Association in Calcutta, the Bombay Women's Work Guild, the All-India Women's Conference at Ahmedabad, as well as many other groups and associations. She also broadcast from a Bombay radio station. On her first tour Mrs. How-Martyn was apparently part of the entourage of Mrs. Sanger when the latter first visited India. Also accompanying Mrs. Sanger was a Miss Phillips. Mrs. Sanger and Miss Phillips travelled over 10,000 miles in the country, visited 18 cities and towns, and addressed 64 meetings. Mrs. Sanger saw scores of government and city officials, and met most of the leaders of Indian public opinion. Mrs. How-Martyn travelled about 6,500 miles and addressed 41 meetings. Altogether the three women also addressed 32 medical organizations. They could not meet all the requests to speak, so great was the interest in their work. At Mrs. Sanger's suggestion, films on contraception were made, and these were subsequently shown to hundreds of doctors in India. On her third visit Mrs. How-Martyn reported that the Bombay Mofussil Maternity Child Welfare and Health council had accepted the birth control programme as a part of its welfare work and had required most of its medical staff to get training in the technique of contraception.[32]

What has been the Indian reaction to birth control? Apparently it has been mixed. Among the intellectuals and in the newspapers, to judge by reports, interest has been great. In certain quarters there has been opposition, especially when it came to policy resolutions and to the initiation of action. As for the Indian masses, they have not been sufficiently exposed to the movement to express their reaction, but one observer, director of the Maternity and Child Welfare Bureau, New Delhi, believes that a common error among birth control enthusiasts in India is in believing "that the women of India are crying out for it."

"Nothing could be further from the truth. In a dumb sort of way thousands, even millions of women do desire release from perpetual child bearing and the misery which so often accompanies it, but that is not synonymous with a desire for birth control. Some of those who want to escape child bearing might be shocked or horrified at the thought of contraception. Moreover the vast majority know nothing whatever about it in the modern sense though they may have a nodding acquaintance with the Indian counterpart of the professional abortionist."[33] This of course is but one person's impression. There has not yet been enough depth-interviewing and public opinion analysis to speak with confidence about the thinking of Indian and Pakistani peasants on birth limitation. One modern study, cited in Chapter 10, throws some light on the matter.[34] It showed the following results [Table 1].

	Attempted contra- ception (per cent)	Desired, but no attempt (per cent)	No desire (per cent)	Total women
Rural section	0.3	4.0	95.7	1,459
Lower middle class city area (Muslim)	3.3	0.3	96.4	1,499
Lower middle class city area (Hindu)	13.2	7.0	80.0	1,265
Upper class city area (Hindu)	38.0	1.0	60.3	1,452

Table 1

	Continence	Safe period	Coitus interruptus	Husband uses	Wife uses	Total cases
Rural section	4	—	—	—	—	4
Lower middle class urban (Muslim)	11	22	10	24	4	50
Lower middle class urban (Hindu)	56	17	32	100	7	167
Upper middle class urban (Hindu)	130	237	235	251	37	551

Table 2

These returns suggest there is a slight desire for birth control in rural areas, but that the higher classes in the large city are the ones who have definitely adopted the practice. As for the methods used, the following are instructive [Table 2].

Obviously a number of couples used more than one method. The most popular methods were those used by the husband (possibly condoms). Next in importance were "coitus interruptus" and "safe period." The frequent use of "continence" is of interest. "The enquiry also collected information as to what was considered the safe period. Although the data have not been tabulated, the feeling among the women seemed to be that the period of risk was either immediately preceding or succeeding the menstrual period – which is contrary to the current scientific view." Among the reasons given for family limitation, inability to look after more children was slightly in the lead, with women's health a close second, and economic reasons a close third. The age group most disposed to practice contraception in the two Hindu areas of the city was the one 25–29. The main reason for not attempting family limitation although desiring it, was given as lack of knowledge of methods. The percentage of women having a "fatalistic" attitude toward family size was as follows for each group:

Rural section	86.8
Lower middle urban (Muslim)	91.0
Lower middle urban (Hindu)	46.3
Upper urban (Hindu)	30.1

In answer to the question, "How many living children should a woman have when she is 40?" the following answers were given [Table 3]

In each case the highest number (in italics) suggests a tendency toward a relatively small family, with the exception of the Muslim group. The rural women show the least concentration on any one family size but, curiously, seem generally to prefer a size smaller than do the urban groups. More studies of this type will reveal, perhaps, that more Indian women use and desire family limitation than has generally been believed.

It has been said that Indian opinion is receptive to birth control education because neither Hinduism nor Mohammedanism has definite tenets against contraception and because the Indians are less prudish in matters of sex than are Europeans. But both of these allegations are offset by other factors. In the first place, there is in Hinduism a strong ascetic element that may be, and already has been turned against birth control. This is pointedly illustrated by Gandhi, who may be taken as the symbol of the modern Hindu ethos.

	nil	1	2	3	4	5	6 & Over
Rural section	3	110	*353*	*353*	299	102	147
Lower middle urban (Muslim)	1	1	21	49	181	*332*	106
Lower middle urban (Hindu)	4	45	289	*431*	217	89	113
Upper urban (Hindu)	1	27	243	*397*	312	81	85

Table 3

Gandhi readily admitted that there were too many people in India for the existing economy. Since he disliked industrialization, there were also too many for his ideal economy. One would think, then, that he might seize upon birth control as the way out, and this he did – but in a peculiar way. "There can be no two opinions about the necessity of birth control. But the only method handed down from ages past is self-control or Brahmacharya. It is an infallible sovereign remedy doing good to those who practice it. The union is meant not for pleasure but for bringing forth progeny. And union is criminal when the desire for progeny is absent."[35] Thus for Gandhi, reflecting Hindu asceticism, sexual pleasure is inherently sinful. It is justified only when it serves a higher purpose – reproduction. It follows that the only permissible form of birth control is abstinence, which implies self-control (i.e. the forgoing of bodily pleasure) and is therefore good. Gandhi's position is the same as that of the Roman Catholic Church, except that he did not get into the complexities of intermittent abstinence and the so-called safe period.

That Gandhi's views are congenial to the Indian people need hardly be said. Up to now birth control has been debated mainly among the Westernized intellectuals. If it is carried to the more orthodox by encompassing a larger circle, it will certainly run the risk of attack from an ascetic point of view. It may be felt that contraception allows people to have sexual pleasure without risking the penalty of having children,[36] that birth control is a materialistic Western innovation, and that it promotes immorality. However, as yet there has been no crystallization of opposition. The main opposition has come from the Roman Catholic Church in India. Certain Hindu scholars, such as Sir S. Radhakrishnan, favor birth control as a help in alleviating India's poverty.

The statement that Indians are less prudish than Westerners in matters of sex is, like the supposed absence of religious attitudes on birth control, open to question. In certain ways the Indians may be less prudish, but in other ways (some bearing directly on birth control) they are not. The seclusion of women (*purdah*) is a Mohammedan institution borrowed by the Hindus; it implies an extreme hypersensitivity in matters pertaining to sex. This hypersensitivity manifests itself in the resistance to having male doctors treat female patients,[37] in the unwillingness of women to enter the medical and nursing professions, and in the hesitancy of female doctors to discuss sexual matters in the presence of male doctors. In 1937, after a tour of South India in behalf of birth control, one M.D. summed up his conclusion on the medical aspects as follows:

"My impressions of the tour are very uncomplimentary to the medical profession. No contraceptive advice whatever is being given except in some of the Mission Hospitals, the male doctors are indifferent or timid to take up the work, asserting that Indian women will not practise birth control methods and that there is no demand for the same. The lady doctors believe just the contrary but are equally timid to seek the required knowledge and are often domineered by anti-birth control superiors. It was distressing to hear from a reliable source that in one station, the lady doctors to whom women were referred for advice told them that birth control methods were harmful, apparently just to cover their ignorance of the subject!

"The doctor to tour the country can be male or female, but I think males would be better. . . . Over 95 per cent of the doctors in this country are males and it will be difficult to get lady doctors in India who would agree to demonstrate and talk on birth control methods to the men doctors and take up the matter with administrative bodies and officials. . . . In a few stations I visited, the lady doctors would not attend the demonstrations *along with* the male doctors."[38]

Opposition to birth control in India and Pakistan may therefore be expected on both religious and moral grounds. To date, this opposition (except for the Catholic community) is not highly organized, but neither are the proponents of birth control. As the birth control movement gains strength and vociferousness, the opposition will doubtless crystallize and gain strength as well. This capacity of a movement to evoke opposition is inevitable. It does not mean that the movement itself will fail. The very controversy itself will tend to spread contraceptive knowledge. In no country of the world has religious opposition been able to stop the diffusion of birth control, any more than it has been able to stop the use of tobacco or alcohol. The practice will eventually come to India in spite of opposition. But as for the government taking the initiative and speeding the diffusion by a vigorous birth control policy, the probability seems remote. "In no country of the world have the politicians openly associated themselves with birth control."[39] But it is precisely this sort of government policy that would be necessary if reliance were placed on birth control alone as the means of quickly reducing the rate of population growth. If it were tried and if it succeeded, it would be (assuming normal economic development) a powerful factor in the rise of the Indian standard of living. But it is safer to predict that it will not be tried than to say that it would not succeed. The speed of birth control diffusion on the Indian subcontinent will probably depend on the enthusiasm and skill that its proponents, in their private capacity, bring to the problem rather than upon official governmental action.

Rapid industrialization

Already in this chapter industrialization has been considered as an alternative to any population policy at all. The conclusion was reached that no conceivable economic development could support indefinitely a steadily growing population. Now we are considering rapid industrialization as a population policy. As such it is really a means of reducing fertility, not directly through officially diffusing contraceptive material and information, but indirectly through changing the conditions of life and thus forcing people in their private capacity to seek the means of family limitation. As a population policy, industrialization obviously means something more than merely allowing social evolution to take its course; if this were all that were implied, it would represent no policy at all. It implies

rather an attempt to speed industrialization beyond what it would otherwise be and to emphasize in the process those elements of modernization that will most likely depress fertility – such as education, urbanization, geographical and class mobility, multi-family dwellings, commercial recreation, and conspicuous consumption. An industrial revolution is so enormous, however, and is instrumental to so many ends, that its feasibility and character are likely to be determined (as is true of birth control too) on grounds other than population alone. The benefits of industrialization are, in fact, among the things *for which* we wish to reduce the rate of population growth. In other words, rapid industrialization has the advantage that it will probably be adopted as an official policy regardless of its connection with population. The only question is whether or not the elements associated most closely with declining fertility will be emphasized. Some of them, such as education, almost certainly will be. But others, such as commercial recreation and class mobility, which are contrary to rural Indian traditions, will probably not be emphasized but will simply arise automatically.

The main disadvantages of quick industrialization as compared to a direct birth control policy (pretending for the moment that one is a substitute for the other) are twofold: first, it is more difficult, and second it is slower. Although economic change seems more acceptable than birth control measures because it interferes less with the mores, the truth is that any policy that rapidly industrialized Pakistan and India would be a far greater shock to the basic social institutions than would any policy that attacked fertility directly. Fast industrialization would sweep both the *ryot* and the *zamindar* from their moorings, transforming them into workers in a collectivized, mechanized agriculture utterly foreign to their habits. The Indian and Pakistani peasants would not undergo this transformation willingly; resisting, they would have to be forced to it. Judged by events in Russia, the cost of this transformation and of the resistance to it would be tremendous in loss of human lives, loss of livestock, and loss of food production. At the same time the existing industrial and commercial system of India and Pakistan would have to be completely overhauled and subjected to strong controls. Production schedules, prices, profits, wages, supply of raw materials, location of industries, flow of capital, and mobility of labor—all would have to be minutely planned and rigidly administered by a central government.[40] The powerful businessmen of the two countries would not necessarily submit willingly to this extreme governmental control. And yet how otherwise could a retarded agricultural region be industrialized rapidly? The Russian example shows that fast industrialization is possible, but it also shows that the cost is heavy. Since the present generation is reared in the customs and institutions that now exist, since it has vested interests (real or fancied) in things as they are, it will not voluntarily approve and bear the costs of a sudden industrial revolution. Its resistance consequently adds to the inherent costs of the economic transformation. Furthermore, its resistance (i.e. the resistance of the people) means that to succeed, the government must become a dictatorship. Many people both outside and inside of the Indian area would rather see a slower pace of industrialization than a dictatorship established.

When, therefore, it is said that rapid industrialization is an easier policy to follow than a policy of direct birth control promotion, all that is meant is that the *statement of the policy* is easier. It cannot mean that the *execution* of the policy is easier. The people will willingly admit that industrial advances are needed; they will not so readily admit that fertility curbs are needed. In the execution of the policy, however, a program of forced industrialization would violate far more taboos and arouse more resistance than would the dissemination of birth control education and propaganda. This fact suggests that a good bit of the talk about rapid industrialization in India and Pakistan is just talk. It sounds good and elicits a favorable reaction. But whether enough official action will be taken to speed the industrial process beyond what its pace would be under ordinary business control is a moot question. If the process is not speeded up, then any announced policy of acceleration will turn out to be a failure.

The economic planners of India, e.g. those who made the famous Bombay plan which calls for a tripling of the national income and a doubling of the per capita income in 15 years, have not faced the true costs.[41] To put down in black and white all the changes, conflicts, privations, and issues involved in rapid industrialization would immediately inform people that not only an economic but a political, social, and religious revolution is in prospect. Except by a handful of Communists, the proposals would be rejected out of hand. So the planners, in order to have their plans even considered, must perforce play down the realistic means and emphasize the goal. Since the goal – a higher standard of living by means of industrial growth – is quite acceptable, it draws popular support for the plan. But when the means become known in detail, they will meet stiff opposition. In Russia the peasantry as a whole, the majority of the population, admittedly opposed collectivization.[42] Only a dictatorship could have forced through such a program.

This does not mean that the costs of forced industrialization will exceed the ultimate gains, but simply that the living generation, reared in the past, is the one that contemplates the proposals. It is this generation that will bear the greatest burdens and reap the fewest benefits. It is this generation that carries an institutional structure in many ways inimical to industrialization. In order for this generation to be willingly seduced, the extent to which the plans call for sacrifice, for social and religious change, and for subjection to political authority must be minimized. In the same way, of course, the advocates of birth control can talk about the advantages of proper child spacing, but straightway the means (contraception) becomes obvious and opposition is aroused.

In addition to its difficulty of execution, a second disadvantage of forced industrialization as a population policy is its slowness. Even granting that industrialization can be greatly accelerated, the time required would nevertheless permit a huge interim growth in numbers. The death rate would for some time continue to fall faster than the birth rate.[43] If the achievement of a Western-type demographic balance were to take 80 years, the interim population growth would be enormous. The same rate of increase as Europe experienced from 1850 to 1933 (a rate smaller than that of India during the last 20 years) would give India and Pakistan in the year 2024 a combined population of 750 million and an average density of 482 per square mile. How fast the modernization process can be speeded up depends on the role of India and Pakistan in the world economy, on the ruthlessness with which industrialization is pushed, and on the absence of chronic internal strife; but it seems hard to believe that it can be done rapidly enough to avoid an enormous growth. Should industrialization be relied on as the sole means of reducing fertility, it could not be successful in time to achieve a marked rise in the standard of living. One can argue, of course, that if India and Pakistan become industrialized, this will automatically raise the standard of living; but it is equally true that the rise will be far below what it would have been had the fertility been lowered at an earlier time. For this reason it seems unwise to rely on rapid industrialization as a substitute for a strong birth control policy, just as it would be unwise to do the opposite.

IDEAL POLICY vs. PROBABLE EVENTS

The conclusion emerges that ideally, in order to maximize real income, the population policy of Pakistan and India would include at least three measures – a program of strategic emigration, a sustained and vigorous birth control campaign, and a scheme for rapid industrialization – because no one of these complex measures can substitute for the others or promise the maximum effect if pursued alone. Emigration would be encouraged with a view to losing as little as possible in terms of skills and capital and gaining as much as possible in terms of remittances. Birth control would be diffused with the help of films, radio, ambulatory clinics, and free services and materials; aided by research on both the techniques of contraception and the methods of mass persuasion; and linked clearly to the public health and child welfare movements. Industrialization would be pushed by central planning and control, by forced capital formation (through rationing and taxation), by intensive training programs, by sweeping agricultural reforms, and by subsidies to heavy industry. The skillful and vigorous pursuit of such a broad policy would probably shorten the rapid growth phase that normally accompanies the industrial revolution. If so, it would mean that control had been deliberately extended to fertility as well as to mortality, that the demographic transition had been achieved quickly by planning. It would mean, as far as demographic factors go, a higher standard of living and a more abundant life for future generations.

But there is little likelihood that such a comprehensive population policy will be adopted. The two governments must pursue other goals in addition to a high standard of living. They are limited as to the means they may use for reaching any particular goal. They find that family behavior is too intertwined with religion and the mores to be manipulated in a purely instrumental way.

The one measure that has the best chance of being pushed is rapid industrialization, but not for demographic reasons. Both Pakistan and the Indian Union will probably try hard to improve their economies without encouraging lower fertility or greater emigration. They will doubtless attempt some reforms on the agricultural side but will place their greatest emphasis on industrial development. It seems likely that they will eventually succeed in industrializing, but because the obstacles are so great (including excessive population) they may not succeed until they have established totalitarian regimes, acquired almost completely planned economies, and experienced sharp temporary rises in mortality. On the other hand they may not succeed in increasing the pace at all beyond what ordinary business evolution would yield. After the economic revolution has been accomplished, the conditions of life for the individual should, as in Europe, North America, Australia, and Japan, be of such a type as to give a powerful personal incentive for limiting births. The birth rate should then drop and a modern demographic balance be achieved. The population will probably be much larger then than it would have been if a full-scale population policy had been carried out in the first place. How great the number will actually be is difficult to say, but it can hardly be double the present figure. How long the process of modernization will take is also hard to say, but it should not take more than a century.

Thus the effect of a full population policy would be not to prevent perpetual population growth (such growth is impossible anyway) but to balance the demographic books at an earlier time. In this way the total number to be cared for would be less and the living standard higher. We have seen that not all industrial peoples enjoy the same real income. The contrast between Japan and Europe, between Europe and America, suggests that the real income in industrial countries is strongly influenced by the point at which demographic growth is stabilized with reference to resources. Even if the whole world becomes industrial, the countries with excessive numbers will still be penalized. Therefore, if India and Pakistan should achieve and hasten industrialization while controlling their population, the resulting level of living would be substantially higher than otherwise.

In short, if we look candidly at the probable future, we must admit that the demographic situation in Pakistan and India will get worse before it gets better. Also, it will get better later than it would if the two governments successfully carried through a comprehensive population policy. The main stumbling block to attempting such a comprehensive policy is birth control; yet if the benefits of civilization are to come increasingly to the people of this region, the birth rate must be brought down. The current discrepancy between the birth and the death rate, which is causing the rapid population growth, is in a sense artificial. The demographic account will have to be balanced sometime. Over any extended period it is impossible to control deaths and not control births. Eventually the birth rate must drop or the death rate rise. Strife, famine, and epidemic disease are an ever-present threat in India, capable of sending the mortality rate higher than ever before, precisely because the population is larger. In order to avoid a catastrophic rise in mortality, the birth rate must eventually fall. With a high density of population in relation to developed resources and with the virtual impossibility of solving the problem quickly by sheer economic measures or by emigration, the two countries must necessarily incorporate planned parenthood as an essential element in any program that actually raises the standard of living to the maximum possible and gives them the greatest national strength. The fact that they will probably not do this does not detract from its advisability. Their unwillingness to do it will not necessarily result in perpetual poverty for their citizens or in absolute catastrophes. But it will result in greater poverty than would otherwise be the case and in greater danger of catastrophes. It is exactly this sort of comparative loss that all policy, including population policy, is designed to avoid.

NOTES

1 From *The Balance of Tomorrow* by Robert Strausz-Hupé, p. 235. Copyright 1945 by Robert Strausz-Hupé. Courtesy of G. P. Putnam's Sons.

2 "It is most unfortunate in the eyes of impartial observers that the new nation should believe it necessary to spend so much of her income for national defense. When one asks against whom Pakistan is preparing to defend her borders, the answer is not Afghanistan, but India. That such a situation should exist does not make sense to foreign observers, since both nations are in the British Commonwealth and their mutual interests inevitably will force them toward joint defense arrangements. . . . The amount is out of all proportion to what is in accordance with reasonably sound accounting principles." Robert Trumbull, "Pakistan Is Facing Economic Pitfall," *New York Times*, May 6, 1949, p. C-9. Other reports cite a slightly smaller military expenditure – around 50 per cent for Pakistan. See Daniel Thorner, "Prospects for Economic Development in Southern Asia," *Foreign Policy Reports*, Vol. 26 (April 15, 1950).

3 For an interesting analysis of the emotional resistances to a rational consideration of population matters see J. C. Flugel, *Population, Psychology, and Peace* (London: Watts, 1947), Ch. 2.

4 *Population and Food Planning in India* (Bombay: Hind Kitabs, 1947), p. 101. Another example of a refusal to consider population policy is Bimal C. Ghose, *Planning for India* (Calcutta: Oxford University Press, 1945), pp. 81–86.

5 Singh, *op. cit.*, pp. 127–28.

6 The frequent habit which people have, when faced with the population question, of saying that agricultural possibilities are not yet exhausted needs serious and immediate challenge. Ghose, *op. cit.*, p. 84, says, for example, that "the crucial problem in this question of overpopulation is the potentialities of Indian agriculture." There are no conceivable potentialities of Indian agriculture that could indefinitely take care of a population expanding at its present rate. Why therefore is not population itself the "crucial question"?

7 Frank W. Notestein, "Problems of Policy in Relation to Areas of Heavy Population Pressure" in *Demographic Studies of Selected Areas of Rapid Growth* (New York: Milbank Memorial Fund, 1944), p. 152.

8 Louis Fischer, *A Week with Gandhi* (New York: Duell, 1942), p. 89.

9 Notestein, *op. cit.*, pp. 148–49.

10 Of course, increased mortality is used as a policy with respect to real or fancied enemies, as in war or genocide. But this simply illustrates its negative evaluation *within* the ethnocentric group. It is a peculiarity of human populations that their growth (apart from migration) must be gradual, but their shrinkage can be sudden. Every group therefore must guard against being wiped out by catastrophe or enemy action. It must place a very high evaluation on keeping alive. For a much fuller exposition of the role of mortality in human culture, see Kingsley Davis, *Human Society* (New York: Macmillan, 1949), Ch. 20.

11 e.g. Warren S. Thompson, *Population and Peace in the Pacific* (Chicago: University of Chicago Press, 1946).

12 Radhakamal Mukerjee, *Races, Lands, and Food* (New York: Dryden Press, 1946), pp. 7, 39. See also p. 82. Similar ideas were expressed in an earlier work, *Migrant Asia* (Rome: Tipografia Failli, 1936).

13 Mukerjee, *Races, Lands, and Food*, p. 28.

14 *ibid.*, p. 68.

15 *ibid.*, p. 66.

16 *ibid.*, pp. 29–30.

17 *ibid.*, p. 79.

18 *ibid.*, pp. 31–32, 35.

19 The major exception is the case of Ireland, where numbers have fallen from 6,548,000 in 1841 to 2,953,000 in 1946. See Wm. Forbes Adams, *Ireland and Irish Emigration to the New World* (New Haven: Yale University Press, 1932) and Conrad M. Arensberg and Solon T. Kimball, *Family and Community in Ireland* (Cambridge: Harvard University Press, 1940), Ch. 6.

20 See, for example, C. Wright Mills, Clarence Senior, and Rose Kohn Goldsen, *The Puerto Rican Journey* (New York: Harper, 1950).

21 For the economic effects of emigration on the home country see Julius Isaac, *Economics of Migration* (New York: Oxford University Press, 1947), Chs. 3–7.

22 For the volume of emigration in relation to population pressure, see Kingsley Davis, "Puerto Rico's Population Problem: Research and Policy" in *International Approaches to Problems of Undeveloped Areas* (New York: Milbank Memorial Fund, 1948), pp. 60–65.

23 Gyan Chand, *India's Teeming Millions* (London: Allen & Unwin, 1939), pp. 291–95; Lanka Sundaram, *India in World Politics* (Delhi: Sultan Chand, 1944), pp. 234–36. Two recent works on Indian population do not discuss emigration as a likely population policy: D. Ghosh, *Pressure of Population and Economic Efficiency in India* (New Delhi: Indian Council of World Affairs) and S. Chandrasekhar, *India's Population, Fact and Policy* (New York: John Day, 1946).

24 On the growth of contraception as a private practice see A. M. Carr-Saunders, *World Population* (Oxford: Clarendon, 1936), Chs. 8, 9, 11, 17. Certain primitive societies, especially those living on small islands, have deliberately limited their numbers for demographic reasons – see, e.g., Raymond Firth, *We, the Tikopia* (London: Allen & Unwin, 1936), pp. 414, 415 – but in most cases the method used has not been contraception but abortion and infanticide. Contemporary examples of government sponsored birth control are the maintenance of health clinics with contraceptive services in Puerto Rico and in some states of the United States. See Christopher Tietze, "Human Fertility in Puerto Rico," *American Journal of Sociology*, Vol. 53 (July, 1947), pp. 34–40, and G. M. Cooper, "Four Years of Contraception as a Public Health Service in North Carolina," *American Journal of Public Health*, Vol. 31 (December, 1941), pp. 1248–52. In the early nineteenth century French public policy seemed to favor limitation of offspring, although at most other times France favored population growth – see D. V. Glass, *Population Policies and Movements* (Oxford: Clarendon, 1940), pp. 146–47.

25 For a fuller analysis of the role of fertility in social organization, see Kingsley Davis, *Human Society, op. cit.*, Ch. 20.

26 Of course, there are plenty of instances in which new methods of preventing death were rejected, but these were nearly all instances in which the people believed that the method did not in fact achieve the result claimed. They distrusted the motive of the physician or the efficacy of the remedy. Once they came to believe that the innovation really promoted health, they generally accepted it.

27 cf. E. F. Penrose, *Population Theories and Their Application* (Stanford University: Food Research Institute, 1934), pp. 115–20.

28 Regine K. Stix and Frank W. Notestein, *Controlled Fertility* (Baltimore: Williams & Wilkins, 1940), Chs. 1, 6, 15.

29 Ruth Young, "Some Aspects of Birth Control in India," *Marriage Hygiene*, First Series, Vol. 2 (August, 1935), p. 40. Gyan Chand, *India's Teeming Millions* (London: Allen & Unwin, 1939), pp. 343–45.

30 See Gilbert W. Beebe, *Contraception and Fertility in the Southern Appalachians* (Baltimore: Williams & Wilkins, 1942). Also Gilbert W. Beebe and Jose S. Belaval, "Fertility and Contraception in Puerto Rico," *Puerto Rico Journal of Public Health and Tropical Medicine*, Vol. 18 (September, 1942), pp. 3–52.

31 Norman E. Himes, *Medical History of Contraception* (Baltimore: Williams & Wilkins, 1936), pp. 131–32.

32 C. V. Drysdale, "The Indian Population Problem," *Marriage Hygiene*, Second Series, Vol. 1 (November, 1947), p. 100; also news notes as follows: "Mrs. Margaret Sanger in India," *ibid.*, First Series, Vol. 2 (May, 1936), pp. 461–64; "Birth Control in India: Mrs. How-Martyn on Her Tour," *ibid.*, Vol. 3 (February, 1937), pp. 241–42. My account is not meant to be definitive and the sources may not be completely accurate in detail; the aim is rather to suggest the kind of birth control work that has been undertaken.

33 Young, *op. cit.*, p. 39.

34 C. Chandra Sekar and Mukta Sen, "Enquiry into the Reproductive Patterns of Bengali Women," under the auspices of the Indian Research Fund Association and the All India Institute of Hygiene and Public Health. The study is not yet published, but some of the tentative results were kindly shown to the writer. The urban groups were all in Calcutta, the rural group about 20 miles from the city.

35 Quoted by Chand, *op. cit.*, pp. 328–29, taken from *Young India*, April 26, 1928.

36 This idea of children as a penalty for pleasure has been criticized by Chand, *op. cit.*, pp. 331–37. Chand also speaks of the utter impracticability of absolute abstinence.

37 This feeling seems to be present to some degree in any peasant society, but it is especially strong in India and Pakistan.

38 A. P. Pillay, "A Birth Control Educative Tour," *Marriage Hygiene*, First Series, Vol. 4 (August, 1937), p. 49.

39 Chand, *op. cit.*, p. 343.

40 Bimal C. Ghose, *Planning for India, op. cit.*, especially pp. 63, 67, 72.

41 Ghose, *op. cit.*, p. 63, says: "The essential weakness of the Bombay Plan is the yawning gap between its far-reaching objectives and the means and machinery proposed to reach them." P. A. Wadia and K. T. Merchant, *Our Economic Problem* (Bombay: New Book, 1946), p. 561, maintain that the Bombay Plan fails to recognize that democracy and economic planning are irreconcilable under capitalism. The authors therefore praise the "People's Plan" of the Radical Democratic Party (by M. N. Roy) as being socialistic but claim that it too is too optimistic with reference to means.

42 Ghose, *op. cit.*, pp. 69–73.

43 It must be recalled that some elements of modernization – e.g. the remarriage of widows, improvement of maternal health, reduction of sterility – will favor high fertility in India and Pakistan.

'Economic Development with Unlimited Supplies of Labour'

Manchester School (1954)

W. A. Lewis

Editors' Introduction

William Arthur Lewis was born in St Lucia in 1915 and died in 1991. He won a scholarship to the London School of Economics, topped his class there, continued to a PhD, and became in 1938 the first black member of staff and later a Reader in Colonial Economics. Arthur Lewis, as he was widely known, moved to Manchester University in 1947 as the Stanley Jevons Professor of Political Economy. It was while he was at Manchester that Lewis became a much-acclaimed 'pioneer' in development economics, first (in 1954) with the paper which we excerpt from here, and one year later with his book-length treatment of *The Theory of Economic Growth* (1955).

It is particularly important in the case of the reading here to note that it is only the first part of a long and at times demanding article. Nevertheless, some of Lewis's key ideas come across very precisely right from the beginning. The first sentence announces that the article is written in the classical tradition (of political economy), and thus not within the neoclassical frameworks which then dominated mainstream (or at any rate, non-Keynesian) economics. In other words, Lewis does not assume that labour is scarce, or that wages, profits and savings must rise together during the process of economic growth. To the contrary, he assumes a 'dual economy', or an economy in which one sector provides a reservoir of cheap labour for another. In this 1954 article the reservoir of cheap labour is provided by the subsistence economy of the agricultural sector in a developing country. Lewis maintains that labour can be decanted from the countryside to more productive jobs in the modern (urban–industrial) economy without significantly raising wages in the cities or dampening productivity in the countryside. People who contribute very little (if anything) in terms of marginal productivity in the countryside – where there is massive disguised unemployment – are redeployed more productively elsewhere.

Paradoxically, however, while the Lewis two-sector model seems to justify a big push for rapid industrialisation, Lewis is careful to argue for balanced economic growth. The growing industrial sector has to sell its output somewhere. If it is not all to go overseas as exports, a good portion must be bought at home, in the countryside. And for that to happen the countryside has to be modernised, said Lewis, in tandem with rapid industrialisation. This would require government planning, but most of all it would need to be financed from the extra savings and investment that capitalist development (in the private and government sectors) would induce. National development would be led, in the long run, by technical change and a growing class of capitalists.

In 1959, Lewis took up a post as the first Principal of the University College of the West Indies, before moving in 1963 to James Madison University in the United States. During the rest of his distinguished career he wrote widely on politics in Africa, on race and economic development, and on the evolution of the international economic order. Readers interested in his mature thoughts on these issues should consult his later book, *The Evolution of the International Economic Order* (1978). W. Arthur Lewis shared the Nobel Prize for Economics in 1979.

Key references

W. A. Lewis (1954) 'Economic development with unlimited supplies of labour', *Manchester School* 22: 131–91.
— (1955) *The Theory of Economic Growth*, London: Allen and Unwin.
— (1965) *Politics in West Africa*, London: Allen and Unwin.
— (1970) *Tropical Development, 1880–1913*, London: Allen and Unwin
— (1978) *The Evolution of the International Economic Order*, Princeton, NJ: Princeton University Press.
— (1984) 'Development economics in the 1950s', in G. Meier and D. Seers, *Pioneers in Development*, Oxford: Oxford University Press.

1. This essay is written in the classical tradition, making the classical assumption, and asking the classical question. The classics, from Smith to Marx, all assumed, or argued, that an unlimited supply of labour was available at subsistence wages. They then enquired how production grows through time. They found the answer in capital accumulation, which they explained in terms of their analysis of the distribution of income. Classical systems thus determined simultaneously income distribution and income growth, with the relative prices of commodities as a minor by-product.

Interest in prices and in income distribution survived into the neo-classical era, but labour ceased to be unlimited in supply, and the formal model of economic analysis was no longer expected to explain the expansion of the system through time. These changes of assumption and of interest served well enough in the European parts of the world, where labour was indeed limited in supply, and where for the next half century it looked as if economic expansion could indeed be assumed to be automatic. On the other hand over the greater part of Asia labour is unlimited in supply, and economic expansion certainly cannot be taken for granted. Asia's problems, however, attracted very few economists during the neo-classical era (even the Asian economists themselves absorbed the assumptions and preoccupations of European economics) and hardly any progress has been made for nearly a century with the kind of economics which would throw light upon the problems. of countries with surplus populations.

When Keynes's *General Theory* appeared, it was thought at first that this was the book which would illuminate the problems of countries with surplus labour, since it assumed an unlimited supply of labour at the current price, and also, in its final pages, made a few remarks on secular economic expansion. Further reflection, however, revealed that Keynes's book assumed not only that labour is unlimited in supply, but also, and more fundamentally, that land and capital are unlimited in supply – more fundamentally both in the short run sense that once the monetary tap is turned the real limit to expansion is not physical resources but the limited supply of labour, and also in the long run sense that secular expansion is embarrassed not by a shortage but by a superfluity of saving. Given the Keynesian remedies the neo-classical system comes into its own again. Hence, from the point of view of countries with surplus labour, Keynesianism is only a footnote to neo-classicism – albeit a long, important and fascinating footnote. The student of such economies has therefore to work right back to the classical economists before he finds an analytical framework into which he can relevantly fit his problems.

The purpose of this essay is thus to see what can be made of the classical framework in solving problems of distribution, accumulation, and growth, first in a closed and then in an open economy. It is not primarily an essay in the history of economic doctrine, and will not therefore spend time on individual writers, enquiring what they meant, or assessing its validity or truth. Our purpose is rather to bring their framework up-to-date, in the light of modern knowledge, and to see how far it then helps us to understand the contemporary problems of large areas of the earth.

1. THE CLOSED ECONOMY

2. We have to begin by elaborating the assumption of an unlimited supply of labour, and by establishing

that it is a useful assumption. We are not arguing, let it be repeated, that this assumption should be made for all areas of the world. It is obviously not true of the United Kingdom, or of North West Europe. It is not true either of some of the countries usually now lumped together as under-developed; for example there is an acute shortage of male labour in some parts of Africa and of Latin America. On the other hand it is obviously a relevant assumption for the economies of Egypt, of India, or of Jamaica. Our present task is not to supersede neo-classical economics, but merely to elaborate a different framework for those countries which the neo-classical (and Keynesian) assumptions do not fit.

In the first place, an unlimited supply of labour may be said to exist in those countries where population is so large relatively to capital and natural resources, that there are large sectors of the economy where the marginal productivity of labour is negligible, zero, or even negative. Several writers have drawn attention to the existence of such "disguised" unemployment in the agricultural sector, demonstrating in each case that the family holding is so small that if some members of the family obtained other employment the remaining members could cultivate the holding just as well (of course they would have to work harder: the argument includes the proposition that they would be willing to work harder in these circumstances). The phenomenon is not, however, by any means confined to the countryside. Another large sector to which it applies is the whole range of casual jobs – the workers on the docks, the young men who rush forward asking to carry your bag as you appear, the jobbing gardener, and the like. These occupations usually have a multiple of the number they need, each of them earning very small sums from occasional employment; frequently their number could be halved without reducing output in this sector. Petty retail trading is also exactly of this type; it is enormously expanded in overpopulated economies; each trader makes only a few sales; markets are crowded with stalls, and if the number of stalls were greatly reduced the consumers would be no whit worse off – they might even be better off, since retail margins might fall. Twenty years ago one could not write these sentences without having to stop and explain why in these circumstances, the casual labourers do not bid their earnings down to zero, or why the farmers' product is not similarly all eaten up in rent, but these propositions present no terrors to contemporary economists.

A little more explanation has to be given of those cases where the workers are not self-employed, but are working for wages, since it is harder to believe that employers will pay wages exceeding marginal productivity. The most important of these sectors is domestic service, which is usually even more inflated in overpopulated countries than is petty trading (in Barbados 16 per cent. of the population is in domestic service). The reason is that in overpopulated countries the code of ethical behaviour so shapes itself that it becomes good form for each person to offer as much employment as he can. The line between employees and dependants is very thinly drawn. Social prestige requires people to have servants; and the grand seigneur may have to keep a whole army of retainers who are really little more than a burden upon his purse. This is found not only in domestic service, but in every sector of employment. Most businesses in under-developed countries employ a large number of "messengers," whose contribution is almost negligible; you see them sitting outside office doors, or hanging around in the courtyard. And even in the severest slump the agricultural or commercial employer is expected to keep his labour force somehow or other – it would be immoral to turn them out, for how would they eat, in countries where the only form of unemployment assistance is the charity of relatives? So it comes about that even in the sectors where people are working for wages, and above all the domestic sector, marginal productivity may be negligible or even zero.

Whether marginal productivity is zero or negligible is not, however, of fundamental importance to our analysis. The price of labour, in these economies, is a wage at the subsistence level (we define this later). The supply of labour is therefore "unlimited" so long as the supply of labour at this price exceeds the demand. In this situation, new industries can be created, or old industries expanded without limit at the existing wage; or, to put it more exactly, shortage of labour is no limit to the creation of new sources of employment. If we cease to ask whether the marginal productivity of labour is negligible and ask instead only the question from what sectors would additional labour be available if new industries were created offering employment at subsistence wages, the answer becomes even more comprehensive. For we have then not only the farmers, the casuals, the petty traders and the retainers (domestic and commercial), but we have also three other classes from which to choose.

First of all, there are the wives and daughters of the household. The employment of women outside the household depends upon a great number of factors, religious and conventional, and is certainly not exclusively a matter of employment opportunities. There are, however, a number of countries where the current limit is for practical purposes only employment opportunities. This is true, for example, even inside the United Kingdom. The proportion of women gainfully employed in the U.K. varies enormously from one region to another according to employment opportunities for women. For example, in 1939 whereas there were 52 women gainfully employed for every 100 men in Lancashire, there were only 15 women gainfully employed for every 100 men in South Wales. Similarly in the Gold Coast, although there is an acute shortage of male labour, any industry which offered good employment to women would be besieged with applications. The transfer of women's work from the household to commercial employment is one of the most notable features of economic development. It is not by any means all gain, but the gain is substantial because most of the things which women otherwise do in the household can in fact be done much better or more cheaply outside, thanks to the large scale economies of specialisation, and also to the use of capital (grinding grain, fetching water from the river, making cloth, making clothes, cooking the midday meal, teaching children, nursing the sick, etc.). One of the surest ways of increasing the national income is therefore to create new sources of employment for women outside the home.

The second source of labour for expanding industries is the increase in the population resulting from the excess of births over deaths. This source is important in any dynamic analysis of how capital accumulation can occur, and employment can increase, without any increase in real wages. It was therefore a cornerstone of Ricardo's system. Strictly speaking, population increase is not relevant either to the classical analysis, or to the analysis which follows in this article, unless it can be shown that the increase of population is caused by economic development and would not otherwise be so large. The proof of this proposition was supplied to the classical economists by the Malthusian law of population.

There is already an enormous literature of the genus: "What Malthus *Really* Meant," into which we need not enter. Modern population theory has advanced a little by analysing separately the effects of economic development upon the birth rate, and its effects on the death rate. Of the former, we know little. There is no evidence that the birth rate ever rises with economic development. In Western Europe it has fallen during the last eighty years. We are not quite sure why; we suspect that it was for reasons associated with development, and we hope that the same thing may happen in the rest of the world as development spreads. Of the death rate we are more certain. It comes down with development from around 40 to around 12 per thousand; in the first stage because better communications and trade eliminate death from local famines; in the second stage because better public health facilities banish the great epidemic diseases of plague, smallpox, cholera, malaria, yellow fever (and eventually tuberculosis); and in the third stage because widespread facilities for treating the sick snatch from the jaws of death many who would otherwise perish in infancy or in their prime. Because the effect of development on the death rate is so swift and certain, while its effect on the birth rate is unsure and retarded, we can say for certain that the immediate effect of economic development is to cause the population to grow; after some decades it begins to grow (we hope) less rapidly. Hence in any society where the death rate is around 40 per thousand, the effect of economic development will be to generate an increase in the supply of labour.

Marx offered a third source of labour to add to the reserve army, namely the unemployment generated by increasing efficiency. Ricardo had admitted that the creation of machinery could reduce employment. Marx seized upon the argument, and in effect generalised it, for into the pit of unemployment he threw not only those displaced by machinery, but also the self-employed and petty capitalists who could not compete with larger capitalists of increasing size, enjoying the benefits of the economies of scale. Nowadays we reject this argument on empirical grounds. It is clear that the effect of capital accumulation in the past has been to reduce the size of the reserve army, and not to increase it, so we have lost interest in arguments about what is "theoretically" possible.

When we take account of all the sources we have now listed – the farmers, the casuals, the petty traders, the retainers (domestic and commercial), women in the household, and population growth – it is clear enough that there can be in an overpopulated economy, an enormous expansion of new industries or new employment opportunities without any shortage of unskilled labour becoming apparent in the labour market. From the point of view of the effect of economic development on wages, the supply of labour is practically unlimited.

This applies only to unskilled labour. There may at any time be a shortage of skilled workers of any grade – ranging from masons, electricians or welders to engineers, biologists or administrators. Skilled labour may be the bottleneck in expansion, just like capital or land. Skilled labour, however, is only what Marshall might have called a "quasi-bottleneck," if he had not had so nice a sense of elegant language. For it is only a very temporary bottleneck, in the sense that if the capital is available for development, the capitalists or their government will soon provide the facilities for training more skilled people. The real bottlenecks to expansion are therefore capital and natural resources, and we can proceed on the assumption that so long as these are available the necessary skills will be provided as well, though perhaps with some time lag.

3. If unlimited labour is available, while capital is scarce, we know from the Law of Variable Proportions that the capital should not be spread thinly over all the labour. Only so much labour should be used with capital as will reduce the marginal productivity of labour to zero. In practice, however, labour is not available at a zero wage. Capital will therefore be applied only up to the point where the marginal productivity of labour equals the current wage. This is illustrated in Figure 1. The horizontal axis measures the quantity of labour, and the vertical axis its marginal product. There is a fixed amount of capital. OW is the current wage. If the marginal product of labour were zero outside the capitalist sector, OR ought to be employed. But it will pay to employ only OM in the capitalist sector, WNP is the capitalists' surplus. OWPM goes in wages to workers in the capitalist sector, while workers outside this sector (i.e. beyond M) earn what they can in the subsistence sector of the economy.

The analysis requires further elaboration. In the first place, after what we have said earlier on about some employers in these economies keeping

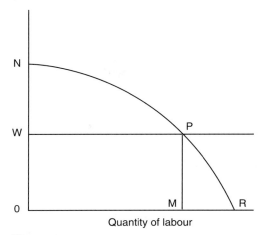

Quantity of labour

Figure 1

retainers, it may seem strange to be arguing now that labour will be employed up to the point where the wage equals the marginal productivity. Nevertheless, this is probably the right assumption to make when we are set upon analysing the expansion of the capitalist sector of the economy. For the type of capitalist who brings about economic expansion is not the same as the type of employer who treats his employees like retainers. He is more commercially minded, and more conscious of efficiency, cost and profitability. Hence, if our interest is in an expanding capitalist sector, the assumption of profit maximisation is probably a fair approximation to the truth.

Next, we note the use of the terms "capitalist" sector and "subsistence" sector. The capitalist sector is that part of the economy which uses reproducible capital, and pays capitalists for the use thereof. (This coincides with Smith's definition of the productive workers, who are those who work with capital and whose product can therefore be sold at a price above their wages.) We can think, if we like, of capitalists hiring out their capital to peasants; in which case, there being by definition an unlimited number of peasants, only some will get capital, and these will have to pay for its use a price which leaves them only subsistence earnings. More usually, however, the use of capital is controlled by capitalists, who hire the services of labour. The classical analysis was therefore conducted on the assumption that capital was used for hiring people. It does not make any difference to the argument, and for convenience we will follow this usage. The subsistence sector is by difference all that part of the economy which is not using reproducible capital. Output per head is lower in this sector than in the capitalist sector, because it is not fructified by capital (this is why it was called "unproductive"; the distinction between productive and unproductive had nothing to do with whether the work yielded utility, as some neo-classicists have scornfully but erroneously asserted). As more capital becomes available more workers can be drawn into the capitalist from the subsistence sector, and their output per head rises as they move from the one sector to the other.

Thirdly we take account of the fact that the capitalist sector, like the subsistence sector, can also be subdivided. What we have is not one island of expanding capitalist employment, surrounded by a vast sea of subsistence workers, but rather a number of such tiny islands. This is very typical of countries in their early stages of development. We find a few industries highly capitalised, such as mining or electric power, side by side with the most primitive techniques; a few high class shops, surrounded by masses of old style traders; a few highly capitalised plantations, surrounded by a sea of peasants. But we find the same contrasts also outside their economic life. There are one or two modern towns, with the finest architecture, water supplies, communications and the like, into which people drift from other towns and villages which might almost belong to another planet. There is the same contrast even between people; between the few highly westernised, trousered, natives, educated in western universities, speaking western languages, and glorying in Beethoven, Mill, Marx or Einstein, and the great mass of their countrymen who live in quite other worlds. Capital and new ideas are not thinly diffused throughout the economy; they are highly concentrated at a number of points, from which they spread outwards.

Though the capitalised sector can be subdivided into islands, it remains a single sector because of the effect of competition in tending to equalise the earnings on capital. The competitive principle does not demand that the same amount of capital per person be employed on each "island," or that average profit per unit of capital be the same, but only that the marginal profit be the same. Thus, even if marginal profits were the same all round, islands which yield diminishing returns may be more profitable than others, the earliest capitalists having cornered the vantage points. But in any case marginal profits are not the same all round. In backward economies knowledge is one of the scarcest goods. Capitalists have experience of certain types of investment, say of trading or plantation agriculture, and not of other types, say of manufacturing, and they stick to what they know. So the economy is frequently lopsided in the sense that there is excessive investment in some parts and under-investment in others. Also, financial institutions are more highly developed for some purposes than for others – capital can be got cheaply for trade, but not for house building or for peasant agriculture, for instance. Even in a very highly developed economy the tendency for capital to flow evenly through the economy is very weak; in a backward economy it hardly exists. Inevitably what one gets are very heavily developed patches of the economy, surrounded by economic darkness.

Next we must say something about the wage level. The wage which the expanding capitalist sector has to pay is determined by what people can earn outside that sector. The classical economists used to think of the wage as being determined by what is required for subsistence consumption, and this may be the right solution in some cases. However, in economies where the majority of the people are peasant farmers, working on their own land, we have a more objective index, for the minimum at which labour can be had is now set by the average product of the farmer; men will not leave the family farm to seek employment if the wage is worth less than they would be able to consume if they remained at home. This objective standard, alas, disappears again if the farmers have to pay rent, for their net earnings will then depend upon the amount of rent they have to pay, and in over-populated countries the rent will probably be adjusted so as to leave them just enough for a conventional level of subsistence. It is not, however, of great importance to the argument whether earnings in the subsistence sector are determined objectively by the level of peasant productivity, or subjectively in terms of a conventional standard of living. Whatever the mechanism, the result is an unlimited supply of labour for which this is the minimum level of earnings.

The fact that the wage level in the capitalist sector depends upon earnings in the subsistence sector is sometimes of immense political importance, since its effect is that capitalists have a direct interest in holding down the productivity of the subsistence workers. Thus, the owners of plantations have no interest in seeing knowledge of new techniques or new seeds conveyed to the peasants, and if they are influential in the government, they will not be found using their influence to expand the facilities for agricultural extension. They will not support proposals for land settlement, and are often instead to be found engaged in turning the peasants off their lands. (Cf. Marx on "Primary Accumulation".) This is one of the worst features of imperialism, for instance. The imperialists invest capital and hire workers; it is to their advantage to keep wages low, and even in those cases where they do not actually go out of their way to impoverish the subsistence economy, they will at least very seldom be found doing anything to make it more productive. In actual fact the record of every imperial power in Africa in modern times is one of impoverishing the subsistence economy, either by taking away the people's land, or by demanding forced labour in the capitalist sector, or by imposing taxes to drive people to work for capitalist employers. Compared with what they have spent on providing facilities for European agriculture or mining, their expenditure on the improvement of African agriculture has been negligible. The failure of imperialism to raise living standards is not wholly to be attributed to self interest, but there are many places where it can be traced directly to the effects of having imperial capital invested in agriculture or in mining.

Earnings in the subsistence sector set a floor to wages in the capitalist sector, but in practice wages have to be higher than this, and there is usually a gap of 30 per cent. or more between capitalist wages and subsistence earnings. This gap may be explained in several ways. Part of the difference is illusory, because of the higher cost of living in the capitalist sector. This may be due to the capitalist sector being concentrated in congested towns, so that rents and transport costs are higher. All the same, there is also usually a substantial difference in real wages. This may be required because of the psychological cost of transferring from the easy going way of life of the subsistence sector to the more regimented and urbanised environment of the capitalist sector. Or it may be a recognition of the fact that even the unskilled worker is of more use to the capitalist sector after he has been there for some time than is the raw recruit from the country. Or it may itself represent a difference in conventional standards, workers in the capitalist sector acquiring tastes and a social prestige which have conventionally to be

recognised by higher real wages. That this last may be

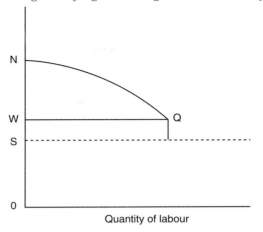

Figure 2

the explanation is suggested by cases where the capitalist workers organise themselves into trade unions and strive to protect or increase their differential. But the differential exists even where there are no unions.

The effect of this gap is shown diagrammatically in Figure 2, which is drawn on the same basis as Figure 1. *OS* now represents subsistence earnings, and *OW* the capitalist wage (real not money). To borrow an analogy from the sea, the frontier of competition between capitalist and subsistence labour now appears not as a beach but as a cliff.

This phenomenon of a gap between the earnings of competing suppliers is found even in the most advanced economics. Much of the difference between the earnings of different classes of the population (grades of skill, of education, of responsibility or of prestige) can be described only in these terms. Neither is the phenomenon confined to labour. We know of course that two firms in a competitive market need not have the same average profits if one has some superiority to the other; we reflect this difference in rents, and ask only that marginal rates of profit should be the same. We know also that marginal rates will not be the same if ignorance prevails – this point we have mentioned earlier. What is often puzzling in a competitive industry is to find a difference in marginal profits, or marginal costs, without ignorance, and yet without the more efficient firm driving its rivals out of business. It is as if the more efficient says: "I could compete with you, but I won't," which is also what subsistence labour says when it does not transfer to capitalist employment unless real wages are substantially higher. The more efficient firm, instead of competing wherever its real costs are marginally less than its rivals, establishes for itself superior standards of remuneration. It pays its workers more and lavishes welfare services, scholarships and pensions upon them. It demands a higher rate on its marginal investments; where its competitors would be satisfied with 10%, it demands 20%, to keep up its average record. It goes in for prestige expenditure, contributing to hospitals, universities, flood relief and such. Its highest executives spend their time sitting on public committees, and have to have deputies to do their work. When all this is taken into account it is not at all surprising to find a competitive equilibrium in which high cost firms survive easily side by side with firms of much greater efficiency.

4. So far we have merely been setting the stage. Now the play begins. For we can now begin to trace the process of economic expansion.

The key to the process is the use which is made of the capitalist surplus. In so far as this is reinvested in creating new capital, the capitalist sector expands, taking more people into capitalist employment out of the subsistence sector. The surplus is then larger still, capital formation is still greater, and so the process continues until the labour surplus disappears.

[In Figure 3] OS is as before average subsistence earnings, and OW the capitalist wage. WN_1Q_1 represents the surplus in the initial stage. Since some of this is reinvested, the amount of fixed capital increases. Hence the schedule of the marginal productivity of labour is now raised throughout, to the level of N_2Q_2. Both the surplus and capitalist employment are now larger. Further reinvestment raises the schedule of the marginal productivity of labour to N_3Q_3. And the process continues so long as there is surplus labour.

Various comments are needed in elaboration. First, as to the relationship between capital, technical progress, and productivity. In theory it should be possible to distinguish between the growth of capital and the growth of technical knowledge, but in practice it is neither possible nor necessary for this analysis. As a matter of statistical analysis, differentiating the effects of capital and of knowledge in any industry is straightforward if the product is homogeneous through time, if the physical inputs are also unchanged (in kind) and if the relative prices of the inputs have remained constant. But when we try to do it for any industry in practice we usually find that the product has changed, the inputs have changed and relative prices have changed, so that we get any number of indices of technical progress from the same data, according to the assumptions and the type of index number which we use. In any case, for the purpose of this analysis it is unnecessary to distinguish between capital formation and the growth of knowledge within the capitalist sector. Growth of technical knowledge outside the capitalist sector would be fundamentally important, since it would raise the level of wages, and so reduce the capitalist surplus. But inside the capitalist sector knowledge and capital work in the same direction, to raise the surplus and to increase employment. They also work together. The application of new technical knowledge usually requires new investment, and whether the new knowledge is capital-saving (and thus equivalent to an increase in capital) or labour-saving (and thus equivalent to an increase in the marginal productivity of labour) makes no difference to our diagram. Capital and technical knowledge also work together in the sense that in economies where techniques are stagnant savings are not so readily applied to increasing productive capital; in such economies it is more usual to use savings for building pyramids, churches, and other such durable consumer goods. Accordingly, in this analysis the growth of productive capital and the growth of technical knowledge are treated as a single phenomenon (just as we earlier decided that

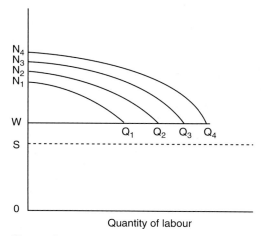

Figure 3

we could treat the growth of the supply of skilled labour and the growth of capital as a single phenomenon in long run analysis).

Next we must consider more closely the capitalist surplus. Malthus wanted to know what the capitalists would do with this ever-growing surplus; surely this would be an embarrassing glut of commodities? Ricardo replied that there would be no glut; what the capitalists did not consume themselves, they would use for paying the wages of workers to create more fixed capital (this is a free interpretation, since the classical economists associated the expansion of employment with an increase of circulating rather than of fixed capital). This new fixed capital would then in the next stage make possible the employment of more people in the capitalist sector. Malthus persisted; why should the capitalists produce more capital to produce a larger surplus which could only be used for producing still more capital and so *ad infinitum?* To this Marx supplied one answer: capitalists have a passion for accumulating capital. Ricardo supplied another: if they don't want to accumulate, they will consume instead of saving; provided there is no propensity to hoard, there will be no glut. Employment in the next stage will not be as big as it would have been if they had created more fixed capital and so brought more workers into the capitalist sector, but so long as there is no hoarding it makes no difference to the current level of employment whether capitalists decide to consume or to save. Malthus then raised another question; suppose that the capitalists do save and invest without hoarding, surely the fact that capital is growing more rapidly than consumption must so lower the rate of profit on capital that there comes a point when they decide that it is not worth while to invest? This, Ricardo replied, is impossible; since the supply of labour is unlimited, you can always find employment for any amount of capital. This is absolutely correct, for his model; in the neo-classical model capital grows faster than labour, and so one has to ask whether the rate of profit will not fall, but in the classical model the unlimited supply of labour means that the capital/labour ratio, and therefore the rate of surplus, can be held constant for any quantity of capital (*i.e.*, unlimited "widening" is possible). The only fly in the ointment is that there may develop a shortage of natural resources, so that though the capitalists get any amount of labour at a constant wage, they have to pay ever-rising rents to landlords.

This was what worried Ricardo; it was important to him to distinguish that part of the surplus which goes to landlords from that part which goes to capitalists, since he believed that economic development inevitably increases the relative scarcity of land. We are not so certain of this as he was. Certainly development increases the rent of urban sites fantastically, but its effect on rural rents depends on the rate of technical progress in agriculture, which Malthus and Ricardo both gravely under-estimated. If we assume technical progress in agriculture, no hoarding, and unlimited labour at a constant wage, the rate of profit on capital cannot fall. On the contrary it must increase, since all the benefit of technical progress in the capitalist sector accrues to the capitalists.

Marx's interest in the surplus was ethical as well as scientific. He regarded it as robbery of the workers. His descendants are less certain of this. The surplus, after all, is only partly consumed; the other part is used for capital formation. As for the part which is consumed, some of it is a genuine payment for service rendered – for managerial or entrepreneurial services, as well as for the services of public administrators, whether these are paid salaries out of taxes, or whether they live off their rents or *rentes* while performing unpaid public duties as magistrates, lord-lieutenants, or the like. Even in the U.S.S.R. all these functionaries are paid out of the surplus, and handsomely paid too. It is arguable that these services are over-paid; this is why we have progressive taxation, and it is also one of the more dubious arguments for nationalisation (more dubious because the functionaries of public corporations have to be paid the market rate if the economy is only partially nationalised). But it is not arguable that all this part of the surplus (*i.e.* the part consumed) morally belongs to the workers, in any sense. As for the part which is used for capital formation, the experience of the U.S.S.R. is that this is increased, and not reduced, by transforming the ownership of capital. Expropriation deprives the capitalists of control over this part of the surplus, and of the right to consume this part at some later date, but it does nothing whatever to transfer this part of the surplus to the workers. Marx's emotional approach was a natural reaction to the classical writers, who sometimes in unguarded moments wrote as if the capitalist surplus and its increase were all that counted in the national income (c.f. Ricardo, who called it "the net revenue" of production). All this, however, is by the way; for our present interest is not in ethical questions, but in how the model works.

5. The central problem in the theory of economic development is to understand the process by which a community which was previously saving and investing 4 or 5 per cent of national income or less, converts itself into an economy where voluntary saving is running at about 12 to 15 per cent. of national income or more. This is the central problem because the central fact of economic development is rapid capital accumulation (including knowledge and skills with capital). We cannot explain any "industrial" revolution (as the economic historians pretend to do) until we can explain why saving increased relatively to national income.

It is possible that the explanation is simply that some psychological change occurs which causes people to be more thrifty. This, however, is not a plausible explanation. We are interested not in the people in general, but only say in the 10 per cent. of them with the largest incomes, who in countries with surplus labour receive up to 40 per cent. of the national income (nearer 30 per cent. in more developed countries). The remaining 90 per cent. of the people never manage to save a significant fraction of their incomes. The important question is why does the top 10 per cent. save more? The reason may be because they decide to consume less, but this reason does not square with the facts. There is no evidence of a fall in personal consumption by the top 10 per cent. at a time when industrial revolutions are occurring. It is also possible that, though they do not save any more, the top 10 per cent. spend less of their income on durable consumer goods (tombs, country houses, temples) and more on productive capital. Certainly, if one compares different civilisations this is a striking difference in the disposition of income. Civilisations in which there is a rapid growth of technical knowledge or expansion of other opportunities present more profitable outlets for investment than do technologically stagnant civilisations, and tempt capital into productive channels rather than into the building of monuments. But if one takes a country only over the course of the hundred years during which it undergoes a revolution in the rate of capital formation, there is no noticeable change in this regard. Certainly, judging by the novels, the top 10 per cent. in England were not spending noticeably less on durable consumer goods in 1800 than they were in 1700.

Much the most plausible explanation is that people save more because they have more to save. This is not to say merely that the national income per head is larger, since there is no clear evidence that the proportion of the national income saved increases with national income per head – at any rate our fragmentary evidence for the United Kingdom and for the United States suggests that this is not so. The explanation is much more likely to be that saving increases relatively to the national income because the incomes of the savers increase relatively to the national income. The central fact of economic development is that the distribution of incomes is altered in favour of the saving class.

Practically all saving is done by people who receive profits or rents. Workers' savings are very small. The middle-classes save a little, but in practically every community the savings of the middle-classes out of their salaries are of little consequence for productive investment. Most members of the middle-class are engaged in the perpetual struggle to keep up with the Jones's; if they manage to save enough to buy the house in which they live, they are doing well. They may save to educate their children, or to subsist in their old age, but this saving is virtually offset by the savings being used up for the same purposes. Insurance is the middle-class's favourite form of saving in modern societies, yet in the U.K., where the habit is extremely well developed, the annual net increase in insurance funds from all classes, rich, middle, and poor is less than 1.5 per cent. of the national income. It is doubtful if the wage and salary classes ever anywhere save as much as 3 per cent. of the national income, net (possible exception: Japan). If we are interested in savings, we must concentrate attention upon profits and rents.

For our purpose it does not matter whether profits are distributed or undistributed; the major source of savings is profits, and if we find that savings are increasing as a proportion of the national income, we may take it for granted that this is because the share of profits in the national income is increasing. (As a refinement, for highly taxed communities, we should say profits net of taxes upon profits, whether personal income or corporate taxes.) Our problem then becomes what are the circumstances in which the share of profits in the national income increases?

The modified classical model which we are using here has the virtue of answering the question. In the beginning, the national income consists almost entirely of subsistence income. Abstracting from population growth and assuming that the marginal product of labour is zero, this subsistence income remains constant throughout the expansion, since by definition labour can be yielded up to the expanding capitalist sector without reducing subsistence output. The process therefore increases the capitalist surplus and the income of capitalist employees, taken together, as a proportion of the national income. It is possible to imagine conditions in which the surplus nevertheless does not increase relatively to national income. This requires that capitalist employment should expand relatively much faster than the surplus, so that within the capitalist sector gross margins or profit plus rent are falling sharply relatively to wages. We know that this does not happen. Even if gross margins were constant, profits in our model would be increasing relatively to national income. But gross margins are not likely to be constant in our model, which assumes that practically the whole benefit of capital accumulation and of technical progress goes into the surplus; because real wages are constant, all that the workers get out of the expansion is that more of them are employed at a wage above the subsistence earnings. The model says, in effect, that if unlimited supplies of labour are available at a constant real wage, and if any part of profits is reinvested in productive capacity, profits will grow continuously relatively to the national income, and capital formation will also grow relatively to the national income.

The model also covers the case of a technical revolution. Some historians have suggested that the capital for the British Industrial Revolution came out of profits made possible by a spate of inventions occurring together. This is extremely hard to fit into the neo-classical model, since it involves the assumption that these inventions raised the marginal productivity of capital more than they raised the marginal productivity of labour, a proposition which it is hard to establish in any economy where labour is scarce. (If we do not make this assumption, other incomes rise just as fast as profits, and investment does not increase relatively to national income.) On the other hand the suggestion fits beautifully into the modified classical model, since in this model practically the whole benefit of inventions goes into the surplus, and becomes available for further capital accumulation.

This model also helps us to face squarely the nature of the economic problem of backward countries. If we ask "why do they save so little," the truthful answer is not "because they are so poor," as we might be tempted to conclude from the path-breaking and praiseworthy correlations of Mr. Colin Clark. The truthful answer is "because their capitalist sector is so small" (remembering that "capitalist" here does not mean private capitalist, but would apply equally to state capitalist). If they had a larger capitalist sector, profits would be a greater part of their national income, and saving and investment would also be relatively larger. (The state capitalist can accumulate capital even faster than the private capitalist, since he can use for the purpose not only the profits of the capitalist sector, but also what he can force or tax out of the subsistence sector.)

Another point which we must note is that though the increase of the capitalist sector involves an increase in the inequality of incomes, as between capitalists and the rest, mere inequality of income is not enough to ensure a high level of saving. In point of fact the inequality of income is *greater* in over-populated under-developed countries than it is in advanced industrial nations, for the simple reason that agricultural rents are so high in the former. Eighteenth century British economists took it for granted that the landlord class is given to prodigal consumption rather than to productive investment, and this is certainly true of landlords in under-developed countries. Hence, given two countries of equal incomes, in which distribution is more unequal in one than in the other, savings may be greater where distribution is more equal if profits are higher relatively to rents. It is the inequality which goes with profits that favours capital formation, and not the inequality which goes with rents. Correspondingly, it is very hard to argue that these countries cannot afford to save more, when 40 per cent. or so of the national income is going to the top 10 per cent., and so much of rent incomes is squandered.

Behind this analysis also lies the sociological problem of the emergence of a capitalist class, that is to say of a group of men who think in terms of investing capital productively. The dominant classes in backward economies – landlords, traders money-lenders, priests, soldiers, princes – do not normally think in these terms. What causes a society to grow a capitalist class is a very difficult question, to which probably there is no general answer. Most countries seem to begin by importing their capitalists from abroad; and in these days many (*e.g.* U.S.S.R., India) are growing a class of state capitalists who, for political reasons of one sort or another, are determined to create capital rapidly on public account. As for indigenous private capitalists, their emergence is probably bound up with the emergence of new opportunities, especially something that widens the market, associated with some new technique which greatly increases the productivity of labour if labour and capital are used together. Once a capitalist sector has emerged, it is only a matter of time before it becomes sizeable. If very little technical progress is occurring, the surplus will grow only slowly. But if for one reason or another the opportunities for using capital productivity increase rapidly, the surplus will also grow rapidly, and the capitalist class with it.

[. . .]

'The Distribution of Gains between Investing and Borrowing Countries'[1]

American Economic Review (1950)

H. W. Singer

Editors' Introduction

Hans Singer was born in Germany in 1910 and died in England in 2006. He fled Hitler's Germany for the UK in 1933 and gained British citizenship in 1946. By that time he had completed a PhD with Maynard Keynes at Cambridge University and was about to begin work in the Economics Department of the United Nations, where he remained until 1969. Singer then joined the Institute of Development Studies at the University of Sussex. He was knighted in 1994 and won the Food for Life Award of the UN World Food Programme in 2001.

Hans Singer wrote widely on development issues, but he is best known for the so-called Singer–Prebisch hypothesis about the costs of international trade. (Singer and Prebisch did not collaborate and appear to have arrived at their conclusions independently.) The Singer–Prebisch hypothesis is often reduced to the insight that countries specialising in primary commodities would generally lose out in the medium and long term to countries able to produce manufactured goods: the net barter terms of trade would move against them. This apparent empirical finding was certainly radical, for it contradicted free trade nostrums based on static comparative advantage theories. It also led Singer to be a strong advocate of foreign aid programmes to the developing world. Developing countries needed financial and technical help to kick-start programmes of industrial development. This would allow them to process domestic foodstuffs and other primary commodities at home, thus adding value. What the reading excerpted here also demonstrates, however, is that Singer had a strong sense of the political structures which had first produced (and were now reproducing) a dependence on primary commodities in 'underdeveloped countries'. Singer pays a good deal of attention in the first part of his classic *American Economic Review* (1950) article to the historical legacies of colonialism, and to continuing foreign ownership of 'the productive facilities for producing export goods in underdeveloped countries'. Singer's sensitivity to questions of ownership and reward, as to power and politics more generally, marked him out as a classic heterodox economist. Throughout his life Singer pressed for greater political voice for the 'Third World'. This would be one means for helping to revamp an international economic system that tended otherwise not to greater equality of outcomes, but to greater inequalities. To this end, Singer spent a lot of his career working for, and helping to set up, a number of key UN bodies, including the United Nations Development Programme, the UN World Food Programme, the UN Industrial Development Organisation and the African Development Bank.

Key references

H. W. Singer (1950) 'The distribution of gains between investing and borrowing countries', *American Economic Review* 40 (2): 473–85.
— (1949) 'Economic progress in under-developed countries', *Social Research* 16: 236–66.
— (1964) *International Development, Growth and Change*, New York: McGraw-Hill.
— (1975) *The Strategy of International Development*, Basingstoke: Macmillan.
— (1998) *Growth, Trade and Development: Selected Essays of Han Singer*, Cheltenham: Edward Elgar.
D. Shaw (2002) *Sir Hans Singer: The Life and Work of a Development Economist*, Basingstoke: Macmillan.

International trade is of very considerable importance to underdeveloped countries, and the benefits which they derive from trade and any variations in their trade affect their national incomes very deeply. The opposite view, which is frequent among economists, namely, that trade is less important to the underdeveloped countries than it is to industrialized countries, may be said to derive from a logical confusion – very easy to slip into – between the absolute amount of foreign trade which is known to be an increasing function of national income, and the ratio of foreign trade to national income. Foreign trade tends to be proportionately most important when incomes are lowest. Secondly, fluctuations in the volume and value of foreign trade tend to be proportionately more violent in that of underdeveloped countries and therefore *a fortiori* also more important in relation to national income. Thirdly, and *a fortissimo*, fluctuations in foreign trade tend to be immensely more important for underdeveloped countries in relation to that small margin of income over subsistence needs which forms the source of capital formation, for which they often depend on export surpluses over consumption goods required from abroad.

In addition to the logical confusion mentioned above, the great importance of foreign trade to underdeveloped countries may also have been obscured by a second factor; namely, by the great discrepancy in the productivity of labor in the underdeveloped countries as between the industries and occupations catering for export and those catering for domestic production. The export industries in underdeveloped countries, whether they be metal mines, plantations, etc., are often highly capital-intensive industries supported by a great deal of imported foreign technology. By contrast, production for domestic use, specially of food and clothing, is often of a very primitive subsistence nature. Thus the economy of the underdeveloped countries often presents the spectacle of a dualistic economic structure: a high productivity sector producing for export coexisting with a low productivity sector producing for the domestic market. Hence employment statistics in underdeveloped countries do not adequately reflect the importance of foreign trade, since the productivity of each person employed in the export sector tends to be a multiple of that of each person employed in the domestic sector. Since, however, employment statistics for underdeveloped countries are notoriously easier to compile than national income statistics, it is again easy to slip, from the fact that the proportion of persons employed in export trade is often lower in underdeveloped countries than in industrialized countries, to the conclusion that foreign trade is less important to them. This conclusion is fallacious, since it implicitly assumes rough equivalence of productivity in the export and domestic sectors. This equivalence may be safely assumed in the industrialized countries but not in the underdeveloped countries.

A third factor which has contributed to the view that foreign trade is unimportant in underdeveloped countries is the indisputable fact that in many underdeveloped countries there are large self-contained groups which are outside the monetary economy altogether and are therefore not affected by any changes in foreign trade. In industrialized countries, by contrast, it is true that repercussions from changes in foreign trade are more widely spread; but they are also more thinly spread.[2]

The previously mentioned fact, namely, the higher productivity of the foreign trade sector in underdeveloped countries might, at first sight, be considered as a cogent argument in favor of the view that foreign trade has been particularly beneficial to underdeveloped countries in raising their general standards of productivity, changing their economies in the direction of a monetary economy, and spreading knowledge of more capital-intensive methods of production and modern technology. That, however, is much less clearly established than might be thought. The question of ownership as well as of opportunity costs enters at this point. The productive facilities for producing export goods in underdeveloped countries are often foreign owned as a result of previous investment in these countries. Again we must beware of hasty conclusions. Our first reaction would be to argue that this fact further enhances the importance and benefits of trade to underdeveloped countries since trade has also led to foreign investment in those countries and has promoted capital formation with its cumulative and multiplier effects. This is also how the matter is looked at in the economic textbooks – certainly those written by nonsocialist economists of the industrialized countries. That view, however, has never been really accepted by the more articulate economists in the underdeveloped countries themselves, not to mention popular opinion in those countries; and it seems to the present writer that there is much more in their view than is allowed for by the economic textbooks.

Can it be possible that we economists have become slaves to the geographers? Could it not be that in many cases the productive facilities for export from underdeveloped countries, which were so largely a result of foreign investment, never became a part of the internal economic structure of those underdeveloped countries themselves, except in the purely geographical and physical sense? Economically speaking, they were really an outpost of the economies of the more developed investing countries. The main secondary multiplier effects, which the textbooks tell us to expect from investment, took place not where the investment was physically or geographically located but (to the extent that the results of these investments returned directly home) they took place where the investment came from.[3] I would suggest that if the proper economic test of investment is the multiplier effect in the form of cumulative additions to income, employment, capital, technical knowledge, and growth of external economies, then a good deal of the investment in underdeveloped countries which we used to consider

as "foreign" should in fact be considered as domestic investment on the part of the industrialized countries.

Where the purpose and effect of the investments was to open up new sources of food for the people and for the machines of industrialized countries, we have strictly domestic investment in the relevant economic sense, although for reasons of physical geography, climate, etc., it had to be made overseas. Thus the fact that the opening up of underdeveloped countries for trade has led to or been made possible by foreign investment in those countries does not seem a generally valid proof that this combination has been of particular benefit to those countries. The very differential in productivity between the export sectors and the domestic sectors of the underdeveloped countries, which was previously mentioned as an indication of the importance of foreign trade to underdeveloped countries, is also itself an indication that the more productive export sectors – often foreign owned – have not become a real part of the economies of underdeveloped countries.

We may go even further. If we apply the principle of opportunity costs to the development of nations, the import of capital into underdeveloped countries for the purpose of making them into providers of food and raw materials for the industrialized countries may have been not only rather ineffective in giving them the normal benefits of investment and trade but may have been positively harmful. The tea plantations of Ceylon, the oil wells of Iran, the copper mines of Chile, and the cocoa industry of the Gold Coast may all be more productive than domestic agriculture in these countries; but they may well be less productive than domestic industries in those countries which might have developed if those countries had not become specialized to the degree in which they now are to the export of food and raw materials, thus providing the means of producing manufactured goods elsewhere with superior efficiency. Admittedly, it is a matter of speculation whether in the absence of such highly specialized "export" development, any other kind of development would have taken its place. But the possibility cannot be assumed away. Could it be that the export development has absorbed what little entrepreneurial initiative and domestic investment there was, and even tempted domestic savings abroad? We must compare, not what is with what was, but what is with what would have been otherwise – a tantalizingly inconclusive business. All we can say is that the process of traditional investment taken by itself seems to have been insufficient to initiate domestic development, unless it appeared in the form of migration of persons.

The principle of specialization along the lines of static comparative advantages has never been generally accepted in the underdeveloped countries, and not even generally intellectually accepted in the industrialized countries themselves. Again it is difficult not to feel that there is more to be said on the subject than most of the textbooks will admit. In the economic life of a country and in its economic history, a most important element is the mechanism by which "one thing leads to another," and the most important contribution of an industry is not its immediate product (as is perforce assumed by economists and statisticians) and not even its effects on other industries and immediate social benefits (thus far economists have been led by Marshall and Pigou to go) but perhaps even further its effect on the general level of education, skill, way of life, inventiveness, habits, store of technology, creation of new demand, etc. And this is perhaps precisely the reason why manufacturing industries are so universally desired by underdeveloped countries; namely, that they provide the growing points for increased technical knowledge, urban education, the dynamism and resilience that goes with urban civilization, as well as the direct Marshallian external economies. No doubt under different circumstances commerce, farming, and plantation agriculture have proved capable of being such "growing points," but manufacturing industry is unmatched in our present age.

By specializing on exports of food and raw materials and thus making the underdeveloped countries further contribute to the concentration of industry in the already industrialized countries, foreign trade and the foreign investment which went with it may have spread present static benefits fairly over both. It may have had very different effects if we think of it not from the point of view of static comparative advantages but of the flow of history of a country. Of this latter school of thought the "infant" argument for protection is but a sickly and often illegitimate offspring.

To summarize, then, the position reached thus far, the specialization of underdeveloped countries on export of food and raw materials to industrialized countries, largely as a result of investment by the latter, has been unfortunate for the underdeveloped

countries for two reasons: (*a*) because it removed most of the secondary and cumulative effects of investment from the country in which the investment took place to the investing country; and (*b*) because it diverted the underdeveloped countries into types of activity offering less scope for technical progress, internal and external economies taken by themselves, and withheld from the course of their economic history a central factor of dynamic radiation which has revolutionized society in the industrialized countries. But there is a third factor of perhaps even greater importance which has reduced the benefits to underdeveloped countries of foreign trade-*cum*-investment based on export specialization on food and raw materials. This third factor relates to terms of trade.

It is a matter of historical fact that ever since the seventies the trend of prices has been heavily against sellers of food and raw materials and in favor of the sellers of manufactured articles. The statistics are open to doubt and to objection in detail, but the general story which they tell is unmistakable.[4] What is the meaning of these changing price relations?

The possibility that these changing price relations simply reflect relative changes in the real costs of the manufactured exports of the industrialized countries to those of the food and primary materials of the underdeveloped countries can be dismissed. All the evidence is that productivity has increased if anything less fast in the production of food and raw materials, even in the industrialized countries[5] but most certainly in the underdeveloped countries, than has productivity in the manufacturing industries of the industrialized countries. The possibility that changing price relations could merely reflect relative trends in productivity may be considered as disposed of by the very fact that standards of living in industrialized countries (largely governed by productivity in manufacturing industries) have risen demonstrably faster than standards of living in underdeveloped countries (generally governed by productivity in agriculture and primary production) over the last sixty or seventy years. However important foreign trade may be to underdeveloped countries, if deteriorated terms of trade (from the point of view of the underdeveloped countries) reflected relative trends of productivity, this could most assuredly not have failed to show in relative levels of internal real incomes as well.

Dismissing, then, changes in productivity as a governing factor in changing terms of trade, the following explanation presents itself: the fruits of technical progress may be distributed either to producers (in the form of rising incomes) or to consumers (in the form of lower prices). In the case of manufactured commodities produced in more developed countries, the former method, i.e., distribution to producers through higher incomes, was much more important relatively to the second method, while the second method prevailed more in the case of food and raw material production in the underdeveloped countries. Generalizing, we may say that technical progress in manufacturing industries showed in a rise in incomes while technical progress in the production of food and raw materials in underdeveloped countries showed in a fall in prices. Now, in the general case, there is no reason why one or the other method should be generally preferable. There may, indeed, be different employment, monetary, or distributive effects of the two methods; but this is not a matter which concerns us in the present argument where we are not concerned with internal income distribution. In a closed economy the general body of producers and the general body of consumers can be considered as identical, and the two methods of distributing the fruits of technical progress appear merely as two formally different ways of increasing real incomes.

When we consider foreign trade, however, the position is fundamentally changed. The producers and the consumers can no longer be considered as the same body of people. The producers are at home; the consumers are abroad. Rising incomes of home producers to the extent that they are in excess of increased productivity are an absolute burden on the foreign consumer. Even if the rise in the income of home producers is offset by increases in productivity so that prices remain constant or even fall by less than the gain in productivity, this is still a relative burden on foreign consumers, in the sense that they lose part or all of the potential fruits of technical progress in the form of lower prices. On the other hand, where the fruits of technical progress are passed on by reduced prices, the foreign consumer benefits alongside with the home consumer. Nor can it be said, in view of the notorious inelasticity of demand for primary commodities, that the fall in their relative prices has been compensated by its total revenue effects.

Other factors have also contributed to the falling long-term trend of prices of primary products in terms of manufactures, apart from the absence of pressure of producers for higher incomes. Technical progress, while it operates unequivocally in favor of manufactures – since the rise in real incomes generates a more than proportionate increase in the demand for manufactures – has not the same effect on the demand for food and raw materials. In the case of food, demand is not very sensitive to rises in real income, and in the case of raw materials, technical progress in manufacturing actually largely consists of a reduction in the amount of raw materials used per unit of output, which may compensate or even overcompensate the increase in the volume of manufacturing output. This lack of an automatic multiplication in demand, coupled with the low price elasticity of demand for both raw materials and food, results in large price falls, not only cyclical but also structural.

Thus it may be said that foreign investment of the traditional type which sought its repayment in the direct stimulation of exports of primary commodities either to the investing country directly or indirectly through multilateral relations, had not only its beneficial cumulative effects in the investing country, but the people of the latter, in their capacity as consumers, also enjoyed the fruits of technical progress in the manufacture of primary commodities thus stimulated, and at the same time in their capacity as producers also enjoyed the fruits of technical progress in the production of manufactured commodities. The industrialized countries have had the best of both worlds, both as consumers of primary commodities and as producers of manufactured articles, whereas the underdeveloped countries had the worst of both worlds, as consumers of manufactures and as producers of raw materials. This perhaps is the legitimate germ of truth in the charge that foreign investment of the traditional type formed part of a system of "economic imperialism" and of "exploitation."

Even if we disregard the theory of deliberately sinister machinations, there may be legitimate grounds in the arguments set out above on which it could be maintained that the benefits of foreign trade and investment have not been equally shared between the two groups of countries. The capital-exporting countries have received their repayment many times over in the following five forms: (*a*) possibility of building up exports of manufactures and thus transferring their population from low-productivity occupations to high-productivity occupations; (*b*) enjoyment of the internal economies of expanded manufacturing industries; (*c*) enjoyment of the general dynamic impulse radiating from industries in a progressive society; (*d*) enjoyment of the fruits of technical progress in primary production as main consumers of primary commodities; (*e*) enjoyment of a contribution from foreign consumers of manufactured articles, representing as it were their contribution to the rising incomes of the producers of manufactured articles.

By contrast, what the underdeveloped countries have to show cannot compare with this formidable list of benefits derived by the industrialized countries from the traditional trading-*cum*-investment system. Perhaps the widespread though inarticulate feeling in the underdeveloped countries that the dice have been loaded against them was not so devoid of foundation after all as the pure theory of exchange might have led one to believe.

[. . .]

It may be useful, rather than end on a wild historical speculation, to summarize the type of economic measures and economic policies which would result from the analysis presented in this paper. The first conclusion would be that in the interest of the underdeveloped countries, of world national income, and perhaps ultimately of the industrialized countries themselves, the purposes of foreign investment and foreign trade ought perhaps to be redefined as producing gradual changes in the structure of comparative advantages and of the comparative endowment of the different countries rather than to develop a world trading system based on existing comparative advantages and existing distribution of endowments. This perhaps is the real significance of the present movement towards giving technical assistance to underdeveloped countries not necessarily linked with actual trade or investment. The emphasis on technical assistance may be interpreted as a recognition that the present structure of comparative advantages and endowments is not such that it should be considered as a permanent basis for a future international division of labor.

Insofar as the underdeveloped countries continue to be the source of food and primary materials and insofar as trade, investment, and technical assistance are working in that direction by expanding primary production, the main requirement of underdeveloped countries would seem to be to provide for some method of income absorption to ensure that the results of technical progress are retained in the underdeveloped countries in a manner analogous to what occurs in the industrialized countries. Perhaps the most important measure required in this field is the reinvestment of profits in the underdeveloped countries themselves, or else the absorption of profits by fiscal measures and their utilization for the finance of economic development, and the absorption of rising productivity in primary production in rising real wages and other real incomes, provided that the increment is utilized for an increase in domestic savings and the growth of markets of a kind suitable for the development of domestic industries. Perhaps this last argument, namely, the necessity of some form of domestic absorption of the fruits of technical progress in primary production, provides the rationale for the concern which the underdeveloped countries show for the introduction of progressive social legislation. Higher standards of wages and social welfare, however, are not a highly commendable cure for bad terms of trade, except where the increment leads to domestic savings and investment. Where higher wages and social services are prematurely introduced and indiscriminately applied to export and domestic industries, they may in the end turn out a retarding factor in economic development and undermine the international bargaining strength of the primary producers. Absorption of the fruits of technical progress in primary production is not enough; what is wanted is absorption for reinvestment.

Finally, the argument put forward in this paper would point the lesson that a flow of international investment into the underdeveloped countries will contribute to their economic development only if it is absorbed into their economic system; i.e., if a good deal of complementary domestic investment is generated and the requisite domestic resources are found.

NOTES

1   The author wishes to acknowledge help and advice received from many friends and colleagues; in particular Mr. Henry G. Aubrey, Dr. Harold Barger, of the National Bureau of Economic Research, Dr. Roberto de Oliveira Campos, of the Brazilian Delegation to the United Nations, Dr. A. G. B. Fisher, of the International Monetary Fund, Professor W. Arthur Lewis, of the University of Manchester (England), and Mr. James Kenny. He also had the inestimable advantage of a discussion of the

subject matter of this paper in the Graduate Seminar at Harvard University, with Professors Haberler, Harris, and others participating.

2   A more statistical factor might be mentioned. Some underdeveloped countries – Iran would be an illustration – exclude important parts of their exports and imports from their foreign trade statistics insofar as the transactions of foreign companies operating in the underdeveloped country are concerned. This is a tangible recognition of the fact that these pieces of foreign investments and their doings are not an integral part of the underdeveloped economy.

3   Often underdeveloped countries had the chance, by the judicious use of royalties or other income from foreign investment, to use them for the transformation of their internal economic structure – a chance more often missed than caught by the forelock!

4   Reference may be made here to the publication by the Economic Affairs Department of the United Nations on "Relative Prices of Exports and Imports of Underdeveloped Countries."

5   According to U.S. data of the WPA research project, output per wage earner in a sample of 54 manufacturing industries increased by 57 per cent during the twenty years, 1919–39; over the same period, agriculture increased only by 23 per cent, anthracite coal mining by 15 per cent, and bituminous coal mining by 35 per cent. In the various fields of mineral mining, however, progress was as fast as in manufacturing. According to data of the National Bureau of Economic Research, the rate of increase in output per worker was 1.8 per cent p.a. in manufacturing industries (1899–1939) but only 1.6 per cent in agriculture (1890–1940) and in mining, excluding petroleum (1902–39). In petroleum production, however, it was faster than in manufacturing.

'Socialism and the Soviet Experience'

from *The Economics of Feasible Socialism* (1983)

Alec Nove

Editors' Introduction

Alexander (Alec) Nove was born in Petrograd, Russia, in 1915 but was moved to London by his father, a prominent Menshevik, in 1923. Nove graduated from the London School of Economics in 1936 and worked for British Intelligence during the Second World War. From 1948 to 1958 Nove again worked for the government, this time for the Board of Trade and the Foreign Office, before taking up a position at the LSE as a Reader. The most productive part of Nove's academic career was played out at the University of Glasgow where in 1963 he was elected Director of its Institute of Soviet and East European Studies. Nove now produced a string of books on the Soviet economy and on Stalinism, one of the most influential of which he produced as an Emeritus Professor, *The Economics of Feasible Socialism*. He died in 1994.

It is important to note that Nove was not a hard-line anti-communist. Far from it: Nove was a left-leaning intellectual who nonetheless recognised, as many of his political colleagues in the 1970s would not, that the Soviet economy was hugely inefficient, in part because it was impossible for managers to measure costs and outputs with any accuracy. In Nove's view, as he explains in this reading, the over-planned and yet badly organised Soviet economy was driven to stop-go cycles and 'Soviet-style inflation'. Overambitious investment plans were also enforced by political clout. The result in the Soviet Union was an unhappy combination of centralised planning and centralised despotism. In this respect, Nove's politics came close to those of Orwell and Koestler. The Soviet Union could only be reformed by concerted efforts at political and economic decentralisation. This would be the 'feasible socialism' of the book's title, and which Nove later celebrated in President Gorbachev's policies of *glasnost* and *perestroika*. Whether Nove thought these policies could be reproduced in the socialist developing world is not clear, although he did visit China in the 1980s and dispensed advice about the dangers of 'partial liberalisation' or market pricing. By the mid-1980s, in any case, the high-watermark of socialism in the 'Third World' had passed. With it went many 1970s-style demands for a more egalitarian (new) international economic order: an order that Nove had once been keen to endorse.

Key references

Alec Nove (1983) *The Economics of Feasible Socialism*, London: George Allen and Unwin.
Alec Nove with Ian Thatcher (1994) *Markets and Socialism*, Aldershot: Edward Elgar.
Alec Nove with Ian Thatcher (1998) *Alec Nove on Communist and Post-Communist Countries*, Aldershot: Edward Elgar.
Alec Nove and D. Muti (eds) (1972) *The Economics of Socialism*, Harmondsworth: Penguin.
R. Bahro (1983) *From Red to Green*, London: New Left Books.

[. . .]

Some of the deficiencies of the Soviet model relate directly to centralised 'non-market' planning, and its many consequences. If the management has no basis for determining what should be produced – and land can produce many different things – then the planners must. Yet the very variety of types of land, and ways in which it could be used, virtually ensure that the plan orders will frequently be unsuitable for the specific conditions of the given locality. Similarly, farms cannot just purchase their inputs, they must apply for them and take what is allocated, which once again leads to predictable errors and omissions. The large size and complexity of farms is due partly to ideological preference, but also to the desire of the overworked planners to reduce the number of the units to which they have to issue instructions. Remedies, to be effective, must surely enlarge the decision-making functions of those on the spot, both farm management and the sub-units within the farm, to reduce the size of the giant farms, to stimulate peasant interest in the outcome of the work they do, and to allow the farms to purchase the inputs they need, as well as the provision of the required infrastructure. It is interesting in this connection to study the much more successful Hungarian experience. The key contrasts are, first, the absence of compulsory deliveries, and therefore the much greater freedom of farms to choose what to produce and what to sell; secondly, almost total freedom to purchase inputs; thirdly, much greater flexibility in internal organisation and in organising the work of peasant members; fourthly, the successful incorporation of the private sector in the production of food, with adequate supplies of fodder, equipment, well-organised marketing facilities, and so on. In my view none of this is inconsistent with a properly understood socialism, but this is not how Soviet (and many 'new left') ideologists see things.

INVESTMENT DECISIONS AND CRITERIA IN THEORY AND PRACTICE

These have several aspects. There is the volume of investments, the share of accumulation in the national income in relation to current consumption.

Then there is the complex question of investment criteria: how to judge which investments are appropriate, rational, efficient. Also of great importance are: who decides what, by reference to whom, and what can and should be the role of the democratic process and/or of interest and pressure groups of various kinds. Finally, there are problems of planning and implementation: how to ensure that the necessary means are available to carry out the intended programme, and then ensure that this (and no other) programme is in fact carried out.

How much should be invested? What balance should be struck between present and future needs? It is only in the more formalistic Western textbooks that this question is 'resolved' by phrases about 'the time preference of the community', since in practice personal savings play a subordinate role as compared with the savings of institutions and the reinvested profits of firms. In any event, in a socialist society a major part of investments must clearly be the responsibility of the planning organs, based on a political decision, taken in the last resort by the political authorities. There is no 'scientific' basis for determining the scale of capital accumulation. Should net investment be 10 per cent, 15 per cent, 25 per cent, or the national income? It must depend on the level of development already achieved, and the perceived urgency of expanding productive capacity, social overhead capital, and so on. The role of the democratic process, in determining how much to invest and in determining broad priorities, seems both important and desirable. Indeed this is fully recognised in the USSR on a formal level: the five-year plan and the annual plan, which incorporates investments, are voted on by the elected Supreme Soviet. The snag, of course, is that the vote is always a unanimous formality, as indeed are the 'elections'.

No historian of economic thought should overlook the remarkable pioneering ideas of the economists of the 1920s on development strategy and investment criteria. Some were political figures, such as Preobrazhensky, others were non-Bolshevik specialists, of which Bazarov was a distinguished example. Indeed an early contributor to the development strategy debate was an anti-Bolshevik, writing in an area occupied by the Whites during the civil war,

Grinevetsky, and Lenin knew of and used his ideas when discussing plans for 'the electrification of Russia' as early as in 1920. War and civil war had interrupted Russia's industrialisation. The triumph of the Bolsheviks had led to the elimination of capitalists and landlords, and therefore meant that savings and investments were to be primarily the direct responsibility of the state. Thus, at a time when Western economics was uninterested in growth and development, Russian thinkers of many political hues or none had to blaze their own trail. The debates coincided with a period in which small-scale private enterprise coexisted with the state sector, and one aspect of the discussions centred on how (and whether) this coexistence could continue – how the resources needed for investment could be 'pumped over' from the private sector, that is, primarily from the peasantry. I can do no more here than underline the importance of the arguments of the period and refer interested readers to the literature.[1]

The dissolution of the so-called New Economic Policy, and the elimination of the private sector (and of nearly all the most talented economists), followed (and was partly caused by) the increase in state investments which was taking place in 1925–7, even before the adoption of the first five-year plan. Even this comparatively modest increase – modest in comparison with what was to follow – caused the gravest strains and shortages and proved incompatible with market equilibrium in a mixed economy, that is, with the principles of NEP. There then occurred, under Stalin's first five-year plan, an investment boom of unprecedented dimensions, which caused massive disproportions and a steep decline in living standards. This tendency to over-invest has been repeated in other communist-ruled countries, and has led to political and economic crises, and also the 'socialist' equivalent of trade cycles, as well as contributing to the chronic sellers' market and shortages which were discussed earlier.

Under Stalin the balance between current needs and investment in the future was tilted very strongly in favour of the latter. In view of the hardships it can cause to the present generation, this strategy is closely connected with the elimination of democratic institutions and the emasculation of the trade unions. It may be argued that Stalin in 1930 had very urgent reasons to press ahead with a huge investment programme with priority for heavy industry, owing to the sense of isolation and external threat. But the same

excesses were committed afterwards, in Poland, Romania, Hungary, Czechoslovakia, not once but repeatedly. Poland's investment boom of 1971–5 led to economic catastrophe and political upheaval. The causes of the Hungarian crisis of 1956 were also connected with the effect of the investment programme on living standards. Much greater sufferings, including real famine, occurred in 1932–3 in the USSR, due partly to the crazy investment tempos and partly to the effects of collectivisation of agriculture; but there the police and the terror were sufficient to maintain order.

What drives the political leadership to over-invest? Various theories have been put forward, some of a psychological nature. A Soviet critic once asserted that Stalin wished the great works of Stalin to be visible from the planet Mars – a kind of pyramids complex. While this is of doubtful validity, it is by no means incorrect to note the predisposition of the leadership to build for the future; indeed, the official aim of 'building communism', requiring, according to orthodox doctrine, a great enlargement of productive capacity, does point to a preference for tomorrow over today. But if this helps to explain the desire for new factories and mines, it cannot be the explanation for the large-scale public buildings, a preference which reminds one of ruling classes of pre-capitalist times.

The large and spectacular has taken precedence over the small, and there was (and is) persistent undervaluation of routine maintenance. Here too one notes the strong desire for 'showmanship', what the Russians frequently refer to as *pokazukha*. This can be seen as an aspect of the success indicator problem, success in this instance meaning an activity which is big, and visible from above. This, and not only a belief in economies of scale, can account for the many instances of 'gigantomania'. The average size of Soviet enterprises is in fact very much larger than is the case in the West, where alongside the giants there are many thousands of small firms. Another reason for the reluctance to set up small units is that it complicates the process of planning: orders have to be issued to larger numbers of managers, so that it seems simpler to have large factories and/or to merge small ones. Yet, as has been pointed out, for instance, by the Soviet economist Kvasha, there is a clear advantage in certain specialised tasks being carried out by small productive units. This is apart from the argument put forward by André Gorz, to the effect that large-scale industry is inherently alienating, that 'small is beautiful', a theme to which it will be necessary to return.

Gigantomania was (and to some extent still is) accompanied by neglect of minor but essential investments, for instance, the persistent lack of what the Russians call *malaya mekhanizatsiya*: auxiliary tasks, such as loading, unloading, repairs, materials handling, storage, are seldom adequately mechanised, and this causes much waste of labour. Prosaic but useful techniques may be neglected for what appear to be irrationally trivial reasons: a Soviet newspaper once reported that an inventor who claimed to have devised an efficient button-holing machine for the clothing industry was met with the reply: 'In an age of Sputniks, you come along with a button-holing machine!'

This brings the discussion closer to the issue of investment criteria. A Western profit-making firm will decide to develop or use whatever it believes to be profitable, whether it is a button-holing machine, a gramophone turn-table or a machine for putting tops on bottles. In the USSR the preference for the grandiose and neglect of 'minor' matters spring from lack of clear criteria for decision-making. Space forbids even a brief analysis of the huge Soviet and non-Soviet literature on investment criteria in the USSR, and the many discussions on the subject. After decades in which the purest 'voluntarism', a 'refusal to calculate', to use the words of Jean-Michel Collette, seemed to prevail, practical planners found that choices had none the less to be made, and that in a choice between techniques the one with the shortest pay-off period should be chosen, other things being equal. It was argued that there should be a cut-off point, beyond which the proposed investment should be adjudged inefficacious. However, several points of great importance remained in dispute, and for good reasons. One was the question of whether the chosen criteria should be identical throughout the economy, or vary by sectors. This was essentially a matter of how to cope with priorities. This in turn had two sub-aspects. One was that *past* priorities had caused great unevenness between the profitabilities (rates of return, pay-off periods) of investments in different industries; to redistribute investment in accordance with a single criterion would cause major imbalances. The second was how to take *present* priorities into account. Example: does the urgent need for energy imply that a lower rate of return should be adopted

for energy, or should the criterion be formally identical and a case made for exceptional treatment for energy? The practical outcome in the two formulations might be the same, but there is a point of principle involved: should objective criteria be recognised, or should one alter the criteria in line with the chosen policy?

Reference to energy should remind us that Western theories on investment criteria are neither satisfactory nor complete. If the world is really to face a shortage of energy towards the end of the 1980s, this in itself constitutes a powerful argument for more investment in this sector, whatever may be the current rate of interest and the actual rate of return on investments in energy at present prices. Of course, we all *ought* to be using the 'shadow' prices which reflect the probable scarcities of seven and more years hence, but these in turn are a consequence of (fallible) calculation as to future supply-and-demand conditions. Calculations about future shortages (input-output, material balances) need not be confined to energy. In so far as, in their major investment decisions, Soviet planners reflect not only general policy priorities but also anticipated shortages (of anything from sheet metal to sports shoes), they should not be regarded as 'irrational' if, in a conflict between rate-of-return and material-balance calculations, they give precedence to the latter.

The above might seem to downgrade, or even ignore, the role of prices. This is not my intention. One must agree with Kornai that, while price information is indeed essential, decisions are not taken on the basis of price information alone (especially bearing in mind the vital role in any major investment decision of uncertainty about the future). To cite Kornai again, while it makes sense to say that if the price of a certain type of cotton cloth has risen, more should be invested in its production, no one is likely to say, in East or West, 'The price of electricity has fallen, let us therefore invest less in power stations'.

Accepting this, it is still necessary to stress the waste occasioned by the nature of Soviet prices. Since – as will be shown – they do not reflect relative scarcities of means, or use-value, or demand, it is clear that all calculations aimed at minimising cost, or maximising effect, are doomed from the start, since the monetary units used in the calculation fail to measure what they ought to be measuring. All this is quite apart from the presence or absence of a rate of interest, and of the rate chosen.

Many a socialist critical of Soviet practice, from Trotsky to Bettelheim, has rightly criticised arbitrariness, has urged the need for rational calculation. But how is one to judge whether a decision is arbitrary or irrational save by reference to some objective criteria? In a choice as to whether to invest in A, B, or C, or whether to produce more of A by methods X, Y, or Z, how can the calculation be made other than in some kind of meaningful prices? How can one avoid having to take time into account too? Evidently, if a project takes two years longer, but economises in the use of some scarce material, there must be some means of comparing the advantages and disadvantages. But the relevance of Soviet experience in the area of prices raises many more issues, and must be put aside for the moment.

Let us pass on to the question of *who* decides what to invest in. One element of Soviet experience must be mentioned first, since it is often overlooked. It may be best illustrated by quoting a conversation with a Soviet specialist, who said: 'There is a large literature on investment choice, but often there is in fact no choice, as the project-makers have only drafted one project.' Other economists, for example, Krasovsky, have strongly criticised the large and bureaucratic *proyektnye organizatsii*.[2] Design bureaux, quantity surveyors, engineers capable of selecting appropriate machines and other such specialists are an important part of the investment process. Whoever chooses can only choose between alternatives that are elaborated by specialists. These also require criteria. If the project-makers have to fulfil plans expressed in millions of roublesworth of projects, this incites them to opt for expensive variants.[3]

The place at which decisions are taken must naturally depend on the relative magnitude and importance of the proposed investment. It may be the top leadership (say, in a decision to develop oil and gas in north-west Siberia, or to launch the virgin lands campaign), or a minister, or more junior officials at the centre or in a locality. The interests and motives of ministerial and other officials below the topmost levels affect the nature of proposals made, the total number of projects started, the speed with which they are completed and much else besides. Here, as in many other instances, Soviet experience underlines the extent to which, in a nominally highly centralised and hierarchically organised system, proposals made and information supplied from below, and the implementation of orders given, are all substantially

affected by subordinates; but also that, given the criteria by which they are judged, the actions of subordinates can have deleterious effects.

One of these is an aspect of the problem of externalities, already discussed. Anyone below the centre can see only a part of the whole picture. A ministry faced with the task of increasing output of, say, farm machinery will tend to save expense and trouble by placing its manufacture in an already-developed area, where the social overhead capital already exists, which may well conflict with the objectives of regional policy and may cause bottlenecks for transport or excess demand for local labour.

A different problem arises out of the commitment of subordinates to large investment programmes of their own. The motives vary. One is the familiar one of 'empire-building'; a party secretary, a minister, a manager, gains in status (and in some cases also salary) if his bailiwick expands. But one need not assume selfishness. As argued in Part 1, people identify sincerely with their area of responsibility: whether it is fertiliser, housing in Tomsk, medical services, artillery, washing-machines, or the development of minerals in Yakutia, genuine reasons exist for more investment. Those in charge of the activities concerned will press for an allocation, authorisation, credits. It is important to note that there is no real cost involved: the 'price' of the investment for the recipient is all but zero.

This is obvious in those instances in which the investment is financed by the centre, by outright grant. It is also the case if the retained profits out of which the investment is financed would otherwise have to be transferred to the state budget. But even if there is a capital charge, or the investment requires an interest-bearing credit from the state bank, this does little to discourage over-application for investment resources. Among several reasons for this, two stand out. The first is what Kornai has called the soft budget constraint: in the end deficits will be covered and no one will be allowed to go bankrupt. The other is that in the end it is not clear who is to be held responsible if the given investment proves to be an error; by the time this becomes known there would probably be a new director. The tendency to over-invest therefore arises *both* because of the centre's own ambitious growth targets *and* because of the efforts of subordinate units to expand, of local bodies to undertake public works. It is, of course, the task of the supreme planning organs to prevent this, to ensure that there is balance between the investment programme and the resources necessary to complete it. But this seldom happens.

In virtually every year since 1930 a Soviet leader has deplored what is called *raspylenie sredstv*, the 'scattering' of investment resources among too many projects. Measures are taken to prevent this, to concentrate on completing what is already started, but the ineffectiveness of these measures is attested by the fact that they have to be repeated, while the percentage of uncompleted investments rises. The result is very long delay in commissioning new factories, houses, and so on. Pressure from above sometimes has paradoxical or even comic results: *Pravda* reported in 1980 on projects certified to be operational but which were not in fact completed, and the excuse given by local officials was that they had strict orders to reduce the percentage of uncompleted construction! It is plainly in the interest of sectoral and local officials to start as much as possible, and to try to divert resources to projects of particular interest to them, and it is equally plain that the central coordinating power is unable to combat this tendency effectively. As is so often the case, the centre is able to ensure that a few key activities are given priority, but cannot cope with the task of controlling everything.[4]

Mention has already been made of fluctuations, 'trade cycles', these being closely connected with over-investment and its consequences. In the smaller countries, heavily dependent on foreign trade, the balance-of-payments constraint (excessive deficits, debt, etc.) compels a downturn in investment, but in due course the upward movement is resumed, with (apparently) no lesson learned from the previous crisis. The case of Poland is a particularly striking instance, and will be discussed in greater detail in Part 3.

In the USSR such a 'classic' cycle-'stop' occurred in 1932–3, but in more recent years it has been more a case of chronic shortage (and construction delays) than a pronounced cyclical trend, at least at macro level. It is quite another matter if one looks at sectors. There one sees grossly excessive investment upswings, for instance, in the chemical industry in the 1958–65 period. These arise for a reason which it is important to analyse, however briefly: the fact that, in a centralised planning system, any important change of structure requires a decision at the very top, but the need for such a decision will only be perceived (by

the very busy men at the summit) if the need is very urgent indeed. Until that moment, working with material-balances or input-output tables, the planning officials naturally tend to preserve the existing pattern, partly because input-output tables are of their nature conservative, reflecting the past, and partly because these officials do not have the authority to make structural changes, which inevitably affect vested interests: one needs a decision at the summit. Let us take the Soviet chemical industry as an example. By the mid-1950s it was woefully backward: production of fertilisers, plastics and detergents was very far below need, and below Western levels. So Khrushchev decided to launch a great campaign to catch up. Over-investment occurred on a massive scale: plans surpassed the possibilities of training personnel, designing (or importing) and installing the machinery, and so on.

It should therefore be a matter of concern to socialist economists that necessary structural changes should not be delayed by the degree of centralisation. One must also assess realistically the time available at the summit to digest information and to take decisions, and then to ensure that they are in fact implemented.

It must also be recognised that in any realistically envisageable situation there will be many conflicting interests involved, each trying to put the best possible case for a larger share of the investment 'cake'. The Soviet economist Maiminas has suggested that one way of limiting the distortions that this can cause is to devise a procedure by which those who draft variants of a project must be distinct from those who select, and a separate body then authorises the necessary financing, all to ensure that information flows and decision-making are kept as clear as possible of prejudice due to interest.[5]

It is also important to appreciate that, in the competition for resources, victory is likely to go to the sector which has the biggest political 'pull', the best hierarchical connections, the most senior and influential boss. As already argued, whoever is in charge of toothbrushes (or an economically backward province, say, Tambov) is bound to have less 'pull' than the official in charge of metallurgy, or of Sverdlovsk province. So instead of bias towards anticipated profitability one has a bias in favour of the already big and powerful, and this helps to explain the chronic under-investment in consumers' goods and services, as well as in 'small' equipment of many kinds (*malaya mekhanizatsiya*), and this despite repeated attempts by the top leadership to redress the balance.

Finally, the tendency to adopt over-ambitious investment plans is an important cause of Soviet-type 'inflation'. Planned demand for producers' goods (building materials, machines and also labour) exceeds actual supply. This would cause prices to rise if they were free to do so. Since they are not, there is shortage, and this gives rise not only to the delays already mentioned but to semi-legal forms of barter, bribery and other unplanned or downright criminal phenomena.

On the positive side, we must note the fact that centralised planning does give to those who plan major structurally significant investments an over-view of the whole economy, and an insight into its probable future needs, which are a source of potential strength, especially if seen against the background of the plain inadequacies of Western theory and practice.

[. . .]

NOTES

1 J.-M. Collette, *Critères d'investissements et calcul économique* (Paris: Cujas, 1964); A. Erlich, *The Soviet Industrialisation Debate* (Cambridge, Mass.: Harvard University Press, 1960); N. Spulber, *Soviet Strategy for Economic Growth* (Bloomington, Ind.: Indiana University Press, 1964); A. Nove, *An Economic History of the USSR* (London: Allen Lane, 1969); E. Preobrazhensky, in *The Crisis of Soviet Industrialisation*, ed. D. Filtzer (White Plains, NY: Sharpe, 1979); and E. H. Carr and R. W. Davies, *Foundations of the Planned Economy* (London: Macmillan, 1969).

2 V. Krasovsky, in *EKO*, no. 1, 1975, pp. 18–19.

3 They were again criticised more recently, on similar lines; see V. Krasovsky in *Voprosy ekonomiki*, no. 1, 1980, p. 110.

4 For a good account of this trend, see G. Markus, 'Planning the crisis: some remarks on the economic system of Soviet-type societies', *Praxis International*, no. 3, 1981.

5 E. Maiminas, *Protsessy planirovaniya v ekonomike*, 2nd edn (Moscow, 1971).

'Man and Nature in China'

Modern Asian Studies (1967)

Rhoads Murphey

Editors' Introduction

Rhoads Murphey was born in 1923. He grew up in Philadelphia where he attended a Quaker school. He later became a distinguished Professor of History and Director of the Asian Studies Programme at the University of Michigan. To say this, however, doesn't begin to convey the extraordinary life that Murphey led in the middle part of the twentieth century. From 1942 to 1946, Murphey was a conscientious objector working for the Friends Ambulance Unit in China, where he helped deliver medical supplies to people in the south-west of the country. During this period, Murphey was able to observe both the Nationalist and Communist armies in China, and got to know several prominent Communist leaders, including Madame Mao and Chou En-Lai.

Murphey's growing expertise on East Asian affairs would later encourage him to write widely about the region, including in the form of undergraduate textbooks. What is remarkable about the article we excerpt from here, however, is its extraordinary sensitivity to the violence of the Maoist project in the 1960s, at least in terms of its ideology of 'man against nature'. The 1950s and 1960s were decades characterised by a more general sense that 'man' might overcome nature. Deserts in Israel and California could be made to bloom, just as malaria could be wiped out, or so it was thought, by DDT. (See Rachel Carson in Part 5 for a contrary view.) But what Murphey shows us here is how far the Chinese Communist Party pushed this ideology. Nature becomes an enemy. Marxism, or Maoism, is conceived in hyper-transformational terms. The aims of the revolution and a perfect society are not to be blocked by internal or external nature. Just as human beings can be made less selfish and more cooperative, so too can mountains be levelled to create new fields for agriculture. All that is required is a body of cheerful conscripts armed with 'the thoughts of Chairman Mao'. And what Murphey also shows us is how quickly this ideology could tip into disaster. Not only does Murphey raise the spectre of some form of environmental backlash, he also notes how the cult of Mao was working to centralise power in China. Reason was undercut and with it empirical experimentation. In James Scott's terms (see Part 6), this was a classic and disturbing case of high modernist development.

Key references

Rhoads Murphey (1967) 'Man and nature in China', *Modern Asian Studies* 1: 313–33.
— (1961) *An Introduction to Geography*, New York: Rand McNally.
— (1980) *The Fading of the Maoist Vision: City and Country in China's Development*, London: Routledge.
K. Buchanan (1970) *The Transformation of the Chinese Earth*, London: G. Bell and Sons.
P. Wheatley (1971) *The Pivot of the Four Quarters*, Chicago, IL: Aldine.

[. . .]

1949 has brought an abrupt about-face.[1] With the dialectic installed as the pre-eminent guide, conflict, contradiction, and struggle are seen not only as the proper condition of human society but as its most important dynamics, the chief means to the progress in which China now fervently, almost mystically, believes, as the West has begun seriously to question it with the collapse of the splendid self-confidence of the nineteenth century under the varied blows of the twentieth. Traditional Chinese patterns of compromise, adjustment, and harmony are specifically attacked. Work has become a good in itself, needing no reward beyond increased production – for someone else's brighter tomorrow. Leisure is condemned as counter-revolutionary sloth. The past is invariably 'the bad old days', and its models are rejected. Nature is dethroned, and man – a confident, pioneering, creating, Communist man – put in its place. 'It is man that counts; the subjective initiative of the masses is a mightly driving force.' And 'It is always the newcomers who outstrip the old'.[2]

Nature is no longer to be accepted but must be 'defied' and 'conquered', words used continually in the contemporary literature to describe the efforts to increase agricultural production and to manipulate the environment. Nature is explicitly seen as an enemy, against which man must fight an unending war, with more conviction and fervour and with a brighter vision of the ultimate results than even the Darwinian-Spencerian West held. Articles in the contemporary press glorify this campaign, with titles such as: 'The Desert Surrenders', 'We Bend Nature to Our Will', 'How We Defeated Nature's Worst', 'Chairman Mao's Thoughts Are our Guide to Scoring Victories in the Struggle Against Nature', 'Hard Work Conquers Nature', and 'The United Will of the People can Transform Nature'. The environment is often personalized, through being seen as an enemy: the 'evil' stream or river which spawns destructive floods, 'the drought devil', or the 'bad old mountain' which stands in the way of a new irrigation channel. Successes are referred to as 'victory reports', and the local secretaries of the Party are said to 'lead the people into battle'. Mountains, the least humanized and most resistant elements in the landscape, have become the leading symbol of hostile nature. The mountains

must be levelled – and in many cases this has literally been done, largely by unaided mass labour. The mountain is the 'contradiction' which Mao urges man to struggle against 'with revolutionary spirit'. 'How the Foolish Old Man Removed the Mountains', an article written by Mao in 1945, is repeatedly quoted. It re-tells the story of Yu Kung, a legendary ancient, who with his two sons and armed only with mattocks levelled two big mountains which blocked their communication with the outside world, despite the scorn of neighbours who said they could never do it because they dug away so little each day. Persistence and hard work pay, and in a socialist China with the boundless power of the people liberated and mobilized, even greater mountains can and must be moved, by their own labour alone. Mao has of course selected a traditional story which suggests the rightness of man's attacking his environment – traditional views were by no means uniform – but has also altered it by removing the supernatural aid from 'heaven' which concluded the original story; for Communist man, the dialectic is enough support.

The transformation of nature is of course a theme familiar from Soviet experience, as is the denial of environmental limits on what man can do.[3] In the Chinese case, however, these convictions seem not only more extreme, held with greater fervour, and acted on with greater dedication and effect, but represent such striking reversals of the long-established and characteristically Chinese traditional view of nature that they call out for notice. In contemporary China, the environment is not ignored as a factor, as in the Soviet Union, where its transformation is left in a few selected spots to huge projects of the central planners; instead, its destructive power and the problems it poses are highlighted.[4] It is directly challenged as an almost living opponent with which every villager must do battle. Part of the reason may be the burning sense of urgency which Communist rule has imposed or infused. The Communists explicitly acknowledge the technological and industrial gap between China and the West (including, as the Chinese do, the Soviet Union as a Western power). Given the ardent national pride and Sinocentric world view which have persisted as strong contemporary characteristics, the closing of that gap is seen as a supreme national task, requiring not only universal mobilization of energies on a gigantic scale, but also a messianic commitment to the idea of progress, and a conviction that it can be achieved by direct attacks, with hands still largely bare, on all of the obstacles which stand in its way. Nature is a convenient target, and can be made an exciting and even inspiring one; its transformation is good Marxist doctrine and offers an uplifting prospect which can be made to capture imagination. The obstacles of a 'degenerate feudal order' and of exploitative colonialism have been destroyed through the leadership of the Party, but the environment remains a roadblock to the revolutionary drive of a united people towards the re-establishment of China's proper place, at the zenith of human achievement. This cannot await the still distant availability of adequate capital investment or machine technology: China is in too great a hurry. Meanwhile it is important not to let these undoubted handicaps lead to discouragement or apathy. The new strength and confidence imputed to a 'liberated people' by the Chinese press is alleged to be so great that it is not only able to move mountains but can scarcely be restrained from doing so.

The rural population may have limited power to improve its immediate economic lot for a long time to come; more capital is extracted from agriculture than is invested in it, so that industrialization may be accelerated. All the more reason therefore to divert attention from circumstances which could yield undesirable sentiments, by preaching commitment to a holy war in which all peasants must be dramatically involved. This is clearly preferable to inaction or to resignation. The emotional crusade against the environment not only gives the peasant something constructive to do but aims to persuade him that he actually can improve his lot by his own efforts in a situation which could otherwise produce despair. Whether nature is significantly or beneficially transformed or not, peasant morale, virtually equivalent to national morale, is bolstered, and there are in fact innumerable concrete accomplishments to point to. Their price, from a cost-accounting point of view, is less important and, given the marked seasonal variation in agricultural employment, the labour which they absorb had limited alternative uses.

As an illustration of what simple determination can accomplish, the story of the villagers of Tachai, in Shansi province, is told and re-told in the press. Tachai was a small and especially indigent mountain village, miserably exploited by landlords and with its

thin, poor arable land scattered in isolated bits over slopes and ravines. The Communist Party freed the villagers from the shackles of the past, and then collectivized agriculture. 'The peasants had stood up. Nature's harshness failed to dishearten the Tachai people; on the contrary, it aroused their revolutionary will to fight. They drew up a grand project to change nature. The ravines were challenged and conquered. In successive battles, they were beaten twice in their campaign to subjugate the Wolf's Den Ravine (but) the undaunted Tachai people marched on the Wolf's Den again.' Finally they 'conquered the most stubborn obstacle nature had confronted them with. Steeled in many battles, they refused to be slaves to nature. They lived deep in the mountains, but they had never had their line of vision blocked by the Taihang peaks. (Contrast this with the traditional view of mountains.) They said "We want to show other people the broad future of socialism and the strength of a collective economy . . ." They did not sit idly by waiting for the mechanization of agriculture, nor did they simply rely on government aid . . . but resolved to maintain a spirit of enterprise based on hard work.' With enormous labour, the villagers converted the ravines into solid fields by constructing terraces and stone retaining walls, and levelled several hundred acres of steeply sloping land to establish a 'rich grain-producing area in place of their poor mountain gullies'.[5] Tachai became a famous model, held up for villagers everywhere to emulate by developing a 'Tachai spirit', becoming 'Tachai men', and constructing 'Tachai fields' in self-reliant attacks on nature. *Mass participation* and *local initiative*, two recurrent themes sounded in the press and exemplified in this story, are the high roads to economic growth. This is of course the spirit of the Great Leap Forward, preached in industry, in education, and even in scholarship, as in agriculture. The urgency, and the radicalism, which it manifests are related to China's sense of damaged pride over the humiliations and revealed backwardness of the century preceding 1949:

> So many deeds cry out to be done,
> And always urgently;
> The world rolls on,
> Time passes.
> Ten thousand years are too long,
> Seize the day, seize the hour!
>   . . .
> Our force is irresistible.[6]

But the continual emphasis on self-reliance, in all fields, while certainly related to national pride, may also result from a semi-mystical belief in the power of a people properly organized and led to 'move mountains', without the material help which is in fact not available to them, through a frontal assault on whatever lies in their path. Self-reliance may be in some respects a counsel of despair, *faute de mieux*, and this may indeed constitute much of the explanation, but the hardships which it imposes are glorified in the same way as those of war. The assault on nature becomes a convenient rallying cry for an economically and technically underdeveloped society striking out, sometimes it would appear blindly, at this among other handy targets (for example, internal and external conspiracy and counter-revolution); the war against nature can be dramatized much more readily than the more prosaic processes of saving, accumulation, and investment.

Achievements in technology, of which there have of course been some brilliant examples, are similarly extolled as self-reliant Chinese successes in outstripping Western technicians. Inadequate Western models are discarded, and hard persevering work from square one cracks the secrets of recalcitrant materials. The series of technical and scientific triumphs culminating in the explosion of a hydrogen bomb, long before Western analysts had assumed this extremely demanding feat was possible, ahead of the French, and since at least 1959 without any outside assistance, was understandably celebrated as a major national victory for self-reliance, of tremendous importance for national morale. The synthesizing of insulin by Chinese chemists in 1966, in a quite different field (and on a problem where Western efforts had not been successful), shows that scientific talent and government support are not all concentrated on the nuclear effort, and there have been many achievements in other scientific fields where research support has been available. With the more mundane problems of actual production techniques, for which foreign experience might be of particular benefit, self-reliance and indeed Chinese superiority are also stressed. Where foreign experts said that something could not be done, or that for example the quality of Chinese ores made a certain product or process impossible, patient and inventive Chinese workers (rarely called engineers), inspired by the thoughts of Chairman Mao to 'proceed from actual conditions', were able to conquer the problem, and

incidentally also to demonstrate that the Chinese environment, seen by Westerners as deficient, is in fact rich when it can be manipulated by dedicated revolutionaries. Such an approach, it is said, is bound to succeed with problems which even in the West the best scientists and technicians have failed to solve, of which continual examples are given in the press. Whether or not this is also whistling in the dark, despite the striking successes in some fields, it is designed further to persuade people that the handicaps and difficulties which confront them need not prevent immediate and successful action, without waiting for the still distant promise of an adequate battery of technological aids. China has to a great extent cut itself off from technical assistance, especially as a result of the Sino-Soviet split and the current anti-foreignism of the Cultural Revolution. The creed of self-reliance may be economically less rational in an immediate sense, but it can be made to command dedication and sacrifice, and provides far more effective experience than dependence on foreign technicians. In the long run, economic growth may be accelerated.

The war against the environment stems in part from two additional factors: the need which the planners see to maintain full mobilization and employment for the entire population, and the need further to dramatize the revolutionary break with the past and its acceptance of things as they are. Revolutionary ardour may cool, and underemployed agricultural workers may become restive if energies and enthusiasms are not continuously committed. One may deduce that an important reason for many projects, for example the drive during the height of the Great Leap to collect wild animal manure – or the Great Leap itself – has been to keep rural labour fully controlled and engaged. The military terminology so commonly used is an understandable part of the campaign. The war against nature, specifically contrasted in official statements with the compromise and adjustment of the bad old days, is also glorified as the gospel of the new age, the dramatic symbol of China's awakening.

As suggested at the beginning of this essay, there is no sharper discontinuity between pre- and post-1949 than in attitudes towards the natural world, and it is not strange that it should be emphasized as the essence of the revolution. The Red Guards are specifically charged with continuing the struggle against 'the four olds': the old ideas, culture, customs, and habits of the degenerate past. Such a confrontation is especially clear and far-reaching between the traditional harmony with nature and new China's assault upon it; the contrast is often used as a hallmark of the revolution, a re-charger of radical fervour, and a goad to greater efforts in all sectors. China's drive for national development will not be deflected by any obstacles, and will never submit to things as they are. From the point of view of the men in power, it is also important to show that something concrete is happening in the less developed countryside where over 70 per cent of the people still live, as well as in the industrializing cities. As the men of Tachai have shown, conflict pays.

One final element in the war against nature is the pronounced anti-urbanism of the Communist Party, and of Mao in particular, something which reflects the Party's early failure in the 1920s and 1930s to create a viable revolutionary base in the cities and its ultimate success in winning power through the peasantry. The official and symbolic strongholds of the traditional Confucian order, of the Western imperialists in the treaty-ports, and of the Kuo Min Tang were also urban, and in opposition the Communists glorified the rural base of the 'real' and 'progressive' peasants – 'the people'.[7] The cities still seem to be distrusted. Industrial or office workers are too easily seduced by what the cultural revolution has labelled 'economism' – concentration on the personal material gain which is clearly more available to them than to the peasants – and by the practical and organizational demands of an industrializing, modernizing society. They may indeed soon cease to see themselves as workers at all and become identified instead with the a-political needs and goals of the management and production system to which they belong. In Mao's phrase, they are soon more 'expert' than 'red', and their stake is not in revolution but in stability, order, and getting on with their jobs. Hence in large part the frenetic efforts of the Cultural Revolution specifically to create conflict, disorder, and even chaos, to throw an increasingly urbanized, organized, stable China back into revolution.[8] By comparison with the counter-revolutionary cities, the unending battles of the war against nature in the countryside and its selfless sacrifices can be portrayed as the heart of the revolution and as the leading model for the whole country to follow. This is the rationale of the movement labelled *hsia fang*,[9] in which urbanites and intellectuals are pressed or required to 'go down' to the countryside and to help not merely in agricultural tasks but specifically in the war against the environment, in order not so much to aid the production processes as to see the real revolution in action and to take part in its self-denying struggles and hardships as the peasants do. Intellectuals are chronically suspect and are especially in need of this kind of re-moulding, through the purifying and regenerating process of physical labour among the peasants, the true proletariat, who exemplify in their fight against nature 'all the truths of Marxism: "to rebel is justified".'[10]

It was not only to recruit students for the Red Guard that all universities and secondary schools were closed for about a year during the height of the Cultural Revolution in 1966–67.[11] The economy will have to pay for this interruption in the training of desperately needed technicians (the overwhelming majority of students are in technical fields), as it must pay for equally serious interferences with the production process during the Cultural Revolution, but at all costs, it seems, the revolution must be kept alive. Mao is fighting a losing battle in this respect, as the frenzy of the Cultural Revolution and its violent accent on youth, in opposition to those with a larger stake in stability, suggest he half realizes. But he is doubtless right to focus it as much as possible on the evils of materialism and economism in the cities and to contrast this with the countryside, where the emotional drama of the war against nature can still be viewed and reacted to as pure and radical revolution, with only scorn for any established order and with the moving of mountains as its battle flag. The army has also been used continuously since the very early days of the Party as shock troops in the transformation of nature, not merely to fill peak labour demands at harvest time, but to reclaim land and to re-shape the environment in a variety of grandiose projects. A special campaign begun in early 1964 at Mao's initiative continues to call on all organizations and individuals to 'Learn from the Peoples' Liberation Army' and its material and ideological model, which can 'arouse the revolutionary millions.'[12]

Despite its focus on agriculture which is a necessary part of the war against nature, contemporary China has also turned on its head the traditional pattern of resource appraisals, of what is important about the environment. The land is now subservient to industry; agricultural development must be given an all-out effort, but as the base for industrialization.

Agriculture will remain for many years the chief source of surplus for investment; it is squeezed hard, and until recently starved of significant capital inputs, for which mass labour must substitute as best it can. But the prime goal of development is industrialization, and the exploitation of non-agricultural resources neglected in the past. Here the Communists are caught in a dilemma, a contradiction of their own making, between anti-urbanism on the one hand and pro-industrialism on the other. At the same time, any form of economic growth must depend on an agricultural system in which the planners have been reluctant to invest capital but from which they must extract surpluses in the face of a rapidly increasing population and of rising demand levels which are hard to hold in check. Industrialization is the symbol and the means of overcoming China's material backwardness; it is also the basis of international political status and security as well as of wealth. Other considerations must be sacrificed to it. This is evidenced not only in the allocation of capital but in the glorification of 'worker' technologists who confound Western experts, increase China's self-reliance, and move the country another step towards industrial leadership. The greatest hero is not the Tachai man but the selfless oil worker or lathe operator who overfulfills his target, improvises an improved technique, or works all night to repair a piece of machinery. The discovery and exploitation of new mineral resources and the development of hydroelectric power are the most enthusiastically praised and admired achievements in the battle with the environment. Special satisfaction is taken in disproving earlier estimates by Western geologists about limitations on China's reserves of important minerals. Enormous teams of students, during their 'vacations', are sent to scour the countryside for iron or coal or oil. The press reports their finds in a tone reminiscent of major sports results, but in the context of revolutionary national achievement. Unfortunately not only these reports, but also the periodic official accounts of the mineral resource inventory are cast in such general terms that they are difficult to use. The precise size, grade or quality, and probable extraction costs – essential data for a meaningful resource inventory – are rarely given. The presumption is not that the data are being withheld but that they are lacking. Mass participation and emotional enthusiasm seem to be as characteristic of the assault on this aspect of the nature as in the agricultural sphere.

For all of the undoubted dividends of the war on the environment, it carries with it the kinds of risks associated with totalitarian planning: decisions tend to have mushrooming consequences. Such risks are augmented by the messianic fervour which is so striking a feature of this Chinese battle, and are increased still further by the largely ineluctable decision to keep capital investment in agriculture to the barest minimum and to substitute for it with mass labour. Nature is now not only to be the servant of man, but agriculture must be the servant of industrialization. There is no denying the truly awesome force generated and channelled by the war against nature, especially in its apparent success in converting individual interests into the sacrifices of an austere campaign whose rewards are predominantly ideological or even mystical. There is more than a shadow of Calvin, Knox, and Cromwell in contemporary China, and also of the frightening *volonté générale* of Robespierre. Mountains are indeed being moved, but there is often a question whether rational economic goals are being served. The present economic margin is perilously thin, and this increases the need for optimum resource allocation. What has sometimes been described as a long-standing characteristic of both Chinese and of Communists generally, fascination with the 'big gamble', may also be involved in many of the planning decisions, not only in the Great Leap. Like most gambles, these may be made out of desperation. There simply is not enough capital to support both the industrial and the agricultural drive on the scale which the planners want, which China obviously needs, and which national ambition dictates. The temptation is strong to take the chance of ordering headlong assaults on both fronts and damn the risks of proceeding with bets uncovered. All of this, in a totalitarian system, multiplies the consequences of any planning errors, which may reach mammoth proportions before they can, at enormous expense, be corrected.

[. . .]

NOTES

1 Many of the ideas and policies of the revolution were of course foreshadowed, not only in the earlier history of the Chinese Communist Party but also by the reformers after 1898, but they were few and powerless. The Kuo Min Tang produced little revolution. Effective change began in 1949.

2 Both statements are taken from Liu Shao-chi, *Report on the Work of the Central Committee*, 2nd Session of the 8th Congress of the CCP, Peking, Foreign Languages Press, 1958. The report was delivered in May 1958. The context makes clear that Liu sees China as a 'newcomer' in its rivalry with the earlier industrialized West, and also that present models are superior to past models. Liu's current disgrace does not invalidate these remarks, which remain, as they were when he made them, basic to Chinese Communist doctrine.

3 For a recent review of this issue in Russian thought, see I. A. Matley, 'The Marxist Approach to the Geographical Environment', *Annals of the Association of American Geographers*, 56 (1966), pp. 97–111.

4 For example, the official explanation that the post-1958 failures in agriculture were due to natural calamities.

5 The quotations are all from Hsin Nung, 'A Great Revolutionary Ambition of the Tachai People', published (in English) in *China Pictorial*, I (1965), pp. 10–15. This article is however typical of scores of others recounting an almost identical story of how other villages have done the same kinds of things.

6 This is the final stanza of a poem by Mao Tse-tung entitled 'Reply to Kuo Mo-jo', dated 5 February 1963. The translation quoted here was published in *China Reconstructs*, xvi (March, 1967), p. 2.

7 Joseph Levenson includes an imaginative analysis of this attitude in his 'The Province, the Nation, and the World', in *Approaches to Modern Chinese History*, ed. A. Feuerwerker, R. Murphey and M. C. Wright, Berkeley, California, 1967, especially pp. 282–86.

8 For a discussion of this issue as applied to Party functionaries, see Ezra Vogel, 'From Revolutionary to Semi-Bureaucrat: the "Regularization of Cadres"', *China Quarterly*, 29 (1967), pp. 36–60.

9 For a survey analysis, see R. W. Lee, 'The Hsia Fang System: Marxism and Modernization', *China Quarterly*, 28 (1966), pp. 40–62. The movement was begun in 1957, but has been much intensified during the Cultural Revolution.

10 *Peking Review*, 16 September 1966, p. 14.

11 The impact of the Cultural Revolution on education was in many respects merely an intensification of the Socialist Education Campaign begun in 1962. This is in turn related to the widening use of the 'half work – half study' plan for education, adopted as a national policy in 1958 – see D. J. Munro, 'Marxism and Realities in China's Educational Policy: The Half Work–Half Study Model', *Asian Survey*, 7 (April, 1967), pp. 254–72.

12 Quoted from Mao in *Hung Ch'i* (Red Flag), 31 March 1964, 'Political Work is the Lifeline of all Work'.

PART FIVE

Challenges to the mainstream

INTRODUCTION TO PART FIVE

Mainstream development thinking in the 1950s and 1960s was dominated for the most part by supremely confident visions of rapid industrial growth and social transformation. The prevailing assumption was that Third World countries had plenty to gain from going second, or from being latecomers instead of pioneers. Backwardness could be turned to their advantage, so long as a planned push to modernity could be confidently diagnosed and followed through. This led to important debates on development strategy, as we have already seen. Should development be more or less balanced or unbalanced, both in sectoral and in spatial terms? Should the process of economic growth be driven mainly by the state or by the private sector? Should import-substitution industrialisation be largely capital-goods based, as in India, or more geared to consumption goods, as in Latin America? And so on.

What did not attract the attention of the mainstream was the very possibility of development as mimicry (with imitation here referring to something rather broader than outright Americanisation). Was it even possible that one group of countries could follow another into an age of high mass consumption? Or might various obstacles be placed in their way? Precisely these questions, however, came to be raised with increasing frequency from the late 1950s by various intellectuals and activists on the Left. Taking the lead in important respects was the American Marxist, Paul Baran.

Paul Baran was a leading light in the *Monthly Review* school of Marxism in the United States. Writing either on his own, as in his classic text from 1957, *The Political Economy of Growth*, or with Paul Sweezy, Baran helped develop a distinctive version of Marxism. Baran did not see 1950s capitalism in the thrusting, competitive terms painted by Karl Marx. Instead, he drew on Lenin to write about the irrationality and contradictions of monopoly capitalism. This was a system dominated by a small number of large concerns, as the term suggests, and in Baran's view it was haunted by the spectre of overproduction. Monopoly capitalists made their Western workers produce far more than they could purchase with their wages. The Third World then became a safety-valve for monopoly capitalists based in the First World. Writing after the Korean War, Baran believed that the capitalist class in the United States was finding it hard to exploit workers at home as freely as it had done in the Depression. At the same time, this class and the government that fronted for it was unwilling to spend heavily on welfare. It thus needed the state to finance expensive military spending abroad. This would help rid US capitalism of surplus production at home, while at the same time securing the super-exploitation of foreign workers. For this reason, Baran suggests, 'the ruling class in the United States (and elsewhere) is bitterly opposed to the industrialisation of the so-called "source countries" and to the emergence of integrated processing economies in the colonial and semi-colonial areas'. The Third World was *required* to be backward, as Baran puts it here. It could not be modernised in the ways that Rostow and others were calling for. Imitation was impossible.

This argument would later be generalised by Gunder Frank and by other theorists of the 'development of underdevelopment'. Frank insisted that the development of the periphery was and always had been impossible under the rule of capital. This became a mantra for many on the Left in the 1960s. In Frank's view, reversing Marx and largely following Baran, capitalism promoted development in the core, but underdevelopment in the periphery. These were two sides of the same coin. Unlike Baran, however, Frank made little of the distinction between competitive capitalism and monopoly capitalism. He argued that the development of the core was linked to the active exploitation of the periphery from the very birth of global capitalism – from the time of Cortés and Pizarro in Mexico and Peru, Clive in India and Rhodes in Africa, as he memorably put it. Modernisation theory assumed that all countries had once been underdeveloped. Frank agreed that all countries had once been undeveloped, in the sense of being pristine. But he insisted that only the Third World had been actively underdeveloped (pillaged, looted, exploited) by outside powers.

Modernisation theory also looked to capitalism as the solution to the problems of underdevelopment. For Frank, this was absurd. In his view, underdevelopment was a necessary consequence of the periphery's unequal integration into a global capitalist system that started to form after 1492. For Frank, then, as for Baran and other Left activists before him, the choice facing countries in the periphery was between 'barbarism or socialism'. What the Third World needed most was less capitalism, or what today would be called less integration into the circuits of globalisation. The sane alternative was to break out of the system. China had done this under Mao, and now Cuba was doing the same under Castro.

Frank's restatement of Baran continued to develop a line of reasoning that owed more to Lenin than to Marx. Marx was prepared to condemn capitalism for many sins, but not for the sin of being unproductive or non-developmental. As the 1960s faded into the 1970s, some on the Left began to worry away at the logic of Frank's model: a logic that made capitalism progressive and regressive in different places at the same time. They also began to observe the rise of a first generation of newly industrialised countries (NICs): Taiwan, for example, and South Korea, along with major industrial regions in Brazil and Mexico. Fernando Henrique Cardoso suggested that Frank had produced his model of the development of underdevelopment just when 'history was preparing a trap for the pessimists'. The Scottish communist, Bill Warren, meanwhile, suggested that the underdevelopment of the periphery was more a function of the territorialised politics of imperialism than of capitalism itself. The European imperial powers had held back capitalism in the periphery for political reasons. Baran notwithstanding, this was not the aim of the new American empire. Making the world safe for US companies to do business – including selling goods to people in the Third World – was more what it was about, even if certain companies feared industrial competition from the South. Following the collapse of the European empires, Warren suggested, capitalism was free to generate economic growth in the periphery, just as it had done in the core. Warren's famous book from 1980, published posthumously, was neatly titled *Imperialism: Pioneer of Capitalism*.

Others on the Left found Warren's analysis unsatisfactory. Some considered it an apologia for Western capitalism and not significantly different from Rostow in its analysis and prognostications. Where was class in this analysis? Where was a sense of the particular pressures facing Southern countries in a world system still dominated by Northern countries and metropolitan capital? Some of the best work to emerge from the broader *Dependencia* school in Latin America had also taken issue with the arch pessimism of Gunder Frank. But it did so while insisting that Latin America's problems had plenty to do with the path-dependencies induced by long histories of Spanish and Portuguese colonialism, as well as by the infernal meddling of Uncle Sam. Scholars including Cardoso (again, with Enzo Faletto), and Osvaldo Sunkel, produced important work in the 1970s on the conditioning relationships that shaped economic and political development in Latin America. The US-supported coup against President Allende in Chile in 1973 lent credence to their view that economic development in Latin America always took shape in relation to the ambitions of Empire.

Elsewhere, analysis moved to the articulation of different modes of production in the global South. This was particularly so among scholars who were influenced by Marxism of a more structuralist bent, borrowing from Louis Althusser in France. This analysis sought to provide a more sophisticated understanding of the often confusing and heterodox landscapes of India, say, or Brazil or South Africa. These were landscapes where small pockets of urban and rural capitalism and economic privilege seemed to coexist with vast hinterlands of pre-capitalist poverty. One mode of production with another. The excerpt we reproduce here from the work of Harold Wolpe seeks precisely to understand these articulations. It is focused on cheap labour flows, race and the reproduction of capitalism and apartheid in South Africa. Wolpe's main beef is with historians who have not grasped the significance of the transition from segregation to apartheid. Wolpe insists that segregation was prompted by the need to maintain a cheap African migrant labour force. African families were required to reproduce their labour power in a set of quasi-independent homelands, or Reserves. Having thus paid for their own, meagre, costs of health care and education, migrant male workers moved from the Reserves to South Africa's mines. There, mining capital paid them only the cost of their daily labour-power. Later on, the capacity to subsidise mining capital through this elaborate articulation of mining and Reserve modes of production came undone. It was then that class and ethno-nationalist politics began to shift to a hardened form of racial and spatial domination. This would become the policy of separate development or *apartheid*.

Wolpe's work contributed significantly to the analytical toolbox of the exiled liberation movement in southern Africa. It also provided crucial empirical energy for the emerging articulation of modes of production debates. It would be fair to say that these debates dominated Left analyses of development and underdevelopment in the late 1970s and early 1980s. This was particularly the case after two devastating critiques of *Monthly Review* (or neo-Smithian) Marxism were published by scholars from the Left in 1977 and 1978, those by Robert Brenner and Gabriel Palma. These authors dismissed as un-Marxist the emphasis on unequal exchange relations in the work of Frank, Samir Amin and others (including Immanuel Wallerstein). Too much emphasis was placed on trade relationships and the exploitative power of multinational corporations. The lineage of these arguments pointed back to Lenin and (more surprisingly, and in upside-down fashion) to Adam Smith. Exchange relationships were either fully good or wholly bad. Not enough attention was directed to Joan Robinson's observation that 'the only thing worse than being exploited by capital was not being exploited at all'. Or to the class struggles through which Europe had escaped feudalism into capitalism, and thereby set in motion its own development. Nor was there sufficient attention to those local structures of class, caste, gender and ethnicity which prevented productive forms of capitalism from deepening their hold in Latin America, Asia and Africa. Too much blame was placed on external agents or on their local stooges, the comprador bourgeoisie.

By the mid-1980s, however, David Booth was not alone in detecting an impasse in Marxist sociologies of development. The rise of the East Asian NICs was by now readily apparent and at one level seemed to offer support for the thesis that state-managed capitalism *could* promote development in the periphery. Meantime, more information was emerging about the failures of some socialist development strategies in the Third World. The inflation of the 1970s and the debt crises of the 1980s also spurred a counter-revolution in development theory and policy, as we shall see in Part 6.

But the Marxist Left was not alone in raising its voice against the development mainstream in the 1960s and 1970s. Our next reading comes from Rachel Carson's classic book, *Silent Spring*, first published in 1962. We include it here not because it is focused on the Third World, but because it prefigures an ecological critique of developmentalism that has grown significantly over the past twenty years. Carson writes of 'the obligation to endure', or what now might be called the imperative of sustainable development. But she also writes about the simplifications that 'development' induces, and the dangers that follow. 'Nature has introduced great variety into the landscape, but man has displayed a passion for simplifying it' (see also Scott, in Part 6).

Carson's reference to man is generic, but some feminists have maintained that development's assault upon nature has been strongly gendered. Women rather than men, Vandana Shiva insists, are more in tune with external nature. As such, they are more likely to protect the environments in which they work (for example, as firewood and vegetable collectors in village forests). Most feminists have refused this line of gender essentialism. Nevertheless, a broad understanding that development is gendered is widely held now in development studies in a way that it was not in the 1950s and 1960s. Considerable challenges remain, not least at the level of practical development outcomes, where women's variegated lives and struggles often differ from those of comparable groups of men, including male kin within households. Challenges also remain within the academy and the policy-making community, where attention to gender issues is sometimes skin-deep and tokenistic, or haunted by Second-Wave Feminism's claims to universal sisterhood. Still, considerable progress has been made and new debates continually open up. And this progress is partly owed to pioneers in the fields of women and development, and gender and development, one of whom in the 1960s and 1970s was Ester Boserup. The reading taken here from her book, *Women's Role in Economic Development*, is dated in several respects, as we should expect. But it still offers insight into the organisation of marital and other relationships, and the status of women, in relation to different systems of farming, notably in West Africa. Boserup ends by noting the dangers of pressing for (modern systems of) birth control in communities where women 'take little part in field work'. 'There is a danger in such a community that the propaganda of birth control, if successful, may further lower the status of women both in the eyes of men and in their own eyes.'

In other communities women were leaving the fields with the aim of joining the urban workforce, whether in the formal or the informal sector. In some countries in Latin America female rural to urban migration exceeded male migration, reversing common patterns in Africa and Asia. The urbanisation of the Third World, however, which had been warmly welcomed in the 1950s by W. Arthur Lewis and others, now began to foster a new set of dissident voices. Bert Hoselitz warned early on that the urbanisation of the Third World might be 'parasitic' rather than 'generative'. It would be very different to the experience of richer countries. Rural to urban migration could be excessive, in the sense of people being attracted to cities by pro-urban ideologies and propaganda rather than by productive jobs and affordable housing. Slums and shanties might then fringe the cities of the developing world. In time they might become breeding grounds for crime and all manner of urban pathologies. The anthropologist Keith Hart coined the term 'informal sector' precisely to attend to these less pleasant aspects of livelihoods in shanty towns.

But if the prevailing tone of discussion darkened in the 1960s and 1970s, the question of what to do about 'premature' urbanisation/urban poverty/urban unemployment provoked many different responses, at least two of which deeply challenged mainstream development thinking. From Michael Lipton came an account of urban bias. In Lipton's view, what mainly caused poor people to remain poor in the global South was a political system that favoured the interests of the urban classes. The urban classes enforced price-twists in Third World economies. They pushed for the underpricing of crops and other rural goods to urban areas and the overpricing of urban goods to rural areas. They did this to buy off dissent among town dwellers. Urban bias was also distributional. More money was spent on urban education and health care, and more taxes, relatively speaking, were taken from rural than from urban populations. Urban bias was thus both unfair and inefficient. Scarce government resources could better be spent in the countryside (on minor irrigation works, say) than on vast projects that pampered the urban middle class. More spending on the countryside would also reduce perverse migration flows from rural to urban areas. The problem of the urban shanty town, Lipton suggested, began and could be dealt with in the countryside.

Lipton's work was later extended by Robert Bates, and by the World Bank in the form of its 1981 report on *Accelerated Development in Sub-Saharan Africa*. It was also linked to determinate policy changes: ideas do have consequences, as Keynes strongly maintained. Most African countries were required to devalue their currencies and remove their marketing boards in the 1980s and 1990s. This was part of the conditionality imposed upon them in return for structural adjustment loans. Overvalued exchange rates had made it difficult for these countries to export crops. Marketing boards and other parastatal agencies were targeted as the main bodies through which price-twists were enforced against farmers.

Another body of academic work which had significant practical consequences is highlighted in the last reading in this Part, by William Mangin. Along with John Turner and others in the 1960s and 1970s, Mangin began to work in the city-edge shanty towns that were attracting so much adverse publicity. This was at a time when many governments were minded to criminalise street hawkers and vendors. Some even sought to bulldoze 'unsightly settlements'. This happened in New Delhi during the Emergency of 1975–77, a period when democracy was suspended in India. Mangin was among the first to challenge these negative and stereotypical portraits. Mangin did not find in Latin American shanty towns a uniform mass of people weighed down in cultures of poverty. What he mainly saw were upwardly mobile families. Like Turner, he discovered people who had first moved from rural areas to city centre slums, and who then had moved to the periphery to protect their growing assets. These people would later try to upgrade properties they had begun to build on illegally occupied land. The solution to Latin America's growing problems of housing supply, Mangin and Turner came to suggest, was not to bulldoze slums and shanties, nor was it to build public housing projects at huge expense. The solution, rather, was to grant legal land tenures to shanty dwellers, to provide sites and services projects (water, electricity, etc.), and then to stand back. The real work of development – of city building – would be done by ordinary families. It would be slow and uneven, but also sure and in accordance with their own needs and capacities. (See also our discussion of William Easterly's distinction between planners and searchers in the Introduction to Part 6.)

Read one way, Mangin seems to be saying that the state needs to know its place. This is a sentiment that became mainstream in development theory and practice in the 1980s, as we shall see in Part 6. But this is to read him out of context. The real thrust of Mangin's argument is about the merits of self-help strategies and the capabilities of poorer people. His work is thus better seen in the context of challenges to mainstream developmentalism that emerged in the 1960s and early 1970s. These challenges sought to critique an earlier body of work that equated development with economic growth, which had little to say about human needs, and which invested the state with sufficient powers of reason and foresight that human agency (gendered human agency at that) became unimportant. What would emerge in the late 1970s and 1980s was something far more far-reaching – even if hinted at by Mangin and Lipton. We refer to a counter-revolution in development studies that was radically anti-dirigiste, on the one hand, and frankly dismissive of development economics, on the other.

REFERENCES AND FURTHER READING

Abrams, C. (1964) *Man's Struggle for Shelter in an Urbanizing World*, Cambridge: MA: MIT Press.

Adams, W. (1990) *Green Development*, London: Routledge.

Amin, S. (1976) *Unequal Development*, New York: Monthly Review Press.

Baran, P. and Sweezy, P. (1966) *Monopoly Capital*, New York: Monthly Review Press.

Bates, R. (1981) *Markets and States in Tropical Africa*, Berkeley: University of California Press.

Brenner, R. (1977) 'The origins of capitalist development: a critique of neo-Smithian Marxism', *New Left Review* 104: 25–92.

Cardoso, F. H. (1977) 'The consumption of dependency theory in the US', *Latin American Research Review* 12: 7–24.

Cardoso, F. H. and Faletto, E. (1979) *Dependency and Development in Latin America*, Berkeley: University of California Press.

Hoselitz, B. (1957) 'Generative and parasitic cities', *Economic Development and Cultural Change* 5: 278–94.

Kabeer, N. (1994) *Reversed Realities: Gender Hierarchies in Development Thought*, London: Verso.

Palma, G. (1978) 'Dependency: a formal theory of underdevelopment or a methodology for the analysis of concrete situations of underdevelopment?', *World Development* 6: 881–924.

Santos, M. (1979) *The Shared Space: The Two Circuits of the Urban Economy in Underdeveloped Countries*, London: Methuen.

Scott, J. (1985) *Weapons of the Weak: Everyday Forms of Peasant Resistance*, New Haven, CT: Yale University Press.

Shiva, V. (1989) *Staying Alive: Women, Ecology and Development,* London: Zed.

Sunkel, O. (1973) 'Transnational capitalism and national disintegration in Latin America', *Social and Economic Studies* 22: 132–76.

Turner, J. F. (1976) *Housing by People*, London: Marion Boyars.

Wallerstein, I. (1974) *The Modern World System*, New York: Academic Press.

Warren, B. (1980) *Imperialism: Pioneer of Capitalism*, London: Verso.

Wolpe, H. (ed.) (1980) *The Articulation of Modes of Production*, London: Routledge and Kegan Paul.

World Bank (1981) *Accelerated Development in Sub-Saharan Africa*, Washington, DC: World Bank.

'The Steep Ascent'

from *The Political Economy of Growth* (1957)

Paul Baran

Editors' Introduction

Paul Baran was born in Russia in 1910 and died in the United States in 1964. He was a major figure in the post-1945 American Left, working closely with Paul Sweezy within the *Monthly Review* school of Marxism (or neo-Marxism). The Monthly Review Press became home in the 1960s to a number of influential accounts of dependency and imperialism, including those of Andre Gunder Frank, Samir Amin and Arghiri Emmanuel, among others. One of the distinctive features of the *Monthly Review* school was its analysis of monopoly capitalism, a concept much explored by Baran and Sweezy, separately and together, and by important figures including Henry Braverman. Out of this analysis, which Paul Baran did much to develop and refine, came a linked analysis of imperialism and uneven development, and thus of the production of the Third World as a site to 'solve' the problems of under-consumption in the metropolitan core. Baran states forcefully in his classic book, *The Political Economy of Growth*, that 'spending on imperialist policy . . . is the one form of [government spending] that is fully acceptable to monopoly capitalists' (pp. 246–7), and that it is for this reason that 'economic development in underdeveloped countries is profoundly inimical to the dominant interests in the advanced countries' (p. 120).

Baran's work was later built upon, popularised and to a considerable degree simplified by Gunder Frank. Frank maintained that capitalism was a system which produced development in the core and underdevelopment in the periphery, and had done so since the time of Cortés, Rhodes and Clive. Baran, for his part, was more sensitive to the different modes of operation of what he called competitive capitalism and monopoly capitalism. Crudely stated, he followed Marx in saluting the productive power of competitive capitalism, and Lenin when he rebuked the inefficiency and imperialist violence associated with monopoly capitalism. (See also his book from 1966 with Paul Sweezy, *Monopoloy Capital: An Essay on the American Economic and Social Order*. It is also worth noting that *The Political Economy of Growth* was addressed to a US audience. Throughout that book Baran is trying to establish the relevance of Marxism in the United States at a time, the 1950s, when Keynesianism was in the ascendancy and when working-class people seemed to be sharing in the bounties of corporate capitalism. Baran's aim was to insist on the inefficiency and ultimate instability of monopoly capitalism at a global scale, and to suggest that welfare spending in the United States was being sacrificed on the altar of militarism and imperialism, an argument that seemed to gain in strength through the Vietnam War.)

In the reading reproduced here, which comes from the beginning of the last chapter of *The Political Economy of Growth*, Baran articulates precisely these themes. The opening sentence maintains that the once 'mighty engine' of capitalist economic development has been re-born as a 'formidable hurdle to human advancement'. Thereafter, the reading develops one of the most important themes in Baran's work, and in that of the *Monthly Review* school more generally; namely, an analysis of the gap which opens up in monopoly capitalism between the actual economic surplus and the potential economic surplus. Socialism is commended as a rational way of bringing these two quantities back into balance, as well as being a means by which the fruits of (full, or potential) economic growth can be more reasonably distributed. What Baran understood to be the achievements of socialist planning in Communist China have to be seen in this broader context – however odd this celebration might strike us today.

Key references

Paul Baran (1973 [1957]) *The Political Economy of Growth*, Harmondsworth: Penguin.
— (1960) *Marxism and Psychoanalysis*, New York: Monthly Review Press.
— (1972) *Longer View: Essays Towards a Critique of Political Economy*, New York: Monthly Review Press.
Paul Baran and P. Sweezy (1968) *Monopoly Capital: An Essay on the American Economic and Social Order*, New York: Monthly Review Press.
Monthly Review website: www.monthlyreview.org

1

It is in the underdeveloped world that the central, overriding fact of our epoch becomes manifest to the naked eye: the capitalist system, once a mighty engine of economic development, has turned into a no less formidable hurdle to human advancement. What Alexis de Tocqueville remarked with reference to political institutions applies on a scale broader than he himself could have visualized: 'The physiognomy of a government may best be judged in its colonies, for there its features are magnified and rendered more conspicuous. When I wish to study the merits of the administration of Louis XIV, I must go to Canada; its deformity is there seen as through a microscope.'[1] Indeed, in the advanced countries the discrepancy between what *could* be accomplished with the forces of production at the disposal of society and what is in fact being attained on the basis of them is incomparably larger than in the backward areas.[2] But while in the advanced countries this discrepancy is obscured by the high *absolute* level of productivity and output that has been reached during the capitalist age, in the underdeveloped countries the gap between the actual and the possible is glaring, and its implications are catastrophic. There the difference is not, as in the advanced countries, between higher and lower degrees of development, between the now reachable final solution of the entire problem of want and the continuation of drudgery, poverty, and cultural degradation; there the difference is between abysmal squalor and decent existence, between the misery of hopelessness and the exhilaration of progress, between life and death for hundreds of millions of people. Therefore, even bourgeois writers occasionally admit that in the underdeveloped countries the transition to a rational economic and social organization is vitally urgent – while holding at the same time that the advanced countries can 'well afford' to remain under the domination of monopoly capitalism and imperialism.[3] Nothing, however, could be more egregiously erroneous. For, as we have seen, the rule of monopoly capitalism and imperialism in the advanced countries and economic and social backwardness in the underdeveloped countries are intimately related, represent merely different aspects of what is in reality a global problem. A socialist transformation of the advanced West would not only open to its own peoples the road to unprecedented economic, social, and cultural progress, it would at the same time enable the peoples of the underdeveloped countries to overcome rapidly their present condition of poverty and stagnation. It would not only put an end to the exploitation of the backward countries; a rational organization and full utilization of the West's enormous productive resources would readily permit the advanced nations to repay at least a part of their historical debt to the backward peoples and to render them generous and unselfish help in their effort to increase speedily their desperately inadequate 'means of employment'.

Yet for reasons that were touched upon earlier,[4] and that would take us beyond the scope of the present discussion to analyse further, this is not the way in which the historical process has unfolded. Far from being aided by the advanced countries, the backward nations' transition to an economic and social order assuring them of a progressive development is taking place against the embittered resistance of the imperialist powers. What Lenin wrote in 1913 about the European countries could well be written today about the entire advanced West:

> In civilized and advanced Europe, with its brilliantly developed machine industry, its rich all-around culture and constitution, a historical moment has been reached when the commanding bourgeoisie, out of fear for the growth and increasing strength of the proletariat, is supporting everything backward, moribund and medieval. The obsolescent bourgeoisie is combining with all obsolete and obsolescent forces in an endeavour to preserve tottering wage slavery.[5]

This support for 'everything backward, moribund and medieval' can be observed everywhere: whether we look at China and Southeast Asia, at the Near East and Latin America, at Eastern and Southeastern Europe, or at Italy, Spain, and Portugal. Its aim is to prevent social revolutions wherever possible, and to obstruct the stabilization and progress of socialist societies wherever such revolutions have taken place.

Little needs to be said at this juncture about the more purely military aspects of the matter. What few traces of genuine humanism still remained in the consciousness of the bourgeoisie from the days of its glorious youth all but vanished under the impact of the intensified class struggle. If the second half of the nineteenth century and the first quarter of the twentieth century were still marked by a series of international agreements directed towards the 'humanization' of warfare, in imperialism's present struggle against the national and social liberation of the peoples inhabiting the underdeveloped countries no holds are barred. 'Operation Killer' is considered to be as legitimate as 'Operation Strangle', and the burning of entire towns and villages as unobjectionable as pouring napalm on civilian populations. This position was epitomized in a statement of President Eisenhower: 'The use of the atom bomb would be on this basis, Does it advantage me or does it not . . .? If I thought the net was on my side I would use it instantly.'[6] Needless to add, this formula does not reflect an exceptional ferocity of particular individuals but the utter moral bankruptcy of a decaying social order.[7]

But since it is far from certain that the 'net' would be on the side of the imperialist camp, the ultimate expedient of war has to be dealt with with the utmost caution and employed only where the very existence of capitalism and imperialism appears to be threatened. Meanwhile everything short of war is used to sabotage the development of the socialist countries. Not that it is not recognized that a great deal is being accomplished and can be accomplished by the nations that have adopted a system of socialist planning. Indeed, the authors of the United Nations report on *Measures for the Economic Development of Under Developed Countries* correctly state that 'if the leaders win the confidence of the country, and prove themselves to be vigorous in eradicating privilege and gross inequalities, they can inspire the masses with an enthusiasm for progress which carries all before it';[8] and John Foster Dulles acknowledges that 'Soviet Communists . . . can and do implement policies with the portrayal of a "great Soviet Communist experiment" with which, during this century, they are catching the imagination of the people of the world, just as we did in the nineteenth century with our "great American experiment".'[9] And while it is generally recognized that the first and foremost need of the underdeveloped countries is a rapid increase of their national income, Professor Mason certifies that 'to the promotion of economic development Communism can bring formidable advantages. . . . Over the long run, given a measure of administrative competence in the investment and use of new capital resources national income is likely to increase at an extremely rapid rate.'[10]

One might expect that under such circumstances the backward nations that have at last managed to emerge from their age-long state of stagnation would receive congratulations and encouragement, if nothing more tangible, from those who are purportedly deeply concerned with their advancement. Such an expectation would reflect, however, a wholly naïve conception of the existing situation. As Lenin asks, 'where, except in the imagination of the sentimental reformists are there any trusts capable of interesting themselves in the condition of the masses instead of the conquest of colonies?'[11] In fact, the progress made in the underdeveloped countries by means of socialist planning is greatly disconcerting to Western official opinion. Although Mr Dulles notes that Communists 'in China have had some success in arousing a sense of social responsibility and in imposing discipline on its supporters' – which is obviously a major step forward in the struggle for economic development – he piously hopes that this advance may come to naught in view of the Chinese 'national character' which he describes, in apparent admiration, as follows: 'The Chinese through their religious and traditional habits of thought have become an individualistic people. The family has been the highest unit of value, and individual loyalty has been to ancestors and descendants. There has been only a little of the broader loyalty to fellow men or to some social or class group or to nation.'[12] Such a 'national character' is, no doubt, a Godsend to imperialists whose sole concern is to dominate the people blessed with it. Accordingly Mr Dulles feels that 'the religious of the East are deeply rooted and have many precious values. Their spiritual beliefs cannot be reconciled with Communist atheism and materialism. That creates a common bond between us, and our task is to find it and develop it.'[13] This sentiment is echoed by Professor Mason who expects religion to be a major obstacle in the way of progress in socialist countries, and who holds that in 'southern Asia as elsewhere religion is a strong bulwark against Communism'.[14] It is hardly surprising that 'everything backward, moribund and medieval' in the underdeveloped countries themselves sees eye to eye with its friends and protectors in the West. Vitally concerned with having the underlying populations form a 'spiritual society of individuals who love God . . . who work hard as a matter of duty and self-satisfaction . . . and for whom life is not merely physical growth and enjoyment, but intellectual and spiritual development',[15] the ruling classes in the underdeveloped countries spare no energy and receive a great deal of American support in their effort to strengthen the sway of religious superstitions over the minds of their starving subjects. What do they or the imperialists care that these superstitions represent a major roadblock on the way to progress? What do they and their Western accomplices care that the cost of maintaining religious obfuscation is increased starvation, multiplied death! As Dr Balogh observed on his trip to India,

the religious revival fostered by the richer classes . . . prevents a rational policy to improve livestock. India has 200,000,000 cattle, many of them quite useless, existing on an extremely scanty food supply. Yet the slaughtering of cattle is banned by

law in many sections and has been stopped *de facto* in most areas. Even monkeys are sacrosanct, though they destroy or eat an estimated one and a quarter million tons of grain annually.[16]

Like the aristocrats at the end of the feudal age, the economic royalists in these latter days of monopoly capitalism and imperialism are not themselves under the sway of obscurantism of this sort. Yet they consider it quite wholesome for their wood-hewers and water-carriers at home and abroad.[17] John Foster Dulles has put the matter in a nutshell: 'We have no affirmative policies beyond, for we cannot go further with material things.'[18]

Indeed, it is capitalism's inability to 'go further with material things', to serve as a framework for economic and social development, that forces its apologists and politicians to rely for its stability on circuses rather than on bread, on ideological claptrap rather than on reason. Thus the campaign for the preservation of capitalism is advertised today more energetically than ever as a crusade for democracy and freedom. In the days of the early struggle against feudalism, when capitalism was a powerful vehicle of progress and when enlightenment and reason were written on the banner of the rising capitalist class, this claim had at least partial historical validity. It had all but lost it in the second half of the nineteenth century, when bourgeois rule was increasingly menaced by the rising socialist movement, and when it became ever more transparent that 'by freedom is meant under the present bourgeois conditions of production free trade, free selling and buying'.[19] And it has turned into an altogether hypocritical sham in the age of imperialism, when capitalism, having lost control over one-third of the globe, is fighting for its very existence. As Engels brilliantly foresaw, 'on the day of the crisis and on the day after the crisis . . . *The whole collective reaction . . . will group itself around pure democracy*'.[20] That the 'whole collective reaction' it is, and that the 'pure democracy' for which it allegedly fights is nothing but pure freedom of exploitation can readily be seen from the membership roster of the so-called free world. Spain and Portugal, Greece and Turkey, South Korea and South Vietnam, Thailand, Pakistan and the sheikhdoms of the Middle East, the military dictatorships of Latin America and the Union of South Africa – all have been promoted by the imperialist crusaders to the status of 'democratic states'. And if Professor Mason, in a passage omitted in a previous quotation, objects to the 'extra-ordinary rapid rate' of increase of national income that can be attained in a socialist society because it would depend on a 'totalitarian régime exercising the weapons of terror [and] . . . squeezing standards of living . . . that no democratic state could possibly accomplish',[21] he does not there note the fact that such terror as has taken place in the course of all social revolutions – frequently excessive, always painful and deplorable – represented the inevitable birth pains of a new society, and that such squeezing of living standards as has occurred has affected primarily, if not solely, the ruling class whose excess consumption, squandering of resources and capital flight had to be 'sacrificed' to economic development. Nor is bourgeois economics in the habit of expressing any such misgivings about the comprador and colonial régimes 'exercising the weapons of terror [and] . . . squeezing standards of living' for the sake of the preservation of the wealth and profits of their supporters and in order to perpetuate misery and stagnation in their countries – as in Formosa or in Greece, in Malaya or in Kenya, in Madagascar or in Algiers, in the Philippines or in Guatemala.

The crude apologetics which identify freedom with freedom of capital, equate the interests of a parasitic minority with the vital needs of the people, and treat imperialism as synonymous with democracy would hardly call for attention were it not for two considerations relating them directly to the problem of future development. The first has to do with the profound impact of this ideology and of the historical circumstances underlying it on the social, political, and cultural evolution of the imperialist nations themselves. This impact is epitomized in Marx' and Engels' trenchant remark that 'no nation can be free if it oppresses other nations'; its tragic importance is manifest beyond possibility of error whether we look at the early history of the 'oppressor nations' or at their most recent record, whether we think of Western Europe or of Tsarist Russia, of Asia or of America. Yet all that is possible at this point is to take note of this terribly important matter; to enlarge upon it would take us too far afield.[22]

2

The other consideration more directly germane to our present problem is the direct effect of the imperi-alist activities reflected and inspired by this 'neo-jingoism' on the course of events in the under-developed countries. This effect is most telling; and its magnitude can be studied with the needed con-creteness. As far as those underdeveloped countries are concerned that still constitute parts of the 'free world', it assumes two principal forms. In the first place, their dominant comprador elements, always supported by the imperialist powers, are now aided more energetically, more systematically, more openly. They not only receive subsidies for the promotion of religion, for the conduct of their political activities, they are also given direct military assistance in their struggle against their increasingly restive people. In an ever-growing number of these countries the régimes based on the reactionary forces owe their existence solely to this help received from the imperialist West.[23]

Secondly, a large number of these governments – if not all of them – are not merely supplied with armaments, they are also compelled to devote con-siderable parts of their countries' national income to the building up and maintenance of large military establishments. The proportion of national income spent on military purposes is over 5 per cent in Pakistan, nearly as large in Turkey, over 3 per cent in Thailand, and much larger in the Philippines, Greece, and some other countries – not to speak of South Vietnam, South Korea, and Formosa, where the per-centage is still greater. It should be recalled that the significance of this burden can be fully appreciated only if it is considered *not* in relation to total national income but as a share of the economic surplus. Indeed, in most if not all of these countries military spending is equal to or exceeds their total productive investment! This wholesale destruction of resources that could by themselves serve as the basis of a massive growth of 'means of employment' is justified by Western imperialists and their agents in the under-developed countries by adducing the supposed danger of Soviet aggression. Yet some who clamour most loudly about the aggressiveness of the Soviet Union do not themselves really believe their own propaganda. They are fully aware that the Soviet Union has no intention of attacking capitalist countries. The accuracy of this is confirmed by many students of Soviet policies not suspect of socialist sympathies. One of the leading United States experts on Soviet problems leaves not the slightest doubt on this question:

The theory of the inevitability of the eventual fall of capitalism has the fortunate connotation that there is no hurry about it. The forces of progress can take their time in preparing the final *coup de grace*. . . . The Kremlin . . . has no right to risk the existing achievements of the revolution for the sake of vain baubles of the future. . . . There is no trace of any feeling in Soviet psychology that . . . the goal must be reached at any given time.[24]

Essentially the same view is held by the man obviously most concerned with the problem, the United States Secretary of Defence, Mr Charles E. Wilson, who 'told a Senate Appropriations subcommittee . . . that the American people should be reassured by Soviet concentration on fighter aircraft production as a sign that the Russians intend to build an Air Force of principally defensive capability.'[25] Innumerable other observers in the United States as well as in Western Europe have expressed their conviction that the socialist camp, preoccupied with internal construction, is utterly unlikely to initiate a war.[26]

Thus what the danger of 'Soviet aggression' really amounts to is the danger of so-called 'subversion' – the now fashionable designation of social revolution. This was clearly expressed by John Foster Dulles: 'The imposition on Southeast Asia of the political system of Communist Russia and its Chinese Communist ally, *by whatever means*, would be a grave threat to the whole free community. The United States feels that that possibility should not be passively accepted, but should be met by united action.'[27] It is, however, either a most fatuous misunderstanding of history or its deliberate misrepresentation to treat social revolutions in individual countries as resulting from 'outside subversion' or as 'imposed' by foreign plots and machinations. Indeed, as the great English historian of the Soviet Union remarks, 'the revolution of 1917, itself the product of the upheaval of 1914, was a turning-point in world history certainly comparable in magnitude with the French revolution a century and a quarter earlier, and perhaps surpassing it'.[28] Was this 'turning-point in world history' the result of skilfully organized 'subversion'? Or was the Chinese revolution, another event of tremendous historical significance, engineered by Soviet specialists in 'subversion'? The answer to this question is provided by the United States Department of State as well as by Mr Kennan, for a long time one of the Department's leading officials.

The unfortunate but inescapable fact is that the ominous result of the civil war in China was beyond the control of the government of the United States. Nothing that this country did or could have done within the reasonable limits of its capabilities could have changed the result; nothing that was left undone by this country has contributed to it. It was the product of internal Chinese forces, forces which this country tried to influence but could not.[29]

And Mr Kennan 'understates' that 'to attribute the revolution which has taken place in China in these recent years primarily to Soviet propaganda or instigation is to underestimate grievously, to say the least, a number of other highly important factors'.[30] The matter is aptly summed up in a remark of Lenin:

The dominance of capitalism becomes subverted not because someone wants to seize power. Such seizure of power would be nonsense. The termination of the dominance of capitalism would be impossible, if the entire economic development of capitalist countries had not led to it. The war has accelerated this process and rendered capitalism impossible. No force would destroy capitalism if it were not undermined and subverted by history.[31]

The conclusion is inescapable that the prodigious waste of the underdeveloped countries' resources on vast military establishments is *not* dictated by the existence of an *external* danger. The atmosphere of such danger is merely created and re-created in order to facilitate the existence of the comprador régimes in these countries, and the armed forces that they maintain are needed primarily, if not exclusively, for the suppression of *internal* popular movements for national and social liberation. The tragedy of the situation has the dimensions of a Greek drama. In Hitler's extermination camps the victims were forced to dig their own graves before being massacred by their Nazi torturers. In the underdeveloped countries of the 'free world', peoples are forced to use a large share of what would enable them to emerge from their present state of squalor and disease to maintain mercenaries whose function it is to provide cannon fodder for their imperialist overlords and to support régimes perpetuating this very state of squalor and disease.[32]

The counter-revolutionary crusade has not merely a crippling effect on the underdeveloped areas under imperialist control; its repercussions are also strongly felt in the countries that belong to the socialist camp. Foremost among them is the inescapable necessity to devote a considerable share of national resources to the maintenance of military establishments. But in the case of the socialist countries those establishments are *defence* establishments. Confronted with implacable hatred on the part of the capitalist class, threatened with programmes of 'liberation' and 'preventive wars', the socialist countries are continually forced to fear an attack from the imperialist powers. David Sarnoff, one of America's leading monopolists, goes a long way towards clarifying the entire issue. 'Though the Soviets want a nuclear war no more than we do,' he writes, 'they accept the risk of it in pushing their political offensive. We, too, cannot avoid risks. (It might become necessary, Mr Dulles said recently, "to forgo peace in order to secure the blessings of liberty"!)'[33] Yet – in remarkable contrast to the anti-socialist propagandists on the highest level – Sarnoff grasps incisively that 'we must realize that world Communism is *not* a tool in the hands of Russia – Russia is a tool in the hands of world Communism. Repeatedly Moscow has sacrificed national interests in deference to world revolutionary needs.' Thus it is obvious that the 'political offensive' General Sarnoff is concerned with has no connection with the absurd notion of 'Russian imperialism' but is simply the spread of social revolution. Indeed, 'that the challenge is global must be kept clearly in view. Red guerrillas in Burma, Communists in France or the U.S., the Huks in the Philippines, Red agents in Central America – these are as much "the enemy" as the Kremlin itself.' As we have seen before, however, it cannot possibly be held that social revolutions are the handiwork of crafty agents or must be attributed to 'Soviet propaganda or instigation'. They are the results of class struggles within capitalist societies that no one can abolish or suspend. What follows from this is that a social revolution in a country that is capitalist today may induce the imperialists to 'forgo peace' and plunge the world into a nuclear war. What follows, furthermore, is that the socialist camp may be faced with such a catastrophe at any time. For it can neither 'regulate' social revolutions so as not unduly to upset the imperialist beneficiaries of the 'blessings of liberty', nor can it actually foresee which social revolution will be considered by the imperialist powers as a *casus belli*, as the signal for starting a general holocaust.

To be sure, this does not mean that a global war may break out 'any minute', that the world lives permanently on the crater of a volcano, and that future developments are altogether unpredictable. What it does mean, however, is that in our age of imperialism and social revolutions the danger of war is continually present, and that the socialist countries have no alternative but to sacrifice considerable parts of their resources to the maintenance of adequate defence.[34] The resulting slowing-down of their advance, the consequent pressure on their standards of living, represent the principal cost of imperialism to the peoples in the socialist countries. The effects of the propaganda campaigns that the imperialist camp launches against them cause an additional strain. These are calculated to create 'a spirit of mutiny, to keep the Kremlin off balance, to deepen existing rifts, to sharpen economic and empire problems', and often consist of 'programs of spiritual and religious character ... [which] preach faith in the Divine, abhorrence of Communist godlessness, resistance to atheism'.[35] And they do provide some succour to the remnants of the former ruling classes in the socialist countries, they strengthen the hold of superstitions on the minds of backward peasants and workers, they increase the difficulties encountered in educating and organizing people for a collective effort to overcome their poverty. Thus they aggravate the internal conditions in those countries, strengthen the hand of those who are most suspicious of Western intentions, and in this way hamper the countries' progress towards democracy and socialism. But to follow General Sarnoff's advice and rename the 'Voice of America' the 'Voice of America – for Freedom and Peace' would not make much difference. 'Facts are stubborn things', and John Foster Dulles has pointed them out with all the necessary accuracy: 'There is no use having more and louder Voices of America unless we have something to say that is more persuasive than anything yet said.'[36]

[. . .]

NOTES

1 Quoted in S. Herbert Frankel, *The Economic Impact on Under-Developed Societies* (Oxford, 1953), p. 17.

2 In that sense Professor Mason is undoubtedly right when he says that 'perhaps the United States is the underdeveloped area rather than the Middle East'. *Promoting Economic Development* (Claremont, California, 1955), p. 9.

3 Thus the authors of the previously cited United Nations report, *Measures for the Economic Development of Under-Developed Countries* (1951), discount for 'a number of under-developed countries ... the prospect of much economic progress until a social revolution has affected a shift in the distribution of income and power'. (Para. 37.)

4 cf. above, [*The Political Economy of Growth*], p. 273.

5 'Backward Europe and Advanced Asia', *Selected Works in Two Volumes* (Moscow, 1950), Vol. 1, Part 2, p. 314.

6 Quoted in the brilliant article by Helen M. Lynd, 'Realism and the Intellectual in a Time of Crisis', *The American Scholar* (Winter 1951–2), p. 26.

7 As Marx observed, speaking of the Paris Commune, 'all this ... only proves that the bourgeois of our days considers himself the legitimate successor to the baron of old, who thought every weapon in his own hand fair against the plebeian, while in the hands of the plebeian a weapon of any kind constituted in itself a crime'. *The Civil War in France*, in Marx and Engels, *Selected Works* (Moscow, 1949–50), Vol. I, p. 489.

8 Para. 38.

9 *War or Peace* (New York, 1950), p. 256.

10 *Promoting Economic Development* (Claremont, California, 1955), p. 6.

11 E. Varga and L. Mendelsohn (eds.), *New Data for Lenin's Imperialism – The Highest Stage of Capitalism* (New York, 1940), p. 184.

12 op. cit., p. 245.

13 ibid., p. 229.

14 op. cit., p. 29.

15 Dulles, op. cit., p. 260.

16 'How Strong Is India?' The *Nation* (12 March 1955), p. 216.

17 Thus while the Rockefeller Foundation has devoted a growing part of its present disbursements to the promotion of divinity schools and other religious pursuits in the United States, the Ford Foundation has been lavishly financing Moslem, Buddhist, and similar enterprises in the underdeveloped countries.

18 op. cit., p. 254.

19 Marx and Engels, *Manifesto of the Communist Party*, in *Selected Works* (Moscow, 1949–50), Vol. I, p. 46.

20 Letter to Bebel, 11 December 1884, in Marx and Engels, *Selected Correspondence* (New York, 1934), p. 434. (Italics in the original.)

21 op. cit., p. 6.

22 cf. above, [*The Political Economy of Growth*], pp. 258–9.

23 This applies to the Philippines no less than to Formosa, to Iran no less than to South Korea, and to Spain no less than to Guatemala.

24 George F. Kennan, *American Diplomacy 1900–1950* (Chicago, 1951), pp. 116, 118.

25 *New York Times*, 20 May 1953.

26 This conviction partly accounts for the pronounced tendency in Western Europe as well as in India – even among people who are most critical of the Soviet Union – to blame the foreign policy of the United States for artificially generating an atmosphere of war danger.

27 Speech to the Overseas Press Club on 29 March 1954, as quoted in *Monthly Review* (May 1954), p. 2. (Italics supplied.)

28 E. H. Carr, *Studies in Revolution* (London, 1950), p. 226.

29 United States Department of State, *United States Relations with China* (Washington, 1949), p. xvi.

30 op. cit., p. 152.

31 *Sochinenya* [Works] (Moscow, 1947), Vol. 24, p. 381.

32 'Brigadier-General W. L. Roberts, U.S. Army, the commander of the Korean Military Advisory Group, told the New York Herald Tribune correspondent on June 5, 1950 ... "KMAG is a living demonstration of how an intelligent and intensive investment of 500 combat-hardened American men and officers can train 100,000 guys who will do the shooting for you. ... In Korea the American taxpayer has an army that is a fine watchdog over investments placed in this country and a force that represents the maximum results at the minimum cost." ' Quoted in Gunther Stein, *The World the Dollar Built* (London, 1952), p. 253.

33 'A New Plan to Defeat Communism', *U.S. News & World Report* (27 May 1955), p. 139. It should be noted, incidentally, that the views of General Sarnoff, then chairman of the Radio Corporation of America, cannot possibly be considered those of an eccentric. As the editors of *U.S. News & World Report* remark in their introductory statement, they were 'discussed thoroughly with President Eisenhower who commended ... the approach in his press conference'.

34 It is here that the political and ideological struggle against imperialism within the advanced capitalist countries, which reduces their willingness to start wars, links up directly with the effort to accelerate and facilitate economic and social progress in the underdeveloped countries, capitalist and socialist alike.

35 Sarnoff, op. cit., pp. 138, 140.

36 *War or Peace* (New York, 1950), p. 261.

'Capitalism and Cheap Labour-Power in South Africa: From Segregation to Apartheid'[1]

from *The Articulation of Modes of Production* (1980)

Harold Wolpe

Editors' Introduction

Harold Wolpe was born in 1926 in Johannesburg, South Africa, to a Lithuanian Jewish family. Wolpe completed a BA in Social Science as well as an LLB from the University of the Witwatersrand. Wolpe became politically active as a student, when he also joined the South African Communist Party (SACP). As a practising attorney, Wolpe represented key anti-apartheid activists, including Nelson Mandela and Walter Sisulu. In 1960, the government opened fire on a crowd of 300 protestors in the black township of Sharpeville, killing 180 and injuring 60. The Sharpeville Massacre was a turning point in state repression and resistance. Despite international condemnation, the South African state became more repressive, and Wolpe was arrested and jailed along with hundreds of other demonstrators. The white electorate voted for South Africa to become a republic, and to quit the British Commonwealth, and the emboldened white supremacist state banned resistance organisations, including the African National Congress (ANC) and the Pan African Congress (PAC). The ANC met in secret to form its new armed wing, *Umkhonto we Sizwe* (Spear of the Nation) in 1961, and in 1963 a large part of the leadership were captured at Lilliesleaf Farm, including Mandela and Sisulu. Wolpe himself tried to escape the country but was arrested at the border, only to engineer a courageous escape with three comrades from Marshall Square Prison in Johannesburg. Wolpe went into exile, initially in Dar-es-Salaam, and from 1964 to 1991 in London, before returning to South Africa after the un-banning of the ANC and SACP. Wolpe's daring escape brought him instant fame or notoriety (depending on one's stance on apartheid). Indeed, the apartheid state called him as a co-conspiritor in the Rivonia Trial, at which Mandela, Sisulu and others were imprisoned for almost 30 years. In England, Wolpe became a sociologist, and he taught at Bradford University, the Polytechnic of North London and Essex University. He remained a part of the exiled anti-apartheid movement, providing an important, independent critical voice. In Britain, Wolpe was part of a revival of Marxist social science on South Africa, including Frederick Johnstone, Shula Marks, Stanley Trapido and Martin Legassick. Never an academic in the narrow or elitist sense, Wolpe saw a role for careful theory that responds to political exigencies but does not provide manifestos. Wolpe also spent time at the University of Dar es Salaam and interacted with other key figures there, including Henry Bernstein and Mahmood Mamdani. From 1977 until his death in 1996, Wolpe also held an abiding interest in what sort of education policy would be suitable to a free South Africa. He worked on education policy within the exiled ANC, and, once in South Africa, at the Education Policy Unit at the University of the Western Cape in Cape Town.

The reading that follows is taken from Wolpe's 1972 article, which would be formative for social scientists and anti-apartheid activists. The article argues against the 'liberal' view that racial policy was 'pre-capitalist' and that it was separable from the capitalist economy. Wolpe demonstrates historically how capitalism was intrinsic to the construction of racial segregation in the early twentieth century, as well as to apartheid by mid-century. This was a 'radical' critique in that it meant that the anti-apartheid movement had to be a struggle against state-sanctioned racism *and* capitalism. Wolpe shows how the migrant labour regime constructed early in the twentieth century relied on 'cheap labour' provided by men who were subsidised by African communities eking out a subsistence in segregated 'Reserves'. Heightening resistance by mineworkers and squatters, and inter-racial political alliances of the oppressed, made it clearer to the state and capital that the capacity of the Reserves to continue to subsidise the system of cheap labour was running out of steam. Apartheid was a policy response to this political economic crisis. Racial ideology, argues Wolpe, 'sustains and reproduces' capitalism in this new form. Stuart Hall elaborates on the notion of 'articulation' as both linking and 'giving expression to', as in the colloquial meaning of articulation as speech. We could ask, for instance, how ideas of race, caste, gender and other ways of expressing difference and inequality link capitalism to other social institutions. There are echoes here of McClintock's (Part 1) use of the 'fetish object' as a means of expressing, and concealing, the mechanisms of power and exploitation.

Key references

Harold Wolpe (ed.) (1980) *The Articulation of Modes of Production*, London: Routledge and Kegan Paul.
— (1985) 'The Liberation Struggle and Research', *Review of African Political Economy*, 32: 72–8.
— (1988) *Race, Class and the Apartheid State*, London: James Currey.
Michael Burawoy (1994) 'From liberation to reconstruction: theory and practice in the life of Harold Wolpe', Wolpe Memorial Lecture, South Africa, July 2004. Online at http://sociology.berkeley.edu/faculty/burawoy/working papers.htm (accessed 5 October 2007).
Stuart Hall (1980) 'Race, articulation and society structured in dominance', in *Sociological Theories: Race and Colonialism*, Paris: Unesco.

■ ■ ■ ■ ■ ■ ■

I. INTRODUCTION

There is undoubtedly a high degree of continuity in the racist ideological foundations of Apartheid[2] and of the policy of Segregation which prevailed in the Union of South Africa prior to the election of the Nationalist Party to power in 1948. It is, perhaps, this continuity which accounts for the widely held view that fundamentally Apartheid is little more than Segregation under a new name. (See, for example, Legassick 1972; Walshe, 1963: 360; Bunting, 1964: 305.) As Legassick expresses it:

> After the Second World War segregation was continued, its premises unchanged, as *apartheid* or 'separate development'.
>
> (*ibid*: 31)

According to this view, such differences as emerged between Segregation and Apartheid are largely differences of degree relating to their common concerns – political domination, the African reserves and African migrant labour. More particularly, the argument continues, in the political sphere, Apartheid entails a considerable increase in White domination through the extension of the repressive powers of the State; the Bantustan policy involves the development of limited local government which, while falling far short of political independence and *leaving unchanged the economic and political functions* of the Reserves, nevertheless, in some ways, goes beyond the previous system in practice as well as in theory; and, in the economic sphere Apartheid 'modernises' the system of cheap *migrant* labour and perfects the instruments of labour coercion:

> *Apartheid*, or separate development, has meant merely tightening the loopholes, ironing out the

informalities, eliminating the evasions, modernizing and rationalizing the inter-war structures of 'segregationist' labour control.

(Legassick, 1972: 47)

[. . .]

. . . *apartheid* has meant an extension to the manufacturing economy of the structure of the gold-mining industry. In the towns, all remnants of African land and property ownership have been removed, and a massive building programme of so-called 'locations' or 'townships' means that the African work force is housed in carefully segregated and police controlled areas that resemble mining compounds on a large scale. All the terms on which Africans could have the right to reside permanently in the towns have been whittled away so that to-day no African . . . has a right to permanent residence except in the 'reserves'. . . .

(1972: 47)

The attempt to relate alterations in policy to changes in social conditions – primarily the development of a class of manufacturing industrialists – unquestionably represents an advance over the simplistic view that Apartheid is the result of ideology. Intense secondary industrialization does have a bearing on the development of Apartheid but the mere fact that it occurs does not explain why it should lead to the attempt to extend the 'structures of gold-mining' to the economy as a whole (just as the mere fact of industrialization of itself does not lead to the opposite result – democratization, elimination of racial discrimination, etc. – as asserted by the 'industrialism thesis'). Legassick is clearly correct in arguing that secondary industrialization intensifies the demand for a *cheap* African labour force at various levels of skill and that this is accompanied by new problems of control for the capitalist state. The problems of control (including the control of wage levels) are *not*, however, simply or primarily a function of the *demand* for labour-power which is cheap, but crucially a function of the conditions of the production and reproduction of that labour-power. It is in this respect that the crucial gap in Legassick's analysis appears, for by focusing largely on the development of secondary industrialization and by assuming that the economic and political functions of the Reserves continue unchanged and, therefore, that the migrant labour system remains what it has always been, he fails to grasp the essential nature of the changes which have occurred in South Africa. The analysis of these essential changes – the virtual destruction of the pre-capitalist mode of production of the African communities in the Reserves and, therefore, of the economic basis of cheap *migrant* labour-power and the consequent changes in and functions of 'tribal' political institutions – will constitute the subject matter of the third and subsequent sections of the paper.

[. . .]

V. APARTHEID: THE NEW BASIS OF CHEAP LABOUR

The focus in the two previous Sections has been largely on the economic foundation of cheap migrant labour-power in the Reserve economy and on the processes which have continuously and to an ever increasing degree undermined this foundation. The immediate result of the decline in the productive capacity of the pre-capitalist economies was a decrease in the agricultural product of the Reserves resulting, therefore, in a decrease of the contribution of the Reserves towards the subsistence necessary for the reproduction of the labour force. This threatened to reduce the rate of surplus value through pressure on wages and posed, for capital, the problem of preventing a fall in the level of profit.

The solution, for capital, to this problem must take account of the complementary effect of the erosion of the economic foundations of cheap migrant labour-power, upon both the African rural societies and the urbanized industrial proletariat. I have already shown that the system of producing a cheap migrant labour force generated rural impoverishment, while at the same time it enabled extremely low wages to be paid to Africans in the capitalist sector. But increasing rural impoverishment, since it removes that portion of the industrial workers' subsistence which is produced and consumed in the Reserves, also intensifies urban poverty. This twofold effect of capitalist development tends to generate conflict, not only about wages, but about all aspects of urban and rural life and to bring into question the structure of the whole society. This broadening and intensification of conflict is met by political measures which in turn lead to an increasingly political reaction. Clearly, the nature, form and extent of the conflicts generated by the structural conditions will depend not only upon the measures of state control but on the complex conjuncture of political ideologies and organization, trade unions, the cohesion of the dominant sector, and so on. Although these may vary, what is continuously present, it must be stressed, is the tendency for the structural conditions to generate conflicts, in one form or another, which centre on the system of cheap labour.

This struggle began long before 1948 when the conditions discussed above began to emerge (and control measures to be taken), but the particularly rapid urbanization and industrialization fostered by the Second World War sharpened and intensified the trends we have been discussing and the resultant conflicts. The 1940s were characterized by the variety and extent of the industrial and political conflicts especially in the urban, but also in the rural areas. In the period 1940–49 1,684,915 (including the massive strike of African mineworkers in 1946) African man-hours were lost as compared with 171,088 in the period 1930–39. Thousands of African workers participated in squatters' movements and bus boycotts. In 1946 the first steps were taken towards an alliance of African, Coloured and Indian political movements and this was followed by mass political demonstrations. Towards the end of the 1940s a new force – militant African intellectuals – appeared on the scene. There were militant rural struggles at Witzieshoek and in the Transkei. These were some of the signs of the growing assault on the whole society (and the structure of cheap labour-power which underpinned it) which confronted the capitalist state in 1948.

For English dominated large-scale capital (particularly mining but also sections of secondary industry), the solution, both to the problem of the level of profit and to the threat to their political control implicit in growing African militance was to somewhat alter the structure of Segregation in favour of Africans. Indeed, the 1948 recommendations of the Native Law Commission (appointed in 1946 by the United Party Government precisely in response to the changing nature of African political struggle) for an alternative mode of control of African labour which included certain restricted reforms and modifications of the racial political-economic structure, were accepted by the United Party as its policy in the 1948 election in which it was defeated by the Nationalist Party. The implementation, had it occurred, of that policy might possibly have had consequences for both the Afrikaner petit-bourgeoisie and, also, the White workers which would have led them into a collision with the State. The point is that reforms which would have resulted in higher real wages and improved economic conditions for Africans could only be introduced without a corresponding fall in the rate of profit provided they were bought at the cost of the White working-class – that is to say, either through a drop in the wages of White workers or the employment of Africans, at lower rates of pay, in occupations monopolized, until then, by White workers. Historically the latter aspect has been at the centre of the conflicts and tension between the White working-class and large-scale capital – conflicts which also reached their peak in terms of strikes in the 1940s.

The alternative for the Afrikaner working-class, resisting competition from African workers, for the growing Afrikaner industrial and financial capitalist class, struggling against the dominance of English monopoly capital, and, perhaps, for a petit-bourgeoisie threatened with proletarianization by the advance of African workers (and the Indian petit-bourgeoisie), was to assert control over the African and other non-White people by whatever means were necessary.[3] For the Afrikaner capitalist class, African labour-power could be maintained as cheap labour-power by repression; for the White worker, this also guaranteed their own position as a 'labour-aristocracy'.

Thus the policy of Apartheid developed as a response to this urban and rural challenge to the system which emerged inexorably from the changed basis of cheap labour-power. What was at stake was nothing less that the reproduction of the labour force, not in general, but in a specific form, in the form of cheap labour-power. Within its framework Apartheid combined both institutionalizing and legitimating mechanisms and, overwhelmingly, coercive measures.

It is beyond the scope of the present paper to set out in any detail the structure of coercive control erected by the Nationalist Government. In a fuller account it would be necessary to do this and to show how Apartheid, as a response to the principal contradiction between capital and cheap African labour, ramifies out and penetrates into the secondary contradictions which in turn have, to some extent, a reciprocal effect on the system. It will be sufficient at this point to refer to three aspects of the mainly coercive mechanisms of Apartheid.

It is true, as was pointed out in the Introduction, that Apartheid differs from Segregation in the degree to which it perfects the mechanisms for ordering the non-White population. This applies particularly to the coercive apparatus of the State.

At the most general level, that of control of the African political challenge, Apartheid entails the removal of the limited rights which Africans and Coloureds had in the Parliamentary institutions of the White State; the revision of old and the introduction of a whole complex of new repressive laws which make illegal militant organized opposition (e.g. Suppression of Communism, Unlawful Organizations and Sabotage Acts, etc.), and the building of all-powerful agencies of control – security police, Bureau of State Security, the army, and police and army civilian reserves, etc.

In the economic sphere measures have been introduced to prevent or contain the accumulation of pressure on the level of wages. Most obvious in this regard is the Natives (Settlement of Disputes) Act which makes it illegal for Africans to strike for higher wages or improved working conditions. This, coupled both with the fact that African trade unions are not legally recognized and that their organization is impeded also by other measures, has effectively prevented the emergence of an African trade union movement capable of having any significant effect on wages. The decline in industrial strikes since 1948 and the tendency of real wages for Africans to fall indicates the success of Government policy.

Less obvious, but having the same purpose of controlling the development of strong African pressure for higher wages, are the important measures introduced by the Nationalist Government relating to African job and geographical mobility. The nature and meaning of these measures has been obscured by the terms of the relevant laws and the Government's policy statements to the effect that Africans were to be regarded only as temporary migrants in the urban areas, there only as long as they ministered to White needs.

The Pass Laws and the Native Urban Areas Act 1925 which regulated the right of residence in urban areas, were, of course, available in 1948. The 'modernization' of the Pass Laws (under the Native (Abolition of Passes and Coordination of Documents) Acts) and the establishment of labour bureaux which serve to direct African workers to where White employers require them has been effected through a battery of amendments to old laws and the introduction of new laws which give the State exceptionally wide powers to order Africans out of one area and into another. There are practically no legal limitations on these powers which can be used to remove 'excess' Africans from areas where their labour is not required or 'troublesome' Africans to outlying, isolated areas where they will be politically harmless. All Africans are, legally, only temporary residents in the urban areas.

These facts have been interpreted as meaning that the Government has elaborated and perfected the migrant labour system. Control over residence and movement is clearly one essential element of a system based on a migrant labour-force, but it is not the only one. Therefore, to treat the increase in the State's legal power to declare Africans temporary sojourners in the urban area and to move them as exigencies demand as constituting the 'modernization' of the system, without taking account of changes in its economic basis, is insufficient. In the present case it results in the failure to grasp the essential changes in the nature of capitalist exploitation in South Africa. It has been the main contention of this paper that in South Africa the migrant labour-force, *properly speaking*, did not mean *merely* a mobile labour-force, or a labour-force that could be made mobile, that is that could be directed and redirected to where it was required. Above all, a *migrant* labour-force is labour-force which is both mobile *and* which has a particular economic basis in the pre-capitalist Reserve economy. With the disappearance of that economic basis, I have argued, the problems of curtailing industrial action and of political control over Africans in the urban areas became extremely acute. That control is exercised, in part, by repressive measures including, importantly, the elaboration of the State's power over the residence and movement of labour. That is to say, the extension of the State's power over the residence and movement of the labour-force, which adds to the State's repressive control over it (precisely, one feature of Apartheid) is a function of the economic changes in the Reserves which generate a threat to the cheapness of labour-power.

The pressure for higher wages has also been combated by largely ideological mechanisms. Two of the most important are, on the one hand, the calculation of the minimum wage required by African workers and, on the other, Bantu Education.

In regard to the former it need only be noted that all the calculations (whether made by the South African Institute of Race Relations, industry or academic investigations) are based on the given, historically determined, 'need' of African workers calculated on the basis of the existing enforced low standards of living. That is why the minimum found to be 'required' by Africans is so many times lower than that required by Whites. In this way low wages for Africans are legitimated.

The pressure on wage levels does not arise only from the economic conditions already discussed but is intensified by the very nature of the changes in the division of labour brought about by the expansion in manufacturing. One of the effects of this expansion is a new and acknowledged demand for African semi-skilled and even skilled labour-power which brings with it higher wages (although still much lower than White wages) and provokes demands for increased wages. The employment of Africans in these occupations results also in the expansion, at the lowest levels, of African education which in the form of 'Bantu Education' reproduces through its ideological content the subjection of Africans.

In its application to the urban areas, Apartheid nevertheless appears predominantly and with ever increasing thoroughness in its coercive form. In its application to the Reserves it has undergone a number of changes in content – culminating in the programme of self-development – in which the attempt both to establish forms of control which Africans would regard as legitimate and to instititutionalize conflict has been an increasingly important ingredient although coercion is never absent. This policy towards the Reserves has been, whatever other purpose it may have had in addition, centrally concerned, as in the past, with the control and supply of a cheap labour-force, *but in a new form.*

The idea of the total separation of the races although an integral element of the Nationalist Party's programme, was not regarded as an attainable objective by the Government. The impossibility of achieving total separation was underlined by the Tomlinson Commission which estimated (or rather, as we now know, grossly over-estimated) that by the turn of the century, if all its recommendations for the reconstruction of the Reserves were implemented, there would be *parity* of Whites and Africans in the 'White' areas.

Nor, in the early years of its regime, did the Government accept the possibility of the Reserves becoming *self-governing and autonomous* areas. In 1951 Verwoerd (then Minister of Native Affairs) told the Upper House of Parliament that the Opposition had tried to create the impression that:

I had announced the forming of an independent Native State . . . a sort of Bantustan with its own leader . . . that is not the policy of the Party. It has never been that, and no leader has ever said it, and most certainly I have not. The Senator wants to know whether the series of self-governing areas will be sovereign. The answer is obvious. How could small scattered states arise? We cannot mean that we intend by that to cut large slices out of South Africa and turn them into independent states.

There is, in fact, little to suggest that, in the first few years of rule, the Nationalist Party had a fully worked out policy in relation to the Reserves or one which differed significantly from that of earlier governments. There are, however, two important points to be noted.

Firstly, the Government already had clearly in mind the establishment of an apparatus of control which would be cheap to run and acceptable to the African people. The 1951 Bantu Authorities Act which strengthened the political authority of the (compliant) chiefs, subject to the control of the State – indirect rule – was the first (and, at the outset, very conflictual) step in that direction. Secondly, political control in the Reserves was obviously recognized to be no solution to the problem of the never-ending enlargement of a working-class totally removed from the Reserves. In this the Government accepted, by implication, the contention, emphasized by the Native Laws Commission (1948), that the flow of people off the land and into the urban areas was an economic process and in 1949 the Tomlinson Commission was established to:

conduct an exhaustive enquiry into and to report on a comprehensive scheme for the rehabilitation of the Native Areas with a view to developing within them a social structure in keeping with the culture of the Native and based on effective socio-economic planning.

(p. xviii)

Nevertheless, the rather modest proposals of the Commission to spend £104 million over ten years for the reconstruction of the Reserves, to end one-man-one-plot in order to create a stable class of farmers and a landless class of workers, and to develop the Reserves economically through White capital investment on the borders and in the Reserves themselves, were not accepted by the Government. There are probably two reasons for this rejection. Firstly, facts brought to light by the Commission showed that to implement the Commission's recommendations relating to agricultural development would have served simply to hasten the ongoing processes which were obviously resulting in the formation of a class of landless rural dwellers and to intensify the migration of workers to the urban centres resulting in a class of workers unable to draw on the Reserves for additional subsistence. Consequently, expenditure on agricultural improvement may have seemed pointless and even dangerous since it would exacerbate the pressures and conflicts in the towns. Secondly, the abolition of restrictions on land-holding and the assisted development of a class of 'kulaks', as recommended by the Commission, also carried with it certain possible dangers. On the one hand, this could lead to a resurgence of African competition to White farmers which it had been one of the purposes of the Native Land Act of 1913 to destroy. On the other hand, the emergence of an economically strong class of large peasants presented a potential political threat to White domination.

Whatever the reasons, by 1959 the Government's policy began to change in significant respects. Without attempting to set out a chronological record, I want to analyse the emergence after 1959 of separate development as the mode of maintaining cheap labour in the Reserves (complementing that in the urban areas) which takes as given the changes in the African 'tribal' economies and erects, under the overarching power of the capitalist state, an institutionalized system of partial political control by Africans. That is to say, the practice and policy of Separate Development must be seen as the attempt to retain, in a modified form, the structure of the 'traditional' societies, not, as in the past, for the purposes of ensuring an economic supplement to the wages of the migrant labour-force, but for the purposes of reproducing and exercising control over a cheap African industrial labour-force in or near the 'homelands', not by means of preserving the pre-capitalist mode of production but by the political, social, economic and ideological enforcement of low levels of subsistence.

In 1959, in the Parliamentary debate on the Promotion of Bantu Self-Government Act, the Prime Minister Dr. Verwoerd stated:

> . . . if it is within the capacity of the Bantu, and if those areas which are allocated to him for his emancipation, or rather, which are already his own, can develop *into full independence*, then it will develop in this way.
>
> (Hansard, 1959, Col. 6520)

This was echoed by Vorster in 1968 (Hansard Col. 3947):

> We have stated very clearly that we shall lead them to independence.

Significantly, the ideological shift from White supremacy to self-determination and independence was accompanied by a parallel alteration in the ideology of race. Thus, whereas in all its essentials Nationalist Party ideology had previously insisted upon the biological inferiority of Africans as the justification for its racialist policies, as the Government was impelled towards the Bantustan policy so it began to abandon certain of its previous ideological positions. Now the stress fell upon ethnic *differences* and the central notion became 'different but equal'. Already in 1959 the Minister of Bantu. Affairs and Development, De Wet Wel stated: (Hansard 1959, Col. 6018.)

> There is something . . . which binds people, and that is their spiritual treasures, the cultural treasure of a people. It is those things which have united other nations in the world. That is why we say that the basis of our approach is that the Bantu, too, will be linked together by traditional and emotional bonds, by their own language, their own culture, their national possessions . . .

More and more the term 'race' gives way to 'nation', 'ethnic group', 'volk'. A fairly recent example appears in the Congress Report (1971), of the National Party's ideological organization, the South African Bureau of Racial Affairs, in which it was stated that the Bureau had from the beginning proceeded upon the principle that separate development was the only basis for good relations between

> . . . the different races as it was previously referred to, but rather different peoples (volk) as we now put it . . .

There is an obvious necessity for this ideological change since a policy of ethnic political independence (for each of the eight ethnic groups identified) was incompatible with an ideology of racial inferiority. Nor would the latter have facilitated the attempt to set up the complex machinery of government and administration intended, in fact, to institutionalize relations between the State and the Reserves *and* to carry out certain administrative functions necessary for economic development in the Reserves. What all this amounts to, as one writer has expressed it, is 'racialism without racism'.

The Transkei Constitution Act was passed in 1963[4] and provided for a legislative assembly to exercise control over finance, justice, interior, education, agriculture and forestry, and roads and works. The Republican Government retains control, *inter alia*, over defence, external affairs, internal security, postal and related services, railways, immigration, currency, banking and customs. It need hardly be stressed that this arrangement in no way approaches political independence. At the same time it must not be overlooked that within limits, set both by the Constitution and the available resources, the Transkeian Government exercises real administrative power. By this means the South African State is able to secure the execution of certain essential social control and administrative functions at low cost particularly as a considerable portion of Government expenditure can be obtained through increased general taxes. Thus in 1971 the Transkeian Government's budget was £18 million of which £3½ million was obtained through taxation of Transkeian citizens.

It is, however, in the sphere of economic development that the emerging role of the Reserves can be seen most clearly. I am not here referring to the rather minor role of the various development corporations (Bantu Development Corporation, Xhosa Development Corporation and so on) in fostering economic development in the Reserves. In fact, up to the present they have largely served to assist small traders and commercial interests by means of loans – that is, they appear to be instruments for the nurturing of a petit-bourgeoisie and have little to do with economic growth in the reserves. Far more important is the State's policy of industrial decentralization.

This policy which has been the subject of government commissions and legislation is also the concern of a Permanent Committee for the Location of Industry. At all times the policy of decentralization has been tied to the Bantustan policy and this meant, at first, the establishment of 'White' industries on the borders of the black 'homelands'. Between 1960 and 1968 some £160 million was invested in industrial plant in the border areas and approximately 100,000 Africans were employed in these industries which were absorbing 30% of Africans entering jobs each year by 1969. By 1971 there were plans for a rapid expansion (including car factories and chemical plants) of industrial development in the border regions. The point has correctly been made that to date most border industries have been established in areas close to the main industrial regions of South Africa including Johannesburg, Pretoria and Durban. This is due to the fact that in remoter border regions the State, in the main, has not provided the necessary infrastructure of transport, communications and so on. But, why decentralize to the borders in any event? One answer has been to suggest that the purpose of border development is to stem the drift of Africans into 'White' South Africa. The question is why? I would suggest that the policy of border industrial development can only be understood if it is seen as an alternative to migration as a mechanism for producing cheap labour-power. There are three aspects of the situation which need to be stressed.

Firstly, neither the provisions of the Industrial Conciliation Act nor Wages Act determinations made for other regions apply to the border industries. This is extremely important in two respects. Since the Industrial Conciliation Act is inapplicable, Section 77 which empowers the Minister of Labour to reserve certain jobs for particular racial groups also does not apply and neither do the provisions of industrial agreements which reserve the higher paid skilled jobs for White workers. This being so it becomes possible to employ Africans in jobs which, in the 'White' areas, are the exclusive preserve of White workers. The effect of this, in conjunction with the inapplicability of wage determinations for other areas, is that a totally different and much lower wage structure becomes possible and has arisen.

Secondly, as elsewhere, African trade unions are not recognized and the provisions of the Natives (Settlement of Disputes) Act apply.

The third, and in some ways perhaps the most important aspect, relates to the conditions of life of the African workers in the border industries. Not only, as has already been indicated, is the level of subsistence extremely low in the 'homelands' but in addition there are virtually no urban areas which might tend to increase this level. The assessment by the State, employers' organizations and so on, of African subsistence requirements in the Reserves is much lower than in the main industrial centres. This fact is not altered (or, at least will not be altered for a considerable period) by the necessity of establishing townships of some kind for the housing of workers employed in industry. It is an interesting index of the State's policy that a major item of expenditure for the so-called development of the Reserves has been for town-planning. A United Nations Report (No. 26, 1970: 15) stated:

> Town planning has throughout been a major portion of expenditure. Thus in 1961 a five-year development plan for the reserves was inaugurated which projected an expenditure of £57 million, but *two-thirds* of this amount was allocated for town-planning, while the next largest item – £7.3 million – was for soil conservation.

The towns planned will be, no doubt, simple in the extreme, supplying little in the way of the complex services and infrastructure of the 'White' urban areas. Despite the State's expenditure all the indications are that what will be established will be rural village slums[5] alleviated marginally, if the Transkei is typical, by the allocation of garden allotments for the purpose of the production of vegetables, etc., which, incidentally, will no doubt provide the rationale for lower wages.

Recently, the Government reversed its previous rejection of the Tomlinson Commission's recommendation that Whites be allowed, under certain conditions, to invest capital in the Reserves. As in the case of the border industries various incentives are held out to induce investment. These include 'tax-holidays', tariff reductions, development loans and so on. All the considerations discussed above in relation to the border industries apply with equal force to industrial development within the Reserves.

It is still too soon to say anything about the likely level of investment inside the Reserves although some investment has already occurred. Nevertheless, the change in policy must be seen as a further significant step towards the establishment of an extensive structure of cheap labour-power in the Reserves.

VI. CONCLUSION

The argument in this paper shows that Apartheid cannot be seen merely as a reflection of racial ideologies and nor can it be reduced to a simple extension of Segregation.

Racial ideology in South Africa must be seen as an ideology which sustains and reproduces capitalist relations of production. This ideology and the political practice in which it is reflected is in a complex, reciprocal (although asymmetrical) relationship with changing social and economic conditions. The response of the dominant classes to the changing conditions, mediated by these ideologies, produces the two faces of domination – Segregation and Apartheid.

The major contradiction of South African society between the capitalist mode of production and African pre-capitalist economies is giving way to a dominant contradiction *within* the capitalist economy. The consequence of this is to integrate race relations with capitalist relations of production to such a degree that the challenge to the one becomes of necessity a challenge to the other. Whether capitalism still has space (or time) for reform in South Africa is an issue which must be left to another occasion.

NOTES

1   In revising an earlier draft of this paper I have benefited from criticisms and comments made by a number of people. I am particularly grateful to S. Feuchtwang, R. Hallam, C. Meillassoux and M. Legassick.

2   Although the term 'Apartheid' has more or less given way to 'self-development' in the language of the Nationalist Party, it remains the term most widely used to characterize the present system in South Africa. In this paper I intend to use 'Apartheid' as the generic term to refer to the period (its policies, practices and ideology) since the Nationalist Party took office in 1948 – in this sense it subsumes the policy of 'separate

development'. 'Racial Segregation' or 'Segregation' is employed throughout, unless the context indicates a contrary intention, to refer to the ideology, policies and practices prior to 1948. It need hardly be added that although 1948 is obviously a year of great importance it is not intended to suggest that in that year Apartheid replaced Segregation.

3   A number of different analyses have been made of the position of different classes in the development of Apartheid. See, for example, Legassick (1972) and Clenaghen (1972). It is not necessary in the present paper, given its concern with the central relationship between White capital and cheap African labour, to pursue this point here.

4   Other Bantustans are in various stages of formation.

5   In a new appendix to the new 1971 edition of his book Mayer (1962) provides an account of a 'dormitory town' which shows that this is exactly what is happening.

REFERENCES

Bunting, B. P. (1964) *The Rise of the South African Reich*, London, Penguin Books.

Clenaghen, W. (1972) 'The State in South Africa', unpublished Seminar Paper.

Legassick, M. (1972) 'South Africa: Forced Labour, Industrialization, and Racial Differentiation', to be published in a forthcoming volume in a series on the political economy of the Third World edited by Richard Harris.

Mayer, P. (1902) *Townsmen or Tribesmen*, London, Oxford University Press.

Report of the Commission for the Socio-Economic Development of the Bantu Areas Within the Union of South Africa (Summary) (U.G. 61/1955) (Tomlinson Commission).

Walshe, A. P. (1963) 'The Changing Content of Apartheid', *Review of Politics*, Vol. XXV, 343–61.

FIVE

'The Obligation to Endure'

From *Silent Spring* (1962)

Rachel Carson

Editors' Introduction

Rachel Carson was born in Pennsylvania, USA, in 1907 and died in 1964. With an academic background in marine biology and zoology, Carson became an inspiration for the global environmental movement with the publication of her book, *Silent Spring*, in 1962. It is important to recognise, however, that *Silent Spring* grew out of a long career of writing, research and advocacy, much of it spent under more trying conditions than commonly faced her (mainly male) peers in the academic world.

In addition to extensive family duties, Rachel Carson pursued her scientific interests, after university, mainly within the US Bureau of Fisheries. Carson joined that agency on a full-time basis in 1936, having come top in the relevant Civil Service exam. She was only the second woman to be appointed to the bureau and took up a position as a junior aquatic biologist. Carson immediately began a parallel career as a science writer, albeit with considerable literary flair. An early article called 'Undersea' was published by *Atlantic Monthly* in 1937. A subsequent work took shape in *Nature* magazine and was published as a book – *The Sea Around Us* (1951). It won a 1952 National Book Award, remained in the bestseller lists for over a year, allowed Carson to work full-time as a writer, and generally made her reputation. The publication of *Silent Spring* built on this fame, which was just as well. Carson had become by the 1950s a major critic of excessive pesticide use in US agriculture and of DDT more specifically. Key figures in the US chemical industry leant on Carson's publisher, Houghton Mifflin, to squash the book – it resisted – and some well-known figures in both the business and academic worlds tried hard to damn the book as the work of an 'unqualified' or even 'hysterical' woman. Carson was accused of wanting all pesticides to be banned, when her position was to ask for less and more responsible use of the most damaging pesticides. *Silent Spring* also strikes a chord – now as then – because of the voice it uses. Carson writes against the God-like pose of some scientists and speaks for what might be called the accumulated wisdom of the ages.

Rachel Carson lived to see the banning of DDT in the United States, but she had by then been diagnosed with breast cancer. Fourteen years after her death she was awarded the Presidential Medal of Freedom. Her work continues to be widely read and discussed.

Key references

Rachel Carson (1962) *Silent Spring*, Boston, MA: Houghton and Mifflin.
— (1951) *The Sea Around Us*, New York: Oxford University Press.
— (1965) *The Sense of Wonder*, New York: Harper and Row.
Rachel Carson with L. Lear (1998) *Lost Woods: The Discovered Writing of Rachel Carson*, Boston, MA: Beacon Press.
L. Lear (1997) *Rachel Carson: Witness for Nature*, New York: Henry Holt.

The history of life on earth has been a history of interaction between living things and their surroundings. To a large extent, the physical form and the habits of the earth's vegetation and its animal life have been moulded by the environment. Considering the whole span of earthly time, the opposite effect, in which life actually modifies its surroundings, has been relatively slight. Only within the moment of time represented by the present century has one species – man – acquired significant power to alter the nature of his world.

During the past quarter-century this power has not only increased to one of disturbing magnitude but it has changed in character. The most alarming of all man's assaults upon the environment is the contamination of air, earth, rivers, and sea with dangerous and even lethal materials. This pollution is for the most part irrecoverable; the chain of evil it initiates not only in the world that must support life but in living tissues is for the most part irreversible. In this now universal contamination of the environment, chemicals are the sinister and little-recognized partners of radiation in changing the very nature of the world – the very nature of its life. Strontium 90, released through nuclear explosions into the air, comes to earth in rain or drifts down as fallout, lodges in soil, enters into the grass or corn or wheat grown there, and in time takes up its abode in the bones of a human being, there to remain until his death. Similarly, chemicals sprayed on croplands or forests or gardens lie long in soil, entering into living organisms, passing from one to another in a chain of poisoning and death. Or they pass mysteriously by underground streams until they emerge and, through the alchemy of air and sunlight, combine into new forms that kill vegetation, sicken cattle, and work unknown harm on those who drink from once-pure wells. As Albert Schweitzer has said, 'Man can hardly even recognize the devils of his own creation.'

It took hundreds of millions of years to produce the life that now inhabits the earth – aeons of time in which that developing and evolving and diversifying life reached a state of adjustment and balance with its surroundings. The environment, rigorously shaping and directing the life it supported, contained elements that were hostile as well as supporting. Certain rocks gave out dangerous radiation; even

within the light of the sun, from which all life draws its energy, there were short-wave radiations with power to injure. Given time – time not in years but in millennia – life adjusts, and a balance has been reached. For time is the essential ingredient; but in the modern world there is no time.

The rapidity of change and the speed with which new situations are created follow the impetuous and heedless pace of man rather than the deliberate pace of nature. Radiation is no longer merely the background radiation of rocks, the bombardment of cosmic rays, the ultra-violet of the sun that have existed before there was any life on earth; radiation is now the unnatural creation of man's tampering with the atom. The chemicals to which life is asked to make its adjustment are no longer merely the calcium and silica and copper and all the rest of the minerals washed out of the rocks and carried in rivers to the sea; they are the synthetic creations of man's inventive mind, brewed in his laboratories, and having no counterparts in nature.

To adjust to these chemicals would require time on the scale that is nature's; it would require not merely the years of a man's life but the life of generations. And even this, were it by some miracle possible, would be futile, for the new chemicals come from our laboratories in an endless stream; almost five hundred annually find their way into actual use in the United States alone. The figure is staggering and its implications are not easily grasped – five hundred new chemicals to which the bodies of men and animals are required somehow to adapt each year, chemicals totally outside the limits of biologic experience.

Among them are many that are used in man's war against nature. Since the mid 1940s over two hundred basic chemicals have been created for use in killing insects, weeds, rodents, and other organisms described in the modern vernacular as 'pests'; and they are sold under several thousand different brand names.

These sprays, dusts and aerosols are now applied almost universally to farms, gardens forests, and homes – non-selective chemicals that have the power to kill every insect, the 'good' and the 'bad', to still the song of birds and the leaping of fish in the streams, to coat the leaves with a deadly film, and to linger on in soil – all this though the intended target may be only a few weeds or insects. Can anyone believe it is possible to lay down such a barrage of poisons on the surface of the earth without making it unfit for all life? They should not be called 'insecticides', but 'biocides'.

The whole process of spraying seems caught up in an endless spiral. Since DDT was released for civilian use, a process of escalation has been going on in which ever more toxic materials must be found. This has happened because insects, in a triumphant vindication of Darwin's principle of the survival of the fittest, have evolved super races immune to the particular insecticide used, hence a deadlier one has always to be developed – and then a deadlier one than that. It has happened also because, for reasons to be described later, destructive insects often undergo a 'flareback', or resurgence, after spraying, in numbers greater than before. Thus the chemical war is never won, and all life is caught in its violent crossfire.

Along with the possibility of the extinction of mankind by nuclear war, the central problem of our age has therefore become the contamination of man's total environment with such substances of incredible potential for harm – substances that accumulate in the tissues of plants and animals and even penetrate the germ cells to shatter or alter the very material of heredity upon which the shape of the future depends.

Some would-be architects of our future look towards a time when it will be possible to alter the human germ plasm by design. But we may easily be doing so now by inadvertence, for many chemicals, like radiation, bring about gene mutations. It is ironic to think that man might determine his own future by something so seemingly trivial as the choice of an insect spray.

All this has been risked – for what? Future historians may well be amazed by our distorted sense of proportion. How could intelligent beings seek to control a few unwanted species by a method that contaminated the entire environment and brought the threat of disease and death even to their own kind? Yet this is precisely what we have done. We have done it, moreover, for reasons that collapse the moment we examine them. We are told that the enormous and expanding use of pesticides is necessary to maintain farm production. Yet is our real problem not one of *over-production*? Our farms, despite measures to remove acreages from

production and to pay farmers *not* to produce, have yielded such a staggering excess of crops that the American taxpayer in 1962 is paying out more than one billion dollars a year as the total carrying cost of the surplus-food storage programme. And the situation is not helped when one branch of the Agriculture Department tries to reduce production while another states, as it did in 1958,

It is believed generally that reduction of crop acreages under provisions of the Soil Bank will stimulate interest in use of chemicals to obtain maximum production on the land retained in crops.

All this is not to say there is no insect problem and no need of control. I am saying, rather, that control must be geared to realities, not to mythical situations, and that the methods employed must be such that they do not destroy us along with the insects.

The problem whose attempted solution has brought such a train of disaster in its wake is an accompaniment of our modern way of life. Long before the age of man, insects inhabited the earth – a group of extraordinarily varied and adaptable beings. Over the course of time since man's advent, a small percentage of the more than half a million species of insects have come into conflict with human welfare in two principal ways: as competitors for the food supply and as carriers of human disease.

Disease-carrying insects become important where human beings are crowded together, especially under conditions where sanitation is poor, as in time of natural disaster or war or in situations of extreme poverty and deprivation. Then control of some sort becomes necessary. It is a sobering fact, however, as we shall presently see, that the method of massive chemical control has had only limited success, and also threatens to worsen the very conditions it is intended to curb.

Under primitive agricultural conditions the farmer had few insect problems. These arose with the intensification of agriculture – the devotion of immense acreages to a single crop. Such a system set the stage for explosive increases in specific insect populations. Single-crop farming does not take advantage of the principles by which nature works; it is agriculture as an engineer might conceive it to be. Nature has introduced great variety into the landscape, but man has displayed a passion for simplifying it. Thus he undoes the built-in checks and balances by which nature holds the species within bounds. One important natural check is a limit on the amount of suitable habitat for each species. Obviously then, an insect that lives on wheat can build up its population to much higher levels on a farm devoted to wheat than on one in which wheat is intermingled with other crops to which the insect is not adapted.

The same thing happens in other situations. A generation or more ago, the towns of large areas of the United States lined their streets with the noble elm tree. Now the beauty they hopefully created is threatened with complete destruction as disease sweeps through the elms, carried by a beetle that would have only a limited chance to build up large populations and to spread from tree to tree if the elms were only occasional trees in a richly diversified planting.

Another factor in the modern insect problem is one that must be viewed against a background of geologic and human history: the spreading of thousands of different kinds of organisms from their native homes to invade new territories. This world-wide migration has been studied and graphically described by the British ecologist Charles Elton in his recent book *The Ecology of Invasions*. During the Cretaceous Period, some hundred million years ago, flooding seas cut many land bridges between continents and living things found themselves confined in what Elton calls 'colossal separate nature reserves'. There, isolated from others of their kind, they developed many new species. When some of the land masses were joined again, about fifteen million years ago, these species began to move out into new territories – a movement that is not only still in progress but is now receiving considerable assistance from man.

The importation of plants is the primary agent in the modern spread of species, for animals have almost invariably gone along with the plants, quarantine being a comparatively recent and not completely effective innovation. The United States Office of Plant Introduction alone has introduced almost 200,000 species and varieties of plants from all over the world. Nearly half of the 180 or so major insect enemies of plants in the United States are accidental imports from abroad, and most of them have come as hitch-hikers on plants.

In new territory, out of reach of the restraining hand of the natural enemies that kept down its numbers in its native land, an invading plant or animal is able to become enormously abundant. Thus it is no accident that our most troublesome insects are introduced species.

These invasions, both the naturally occurring and those dependent on human assistance, are likely to continue indefinitely. Quarantine and massive chemical campaigns are only extremely expensive ways of buying time. We are faced, according to Dr Elton, 'with a life-and-death need not just to find new technological means of suppressing this plant or that animal'; instead we need the basic knowledge of animal populations and their relations to their surroundings that will 'promote an even balance and damp down the explosive power of outbreaks and new invasions'.

Much of the necessary knowledge is now avail-able but we do not use it. We train ecologists in our universities and even employ them in our govern-mental agencies but we seldom take their advice. We allow the chemical death rain to fall as though there were no alternative, whereas in fact there are many, and our ingenuity could soon discover many more if given opportunity.

Have we fallen into a mesmerized state that makes us accept as inevitable that which is inferior or detrimental, as though having lost the will or the vision to demand that which is good? Such thinking, in the words of the ecologist Paul Shepard,

idealizes life with only its head out of water, inches above the limits of toleration of the corruption of its own environment. . . . Why should we tolerate a diet of weak poisons, a home in insipid surround-ings, a circle of acquaintances who are not quite our enemies, the noise of motors with just enough relief to prevent insanity? Who would want to live in a world which is just not quite fatal?

Yet such a world is pressed upon us. The crusade to create a chemically sterile, insect-free world seems to have engendered a fanatic zeal on the part of many specialists and most of the so-called control agencies. On every hand there is evidence that those engaged in spraying operations exercise a ruthless power. 'The regulatory entomologists . . . function as prosecutor, judge and jury, tax assessor and collector and sheriff to enforce their own orders,' said

Connecticut entomologist Neely Turner. The most flagrant abuses go unchecked in both state and federal agencies.

It is not my contention that chemical insecticides must never be used. I do contend that we have put poisonous and biologically potent chemicals indiscriminately into the hands of persons largely or wholly ignorant of their potentials for harm. We have subjected enormous numbers of people to contact with these poisons, without their consent and often without their knowledge. If the Bill of Rights contains no guarantee that a citizen shall be secure against lethal poisons distributed either by private individuals or by public officials, it is surely only because our forefathers, despite their considerable wisdom and foresight, could conceive of no such problem.

I contend, furthermore, that we have allowed these chemicals to be used with little or no advance investigation of their effect on soil, water, wildlife, and man himself. Future generations are unlikely to condone our lack of prudent concern for the integrity of the natural world that supports all life.

There is still very limited awareness of the nature of the threat. This is an era of specialists, each of whom sees his own problem and is unaware of or intolerant of the larger frame into which it fits. It is also an era dominated by industry, in which the right to make a dollar at whatever cost is seldom challenged. When the public protests, confronted with some obvious evidence of damaging results of pesticide applications, it is fed little tranquillizing pills of half truth. We urgently need an end to these false assurances, to the sugar coating of unpalatable facts. It is the public that is being asked to assume the risks that the insect controllers calculate. The public must decide whether it wishes to continue on the present road, and it can do so only when in full possession of the facts. In the words of Jean Rostand, 'The obligation to endure gives us the right to know.'

LIST OF PRINCIPAL SOURCES

'Report on Environmental Health Problems', *Hearings*, 86th Congress, Subcom, of Com. on Appropriations, March 1960, p. 170.

The Pesticide Situation for 1957–58, U.S. Dept of Agric., Commodity Stabilization Service, April 1958, p. 10.

Elton, Charles S., *The Ecology of Invasions by Animals and Plants*, New York, Wiley, 1958; London, Methuen, 1958.

Shepard, Paul, 'The Place of Nature in Man's World', *Atlantic Naturalist*, Vol. 13 (April–June 1958), pp. 85–9.

'The Economics of Polygamy'

from *Women's Role in Economic Development* (1970)

Ester Boserup

Editors' Introduction

Ester Boserup was born in 1910 in Copenhagen. Boserup attended the University of Copenhagen between 1929 and 1935, where she studied theoretical economics. Boserup distinguished herself as a writer during her studies, and was also involved in an independent socialist group. For the next ten years, Boserup worked in the Danish economic administration, after which she spent another ten years with the United Nations Economic Commission for Europe. On the invitation of the great Swedish economist Gunnar Myrdal, Boserup went to India to participate in a study of South and South-East Asian agriculture. Boserup's engagement with agrarian livelihoods and with experts across India led her to question some of the prevailing foundations of economic thought, as for example on the existence of a surplus population in densely populated countries like India. Continuing various consultancies, Boserup wrote from her experience in South Asia the controversial book, *The Conditions of Agricultural Growth* (1965). Countering Malthusian arguments, Boserup argued that population pressure spurred agricultural innovation, particularly in increasing the frequency of cropping. Population pressure provides a disincentive to fallowing, and the decline in soil fertility leads to increased agricultural innovation, but also to an increased marginal cost of farming labour. Boserup called this process of innovation through more labour and lower efficiency a process of agricultural intensification.

After spending a year in Dakar in 1964–65, Boserup began to work on agriculture–industry linkages, and she became more interested in the gender division of labour through which male migrants left women in charge of subsistence for long periods. An interest in gender relations at work, and the centrality of women's labour in agrarian livelihoods, led to Boserup's landmark *Woman's Role in Economic Development* (1970), from which the reading that follows is excerpted.

Boserup's main argument in this book is that, since 'sex roles' vary across contexts, and since women are always primarily involved in 'reproduction', variation in their involvement in 'production' is incredibly important. In what she calls the 'female farming systems' prevalent in sub-Saharan Africa and South-East Asia, Boserup argued that involvement in production afforded women considerable independence. In contrast, 'male farming systems' in South and West Asia restricted women to 'reproductive' rather than 'productive' work, curtailing their liberties significantly. However, even in 'female farming systems', in Boserup's view, 'modernisation' discriminates against women in various ways – biasing land rights, education, technology and training to male farmers. The excerpted reading looks at the economic rationalities of polygamy, and to a lesser extent dowry, seclusion and veiling, as linked to the broad requirements of each context. The main policy recommendation Boserup makes in the book is for appropriate forms of female education to break gender barriers to modernisation. The reader should be able to situate Boserup's views on modernisation and the market, while also noting her careful attention to variance and practical outcomes. Boserup's work would become foundational to the 'Women in Development' approach in development institutions, which pressed for incorporating women in development projects. Boserup went on to publish *Population and Technological Change* (1981), which extends her earlier argument concerning population growth on rural and urban change. Ester Boserup died in 1999, a pioneering thinker on the centrality of gender relations to development processes.

Key references

Ester Boserup (1970) *Women's Role in Economic Development*, London: George Allen and Unwin.
— (1965) *The Conditions of Agricultural Growth: The Economics of Agrarian Change under Population Pressure*, Chicago, IL: Aldine Publishing.
— (1981) *Population and Technological Change: A Study of Long-Term Trends*, Chicago, IL: University of Chicago Press.
Naila Kabeer (2003 [1994]) *Reversed Realities: Gender Hierarchies in Development Thought*, London: Verso.
Sylvia Chant (2007) *Gender, Generation and Poverty: Exploring the 'Feminisation of Poverty' in Africa, Asia and Latin America*, Cheltenham: Edward Elgar.

Some years ago, UNESCO held a seminar on the status of women in South Asia. The seminar made this concluding statement after a discussion of the problem of polygamy: 'Polygamy might be due to economic reasons, that is to say, the nature of the principal source of livelihood of the social group concerned, e.g. agriculture, but data available to the Seminar would not permit any conclusions to be drawn on this point.'[1]

It is understandable that such a cautious conclusion should be drawn in Asia where the incidence of polygamy is low and diminishing. In Africa, however, polygamy is widespread, and nobody seems to doubt that its occurrence is closely related to economic conditions. A report by the secretariat of the UN Economic Commission for Africa (ECA) affirms this point: 'One of the strongest appeals of polygyny to men in Africa is precisely its economic aspect, for a man with several wives commands more land, can produce more food for his household and can achieve a high status due to the wealth which he can command.'[2]

It is self explanatory, after our discussion in the preceding chapter of female work input in African farming, that a man can get more food if he has more land and more wives to cultivate it. But why is it that the more wives he has got, the more land he can command, as the ECA statement says? The explanation lies in the fact that individual property in

land is far from being the only system of land tenure in Africa. Over much of the continent, tribal rules of land tenure are still in force. This implies that members of a tribe which commands a certain territory have a native right to take land under cultivation for food production and in many cases also for the cultivation of cash crops. Under this tenure system, an additional wife is an additional economic asset which helps the family to expand its production.

In regions of shifting cultivation, where women do all or most of the work of growing food crops, the task of felling the trees in preparation of new plots is usually done by older boys and very young men, as already mentioned. An elderly cultivator with several wives is likely to have a number of such boys who can be used for this purpose. By the combined efforts of young sons and young wives he may gradually expand his cultivation and become more and more prosperous, while a man with a single wife has less help in cultivation and is likely to have little or no help for felling. Hence, there is a direct relationship between the size of the area cultivated by a family and the number of wives in the family. For instance, in the Bwamba region of Uganda, in East Africa, it appeared from a sample study that men with one wife cultivated an average of 1·67 acres of land, while men with two wives cultivated 2·94 acres, or nearly twice as much. The author of the study describes women in this region as 'the cornerstone and the limiting factor in the sphere of agricultural production' and notes that almost all the men desire to have additional wives. A polygamic family is 'the ideal family organization from the man's point of view'.[3]

In female farming communities, a man with more than one wife can cultivate more land than a man with only one wife. Hence, the institution of polygamy is a significant element in the process of economic development in regions where additional land is available for cultivation under the long fallow system. In Chapter 1 we found an inverse correlation between the use of female family labour and the use of hired labour. It seems that farmers usually either have a great deal of help from their wives, or else they hire labour. Thus farmers in polygamic communities have a wider choice in this than have farmers in monogamic communities. In the former community, the use of additional female family labour is not limited to the amount of work that one wife and her children can perform; the total input of labour can be expanded by the acquisition of one or more additional wives.

This economic significance of polygamy is not restricted to the long fallow system of cultivation. In many regions, farmers have a choice between an expansion of cultivation by the use of more labour in long fallow cultivation, with a hoe, or an expansion by the transition to shorter fallow with ploughs drawn by animals.[4] In such cases, three possible ways of development present themselves to the farmer: expansion by technical change (the plough); expansion by hierarchization of the community (hired labour); or expansion by the traditional method of acquiring additional wives. In a study of economic development in Uganda, Audrey Richards pointed to this crucial role of polygamy as one of the possible ways to agricultural expansion: 'It is rare to find Africans passing out of the subsistence farm level without either the use of additional labour (read: hired labour E.B.), the introduction of the plough, which is not a practical proposition in Buganda; or by the maintenance of a large family unit, which is not a feature of Ganda social structure at the moment.'[5]

In the same vein, Little's classical study of the Mende in the West African state of Sierra Leone concluded that 'a plurality of wives is an agricultural asset, since a large number of women makes it unnecessary to employ much wage labour'.[6] At the time of Little's study (i.e. in the 1930s), it was accepted in the more rural areas that nobody could

run a proper farm unless he had at least four wives. Little found sixty-seven wives to the twenty-three cultivators included in his sample and an average of 2·3 wives per married man in a sample of 842 households. He describes how the work of one wife enables him to acquire an additional one: 'He says to his first wife, "I like such and such a girl. Let us make a bigger farm this year." As soon as the harvest is over for that year, he sells the rice and so acquires the fourth wife'.[7]

Little's study is thirty years old, and the incidence of polygamy has declined since then. But, although households with large numbers of wives seem to have more or less disappeared in most of Africa, polygamy is still extremely widespread and is considered an economic advantage in many rural areas. The present situation can be gleaned from Table 1, which brings together the results of a number of sample studies about the incidence of polygamy. It is seen that none of the more recent studies shows such a high incidence of polygamic marriages as in the period of Little's old study. Most of the studies show an average number of around 1·3 wives per married man.[8]

In most cases over one-fifth of all married men were found to have more than one wife at the time of enquiry.[9]

The acquisition of an additional wife is not always used as a means of becoming richer through the expansion of cultivation. In some cases, the economic role of the additional wife enables the husband to enjoy more leisure. The village study from Gambia mentioned in Chapter 1 showed that in this village, where rice is produced by women, men who had several wives to produce rice for them produced less millet (which is a crop produced by men) than did men with only one wife.[10] Likewise, in the villages in the Central African Republic men with two wives worked less than men with one wife, and they found more time for hunting, the most cherished spare time occupation for the male members of the village population.[11]

Undoubtedly, future changes in marriage patterns in rural Africa will be closely linked to future changes in farming systems which may lessen (or enhance) the economic incentive for polygamic marriages. Of course, motives other than purely economic considerations are behind a man's decision to acquire an additional wife. The desire for numerous progeny is no doubt often the main incentive. Where both the desire for children and the economic considerations are at work, the incentives for polygamy are likely to

Country in which sample areas are located		Average number of wives per married man	Polygamic marriages as percentage of all existing marriages
Senegal	A	1.1	24
	B	1.3	23
	C	1.3	21
Sierra Leone		2.3	51
Ivory Coast		1.3	27
Nigeria	A	2.1[a]	63
	B	1.5	
Cameroon		1.0–1.3[b]	
Congo	A	1.3	11
	B	1.2	17
South Africa			14
Uganda	A	1.7	45
	B	1.2	

[a] The figures refer to male head of families, while married sons living with these seem to be excluded.
[b] The lowest ratio refers to unskilled workers, the highest ratio to own-account workers.

Table 1 Incidence of polygamy in Africa

be so powerful that religious or legal prohibition avails little.

The study of the Yoruba farmers of Nigeria mentioned in Chapter 1 has this to say: 'There are no doubt other reasons why polygamy prevails in the Yoruba country as in other regions of the world; but the two which seem to be most prominent in the minds of Yoruba farmers are that wives contribute much more to the family income than the value of their keep and that the dignity and standing of the family is enhanced by an increase of progeny. While these beliefs persist the institution of polygamy will be enduring, even in families which have otherwise accepted Christian doctrine. The Yoruba farmer argues that the increased output from his farms obtainable without cash expense when he has wives to help him outweighs the economic burden of providing more food, more clothing and larger houses.'[12]

THE STATUS OF YOUNGER WIVES

It is easy to understand the point of view of the Yoruba farmers quoted above when one considers the contribution to family support which women make in this region. Economic relations between

husband and wife among the Yoruba differ widely from the common practise of countries where wives are normally supported by their husbands. Only 5 per cent of the Yoruba women in the sample reproduced in Table 2 received from their husbands everything they needed – food, clothing and some cash – and only 2 per cent of them did no work other than domestic activities. A large majority were self-employed (in agriculture, trade or crafts) and many helped a husband on his farm in addition to their self-employment and their domestic duties. Most of these self-employed women had to provide at least part of the food for the family as well as clothing and cash out of their own earnings. Nearly one-fifth of the women received nothing from their husband and had to provide everything out of their own earnings; nevertheless they performed domestic duties for the husband and half of them also helped him on his farm.

There may not be many tribes in Africa where women contribute as much as the Yorubas to the upkeep of the family, but it is normal in traditional African marriages for women to support themselves and their children and to cook for the husband, often using food they produce themselves. A small sample from Bamenda in the West African Cameroons

Percentage of women with the following rights and duties:

Wife receives from husband	Wife contributes to household:				Total
	as self-employed, family aid and housewife	as self-employed and housewife	as family aid and housewife	as housewife	
Nothing	8	11			19
Part of food	32	16			48
All food	15	11	1	1	28
Food, clothing and cash	1		3	1	5
Total	56	38	4	2	100

Table 2 Rights and duties of Yoruba women

showed that the women contributed 44 per cent of the gross income of the family.[13] Many women of pastoral tribes, for instance the Fulani tribe of Northern Nigeria and Niger, are expected to provide a large part of the cash expenses of the family out of their own earnings from the sale of the milk and butter they produce. They cover the expenditure on clothing for their children and themselves as well as buying food for the family.[14] In many regions of East Africa, women are traditionally expected to support themselves and many women are said to prefer to marry Moslems because a Moslem has a religious duty to support his wife.

In a family system where wives are supposed both to provide food for the family – or a large part of it – and to perform the usual domestic duties for the husband, a wife will naturally welcome one or more co-wives to share with them the burden of daily work. Therefore, educated girls in Africa who support the cause of monogamous marriage as part of a modern outlook are unable to rally the majority of women behind them.[15] In the Ivory Coast, an opinion study indicated that 85 per cent of the women preferred to live in polygamous rather than monogamous marriage. Most of them mentioned domestic and economic reasons for their choice.[16]

In many cases, the first wife takes the initiative in suggesting that a second wife, who can take over the most tiresome jobs in the household, should be procured. A woman marrying a man who already has a number of wives often joins the household more or less in the capacity of a servant for the first wife, unless it happens to be a love match.[17] It was said above that in most parts of the world there seems to be an inverse correlation between the use of female labour and the use of hired labour in agriculture, i.e. that most farmers have some help either from their wives or from hired labour. However, in some regions with widespread polygamy, hired labour is a supplement to the labour provided by several wives, in the sense that the tasks for which male strength is needed are done by hired labour, while the other tasks are done by wives. In such cases the husband or his adult sons act only as supervisors.

Reports from different parts of Africa, ranging from the Sudan to Nigeria and the Ivory Coast, have drawn attention to this frequent combination of male labourers and wives of polygamous cultivators working together in the fields under the supervision of one or more male family members.[18] In such cases, the availability of male labour for hire is not a factor which lessens the incentive to polygamous marriages. On the contrary, it provides an additional incentive to polygamous marriages as a means of expanding the family business without changing the customary division of labour between the two sexes. Little reported that in Sierra Leone men with several wives sometimes used them to ensnare male agricultural labourers and get them to work for them without pay.[19]

In regions where polygamy is the rule, it is likely, for obvious demographic reasons, that many males will have to postpone marriage, or even forgo it. Widespread prostitution or adultery is therefore likely to accompany widespread polygamy, marriage payments are likely to be insignificant or non-existent for the bride's family and high for the bridegroom's family, sometimes amounting to several years'

Age group	Percentages		
	First wives	Later wives	Husbands
Below 25 years	12	35	
25–34 years	49	44	10
35–49 years	35	19	59
50 years and over	4	2	31
All ages	100	100	100

Table 3 Age distribution of married Moslem population of Dakar in Senegal

earnings of a seasonal labourer.[20] This will induce parents to marry off their daughters rather young, but in a period like the present, where each generation of girls is numerically larger than the previous one, the difference in age between the spouses will be narrower than it was previously.

Figures from Dakar, the capital of Senegal, shown in Table 3, illustrate the importance of the age difference between the spouses. Here, the average marriage age for women is 18 years, and the average age of first marriage for men is between 27 and 28

years. The average age difference between men and their second wives is over 15 years, and nearly all wives belong to age groups which are larger than those to which their husbands belong.[21] No less than 90 per cent of married men belong to the relatively small generations over 35, as can be seen from the table, while only 39 per cent of their first wives and 21 per cent of their second wives belong to these generations.

Economic policy during the period of colonial rule in Africa contributed to the introduction or reinforcement of the customary wide difference in marriage age of young men and girls. In order to obtain labour for head transport, construction works, mines and plantations, the Europeans recruited young villagers at an age where they might have married had they stayed on in the village. Instead they married after their return several years later. The result was an age structure in the villages with very few young men in the age group between 20 and 35 and the need to marry young girls to much older men who had returned from wage labour.

The difference between the numbers of boys and girls in villages where the custom of taking away wage labourers before marriage persists, can be seen from Figure 1 which gives the age distribution in Rhodesian villages as reported in a study by J. Clyde

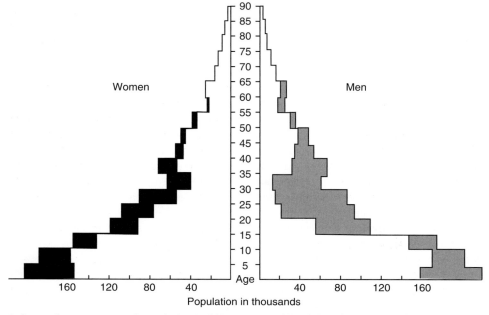

Figure 1 Sex and age structure of population in African areas of South Rhodesia in 1956 (shaded portion represents persons absent from the African areas)

Mitchell.[22] In the age groups 20–35 nearly all the men are away and the number of women in these age groups is several times higher than that of the men. In many other parts of Africa, recruitment for mines, plantations and urban industries results in similarly abnormal age distributions in the villages where the labourers are recruited.

Normally, the status of the younger wife is inferior as befits the assistant or even servant to the first wife. This can be explained partly as a result of the wide age difference between husband and wife and between first and younger wife, but the historical background of the institution of polygamy must also be kept in mind. Domestic slavery survived until fairly recently in many parts of Africa, and the legal ban on slavery introduced by European colonial powers provided an incentive for men to marry girls whom otherwise they might have kept as slaves. In a paper published as recently as 1959, it is mentioned that in the Ivory Coast women were still being pawned by husbands or fathers to work in their creditor's fields, together with his own wives and daughters and without pay until the debt was paid off, when they were free to return to their own families.[23] Today, such arrangements may be rare in Africa, but it is probable that the bride price for an additional wife is sometimes settled by the cancellation of a debt from the girl's family to the future husband, which would come to much the same thing in terms of real economic relationships.

Embodied in Moslem law is the well-known rule that all wives must be treated equally, which implies that the younger wives must not be used as servants for the senior wives. Moreover, a limit is set to the use of wives for expansion of the family business, partly by limiting the allowable number of wives to four, and partly by making the husband responsible for the support of his wives. We have already mentioned that this serves to make Moslem men desirable marriage partners for many African girls in regions where girls married to non-Moslems are expected to support themselves and their children by hard work in the fields. Because of this principle of equal treatment, first wives in orthodox Moslem marriages may desist from making the younger wives perform the most unpleasant tasks. Often in African families – Moslem and non-Moslem – each wife has her own hut or house and cooks independently, while the husband in regular succession will live and eat with each of his wives. Even so, the wife gains by not having to feed

her husband all the time, and we sometimes find that women prefer polygamy even where the wives are treated equally.

In most of Africa the rule is that a wife may leave her husband provided that she pays back the bride price. In regions where wives must do hard agricultural work, many young girls wish to find money to enable them to leave a much older husband, and many husbands fear that their young wives will be able to do so.[24] This makes older men take an interest on one hand in keeping bride prices at a level which makes it difficult for women to earn enough to pay them back and on the other hand in preventing their young wives from obtaining money incomes. Later, we shall see what role these conflicting interests between men and women are playing in development policy.

WORK INPUT AND WOMEN'S STATUS

Polygamy offers fewer incentives in those parts of the world where, because they are more densely populated than Africa, the system of shifting cultivation has been replaced by the permanent cultivation of fields ploughed before sowing. However, in some regions where the latter system prevails, polygamy may have advantages. This is true particularly where the main crop is cotton, since women and children are of great help in the plucking season.[25] But in farming systems where men do most of the agricultural work, a second wife can be an economic burden rather than an asset. In order to feed an additional wife the husband must either work harder himself or he must hire labourers to do part of the work. In such regions, polygamy is either non-existent or is a luxury in which only a small minority of rich farmers can indulge. The proportion of polygamic marriages is reported to be below 4 per cent in Egypt, 2 per cent in Algeria, 3 per cent in Pakistan and Indonesia.[26] There is a striking contrast between this low incidence of polygamy and the fact that in many parts of Africa South of the Sahara one-third to one-fourth of all married men have more than one wife.

In regions where women do most of the agricultural work it is the bridegroom who must pay bride wealth, as already mentioned, but where women are less actively engaged in agriculture, marriage payments come usually from the girl's family. In South and East Asia the connection between the work of women and the direction of marriage payments is close and unmistakable. For instance, in Burma, Malaya and Laos women seem to do most of the agricultural work and bride prices are customary.[27] The same is true of Indian tribal people, and of low-caste peoples whose women work. By contrast, in the Hindu communities, women are less active in agriculture, and instead of a bride price being paid by the bridegroom, a dowry has to be paid by the bride's family.[28] A dowry paid by the girl's family is a means of securing for her a good position in her husband's family. In the middle of the nineteenth century it was legal for a husband in Thailand to sell a wife for whom he had paid a bride price, but not a wife whose parents had paid a dowry to the husband.[29]

Not only the payment of a dowry but also the use of the veil is a means of distinguishing the status of the upper class wife from that of the 'servant wife'. In ancient Arab society, the use of the veil and the retirement into seclusion were means of distinguishing the honoured wife from the slave girl who was exposed to the public gaze in the slave market.[30] In the Sudan even today it appears to be a mark of distinction and sophistication for an educated girl to retire into seclusion when she has finished her education.[31]

In communities where girls live in seclusion, and a large dowry must be paid when they marry, parents naturally come to dread the burden of having daughters. In some of the farming communities in Northern India, where women do little work in agriculture and the parents know that a daughter will in due course cost them the payment of a dowry, it was customary in earlier times to limit the number of surviving daughters by infanticide. This practise has disappeared, in its outward forms, but nevertheless the ratio of female to male population in these districts continues to be abnormal compared to other regions of India and to tribes with working women living in the same region. A recent study of regional variations in the sex ratio of population in India[32] reached the conclusion that the small number of women in the Northern districts could not be explained either by undernumeration of females, or by migration, or by a low female birthrate. The only plausible hypothesis would be that mortality among girls was higher than among boys. The conclusion drawn was that 'the persistence of socio-cultural factors are believed to be largely responsible for the excess of female mortality over the male'.[33] One of these socio-cultural factors seems to be a widespread supposition that milk is not good for girls, but is good for boys. There is also a tendency to care more for sick boys than for sick girls.[34]

In a study from a district in Central India with a deficit of women, the author is very outspoken about the neglect of girls: 'The Rajputs always preferred male children. . . . Female infanticide, therefore, was a tolerated practise. . . . Although in the past 80 years the proportion of the females to males has steadily risen yet there was always a shortage of women in the region. . . . When interrogated about the possibility of existence of female infanticide, the villagers emphatically deny its existence. . . . It was admitted on all hands that if a female child fell ill, then the care taken was very cursory and if she died there was little sorrow. In fact, in a nearby village a cultivator had twelve children – six sons and six daughters. All the daughters fell ill from time to time and died. The sons also fell ill but they survived. The villagers know that it was by omissions that these children had died. Perhaps there has been a transition from violence to non-violence in keeping with the spirit of the times.[35] The report adds that 'no records of birth or deaths are kept . . . it was enjoined upon the Panchayat (village council) to keep these statistics, but they were never able to fulfil the task.'[36] It is explicitly said in the study that the district is one where wives and daughters of cultivators take no part in field work.[37]

To summarize the analysis of the position of women in rural communities, two broad groups may be identified: the first type is found in regions where shifting cultivation predominates and the major part of agricultural work is done by women. In such communities, we can expect to find a high incidence of polygamy, and bride wealth being paid by the future husband or his family. The women are hard working and have only a limited right of support from their husbands, but they often enjoy considerable freedom of movement and some economic independence from the sale of their own crops.

The second group is found where plough cultivation predominates and where women do less agricultural work than men. In such communities we may expect to find that only a tiny minority of marriages, if any, are polygamous; that a dowry is usually paid by the girl's family; that a wife is entirely dependent upon her husband for economic support; and that the husband has an obligation to support his wife and children, at least as long as the marriage is in force.

We find the first type of rural community in Africa South of the Sahara, in many parts of South East Asia and in tribal regions in many parts of the world. We also find this type among descendants of negro slaves in certain parts of America.[38] The second type predominates in regions influenced by Arab, Hindu and Chinese culture.

Of course, this distinction between two major types of community is a simplification, like any other generalization about social and economic matters. This must be so because many rural communities are already in transition from one type of technical and cultural system to another, and in this process of change some elements in a culture lag behind others to a varying degree. For example, some communities may continue to have a fairly high incidence of polygamy or continue to follow the custom of paying bride price long after the economic incentive for such customs has disappeared as agricultural techniques changed.

It was mentioned in Chapter 1 that in many rural communities hired labour is replacing the work of women belonging to the cultivator family. Where this happens, the economic incentive for polygamy may disappear, since additional wives are liable to become an economic burden. This, of course, is true only if it is assumed that the women who give up farm work retire into the purely domestic sphere. It is another matter if the women substitute farming by another economic activity such as trade. We shall revert to the problems of trading women in a later chapter.

In the type of rural community where women work hard, it is a characteristic that they are valued both as workers and as mothers of the next generation and, therefore, that the men keenly desire to have more than one wife. On the other hand, in a rural community where women take little part in field work, they are valued as mothers only and the status of the barren woman is very low in comparison with that of the mother of numerous male children. There is a danger in such a community that the propaganda for birth control, if successful, may further lower the status of women both in the eyes of men and in their own eyes. This risk is less in communities where women are valued because they contribute to the well-being of the family in other ways, as well as breeding sons.

NOTES

1 Appadorai, 19.
2 UN. ECA., *Wom. Trad. Soc.*, 5.
3 Winter, 24.
4 Simons, 79–80.
5 Richards 1952, 204.
6 Little 1951, 141–2.
7 Little 1951, 141–2, 145.
8 Some of the samples were taken in urban areas, where the incidence of polygamy is often, though not always, lower than in rural areas.
9 To evaluate correctly this figure for the incidence of polygamy it must be taken that some of the married men, at the time of the enquiry, had one wife only because they were at an early stage of their married life, while others were older men living in monogamous marriage because they had lost other wives by death or divorce. Therefore, the figure for the incidence of polygamy would have been considerably higher if it were to show the proportion of men who have more than one wife at some stage of their married life.
10 Haswell, 10.
11 Georges, 18, 25, 31.
12 Galletti, 77.
13 Kaberry, 141.
14 Forde, 203; Dupire 1960, 79.
15 UN. ECA., *Polygamy*, 32.
16 Boutillier 1960, 120.
17 Little 1951, 133.
18 Baumann, 307; Forde, 45; Boutillier 1960, 97; Gosselin, 521.
19 Little 1951, 141; 1948; 11.
20 Forde, 75n.
21 UN. ECA., *Polygamy*, 24.
22 Mitchell, *Soc. Backgr.*, 80.
23 d'Aby, 49.
24 Winter, 23.
25 Arnaldez, 50.
26 UN. ECA., *Wom. N. Afr.*, 41; Appadorai, 18.
27 MiMi Khaing, 109; Swift, 271; Lévy, 264.
28 Mitham, 283–4.
29 Purcell, 295.
30 Izzedin, 299.
31 Tothill, 245.
32 Visaria, 334–71.
33 Visaria, 370.
34 Karve, 103–4.
35 Bhatnagar, 61–2.
36 Bhatnagar, 65.
37 In some cases, the shortage of women in rural communities in North India induces the cultivators to acquire low caste women from other districts, or from other Indian States, against the payment of a bride price (Nath, May 1965, 816). This need not be an infringement on caste rules. Although it is usually forbidden for a man to marry a woman of a higher caste, men of higher caste may have the right to marry women of lower castes (Majundar, 61). This may then entail the

payment of a bride price instead of the receipt of a dowry as would be customary in the husband's own subcaste.

38 Bastide, 37ff.

REFERENCES

Appadorai, A., *The Status of Women in South Asia*, Bombay 1954.

Arnaldez, Roger, 'Le Coran et L'emancipation de la Femme', in Mury, Gilbert (ed.) *La Femme à la Recherche d'elle-même*, Paris 1966.

Bastide, Roger, *Les Amériques Noires*, Paris 1967.

Baumann, Hermann, 'The Division of Work according to Sex in African Hoe Culture', in *Africa*, Vol. I, 1928.

Bhatnagar, K. S., *Dikpatura, Village Survey*, Monographs No. 4, Madhya Pradesh, Part VI, Census of India 1961, Delhi 1964, processed.

Boutillier, J. L., *Bongouanou Côte d'Ivoire*, Paris 1960.

d'Aby, F. J. Amon, 'Report on Côte d'Ivoire', in International Institute of Differing Civilizations, *Women's Role in the Development of Tropical and Sub-Tropical Countries*, Brussels 1959.

Dupire, Marguerite, 'Situation de la Femme dans une Société Pastorale (Peul Wo Da Be-Nomades du Niger)', in Paulme, Denise (ed.) *Femmes d'Afrique Noire*, Paris 1960.

Forde, Daryll, 'The Rural Economies', in Perham, Margery (ed.) *The Native Economics of Nigeria*, London 1946.

Galletti, R., Baldwin, K. D. S. and Dina, I. O., *Nigerian Cocoa Farmers*, London 1956.

Georges, M. M. and Guet, Gabriel, *L'emploi du Temps du Paysan dans une Zone de L'oubangui Central 1959–60*, Bureau pour le Developpement de la Production Agricole, Paris 1961, processed.

Gosselin, Gabriel, 'Pour une Anthropologie du Travail Rural en Afrique Noire', in *Cahiers d'Études Africaines*, Vol. III, 1962–3.

Haswell, M. R., *The Changing Pattern of Economic Activity in a Gambia Village*, HMS London 1963, processed.

Izzeddin, Nejla, *The Arab World*, Chicago 1953.

Kaberry, Phyllis M., *Women of the Grassfields. A Study of the Economic Position of Women in Bamenda, British Cameroons*. Colonial Office, Research Publication No. 14, London 1952.

Karve, Irawati, 'The Indian Woman in 1975', in *Perspectives, Supplement to the Indian Journal of Public Administration*, January–March 1966.

Lévy, Banyen Phimmasone, 'Yesterday and today in Laos: a Girl's Autobiographical Notes', in Ward, Barbara E. (ed.) *Women in the New Asia, The Changing Social Roles of Men and Women in South and South-East Asia*, UNESCO 1963.

Little, K. L., 'The Changing Position of Women in the Sierra Leone Protectorate', in *Africa*, Vol. XVIII, 1948.

Little, K. L., *The Mende of Sierra Leone. A West African People in Transition*, London 1951.

Majumdar, D. N., 'About Women in Patrilocal Societies in South Asia', in Appadorai, A. (ed.) *The Status of Women in South Asia*, Bombay 1954.

Mi Mi Khaing, 'Burma: Balance and Harmony', in Ward, Barbara E. (ed.) *Women in the New Asia. The Changing Social Roles of Men and Women in South and South-East Asia*, UNESCO 1963.

Mitchell, J. Clyde, *Sociological Background to African Labour*, Salisbury 1961.

Nath, Kamla, 'Women in the New Village', in *The Economic Weekly*. May 1965.

Purcell, V., 'Report on Burma, Thailand and Malaya', in International Institute of Differing Civilizations, *Women's Role in the Development of Tropical and Sub-Tropical Countries*, Brussels 1959.

Richards, Audrey I., *Economic Development and Tribal Change*, Cambridge 1952.

Simons, H. J., *African Women. Their Legal Status in South Africa*, Evanston 1968.

Swift, Michael, 'Men and Women in Malay Society', in Ward, Barbara E. (ed.) *Women in the New Asia. The Changing Social Roles of Men and Women in South and South-East Asia*, UNESCO 1963.

Tothill, J. D., *Agriculture in the Sudan*, London 1954.

United Nations Economic Commission for Africa, 'Polygamie, Famille et Fait Urbain (Essai sur le Sénégal)', *Workshop on Urban Problems*, Addis Ababa 1963, processed.

United Nations Economic Commission for Africa, 'The Employment and Socio-Economic Situation of Women in some North African Countries', *Workshop on Urban Problems*, Addis Ababa 1963, processed.

United Nations Economic Commission for Africa, 'Women in the Traditional African Societies', *Workshop on Urban Problems*, Addis Ababa 1963, processed.

Visaria, Pravin M., 'The Sex Ratio of the Population of India and Pakistan and Regional Variations During 1901–61', in Bose, Ashish (ed.) *Pattern of Population Change in India 1951–61*, Bombay 1967.

Winter, E. H., 'Bwamba Economy', *East African Studies*, No. 5, Kampala 1955.

'What is "Urban Bias", and is it to Blame?'

from *Why Poor People Stay Poor: Urban Bias in World Development* (1977)

Michael Lipton

Editors' Introduction

Michael Lipton was born in 1937 and read Philosophy, Politics and Economics at Oxford University, where he also won a Fellowship at All Souls' College. While still at Oxford, Lipton was directed by his mentor, Paul Streeten, to work for Gunnar Myrdal on what became a monumental study of the *Asian Drama* (1968). Partly through this commission, and later as a young lecturer at Sussex University, Lipton gained expertise in the still young field of development economics. He was somewhat unusual among his peers, however, by virtue of the fact that he spent more than half a year in 1965 living in a village in Maharashtra, India. Lipton used this experience to think about how ordinary peasant farmers were making economic decisions and how they were being treated under a planning regime that prioritised urban–industrial growth. Out of this work, and following the near-famine conditions that hit parts of India in 1966–67, Lipton began to fashion an account of 'urban bias' in the development process. This work gave rise to his book-length treatment of the same issue in 1977, *Why Poor People Stay Poor*. Lipton famously began the book by declaring that

> The most important class conflict in the poor countries of the world today is not between labour and capital. Nor is it between foreign and national interests. It is between the rural classes and the urban classes. The rural sector contains most of the poverty, and most of the low-cost sources of potential advance; but the urban sector contains most of the articulateness, organisation and power. So the urban classes have been able to 'win' most of the rounds of the struggle with the countryside; but in so doing they have made the development process needlessly slow and unfair.
>
> (1977: 13)

Lipton's account of urban bias in the development process was a major challenge to mainstream views in the 1960s and 1970s. Interestingly, too, his urban bias thesis (UBT) was intended not so much as a moral or normative argument – this is not a Gandhian endorsement of the authenticity of rural virtues – but as a scientific account of the inefficiencies and inequities induced by urban biased development. In his view, urban biased policies led to lower levels of food production than would otherwise prevail. This was because farmers saw prices 'twisted' against them by marketing boards and other parastatals that pandered to the short-run interests of the urban middle classes. Urban biased policies also led to unfair (that is to say, too low) allocations of funds to the countryside to support local education and health care; excessive rural–urban migration was one result of this bias.

In the reading which follows we see Lipton elaborating on the UBT specifically with reference to food prices. He makes the controversial argument that 'the whole interest of the rural argument is against cheap food' – controversial, because for many of his critics this ignores the sharp division of interests between food-selling farmers and food-purchasing agricultural labourers. Lipton also examines why urban bias is overlooked and why it leads to the persistence of mass poverty. In the rest of *Why Poor People Stay Poor*, a long book running to more than four hundred pages, Lipton tries to substantiate these and other key arguments.

Since 1977, we should note, many countries in the global South, and perhaps most especially in sub-Saharan Africa, have removed the main price-twists that Lipton saw as so damaging in the 1960s and 1970s, and which were linked to the damaging crises of food production that hit Africa in the 1970s and 1980s. Here is a case of a development economist making a difference. Lipton maintains that World Bank conditionality policies have dealt more with the 'economics' than with the underlying 'politics' of urban bias, but he concedes that progress has been made in terms of currency devaluations and the abolition of marketing boards. Lipton does not accept, however, that urban bias is no longer a problem in the developing world. In more recent work with Robert Eastwood, he has maintained that allocational urban bias is actually worsening in many parts of the global South, including in China. The urban–rural gap remains a major issue for policy makers, as well as being a continuing source of acute (and in Lipton's view unnecessary and unwarranted) political tensions. Lipton's critics, for their part, argue that his view of urban growth is unduly pessimistic. Cities are seen as predatory rather than as sites of innovation. Liptonian policy, accordingly, is less attuned than it should be to the benefits that cities provide in terms of agglomeration economies and simple 'buzz'.

Key references

Michael Lipton (1977) *Why Poor People Stay Poor: Urban Bias in World Development*, London: Temple Smith.
— (2005) 'Urban bias', in T. Forsyth (ed.) *Encyclopedia of International Development*, London: Routledge.
G. Jones and S. Corbridge (2008) 'The urban bias thesis revisited', *Progress in Development Studies* (forthcoming).
R. Eastwood and M. Lipton (2000) 'Pro-poor growth and pro-growth poverty reduction: meaning, evidence, and policy implications', *Asian Development Review* 18 (2): 22–58.
World Bank (1981) *Accelerated Development in Sub-Saharan Africa,* Washington, DC: World Bank.

[. . .]

URBAN BIAS AND CLASS STRUCTURES: THE EXAMPLE OF FOOD PRICES

A glance at the real class structure of most poor countries casts doubt on the usual analyses of classes defined by their relationship to the means of production – land, capital and labour.[1] One sees rural rentiers – big landlords – certainly, but also many thousand small farmers, tenants or proprietors. They live in no communal idyll, and are poor and exploited by local monopolists – suppliers of credit, marketing facilities or land. Nor are these peasants egalitarian; they are highly differentiated. But, on the whole, they are consumer-producers for whom the separation of capital and labour, profit and wage,[2] process of production and use of end-product, is meaningless.

Nor is the urban sector inviting to the classical Marxian analysis. One sees a mass of urban jobless, but they are often in reality *fringe villagers*, waiting until penury forces them back to the land and meanwhile living on casual work – or on their rural relatives. They hang around the city slums more in temporary hope than in expectation of work. They are kept half-employed partly by employers' preferences for machines and small workforces: preferences due to subsidised imports and high wages, both maintained in part by unions or other skilled-labour pressures. The wage levels (and less obviously the import subsidies) are paid for by villagers, who could otherwise find urban jobs (or get more imports themselves). They are enforced through naked urban power; they lead to situations where employers prefer heavy capital equipment worked well below capacity to more employment-generating strategies of production. The existing urban labour aristocracy enjoys high wages largely because it is small: it is too costly for the employer to do without their skills, cheap to pay them off, and easy to acquire capital subsidies to keep employment levels low.

Yet if he did go for more labour-intensive methods, he would advance equality. More people would be employed at a slightly lower urban wage. Despite some vulgar *Marxisants*, the basic conflict in the Third World is not between capital and labour, but between capital and countryside, farmer and townsman, villager (including temporarily urban 'fringe villager') and urban industrial employer-cum-proletarian elite, gainers from dear food and gainers from cheap food.[3] While the urban centres of power and government remain able and willing to steer development overwhelmingly towards urban interests, development will remain unequalising.

The systematic action by most governments in poor countries to keep down food prices (chapter 13) clarifies the operation of class interests in urban bias. Town and country are polarised, yet the powerful country interests are bought off (by subsidies for inputs, such as tractors and tubewells, that they are almost alone in using). The urban employer wants food to be cheap, so that his workforce will be well fed and productive. The urban employee wants cheap food too; it makes whatever wages he can extract from the boss go further.

Less obviously, the *whole* interest of the rural community is against cheap food. This is clear enough for the farmers who sell food to the towns (largely big farmers, bought off by input subsidies); but even the 'deficit farmer' or net food buyer (who grows too little to feed himself from his land alone) often gains when food is dear, except perhaps in the very short term. Deficit farmers cannot make ends meet on their land alone, and to buy enough food must work for others.[4] Often they work on farms for a fixed share of the crop, which is worth more when food prices are high. Whether they work for crop wages or for cash, it pays the big farmer to hire more labour when food is dearer, and this bids up farm wages as well as rural employment. The rural

craftsmen who serve the big farmers' production and consumption needs – carpenters, ropemakers, gold-smiths – receive more offers of work, at higher wages, when their patrons are enriched because food is dearer; and many poor agriculturists eke out their income by traditional craft activities. Moreover, the richer farmers have more cash to lend out when food is dear and their income high, so the interest rate to the poor borrower is reduced as lenders compete. Even the people on the fringe of the countryside, the recently migrant urban unemployed, find their remittances from the village increasing when their farming fathers and brothers benefit from high food prices.

There is a 'deep' reason why an issue such as the price of food polarises city and country into opposing classes, each fairly homogeneous. The reason is that within each rural community (though hardly one is nowadays completely closed) extra income generated tends to circulate. The big farmer, when he gets a good price for his output, can buy a new seed drill from the village carpenter, who goes more often to the barber and the laundryman, who place more orders with the village tailor and blacksmith. When food becomes cheap, this sort of circulation of income is transferred from the village to the city, because it is in the city that the urban worker will spend most of the money he need no longer use to buy food.

We shall see in chapter 13 how urban power, urban government in the urban interest, has made farm products artificially cheap and farmers' requirements artificially dear in most poor countries. In Pakistan, in the early 1960s, the total effect of State action and private power balances was to *triple* the number of hours the farm sector had to work to get a typical bundle of urban goods.[5] By severe restrictions on cheap imports of industrial consumer goods, by cheap imported raw materials for factory-owners but not farmers, and by many other means, the ratio of industrial prices to agricultural prices was trebled! This is by no means unusual in poor countries. It shows a degree of exploitation, of unequal dealing, next to which intra-urban conflict between capitalist and proletariat is almost negligible.

There is nothing wicked or conspiratorial about this. It is the natural play of self-interest and power, only obfuscated by moralising from outside – *cosi fan tutti*, moralisers as well. And it is only one of many ways in which the city (where most government is) screws the village (where most people are) in poor countries. In tax incidence, in investment allocation, in the provision of incentive, in education and research: everywhere it is government by the city, from the city, for the city.

In isolated moments of war or revolution, a nation may develop a sense of shared interest between land-less agricultural and casual industrial labourer, or between city capitalist and village landlord; even when things are quiet, the urban elite pacifies the big farmer by allocating him most of the few resources that can be spared for agriculture. But usually the contradictions between city capitalist and city pro-letariat can be resolved by negotiation – are not 'antagonistic'[6] – because they can be settled at the expense of the rural interest.[7]

Poverty persists alongside development largely because poor countries are developed from, by and for people in cities: people who, acting under normal human pressures, deny the fruits of development to the pressure-less village poor. Few of these can escape the trap by joining the exploitative city elite, because high urban wages (and subsidised capital imports) deter employers from using extra labour.[8] Many villagers, once migration has failed to secure entry to the urban labour aristocracy, return to an increasingly land-scarce village: a village that is by policy denied the high food prices that would normally be linked to land scarcity, by policy starved of public investment allocations, and hence by policy prevented from sharing in development and thus from curing its own poverty.

This is one reason why, as suggested in chapter 1, we are dealing not with temporary inequality caused by a passing weakness of the impoverished prole-tariat, but with self-confirming inequality caused by the alliance of the urban employer and proletariat against the rural poor. Population growth, moreover, makes it unlikely that the rural poor will be sucked out, and up, by a labour shortage for a very long time. But what is the structure of power that prevents poor villagers, even in democratic countries, from calling the tune? Why is that power structure not generally recognised for what it is?

WHY URBAN BIAS IS OVERLOOKED

In almost all countries now poor, but in very few now rich, most people live on the land and off the land;

but they are governed by, from and (we argue) for townsmen. This fact is so 'set up' as to obscure it from most people likely to analyse its consequences intelligently. Teachers, planners, politicians – one would imagine – can succeed only by knowing their own world. If indigenous, they *have* succeeded, so how *can* they have misdefined that world? If foreign, they find it hard to accept that an elite can effectively remove its peasant masses from its world; when I taught a final-year class of honours students of economics in Khartoum University in 1962, I was amazed to discover that many of them had never set foot in any of the villages where over 80 per cent of their compatriots lived and worked; and even now that I know how common this situation is, I can hardly believe it. In most poor countries the politician, the civil servant, the university teacher, the businessman, or the trade-union leader is selected by townsmen, caters largely to an urban audience,[9] and in pursuing his interests or his career has every incentive to spend his time almost wholly in big cities. The potential member of the elite is 'set up' to define rural life out of his world. Even if village-born, he has reason to regard his relatives as a burden, the prospect of reabsorption into rurality as the ultimate threat, and the whole rural episode as best forgotten.

The foreign observer from a rich country is even worse placed. He may virtuously stir from his air-conditioned hotel to inspect city slums. He is almost certain never to spend long enough in any village to learn what happens. If he is an egalitarian, the gross-ness of intra-urban inequality – worse than within most villages – leads him to 'reward' the cities for the obviousness of their poverty, with recommended resource allocations that leave the less obvious mass rural poverty untouched.

The opinions and actions of people who make and influence policy are bound to be affected by what they see, hear and fear. They hear the conversations of the city employers and city trade unionists. They see *urban* poverty, especially noticeable and squalid (and severe) because of the adjacent extremes of unconcealed wealth – it is in this sense that high allo-cations for housing reward the city for its inequality – and because of the glamorous prospects of modern industrial technology in an urban setting. They fear the pressures of a city elite, the riots of a unionised city proletariat. Naturally they allocate an economic-ally indefensible share of resources to the cities. So would you or I in their place.

URBAN BIAS AND THE PERSISTENCE OF POVERTY

Even if urban bias is significant in many countries, what has it to do with the persistence of poverty? After all, we have shown that most poor countries have greatly accelerated their growth and development since 1945 – at the same time as centralising and independent governments emerged. Indeed, the earlier advocacy of agricultural primacy by colonial regimes – for the wrong reasons – has helped to produce the urban bias of their independent successors. Typically, in Burma, the planners' 'main emphasis was to shift the economy away from a pattern of primary production ... established in colonial times'; negligible farm investment, prices turned against farmers, and a cover-up of 'repeated assurances ... that agriculture merited priority' followed naturally.[10]

Poor countries could have raised income per person since 1945 much faster than they did, if allocative urban bias had been reduced. I shall show that many of the resources allocated by state action to city-dwellers would have earned a higher return in rural areas; that private individuals, furthermore, were indirectly induced by administrative decisions and price distortions to transfer from countryside to town their own resources, thereby reducing the social (but increasing the private) rate of return upon those as well; and that, ultimately, inadequate inputs of rural resources substantially reduced even the efficient use of urban resources. But so what? Unprecedented growth in poor countries has proved unable to make a major impact on the conditions of the poor people who live there. A reduction of urban bias might speed growth even further; but why should that help the poor any more than past accelerations of growth?

The reason is that allocations biased townwards with respect to the efficiency norm are almost certainly heavily biased townwards with respect to the equity norm. By reallocating capital, skills and administrative attention from city to countryside, we can hardly help reducing the inequality of incomes; countrymen start much poorer than townsmen, share their income somewhat more equally, and are likelier to owe their poverty to conditions curable by income from work.

By shifting resources from city to country, a poor nation almost certainly relieves poverty in the short term. If such shifts also raise output in the sense defined above, and if the present level of urban allocations is kept up by urban bias (dispositional and hence allocative), such bias is a main cause of the persistence of poverty. So far, I have merely outlined the case; before providing proof, I look at alternative explanations for the persistence of poverty. If such explanations are in some sense 'right' *as well as* explanations of urban bias, we must ask (for instance) which, of urban bias and 'imperialism', is cause and which effect; or else which maintains poverty where – and why; or else we must admit that poverty is overdetermined: remove one cause, say urban bias, and another component of the eternal pattern of exploitation, say 'imperialism', still keeps the poor down.

Need we be so gloomy?

NOTES

1 See, however, K. Marx, *Pre-Capitalist Formations* (ed. E. Hobsbawm), Lawrence & Wishart, 1964, and, indeed, the subtle class analysis of Marx's *The Eighteenth Brumaire of Louis Bonaparte*.

2 A. V. Chayanov, *Theory of Peasant Economy* (tr. B. Kerblay, ed. D. Thorner), Irwin, Homewood (Illinois), 1966.

3 Nowhere has this conflict been more brilliantly described and analysed than in K. Kautsky, *Die Agrarbrage*, Dietz (Stuttgart), 1899 (hereafter cited as Kautsky), especially pp. 208–21; see below, chapter 4.

4 A landless labourer is just a 100% deficit farmer.

5 As compared with the relative world prices prevailing, the price of industrial products was kept thrice as high relative to farm products. S. R. Lewis, *Pakistan: Industrialisation and Trade Policies*, OECD Oxford, 1972 (hereafter cited as Lewis), p. 65; see below, pp. 306–7, and pp. 124–30 for comparison with the policies subtly advocated in the USSR by Preobrazhensky, and crudely brutalised by Stalin.

6 In other words, 'antagonistic' and 'non-antagonistic' contradictions in both capitalist and socialist economies – while rightly *distinguished* by Mao Tse-tung ('On Contradiction', in *Collected Works*, Foreign Languages Press (Peking), vol. 1, pp. 343–5) – are wrongly *identified*.

7 Conversely, however, intra-village conflict and faction are more serious, because the protagonists are weak *vis-à-vis* the city, and cannot settle their disputes at its expense. For an extreme but telling statement of the high barriers presented by factionalism to village development, see Baljit Singh, *Next Step in Village India*, University of Lucknow (Department of Economics), 1959.

8 See below, pp. 223–30. Another confirmation of the smallness of true, *permanent* rural–urban migration comes from Tanzania; see M. A. Bienefeld and R. H. Sabot, *The National Urban Mobility, Employment and Income Survey of Tanzania, 1971*, Economic Research Bureau and Ministry of Economic Affairs and Development Planning, Dar es Salaam, 1972.

9 Even when there are quinquennial elections and massively rural electorates, it is the townsmen who write the papers and put out the radio programmes (even a largely illiterate village usually has access to a radio), fund political parties, pick candidates, own or frequent 'political' coffee-houses and bars.

10 H. V. Richter, 'The Union of Burma', in R. T. Shand (ed.) *Agricultural Development in Asia*, Australian National University Press and Allen & Unwin, 1969, pp. 159–63.

'Latin American Squatter Settlements: A Problem and a Solution'

from *Latin American Research Review* (1967)

William Mangin

Editors' Introduction

William Mangin is an Emeritus Professor of Anthropology at the Maxwell School of Syracuse University. Mangin's main area of research since 1951 has been on the anthropology of Peru. His work has covered a range of issues, including urbanisation, ethnicity, education, health, politics and alcohol use. More recently, Mangin has also begun anthropological research in Spain.

Mangin's essay here addresses an important issue researched in particular by Latin Americanists in the 1960 and 1970s. These scholars set out to disprove some of the reigning assumptions of the Chicago School of Urban Sociology, to show how urban poor communities were not passive subjects of domination. They showed how the urban poor participated in making livelihoods, communities and neighbourhoods in ways that most urban scholars in the West were failing to understand. Mangin's review, which we excerpt in the reading, shows that squatter settlements – whether Mexico's *colonias proletarias*, Peru's *barriadas*, Chile's *callampas*, Uruguay's *cantegriles* or Brazil's *favelas* – are not what they seem. Scholars and elites have seen squatter settlements as collections of peasants in a state of rural disorientation, disorganisation, crime, decay and lack of knowledge: the ingredients that would force people to militancy and communism. These were Cold War myths and elite stereotypes, rather than the products of actual research. The scholar Janice Perlman (1976), for instance, went to the *favelas* of Rio to find that people were not at all as marginal as scholars sitting in Chicago might presume; rather, they were active members of society, organising their livelihoods and also pressing for rights through political organisation. In the reading here, Mangin also takes apart these myths by looking at how people actually make informal settlements, how they organise, and how they press for a better future. Squatters do not simply sit in a state of abjection and resignation.

These questions could not be more vital today. Mike Davis (Part 1) has written a book called *The Planet of Slums* which reproduces Northern and elite views of 'slums' as places of irrationality and domination. Intellectuals of the south often disagree. South Africa today has been the site of the largest social movement since the uprising to bring down the white-supremacist Apartheid regime in the 1980s. This movement, called *Abahlali base Mjondolo*, or 'those who stay in shacks', has emerged autonomously from informal settlements. What is striking is that in this Internet age, they also imagine and write about what they do; the work of these grassroots intellectuals can be found on their website, at the 'University of Abahlali base Mjondolo'. They are not the irrational Africans of Davis's *Planet of Slums*, and they refuse what McClintock calls the 'porno-tropical gaze' which sees Africans as irrational sexual predators. They are fighting for adequate housing, jobs and a new understanding of democratic politics.

The edited collection by Al Sayad and Roy, *Urban Informality* (2005), shows that Latin Americanists early on saw informal settlements as sites of politics. This was partly because Latin American states were engaging in populist politics that required them to reach out to vast squatter settlements who might use their vote against the party in power. In more authoritarian regimes, the poor have often encroached in subtle ways. Asef Bayat calls this 'the quiet encroachment of the ordinary', as poor people slowly colonise urban space, as in Bombay, Cairo or Tehran. Periodic slum clearance is used against the poor, but the state often allows informal settlements to persist because poor people are forced to fend for themselves rather than rely on the state. Networked activism such as that of the *Abahlali base Mjondolo* stands to threaten this assumption that the poor will always subsidise the state in this manner.

Key references

William Mangin (1967) 'Latin American squatter settlements: a problem and a solution', *Latin American Research Review* 2: 65–98 .
— (ed.) (1970) *Peasants in Cities: Readings in the Anthropology of Urbanization*, Boston, MA: Houghton-Mifflin.
Janice Perlman (1976) *Myth of Marginality: Urban Poverty and Politics in Rio de Janeiro*, Berkeley: University of California Press.
Nezar Al Sayad and Ananya Roy (eds) (2005) *Urban Informality: Transnational Perspectives from the Middle East, Latin America, and South Asia*, Lanham, MD: Lexington Books.
The University of Abahlali base Mjondolo, http://www.abahlali.org/node/237 (accessed 6 October 2007).

Squatter settlements have formed around large cities throughout the world, mushrooming particularly since the end of World War II. In an excellent preview to a forthcoming book, Turner (1966) has discussed some common features among squatter settlements in Latin America, Asia, Africa, and Europe. Morse (1965a) has also referred to squatter communities and to general characteristics of Latin American urbanization in an article in this journal.[1] Without repeating the work of Turner and Morse I would like to present a preliminary survey of Latin American squatter settlements with a model of their formation, growth, and social development that contradicts many views held by planners, politicians, newspapermen, and much of the general population, including many residents of the settlements themselves.

Several writers have referred to various local names given to squatter communities: *colonias proletarias* in Mexico, *barriadas brujas* in Panama, *ranchos* in Venezuela, *barriadas* in Peru, *callampas* in Chile, *cantegriles* in Uruguay, *favelas* in Brazil, and, in other places, marginal areas, clandestine urbanizations, *barrios* of invasion, parachutists, phantom towns, etc. There are no general works on the subject, but some good descriptions of local conditions do exist. The major sources used are listed by country in note 2.[2] Not many sources are available and, unfortunately, several appear only in mimeographed form. The reports point out that squatter populations consist mainly of low income families but all of the authors distinguish between squatter settlements and other types of lower class housing in tenements, alleys (*callejones*), shack yards (*coralones, jacales*), and rented slum buildings. They agree, sometimes to their own surprise, that it is difficult to describe squatter settlements as slums. The differentiation of squatter settlements from inner-city slums is, in fact, one of the first breaks from the widely shared mythology about them. (See, for example, Patch, 1961; Mangin, 1965.) The purpose in noting this mythology is not merely to set up a straw man for the paper. A review of the "Chaos, Crisis, Revolution and Wither Now Latin America" literature, or, of most governmental, United Nations, or AID reports, or, of most newspapers and magazines in Latin America will show that it is the predominant position on squatter settlements.

The standard myths, not all incorrect and by no means mutually consistent, are, with some variation among countries, as follows:

1 The squatter settlements are formed by rural people (Indians where possible) coming directly from "their" farms.

2 They are chaotic and unorganized.

3 They are slums with the accompanying crime, juvenile delinquency, prostitution, family breakdown, illegitimacy, etc.

4 They represent an economic drain on the nation since unemployment is high and they are the lowest class economically, the hungriest and most poorly housed, and their labor might better be used back on the farms.

5 They do not participate in the life of the city, illiteracy is high and the education level low.

6 They are rural peasant villages (or Indian communities) reconstituted in the cities.

7 They are "breeding grounds for" or "festering sores of" radical political activity, particularly communism, because of resentment, ignorance, and a longing to be led.

8 There are two solutions to the problem: a) prevent migration by law or by making life in the provinces more attractive; or b) prevent the formation of new squatter settlements by law and "eradicate" (a favorite word among architects and planners) the existing ones, replacing them with housing projects.

The myths are embodied in writings ranging from the ridiculous – for example, a North American M.D. from the ship *Hope* (Walsh, 1966), reflecting the views of many of his Peruvian medical colleagues, says of a barriada near the city of Trujillo,

> In this enormous slum lived some 15,000 people many of whom had come down from the mountains, lured by communist agitators. Why starve on a farm the agitators asked, when well-paid jobs, good food, housing and education were waiting in Trujillo? This technique for spreading chaos and unrest has brought as many as 3,000 farmers and their families to the barriadas in a month. Once they arrive (on a one-way ride in communist-provided trucks) they are trapped in the festering slums with no money to return to their farms.

to serious observations by responsible people – for example, a North American sociologist (Schulman, 1966) who worked in a Colombian *tugurio* for nine months writes of squatter setlements,

> It is the rudest kind of slum, clustering like a dirty beehive around the edges of any principal city in Latin America. In the past two decades poor rural people have flocked to the cities, found no opportunities but stayed on in urban fringe shanty-towns squatting squalidly on the land. . . . Living almost like animals, the tugurio's residents are overwhelmed by animality. Religion, social control, education, domestic life are warped and disfigured.

A familiar theme of anti-city feeling and rural, small town bias runs through much of the European and North American commentary on squatter settlements.[3] Latin American academics and politicians, on the other hand, tend to be anti-countryside and pro-city, considering the latter to be the repository of all that is good and beautiful in Spanish and Portuguese culture. In a strange way, both points of view reinforce each other in condemning squatter settlements as disorganized products of outside agitation, and in suggesting eradication and shipment back to the rural areas as a solution.

It is probably apparent from the tone of the above remarks that my own views differ from those mentioned. In discussing the eight myths, I believe that I can portray a very different picture. Other myths, however, influence my own thinking so not all of the observations to which I refer are mistaken. They are biased, but they do reflect an aspect of reality. In presenting a different, and I believe a more hopeful and realistic, view, I do not mean to minimize the problems of overpopulation, rapid urbanization, poverty, prejudice, and lack of elementary health and social services that play such an important part in squatter setlement life.

William F. Whyte (1943) provided an effective counter to the anti-city literature by revealing strong informal institutions created by people in a Boston slum to serve their own needs. Since then many studies have drawn similar conclusions, including a recent study of the same area (Gans, 1962) that shows stability through change. Turner (1963, 1966) and I (1963, 1964, 1965) have noted similar kinds of institutions in squatter settlements, viewing them as solutions to difficult social problems rather than as problems in themselves.[4] The formation of squatter settlements is a popular response to rapid urbanization in countries that cannot or will not provide services for the increasing urban population.[5] In their study of the

Polish peasant in Europe and America, Thomas and Znaniecki (1920) viewed situations of great change and disorganization not as "a mere reinforcement of the decaying organization," but as ". . . a production of new schemes of behavior and new institutions better adapted to the changed demands of the group; we call this production of new schemes and institutions social reconstruction." I, too, see the squatter settlements as a process of social reconstruction through popular initiative.

FORMATION OF SQUATTER SETTLEMENTS

In two studies I have described the planning of a barriada invasion (Mangin, 1963) and constructed a composite history of a barriada in Peru (1960). The process is also discussed and photographed in a UN film, *A Roof of My Own*, advised by John Turner in the International Zone series. Subsequent investigation has indicated that strikingly similar sequences have occurred in other countries.

As a general pattern the majority of residents of a settlement have been born in the provinces and have migrated from farms or small towns. They have also come largely from tenements, alleys, and other slums within city limits where they settled upon arrival. According to a census of a typical Lima barriada in 1959, the average time of residence in Lima for heads of families originally from the provinces was nine years, and practically none of them had been in Lima less than three years. [Barriadas may now contain 400,000 of Lima's population of two million as opposed to 45,000 in 1940 (Abrams, 1965).] Frieden (1964) writes that in a 1958 census in Mexico City more than half of the residents of metropolitan tenements were of rural origin and that they were the squatters in the colonias proletarias. The colonias have two million inhabitants and may occupy 40 per cent of the land area of the Federal District by 1970 (Harth Deneke, 1966). Cortén (1965) reports that the majority of squatters in Santo Domingo are from the provinces and that not a single rural immigrant in his sample settled directly in a newly formed squatter settlement. Cuevas (1965) notes that in Guatemala City, where the squatters comprise 10 per cent of the population, "the great majority of the inhabitants (73 per cent) are economically displaced from the city itself," and that they are originally from the

provinces or the rural areas of the Department of Guatemala. Lopez et al. (1965) corroborate this finding. In Venezuela, where squatters in ranchos comprise about 35 per cent of the population of Caracas and 50 per cent of that of Maracaibo (Abrams, 1962), Talton Ray's study (1966) shows that most rancho residents are migrants but that "close to 100 per cent come from barrios within the city, not from the countryside." C. B. Turner (1964) and Usandizaga and Havens (1966) describe the same circumstances in Bogotá and Barranquilla, Colombia. Lutz (1966) found the same situation in Panama City, Bon (1963) in Montevideo, and Leeds (1966) in Rio de Janeiro, where favelas contain more than 500,000 people. Rosenbluth found that this relationship obtained for the callampas in Chile (1963) (6 per cent had come directly from the provinces only since 1960, 12 per cent before 1930). In the callampa sample of Goldrich, Pratt, and Schuller (1966) the residents are from rural areas but only 6 per cent had been in Santiago less than three years, and 80 per cent had been there 12 years or more. They estimate that about one-tenth of Santiago's population of two million are in callampas; Abrams (1962) uses the figure of 25 per cent.

The establishment of the community varies with local conditions, geographic and political. They arise on vacant land, usually uncultivated and owned by a governmental entity, on the outskirts of cities, or on undesirable land within the city, e.g., on steep hillsides, swamps, river beds, and dumps. Where there has been no active opposition from governments, settlements have been formed in an unorganized fashion by a few families drifting onto a site or, as in Guatemala, Guayaquil (Vitale, which I have personally observed), and parts of Lima, augmenting a group of families relocated by the government after a natural disaster. More common, however, is an organized invasion in the face of active opposition from the police. In countries not noted for governmental efficiency some of the invasions have been remarkably well organized. In Lima, after months of planning, thousands of people moved during one night to a site that had been secretly surveyed and laid out. They arrived with the materials to build a straw house, all their belongings, and a Peruvian flag. They were determined and, in several cases, returned to sites two and three times after police burned their belongings and beat and killed their fellows (reported in the Peruvian bimonthlies, *Caretas* and *Oiga*, 1963, for example). Police opposition to organized occupations has also been encountered in Brazil, Santiago de Chile (Clark, 1962), Cali (Powelson, 1964), Bogotá, Acapulco, and other places. Very little data is available on the organizers of invasions outside Peru. In Peru, evidence from my own experience and that of many Peace Corps volunteers and other observers indicates that the organizers were generally residents of the barriada of invasion. They often sought help from outsiders such as engineering students, lawyers, and, at least in two cases, army officers. The situation most closely approximating the often alleged agitation by outsiders involved men who helped form invasion groups on several occasions apparently because they thoroughly enjoyed and gained status from the meetings and the excitement. Land speculators have been involved in several countries, but they generally appear after the invasion.

The migrants come from rural provinces but many of them are from towns rather than farms. In Peru they come from a variety of settings and economic levels (Matos, 1961). In Chile 65 per cent of the migrants to Santiago had lived in towns of more than 5,000 as of 1946 (Herrick, 1965). Sixty per cent of the migrants to Buenos Aires studied by Germani (1961) had fathers who did manual labor, but many had shopkeeper fathers. Many of the migrants in countries where the data are available have lived for a time somewhere between the rural setting and the metropolis. Of migrants studied in villas miserias, Germani (1961) notes that 15 per cent had migrated from areas of less than 2,000 population, more than a third from settings with 2,000 to 20,000 inhabitants, and 50 per cent from large towns.

INTERNAL ORGANIZATION OF THE SQUATTER SETTLEMENTS

In the barrios formed by organized invasion, the original organization is strong at first. In those formed by other means, organizations are established to defend the community and advocate its cause to the government. The only exception is the Dominican Republic in which Cortén (1965) suggests that the lack of organizations was due to the dictator Trujillo and that since his regime they have been formed in new squatter settlements. In Lima the invasion organizations lose their power as the barriada becomes integrated with the city, but in some places

they remain important as intermediaries with the government and as organizers of mutual aid public works. In most of the barriadas studied, orderly, unofficial elections were held annually, and the importance of the organization depended largely on the personality of the elected leader. National politics had a part in many of the elections, but regionalism and personal charisma played a greater role in the elections I observed. Goldrich et al. (1966) compare political organizations in barriadas and callampas in Lima and Santiago and find the Chilean callampa organizations to be more permanent because they are able to change their function from invasion and defense to that of "output agencies mediating between the community and the government." Lutz (1966) emphasizes that practically the only political activity in the barriadas of Panama is the barriada organization. In the papers by Leeds, Hoenack, Wygand, and Morocco (1966) on Rio's favelas, the degree of organization in the favela associations is striking. They have organized everything from private water systems, markets, labor division, and groups to raise money to buy the land on which they live, to Carnaval dance groups essential to the famous Rio festival. In Lima and Rio, national political parties figure somewhat in squatter setlement politics, while in Venezuela, according to Ray (1966), they play a larger role. Nevertheless, the internal organization of a Venezuelan rancho is strong and, despite the struggles over plots of land during the original occupation of the rancho, the residents show great respect for each other's lot once they begin construction.

Membership in the associations varies but there are always nonmembers. In most cases membership doesn't imply participation but it appears that in squatter settlements there is at least as much and probably more participation in local politics than in other parts of the countries. The associations have trouble enforcing rules and often fail in their dealings with the bureaucracy. They do seem to be able to control, to a certain extent, who will be the members of the invasion group and the new residents; they favor nuclear families with an employed male as head. Associations frequently manage to get some assistance in installing sewers, water, roads, etc., and they provide a low-level, unofficial court for minor disputes. Most important, associations often give people a feeling of controlling their own destinies. This is frequently illusory, and distrust of local leadership is present almost from the start. The association does, however, have a major accomplishment that is visible at all times, namely, the invasion and successful retention of a piece of land.

SQUATTER SETTLEMENTS AND SOCIAL DISORGANIZATION

In Rio and Lima, many people warned me not to go into barriadas or favelas because they were full of criminals. Other scholars have had the same experience in other countries. In fact, however, squatter settlements are overwhelmingly composed of poor families who work hard and aspire to get ahead legitimately. Petty thievery is common, low-level tax evasion a national pastime; disputes occur over land titles, children annoying or damaging others' property, small debts, dogs, etc. Wife and child beating are frequent, at least in Lima and Rio, and drunkenness is common. Assault outside the family is rare. Organized crime is practically nonexistent.

The favelas of Rio de Janeiro seem to present something of an exception. The situation is not quite that described by *Time* (September 23, 1957),

> Squeezed by belt-cinching inflation and an influx of some 3,700 newcomers a month, the favelas gangsters have moved into the city's streets, boosted the crime rate alarmingly in recent weeks. In previously safe lovers' lanes, girls were raped, and their boy friends robbed, beaten or murdered. . . . In the favelas, where policemen rarely tread, gang lords built up Robin Hood reputations, casually rubbed out rivals and stool pigeons, and treated attempts to catch them with growing contempt.

Studies such as the SAGMACS (1960) report, however, indicate a higher rate of crime in the Rio favelas than in those of most of the other countries, and the existence of criminal gangs. A BEMDOC official in Rio told me last year that the vast majority of favelas were safe and several Peace Corps volunteers confirmed this assertion. According to the same official, one of BEMDOC's sociologists found that a gang of thieves living in a favela enforced very strict rules of honesty among their members within the favela to avoid attracting police attention.

Empirical evidence does not indicate that crimes occur with more frequency within squatter settle-

ments than outside. In the only comparative study I could find (Rotondo et al., 1963) the rate of delinquency in a central city slum of Lima was high and varied, while that of a barriada was low and unvaried – primarily complaints of wives against husbands and petty thievery.

The traffic in most squatter settlements doesn't warrant serious prostitution efforts but there are part-time prostitutes in Peru and Brazil. Some prostitutes lived in a barriada in which I worked in Lima, but they plied their trade elsewhere. Gambling is also on a low level because of the lack of money and traffic. Goldrich (1966) writes of callampas in Chile and barriadas in Peru, "In the face of their reputation for social disorganization, these settlers reveal remarkably little of it; crime, promiscuity, broken homes occur infrequently, particularly in comparison to the bridgehead settlements and traditional city slums. Though born as illegal invasions, these communities display a prevailing orientation toward law and order."

Family and kinship relationships are strong and provide a degree of crisis insurance. Rotondo, referring to a Lima barriada (1965) and a slum (1961), notes greater involvement with kin and much less social pathology in the barriada. Pearse (1961) discusses kin group controls on behavior in favelas. The importance of the family cannot be overestimated. The information on all the squatter settlements indicates that by far the greatest number of households consist of nuclear, bilateral families with resident fathers. The exceptions are Buenaventura, Colombia, where Mallol (1963) found a large number of female-based households, and Puerto Rico, where Safa (1965) encountered a similar situation. Grandparents and other relatives are also prevalent but the populations are generally younger than the already young national population averages. Since some squatter settlements have been in existence for 20 years there is variation in the age levels and one can see a family cycle (see Hammel, 1961, 1964). As Turner (1966) points out, the invader spends several years consolidating his building lot and constructing a house. This provides a start for his children. Desertion and early death of males are probably about the same as outside the settlement, but the widow or abandoned wife has the assets of a house and lot to make her an attractive marriage or common law partner.

The birth rate is higher than the national average in all of the squatter settlements. Rosenbluth (1963) notes that there are 4.5 children per mother in callampas as opposed to 2.38 in the rest of Santiago. The SAGMACS report (1960) indicates the same ratio as does a more recent publication of BEMDOC in Rio (BEMDOC, 1965). The Colombian and Peruvian data also give similar statistics. Illegitimacy figures don't have great significance in lower class Latin American populations because of the large number of stable consensual unions. I could find no solid information on the subject in squatter settlements.

Family relationships are as ambivalent in the squatter settlements as outside. Gordon Parks' account of violence and bitterness within a favela family (1961) could be duplicated inside and outside squatter settlements throughout Latin America and inside and outside slums throughout the world.[6] Padilla (1958) and Wakefield (1957) report social disorganization among Puerto Rican migrants to New York but emphasize that parents integrate and sacrifice for their children. Lewis (1966a), in a much more intensive study of fewer people, reports some integration and sacrifice for children but emphasizes disorganization, bitterness, and violence. Lewis notes the same conditions among rural migrants to Mexico City (1960) but many observers have taken issue with him (for example, Butterworth, 1962; Paddock, 1961; Bendiner, 1967), and in other studies (Lewis, 1952, 1959, 1965) he, too, emphasizes very different sorts of migrant behavior in Mexico City. Fried (1959) and I (1961, 1964) describe family attitudes in a Peruvian barriada in very different terms. He found and stressed pity and pessimism; I encountered and stressed self-help and optimism. As I have pointed out (1967), we are both correct in stressing apparently contradictory aspects of reality that exist in the same population. Lewis and Butterworth are both right. Lewis and Wakefield are both right about Puerto Rico. Bonilla (1961) says that Camus, in his film *Black Orpheus*, and Parks in his *Life* photographic essay, "each fasten on separate phases of the emotional cycle of the favela." The film stresses favelados as ". . . carefree, full of boundless energy and rhythm, with an unfettered zest for life." Parks' family is ". . . a proto-human band, weakened by chronic illness, callously indifferent to each other's suffering, living in a monotone degradation punctured only by flashes of violence." Both exist in favelas, barriadas, colonias proletarias, and outside.

The differences, pessimism and optimism, exist between families. Contrast, for example, the bitter complaining of Carolina María de Jesús in her diary of favela life (1962) with the enterprising hopefulness of Achilles and María (Leeds, 1966), a favela couple. Contradictions also exist within the same family and within individuals. In a Lima barriada we administered a section of an MMPI to a randomly selected sample of 74 adults and found apparent contradictions:

Q. The future looks blacker every day.
    Yes 61   No 13
Q. A young man of today can have much hope for the future.
    Yes 69   No 5
Q. Many children are a burden.
    Yes 50   No 23
Q. One should sacrifice all for one's children.
    Yes 71   No 3

The same people who see the future becoming blacker every day have hope for the future. The same people who view their children as a burden believe that they should sacrifice for them. The contradictions are more apparent than real. This kind of ambivalence characterizes the human condition.

Discussions of alienation and the quality of urban life probably digress too far from the purpose of a survey article, but I believe existing evidence indicates that inhabitants of squatter settlements are less alienated from the national state and more involved with each other than are residents of central city slums. Although reports are not explicit on the subject, one of the reasons for this difference is that squatters can constantly see around them a major accomplishment of their own, i.e., the seizure of land and the creation of a community.

THE ECONOMIC CONDITION OF SQUATTER SETTLEMENTS

Just as traditional economists have difficulty with peasant economics, they have difficulty in evaluating the contribution (or lack of it) of squatter settlements to national and city economies. Obviously one major contribution is that millions of people have solved their own housing problem in situations where the national governments were practically unable to move. On the other hand, as Abrams (1965), Turner (1963), and many governmental planners have pointed out, by so doing they have occupied land that might have a more "logical" use in a city, and they have made the provision of services such as water, sewage disposal, electricity, and paved streets much more expensive than if the land were empty. This dilemma, unfortunately, is only a dilemma for planners. People who need housing can't be kept in the pipeline for years as a plan can. Clearly the land invaders want services and, in most cases, have shown themselves to be quite willing to pay for them. They are not, however, willing to wait for them on the basis of a governmental promise. The result has been the creation of many unsanitary communities with expensive private water and electric arrangements and poor internal transportation. One of the rational considerations weighed by potential residents in established squatter settlements as well as by invaders is the balance between the economic advantages of paying no rent, "owning" your own home, and joining a community, on the one hand, and having few services, building your own house, risking great loss if dislocated, and taking unpopular and often illegal political action, on the other. Since some squatter settlements are located close to the dwellers' places of employment, this is also a consideration. The central city slums are also close to employment but they are expensive and undesirable. The few governmental attempts to build new projects (Peru, Brazil, Venezuela, Argentina) have chosen sites that are much too far from work and have proved to be much too expensive for the squatters.

[. . .]

NOTES

1   There are several good sources on Latin American urbanization. The sources I have consulted as background are Hauser (1961), a report on a UN seminar on urbanization in Latin America; Dorselaer and Gregory (1962); Union Panamericana (1965); and T. Lynn Smith (1960) on population studies; Rycoft and Clemmer (1963); the special issue on urbanization of *Scientific American* (Sept. 1965); and Charles Abrams' book (1964) on world housing problems.

2   I refer specifically to studies in the text when I discuss matters in detail. When I refer to Peru without giving a specific source I am using my own experience based on more than six years of field work. I do, on occasion, refer to "the studies in the survey," "all the reports," or

say "the sources all agree," etc. Following is a list of the major sources I consulted by country. Mexico: Frieden, 1964, 1965; Lewis, 1959, 1960; Harth, 1966; Butterworth, 1962. Guatemala: Cuevas, 1965; Lopez y otros, 1965. Dominican Republic: Cortén, 1965. Venezuela: Ray, 1966; Peattie, 1962, 1966; Jones, 1964. Colombia: C. B. Turner, 1964; Usandizaga and Havens, 1966; Pinto, 1966; Mallol, 1963; Bernal, 1963. Panama: Guiterrez, 1961; Lutz, 1966. Puerto Rico: Lewis, 1966a; Safa, 1964, 1966. Uruguay: Bon, 1963. Argentina: Germani, 1961; Wilson, 1965. Brazil: SAGMACS, 1960; Pearse, 1961; BEMDOC, 1965; and an invaluable set of mimeographed reports of chapters in a forthcoming book on favelas by Hoenack, Leeds, Modesto, Morocco, O'Neil, Smith, and Wygand, 1966. Peru: Turner, 1963, 1966; Matos, 1961; Hammel, 1961, 1964; Rotondo y otros, 1961, 1963; Rotondo, 1965; Mangin, 1960, 1963, 1964.

3   For an analysis of some of the anti-city literature in U.S. social science, see Mills, 1943. For a discussion of urbanism in Latin America, and elsewhere, see Morse, 1965b.

4   Lima and other cities have always grown through the formation of squatter settlements. In *Relación del virrey Conde de Superinda*, Documento C1312, Biblioteca Nacional del Peru, a report from 1746 denounces the formation of barriadas in present-day Rimac, notes that they are unsanitary, and that the people should be sent back to the mountains (I am indebted to John Te Paske for calling my attention to the document). Schaedel (1966) points out that pre-Spanish, and pre-Incaic cities on the Peruvian coast used the same sorts of land for housing that are presently being used by barriadas, and he refers to this type of use as "characteristic of indigenous America at its highest point of development," and "resolving urban living problems better than the imported Spanish variety."

5   As Mack and McElrath (1964) point out, "The fastest rates of urbanization now are occurring in societies with relatively low levels of urbanization." Wingo (1966) has made the same statement about Latin America, and he separates Argentina, Uruguay, and Cuba from other countries on the basis of their more advanced urbanization. Migration to cities has slowed down, and squatter settlements reflect a pushing out of the least competitive members of the society rather than an outlet for the pressures of rapid urbanization. Silvert (1966) has made the same division, adding Costa Rica, and making a similar point. The Argentine material does suggest a difference from the others. Germani (1961) points out that delinquency is higher in squatter settlements than in tenements and that recent migrants are more apt to be found in squatter settlements. This contradicts the information from other countries and may be due to the distinction made by Wingo and Silvert.

Wilson (1965), however, says that squatters are very much like other Argentines and are in the settlements because of the housing shortage that affects the total population of Buenos Aires. The Uruguayan situation is not described in sufficient detail, but it appears to fit the general, rather than the Argentine, pattern. I could find no information on squatter settlements in Cuba or Costa Rica.

6　The concept of the culture of poverty developed by Lewis, as described in the introduction to *Children of Sanchez* and revised in *Scientific American*, October, 1966, is an important idea that is applicable to certain older squatter settlements, but more so to central city slums. I would take issue with the term "culture," since a change in employment or sudden acquisition of wealth *may* change an individual's "culture" if it is the culture of poverty. Carolina María de Jesús (1962), from a Rio favela, and the people Lewis describes in La Esmeralda, a Puerto Rican squatter settlement, fit the concept perfectly, and sudden wealth does not necessarily change their culture. I do feel that the concept has no particular application to the majority of the residents of the majority of squatter settlements. Despite their poverty in relation to that of poor people in the United States, the fantastic disparity in wealth and power between the squatters and their own upper and middle classes, and the ambivalent attitude of the national governments reflected in the violence of the army and the police toward the squatters coupled with half-hearted attempts to assist them with housing, they are not alienated, hopeless people caught in a vicious circle of poverty. For most of the adults their condition in the squatter settlements is the best of their lives and a marked improvement on their previous two or three houses. I am always suspicious of the characterization of any population as apathetic, and it is certainly an inappropriate term for squatter settlements.

BIBLIOGRAPHY

ABRAMS, CHARLES
　　1964　Man's Struggle for Shelter in an Urbanizing World. Cambridge, Mass.
　　1965　The use of land in cities. Scientific American. September. 151–161.

ARIAS B., JORGE
　　1965　La concentración urbana y las migraciones internas. In: Problemas de la Urbanización en Guatemala. 19–45. Ministerio de Educación.

BEMDOC (BRASIL ESTADOS UNIDOS MOVIMENTO DE DES-
　　ENVOLVIMENTO E ORGANIZAÇÃO DE COMUNIDADE)
　　1965　Vial Proletaria da Penha. Rio de Janeiro.

BENDINER, ELMER
　　1967　Outside the kingdom of the middle class: La Vida by Oscar Lewis. The Nation. January 2.

BERNAL A., HERNANDO
　　1963　Ritmos de vida en Buenaventura. Revista Colombiana de Antropología. 12: 331–355.

BON ESPASANDIN, MARIO
　　1963　Cantegriles: Familia, Educación, Niveles Económicos—Laborales, Vivienda y Aspectos Generales que Componen el "Collar de Miserias" de Montevideo. Monte-video.

BONILLA, FRANK
　　1961　Rio's favelas: the rural slum within the city. American Universities Field Staff Report. August.

BOURRICAUD, FRANCOIS
　　1964　Lima en la vida política peruana. América Latina. October–December.

BRADLEY, JOHN
　　1966　The market system and urbanization in Lima. (Mimeographed).

BUTTERWORTH, DOUGLAS
　　1962　A study of the urbanization process among the Mixtec migrants from Tilantongo in Mexico City. América Indígena. 22: 257–274.

CARETAS
　　1963　Enero-Febrero.

CLARK, GERALD
　　1963　The Coming Explosion in Latin America. New York City.

CORTÉN, ANDRE
　　1965　Como vive la otra mitad de Santo Domingo: estudio de dualismo estructural. Caribbean Studies. 4: 3–19.

CUEVAS, MARCO ANTONIO
　　1965　Análisis de tres áreas marginales de la ciudad de Guatemala. In: Problemas de la Urbanización en Guatemala. Ministerio de Educación.

DAVIS, KINGSLEY
　　1965　The urbanization of the human population. Scientific American. September. 41–50.

DORSELAER, JAIME AND GREGORY, ALFONSO
　　1962　La Urbanización en América Latina. Oficina Internacional de Investigaciones Sociales de FERES. Bogotá.

FRIED, JACOB
　　1959　Acculturation and mental health among migrants in Peru. In: Culture and Mental Health. Marvin Opler ed. New York City.

FRIEDEN, BERNARD
　　1964　A program for housing and urban develop-ment in Mexico City. (Mimeographed.) AID. Washington, D.C.
　　1965　The search for housing policy in Mexico City. Town Planning Review. 35: July.

FOSTER, GEORGE
  1965 Peasant society and the image of limited good. American Anthropologist. 67: 293–315.

GANS, HERBERT
  1962 The Urban Villagers. New York City.

GERMANI, GINO
  1961 Inquiry into the social effects of urbanization in a working-class sector of Buenos Aires. In: Urbanization in Latin America. Philip Hauser ed. 206–233. UNESCO. New York City.

GOLDRICH, DANIEL
  1965 Toward the comparative study of politicization in Latin America. In: Contemporary cultures and societies of Latin America. Dwight Heath and Richard Adams eds. 361–378. New York City.

GOLDRICH, DANIEL, PRATT, R. B., AND SCHULLER, C. R.
  1966 The political integration of lower class urban settlements in Chile and Peru. (Mimeographed.) Annual Meeting of Aemrican Political Science Association. New York City.

GUTIERREZ, SAMUEL
  1961 El problema de las barriadas brujas en la ciudad de Panama. Instituto de Vivienda y Urbanismo.

HALPERIN, ERNST
  1965 The decline of communism in Latin America. Atlantic Monthly. May. 65–70.

HAMMEL, EUGENE A.
  1961 The family cycle in a coastal Peruvian slum and village. American Anthropologist. 63: 989–1005.
  1964 Some characteristics of rural village and urban slum populations on the coast of Peru. Southwestern Journal of Anthropology. 20: 346–358.

HARTH DENEKE, JORGE
  1966 The colonias proletarias of Mexico City, low income settlements on the urban fringe. Master's thesis in City Planning. Massachusetts Institute of Technology. Cambridge, Mass.

HAUSER, PHILIP M. ed.
  1961 Urbanization in Latin America. UNESCO. New York City.

HEATH, DWIGHT AND ADAMS, RICHARD eds.
  1965 Contemporary cultures and societies of Latin America. New York City.

HERRICK, BRUCE H.
  1966 Urban Migration and Economic Development in Chile. Cambridge, Mass.

HOENACK, JUDITH
  1966 Marketing, supply and their social ties in Rio favelas. (Mimeographed.)*

JONES, EMRYS
  1964 Aspects of urbanization in Venezuela. Ekistics. 18: 109.

KOTH, MARCIA, SILVA, JULIA AND DIETZ, ALBERT
  1964 Housing in Latin America. Cambridge, Mass.

LEEDS, ANTHONY
  1964 Brazilian careers and social structure: an evolutionary model and case history. American Anthropologist. 66: 1321–1347.[†]
  1966 The investment climate in Rio favelas. (Mimeographed.)*

LEWIS, OSCAR
  1952 Urbanization without breakdown. Scientific Monthly. 75: 31–41.[†]
  1959 Five Families. New York City.
  1960 The Children of Sanchez. New York City.
  1965 The folk urban ideal types. In: The Study of Urbanization. P. Hauser and L. Schnore eds. New York City.
  1966a La Vida: A Puerto Rican Family in the Culture of Poverty. New York City.
  1966b Even the saints cry. Trans-Action. November. 18–23.
  1966c The culture of poverty. Scientific American. October. 19–25.

LOPEZ T., JOSE AND OTHERS
  1965 Barrios Marginales. Informe Sobre La. Colonia "La Verbena," Ciudad de Guatemala. Dirección de Obras Públicas.

LUDWIG, ARMIN K.
  1966 The planning and creation of Brasilia. In: New Perspectives of Brazil. E. Baklanoff ed. Nashville.

LUTZ, THOMAS
  1966 Some aspects of community organization and activity in the squatter settlements of Panama City. (Mimeographed.)

MALLOL DE RECASSENS, MARIA ROSA Y RECASSENS, JOSE
  1963 Estudios comparativos de los niveles de vivienda en Buenaventura y Puerto Colombia. Revista Colombiana de Antropología. 12: 295–328.

MANGIN, WILLIAM
  1959 The role of regional associations in the adaptation of rural population in Peru. Sociologus. 9: 21–36.[†]
  1960 Mental health and migration to cities. Annals of the NY Academy of Sciences. 84: 17: 911–917.[†]
  1963 Urbanization case history in Peru. Architectural Design. August.
  1964 Sociological, cultural and political characteristics of some rural Indians and urban migrants in Peru. Wenner-Gren. Symposium

on Cross-Cultural Similarities in the Urbanization Process. (Mimeographed.)

1965 The role of social organization in improving the environment. In: Environmental Determinants of Community Well-Being. Pan-American Health Organization.

1967 Comunidades de la sierra de América Indígena. (In press.)

MARIA DE JESUS, CAROLINA

1962 Child of the Dark. Translated from Portuguese by St. Clair Drake. New York City.

MATOS MAR, JOSE

1961 Migration and urbanization: the barriadas of Lima. In: Urbanization in Latin America. Phillip M. Hauser ed. 170–190. New York City.

MILLER, S. M. AND REIN, MARTIN

1964 Poverty and social change. The American Child. March.

MILLS, C. WRIGHT

1943 The professional ideology of social pathologists. American Journal of Sociology. 49: 165–180.

MODESTO, HELIO

1966 Favelas-reflexoes sôbre o problema. (Mimeographed.)*

MOROCCO, DAVID

1966 Carnaval groups-maintainers and intensifiers of the favela phenomenon in Rio. (Mimeographed.)*

MORSE, RICHARD M.

1965a Urbanization in Latin America. Latin American Research Review. 1: 1: 35–74.

1965b The sociology of San Juan: an exegesis of urban mythology. Caribbean Studies. 5: 45–55.

NEHEMKIS, PETER

1964 Latin America: Myth and Reality. New York City.

OIGA

1963 Enero.

O'NEILL, CHARLES

1966 Some problems of urbanization and removal of Rio favelas. (Mimeographed.)*

PADDOCK, JOHN

1961 Oscar Lewis's Mexico. Anthropological Quarterly. 34: 129–149.

PADILLA, ELENA

1958 Up From Puerto Rico. New York City.

PAREDES, ERNESTO

1963 Fuentes de la población de la barriada Fray Martín de Porres. In: Migración e Integración en el Perú. H. Dobyns and M. Vasquez eds. Lima.

PARKS, GORDON

1961 Freedom's fearful foe: poverty. Life. June 16.

PATCH, RICHARD W.

1961 Life in a callejón. American Universities Field Staff Report. June.

PEARSE, ANDREW

1961 Some characteristics of urbanization in the city of Rio de Janeiro. In: Urbanization in Latin America. Philip M. Hauser ed. 191–205. New York City.

PEATTIE, LISA

1962 A short ethnography of La Laja. (Mimeographed.)

1966 Social issues in housing. Joint Center for Urban Studies. (Mimeographed.) Cambridge, Mass.

PINTO BARAJAS, EUGENIO ed.

1966 Control y Erradicación de Tugurios en Bucaramanga. Santander, Colombia.

POWELSON, J. P.

1964 The land-grabbers of Cali. The Reporter. January 16.

RAY, TALTON

1966 The political life of a Venezuelan barrio. (Mimeographed.)

REINA, RUBEN E.

1964 The urban world view of a tropical forest community in the absence of a city, Petén, Guatemala. Human Organization. 23: 265–277.

RODWIN, LLOYD

1965 Ciudad Guayana: a new city. Scientific American. September. 122–132.

ROSENBLUTH L., GUILLERMO

1963 Problemas Socio-Económicos de la Marginalidad y la Integración Urbana: El Caso de las Poblaciones Callampas en el Gran Santiago. Santiago, Chile.

ROTONDO, HUMBERTO

1965 Adaptability of human behavior. In: Environmental Determinants of Community Well-Being. Pan-American Health Organization.

ROTONDO, HUMBERTO AND OTHERS

1963 Un estudio comparativo de la conducta antisocial de menores en áreas urbanas y rurales. In: Estudios de Psiquiatría Social en el Perú. Lima.

RYCROFT, W. STANLEY AND CLEMMER, MYRTLE

1963 A Study of Urbanization in Latin America. United Presbyterian Church. New York City.

SAFA, HELEN

1964 From shantytown to public housing: a comparison of family structure in two urban neighborhoods in Puerto Rico. Caribbean Studies. 4: 3–12.

1965   The female-based household in public housing: a case study in Puerto Rico. Human Organization. 24: 135–139.

SAGMACS

1960   Aspectos Humanos de Favela Carioca. Report of Mission of Father Lebret. Special Supplement to O Estado de São Paulo. April 13.

SALMON, LAWRENCE

1966   Report on Vila Kennedy and Vila Esperança. (Mimeographed.) Cooperativa Habitacional (COHAB).

SCHAEDEL, RICHARD P.

1966   Urban growth and ekistics on the Peruvian coast. 36th International Congress of Americanists. 1: 531–539. Sevilla.

SCHMITT, KARL AND BURKS, DAVID

1963   Evolution or Chaos: Dynamics of Latin American Government and Politics. New York City.

SCHULMAN, SAM

1966   Latin American shantytown. N.Y. Times Magazine. January 16.

SILVERT, KALMAN H.

1966   The Conflict Society: Reaction and Revolution in Latin America. American Universities Field Staff. New York City.

SIMMONS, OZZIE G.

1952   El uso de los conceptos de aculturación y asimilación en el estudio del cambio cultural en el Perú. Perú Indígena. 2: 40–47. Lima.

SMITH, NANCY

1966   Eviction in a Rio favela-leadership, land tenure and legal aspects. (Mimeographed.)*

SMITH, T. LYNN

1960   Latin American Population Studies. University of Florida. Gainesville.

STOKES, CHARLES J.

1962   A theory of slums. Land Economics. 38: 3.

THOMAS, WILLIAM I. AND ZNANIECKI, FLORIAN

1920   The Polish Peasant in Europe and America. Chicago.

TURNER, CHARLES BARTON

1964   Squatter settlements in Bogotá. (Mimeographed.) CINVA. Bogotá.

TURNER, JOHN F. C.

1963   Dwelling resources in South America. Architectural Design. August.

1966   Asentamientos Urbanos No Regulados. Cuadermos de la Sociedad Venezolana de Planificación. 36.

USANDIZAGA, ELSA AND HAVENS, EUGENE

1966   Tres Barrios de Invasión. Facultad de Sociología. Bogotá.

WAGNER, BERNARD, MCVOY, DAVID, AND EDWARDS, GORDON

1966   Guanabara Housing and Urban Development. AID Housing Report. July 1.

WAKEFIELD, DAN

1959   Island in the City. Boston.

WALSH, WILLIAM, MD.

1965   Class, Kinship and Power in an Ecuadorean Town: The Negroes of San Lorenzo. Stanford.

WHITTEN, NORMAN E., JR.

1962   Urbanization and Political Extremism. (Mimeographed.) Cambridge, Mass.

WHYTE, WILLIAM FOOTE

1943   Street Corner Society. Chicago.

WIENER, MYRON

1966   Yanqui, Come Back! The Voyage of the U.S. Hope to Peru. New York City.

WILLEMS, EMILIO

1966   Religious mass movements and social change in Brazil. In: New Perspectives of Brazil. E. Baklanoff ed. Nashville.

WILSON, L. ALBERT

1965   Voice of the Villas. Foundation for Cooperative Housing. Washington, D.C.

WINGO, LOWDON, JR.

1966   Some aspects of recent urbanization in Latin America. Resources for the Future. (Mimeographed.) Washington, D.C.

WYGAND, JAMES

1966   Water networks: their technology and sociology in Rio favelas. (Mimeographed.)*

* Papers presented at 36th International Congress of Americanists. September 1966. Mar del Plata.
† Reprinted in Contemporary cultures and societies of Latin America. Dwight Heath and Richard Adams eds. New York City.

PART SIX

The hubris of development

INTRODUCTION TO PART SIX

Attempts to understand development through the lens of Development Studies flourished in the 1950s, 1960s and 1970s. Major centres of Development Studies were opened during this period in Western Europe, including the Institute for Social Studies at The Hague, the Institute for Development Studies at the University of Sussex, and the School of Development Studies at the University of East Anglia. Other centres flourished in Delhi, Dar-es-Salaam and across Latin America. The oldest journals in the field (including *Economic Development and Cultural Change*, from Chicago since 1952) were also added to by the *Journal of Development Studies* (UK, 1964), *World Development* (US, 1970) and *Development and Change* (Netherlands, 1971), among several others. The professionalisation of studies of development continued apace through the 1980s and 1990s and into the present decade. New journals continued to emerge (for example, *Third World Quarterly* and *Journal of International Development*), and alongside them large numbers of degree courses (mainly at graduate levels, and well beyond Western Europe). Development Studies became a more contested and hybrid discipline.

In the second half of the 1970s, however, and through much of the 1980s, the enterprise of development itself stalled in important respects. Some countries in Africa were struck by famines. Critics in India bemoaned the 'Hindu rate of growth' that afflicted the country. About half the population there remained in absolute poverty by 1980. They were not helped by per capita rates of GDP growth of 1 or 2 per cent a year. And much of Latin America was damaged by the debt crisis that exploded in 1982–83 and which led to a lost decade of development.

The debt crisis followed the collapse of the Bretton Woods system in 1971/73 and a shift to floating exchange rates. It also took shape with reference to a regime of international financial transfers that had been privatising since the 1960s. Loans from money centre banks in the United States, Western Europe and Japan were fast replacing aid flows to many middle-income countries. Borrowing countries generally liked this arrangement since it removed the conditionalities that aid brought with it. Substantial loans were also required in the 1970s to pay for more expensive oil imports, as well as 'development'. (The price of oil increased fourfold in 1973–74, partly in response to the actions of the Organisation of Petroleum Exporting Countries.) The loans advanced by the money centre banks were made at floating interest rates. This was done to help protect the banks against endemic risk, but in the 1970s it also worked to the advantage of borrowing countries. Real interest rates then were generally very low – that is after adjusting for inflation. But when repayments fell due in the early 1980s, typically after a 5–7-year grace period, global interest rates had shot up to a record post-war high. The United States was now squeezing inflation out of the system by means of tight money policies. These had been introduced by the Federal Reserve Bank Chairman Paul Volcker in 1978–79. Latin American borrowers, especially, paid the price of these tight money policies in 1981–83, when the interest rates on their outstanding loans rose in tandem with the prime (interest) rate in the United States. In August 1982, Mexico defaulted on its debt repayments, which then had to be rescheduled. Many more Latin American countries followed suit in 1983, prompting dire warnings about the stability of the international banking system.

These events, along with the aforementioned inflation that hit the world economy in the 1970s, and alongside a parallel crisis of public indebtedness in Africa, threw up fresh problems for analysis within

Development Studies. They also caused some critics to take aim at the intellectual standing of the new discipline. Major attacks came first from the economic and political Right. They came from what John Toye (1987) called 'the counter-revolution in development theory and policy'. But other critiques soon followed, and in some cases they joined hands with the neo-liberals, as the counter-revolutionaries were also called. The relevance of development studies was called into question, along with the humility of those who thought that development could be spirited quickly into existence by planners, experts and technocrats.

Some free-market economists had never cared much for what they called the *dirigiste*, or state-centred, nature of post-1950 developmentalism. Milton Friedman, for example, often acclaimed as the intellectual father of monetarism, wrote strongly against foreign aid programmes in the late 1950s. Peter Bauer wrote in similar terms in the UK in the 1960s and 1970s. He was rewarded with a peerage by Margaret Thatcher in the 1980s. Both men argued that aid created dependency and that the already rich countries had developed without aid. Trade was preferable to aid. At its worst, it was argued, foreign aid amounted to a tax on poorer people in rich countries in order to transfer funds to rich people in poor countries.

Until the late 1970s these voices stood outside the mainstream. The international spread of inflation in the 1970s changed things, however, as did the debt crisis and rising labour militancy in many industrialised countries. The New Right was now encouraged to mount a sustained challenge to the damaging effects of Keynesianism. Arguably, this began in the UK, the United States and Canada before it was extended to the developing world. But the diffusion of what some have called Thatcherite ideas was not a long time coming. In a famous pamphlet from 1983, the economist Deepak Lal saw 'development economics' as the scourge of the Third World. Lal could not bring himself to combine these two words except in scare quotes. In his view, there was only one body of reputable economic theory. This corpus of mono-economic theory was based around propositions that signalled the efficiency of the market as a resource allocation mechanism. It also embraced comparative advantage theory and free trade.

In contrast, Lal argued, developing countries had been steered in the wrong direction by economists who peddled the false doctrines of development economics. These doctrines held that free markets did not or could not work (yet) in poorer countries. Weak markets had to be supplemented by strong state actions. Savings and investment could then be directed to the commanding heights of the economy: usually a set of import-substituting manufacturing industries, as we have seen. Foreign aid would supplement domestic resources. Lal further accused development economists of wanting to address poverty and inequality by means of large-scale programmes of social engineering, rather than by rapid and generalised economic growth. The result of all this, for Lal, was exactly the opposite of what the planners and dual economists had aimed for. The centrality of the state, combined with a lack of concern for microeconomic efficiency, led to slow growth, high levels of rent-seeking and corruption. It also led to the manufacturing of products – those cars in India, again – that few people outside the country would be willing to buy.

Lal's pamphlet was widely read in the early 1980s. He also wrote columns for leading newspapers and magazines on the perils of development economics. By the mid-1980s some of his ideas were being mainstreamed in the World Bank. The neo-liberal economist Anne Krueger was appointed as that institution's Vice President for Economics and Research half way through the first Reagan Presidency. The World Bank and the International Monetary Fund (IMF) now began strenuously to promote structural adjustment policies in countries in default on their debts. They exploited the greater bargaining power of Northern countries following the debt crisis and called for less public spending, tighter monetary policies, and devaluation in the South. Developing countries were to be made leaner. Their economies were to be weaned away from the domestic public sector. They would be oriented more to the global economy, where hard currencies could be earned.

The so-called Washington Consensus became a codification of these ideas at the start of the 1990s. Its principal author, however, John Williamson, was rather less of a true believer than some at the World Bank and the IMF. As is evident from the reading reproduced here, Williamson was not so much a market fundamentalist as a believer in economic fundamentals. Williamson took the view that certain forms of development policy since 1950 had been plain bad. We might not know what works most efficiently to produce development, he maintained. But we do know what doesn't work, and we should avoid these policies for the economic health of all.

James Scott makes a parallel argument in the reading reproduced here, which comes from the Conclusion to his book, *Seeing Like a State: Why Certain Schemes to Improve the Human Condition Have Failed.* Scott is less concerned than Williamson or Lal with development economics. His book is focused on some of the worst tragedies of the twentieth century: forced collectivisations, for example, or agricultural monocropping. But Scott is driven to the conclusion that these tragedies emerged as much from arrogance and plain bad ideas as from brute power. Planning, Scott notes, was not only the hand-maiden of a first generation of Developmentalism, it also represented a damaging moment of simplification. As he puts it so crisply: 'If I were asked to condense the reasons behind these failures into a single sentence, I would say that the progenitors of such plans regarded themselves as far smarter and farseeing than they really were and, at the same time, regarded their subjects as far more stupid and incompetent than *they* really were' (1998: 343, emphasis in the original; and see also Rhoads Murphey in Part 4).

This is not an exceptionally novel observation – there are hints of Burke here, and of Hayek in a different way – but it is a telling one, and it was brought back to the boil in 2006 in a very public debate between two development economists based in New York, Jeffrey Sachs and William Easterly. We have met Sachs before, of course (see Part 1). Sachs caught the public eye in 2005 when he published his book, *The End of Poverty*, with a foreword by the rock star Bono. Sachs also worked closely with Tony Blair and Gordon Brown in the UK government. The challenge of ending poverty in Africa was made the key moment of the UK's presidency of the G8 countries in 2005. Trade reform, debt relief and a $75 billion 'Marshall Plan for Africa' were widely canvassed as the ways to ensure that most African countries would meet their Millennium Development Goal (MDG) targets for poverty reduction by 2015, or perhaps in 2025. Sachs also worked on the MDGs as Special Adviser to UN Secretary General Kofi Annan.

For William Easterly (2006), however, this was all so much bunkum. It was further evidence, he said, of the mentality of the White Man's Burden, or the crass idea that 'we' could pull 'them' out of poverty with vast infusions of money, infrastructure and expertise over a short period of time. Another case of smart and farsighted planners – Easterly's preferred word of abuse – feeling and acting smarter than they really are. In contrast, Easterly said, we should go with the searchers. Work with local people solving local problems innovatively and in an accountable fashion. Support them with limited funds and stand back. And don't expect miraculous changes any time soon. Be sceptical. 'The unaccountable foreign experts', Easterly wrote, 'who promise to comprehensively end poverty "at an amazingly low cost" [Sachs and others], a claim that bears stronger intellectual kinship to late-night TV commercials than to African realities, will accomplish very little' (2007: 66).

Michael Edwards was also in New York City at this time, working for the Ford Foundation's Governance and Civil Society Unit. It is possible that Edwards would have agreed with large parts of Easterly's argument, although he has been a more consistent advocate for civil society than has Easterly. (Easterly trusts more to markets – albeit not market reform planned from above – and democracy.) Certainly, Edwards picked up on a commonly expressed sentiment when he published an essay in 1989 on the 'irrelevance of development studies' (reproduced here). By this time, just before the end of the Cold War and the so-called End of History, a renewed pessimism, or realism, was dawning in development studies. The optimism of the 1950s and 1960s had long since dissipated. Nevertheless, the lost decade of development in Latin America, and famine in the Horn of Africa, both of which could be set against the rise of South-East Asia, forced a more searching analysis of where development had got to in the 1980s, and with it Development Studies. Edwards provided one well-known contribution to the mix.

Two more came from Diane Elson and James Ferguson with Larry Lohmann. Elson builds on the work of Ester Boserup and others to spotlight continuing male bias in the development process – another form of hubris, surely. This is discussed here in respect of structural adjustment's damaging effects on the lives of women in sub-Saharan Africa. Ferguson and Lohmann, for their part, accuse the development industry in Lesotho of being an anti-politics machine. They argue that the cumulative effect of one failing develop-ment project after another works, curiously, to embed the idea that still more projects are required, and still more expertise from outside. It is in this manner that the technicalisation of development proceeds. And what is silenced, Ferguson and Lohmann suggest, are not only the voices of local searchers, but also the regional political issues that most need attention: questions of gender relations, for example, access to and control over land, and unequal development across southern Africa.

Many more contributions along these lines would emerge throughout the 1990s. They did so once the fallout from the break-up of the Soviet Union became more widely recognised and understood. Ironically, too, they also did so once the failures of neo-liberalism became more apparent. A discourse on development that was first advanced as an antidote to the overweening vanities of Developmentalism became in the 1980s and 1990s the most hubristic form of development of all. Particularly within the Bretton Woods institutions, it has only been in the past ten years that serious efforts have been made to look past neo-liberalism's obsession with market economics. It has been a long and painful process to restore to the development agenda a broader – and more traditional – concern for such matters as politics (or, more thinly, institutions and governance), the environment, vital path-dependencies, and questions of human agency, rights-based initiatives and cultural difference.

REFERENCES AND FURTHER READING

Burnside, C. and Dollar, D. (1997) 'Aid, policies and growth', *World Bank Policy Research Working Paper* No. 596252.

Corbridge, S. (2007) 'The (im)possibility of development studies', *Economy and Society* 36(2): 179–211.

Easterly, W. (2006) *The White Man's Burden,* New York: Penguin Press.

Easterly, W. (2007) 'The white man's burden: letter to the editor', *New York Review of Books*, 11 January, p. 66.

Ferguson, J. (1990) *The Anti-Politics Machine: 'Development', Depoliticisation and Bureaucratic Power in Lesotho*, Cambridge: Cambridge University Press.

Forbes, K. (2000) 'A reassessment of the relationship between inequality and growth', *American Economic Review* 90: 869–87.

Goldsmith, A. (1995) 'The state, the market and economic development: a second look at Adam Smith in theory and practice', *Development and Change* 26: 633–50.

Gray, J. (2001) *False Dawn: The Delusions of Global Capitalism*, Cambridge: Polity.

Harriss-White, B. (1996) 'Free market romanticism in an age of deregulation', *Oxford Development Studies* 24: 27–45.

Krueger, A. (1974) 'The political economy of the rent-seeking society', *American Economic Review* 64: 291–303.

Kwon, J. (1994) 'The East Asian challenge to neo-classical orthodoxy', *World Development*, 22: 635–44.

Lindert, P. (2004) *Growing Public: Social Spending and Economic Growth Since the Eighteenth Century*, Cambridge: Cambridge University Press.

Mrázek, R. (2002) *Engineers of Happy Land: Technology and Nationalism in a Colony*, Princeton, NJ: Princeton University Press.

Sachs, J. (2005) *The End of Poverty: How We Can Make It Happen In Our Lifetime*, London: Penguin.

Toye, J. (1987) *Dilemmas of Development: Reflections on the Counter-Revolution in Development Theory and Policy*, Oxford: Blackwell.

'Foreign Aid, Forever?'

Encounter (1974)

P. T. Bauer

Editors' Introduction

Peter Thomas Bauer was born Pieter Tamàs Bauer in Budapest in 1915; he died in London in 2002, having been made a Life Peer (Baran Bauer of Market Ward, Cambridge) by Margaret Thatcher in 1983. Bauer studied Economics at Cambridge University, graduating in 1937, but spent most of his professional career at the London School of Economics. He was a friend of Friedrich von Hayek and a member of the Mont Pelerin Society that Hayek founded. The left-of-centre *Observer* newspaper dubbed Bauer 'Lord Anti-Aid' for his vigorous and sustained critiques of foreign aid programmes.

For much of his career Peter Bauer was a maverick in the field of development economics. Strongly opposed to big government and centralised economic planning, Bauer was ahead of his time in proposing 'the extension of the range of choice, that is, an increase in the range of effective alternatives open to people, as the principal objective and criterion of economic development' (1957, in *Economic Analysis and Policy in Under-Developed Countries*). His field experience in Malaysia and West Africa cautioned him against the view that farmers did not respond rationally to price incentives, or that markets were poorly embedded in the tropics. Bauer was also deeply suspicious of the claim that developing countries needed to be gifted (or burdened) with huge infrastructure projects, or that they should be sheltered from the competitive forces he associated with free markets and free trade.

Bauer's trademark critique of foreign aid programmes was sharpened as early as 1959 in a review of US Aid and Indian Economic Development that he produced for the American Enterprise Association. To some degree, perhaps, Bauer's often repeated critique of foreign aid programmes has detracted from a broader corpus of work on markets and marketing boards or the rubber industry. Bauer also argued strongly in the 1980s against debt write-downs for Latin America. He argued that such policies would reward 'dishonourable behaviour' and would increase moral hazard in the international financial system.

Nevertheless, it is one of Bauer's several critiques of foreign aid that we reproduce here. Written not so much to celebrate as to debunk twenty-five years of foreign aid (its Silver Jubilee), Bauer denounces its 'far reaching and brutal consequences'. The article remains of interest because it takes the reader quickly through the usual reasons for giving aid – including what Bauer calls Western feelings of guilt – and subjects each one of them to withering critique. One-sided though it clearly is, Bauer's paper retains its force as a principled critique of those parts of the aid business which mainly work to the advantage of political and bureaucratic elites in donor and recipient countries alike. (Readers might like to compare the target of Bauer's wrath in 1974 with the stated aim and practices of one of the more admired aid or development agencies working today, the UK's Department for International Development. Check out DFID's 2006 publication, *Making Governance Work for the Poor* available online at www.dfid.gov.org).

Key references

P. T. Bauer (1974) 'Foreign aid, forever?', *Encounter* 42: 15–28.
— (1948) *The Rubber Industry: A Study in Competition and Monopoly*, London: Longman, Green and Co.
— (1957) *Economic Analysis and Policy in Under-Developed Countries*, London: Cambridge University Press.
— (1972) *Dissent on Development*, London: Weidenfeld and Nicolson.
— (1981) *Equality, the Third World and Economic Delusion*, London: Methuen.
P. T. Bauer and B. Yamey (1957) *The Economics of Under-Developed Countries*, London: Cambridge University Press.

"It is obvious that official aid is vitally important because it has special characteristics which make it indispensable to the developing countries."

Richard Wood,
Minister for Overseas Development,
Hansard (9 June 1971)

"Party political battle is able to rage to the right and left, but on aid and development there is a bilateral agreement. We tend to agree with each other and to express our mutual concern about the third world."

Judith Hart,
Shadow Minister for Overseas Development,
Hansard (27 April 1972)

Foreign aid celebrated its Silver Jubilee in January. Point Four of President Truman's message to Congress of January 1949 which proposed aid to under-developed countries inaugurated a new political, economic and ideological movement in our times. It has had far-reaching and sometimes brutal consequences, enormous costs, little success, and virtually no adverse criticism.

The early days of foreign aid were marked by high hopes and what must now seem absurd optimism. In the 1950s prominent American economists argued that limited aid over a few years would ensure the self-sustaining economic development of the recipients, and thus remove the need for external assistance. But now aid is envisaged as a practically open-ended commitment extending into the indefinite future. Thus, the Pearson Report envisages aid as extending into the 21st century; and Mr Reginald Prentice, former Minister for Overseas Development, has even said that we must expect it to extend beyond the 21st century. Yet there is little talk of the failure of foreign aid. In fact foreign aid continues to increase. Even

President Amin of Uganda still receives aid from Britain (besides new and additional UN aid).

Incredible as it may seem, British and multi-national aid even goes to the oil states of the Middle East and North Africa including Bahrein, Iraq, Kuwait, Libya and Saudi Arabia, although they have vast oil reserves and oil revenues which they are often unable to spend, and whose per capita incomes in some cases much exceed those of many donors including Britain. The governments of several aid-recipient oil states themselves grant subsidies to other countries.

The explanation is, of course, that foreign aid is an act of faith.

1. THE ACT OF FAITH

"As with other articles of faith, however, once accepted, certain consequences, which must also be accepted, flow from it. The most important is the obligation of every government to play its part in cooperation with all others to ensure that all people have a reasonable chance to share in the resources of the world. . . . We live at a time when the ability to transform the world is only limited by faintness of heart or narrowness of vision."

Partners in Development, Report of the
Commission on International Development
(Pearson Report, N.Y., 1969)

"If one per cent, in total, is the present target, what is wrong with two per cent? And why should not the developed countries be compelled not merely by shame but by enlightened self-interest to channel increasing proportions of their foreign aid through the multi-national institutions now operating in this field?"

Nature (26 May 1972)

The case for foreign aid is usually taken for granted. The results of aid and the conduct of recipient governments are rarely discussed, and then only in the most cursory fashion. The case for foreign aid is regarded as axiomatic, so that either progress or lack of progress can be used to argue for more aid. Progress is evidence of success, and lack of progress is evidence that more must be done. It is usually assumed that giving more aid means doing better, that more aid is more meritorious. In the literature "increase" and "improvement" are interchangeable terms, as in the following passage from an official OECD report:

> In these GNP terms, the biggest *increases* were made by Sweden, Portugal and Denmark, France, with the United Kingdom also doing quite well. Belgium, Germany, Italy, Norway and the United States also *improved* their performance, the latter despite their well-publicised difficulties in increasing aid appropriations. . . . What are the prospects for a continuation of this modest *improvement* in the levels of official development assistance?[1]

Whatever happens, progress or failure becomes an argument for more aid. But then evidence has always been irrelevant to an act of faith.

Furthermore, the burden of proof has somehow come to be placed on the critics instead of the advocates. As aid involves the transfer of money to governments and international organisations over whom the representatives elected by the taxpayers in the donor countries have no control, this seems particularly anomalous. Until well after World War II it was an important principle in the allocation of even modest grants to British colonial governments that they should go to governments for whose conduct the Secretary of State for the Colonies was answerable to Parliament.[2]

At about £275 million per year British aid now exceeds by more than one third the total British contribution to the European Economic Community plus all other expenditure on external relations; it is equal to the yield of surtax and twice that of betting and gaming taxes; and it is about one quarter of the expected balance of payments deficit for 1973.

Aid, nevertheless, remains the only category of government spending which goes unquestioned (except for suggestions that it should be higher). Spending on such traditional, necessary, or popular items as the monarchy, defence, education, national health or school milk has at times been criticised, but not foreign aid. This unquestioning attitude has prevailed for well over a decade, through periods of substantial restriction on state and private spending in general, and on overseas investment and travel in particular.

[. . .]

4. INFRASTRUCTURE AND PRIVATE INVESTMENT

In the 1972 House of Commons debate on aid Mr Wood asserted that official aid obviously had certain characteristics which made it indispensable for development; but it was Mrs Judith Hart, his opposite number, who attempted to spell out the reasons. Aid was indispensable, she said, because private capital could not finance the "infrastructure" necessary for development such as roads, power supplies, and schools ("the private investors cannot finance infrastructure"); because many poor countries were too small for profitable investment ("private investment demands markets for what it produces and this necessarily means that it does not go to the poorest countries"); because private investment was unpredictable ("private investment is unpredictable and cannot be taken into account in formulating development plans"). Although these frequently-heard arguments are rather more specific than Mr Wood's empty assertion, they are quite invalid.

Private investment may indeed fail to finance some parts of the infrastructure.[3] But the government can always borrow abroad for this purpose and service the loans from higher tax revenues, as used to be standard practice. Moreover, infrastructure is not a pre-condition of development but a collection of facilities which emerge in the course of material progress.

Contrary to Mrs Hart's assertion, the small size of the domestic market does not preclude private investment. Whatever the size of the market, if the required conditions (other than capital) are present investment will be generated locally or supplied commercially from abroad. Hong Kong, Taiwan, Malaysia, Israel, and a number of Latin American countries are among the most obvious examples. Mrs Hart gave Uganda as an example of a small country unable to secure private capital. In fact the country had no such difficulties until recently when its governments became increasingly hostile to private capital.[4]

The flow of private investment may well be unpredictable, notably so when the political climate is hostile to private capital, especially foreign capital. But the flow of aid is also uncertain; and development planning (which is in any case not a precondition of progress) should take account of these fluctuations.

5. REPERCUSSIONS

Aid is an inflow of subsidised or *gratis* resources, and in this sense enriches the recipients. But unlike manna from heaven, it does not descend indiscriminately on everybody. It sets up a whole variety of damaging repercussions, and these have to be taken into account.

1. Official aid reinforces the disastrous tendency to politicise life in poor countries, the trend towards *politique d'abord*. This tendency operates even without aid, but is buttressed and intensified by it. The hand-outs increase the power, resources, and patronage of governments compared to the rest of society; and this is exacerbated by the more favourable treatment of governments which try to establish state-controlled economies. Many Third World governments, especially in Africa, could not operate their close economic controls without expatriate staffs, employed under various aid programmes. Politicisation of life provokes political tension because it becomes supremely important, often a matter of life and death, who has the power – witness Indonesia, Nigeria, Pakistan, Tanzania, Uganda, and Zaire (the Congo). This politicisation diverts the attention, energies, and activities of able and enterprising people from economic activity to politics and administration, sometimes from choice (because this diversion is profitable), but quite often from necessity (because economic or even physical survival comes to depend on political developments and administrative decisions).

2. Besides politicising life and thereby contributing to social and political tension, official aid similarly reinforces the pursuit of policies damaging to material progress (and often also inhuman). Many recipient governments restrict the activities of highly productive and economically successful minorities – Chinese in Indonesia, Asians in Africa, Indians in Burma, Europeans everywhere. And many have also maltreated and persecuted politically ineffective groups, especially ethnic minorities. Such policies reduce current and prospective incomes in these countries, and widen income differences between them and the West. Even where it does not promote economic advance, aid represents current resources accruing to the governments, resources which are available for the purchase of imports or for the distribution of largesse, so that it helps governments temporarily to conceal from their own people some of the economic consequences of their actions.

3. Aid often supports extremely wasteful projects which make large losses year after year, and which can absorb more local resources than the value of their output. For political reasons such hopeless projects often have to be continued for years after it has become plain that they are thoroughly wasteful.[5]

4. Aid is often linked to balance-of-payments deficits of the recipients, especially when these deficits are considered as the results of laudable official efforts to speed up progress. Balance-of-payments crises in the course of development planning are especially useful in supporting appeals for aid; governments of poor countries are understandably encouraged to embark on ambitious plans involving large expenditures financed by inflationary monetary and fiscal policies, and also to run down their foreign-exchange reserves. Inflationary policies, payments difficulties, and the detailed economic controls which they promote, all engender a widespread feeling of insecurity or even a crisis atmosphere, a sequence which inhibits domestic saving and investment and even promotes a flight of capital. These sequences, in turn, serve as arguments for further external assistance.

5. The insistence on the need for external assistance obscures the necessity for the people of poor countries themselves to develop the facilities, attitudes, and institutions which are required if these societies are to achieve sustained substantial material progress. Indeed, this insistence on external aid helps to perpetuate the ideas and attitudes widespread in these countries, which are damaging to economic progress: that opportunities and resources for advance of oneself and one's family must come from someone else – the state, the rulers, one's superiors, richer people or foreigners. In this sense aid pauperises those it purports to assist.

6. Aid frequently influences policies into inappropriate directions by promoting unsuitable external models, such as Western-type universities whose graduates cannot get jobs, Western-style trade unions which are only vehicles for the self-advancement of politicians, and a Western pattern of industry even where quite inappropriate (*e.g.* national airlines and steelworks).[6]

7. Aid is money which could be used more productively in the donor countries (or, at any rate, by their citizens), so that it reduces the combined national incomes of donors and recipients. It also retards the material advance of the donors by affecting adversely their own balance-of-payments; and payments difficulties in donor countries have often brought about restrictive domestic economic policies – with adverse effects on the Third World's export markets.

8. As a result of the operation of various sectional interests, the most important donor countries erect substantial barriers against the exports from the less developed countries to whom they are giving aid. Foreign aid unfortunately diminishes the political resistance in the recipient countries to the erection of these barriers, both within the donor countries and by spokesmen of recipient countries.

All these adverse repercussions (and the list could easily be extended), even when taken together, do not mean that foreign aid cannot promote development. But it is certainly unwarranted to assume that it must do so simply because it represents an inflow of subsidised resources. Aid may well improve current economic conditions in the recipient countries.

But the contribution of aid to long-term development is at best marginal. It means that some investible funds are available more cheaply than would be the case otherwise. These funds are likely to be less productive than capital supplied commercially since their use cannot be adjusted so readily to market conditions including supplies of complementary factors. And when capital is productive, it will usually be generated locally or be readily available from abroad on commercial terms.

Moreover, the crucial personal and social determinants of development are apt to be affected adversely by the repercussions of the inflow of aid. These adverse repercussions are likely to offset, or more than offset, the benefits from the inflow of subsidised investible funds in otherwise propitious circumstances. There is, therefore, not even a general presumption that aid is more likely to promote development rather than retard it. As it has operated in the past and as it is likely to operate in the foreseeable future, any general presumption would tend to be in the very opposite direction.[7]

6. AID AS AN AXIOM

The axiomatic treatment of aid is also reflected in its terminology. Thus the very name of the Overseas Development Administration prejudges the results of its operations. It should be called the Overseas Donation Administration, which would appropriately describe its activities. A terminology directed towards description rather than prediction (or hope) would also more nearly conform to the usual practice of government departments. The same comment goes for the Development Assistance Committee of the Organisation for Economic Cooperation and Development. In the last year or two the term "foreign aid" has come to be increasingly replaced by the term "development finance" to describe what is plainly a system of doles.

Aid lobbyists often insist that neither the culture, the social institutions, nor the policies of the recipient governments should be criticised. But what if these are incompatible with substantial material progress? Material advance requires a modernisation of the mind, and this is inhibited by many of the mores and institutions of the poor countries.

Some aid lobbyists do suggest at times that recipient governments should be persuaded to adopt different policies, such as the adoption of "comprehensive planning" or the expropriation of politically unpopular and ineffective groups, or both. But such measures are not calculated to promote development. The success of specific projects financed by aid is often publicised without enquiry into the cost in terms of alternative use of the resources, or into the possibility of financing the project from sources other than aid.[8] And when changes in the national income of aid-recipients are occasionally discussed, the enquiry does not extend to the meaning or limitations of the underlying statistcs (an issue to which I shall return shortly) or to the relation between statistical growth and general living standards, or to the problem of establishing a causal relationship between the inflow of aid and the rate of development, or to the question why aid should be necessary if investment in the recipient countries is indeed genuinely productive.

7. THE RELIEF OF POVERTY

In the last few years the aid lobbies have changed their emphasis from development to the relief of poverty. If the advocacy has become much more strident, it is perhaps because the new emphasis on "need" itself implicitly recognises the previous failure of aid as an instrument for "development."

Poverty is said to be indicated by comparison of per capita incomes, coupled with references to hunger, famine, and starvation and the alleged "ever-widening gap between the haves and have-nots." These arguments are even less substantial than earlier ones for aid as an instrument for development. Here are some of the reasons why this is so:

1. The statistics of income per head in less developed countries are quite useless for international comparison and for the measurement of need. Professor Dan Usher (a Canadian statistician who lived for years in South-east Asia and who has a Thai wife), has written pertinently on this subject. The following passages from the introduction and summary of his book *The Price Mechanism and the Meaning of National Income Statistics* (Oxford, 1968) epitomise his conclusions.

Using Thailand as an example this book shows that statistics like these may contain errors of several hundred per cent . . . the discrepancy is not due primarily to errors in data . . . the fault lies with the rules [of national income comparisons] themselves . . . [which] generate numbers that fail to carry the implications expected of them.

In Thailand I saw a people not prosperous by European standards but obviously enjoying a standard of living well above the bare requirements of subsistence. Many village communities seemed to have attained a standard of material comfort at least as high as that of slum dwellers in England or America. But at my desk I computed statistics of real national income showing people of underdeveloped countries including Thailand to be desperately if not impossibly poor. The contrast between what I saw and what I measured was so great that I came to believe that there must be

some large and fundamental bias in the way income statistics are compiled. Something is very wrong with these statistics. For instance, if the figure of $40 for Ethiopia means what it appears to mean, namely that Ethiopians are consuming per year an amount of goods and services no larger than could be bought in the United States for $40 then most Ethiopians are so poor that they could not possibly survive let alone increase in numbers.

These observations are clearly pertinent, not only to the case for foreign aid as an instrument for the relief of poverty but to a much wider range of problems. Conceptual and statistical problems apart, there are also frequent instances of deliberate manipulation of national income statistics and similar data for political purposes (as has been noted by Professor Oskar Morgenstern to whom I will refer later). Even the population statistics of the Third World (which underlie per capita incomes) are subject to wide margins of error. Thus, the 1963 census put the population of Nigeria at 55.6 million while a prominent American scholar of Nigerian affairs estimated it at 37.1 million for the same year. All this should put into perspective the value of statistics which purport to estimate changes in Afro-Asian per capita incomes to within one or two per cent. Yet such estimates feature prominently in the advocacy of aid.

2. Foreign aid is paid by governments to governments. Unlike voluntary charity (or for that matter domestic progressive taxation), it cannot take into account differences in incomes and conditions of families and persons. Indeed, it operates perversely. The poorest are largely untouched by aid, as for instance desert people in Africa and Asia, aborigines in South-east Asia, Africa and Latin America, and various other backward groups, categories which total many millions. (Even the presence of these groups is rarely acknowledged in aid discussions which all too often ignore the diversity of the less developed world and tend to envisage it as a largely undifferentiated, uniform mass.) Much of aid only benefits the urban population, among whom the more articulate, influential, educated, skilled and enterprising people are disproportionately represented (notably, politicians, civil servants, academics and urban businessmen).

There are also substantial classes of beneficiaries

of official aid in the donor countries (exporters, consultants, civil servants, and academics). An important class of beneficiary is the staff members of the international agencies, both from rich and poor countries, who alone pay no taxes and thus contribute nothing to the foreign aid they advocate so warmly. The taxes which go to finance aid are levied on the whole population in the donor countries, including the poorest.

3. Quite apart from these various statistical and conceptual problems and from the uncertain distributive effects of aid, evident anomalies or absurdities result from the uncritical adoption of need as a ground for aid, especially need as conventionally measured by per capita income. The maltreatment and expulsion of minorities by aid recipients, already mentioned, provides an instructive example. These minorities have been the main agents of economic progress in these countries, and their incomes were above the average. Hence their elimination necessarily reduces per capita incomes. Should aid be increased because these policies have directly reduced per capita incomes and thus aggravated need? And should it be increased yet again if this encourages further maltreatment and expulsions?[9]

There are also many societies where the local attitudes, aptitudes, and institutions are such that aid (especially government-to-government aid) can do little or nothing to relieve need – either because the official machinery will ensure that it does not reach the needy, or because aid promotes policies which aggravate poverty, or because the attitudes of the people themselves may largely resist improvements in their material conditions.

8. THE WIDENING GAP

"Over 1,500,000,000 people, something like two-thirds of the world's population, are living in conditions of acute hunger, defined in terms of identifiable nutritional disease. This hunger is at the same time the effect and the cause of the poverty, squalor, and misery in which they live."

HAROLD WILSON, *The War on World Poverty* (1953)

"It is on the ethical plane that the present situation is scandalous. One third of the world lives in comfort and two thirds in misery."

JOHN PINCUS, *Trade, Aid and Development* (1967)

"The widening gap between the developed and developing countries has become a central issue of our time."

Partners in Development (Pearson Report, 1969)

Three further variants of the general argument for aid deserve to be noted: the population explosion; large-scale starvation; and the widening gap.

The rapid growth of population in the Third World (the so-called population explosion) does not rescue the conventional arguments for aid. For it reflects a fall in mortality, a longer life-expectation of both infants and adults, *i.e.* some improvement in basic conditions. This improvement is omitted in conventional statistics of per capita incomes, as they do not recognise health, life-expectation and the possession of children as components of living standards.

It is often asserted that without foreign aid there would be widespread starvation in the Third World. These countries are supposed to be living at below-subsistence levels and under persistent threat of starvation, while at the same time alarm is expressed at the growth of their populations – a paradox which is seldom noticed. If there is not enough food for the existing population, there could be no large-scale increase in numbers. Much aid directly or indirectly finances uneconomic enterprises or activities which produce neither food nor exports to purchase it.

The widening gap is another plausible but insubstantial argument for aid. To begin with, the argument again prejudges the effects of aid by implying that aid promotes the long-term improvement in living standards in poor countries. Aid certainly removes resources from the donors; but this does not mean that it improves incomes or living standards of the recipients. To make the rich poorer does not make the poor richer. The impact of aid on differences in income and living standards cannot be judged without examining the factors behind these differences and also the likely effects of aid in specific instances.

The concept of the so-called gap is vague to a degree. There is continuous gradation and no clear gap or discontinuity in the international range of per capita incomes. The distinction between what are called rich and poor countries on the basis of per capita incomes depends simply on where the line between the two categories is drawn; and in the absence of clear discontinuity this is quite arbitrary. Consequently the difference in per capita incomes between the two categories is similarly arbitrary. Moreover, many Afro-Asian lands in recent years have developed much faster than most rich countries – South Korea, Thailand, Hong Kong, Malaysia, Kenya, the Ivory Coast, Nigeria, as well as some Latin American countries. The oil states of the Middle East (usually included in the less developed world) have per capita incomes among the highest in the world. Thus the difference in conventionally measured per capita incomes between some of these countries and many developed countries has narrowed and not widened in recent years or decades. The arbitrary and crude aggregation of the developed and less developed world conceals far-reaching differences within these aggregates, including differences in material prosperity and rates of progress.

When referring to international income differences and to the widening gap, aid advocates often suggest that the higher income of the West has been somehow secured at the expense of the Third World. This has been a familiar theme of Marxist and Leninist pamphlets, and of the arguments of so-called Third World politicians. In recent years it has increasingly found its way into the advocacy of aid.

These allegations are quite without substance. The poorest societies in the less developed world are those with the fewest (or no) external economic contacts, so that their poverty cannot be the result of deprivation by external powers. Conversely, those involved in extensive foreign trade are the most prosperous. Obvious examples are Malaysia and Ghana, not to speak of the oil-producing countries. The material prosperity of Western societies and of Japan is the achievement of their own people, whose activities have also promoted such economic advance as has taken place in the Third World. But the advocacy of aid is presumably made easier by suggesting the contrary.

9. INDEBTEDNESS AND INTERNATIONAL TRADE

"We shall produce any statistic that we think will help us to get as much money out of the United States as we possibly can. Statistics which we do not have, but which we need to justify our demands, we will simply fabricate."

Unnamed civil servant, quoted in O. Morgenstern, *On the Accuracy of Economic Observations* (1963)

"The foreign earnings of the developing countries have suffered severely from the deterioration in terms of trade. Unless these countries succeed in obtaining additional resources, they will be unable to achieve the reasonable rate of growth set as a target in their plans. The situation will be worse still if the terms of trade deteriorate further in the future."

Towards A New Trade Policy For Economic Development, United Nations (1964)

"Between 1960 and 1969 exports from developing countries grew at the rate of 9·5 per cent. If we look in the same period at the primary commodities on which many of the developing countries, and particularly the poorest, depend completely – if we exclude oil, which benefits only a tiny number of highly privileged countries – we find that the export of primary commodities grew at the rate of only 2.5 per cent."

Judith Hart, *Hansard* (27 April 1972)

The burden of indebtedness and the alleged deterioration of the position of the Third World in world trade are often-heard subsidiary arguments for aid, straddling the promotion of development and the relief of need. They are major themes of the proceedings of international organisations, and are widely reported and discussed in the so-called quality press. These notions, though insubstantial, are worth examining both because they are so widely canvassed and because they show up the regrettable lack of understanding (or scruple) in the advocacy of aid.

The indebtedness of many less developed countries is often instanced as an argument for additional aid. Most of these debts reflect soft loans, usually with large grant elements, incurred under earlier aid agreements, often supplemented by outright grants. With the worldwide rise in prices (especially of aid-recipients' export prices, particularly in 1972–73), the real burden even of these soft loans had diminished greatly. The inability to service these loans suggests that they have been wasted, and have not increased incomes or taxable capacity.[10] This waste of earlier aid now becomes an argument for more of it. It is the donors who are held responsible for failures to service even heavily subsidised loans, and who are made to feel guilty for "draining away" the foreign exchange resources of the Third World.

The so-called deterioration of the external trade position of less developed countries as an argument for aid takes two forms: a decline in their share in world trade, and deterioration in their terms of trade. The allegations are either untrue, or irrelevant, or both.

Why does the whole discussion usually ignore exports of mineral oil? Oil was excluded in the oft-quoted U.N. publication *Towards a New Trade Policy for Economic Development* (the Prebisch Report) which set the tone of the UN Conference on Trade and Development (UNCTAD) of which Dr Prebisch became Secretary-General. Yet oil is by far the most important export from the Third World, worth well over double that of any other export, and the value of mineral oil exports has increased phenomenally since World War II. The omission is usually justified on the ground that oil exports benefit only a handful of countries with tiny populations and rulers who are not interested in the benefit of their peoples. This was specifically stated by Mrs Hart in a House of Commons debate on 27 April 1972, in the course of which she complained about Third World difficulties allegedly caused by external factors; and she instanced the damage to exporters of copper, especially Chile, resulting from a fall in the copper price (from a very high level). In fact, some of the oil exporters are rather large countries: Nigeria, Iran, and Iraq.[11] The population of Nigeria is about five times that of Chile and substantially exceeds the combined total of all copper-exporting Third World countries. The population of Iran is also much larger than that of any copper-exporting poor country. In any case, the character of the governments of exporting countries is irrelevant to a discussion of external trade (though as it happens the governments of several oil-exporting countries pursue policies which are nowadays often regarded as progressive, *viz.* expropriation of property, hostility towards the prosperous, and maltreatment of ethnic minorities).

The decline in the share of a country or group of countries in world trade has no adverse welfare implications and is not evidence of adverse external conditions. For instance, since the 1950s the large increase in the foreign trade of Japan, the reconstruction of Europe and the liberalisation of intra-European trade have brought about a decline in the share of other groups in world trade, including that of the United States and the United Kingdom (the external trade of which has grown proportionately less than that of western Europe and Japan). Exports from a country or group of countries often decline as a result of increased domestic use of previously exported products (Nigerian groundnuts and palm oil, Indian groundnuts); or the intensification of protectionist policies of their governments; or domestic inflation; or special taxation of exporters (Burma rice). A decline in the share of a group of countries in external trade is no evidence whatever of unfavourable external developments. As it happens, the share of less developed countries in world trade has not decreased, but has increased in recent decades, notably since before World War I.

Allegations about the persistent deterioration in the terms of trade of less developed countries or primary producers (these categories are often treated as if they were interchangeable) are, again, meaningless, or untrue, and usually both. The lack of similarity in the trading patterns of the Third World (which compromises well over half of the world) makes aggregation of their terms of trade practically meaningless. The terms of trade of individual countries and even of groups of countries often move in opposite directions. In any case the terms of trade (the ratio of import and export prices) even of individual poor countries are of practically no interest if they relate to more than a very short period – unless changes in the cost of production of exports, and in the range and quality of imports and in the volume of trade, are also considered. What matters is the amount of imports which can be purchased with a unit of domestic resources or a specified volume of resources; and this cannot be inferred from the ratio of import and export prices. (In technical language: the comparisons relevant to economic welfare and development are the factoral terms or the income terms of trade, not the commodity terms of trade.) Moreover, the proportion of export earnings retained in poor countries has risen substantially in recent years as a result of major increases in corporation taxes and royalty rates. Even the crude commodity terms of these countries have been exceptionally favourable in recent years. When changes in the cost of production, the great improvement in the range and quality of imports available and the huge increase in the volume of trade are taken into account, the external purchasing power of their exports, of *ldcs* whether per unit of resources or in the aggregate, is now extremely favourable, probably more so than ever before.[12]

These observations should not come as a surprise. It was Sir Arthur Lewis who said this a few years ago:

> The terms of trade for primary as against manu-factured products averaged higher in the 1950s than at any time in the preceding eighty years. The first half of the 1950s was especially good because of the Korean war and heavy stock-piling in the United States and elsewhere. The terms of trade deteriorated in the second half of the decade and on till 1962, since when they have moved upwards. However, even in 1962 they were 5 per cent above 1929, which preceded the Great Depression.[13]

This was written before the recent upsurge in primary product prices and does not take into account the various favourable factors noted in the previous paragraph.

Moreover, because of the certain further increase in world population, the virtually certain technical progress in manufacturing, and the virtually certain increase in total world income, it is probable that the terms of trade of countries which produce mineral and agricultural products for export will improve. This probability, incidentally, underlies the irony of much current discussion – which predicts a world shortage of food and simultaneously urges developing countries to divert resources away from agriculture and towards manufacturing.

10. THE MORALITY OF AID

> "Could the moral and social foundations of their own societies remain firm and steady if they washed their hands of the plight of others?"
>
> *Partners in Development* (Pearson Report, 1969)

The belief that foreign aid is the discharge of a moral obligation to help the poor is perhaps the most influential argument – or, rather, emotion – behind the advocacy of aid in popular discussion. I thought it appropriate to postpone discussion of this argument until after consideration of the more technical or systematic arguments for aid and of its major results and implications, since a worthwhile assessment of the morality of aid must be affected by its results.

The suggestion that aid represents the discharge of moral duty is usually based on the poverty of the Third World as compared to the West. However, this argument prejudges the effects of aid by taking it for granted that aid is bound to raise living standards in the recipient countries. I have shown that this belief is quite unwarranted. I have noted also that the standard international income comparisons bandied about in this context are worthless, and further that much of aid benefits the relatively well-off in the recipient countries and leaves the poorest untouched or even affects them adversely. Nor is this all. Foreign aid often facilitates the pursuit of measures which provoke or exacerbate political tension, increase the flow of refugees, and thus promote much suffering and human misery.

The conduct of many aid-recipient governments in the Third World clearly offends the most elementary moral principles. The expulsion of tens of thousands of Asians from Uganda is only the most widely publicised and, therefore, the most familiar instance. Large-scale maltreatment of minorities, including expropriation and expulsion, has taken place since World War II in many aid-receiving countries, including Burma, Burundi, Sir Lanka, Egypt, Indonesia, Iraq, Nigeria, Pakistan, the Sudan, Tanzania and Zaire. In some of these countries there have been large-scale massacres.

Many of the aid lobbyists seem to be sublimely uninterested in the social results of aid, or in the policies it buttresses. Morality appears to be satisfied as long as the donors are made to feel guilty and are divested of a goodly portion of their resources.

The argument that aid is the discharge of moral duty to help the poor is open to a further objection; and to some it may even be more fundamental. The implicit analogy between foreign aid and the morality of voluntary charity fails. Foreign aid is taxpayers' money – the donors have to pay whether they like it or not. By and large they do not even know that they do in fact contribute to aid. Aid lobbyists do not give away their own money. They propose taxes on others. The moral obligation to help one's fellow man rests on persons who are prepared to make sacrifices. It cannot be discharged by entities such as governments.

Foreign aid also differs from voluntary charity in various other ways. For instance, unlike voluntary action it cannot easily be directed to the specific needs of groups or persons, since it is distributed to governments and not to voluntary organisations or to individuals.

The few people in poor countries who know about aid sense that it differs radically from voluntary charity. This is one reason why they suspect statements about its alleged humanitarian quality.

11. ENLIGHTENED SELF-INTEREST

The arguments and sentiments in favour of aid usually focus on the needs of the recipients. But aid is often said to benefit donors on the ground that it promotes political strategy or develops export markets.

1. In spite of these suggestions, the Western donors no longer envisage aid as an instrument of political strategy or for the promotion of a more secure world. In this they are quite right. Aid can hardly serve to promote the security of the donors. It reduces their resources and, to that extent, weakens their position. Moreover, it is only materially developed countries that can imperil the security of the donors. In the unlikely event of aid substantially promoting the material development of the recipient countries, this would improve their military potential. This outcome would enable them to become a threat to the donors, which at present they are not. The military and technical resources of less developed countries are too meagre to enable them to endanger the security of the donors (with the exception of mainland China, a country not usually considered in this context).

Does one still have to cite any specific examples to show that the granting of aid has not actually promoted the political interests of the Western donors? The recipients usually resent the Western donors, and find it satisfying to assert their independence, a stance which is often useful for both domestic and external political reasons. The indiscriminate distribution of Western millions understandably arouses suspicion, especially of political domination. Again, the feelings of guilt in the West make the recipients believe that aid is at best a partial restitution for past wrongs. It is, therefore, not surprising that the recipients find regular occasions to abuse and embarrass the Western donors. The recent statements of General Amin are only one example. Similar sentiments have come from the present and past leaders of Algeria, Ghana, India, Indonesia, Iraq, Tanzania, the United Arab Republic, and Zambia. And when the governments of India and Pakistan sought external mediation in the Kashmir dispute (1965), they turned to the Soviet Union although they had for years been receiving much US and UK aid.

2. Aid cannot provide markets for exports, since exports bought with aid are given away. This benefits donors as little as a shopkeeper would benefit from being burgled because the thief may spend part of the proceeds in his shop. The suggestion that aid benefits donors through the long-term development of the recipients anticipates the success of aid. It also ignores that alternative and more productive uses of resources exist within the donor countries and elsewhere. This consideration is reinforced by the inability of the recipients to service even very soft loans. The low productivity of aid compared to alternative uses implies a reduction in the rate of growth of the combined incomes of donors and recipients, so that in the long run the supply of investible funds will be less than it would otherwise have been.

12. CURIOUSER AND CURIOUSER

The behaviour in recent years of both donor and recipient countries has thrown into relief many anomalies and paradoxes.

1. President Amin, having expelled tens of thousands of people including the most productive East African traders and industrialists, has inflicted much hardship, and damaged the development prospects of the country; he has also aggravated a social problem in Britain, a major aid-donor to Uganda. Amin's policies have further endangered the already precarious position of the remaining minorities in other aid-recipient African countries (including Kenya, Nigeria, Tanzania and Zambia). Yet Amin continues to receive appreciable Western aid. British aid contributed substantially to Amin's rise to power. A quasi-totalitarian régime in Uganda preceded Amin. In 1966, Dr Milton Obote (Amin's predecessor) destroyed the opposition in a full-scale and bloody civil war with the help of money supplied by British aid.

In Zanzibar the government (which came to power after a coup in 1964) massacred hundreds and probably thousands of Arabs and Indians, as well as many African opponents. British aid was resumed after a brief suspension. Tanzania, the merger of Tanganyika and Zanzibar, has consistently received Western aid, including British aid (with the exception of the period when President Nyerere turned down British cash), in the face of continued political executions in Zanzibar, expulsions and expropriations throughout the country, large-scale uprooting of people, and compulsory collectivisation of farming.

Such instances of large-scale suffering and hardship are rarely noted by aid lobbyists.

2. It is now widely proposed that additional special drawing rights (SDRs) should be created by the international monetary authorities to be issued to poor countries as foreign aid. The creation of this additional international money would be inflationary. Such a proposal is anomalous when inflation causes widespread hardship and its control is (or so we are told) the major economic problem of Western governments. Moreover, some concomitants of inflation (notably, international differences in rates of inflation) are apt to lead to the imposition of import controls, which damage the economies of less developed countries.

Even apart from the creation of SDRs as a form of multi-national aid, foreign aid adds to inflationary pressures. As I have already noted, the prospect of aid often encourages Third World inflationary policies because they promote balance-of-payments difficulties, which in turn make appeals for more aid more effective. As aid represents resources given away, it adds to inflationary pressure in the donor countries (unless offset by deliberate restrictions on domestic spending). And in this way it also weakens the balance-of-payments position of the donors. In Britain some of these adverse effects of aid have contributed to the loss of the value of the sterling balance of overseas countries including less developed countries, among which Hong Kong and Malaysia have been conspicuous sufferers.

3. Commodity agreements provide another rich field of paradoxes. They are often intended primarily as a form of aid by raising or maintaining the already high prices of major foodstuffs and raw materials exported (*e.g.* cocoa, coffee, sugar, and tin).

These agreements benefit countries which are relatively prosperous, such as Malaysia, Ghana, Colombia, the Ivory Coast, Nigeria, Thailand, since it is in such countries that external trade is relatively important. And within these countries it is the relatively well-off who benefit most.

4. Many aid-recipient governments have levied specially heavy taxes for many years on the producers of food export crops (Thai and Burma rice, West African cocoa), often through state export monopolies which pay the farmers much less than the world price. Many governments have prohibited, or greatly restricted, private trade in local foodstuffs, raw materials, and imported merchandise, in favour of state trading corporations and state-subsidised or -operated co-operatives. These measures reduce agricultural output and raise the prices of agricultural products. Yet they are often encouraged by aid donors who frequently supply the required administrative personnel. The same applies to the operation of state-sponsored industries and the construction of prestige projects financed from aid, which divert resources from food production.

5. Although the case for aid rests supposedly on the need of the recipients for more capital, practically all of them restrict the inflow and deployment of often highly productive private capital. In India, for instance, foreign investment is strictly controlled and is barred from a wide range of industry and commerce. The expansion and even the current operation of both domestic and foreign enterprises in India are often severely restricted, so much so that factories and businesses are forced to work below capacity – while the government receives aid to expand its own enterprises, using both foreign and domestic resources. Similar policies are pursued by many other aid-recipients. The inflow of subsidised or *gratis* funds from abroad, accruing to governments in various ways, facilitates and encourages the pursuit of such restrictive and economically damaging policies.

6. Aid is supposedly necessary to supplement the resources of the recipients. Many African and Arab governments regularly pass on substantial aid to their actual or potential allies. The Nigerian government has recently granted aid to several neighbouring countries, notably Togo and Dahomey.

7. Aid continues regardless of improvements in the fortunes of recipients (*e.g.* through the discovery of minerals or a rise in export prices). In the early 1960s Nigerian exports amounted to around £160 million a year. In 1973 they will exceed £800 million, and for 1974 they are estimated to exceed £1000 million. This huge increase in export earnings was the result partly of the discovery and rapid development of oil deposits by foreign companies, and partly of large rises in the prices of the country's traditional exports. According to press reports in July 1973, the Nigerian government proposes to spend £120 million on facilities required for an "African & Black Festival" in 1974. However, the British aid programme of several million a year goes on. Similarly, some oil-producing countries in the Middle East, with huge oil royalties, still receive Western aid.

8. Some ideas and proposals, advanced in all seriousness in the advocacy of aid, are nothing less than bizarre. One example is a long and prominent leading article in *The Times* (25 June 1973) which drew on an article by Mr Escott Reid in *Foreign Affairs*. The *Times* article chastises the World Bank and the Western donors for occasional hesitation in granting aid to governments which have expropriated the property of their citizens and that of foreign investors. This hesitation, or passing reluctance, is described as arm-twisting. Both articles advocate that voting on World Bank's aid decisions should be substantially based on the population of the member countries. They were concerned with the inadequate representation of India, Indonesia, Brazil, Bangladesh, Pakistan and Nigeria. This suggestion is analogous to proposing that charities should be governed by recipients of their alms, or that banks should be controlled by their borrowers.

13. FORCES BEHIND AID

Supporters of aid generally seem to be well-intentioned. But they appear to be interested not so much in the effects of aid or in the welfare of the recipients, as in giving away money, sometimes their own, more often that of other people. There is a widespread if vague feeling of guilt in the West, often unaccompanied by any sense of responsibility for the consequences of the actions that issue from it.

Foreign aid may not do much for its supposed beneficiaries, and often contributes to their suffering and hardship. But it does demonstrably benefit influential and articulate sectional interests in the West. These include the staffs of international agencies and of government departments; bored, power- and money-hungry academics; the churches, which increasingly look upon themselves as secular welfare agencies; and exporters who benefit from sheltered markets. There are also supporters of aid who favour it on various political grounds, such as its use as an instrument for promoting what they call social change in the recipient countries, that is the establishment of socialist societies there.

There are, thus, reasons why both the actual effects of aid and the shortcomings of the arguments in its favour will continue to be ignored. The substantial flow both of aid and of the familiar arguments for it is therefore almost certain to continue.

NOTES

1 Organisation for Economic Cooperation and Development, *Development Cooperation 1972: efforts and policies of the members of the Development Assistance Committee* (Paris, 1972), italics added.

2 The increasing amount of aid channelled through the international agencies gradually severs all links between donors and recipients. It also means that, at one remove, British aid supports governments which do not receive British aid directly. For instance, in 1973 new UN aid projects were approved for Uganda – after the mass expulsion of Asians. Analogous considerations apply even to bilateral aid. Britain may come eventually not to give any aid to Uganda – but then British aid to other countries sets free resources of other donors to grant aid to Uganda.

3 As a matter of fact much infrastructure in less developed countries (*ldcs*) has been financed and operated privately until recently and often still is, as for instance, transport enterprises and utilities, and sometimes even roads, schools and hospitals established by plantation and mining enterprises.

4 Mrs Hart quoted a Uganda spokesman in support of her allegations: "Yet we do not succeed in attracting private investors. Why don't we succeed? Because we cannot offer the markets. We are too small a country to be interesting to the private investor. . . ."

This is untrue: there was, until recently, substantial Asian and European investment in that country.

5 This consideration applies not only to industrial ventures and other familiar prestige projects, but often also to agricultural schemes. For instance, some years ago Tanzania received substantial amounts of bush clearing equipment under a Yugoslav aid scheme. The equipment was designed for use in temperate climates. The attempt to keep it going rather ineffectively in Tanzania absorbed large amounts of labour and scarce water for cooling. Protracted pressure by external advisers was required before the government agreed to abandon the equipment. (The episode bears some resemblance to the celebrated "groundnut scheme" of the late 1940s in the same area.)

6 See Harry Johnson's critical account of the general Western intellectual misconceptions which helped to form the "Afro-Asian" ideology: "A Word to the Third World", Encounter, October 1971.

7 "Marshall Aid" is often quoted as a precedent for the development possibilities of aid. But the analogy is false. The economies of Western Europe had to be restored, while those of the present recipients have to be developed. The peoples of Western Europe had the faculties, motivations, and institutions for centuries before the second World War. The rapid return to prosperity in Western Europe and the ending of Marshall Aid after four years – even though West Germany had to absorb many millions of refugees (among whom old people and children were disproportionately represented) and also had to make substantial reparation payments to the Soviet Union – contrast sharply with the quasi-permanent flow of aid to the Third World.

8 If "development" has any meaning as a desirable process, it must mean an increase in the volume of goods and services desired by the population. It should not be related simply to the growth of particular sectors and activities (say, the public sector, or manufacturing output, or the output of capital goods), or simply to the output of arbitrarily chosen goods and services unrelated to present or prospective consumer demand. The neglect of these considerations explains the paradox of the acute shortage of simple consumer goods (razor blades, stockings, clothing, not to speak of consumer durables) in countries whose official statistics record impressive growth rates.

9 It is sometimes suggested that these expulsions need not greatly obstruct development because technical assistance personnel can replace the groups expelled from the aid-recipient countries. Quite apart from the moral and political implications of this kind of reasoning, the suggestion is unfounded.

   Technical assistance personnel do not usually possess the relevant commercial, administrative or even technical skills; and even when they do, they lack the experience and local knowledge. A large proportion of technical assistance personnel is quite inexperienced. They are certainly no effective replacement for the thousands of traders, industrialists and farmers with decades of local experience.

10 Ghana, a substantial aid beneficiary of many years standing, provides an instructive example. Dr Nkrumah had at his disposal huge sums of money, largely from the operation of the state's export monopoly of cocoa. These funds, amounting to hundreds of millions of pounds, served as the major financial base for his political operations as well as for large-scale personal spending by himself and his political allies, and also for indiscriminate public spending. The economic usefulness of the huge expenditure of the Nkrumah government can be gauged from the fact that after the rapid expenditure of the large reserves of the Gold Coast–Ghana export monopolies and of current cocoa revenues, and in spite of many years of acute death of consumer goods in the country, the government incurred debts on which it had to default.

11 Incidentally, the oil states derided by Mrs Hart do receive Western aid, including British aid.

12 For political reasons many Third World governments often rely on high-cost imports; but this does not affect the argument although it shows that the terms of trade are not outside the control of these governments.

13 W. Arthur Lewis, "A Review of Economic Development" (Richard T. Ely Lecture at the 75th Annual Meeting of the American Economic Association 1964), published in the *American Economic Review* (May 1965).

'The Dirigiste Dogma'

from *The Poverty of 'Development Economics'* (1983)

Deepak Lal

Editors' Introduction

Deepak Lal was born in Lahore in 1940 and read Philosophy, Politics and Economics at Oxford University before returning to India (where his family had mainly settled after Partition in 1947) and the Indian Foreign Service. His subsequent career saw him move in and out of academic jobs (at Oxford, University College London and since 1990 as the James Coleman Professor of International Development Studies at the University of California, Los Angeles (UCLA) and positions with the Planning Commission in India, the International Labour Organisation and the World Bank (where he worked during 1978–80 and 1983–87). On his website, Lal describes how he broke with the Fabianism of his youth in the second half of the 1970s. By that time, too, he was beginning to develop his influential account of the 'repressed economy'. This refers to the damage that was done to developing economies by the well-meaning but in his view wrong-headed nostrums of 'development economics'.

The reading that we reproduce here is the first chapter of the damning pamphlet that Lal produced in 1983 for the UK think-tank, the Institute of Economic Affairs. Lal outlines what he considers to be the core beliefs of the 'dirigiste dogma' that stood behind conventional thinking on development in the 1970s. In a sense, these beliefs begin and end with the proposition that, 'the price mechanism, or the working of a market economy, needs to be supplanted (and not merely supplemented) by various forms of direct government control, both national and international, to promote economic development'. Everything else in the pamphlet takes shape with reference to this first proposition. What is important to recognise now, some twenty-five years later, is that Lal's pamphlet had significant force when it was published. In part, this was because it hit a nerve. From 1983 to 1987 Deepak Lal worked closely with Anne Krueger to reform the World Bank's research department. Prior to that, in the 1970s, he had been involved in the polite battles that were gathering pace in Oxford between Paul Streeten's heterodox/mainstream/dirigiste centre for development studies at Queen Elizabeth House and Ian Little's neo-classical take on the economics of development at Nuffield College. In part, too, it was because Lal's crisp prose brought the counter-revolution in development studies into the worlds of students long used to reading about dependency theories and unequal exchange, but less often about the virtues of market pricing in poorer countries. Although he has since published a number of books and papers on more focused topics, including the so-called Hindu equilibrium in India, Lal remains best known for this widely read and much discussed pamphlet. What he was calling for, of course, was nothing less than the death of 'development economics'. Similar views are apparent in his most recent book (2006), which is a paean to the virtues of classical liberalism.

Key references

Deepak Lal (1983) *The Poverty of 'Development Economics'*, London: Institute of Economic Affairs.
— (1989) *The Hindu Equilibrium, Vol. 1: Cultural Stability and Economic Stagnation*, Oxford: Oxford University Press.
— (2004) *In Praise of Empires: Globalization and Order*, London: Macmillan.
— (2006) *Reviving the Invisible Hand: The Case for Classical Liberalism in the Twenty-first Century*, Princeton, NJ: Princeton University Press.
J. Toye (1993) *Dilemmas of Development: The Counter-revolution in Development Theory and Policy*, 2nd edition, Oxford: Blackwell.

INTRODUCTION

The essential elements of the *Dirigiste Dogma*, as I see them, can be briefly stated. The major one is the belief that the price mechanism, or the working of a market economy, needs to be supplanted (and not merely supplemented) by various forms of direct government control, both national and international, to promote economic development. A complementary element is the belief that the traditional concern of orthodox micro-economics with the allocation of given (though changing) resources is at best of minor importance in the design of public policies. The essential task of governments is seen as charting and implementing a 'strategy' for rapid and equitable growth which attaches prime importance to macro-economic accounting aggregates such as savings, the balance of payments, and the relative balance between broadly defined 'sectors' such as 'industry' and 'agriculture'.

The third element is the belief that the classical 19th-century liberal case for free trade is invalid for developing countries, and thus government restriction of international trade and payments is necessary for development. Finally, it is believed that, to alleviate poverty and improve domestic income distribution, massive and continuing government intervention is required to re-distribute assets and to manipulate the returns to different types of labour and capital through pervasive price and (if possible) wage controls — and through controls which influence the composition of commodities produced and imported — so that scarce resources are used to meet the so-called 'basic needs' of the poor rather than the luxurious 'wants' of the rich.[1]

In arguing against the *Dirigiste Dogma*, I do not want to question the objectives it ostensibly seeks to serve, namely, equitable and rapid growth to make an appreciable dent, as quickly as possible, in poverty in the Third World. My case is that the means proposed are of dubious merit. Nor, more importantly, am I arguing for *laissez-faire*. That doctrine, as Keynes noted in his famous book, *The End of Laissez-Faire* — better known, alas, for its title than its contents — has been under attack by orthodox economics since John Stuart Mill.[2]

Sadly, many *dirigistes* implicitly contrast their set of beliefs as an alternative to one based on *laissez-faire*. The real issue between them and orthodox economists, however, is the form and extent of government intervention, not its complete absence. Just as Keynes noted that it was not the economists but 'the popularisers and vulgarisers'[3] who spread the *laissez-faire* doctrine, so it cannot be assumed that many distinguished contemporary economists whose views have fed the modern-day *Dirigiste Dogma* thereby necessarily subscribe to it themselves. It has been argued that Marx was not a Marxist, nor Keynes a Keynesian, and many a thinker who has nourished the *dirigiste* stream is not a *dirigiste*. This *Hobart Paperback*, therefore, is concerned with correctly interpreting not so much what particular economists meant as what they have been taken to mean by a wider lay public. For it is the latter which ultimately determines the climate of opinion in which alternative policies are judged and implemented.

1. THE ALLEGED IRRELEVANCE OF ORTHODOX ECONOMICS

Before we enter the more important debates on some of the specific beliefs of the *dirigistes*, it remains to chart the major intellectual foundations of the broad claim that *dirigisme* is required to promote development. Fortunately, an important contributor to this set of beliefs has recently characterised the major underlying assumptions which distinguish what he labels 'development economics' from both orthodox economics and various Marxist and neo-Marxist schools of thought on the economics of developing countries. Albert Hirschman distinguishes the various schools in terms of what he calls the 'mono-economics' claim and the 'mutual-benefit' claim.[4] According to Hirschman, the mono-economics claim asserts that traditional economics is applicable to developing countries in the same way as it is to developed ones; the mutual-benefit claim asserts that 'economic relations between these two groups of countries could be shaped in such a way as to yield gains for both'.[5] Whilst orthodox economics accepts both claims and neo-Marxists are presumed to reject both, Hirschman argues that development economics rejects the mono-economics but accepts the mutual-benefit claim – unlike Marx himself who would have accepted the mono-economics but rejected the mutual-benefit claim!

It is chiefly the influence of Hirschman's 'development economics' that I wish to counter in this Paper – though, to the extent there are many neo-Marxist influences on policies for and towards the Third World, I shall be dealing briefly with these too. Despite Hirschman's categorisation, development economics is closer to the neo-Marxists than to orthodox economics in its view of the mutual-benefit claim. For development economics, mutual gains can be realised only after legitimate departures from the orthodox case for free trade which must be enforced by government action both nationally and internationally. In practice, therefore, whilst not going as far as the neo-Marxists in their desire to smash the whole world capitalist system based on 'unequal exchange',[6] development economists nevertheless accept that developing countries are 'unequal partners'[7] in the current world trading and payments system, and that the rules of the game of the liberal international economic order must be changed to serve their interests.

2. THE KEYNESIAN HERITAGE

The analytical and empirical bases of development economics were provided by the Keynesian 'revolution' in economic thought and the experience of the developing countries during the Great Depression of the 1930s. While the next Section will consider the lessons that were drawn from the latter, a few remarks are required here about the Keynesian lineage of development economics and the revolt against orthodox economics that it was supposed to represent.

The specific Keynesian remedy for curing mass unemployment during a depression was soon seen to be irrelevant to developing countries which, unlike developed ones, did not face unemployment of both men and machines. Rather, their problem was too few 'machines' adequately to employ the existing 'men'.[8] All the same, in contrast with the orthodox economics castigated by Keynes, Keynesian modes of thought were seen as relevant to the problems of development. Both the central theoretical concern of Keynesian economics – namely, the determinants of the level of economic activity rather than the relative prices of commodities and factors of production – and its distinctive method – namely, national income-expenditure analysis – were enthusiastically adopted by development economics. The allocation of given resources, a major concern of orthodox economics, was considered of minor importance compared with the problems of increasing material resources – subsumed in the portmanteau term 'capital' – and of ensuring their fullest utilisation.

These Keynesian modes of thought also led to an implicit or explicit rejection of the primary rôle assigned by orthodox economics to changes in relative prices in mediating imbalances in the supply and demand for different 'commodities' – including not merely such obvious commodities as carrots and clothes, but also hypothetical composite 'commodities' such as 'savings', 'investment' and 'foreign exchange'. Changes in income were substituted as the major adjustment mechanism for bringing supply and demand into balance. This neglect of the role of the price mechanism was usually justified by assumptions based on casual empiricism: that there were limited possibilities for consumers in developing countries to substitute different commodities as their relative prices changed since their consumption consisted of bare essentials, for which no substitutes existed; and that producers could not substitute cheaper inputs for more expensive ones because, by assumption, their production techniques required inputs to be used in fixed proportions. The implicit or explicit assumption of what economists call 'limited substitutability' in both consumption and production meant the downgrading of a large part of the rôle played by relative price changes in adjusting the demand and supply of different commodities and factors of production to each other.

Moreover, the concentration on macro-economics, flowing from Keynesian modes of thought, required thinking in terms of aggregates of different 'commodities'. At its simplest, this conceptual aggregation necessitates an assumption that the relative prices of real-world commodities which constitute the aggregate composite 'commodity' remain unchanged during the period of analysis. As a result, the neglect of the price mechanism, except for the relative 'prices' of these composite 'commodities', is almost inbuilt into macro-economic thinking. Though undoubtedly useful for certain analytical and policy purposes, there is a consequent temptation – not often resisted in development economics – to ignore micro-economic problems altogether in the design of public policies.

The concentration on macro-economics was further aided by the spread of national income accounting and the establishment of statistical offices in most developing countries to provide the necessary data. Though the resulting information has considerably improved our quantitative knowledge of developing countries, it has also given a fillip to a particular type of applied economics research in both developed and developing countries which can be termed 'mathematical planning'. Building on the work of Tinbergen and his associates[9] in estimating statistical macro-economic relationships (from the 'time series' data supplied by the national income statisticians), and on the work of Leontief in refining 'input–output analysis' to describe the interrelationships in the production structure of an economy, development planning seemingly acquired a hard scientific and quantifiable character.

The Leontief input–output system,[10] building as it did on the Soviet practice of 'material balance planning', ignored relative price changes by assuming that the inputs for producing particular real-world commodities were required, for technological reasons, in fixed proportions. The typical development plan first laid down a desired rate of growth of aggregate consumption. Then the quantities of different commodities required in fixed proportions, either as inputs into production or outputs for consumption, were derived from an input–output table for the economy. Since such plans were presented in terms of desired quantities of production of various goods, their implementation most often entailed direct controls on production, including state provision of some goods considered either too important to be supplied by the private sector or unlikely to be produced by the private sector in the planned amounts.

3. THE NEGLECT OF WELFARE ECONOMICS

The final intellectual strand in the making of development economics was a neglect of the one branch of economic theory which provides the logic to assess the desirability of alternative economic policies, namely, welfare economics. This was due partly to its rejection of much of micro-economics, and partly to what was seen as the inherently limited applicability of conventional welfare economics, whether of the classical sort as systematised by Pigou or the 'new welfare economics' of Hicks and Kaldor. Broadly two types of objections were raised against this branch of economics, and they continue to be echoed in contemporary development economics. The first concerned its ethical foundations, the second the real-world relevance of its assumptions about consumers' tastes and producers' technology.

It is important to assess these objections, and the current status and scope of welfare economics, for three reasons. First, because welfare economics provides 'the grammar of arguments about policy':[11] those seeking to argue the case for increased government intervention might have been expected to use it to bolster their claims. Secondly, the development of what is labelled 'second-best' welfare economics (below, pp. 14–16) was stimulated in part by the problems and debates about developing countries.[12] Thirdly, and equally important, the analytical framework for assessing the claims of the *Dirigiste Dogma* (as of *laissez-faire*) is provided in large part by welfare economics, and it is therefore necessary to outline briefly the logic of this important branch of economics.

Welfare economics is concerned with two general classes of practical questions: (*a*) the measurement of real national income,[13] and (*b*) the efficiency and equity of particular economic outcomes, including the scope for improving them through various instruments of public policy. These are the very issues of assessing economic performance and designing policies to improve it which lie at the heart of the practical debates on development taken up in later sections of this Paper. I turn, therefore, to outlining the development of modern welfare economics, albeit very cursorily, to show how it lends *prima facie* support to the *Dirigiste Dogma*, but also to show why this support is deceptive.

One major strand of objections to welfare economics concerns its ethical foundations. For our purpose, it is sufficient to note that such objections are related to questions about the distribution of income – whether and how the distributional effects of economic change should be accounted for in measuring changes in aggregate economic welfare.[14] Not surprisingly, there is no consensus to date on these normative issues since the ethics of income distribution and other political aspects of the good society remain controversial. But does that invalidate the 'positive' welfarist conclusions about the so-called optimum conditions for production and exchange required for an efficient allocation of resources? There are some development economists who believe so.[15] This is to misunderstand the logic of modern welfare economics, however, and to derive illegitimate inferences from the legitimate criticism that it might be ethically blinkered. For, as we shall see, the most useful results of modern applied welfare economics do not depend upon accepting a particular ethical viewpoint. They are 'the logical conclusions of a set of consistent value axioms which are laid down for the welfare economist, by some priest, parliament or dictator',[16] whilst, as far as the *Dirigiste Dogma* is concerned, the policies *it* has engendered have aided neither efficiency nor equity nor liberty in the Third World.

4. THE THEORETICAL ATTACK ON *LAISSEZ-FAIRE*

It remains to chart the second set of objections to welfare economics which have also led to its neglect in development economics. These concern its 'positive' aspects. The basic theorems of welfare economics – rigorously derived in the 1950s by Professors Arrow and Debreu – show that, in a perfectly-competitive economy with universal markets for all commodities distinguishable not only by their spatial and temporal characteristics but also by the various conceivable future 'states of nature' under which they could be traded (that is, there is a 'complete' set of futures markets for so-called 'contingent' commodities[17]), a *laissez-faire* equilibrium will be Pareto-efficient in the sense that, with given resources and available technology, no individual can be made better-off without someone else being made worse-off. Since, however, this competitive, Pareto-efficient equilibrium may not yield the distribution of income considered socially desirable according to the prevailing ethics, government intervention may be necessary to legislate the optimum income distribution even in a perfectly competitive economy with complete markets. If government can levy lump-sum taxes and disburse lump-sum subsidies, the perfectly-competitive economy can attain a full 'welfare optimum'.

It is, however, premature to cheer this rigorous establishment of the case for a *laissez-faire* economy in which the government's rôle (apart from providing a legal framework to enforce property rights and maintain law and order) is confined to *lump-sum* redistributive measures (assuming the unmodified distribution conflicts with prevailing ethical norms). For, as many critics of the price mechanism have been only too ready to point out, the conditions (or assumptions) for establishing it are extremely unrealistic. Broadly, the assumptions fall into those required for (a) perfect competition and (b) universal markets.

Perfect competition depends on stringent assumptions about the tastes of consumers and the nature of producers' technology. First, there must be no interdependencies in either consumption or production not mediated through markets (that is, there must be no so-called 'external effects', such as keeping up with the Jones's or emitting smoke which damages the output of a nearby laundry). And, secondly, there must not be too many industries with decreasing costs of production (that is, 'increasing returns' in production must not be large relative to the size of the economy) since these are likely to lead to monopoly. Development economists have emphasised the importance of 'externalities' and 'increasing returns' in developing countries.[18] Though usually asserted rather than empirically demonstrated, it has led them to reject the argument for a market economy implicit in the notion of a perfectly-competitive Utopia.

These two assumptions are not nearly as

unrealistic,[19] however, as the other major one required to show the Pareto-efficiency of a *laissez-faire* competitive economy, namely, the existence of universal markets. The lack of markets for all current and future 'contingent' commodities is likely to be the fundamental cause of so-called 'market failure'. Externalities pose problems essentially because of the difficulty (if not impossibility) of creating a market for them even though, conceptually, they can be readily identified as 'commodities' (factory smoke, for example, is a commodity, but also a 'bad' for which there is no market). The reason is that, to create a market in any commodity, non-buyers must be excluded from obtaining it. Exclusion may be technically impossible or prohibitively expensive, in terms of resource costs, for most 'externalities'. Where exclusion is possible, there may be so few buyers and sellers in the market for the externality that it cannot be perfectly competitive.[20]

The difficulty of establishing markets in these commodities reflects what are broadly termed the costs of making transactions attached to any market or indeed any mode of resource allocation. Transaction costs include the costs of exclusion as well as those of acquiring and transmitting information by and to market participants. They drive a wedge, in effect, between the buyer's and the seller's price. The market for a particular good will cease to exist if the wedge is so large as to push the lowest price at which anyone is willing to sell above the highest price anyone is willing to pay.

Apart from making it difficult to deal through an unfettered market with externalities, these transactions costs will also limit the development of futures markets for all commodities. Thus, far from being an apologia for the *laissez-faire* doctrine, as many suppose, modern welfare economics provides the precise reason why, even in the absence of distributional considerations, a real-world *laissez-faire* economy is not likely to be Pareto-efficient – because (a) it is unlikely to be perfectly competitive, and (b) it will certainly lack universal markets.

5. THE LIMITS OF RATIONAL DIRIGISME

Thus, even if income distribution is disregarded, there would seem to be a *prima facie* case for government intervention. It would be absurd, however, to jump to the conclusion that, because *laissez-faire* may be inefficient and inequitable, any form of government intervention thereby entails a welfare improvement. For transactions costs are also incurred in acquiring, processing and transmitting the relevant information to design public policies, as well as in enforcing compliance. There may consequently be as many instances of 'bureaucratic failure' as of 'market failure', making it impossible to attain a Pareto-efficient outcome.

Let us consider the question of legislating for the optimal income distribution in an otherwise competitive, Pareto-efficient economy. If government could levy *lump-sum* taxes which were inescapable and could not be avoided by economic agents altering their otherwise efficient choices, it could achieve the full welfare optimum. If, for example, income differences were related to the inescapable abilities of individuals, and there was an unambiguous and readily available (at low cost) index of these abilities, a lump-sum tax/subsidy system based on differential abilities would allow the full welfare optimum to be achieved (in a perfectly-competitive economy with complete markets). Clearly, such a system is not feasible because of the costs of acquiring the necessary information.

By contrast, a tax/subsidy system based on *income* differences which aimed at legislating for a desired income distribution would not be lump-sum because it would affect the choices individuals make at the margin between work and leisure. By distorting the initial, *ex hypothesi*, efficient allocation, the income-based tax/subsidy system, though improving the distribution of income, would impair the productive efficiency of the economy. The feasible instrument of government intervention would mean that the welfare gain from an improved distribution could only be obtained by inflicting a welfare loss in the form of lower productive efficiency. Because of the 'bureaucratic failure' inherent in the inability of government to introduce a lump-sum tax/subsidy system, a full welfare optimum is not attainable even with government intervention. All that can be achieved is a 'second-best' optimum where the *net* gain from the distributional gain and efficiency loss are at a maximum.

The same argument applies to government intervention to correct market failures in any real-world economy which is not perfectly competitive or which lacks complete futures markets. There are few, if any, instruments of government policy which are non-distortionary, in the sense of not inducing economic agents to behave less efficiently in some respects. Neither markets nor bureaucrats as they exist can therefore be expected to lead an economy to a full welfare optimum. The best that can be expected is a second-best.

Given that the optimum is unattainable, the relevant policy problem becomes that of assessing to what extent particular government interventions may raise welfare in an inherently and inescapably imperfect economy. The Utopian theoretical construct of perfect competition then becomes relevant as a reference point by which to judge the health of an economy, as well as the remedies suggested for its amelioration. Since improvements will not necessarily entail a movement towards the perfectly-competitive theoretical norm, evaluating the likely consequences of alternative policies in an imperfect economy becomes a subtle exercise in what is nowadays termed 'second-best welfare economics'.

An early theoretical contribution by Lipsey and Lancaster correctly argued that, in an imperfect economy, the restoration of some of the conditions which would exist under perfect competition would not necessarily result in an improvement in welfare.[21] This insight was unfortunately taken to mean that there was no way in which the effects on economic welfare of alternative piecemeal policies to improve the working of the price mechanism or to alter the distribution of income could be judged. Many took it to imply that, since every economy is imperfect, welfare economics (and by implication microeconomics) was irrelevant in the design of public policy. One of the major analytical advances of the last two decades, prompted by the problems of developing countries, has been to show that this is not so.[22] Specific examples of the application of modern second-best welfare economics are given in the next sections, and Appendix 1 (p. 111) provides an outline of the logic of the exercise.

The major point to note is that no general rule of second-best welfare economics permits the deduction that, in a necessarily imperfect market economy, particular *dirigiste* policies will increase welfare. They may not; and they may even be worse than *laissez-faire*. Moreover, any economic justification for a *dirigiste* policy not based on the logic of second-best welfare economics must be incoherent, and akin to the miracle cures peddled by quacks which are adopted because of faith rather than reason. The burden of the case against the *Dirigiste Dogma* in its application to developing countries is that, though in many instances some forms of *dirigisme* may have been beneficial had they been feasible, the *dirigiste* policies actually adopted (either because they were considered the only feasible ones, or else because the relative costs and benefits of alternative policies were never examined) have often led to outcomes which, by the canons of second-best welfare economics, may have been even worse than *laissez-faire*. The conclusion, therefore, of this theoretical tour is that the very analysis which seemingly establishes a *prima facie* intellectual justification for the *Dirigiste Dogma* provides, in its fullness, the antidote!

NOTES

1 Nurkse [167, 168], Myrdal [162], Hirschman [78], Balogh [14], Rosenstein-Rodan [177], Chenery [39], Prebisch [171], Singer [192], and Streeten [202] are notable amongst many others who would consider themselves to be non-neo-classicals and whose writings have been influential in providing various elements of the *Dirigiste Dogma*. There has, however, always been some opposition to these views: Haberler [65, 66], Viner [210], Bauer and Yamey [18], Schultz [181].

2 Keynes [91], p. 26. Nor is 'the phrase *laissez faire* to be found in the works of Adam Smith, of Ricardo or of Malthus. Even the idea is not present in a dogmatic form in any of these authors'. (p. 20) 'This is what the economists are *supposed* to have said. No such idea is to be found in the writings of the greatest authorities.' (p. 17) 'Some of the most important work of Alfred Marshall – to take one instance – was directed to the elucidation of the leading cases in which private interest and social interest are *not* harmonious. Nevertheless, the guarded and undogmatic attitude of the best economists has not prevailed against the general opinion that an individualistic *laissez-faire* is both what they ought to teach and what in fact they do teach.' (p. 27)

3 *Ibid.*, p. 17.

4 A. Hirschman [79].

5 *Ibid.*

6 The title of an influential neo-Marxist work by A. Emmanuel [55].

7 The title of a well-known collection of papers by Lord Balogh [14].

8 V. K. R. V. Rao [173].

9 J. Tinbergen [208].

10 W. Leontief [124], H. Chenery *et al.* [41].

11 F. H. Hahn [67].

12 I. M. D. Little and J. A. Mirrlees [142, 143], P. Dasgupta–
    S. Marglin–A. Sen [49], A. H. Harberger [70], Little,
    Scitovsky, Scott [145].
13 Sen [190] provides a lucid survey of the issues.
14 A. K. Sen [189].
15 P. Streeten and S. Lall [206], for instance.
16 Little [135], p. 80.
17 A 'contingent commodity', for instance, would be 'ball-
    bearings' for delivery on 16 September 1995, where the
    future price is conditional upon whether or not industrial
    production in Indonesia is 20 per cent above average
    that month.
18 Rosenstein-Rodan [177], T. Scitovsky [184], Hirschman
    [78], Chenery [39].
19 Many externalities can be dealt with by suitable taxes
    and subsidies; and, in an open economy, the potential
    danger of decreasing-cost industries becoming mon-
    opolistic is reduced by foreign competition.
20 Arrow [7] cites the example of the lighthouse keeper
    who knows exactly 'when each ship will need its ser-
    vices, and . . . abstract from indivisibility (since the light
    is either on or not). Assume further that only one ship
    will be within range of the lighthouse at any moment.
    Then exclusion is perfectly possible; the keeper need
    only shut off the light when a non-paying ship is coming
    into range. But there would be only one buyer and
    one seller and no competitive forces to drive the two
    into a competitive equilibrium. If in addition the costs
    of bargaining are high it may be most efficient to offer
    the service free.' (p. 15)
21 Lipsey and Lancaster [133].
22 Two sets of applications of this theory are in Little and
    Mirrlees [142], and W. M. Corden [46].

REFERENCES

[1] K. J. Arrow: 'Political and Economic Evaluation of
     Social Effects and Externalities', in J. Margolis (ed.), *The
     Analysis of Public Output*, NBER, Columbia, New York,
     1970.
[2] T. Balogh: *Unequal Partners*, 2 vols. Blackwells, Oxford,
     1963.
[3] P. T. Bauer and B. S. Yamey: *The Economics of Under-
     developed Countries*, Cambridge, 1957.
[4] H. Chenery: 'The Interdependence of Investment
     Decisions', in A. Abramowitz (ed.), *The Allocation of
     Economic Resources*, Stanford, 1959.
[5] H. Chenery *et al.*: *Structural Change and Development
     Policy*, Oxford, 1979.
[6] W. M. Corden: *Trade Policy and Economic Welfare*,
     Oxford, 1974.
[7] P. Dasgupta, S. Marglin and A. K. Sen: *Guidelines for
     Project Evaluation*, UNIDO, New York, 1972.

[8] A. Emmanuel: *Unequal Exchange*, Monthly Review
     Press, New York, 1972.
[9] G. Haberler: 'Critical Observations on Some Current
     Notions in the Theory of Economic Development',
     L'Industria, No. 2, 1957, reprinted in G. Meier (ed.)
     [153].
[10] G. Haberler: *A Survey of International Trade Theory*,
     Princeton Special Papers in International Economics,
     1961.
[11] F. H. Hahn: 'On Optimum Taxation', *Journal of
     Economic Theory*, February 1973.
[12] A. C. Harberger: *Project Evaluation – Selected Essays*,
     Chicago, 1972.
[13] A. O. Hirschman: *The Strategy of Economic Develop-
     ment*, Yale, 1958.
[14] A. O. Hirschman: *Essays in Trespassing – Economics to
     Politics to Beyond*, Cambridge, 1981.
[15] J. M. Keynes: *The End of Laissez-Faire*, Hogarth Press,
     London, 1926.
[16] W. Leontief: *The Structure of the American Economy*,
     Oxford, 1941.
[17] R. G. Lipsey and K. Lancaster: 'The General Theory of
     the Second Best', *Review of Economic Studies*, Vol. 26,
     1956–57.
[18] I. M. D. Little: *A Critique of Welfare Economics*, Oxford,
     1950.
[19] I. M. D. Little and J. A. Mirrlees: *Manual of Industrial
     Project Analysis, Vol. II: Social Cost–Benefit Analysis*,
     OECD Development Centre, Paris, 1969.
[20] I. M. D. Little and J. A. Mirrlees: *Project Appraisal and
     Planning for Developing Countries*, Heinemann, London,
     1974.
[21] I. M. D. Little, Tibor Scitovsky, and Maurice Scott:
     Industry and Trade in Some Developing Countries, Oxford
     University Press, 1970. There are separate case studies
     of India, Pakistan, Brazil, Mexico, the Philippines, and
     Taiwan in this OECD series, all published by the
     Oxford University Press.
[22] G. Myrdal: *Economic Theory and Underdeveloped
     Regions*, Vora and Co., Bombay, 1958. (This is an
     expanded version of his 1956 Bank of Cairo lectures
     entitled: *Development and Underdevelopment – A Note
     on the Mechanism of National and International
     Economic Inequality*.)
[23] R. Nurkse: *Problems of Capital Formation in Under-
     developed Countries*, Oxford, 1953.
[24] R. Nurkse: *Equilibrium and Growth in the World Econ-
     omy*, Harvard University Press, 1961.
[25] R. Prebisch: *The Economic Development of Latin
     America and Its Principal Problems*, United Nations,
     New York, 1950.
[26] V. K. R. V. Rao: 'Investment, Income and the Multiplier
     in an Underdeveloped Economy', *Indian Economic
     Review*, February 1952; reprinted in Agarwala and
     Singh (eds.) [3].

[27] P. N. Rosenstein-Rodan: 'Problems of Industrialisation of Eastern and South-Eastern Europe', *Economic Journal*, June–September 1943.

[28] T. Schultz: *Transforming Traditional Agriculture*, Yale, 1964.

[29] T. Scitovsky: 'Two Concepts of External Economies', *Journal of Political Economy*, April 1954.

[30] A. K. Sen: 'Personal Utilities and Public Judgments: Or What's Wrong With Welfare Economics?', *Economic Journal*, September 1979; Comment by Y. K. Ng and Reply by Sen in *Economic Journal*, June 1981.

[31] A. K. Sen: 'The Welfare Basis of Real Income Comparisons – A Survey', *Journal of Economic Literature*, March 1979.

[32] H. Singer: 'The Distribution of Gains Between Borrowing and Investing Countries', *American Economic Review*, May 1950.

[33] P. Streeten: *The Frontiers of Development Studies*, Macmillan, 1972.

[34] P. Streeten and S. Lall: *Foreign Investment, Transnationals and Developing Countries*, Macmillan, 1977.

[35] J. Tinbergen: *Statistical Testing of Business-Cycle Theories I: A Method and Its Application to Investment Activity*, League of Nations, Geneva, 1939.

[36] J. Viner: *International Trade and Economic Development*, Oxford, 1953.

'Democracy and the "Washington Consensus" '

World Development (1993)

John Williamson

Editors' Introduction

John Williamson has been a Senior Fellow of the Institute for International Economics in Washington, DC, since 1981. He was an economic consultant to the UK Treasury from 1968 to 1970 and an adviser to the International Monetary Fund from 1972 to 1974. Williamson has also worked as an economics professor in the UK, the United States and Brazil, and was the project director for the UN's High-Level Panel on Financing for Development (the so-called Zedillo Report of 2001). He also took leave from his Washington think-tank to serve as the Chief Economist for South Asia at the World Bank (1996–99), although, perhaps significantly, most of his academic writings have dealt with economic issues (aid, finance, debt, adjustment) in Latin America, Eastern Europe and the ex-Soviet Union. Williamson has also written extensively on the changing position of the US economy within the world economy, and, relatedly, on the changing position of the US dollar as the world's leading international unit of account.

For better or worse, however, John Williamson is best known today as the scholar who sought to define the Washington Consensus of the late 1980s and early 1990s (see the reading published here). Williamson maintains that his account of the Washington Consensus was never meant to describe a philosophy of extreme neo-liberalism. His original thoughts from 1989 and 1993 did not suggest that policy makers in Washington were committed to promoting monetarism, supply-side economics or a minimal state in developing countries. Those policies, he later remarked, were more suited to the Mont Pelerin society than to Washington's elites. (The Mont Pelerin society was founded by von Hayek and promotes a classically liberal, indeed at times libertarian, world view.) Nor was Williamson commending capital account liberalisation in 1989/93. Williamson rather intended the Washington Consensus to define ten main areas of reform in Latin America that were being widely talked about by the IMF and the World Bank, and by other Washington-based think-tanks (presumably from the Brookings Institution to the Institute for International Economics). These initiatives are clearly spelled out here, although Williamson has conceded that some aspects of the Consensus – notably those linked to privatisation, which sometimes opened the way for the privatisation of corruption and elite looting of state assets – have since become contentious. Williamson continues to believe that most parts of the Washington Consensus describe a common-sense approach to economic policy in Latin America, a view that has been strongly challenged by his critics. Events have forced him to acknowledge, however, that the term 'Washington Consensus' has taken on a life beyond his original summary. Even in the work of scholars who might be expected to be more precise (Williamson has mentioned Joseph Stiglitz in this regard), the Washington Consensus is identified exactly as that hard-line 'neo-liberalism' that Williamson disdains, and which in his view did not dominate policy circles in Washington even in the late 1980s. As ever, ideas escape their owners.

Key references

John Williamson (1993) 'Democracy and the "Washington Consensus"', *World Development* 21(8): 1329–36.

— (2000) *Exchange Rate Regimes for Emerging Markets*, Washington: Institute for International Economics.

— (2004/5) 'The strange history of the Washington Consensus', *Journal of Post-Keynesian Economics* 27: 195–206.

Paul Mosley (2001) 'Attacking poverty and the "post-Washington consensus"', *Journal of International Development* 13: 307–13.

Z. Onis and F. Senses, 'Rethinking the emerging Post-Washington Consensus', *Development and Change* 36 (2): 263–90.

1. INTRODUCTION

In mid-1989 I prepared a list of the principal economic reforms that were being urged on Latin American countries by the powers-that-be in Washington (see Williamson, 1990). I entitled the paper "What Washington means by policy reform" and rashly dubbed it the "Washington consensus," a term immediately challenged by one of my discussants, Richard Feinberg, on the grounds that it was not universally endorsed in Washington, and was therefore not a consensus, while the geographical scope of its acceptance went far beyond the Beltway. He suggested that it would have been more apt to name it the "universal convergence." My Latin friends in due course let me know that many of them also took a poor view of the term, since it suggested that Washington had figured out what they should be doing and was engaged in imposing "economic correctness" on them. Their latest barb (due to Carlos Prima Braga), which I endorse in substance, is that my term is a misnomer because the "Washington consensus" is being implemented in every capital of the hemisphere bar Washington. (This seems to overlook Havana.)

My view is in fact that the "Washington consensus" is the outcome of worldwide intellectual trends to which Latin America contributed (principally through the work of Hernando de Soto) and which have had their most dramatic manifestation in Eastern Europe. It got its name simply because I tried to ask myself what was the conventional wisdom of the day among the economically influential bits of Washington, meaning the US government and the international financial institutions. I did not intend to imply that they could claim any particular priority in having nurtured the conventional wisdom.

I tried to describe what was conventionally thought to be wise rather than what I thought was wise: that is, it was intended as a positive rather than a normative list. It happens, however, that I endorse everything on the list, which is hardly surprising since I live in Washington and like to think of myself as reasonably eclectic rather than temperamentally rebellious. The list is nonetheless materially different from the list that I would have drawn up had I been aiming to produce a normative list, since the latter would have had a substantially bigger equity-oriented component.[1] I deliberately excluded from the list anything which was primarily redistributive, as opposed to having equitable consequences as a byproduct of seeking efficiency objectives, because I felt the Washington of the 1980s to be a city that was essentially contemptuous of equity concerns.[2]

The questions addressed in this paper concern the political status of the "Washington consensus," alias the "universal convergence." Would political agreement that this is good economics be desirable? Or would democratic politicians have nothing worthwhile left to decide in the field of economics if they all subscribed to the universal convergence? How much consensus on economic policy would be desirable, and how much is it reasonable to expect might come about?

2. THE CASE FOR CONSENSUS

I can see no advantage to democracy in having major parties espousing economic nonsense. If they win an election, the economy will suffer; but to the extent that the electorate has a certain amount of earthy common sense, the chances are that they will lose and thus that the range of effective electoral choice will be reduced. To those of us who regard ourselves as left of center, it ought to seem particularly unfortunate when left-wing parties espouse economic rubbish that jeopardizes their prospects of being able to further egalitarian causes.

Consensus on good economics is important if economic reform is to succeed. Continual policy reversals are obviously disruptive. The pressures to manufacture policy differences produced by a two-party system, and then to lurch between the one position and the other as the parties alternate in power, help to explain the inferiority of the postwar economic performance of the United Kingdom in comparison to the continental countries where coalition government is the norm and the pressures are therefore to seek convergence rather than confrontation.

In many cases the benefits of reform depend not just on sensible policies being enacted but also upon confidence that those policies will remain invariant to political change. Perhaps the most obvious case in point, at least so far as Latin America is concerned, relates to capital flight. The prime cause of capital flight was populist macroeconomic policies, and hence the return of flight capital demanded restoration of macroeconomic discipline. But simply restoring macro discipline may not suffice, because wealth-holders look to the future and may repatriate their funds only when they are confident that the reformed macro policies are here to stay. The same thing is true with any decision that has long-term implications, notably the decision to invest.

A further benefit of broad political consensus in favor of the universal convergence is that it will help to limit the damage that can be done by the political mafia hypothesized by public choice theory. Rules that make state help the exception rather than the rule, and require it to be transparent and based on agreed principles, will cramp the style of those politicians whose main objective is to feather their own nests rather than to further any concept of the public good that can be represented in a plausible social welfare function.

3. CONSENSUS AND DEMOCRACY

It would be ridiculous to argue that as a matter of principle every conceivable point of view should be represented by a mainstream political party. No one feels that political debate is constrained because no party insists that the Earth is flat. No one demands that one or other of the principal parties should advocate racism or the denial of human rights. Until those battles are won, of course, the establishment of human rights and the elimination of racism deserve to be high on the political agenda. But one hopes that in due course a consensus will be established in favor of human rights and against racism, and at that time those very important issues will cease to be subjects of political controversy between the mainstream parties.

The universal convergence seems to me to be in some sense the economic equivalent of these (hopefully) no-longer-political issues. Until such economic good sense is generally accepted, then its promotion must be a political priority. But the sooner it wins general acceptance and can be removed from mainstream political debate, the better for all concerned. Indeed, the chances of removing these basic economic issues from the political agenda should be better than those of keeping human rights and racism off the agenda, inasmuch as the latter depend solely upon value judgments whereas the superior economic performance of countries that establish and maintain outward-oriented market economies subject to macroeconomic discipline is essentially a positive question. The proof may not be quite as conclusive as the proof that the Earth is not flat, but it is sufficiently well established as to give sensible people better things to do with their time than to challenge its veracity.

Since the world has its share of cranks, however, we cannot expect unanimous endorsement of the universal convergence. A democratic political system needs a way of allowing dissent to be expressed. This is important both because suppression is the best way to convince the conspiratorially minded that there is something to hide, and because there must always remain some doubt as to whether what deserves to pass for conventional wisdom today will remain valid for the indefinite future. The system needs some mechanism whereby orthodoxy[3] can be challenged, for truth is never absolute in the social sciences – and it also changes in a way that truth in the natural sciences does not.

I have no view on whether this escape channel is best provided by the existence of fringe parties or by the mainstream parties tolerating eccentric minorities. I see no reason to try and limit the expression of dissenting viewpoints to one mechanism rather than the other.

Neither do I see any problem analogous to the "democrats' dilemma" – the question as to whether or not democrats should tolerate use of the democratic system by totalitarian parties that would not allow themselves to be democratically replaced if they ever gained power. Parties with crazy economic programs must obviously be expected to do economic damage if they gain power, but as long as they can be replaced democratically there is no case for limiting their political freedom. Thus my position is not that democracy should be in any way circumscribed so as to promote good economic policy, but rather that both economic policy and democracy will benefit if all mainstream politicians endorse the universal convergence and the scope of political debate on economic issues is *de facto* circumscribed in consequence.

4. THE SCOPE OF CONSENSUS

How much consensus on economic policy should one hope for? As much as is justified by the state of economic knowledge and the convergence in fundamental (ultimately political) values.

The hope that we can now develop far more consensus than would have been conceivable or appropriate in the 1950s is based ultimately on the fact that we now know much more about what types of economic policy work. At that time it looked as though socialism was a viable alternative to a market economy; now we know that it is not. At that time we had not discovered that pushing import substitution beyond the first ("easy") stage was vastly inferior to a policy of outward orientation that allowed non-traditional exports to develop; now we know better. At that time we thought that more expansionary macro policies produced more output rather than just more inflation even in the long term; there is now

evidence that this is the opposite of the truth (De Gregorio, forthcoming). At that time it was still possible to hope that greed could be displaced by altruism as effectively as it can be made to work in the public interest by a competitive market economy; after the collapse of communism that just looks naive.

Consensus, however, will and should be circumscribed by two types of consideration: the limitations of our knowledge and differences in our values. We know from elementary economic theory that markets will not work satisfactorily where externalities are important unless some deliberate action is taken to internalize them. We have also been learning about the sorts of situations where externalities actually are sufficiently important as to demand countervailing action: where spillovers affect the quality of the environment, with regard to invention, where the safety of financial institutions is at stake, where the labor training provided by one firm can be poached by another, and so on. At present many of these issues are still too novel to have established any robust knowledge of the best way of internalizing the externalities, and hence they will remain controversial for the time being. But over time experience will accumulate and controversy should accordingly diminish, except in countries such as the United Kingdom where the political system is so structured as to manufacture controversy artificially.

The other reason why consensus will remain limited is that normative values differ. The classical view of politics is that it provides the mechanism whereby society decides whose normative values will prevail. The winning party or coalition specifies the social welfare function to be maximized by the choice of economic policies. Since I cannot conceive of a social welfare function that would specify means (such as the choice of economic system or the extent of the role of the state) rather than ends (notably the degree to which equity is to be traded off against efficiency, as sketched in the appendix), this suggests that the fundamental political divide (on economic issues) is not between capitalism and socialism, or between free markets and state intervention, but between those concerned to promote an equitable income distribution ("the left") and those concerned to defend established privileges ("the right"). Note that this dividing line helps explain why it seemed natural for communist hardliners to be called right-wing in recent Soviet debates while those seeking the transition to a market economy were called the left. (But there can be circumstances in which the distinction is less clear, notably where a privileged elite has been partially replaced by another – as in 1979 Britain, where the old elite had been partially replaced by the trade union barons and the displacement of the latter gave the Thatcher government a claim to being considered radical.)

Thus economic policies can be classified into three categories:

(a) those where consensus has been established;
(b) those where controversy still reigns, but where the controversy is nonideological in the sense that there is no inherent reason for an egalitarian to be driven to favor one type of outcome rather than another;
(c) those where controversy reigns now and should be expected to remain in the future, since the appropriate choice depends on political (normative) values.

Let me take the 10 headings of the Washington consensus and suggest the degree to which policies in those fields ought to fall in one or another of these three categories.[4] While the list was compiled with reference to Latin America, it seems to me to be generally applicable.

Fiscal Discipline. Budget deficits, properly measured to include provincial governments, state enterprises, and the central bank, should be small enough to be financed without recourse to the inflation tax. This typically implies a primary surplus (i.e., before adding debt service to expenditure) of several percent of GDP, and an operational deficit (i.e., the deficit disregarding that part of the interest bill that simply compensates for inflation) of no more than about 2 percent of GDP.

This I place unreservedly in category (a). The only scope I see for sensible debate concerns the extent to which it may be rational to allow cyclical budget deficits in the interest of stabilization policy (something not precluded by my wording), which I place in category (b).

Public Expenditure Priorities. Policy reform consists in redirecting expenditure from politically sensitive areas which typically receive more resources than their economic returns can justify, like administration, defense, indiscriminate subsidies, and white elephants, toward neglected fields with high economic returns and the potential to improve income distribution, like primary health and education, and infrastructure.

I phrased this to cover as much as I felt I could have commanded a consensus in the Washington of the 1980s, and I suspect this is not that different from what can be expected to command a consensus, i.e. fall in category (a), in Latin America or elsewhere. But this is the area par excellence where I would want and expect political debate, category (c), to focus. The left will want expenditures aimed at improving income distribution even where they do not have a particularly favorable impact on growth, while the right will not. In another paper (Williamson, 1991) I have discussed the sorts of expenditure that seem to me to deserve priority in Latin America if something is to be done about the region's appalling income distribution.

Tax Reform involves broadening the tax base and cutting marginal tax rates. The aim is to sharpen incentives and improve horizontal equity without lowering realized progressivity. Improved tax administration is an important aspect of broadening the base in the Latin context. Taxing interest on assets held abroad ("flight capital") should be another priority in the coming decade.

Once again that was worded with the intention of making it acceptable across the political spectrum (category a), but not with the expectation that this would remove tax policy from the political arena. If the left is to achieve higher public expenditure without compromising fiscal discipline then it will have to raise more revenue by taxes, including higher marginal tax rates, although perhaps still recognizing the tradeoff with efficiency that this implies. In Williamson (1991) I suggest the desirability of land value taxation, as well as taxing interest on flight capital, as tools to improve equity with no efficiency cost in the Latin context.

Financial Liberalization. The ultimate objective is market-determined interest rates, but experience has shown that, under conditions of a chronic lack of confidence, market-determined rates can be so *high* as to threaten the financial solvency of productive enterprises and government. Under that circumstance a sensible interim objective is the abolition of preferential interest rates for privileged borrowers and achievement of a moderately positive real interest rate.

Once again this was carefully worded in the hope of capturing common ground, but it is probably true that among economists this recommendation would be more controversial than those concerning fiscal policy. Some successful modernizing countries (France, Korea) retained state direction of credit till they were well beyond the stage where Latin America is today: thus this recommendation may fall in category (b) rather than (a). Williamson (1991) suggests that the Bangladeshi Grameen Bank is a social innovation that makes it possible to channel credit to micro enterprises where it has an unusually high rate of return as well as benefiting the poor, which should place it in category (b) rather than (c) although one would hope the left would give particular priority to the transplantation of such institutions.

Exchange Rates. Countries need a unified (at least for trade transactions) exchange rate set at a level sufficiently competitive to induce a rapid growth in non-traditional exports, and managed so as to assure exporters that this competitiveness will be maintained in the future.

Given the succes of export-led growth and the evidence that a competitive exchange rate is the key to such success, this recommendation falls in category (a). Given also the compelling evidence that markets cannot be relied on to take exchange rates to competitive levels that will support prudent macroeconomic policies and export-led growth, I do not see that this is consistent with allowing the exchange rate to float (but I am not sure that this is yet universally conceded). Of course, there is always a tradeoff between securing a competitive exchange rate and restraining inflation, and there may be times when a government will choose to give a greater weight to the latter, e.g., by joining an institution such as the EMS exchange rate mechanism. But I would argue that there probably would be (and certainly ought to be) a consensus that this should be done only if the prospective cost in terms of diminished competitiveness is tolerable.

Trade Liberalization. Quantitative trade restrictions should be rapidly replaced by tariffs, and these should be progressively reduced until a uniform low tariff in the range of 10 percent (or at most around 20 percent) is achieved. There is, however, some disagreement about the speed with which tariffs should be phased out (with recommendations falling in a band between 3 and 10 years), and about whether it is advisable to slow down the process when macroeconomic conditions are adverse (recession and payments deficit).

This is the one topic where I explicitly recognized the existence of a significant difference of opinion in Washington, and there is no question but that a similar disagreement exists in the profession at large. Perhaps one could get consensus on my wording given that it includes this caveat, but the need for that caveat obviously means that we are closer to category (b) rather than (a).

Foreign Direct Investment. Barriers impeding the entry of foreign firms should be abolished; foreign and domestic firms should be allowed to compete on equal terms.

This would presumably provoke political disagreement, not from the left or right, but from nationalists. It would be nice to be able to assign it to category (a), but I am not sure it would be justified.

Privatization. State enterprises should be privatized.

One frequent criticism of my list as an exercise in reporting was that I should have included the restructuring of public enterprises as an alternative way of handling the problem of poor public-sector performance that would in some instances be more practical than privatization. With that addition, perhaps even this entry may fall in category (a).

Deregulation. Governments should abolish regulations that impede the entry of new firms or restrict competition, and ensure that all regulations are justified by such criteria as safety, environmental protection, or prudential supervision of financial institutions.

Perhaps the general form of this recommendation would be acceptable in category (a), but there are still lots of category (b) debates about how it should be implemented.

> *Property Rights*. The legal system should provide secure property rights without excessive costs, and make these available to the informal sector.

I would hope that this recommendation would fall in category (a), but I urged in Williamson (1991) that it be supplemented by land reform, which is surely a category (c) topic.

Thus most of the universal convergence falls in category (a). It is drawn from that body of robust empirical generalizations that forms the core of economics. I do not for a moment doubt that it is possible to conceive of pathological circumstances under which all of the advice listed in the universal convergence would be inopportune. The interesting question is whether these pathological circumstances actually occur with sufficient frequency to make it important to be alert to their possible presence. My judgment is that they in fact arise sufficiently infrequently to make it likely that far more mischief will be done by making a fetish of their possible presence than by advising policy makers to ignore their possible occurrence. Of course, good economists will still try to stay alert to the possibility that pathological circumstances will sometimes arise, and will advise a government (or opposition) to modify its policies appropriately. But they should be pressed to explain their reasons for deviating from orthodoxy, and certainly not allowed to get away with a Baloghian inference that it is justification in itself that advice be unorthodox.

5. CONCLUDING REMARKS

Acceptance of the proposition that there is a substantial body of economic advice, roughly that summarized in the "Washington consensus/universal convergence," that deserves to be endorsed across the political spectrum does not mean the "end of politics," although it may help curb the anti-social style of politics described by the "new political economy." But civilized politics, meaning the use of the electoral and parliamentary systems in order to determine the specification of the social welfare function that economic policy should seek to maximize, still has a crucial role to play, because there will be a

tradeoff between equity and efficiency when Latin America finally gets its policies sufficiently in order to reach the frontier. Doing something to remedy the region's disgraceful inequalities is going to be a major challenge to the left, though it is one that the left will be in a position to take on only when it liberates itself from its historical legacies of antipathy to the market, populism, and inward-looking nationalism. This is not a hopeless cause, since in some countries (notably Australia, New Zealand and Spain) the universal convergence was pioneered by the left, while in many others it has now been endorsed by the left as much as by the right.

Wide political endorsement of mainstream economics will also leave room for another type of political contest, one in which the political parties search for policy proposals that will better correct for externalities. This search is essentially non-ideological, i.e. there is no inherent reason why an egalitarian should be more or less interested in devising a scheme that will (for example) encourage companies to provide the optimal amount and form of training than a conservative. But the fact that all parties believe that they can gain votes by developing proposals for better management of the system will improve the prospects for constructive reform, especially in comparison with a situation in which left-wing parties feel themselves under an ideological obligation to force any reform into the mould of an extension of the state's power to direct economic activity.

6. POSTSCRIPT

One question that arose during the conference at Forlí was the relation of the Washington consensus to what was called "neoliberalism." When I asked how I should understand this term, I was told that it encompassed Austrian economics, monetarism, new classical macroeconomics, and public choice theory. (Since these are more or less the bits of economics that are customarily dressed up as positive economics even though only conservatives subscribe to them, I must say that I would find the alternative term "neoconservatism" more apt.) If interpreted in that sense I would certainly deny that the Washington consensus is neoliberal (or neoconservative).

I suspect that the discovery that it was being interpreted this way may have been the stimulus to a list that I subsequently drew up of issues on which the Washington consensus does *not* imply a particular view (Williamson, 1992). This list is as follows:

— the desirability of maintaining capital controls
— the need to target the current account
— how rapidly and how far inflation should be reduced
— the advisability of attempting to stabilize the business cycle
— the usefulness of incomes policy and wage/price freezes (sometimes called "heterodox shocks")
— the need to eliminate indexation
— the propriety of attempting to correct market failures through such techniques as compensatory taxation
— the proportion of GDP to be taken in taxation and spent by the public sector
— whether and to what extent .income should be deliberately redistributed in the interest of equity
— the role of industrial policy
— the model of the market economy to be sought (Anglo-Saxon *laissez-faire*, the European social market economy, or Japanese-style responsibility of the corporation to multiple stakeholders)
— the priority to be given to population control and environmental preservation.

These topics were not included under the Washington consensus because, chronic consensus-seeker as I am, I did not perceive that any particular view could come close to commanding a consensus in Washington. In some cases (e.g., as regards population control) I thought this was scandalous, while in other cases it struck me as quite natural. In most cases my personal views on these controversial issues are far removed from those of neoconservatives, so I find it ironic that some critics have condemned the Washington consensus as a neoconservative tract. I regard it rather – as Luiz Carlos Bresser Pereira said at Forlí – as an attempt to summarize the common core of wisdom embraced by all serious economists.

NOTES

Revised version of a paper presented to a conference organized by the Social Science Research Council (SSRC) and the European Center for the Study of Democratization (CESDE) on "Economic Liberalization and Democratic Consolidation" at the University of Bologna – Forlí on April 2–4, 1992. The author acknowledges the helpful comments of participants at that conference.

Reprint and permission requests should be sent to: Institute for International Economics, 11 Dupont Circle NW, Washington, DC 20036–1207, U.S.A.

1 I tried to complete the list from a normative standpoint in a recent paper: Williamson (1991).
2 Another of my original discussants, Stanley Fischer, challenged me on that, and argued that at least so far as the World Bank was concerned I should have added cost-effective social expenditures.
3 Perhaps Keynes's most unfortunate legacy to the economics profession was his habit of parodying orthodoxy. It was actually rather funny when he did it, and doubtless some of his targets deserved the treatment, but lesser followers reduced it to a cheap routine. The fact is that in most cases things become orthodox because they contain a large element of sense, so to dismiss something because it is orthodox is silly. I hope that this banal observation will not be interpreted as my saying that orthodoxy is always right or should never be challenged.
4 The summary under each heading is taken from the appendix to Williamson (1991).

REFERENCES

De Gregorio, Jose, "Economic growth in Latin America," *Journal of Development Economics* (forthcoming).

de Soto, Hernando, *The Other Path: The Invisible Revolution in the Third World* (New York: Harper and Row, 1989).

Okun, Arthur, *The Big Trade-off: Equity versus Efficiency* (Washington, DC: Brookings Institution, 1975).

Wiliamson, John, "In search of a manual for technopols" (Washington, DC: Institute for International Economics, mimeo, 1992).

Williamson, John, "Development strategy for Latin America in the 1990s" Paper presented at the Raul Prebisch Conference (Washington, DC: Inter-American Development Bank, November 1991).

Williamson, John, *Latin American Adjustment: How Much Has Happened?* (Washington, DC: Institute for International Economics 1990).

APPENDIX: THE SOCIAL WELFARE FUNCTION OF THE LEFT

Any civilized person would like to see social arrangements chosen in a way that would maximize a social welfare function of the general form

$$W = W(U_1, U_2, \ldots \ldots U_n),$$

with $W_i > 0$ for $i = 1 \ldots n$, the n individuals recognized as composing society in the sense that their welfare is relevant to the evaluation of social welfare, and where U_i is the *i*th individual's assessment of his or her own utility. In the past the left has devoted a lot of its energy to enlarging *n*, the domain of individuals recognized as being relevant in evaluating social welfare (from kings and lords to include the male propertied middle classes, the working class, serfs, women, slaves, foreigners, gypsies, natives . . . and perhaps in the future the effort will be to extend the domain of the welfare function to include species other than *Homo sapiens*).

The economic tradition has been to seek to avoid imposing more conditions on $W(\ldots)$ than that it be an increasing function of each of its arguments, on the ground that the latter requires only a very weak value judgment that will be accepted by all civilized people. This gives rise to the criterion of Pareto optimality, which defines a social improvement as occurring when the welfare of someone improves without a diminution in the welfare of anyone else. Since almost no social change could realistically be expected to pass this test, however, an attempt was made a half century ago to extend the criterion to be able to deal with realistic cases where the gains of some come at the expense of others. If the winners were to compensate the losers, the Pareto criterion would be satisfied. The "compensation test" suggested that an improvement could be defined to have occurred if the winners *could have* compensated the losers, even though no actual compensation was paid.

This leads to a very conservative, but commonly used, welfare criterion, in which an extra dollar is evaluated as being equally valuable to society no matter whether it accrues to a pauper or a millionaire. Society wants to maximize Σy_i. Paradoxically this criterion was developed by two left-of-center economists, Tibor Scitovsky and Nicholas Kaldor.

Let us overlook the technical objections to the compensation principle (notably its failure to pass the reversibility test) and assume that economic efficiency can be measured by the total value of income, Σy_i. Let us also agree that it is possible to measure the equity of income distribution by some measure such as the inverse of the variance of income (when income has been made comparable by transforming it to household per capita real income

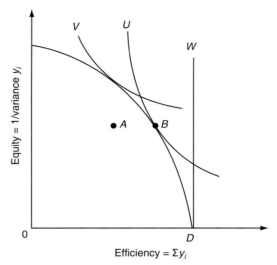

Figure 1 The Big Tradeoff.

by allowing for the differing real income needs of family members at different stages of the life cycle), 1/variance (y_i). Then the "big tradeoff" of Okun (1975) is the curve shown in Figure 1.

Political values can be characterized by the social welfare function used to evaluate alternative feasible outcomes. Assuming that we have resolved the issue of the domain of the welfare function, the most conservative civilized welfare function is that implied by the compensation principle and shown in Figure 1 by the vertical lines such as *W*. More egalitarian welfare functions are those shown by curves such as *U* or *V*, with *V* representing a more egalitarian set of preferences than *U*.

So long as left-wing parties are so misguided as to adopt policies that are populist, socialist, or protectionist they will tend to lead the economy to inefficient points such as *A*. Endorsement of the Washington consensus/universal convergence will lead them toward the frontier, say at *B*. Since fewer people suffer lower welfare at *B* than at *A*, relative to the conservative choice *D*, their electoral prospects will be enhanced to the extent that the electorate understands the implications of the parties' policy programs. Hence the conclusion that the jettisoning of the Marxist baggage that the left has been saddled with for the last 140 years or so should enhance the prospects of the left and open the way for a more vigorous confrontation on the fundamental issue of income distribution.

'Seeing Like a State: Conclusion'

from *Seeing Like a State: How Certain Schemes to Improve the Human Condition Have Failed* (1998)

James C. Scott

Editors' Introduction

James C. Scott was born in 1936 and is Sterling Professor of Political Science at Yale University. A past president of the (American) Association of Asian Studies, Scott's early publications were concerned with elite attitudes in Malaysia (the home of his principal field sites) and comparative political corruption. Scott came to wider attention in 1976 with his book, *The Moral Economy of the Peasant: Subsistence and Rebellion in Southeast Asia.* In this work, Scott insisted, in opposition to many Marxist and rational choice accounts, that the peasantry in the 'Third World' shared certain moral precepts about the world – including linked norms of hierarchy and reciprocity – that were distinctive and which signalled their less than full subordination to capitalism. In this book, and in *Weapons of the Weak: Everyday Forms of Peasant Resistance* (1985), Scott also began to set out his well-known account of the nature of peasant resistance/rebellion. Scott asked himself a seemingly simple question: if the linked processes of agrarian capitalism and the Green Revolution increasingly were promoting economic and social differentiation among the Malaysian peasantry, why were organised peasant rebellions so thin on the ground? The question had particular force because many Marxists in the 1970s and 1980s expected to see – and thought they were seeing – organised peasant revolts in South and South-East Asia. Writing of India, the activist-scholar Hari Sharma likened the Green Revolution there to a water melon: green on the outside but red inside. Scott came to disagree. In his view, peasants were wise not to protest too visibly or to confront their more heavily armed antagonists (landlords, state officials) too directly. They tended to lose on such occasions. Their rebellions were often put down violently by superior force. But this did not mean that the peasantry was quiescent. The real message of *Weapons of the Weak* is that poor men and women fight back by foot-dragging, joke-telling, cheating and dissimulation. These are the classical politics, Scott insists, of subaltern groups, and we have to learn to read these 'hidden transcripts' of resistance.

In the late 1990s, Scott returned to the question of 'seeing political actions', but this time at a grander scale and in regard to state actions to force visibility on their subjects-citizens. In *Seeing Like a State: How Certain Schemes to Improve the Human Condition Have Failed* (1998), Scott develops a broader argument that will be familiar to readers of Burke, Orwell and Koestler. He notes, in particular, that various high modernist states in the twentieth century caused enormous damage to their populations and environments by misguidedly pursuing ideologies of simplification and perfection. In pursuit of the perfectly planned city they created anomie and alienation; in pursuit of scientific forestry systems and a perfectly tamed nature they produced biodiversity loss and an ecological backlash. Scott is not against 'development' per se, but his work is marvellously instructive on the sometimes catastrophic effects of projects of social and environmental engineering that ride roughshod over the complexity of the 'real world' and the organic local knowledge (*mētis*) that ordinary people (peasants, city dwellers) come to have of it. In this reading we see Scott summing up this argument. Quite memorably, we think, he berates the arrogance of those who promise us a future of conforming order untempered by nonconforming practice.

Key references

James C. Scott (1998) *Seeing Like a State: How Certain Schemes to Improve the Human Condition Have Failed*, New Haven, CT: Yale University Press.
— (1968) *Political Ideology in Malaysia: Reality and Beliefs of an Elite*, New Haven, CT: Yale University Press.
— (1976) *The Moral Economy of the Peasant: Subsistence and Rebellion in Southeast Asia*, New Haven, CT: Yale University Press.
— (1985) *Weapons of the Weak: Everyday Forms of Peasant Resistance*, New Haven, CT: Yale University Press.
— (1990) *Domination and the Arts of Resistance: Hidden Transcripts*, New Haven, CT: Yale University Press.

They would reconstruct society on an imaginary plan, much like the astronomers for their own calculation would make over the system of the universe.

Pierre-Joseph Proudhon, *on the utopian socialists*

Yet a man who uses an imaginary map, thinking that it is a true one, is likely to be worse off than someone with no map at all; for he will fail to inquire whenever he can, to observe every detail on his way, and to search continuously with all his senses and all his intelligence for indications of where he should go.

E. F. Schumacher, *Small Is Beautiful*

The great high-modernist episodes that we have examined qualify as tragedies in at least two respects. First, the visionary intellectuals and planners behind them were guilty of hubris, of forgetting that they were mortals and acting as if they were gods. Second, their actions, far from being cynical grabs for power and wealth, were animated by a genuine desire to improve the human condition – a desire with a fatal flaw. That these tragedies could be so intimately associated with optimistic views of progress and rational order is in itself a reason for a searching diagnosis. Another reason lies in the completely ecumenical character of the high-modernist faith. We encounter it in various guises in colonial development schemes, planned urban centers in both the East and the West, collectivized farms, the large development plans of the World Bank, the resettlement of nomadic populations, and the management of workers on factory floors.

If such schemes have typically taken their most destructive human and natural toll in the states of the former socialist bloc and in revolutionary Third World settings, that is surely because there authoritarian state power, unimpeded by representative institutions, could nullify resistance and push ahead. The ideas behind them, however, on which their legitimacy and appeal depended, were thoroughly Western. Order and harmony that once seemed the function of a unitary God had been replaced by a similar faith in the idea of progress vouchsafed by scientists, engineers, and planners. Their power, it is worth remembering, was least contested at those moments when other forms of coordination had failed or seemed utterly inadequate to the great tasks at hand: in times of war, revolution, economic collapse, or newly won independence. The plans that they hatched bore a family resemblance to the schemes of legibility and standardization devised by the absolutist kings of the seventeenth and eighteenth centuries. What was wholly new, however, was the magnitude of both the plans for the wholesale transformation of society and the instruments of statecraft – censuses, cadastral maps, identity cards, statistical bureaus, schools, mass media, internal security apparatuses – that could take them farther along this road than any seventeenth-century monarch would have dreamed. Thus it has happened that so many of the twentieth century's political tragedies have flown the banner of progress, emancipation, and reform.

We have examined in considerable detail how these schemes have failed their intended beneficiaries. If I were asked to condense the reasons behind these failures into a single sentence, I would say that the progenitors of such plans regarded themselves as far smarter and farseeing than they really were and, at the same time, regarded their subjects as far more stupid and incompetent than *they* really were. The remainder of this chapter is devoted to expanding on this cursory judgment and advancing a few modest lessons.

"IT'S IGNORANCE, STUPID!"

> The mistake of our ancestors was to think that they were "the last number," but since numbers are infinite, they could not be the last number.
>
> Eugene Zamiatin, *We*

The maxim that serves as the heading for this section is not simply suitable for bumper stickers mimicking the insider slogan of Bill Clinton's 1992 presidential campaign, "It's the economy, stupid!" It is meant to call attention to how routinely planners ignore the radical contingency of the future. How rare it is to encounter advice about the future which *begins* from a premise of incomplete knowledge. One small exception – a circular on nutrition published by the health clinic at Yale University, where I teach – will underscore its rarity. Normally, such circulars explain the major food groups, vitamins, and minerals known to be essential for balanced nutrition and advise a diet based on these categories. This circular, however, noted that many new, essential elements of proper nutrition had been discovered in the past two decades and that many more elements will presumably be identified by researchers in the decades ahead. Therefore, *on the basis of what they did not know*, the writers of this piece recommended that one's diet be as varied as possible, on the prudent assumption that it would contain many of these yet unidentified essentials.

Social and historical analyses have, almost inevitably, the effect of diminishing the contingency of human affairs. A historical event or state of affairs simply *is* the way it is, often appearing determined and necessary when in fact it might easily have turned out to be otherwise. Even a probabilistic social science, however careful it may be about establishing ranges of outcomes, is apt to treat these probabilities, for the sake of analysis, as solid facts. When it comes to betting on the future, the contingency is obvious, but so is the capacity of human actors to influence this contingency and help to shape the future. And in those cases where the bettors thought that they knew the shape of the future by virtue of their grasp of historical laws of progress or scientific truth, whatever awareness they retained of the contingency seemed to dissolve before their faith.

And yet each of these schemes, as might also have been predicted, was largely undone by a host of contingencies beyond the planners' grasp. The scope and comprehensiveness of their plans were such that they would have had indeterminate outcomes even if their historical laws and the attendant specification of variables and calculations had been correct. Their temporal ambitions meant that although they might, with some confidence, guess the immediate consequences of their moves, no one could specify, let alone calculate, the second- or third-order consequences or their interaction effects. The wild cards in their deck, however, were the human and natural events outside their models – droughts, wars, revolts, epidemics, interest rates, world consumer prices, oil embargoes. They could and did, of course, attempt to adjust and improvise in the face of these contingencies. But the magnitude of their initial intervention was so great that many of their missteps could not be righted. Stephen Marglin has put, their problem succinctly: If "the only certainty about the future is that the future is uncertain, if the only sure thing is that we are in for surprises, then no amount of planning, no amount of prescription, can deal with the contingencies that the future will reveal."[1]

There is a curiously resounding unanimity on this point, and on no others, between such right-wing critics of the command economy as Friedrich Hayek and such left-wing critics of Communist authoritarianism as Prince Peter Kropotkin, who declared, "It is impossible to legislate for the future." Both had a great deal of respect for the diversity of human actions and the insurmountable difficulties in successfully coordinating millions of transactions. In a blistering critique of failed development paradigms, Albert Hirschman made a comparable case, calling for "a little more 'reverence for life,' a little less straitjacketing of the future, a little more allowance for the unexpected – and a little less wishful thinking."[2]

One might, on the basis of experience, derive a few rules of thumb that, if observed, could make development planning less prone to disaster. While my main goal is hardly a point-by-point reform of development practice, such rules would surely include something along the following lines.

Take small steps. In an experimental approach to social change, presume that we cannot know the consequences of our interventions in advance. Given this postulate of ignorance, prefer wherever possible to take a small step, stand back, observe, and then plan the next small move. As the biologist J. B. S. Haldane metaphorically described the advantages of smallness: "You can drop a mouse down a thousand-yard mineshaft; and on arriving at the bottom, it gets a slight shock and walks away. A rat is killed, a man broken, a horse splashes."[3]

Favor reversibility. Prefer interventions that can easily be undone if they turn out to be mistakes.[4] Irreversible interventions have irreversible consequences.[5] Interventions into ecosystems require particular care in this respect, given our great ignorance about how they interact. Aldo Leopold captured the spirit of caution required: "The first rule of intelligent tinkering is to keep all the parts."[6]

Plan on surprises. Choose plans that allow the largest accommodation to the unforeseen. In agricultural schemes this may mean choosing and preparing land so that it can grow any of several crops. In planning housing, it would mean "designing in" flexibility for accommodating changes in family structures or living styles. In a factory it may mean selecting a location, layout, or piece of machinery that allows for new processes, materials, or product lines down the road.

Plan on human inventiveness. Always plan under the assumption that those who become involved in the project later will have or will develop the experience and insight to improve on the design.

PLANNING FOR ABSTRACT CITIZENS

The power and precision of high-modernist schemes depended not only on bracketing contingency but also on standardizing the subjects of development. Some standardization was implicit even in the noblest goals of the planners. The great majority of them were strongly committed to a more egalitarian society, to meeting the basic needs of its citizens (especially the working class), and to making the amenities of a modern society available to all.

Let us pause, however, to consider the kind of human subject for whom all these benefits were being provided. This subject was singularly abstract. Figures as diverse as Le Corbusier, Walther Rathenau, the collectivizers of the Soviet Union, and even Julius Nyerere (for all his rhetorical attention to African traditions) were planning for generic subjects who needed so many square feet of housing space, acres of farmland, liters of clean water, and units of transportation and so much food, fresh air, and recreational space. Standardized citizens were uniform in their needs and even interchangeable. What is striking, of course, is that such subjects – like the "unmarked citizens" of liberal theory – have, for the purposes of the planning exercise, no gender, no tastes, no history, no values, no opinions or original ideas, no traditions, and no distinctive personalities to contribute to the enterprise. They have none of the particular, situated, and contextual attributes that one would expect of any population and that we, as a matter of course, always attribute to elites.

The lack of context and particularity is not an oversight; it is the necessary first premise of any large-scale planning exercise. To the degree that the subjects can be treated as standardized units, the power of resolution in the planning exercise is enhanced. Questions posed within these strict confines can have definitive, quantitative answers. The same logic applies to the transformation of the natural world. Questions about the volume of commercial wood or the yield of wheat in bushels permit more precise calculations than questions about, say, the quality of the soil, the versatility and taste of the grain, or the well-being of the community.[7] The discipline of economics achieves its formidable resolving power by transforming what might otherwise be considered qualitative matters into quantitative issues with a single metric and, as it were, a bottom line: profit or loss.[8] Providing one understands the heroic assumptions required to achieve this precision and the question that it cannot answer, the single metric is an invaluable tool. Problems arise only when it becomes hegemonic.

What is perhaps most striking about high-modernist schemes, despite their quite genuine egalitarian and often socialist impulses, is how little confidence they repose in the skills, intelligence, and experience of ordinary people. This is clear enough in the Taylorist factory, where the logic of work organization is to reduce the factory hands' contribution to a series of repetitive, if practiced, movements – operations as machinelike as possible. But it is also clear in collectivized farms, ujamaa villages, and planned cities, where the movements of the populace have been to a large degree inscribed in the designs of these communities. If Nyerere's aspirations for cooperative state farming were frustrated, it was not because the plans had failed to integrate a scheme of cooperative labor. The more ambitious and meticulous the plan, the less is left, theoretically, to chance and to local initiative and experience.

STRIPPING REALITY TO ITS ESSENTIALS

The quantitative technologies used to investigate social and economic life work best if the world they aim to describe can be remade in their image.
Theodore M. Porter, *Trust in Numbers*

If the facts – that is, the behavior of living human beings – are recalcitrant to such an experiment, the experimenter becomes annoyed and tries to alter the facts to fit the theory, which, in practice, means a kind of vivisection of societies until they become what the theory originally declared that the experiment should have caused them to be.
Isaiah Berlin, "*On Political Judgment*"

The clarity of the high-modernist optic is due to its resolute singularity. Its simplifying fiction is that, for any activity or process that comes under its scrutiny, there is only one thing going on. In the scientific forest there is only commercial wood being grown; in the planned city there is only the efficient movement of goods and people; in the housing estate there is only the effective delivery of shelter, heat, sewage, and water; in the planned hospital there is only the swift provision of professional medical services. And yet both we and the planners know that each of these sites is the intersection of a host of interconnected activities that defy such simple descriptions. Even something as apparently monofunctional as a road from *A* to *B* can at the same time function as a site for leisure, social intercourse, exciting diversions, and enjoying the view *between A* and *B*.[9]

For any such site, it is helpful to imagine two different maps of activity. In the case of a planned urban neighborhood, the first map consists of a representation of the streets and buildings, tracing the routes that the planners have provided for the movements between workplaces and residences, the delivery of goods, access to shopping, and so on. The second map consists of tracings, as in a time-lapse photograph, of all the *unplanned* movements – pushing a baby carriage, window shopping, strolling, going to see a friend, playing hopscotch on the sidewalk, walking the dog, watching the passing scene, taking short-cuts between work and home, and so on. This second map, far more complex than the first, reveals very different patterns of circulation. The older the neighborhood, the more likely that the second map will have nearly superseded the first, in roughly the same way that planned, suburban Levittowns have, after fifty years, become thoroughly different settings from what their designers envisioned.

If our inquiry has taught us anything, it is that the first map, taken alone, is misrepresentative and indeed nonsustainable. A same-age, monocropped forest with all the debris cleared is in the long run an ecological disaster. No Taylorist factory can sustain production without the unplanned improvisations of an experienced workforce. Planned Brasília is, in a thousand ways, underwritten by unplanned Brasília. Without at least some of the diversity identified by Jacobs, a stripped-down public housing project (like Pruitt-Igoe in Saint Louis or Cabrini Green in Chicago) will fail its residents. Even for the limited purposes of a myopic plan – commercial timber, factory output – the one-dimensional map will simply not do. As with industrial agriculture and its dependency on landraces, the first map is possible only because of processes lying outside its parameters, which it ignores at its peril.

Our inquiry has also taught us that such maps of legibility and control, especially when they are backed by an authoritarian state, *do* partly succeed in shaping the natural and social environment after their image. To the degree that such thin maps do manage to impress themselves on social life, what kind of people do they foster? Here I would argue that just as the monocropped, same-age forest represents an impoverished and unsustainable ecosystem, so the high-modernist urban complex represents an impoverished and unsustainable social system.

Human resistance to the more severe forms of social straitjacketing prevents monotonic schemes of centralized rationality from ever being realized.

Had they been realized in their austere forms, they would have represented a very bleak human prospect. One of Le Corbusier's plans, for example, called for the segregation of factory workers and their families in barracks along the major transportation arteries. It was a theoretically efficient solution to transportation and production problems. If it had been imposed, the result would have been a dispiriting environment of regimented work and residence without any of the animation of town life. This plan had all the charm of a Taylorist scheme where, using a comparable logic, the efficient organization of work was achieved by confining the workers' movements to a few repetitive gestures. The cookie-cutter design principles behind the layout of the Soviet collective farm, the ujamaa village, or the Ethiopian resettlement betray the same narrowness of vision. They were designed, above all, to facilitate the central administration of production and the control of public life.

Almost all strictly functional, single-purpose institutions have some of the qualities of sensory-deprivation tanks used for experimental purposes. At the limit, they approach the great social control institutions of the eighteenth and nineteenth centuries: asylums, workhouses, prisons, and reformatories. We have learned enough of such settings to know that over time they can produce among their inmates a characteristic institutional neurosis marked by apathy, withdrawal, lack of initiative and spontaneity, uncommunicativeness, and intractability. The neurosis is an accommodation to a deprived, bland, monotonous, controlled environment that is ultimately stupifying.[10]

The point is simply that high-modernist designs for life and production tend to diminish the skills, agility, initiative, and morale of their intended beneficiaries. They bring about a mild form of this institutional neurosis. Or, to put it in the utilitarian terms that many of their partisans would recognize, these designs tend to reduce the "human capital" of the workforce. Complex, diverse, animated environments contribute, as Jacobs saw, to producing a resilient, flexible, adept population that has more experience in confronting novel challenges and taking initiative. Narrow, planned environments, by contrast, foster a less skilled, less innovative, less resourceful population. This population, once created, would ironically have been exactly the kind of human material that would in fact have needed close supervision from above. In other words, the logic of social engineering on this scale was to produce the sort of subjects that its plans had assumed at the outset.

That authoritarian social engineering failed to create a world after its own image should not blind us to the fact that it did, at the very least, damage many of the earlier structures of mutuality and practice that were essential to metis. The Soviet kolkhoz hardly lived up to its expectations, but by treating its workforce more like factory hands than farmers, it did destroy many of the agricultural skills the peasantry had possessed on the eve of collectivization. Even if there was much in the earlier arrangements that ought to have been abolished (local tyrannies based on class, gender, age, and lineage), a certain institutional autonomy was abolished as well. Here, I believe, there is something to the classical anarchist claim – that the state, with its positive law and central institutions, undermines individuals' capacities for autonomous self-governance – that might apply to the planning grids of high modernism as well. Their own institutional legacy may be frail and evanescent, but they may impoverish the local wellsprings of economic, social, and cultural self-expression.

THE FAILURE OF SCHEMATICS AND THE ROLE OF MĒTIS

> Everything is said to be under the leadership of the Party. No one is in charge of the crab or the fish, but they are all alive.
>
> Vietnamese villager, *Xuan Huy village*

Not long after the decisive political opening in 1989, in what was then still the Soviet Union, a congress of agricultural specialists was convened to consider reforms in agriculture. Most participants were in favor of breaking up the collectives and privatizing the land in the hope of recreating a modern version of the private sector that had thrived in the 1920s and that Stalin had destroyed in 1930. And yet they were nearly unanimous in their despair over what three generations had done to the skills, initiative, and knowledge of the kolkhozniki. They compared their situation unfavorably to that of China, where a mere twenty-five years of collectivization had, they imagined, left much of the entrepreneurial skill of the peasantry intact. Suddenly a woman from Novosibirsk scolded them: "How do you think the rural people survived sixty years of collectivization in the first place? If they hadn't used their initiative and wits, they wouldn't have made it through! They may need credit and supplies, but there's nothing wrong with their initiative."[11]

Despite the manifold failures of collectivization, it seems, the kolkhozniki had found ways and means to at least get by. We should not forget in this context that the first response to collectivization in 1930 was determined resistance and even rebellion. Once that resistance was broken, the survivors had little choice but to comply outwardly. They could hardly make the rural command economy a success, but they could do what was necessary to meet minimal quotas and ensure their own economic survival.

An indication of the kinds of improvisations both tolerated and required may be inferred from an astute case study of two East German factories before the Wall came down in 1989.[12] Each factory was under great pressure to meet production quotas – on which their all-important bonuses depended – in spite of old machinery, inferior raw materials, and a lack of spare parts. Under these draconian conditions, two employees were indispensable to the firm, despite their modest place in the official hierarchy. One was the jack-of-all-trades who improvised short-term solutions to keep machinery running, to correct or disguise production flaws, and to make raw materials stretch further. The second was a wheeler-dealer who located and bought or bartered for spare parts, machinery, and raw material that could not be obtained through official channels in time. To facilitate the wheeler-dealer's work, the factory routinely used its funds to stock up on such valued nonperishable goods as soap powder, cosmetics, quality paper, yarn, good wine and champagne, medicines, and fashionable clothes. When it seemed that the plant would fall short of the quota because it lacked a key valve or machine tool, these knowledgeable dealers would set off across the country, their small Trabant autos jammed with barter goods, to secure what was needed. Neither of these roles was provided for in the official table of organization, and yet the survival of the factory depended more on their skills, wisdom, and experience than on those of any other employee. A key element in the centrally planned economy was underwritten, always unofficially, by mētis.

Cases like the one just described are the rule, not the exception. They serve to illustrate that the formal order encoded in social-engineering designs inevitably leaves out elements that are essential to their actual functioning. If the factory were forced to operate only within the confines of the roles and functions specified in the simplified design, it would quickly grind to a halt. Collectivized command economies virtually everywhere have limped along thanks to the often desperate improvisation of an informal economy wholly outside its schemata.

Stated somewhat differently, all socially engineered systems of formal order are in fact subsystems of a larger system on which they are ultimately dependent, not to say parasitic. The subsystem relies on a variety of processes – frequently informal or antecedent – which alone it cannot create or maintain. The more schematic, thin, and simplified the formal order, the less resilient and the more vulnerable it is to disturbances outside its narrow parameters. This analysis of high-modernism, then, may appear to be a case for the invisible hand of market coordination as opposed to centralized economies. An important caution, however, is in order. The market is itself an instituted, formal system of coordination, despite the elbow room that it provides to its participants, and it is therefore similarly dependent on a larger system of social relations which its own calculus does not acknowledge and which it can neither create nor maintain. Here I have in mind not only the obvious elements of contract and property law, as well as the state's coercive power to enforce them, but antecedent patterns and norms of social trust, community, and cooperation, without which market exchange is inconceivable. Finally, and most important, the economy is "a subsystem of a finite and nongrowing eco-system," whose carrying capacity and interactions it must respect as a condition of its persistence.[13]

It is, I think, a characteristic of large, formal systems of coordination that they are accompanied by what appear to be anomalies but on closer inspection turn out to be integral to that formal order. Much of this might be termed "mētis to the rescue," although for people ensnared in schemes of authoritarian social engineering that threaten to do them in, such improvisations bear the mark of scrambling and desperation. Many modern cities, and not just those in the Third World, function and survive by virtue of slums and squatter settlements whose residents provide essential services. A formal command economy, as we have seen, is contingent on petty trade, bartering, and deals that are typically illegal. A formal economy of pension systems, social security, and medical benefits is underwritten by a mobile, floating population with few of these protections. Similarly, hybrid crops in mechanized farm operations persist only because of the diversity and immunities of antecedent landraces. In each case, the nonconforming practice is an indispensable condition for formal order.

[. . .]

NOTES

1 Stephen A. Marglin, "Economics and the Social Construction of the Economy," in Stephen Gudeman and Stephen Marglin, eds., *People's Ecology, People's Economy* (forthcoming).

2 Albert O. Hirschman, "The Search for Paradigms as a Hindrance to Understanding," *World Politics* 22 (April 1970): 239. Elsewhere Hirschman takes social science in general to task in much the same fashion: "But after so many failed prophecies, is it not in the interest of social science to embrace complexity, be it at some sacrifice of its claim to predictive power?" ("Rival Interpretations of Market Society: Civilizing, Destructive, or Feeble?" *Journal of Economic Literature* 20 [December 1982]: 1463–84).

3 Quoted in Roger Penrose, "The Great Diversifier," a review of Freeman Dyson, *From Eros to Gaia*, in the *New York Review of Books*, March 4, 1993, p. 5.

4 Like all rules of thumb, this rule is not absolute. It could be waived, for example, if catastrophe seems imminent and quick decisions are essential.

5 This is, I believe, the strongest argument against capital punishment for those who are not opposed to it on other grounds.

6 Aldo Leopold, quoted in Donald Worster, *Nature's Economy*, 2nd ed. (New York: Cambridge University Press, 1994), p. 289.

7 The typical social science solution to this sort of issue is to turn it into a quantitative exercise by, say, asking citizens to assess the well-being of the community on a predetermined scale.

8 "Everything becomes crystal clear after you have reduced reality to one – one only – of its thousand aspects. You know what to do. . . . There is at the same time the perfect measuring rod for the degree of success or failure. . . . The point is that the real strength of the theory of private enterprise lies in its ruthless simplification, which fits so admirably into the mental patterns created by the phenomenal successes of science. The strength of science too derives from its 'reduction' of reality to one or another of its many aspects, primarily the reduction of quality to quantity" (E. F. Schumacher, *Small Is Beautiful: A Study of Economics as if People Mattered* [London: Blond and Briggs, 1973], pp. 272–73).

9 See John Brinckerhoff Jackson, *A Sense of Place, a Sense of Time* (New Haven: Yale University Press, 1994), p. 190.

10 For this insight I am much indebted to Colin Ward's *Anarchy in Action* (London: Freedom Press, 1988), pp. 110–25.

11 Personal notes from the first congress of the Agrarian Scientists' Association, "Agrarian Reform in the USSR," held in Moscow, June 24–28, 1991.

12 Birgit Müller, *Toward an Alternative Culture of Work: Political Idealism and Economic Practices in a Berlin Collective Enterprise* (Boulder: Westview Press, 1991), pp. 51–82.

13 Herman E. Daly, "Policies for Sustainable Development," paper presented at the Program in Agrarian Studies, Yale University, New Haven, February 9, 1996, p. 4.

'The Irrelevance of Development Studies'

Third World Quarterly (1989)

Michael Edwards

Editors' Introduction

Michael Edwards, a geographer by training, is currently Director of the Ford Foundation's Governance and Civil Society Unit in New York City. Before moving to the United States from the UK, Edwards worked for many years for Oxfam, including in Zambia, and for Save the Children. He is widely recognised as a leading authority on non-governmental organisations (NGOs) and civil society (on which topic he worked for the World Bank in the 1990s). Edwards is a prolific writer and his book from 1992 with David Hulme – *Making a Difference: NGOs and Development in a Changing World* – is still widely cited. Other books include the edited collections *Beyond the Bullet: NGO Performance and Accountability in the Post-Cold War World* and *NGOs, States and Donors: Too Close for Comfort?* (1997) (both with David Hulme as his co-editor), and two monographs from 2004, *Civil Society* and *Future Positive: International Cooperation in the Twenty-First Century*.

Edwards' recent work has struck an upbeat note. If one of the main arguments of *Future Positive* is that international cooperation was a feature of the global system c.1945–50 and has since been corrupted by greed and self-interest, a second is that international cooperation can be rebuilt by well-meaning actors and institutions. This optimism, however, is underpinned by the streak of pragmatism that runs through Edwards' work and which is well to the fore in the reading here, from his well-known essay on 'The irrelevance of development studies'. Edwards is clear that NGOs and civil societies have to be rigorously self-critical. They stand to benefit from the sorts of protocols that dominate (or that are meant to dominate) in the academy. But he also insists on a two-way street. Edwards has little time for academic theorising which can too often seem like navel-gazing. Although his call for practical, 'relevant' academic work has been much criticised since the time of its publication – relevant to whom and how, his critics ask – this 1989 essay continues to be widely referred to in the broad development community.

Key references

Michael Edwards (1989) 'The irrelevance of development studies', *Third World Quarterly* 11: 116–35.

— (1996) *Beyond the Bullet: NGO Performance and Accountability in the Post-Cold War World*, Bloomfield, CT: Kumarian Press.

— (2004) *Civil Society*, Cambridge: Polity.

— (2004) *Future Positive: International Cooperation in the Twenty-First Century*, London: Earthscan.

Michael Edwards and D. Hulme (1992) *Making a Difference: NGOs and Development in a Changing World*, London: Earthscan.

D. Hulme and Michael Edwards (eds) (1997) *NGOs, States and Donors: Too Close for Comfort?* London: Macmillan.

▨ ▨ ▨ ▪ ▨ ▨

Development, however defined, is a slow and uneven process. Indeed, in some areas we seem to be going backwards rather than forwards. In material terms, food production per capita may be declining throughout much of sub-Saharan Africa, while real incomes stagnate and the provision of basic services continues to lag behind population growth.[1] In terms of people's control over the forces which shape their lives (a definition of development that I prefer), progress seems equally fragile. The world's poor remain very much in the grip of national and international forces over which they have little influence. Yet, at the same time, we face an unprecedented expansion in the quantity of development research being undertaken, advice being proffered, and projects being financed. As Lloyd Timberlake has pointed out, advising Africa has become a major industry, with at least 80,000 expatriates at work south of the Sahara at a cost of more than US$4 billion a year.[2] The volume of magazines, books, periodicals and papers concerning development continues to increase, while there are so many conferences on development that attendance has become almost a profession in itself.

Clearly, the links between research and development are complex, and few would posit a simple, linear relationship between the two. However, the fact that this immense outpouring of information and advice is having demonstrably little effect on the problems it seeks to address should at least give us cause for concern. Why is it that our increasing knowledge of the Third World does not enable solutions to be found? Is this because practitioners refuse to listen? Is 'development' a matter to be left to practitioners anyway? Are there other, stronger forces that prevent the right action being taken? Or could it be, in Paul Devitt's words, that 'our kind of knowledge is simply not enough'?[3]

This article looks at one aspect of the relationship between research and development which appears to me to underlie our current predicament: the absence of strong links between *understanding* and *action*. The weakness (indeed the non-existence) of these links makes much current thinking irrelevant, in the literal sense, to the problems it pretends to address. As a result, many of us who are involved in the field of development studies have become part of the problems of underdevelopment, rather than being part of the solutions to these problems.

Pleas for change in the methodology of development studies are not new. Indeed, much of the argument contained in this article has already been stated elsewhere.[4] However, these new approaches appear to be making little headway, perhaps because the politics of development studies are against them. There seems to me to be a need to reinforce constantly the message that fundamental changes are essential. The first section of this article looks at conventional approaches to development studies and outlines some of the factors which underlie their failure to come to grips with the problems they seek to address. In the next section I look at how development studies might be restructured so as to increase their usefulness, focusing on the methodology of 'participatory' or 'action' research and the practical issues surrounding its use. By way of conclusion, I question whether such a restructuring is possible without much deeper changes in attitudes and values among academics and practitioners alike. Throughout, the term 'development studies' is used to cover all forms of writing and talking about development, as well as forms of action (such as training and even project work) which grow out of these processes. When I use the terms 'the poor', 'poor people', and 'popular participation', I am referring specifically to the most vulnerable and exploited groups in communities and societies.

CONVENTIONAL APPROACHES TO DEVELOPMENT STUDIES

Although some progress has been made in exploring alternative approaches, inspired particularly by the early work of Paulo Freire, development studies are still based largely on traditional 'banking' concepts of education. These traditional concepts embody a series of attitudes that contribute to the irrelevance of much of their output to the problems of the world in which we live. Most importantly, people are treated as objects to be studied rather than as subjects of their own development; there is therefore a separation between the researcher and the object of research, and between understanding and action. Research and education come to be dominated by content rather than form or method; they become processes which focus on the transmission of information, usually of a technical kind, from one person to another. The 'transmitter' and the 'receiver' of information are distanced from each other by a basic inequality in the amount of technical knowledge they each possess. The most extreme example of this process is the 'empirical' questionnaire-based survey designed, analysed and controlled by people outside the community which is being studied. However, the same attitudes pervade most forms of writing and talking about development to one degree or another. As Nicholas Maxwell puts it, 'Insofar as academic inquiry does try to help to promote human welfare, it does so . . . by seeking to improve knowledge of various aspects of the world.'[5] Let us look at five crucial types of problem which arise when such an approach is adopted.

Experts, 'expats', and the devaluation of popular knowledge

The natural consequence of a concern for technical interpretations of reality is that knowledge, and the power to control it, become concentrated in the hands of those with the technical skills necessary to understand the language and methods being used.[6] The idea that development consists of a transfer of skills or information creates a role for the expert as the only person capable of mediating the transfer of these skills from one person or society to another. Herein lies the justification, if justification it is, for the 80,000 expatriate 'experts' at work south of the Sahara today. They are there to promote 'development', defined implicitly as a transfer of knowledge from 'developed' to 'underdeveloped' societies. Yet this 'expert' status is usually quite spurious. As Adrian Adams has pointed out, 'In Britain a doctor is a doctor; he'll be a medical expert if he goes to help halve the birthrate in Bangladesh . . . what matters is the halo of impartial prestige his skills lend him,

allowing him to neutralise conflict-laden encounters . . . and disguise political issues, for a time, as technical ones.'[7]

Of course, this is not simply a problem of the North. The influence of such ideas also permeates much official thinking about development in the Third World, through the elites that mimic the behaviour of their counterparts in Europe and North America. Anyone who has discussed development issues with African extension workers in health or agriculture will recognise immediately the deadening effect of conventional approaches to education and training, passed down to the grassroots by successive levels of a hierarchy schooled in the language and methods of the expert. Yet, as a report from the Massachusetts Institute of Technology points out, 'The fundamental problem confronting agriculture is not the adoption . . . of any particular set of inputs or economic arrangements or of organisational patterns . . . . Rather, it is to build in an attitude of experiment, trial and error, innovation and the adoption of new ideas.'[8] This conclusion is not unique to agriculture. In all sectors of development, the adoption of problem-solving approaches is much more important than communicating particular packages of technical information. If people can analyse, design, implement and evaluate their work in a critical fashion, they stand a good chance of achieving their objectives. However, a system of education and training that relies on experts will never be able to do this, because the attitudes of the expert prevent people from thinking for themselves. To quote Nicholas Maxwell again, 'Whereas for the philosophy of knowledge, the fundamental kind of rational learning is acquiring knowledge, for the philosophy of wisdom the fundamental kind of rational learning is learning how to live, how to see, to experience, to participate in and to create what is of value in existence.'[9] I shall return to this theme later in my argument.

This is not a problem unique to research, education and training. It also underlies the dangerous obsession with 'projects' that characterises the work of most development agencies. The logical corollary of a world-view which sees development as a series of technical transfers mediated by experts is that, given a sufficient number of situations, or projects, in which these transfers can be made, 'development' will occur. But, as Sithembiso Nyoni has pointed out, no country in the world has ever developed itself through projects;[10] development results from a long process of experiment and innovation through which people build up the skills, knowledge and self-confidence necessary to shape their environment in ways which foster progress toward goals such as economic growth, equity in income distribution, and political freedom. At root then, development is about processes of enrichment, empowerment and participation, which the technocratic, project-oriented view of the world simply cannot accommodate.[11]

A further consequence of the predominant technical view of development is the devaluation of indigenous knowledge (which grows out of the direct experience of poor people) in the search for solutions to the problems that face us. This is inevitable if knowledge is associated with formal education and training. The result is that general solutions manufactured from the outside are offered to specific problems which are highly localised. The practice of development work teaches us that problems are often specific in their complexity to particular times and places. A number of recent studies have recognised this in their exploration of the complex relationships which evolve over time between people and their environment within geographically-restricted areas.[12] These relationships are dynamic, seasonal and often unpredictable. They cannot be subjected to general models and it is therefore difficult for conventional approaches to knowledge to accommodate them. What is required, as Paul Richards has argued, is a 'people's science' which uses local knowledge to explore local solutions to local problems.[13] Such a 'science' differs radically in approach from the traditional one, which ignores indigenous knowledge or relegates it to a subordinate position. To take just one example: in his investigations of the relationship between nutrition and cash-cropping in Zambia's Northern Province, Barrie Sharpe has found tremendous diversity in the informal networks of exchange and innovation that have evolved over hundreds of years to ensure peoples' survival in the face of hostile environmental conditions.[14] These networks are highly localised. Indigenous knowledge of this kind could have been used as the basis of a successful development policy for the region. Instead, government and aid agencies have applied pressure to commercialise the cultivation of maize on a large scale. This is threatening the survival of these informal networks, and with it, the nutrition of children in families which had previously found a better balance between the conflicting demands of food, cash and welfare.

A final consequence of the technocratic approach to knowledge is a refusal to accept the role of *emotion* in understanding the problems of development. Development research is full of a spurious objectivity; this is a natural consequence of divorcing subject from object in the process of education. Any hint of 'subjectivity' is seized upon immediately as 'unscientific' and therefore not worthy of inclusion in 'serious' studies of development. Yet it is impossible to understand real-life problems fully unless we can grasp the multitude of constraints, imperfections and emotions that shape the actions and decisions of real, living people. People act on issues about which they have strong feelings, 'so all education and development projects should start by identifying the issues which local people speak about with excitement, hope, fear, anxiety or anger'.[15] This is precisely what conventional research does not do, because it divorces itself from the everyday context within which an understanding of these emotions can develop. Unless, as Raymond Williams puts it, we move beyond the current separation of emotion and intelligence and accept emotion as 'a direct concern with people, the key to the new order', we will never develop a better understanding of the problems that face us.[16]

The net effect of the characteristics briefly explored above is to render technocratic approaches to development research, training and practice irrelevant to the problems they seek to address. Expert-oriented views of development distance the researcher from reality, create barriers which promote ignorance, and perpetuate inappropriate models based on the views of outsiders. The crucial issue of 'relevance' is explored in more detail below.

'First' and 'last' in values and priorities

Conventional approaches to development studies embody certain values, or 'mind sets', which act as a barrier to the genuine understanding of issues and problems; as Robert Chambers has written in detail about these mind sets, a brief summary of his conclusions will suffice here. Chambers emphasies that the 'values and preferences of first professionals are typically polar opposites of last realities'.[17] That is, whereas researchers and advisers tend to prefer the qualities of modernity, quantification, prediction and tidiness (to name but four), the reality of those being 'researched' reflects preferences for the traditional, non-quantifiable, unpredictable and messy. The mismatch between these two sets of preferences results in a series of biases in the perceiver that obscure a real understanding of the situation at hand. The 'gender blindness' of much current research and practice is another component of the same process. This is an inevitable consequence of the values and attitudes which we adopt as a result of long exposure to conventional approaches to education and training, and is reinforced by our inability (or unwillingness) to change these approaches. The 'rural development tourism' practised by many academics and development agencies is a classic example of such attitudes at work.

The values and attitudes embodied in the conventional approach are essentially selfish. As Odhiambo Anacleti has pointed out, much conventional research is useless 'because it is for the satisfaction of the researcher rather than the researched'.[18] Or, as a Senegalese villager put it to Adrian Adams, 'Les chercheurs ne cherchent pas la vérité'.[19] Yet the one characteristic which is essential if a genuine dialogue is to be established with poor people is humility; humility on the part of the researcher or practitioner towards the subject of his or her attention. This is impossible if people are treated as objects to be studied. It is the absence of humility that places many academics in attitudes of self-appointed superiority over people who are more directly involved in practical development work. Instead of cooperating, academics and practitioners often see themselves as adversaries rather than collaborators. This is one reason why links between research and practice are so weak (see below). We will continue to speak past each other instead of to each other until this sense of inequality is eradicated.

Lest development workers feel themselves immune to this criticism, we should remind ourselves that the same attitudes pervade the work of many governmental, multilateral, and even non-governmental organisations (NGOs). How many of us really possess the humility to learn from the poor? Working for an NGO is no guarantee that the right approach will be adopted. Many NGOs are subject to the same prejudices as elitist researchers. The first essential step toward greater relevance in

development studies is to change the way we think and act, so that we become able to listen and to learn 'from below'.

The monopoly of knowledge and the control of power

The field of development studies is dominated by the North and, to a lesser extent, by the Third World elites whom we have trained and sponsored. The prodigious output of books and journals from Northern universities rarely has any positive influence over the lives of poor people. This is partly because materials are priced so highly that even libraries in the Third World cannot afford to buy them. More importantly, even when it is accessible such material is often irrelevant because of the biases and misperceptions that form the subject-matter of this paper. The most complete divorce of all lies between research output and the subjects of this research – poor people themselves. The barriers created by jargon, language, literacy, price, availability and method create a situation where people are denied access to the information which is supposed to concern them. The usefulness of research then depends upon its effectiveness in changing attitudes among elite groups in a direction that will ultimately enable poor people to think and act for themselves. If research does not do this it serves only to perpetuate the monopoly over knowledge which lies at the root of the problem in the first place.

Northern domination over education and training is now being challenged by the growth of Third World centres of knowledge. Unfortunately, these institutions often take on exactly those characteristics which render their Northern counterparts ineffective as catalysts for development. They over-emphasise the acquisition of technical skills and fail to challenge the prejudices which prevent people from 'learning from below'. Thankfully, this situation is beginning to change as more institutions emphasising 'participatory' approaches to problem-solving find their voice. The growth of participatory research networks throughout the Third World is particularly encouraging.

Increasingly, 'popular participation' is accepted as the only real basis for successful development. In reality, however, the practice of development studies continues to be anti-participatory. This contradiction shows itself in the advocacy of participation by writers who do not allow the subjects of their research to participate, and by development agencies who parrot the virtues of participation while telling their partners in the Third World what to do and how to do it. Recipients of Northern NGO aid must demonstrate a high degree of internal democracy in order to qualify for assistance. Yet the donor agencies themselves would never qualify on these grounds; they consistently refuse to share their own power while insisting on 'popular participation' as the fundamental criterion for development work overseas. This is the same form of hypocrisy that allows Northern thinkers to use the Third World as a laboratory for social, economic, and political experiments which they are unable (or unwilling) to conduct in their own societies. Development has become a spectator sport, with a vast array of experts and others looking into the 'fishbowl' of the Third World from the safety and comfort of their armchairs. We need to remind ourselves that the principles underlying participatory approaches to development are universal, and any attempt to restrict them to the developing world is meaningless.

Underlying all these problems is a simple inequality of power between North and South. Northern academics are able to monopolise control over the process of research because they have the resources to do so; Northern agencies are able to control the funding of development work because they also have the resources to do so. Knowledge is power, and the control of knowledge is the control of power. Power is the central component of development and without it there is little that the poor can do to change their circumstances. Centralised control over development studies is therefore directly anti-developmental in its effects. It undermines local self-confidence and prevents people at grassroots level from acquiring the skills and abilities which they need to analyse and solve problems for themselves. It prevents the transformation of people into agents of their own development by retarding the sharing of knowledge and information. This monopoly needs to be broken so that people become able to participate fully in the creation and use of their own knowledge. If this is not done, research and information will continue to circulate in a closed circuit from which poor people will always be excluded.

Understanding and changing the world: which comes first?

The famous inscription on Karl Marx's gravestone in Highgate Cemetery, London, poses a dichotomy which is central to the irrelevance of much development thinking today. Marx emphasised that changing the world, rather than understanding or interpreting it, is the prime task of the revolutionary. Conventional approaches to development studies posit the opposite view: that understanding the world must precede the ability successfully to change it, if indeed the link between understanding and change is made at all. In fact, there is no real dichotomy here since the processes of change and understanding should be synchronous. They must occur together if either is to be effective. We cannot change the world successfully unless we understand the way it works; but neither can we understand the world fully unless we are involved in some way with the processes that change it. The problem with much in development studies today is that they are divorced completely from these practical processes of change, as I have sought to show.

Listen to the following quotation from Canaan Banana, the former President of Zimbabwe:

> Whereas an armchair intellectual of rural development, lost in the labyrinth of misty theories and postulations, can afford to oversimplify matters and get away with it, a practitioner of rural development, that man or woman in the constant glare of various vicious and different shades of rural poverty and suffering, cannot. Time and again, now and in the future, they face the bleak disjuncture and mismatch between lengthy and laborious theories, decked out in figures and ornate expressions, and the ugly, undercorated and sordid reality of rural poverty.[20]

Banana's point is that a proper understanding of the problems of development requires a measure of involvement in the process of development itself. To this extent, development cannot be 'studied' at all; we can participate in the processes that underlie development and observe, record and analyse what we see, but we can never be relevant to problems in the abstract. However, this is precisely the position of much development research today. The reality of development studies bears little or no relation to the reality it seeks to address.

Researchers from the political left have been no more successful in this respect than commentators from the right, who form perhaps a more convenient target for criticism. As Adrian Adams points out, 'radical' critiques of conventional approaches to development have done little to change the way in which we look at this central relationship between understanding and action. 'All they have done is to create, alongside the activities of development experts, a body of ideas which cannot embody themselves in action and so proliferate in helpless parasitic symptoms with that which they criticise.'[21] The usual conclusion of these Marxist and Neo-Marxist critiques is that 'progress' is impossible without revolutionary changes in the structure of (capitalist) society. This is hardly an original or a useful conclusion to those who are actively working for change, by definition and inevitably within the social and political structures in which they live.

Our tendency to separate the processes of understanding and change leads naturally on to irrelevance because, while abstract research cannot be applied in practice, practice is often deficient because it fails to understand the real causes and character of the problems it seeks to address! Research and action become two parallel lines that never meet. Unless and until they do, we will neither understand nor change the world successfully.

[. . .]

NOTES

* The views expressed in this paper are the author's own views and not necessarily those of OXFAM.

1 For example, the FAO's indices of food production per capita (each country's average for 1979–81 = 100) show a fall from 94 in 1974 to 88 in 1986 for Ethiopia, from 116 to 93 for Kenya, and from 133 to 85 for Mozambique. Figures from *Monthly Bulletin of Statistics*, April 1986 and April 1987, Rome: FAO, 1987.

2 L Timberlake, *Africa in Crisis*, London: Earthscan, 1985, pp 8–10.

3 P Devitt, *Good Local Practice* (mimeographed), Oxford: OXFAM, 1986.

4 See especially the recent work of Robert Chambers: *Rural Development: putting the last first*, London: Long-man, 1983; 'Putting Last Thinking First; a professional revolution', *Third World Affairs* 1985, pp 78–94; 'Putting The Last First' in P Ekins (ed), *The Living Economy: a new economics in the making*, London: Routledge & Kegan Paul, 1986, pp 305–22. For a more general critique, see N Maxwell, *From Knowledge to Wisdom*, Oxford: Basil Blackwell, 1984.

5 N Maxwell, *From Knowledge to Wisdom*, p 2.

6 See D Hall, A Gillette and R Tandon, *Creating Knowledge: a monopoly? Participatory research in development*, Toronto: International Council for Adult Education, 1982.

7 A Adams, 'An Open Letter to a Young Researcher', *African Affairs* 78 (313) 1979, pp 453–79.

8 Quoted in R Bunch, *Two Ears of Corn: a guide to people-centred agricultural improvement*, Oklahoma City: World Neighbors, 1982, p 138.

9 N Maxwell, *From Knowledge to Wisdom*, p 66.

10 S Nyoni, the Director of the Organisation of Rural Associations For Progress (ORAP) in Zimbabwe, made this observation in an address given to OXFAM staff and supporters in Oxford in 1985.

11 See C Elliot, *Comfortable Compassion? Poverty, power and the church*, London: Hodder & Stoughton, 1987.

12 See P Richards, *Indigenous Agricultural Revolution*, London: Hutchinson, 1985; R Chambers, *Rural Development*; P Devitt, *Good Local Practice*; M Vaughan, *The Story of an African Famine: Gender and famine in twentieth-century Malawi*, Cambridge: Cambridge University Press, 1987.

13 P Richards, *Indigenous Agricultural Revolution*.

14 B Sharpe, 'Interim Report on Nutritional Anthropology Investigation' (mimeographed), Mpika: ODA/IRDP, 1987.

15 A Hope, S Timmel and C Hodzi, *Training for Transform-ation: a handbook for community workers*, Gweru: Mambo Press, 1984, p 8.

16 R Williams, *Towards 2000*, London: Penguin, 1983, p 266.

17 R Chambers, 'Putting the Last First', p 307.

18 O Anacleti, quoted in Y Kassam and K Mustafa, *Partici-patory Research: an emerging alternative methodology in social science research*, Nairobi: African Adult Education Association, 1982, p 110.

19 Quoted in A Adams, 'An Open Letter', p 473.

20 C Banana, quoted in M de Graaf, *The Importance of People: experiences, lessons and ideas on rural develop-ment training in Zimbabwe – Hlekweni and beyond*, Bulawayo: Hlekweni Friends Rural Service Centre, 1987, p 9.

21 A Adams, 'An Open Letter', p 477.

'Male Bias in the Development Process: An Overview'

from *Male Bias in the Development Process* (1995)

Diane Elson

Editors' Introduction

Diane Elson is from the English Midlands, and was the first in her family to attend university where she studied Philosophy, Politics and Economics at Oxford University. Between 1968 and 1985, Elson held research and teaching positions at Oxford, Sussex and York universities. In 1985, Elson became lecturer in Development Economics at the University of Manchester, after which she earned her PhD, advanced to Professor of Development Studies, and published a series of important essays and the 1991 landmark *Male Bias in the Development Process*. Since 2000, she has been Chair in Sociology at the University of Essex. Diane Elson's distinction has been to provide a series of ground-breaking feminist critiques of development processes and institutional practices, the latter of which has fed into important feminist interventions, through 'Gender and Development' (GAD) networks and initiatives. It is this combination of scholarly research, intervention in development theory, critique of institutions, and work on the nuts-and-bolts of gender transformation in development practices that makes Elson quite a unique figure in the history of development theory and practice.

Recall that Ester Boserup's work was foundational to the earlier development work on 'Women in Development' (WID), in order to shift development initiatives to incorporate 'women' as a constituency. GAD has emerged as a response to some of the liberal assumptions of WID, with respect to self-interested action, or the separation of 'productive' and 'reproductive' spheres, or the sense that women were at the centre of certain kinds of development (agriculture) while being marginalised in new spheres (industry). Elson, along with others like Ruth Pearson and Nancy Folbre, began to think of gender in a more dynamic sense, in relation to a radical critique of capitalist development as well as of development institutions themselves. For instance, in 1981, Elson and Pearson took issue with Boserup and dependency theorists on the question of whether and how industrialisation 'marginalises' women. Turning to export-led, labour-intensive industries in the South, Elson and Pearson show that when capital's preference for 'cheap labour' is decomposed to include various elements of the social wage, industries prefer to hire women, and more precisely young, unmarried women 'freed' from domestic work. Along with this kind of research, GAD scholars like Elson, Pearson and Sylvia Chant have also reflected on the practical side of the broadening of feminist concerns into a wider critique of gender, outside the domains of 'reproduction' and domesticity, for instance, in the widespread use of gender budgeting involving civic and governmental groups across global contexts.

The reading is an exemplary text in this regard, as it demonstrates Elson's pioneering feminist insights applied both to the gendered underpinnings of development economics and to the outcomes of structural adjustment programmes (SAPs), or neo-liberal reform packages. What Elson shows is that SAPs not only increase women's obligations to engage in waged labour, but that cutbacks in welfare also mean that women bear the brunt of increases in unpaid labour. What is important in the reading is that it sets out an approach to understand 'male bias' beyond proximal attitudes, and beyond the simple notions of male/ female preference, the legacy of WID. Rather, Elson's feminism extends across sites and scales, to the effects of analytical categories, institutional practices, social and economic relations, as well as to feminist interventions to transform male bias.

Key references

Diane Elson (ed.) (1995) *Male Bias in the Development Process*, Manchester: Manchester University Press.
— (1980) 'The Value Theory of labour', in Diane Elson (ed.) *Value: The Representation of Labour in Capitalism*, London: CSE Books.
Diane Elson and Ruth Pearson (1981) ' "Nimble fingers make cheap workers": an analysis of women's employment in Third World export manufacturing', *Feminist Review* 7.
Cecile Jackson and Ruth Pearson (eds) (1998) *Feminist Visions of Development: Gender Analysis and Policy*, London: Routledge.

This is not another book about women in development. Books about women in development have been a necessary stage in making gender relations visible in the development process, but posing the issue in terms of women in development has several limitations. It facilitates the view that 'women', as a general category, can be added to an existing approach to analysis and policy, and that this will be sufficient to change development outcomes so as to improve women's position. It facilitates the view that 'women's issues' can be tackled in isolation from women's relation to men. It may even give rise to the feeling that the problem is women rather than the disadvantages women face; and that women are unreasonably asking for special treatment rather than for redress for injustices and for removal of distortions which limit their capacities. It tends to encourage the treatment of women as a homogeneous group with the same interests and viewpoint everywhere. It is necessary to move on from 'women in development' to approaches that emphasise gender relations.

Gender relations are the socially determined relations that differentiate male and female situations. People are born biologically female or male, but have to acquire a gender identity. Gender relations refer to the gender dimension of the social relations structuring the lives of individual men and women, such as the gender division of labour and the gender division of access to and control over resources. An emphasis on gender highlights the fact that work is gendered; that some tasks are seen as 'women's work', to do which is demeaning for men; while other tasks are 'men's work', to do which unsexes women. An emphasis on gender relations encourages a questioning of the supposed unity of the household and facilitates the posing of questions about the relative power of women and men.

There is a wealth of evidence demonstrating the differences in power between women and men throughout the world. It is not that women are powerless victims or that no women are in positions of power over men, but rather that, relatively speaking, women are less powerful than men of similar economic and social position. A graphic example is the risk of sexual violence faced by any woman who finds herself alone in a public place at night. The rich woman whose car has broken down is in the same position as the poor woman waiting for a bus. They are both at risk because they are breaking a gender norm, the norm that 'respectable' women should not be alone at night in public places. In breaking this norm they can be perceived as legitimate targets, as 'asking for it'. Men too may risk violence on the streets, but mugging has a quite different significance from rape.

A gender approach has greater flexibility than a women-in-development approach. For instance, an emphasis on gender relations tends to permit greater awareness of the different ways that different women experience gender. Though rich and poor women both face a common danger of rape if they are alone in a public place after dark, poor women have more of an interest in improvements in public transport than do rich women.

The asymmetry between male and female gender can be expressed in terms of the language of gender subordination: the idea that women as a gender are subordinate to men as a gender. But this language focuses on structures rather than agents. It can obscure individual responsibility and suggest the presence of immovable social forces in whose operation we can only acquiesce. It can even be used to justify the denial of equal opportunities to women, as in the case of Equal Opportunity Commission *v.* Sears, Roebuck & Co. in the USA (Kessler-Harris, 1987).[1] Women active in grass-roots feminist activities, such as Women's Aid and Rape Crisis Centres, have suggested to me that it is also a language which is too academic, too sanitised, too polite. It is time to stop talking about gender subordination and start talking about male bias.

MALE BIAS

Talk of bias can be simply emotive, so it is certainly necessary to think carefully about criteria for use of the term. There are some precedents for using it in examining development issues, most notably the term 'urban bias' (Lipton, 1977). Whatever reservations one has about the explanatory power of 'urban bias', there is no doubt that it served to mobilise analysis and policy to address the important question of rural–urban inequality. An essential contribution to its mobilising ability was the way it combined the flavour of condemnation of the word 'bias' with appeals to objective criteria and empirical evidence. It is in the same spirit that I shall use the term 'male bias'.

By *male* bias I mean a bias that operates in favour of men as a gender, and against women as a gender, not that all men are biased against women. Some men have contributed substantially to the diagnosis and understanding of male bias and have campaigned to overcome it. Some women show little understanding of the operation of male bias and do much to perpetuate it. To emphasise this point, in

what follows I shall draw on the work of a male economist, A. K. Sen, who has provided some useful conceptual tools for the elucidation of male bias. Nevertheless, on the whole women are more likely to recognise the significance of male bias, and to wish to combat it, than are men. But this is a matter of differences in the experience of women and men, not of differences in some essential femininity or masculinity.

What is bias? It is asymmetry that is ill-founded or unjustified. There is no problem in demonstrating gender asymmetry in the outcomes of development processes, in the lived experience of women and men throughout the world; the arguments are about the extent to which such asymmetry is ill-founded and unjustified. No attempt will be made here to review the enormous literature depicting gender asymmetry in developing countries. A useful overview of the literature and summary of key features of gender relations in the major regions of the Third World is provided by Brydon and Chant (1989). Compilations of statistical evidence can be found in United Nations publications, from organisations like the International Labour Office (ILO/INSTRAW, 1985) and the Department of International Economic and Social Affairs (UN, 1986). But there remains a question of interpretation. How far do the asymmetries represent male bias, and how far difference and complementarity?

MALE BIAS IN DEVELOPMENT OUTCOMES

The first point that must be tackled is the issue of the benchmark against which bias in development outcomes is to be judged. What counts as lack of bias? Equal treatment of equals? But equal in what respects? Different people in different situations have different needs and different talents. Removing bias does not mean complete standardisation and removal of all differences. One approach to defining bias is in terms of differences which are not the result of differences in endowments and preferences. This is the procedure favoured by neo-classical economists setting up models of the labour market and the household (for example, Becker, 1981). Such models tends to downplay the prevalence of bias through using oversimple, uncritical notions of endowments and preferences.

Aptitudes that are often ascribed to endowments, such as women's supposedly 'nimble fingers', may be due to the upbringing women have received at home and at school, which trains them in sewing and in repetitive sorting tasks (like tidying up and separating grains of rice from stones and husks), and emphasises the virtues of patience and endurance of routine (Elson and Pearson, 1981). It is virtually impossible to separate out endowments from acquired characteristics for a wide range of attributes (Block and Dworkin (eds), 1977). Levels of nutrition before birth can have an impact on subsequent achievements. Characteristics that are unproblematically genetic endowments, such as eye colour, are those which are least interesting from the point of view of explaining social outcomes. Part of the problem of male bias is that it tends to hamper women from acquiring those characteristics which are well-rewarded in the market; and that it tends to hamper social scientists from understanding the limitations of notions of male and female endowments of aptitudes or talents.

If innate aptitudes cannot be taken for granted, neither can well-defined individual preferences. Sen argues that family identity may exert such a strong influence on the perceptions of rural Indian women that they find it unintelligible to think in terms of their own preferences and welfare. Instead, they think in terms of the welfare of their families (Sen, 1987). This is a theme which has also run through much feminist literature on women's consciousness in developed countries. Part of the problem of male bias is that it tends to hamper women from forming well-defined notions of what they want; women submerge their own interests beneath those of men and children.

Instead of judging bias against endowments and preferences, it may be judged against rights and capabilities (Sen, 1984, ch. 13). Equal rights have been a rallying cry for many women's movements, and in many countries women have won a substantial measure of legal equality. But even in countries where equal rights for women are enshrined in the constitution, women find enormous difficulties in exercising those rights. Key rights for poor rural women, such as rights to land, have no practical purchase because land rights are vested in household heads: that is, in men, unless there are no adult males in the household (Jiggins, 1988). Key rights for poor urban women, such as equal pay, have no purchase

because women are concentrated in the informal sector where legislation does not reach, or in female ghettos in the formal sector where there is no male standard with which to establish equality (Joekes, 1987). Moreover, entitlement systems, governing who can have use of what, which regulate market transactions, typically do not regulate intra-household resource distribution (Sen, 1987). Thus, an emphasis on rights has to be supplemented by an emphasis on socially conferred capabilities – what are women in practice able to do? Are they able to be well-nourished; to enjoy good health and long lives; to read and write; to participate freely in the public sphere; to have some time to themselves; to enjoy dignity and self-esteem? How does women's enjoyment of these capabilities compare with that of men? Do women face constraints which are not faced by men? In so far as women enjoy fewer and more circumscribed capabilities than do men, then there is male bias in development outcomes. Constraints on women operate to men's advantage in the short run, as in bargaining within the household (discussed below). Male bias exists even if women do not manifest any lesser satisfaction with their lot in life than do men. As Sen points out:

> There is much evidence in history that acute inequalities often survive precisely by making allies out of the deprived. The underdog comes to accept the legitimacy of the unequal order and becomes an implicit accomplice. It can be a serious error to take the absence of protests and questioning of inequality as evidence of the absence of that inequality.
>
> (Sen, 1987, p. 3)

Sen's argument has force with respect to any kind of inequality, and in emphasising male bias I do not intend to imply that it is the only important form of bias. Class bias, regional bias, urban bias, racial and ethnic bias, are all important; and the different kinds of bias are imbricated with one another, forming differentiable but not separated aspects of a whole lived situation for any individual. Thus all women do not face the same kind and same degree of male bias; and they may enjoy the fruits of other kinds of bias, or share the deprivations, with men in the same class, region, ethnic group.

What perhaps is unique about male bias is that those who are disadvantaged by it live daily in intimate personal relationships with those who are advantaged by it. The relationship between women and men living together in households has been usefully depicted by Sen (1985, 1987) in terms of co-operative conflicts. Women and men gain from co-operating with one another in joint living arrangements in so far as this increases the capabilities of the household as a whole; but the division of the fruits of co-operation is a source of conflict. Women are at a disadvantage in bargaining over the division of the fruits of co-operation because their fall-back position tends to be worse. That is, if no bargain is struck, and women are on their own, without husband, father, brother or other male relative to co-operate with, they tend to be worse off, in terms of capabilities, than if they agree to strike a bargain and enter into some kind of co-operation with men. The evidence of the poverty of female-headed households is overwhelming testimony to the weakness of women's fall-back position. However, co-operative conflicts between people of different genders are more than simple bargaining problems because of the gender differentiation in the specification of preferences discussed earlier. As a result of an upbringing shaped by male bias, women tend not to have such sharp perceptions as men of their own interests, needs, rights or deserts. And this perpetuates male bias, because the co-operative arrangement arrived at is likely to be less favourable to those individuals with less well-defined perceptions of their own interests.

Male bias is contradictory in that while it preserves the subordination of women as a gender to men, it also has costs for society considered as a whole. For instance, male bias distorts resource allocation by denying women adequate access to productive inputs. This lowers women's productivity and reduces total output in comparison with what could be achieved if resource allocation were free of gender distortion (Palmer, 1988). Thus male bias is a barrier to the achievement of development objectives such as growth of output. So why don't more men show eagerness to overcome male bias? Perhaps it is because the disadvantages of relinquishing male power are more immediately apparent, while the distribution of the gains is uncertain and the transition period may be painful. If women's productivity is enhanced because male bias in resource allocation is reduced, total output may rise, but so may women's bargaining power. The total size of the cake may increase, but men's share of it may fall.

THE PROXIMATE CAUSES OF MALE BIAS IN DEVELOPMENT OUTCOMES

The proximate causes of male bias in development outcomes can be analysed in terms of male bias in everyday attitudes and actions, in theoretical reasoning, and in public policy. The underlying supports of male bias are to be found in the particular ways in which getting a living is integrated with raising children.

Male bias in everyday attitudes and actions may be the result of prejudice and discrimination at the conscious level, but this is not necessarily the case. Bias may be deeply embedded in unconscious perceptions and habits, the result of oversight, faulty assumptions, a failure to ask questions. For instance, women's contribution to family income tends to be overlooked because much of it is unpaid or takes the form of repetitive services rather than products that can be massed together in an unmistakable sign of contribution made. As a result, women tend to be regarded as less deserving than men when it comes to intra-household distribution (Sen, 1987). Such unconscious bias is not unreachable and unchangeable. People can be brought to recognise it through education, consciousness-raising groups, politicisation, social change. Domitila Barrios deChungara, a women's leader and miner's wife in a tin-mining community in Bolivia, explains how she went about doing this:

> But in spite of everything we do, there's still the idea that women don't work, because they don't contribute economically to the home, that only the husband works because he gets a wage. We've often come across that difficulty.
>
> One day I got the idea of making a chart. We put as an example the price of washing clothes per dozen pieces and we figured out how many dozens of items we washed a month. Then the cook's wage, the babysitter's, the servant's. We figured out everything that we miners' wives do every day. Adding it all up, the wage needed to pay us for what we do in the home, compared to the wages of a cook, a washerwoman, a babysitter, a servant, was much higher than what the men earned in the mine for a month. So that way we made our companeros [husbands] understand that we really work, and even more than they do in a certain sense.
>
> (Johnson and Bernstein (eds), 1982, p. 235)

However, conscious and unconscious male bias in thought and action is frequently buttressed by economic and social structures which make such practices seem rational, even to those who are disadvantaged by them. Thus it can seem entirely rational for mothers to allocate more food to sons than to daughters when food is short, in circumstances where sons are more valuable, socially and materially, than daughters; and where future survival of the household depends crucially on survival of sons to adulthood, then it can seem entirely rational to prioritise their needs and neglect those of daughters. Such behaviour is acclaimed by neo-classical economists as evidence that their harmonious 'joint utility' models of the household are correct (Rosenzweig, 1986), though it can equally well be explained in terms of a co-operative conflicts model (Folbre, 1986). But the important point is that although actions that perpetuate male bias are rational from the point of view of a highly constrained individual, they do not testify to the overall rationality, much less desirability, of the social system. Rather, they suggest that the constraints on individuals need to be changed through some collective process. In the absence of such a process, individual women will certainly find it rational to do things that perpetuate male bias. This has been recognised by careful thinkers about individual choice and well-being, such as Sen, who points out that:

> Deprived groups may be habituated to inequality, may be unaware of possibilities of social change, may be hopeless about upliftment of objective circumstances of misery, may be resigned to one's fate, and may be willing to accept the legitimacy of the established order. The tendency to take pleasure in small mercies would make good sense given these perceptions, and cutting desires to shape (in line with perceived feasibility) can help to save one from serious disappointment and frustration.
>
> (Sen, 1987, p. 9)

Male bias in theoretical reasoning may often not be so immediately apparent because the reasoning is presented in terms which appear to be gender neutral. Rather than talking about women and men, and sons and daughters, use is made of abstract concepts like the economy, the formal sector, the informal sector, the labour force, the household. Or the argument is conducted in terms of socio-economic categories which, on the face of it, include both women and men, such as 'farmer', and 'worker'. It is only on closer analysis that it becomes apparent that these supposedly neutral terms are in fact imbued with male bias, presenting a view of the world that both obscures and legitimates ill-founded gender asymmetry, in which to be male is normal, but to be female is deviant. This is more immediately apparent in analysis conducted in terms of socio-economic groups, where we soon read of 'farmers and their wives' and 'workers and their wives' but never of 'farmers and their husbands' and 'workers and their husbands', despite the fact that large numbers of women are farmers or wage-earners in their own right.

Let us examine this more closely using the example of the category 'farmer'. Though this appears to be a gender-neutral category, it is used in a way that implies farmers are men; this suggests that major decision-making and farm management is undertaken by men, while women serve as unpaid family labour, helping their husbands. While this may be true of some areas, and some types of farming, we have enough case study data to know that it is certainly not universally true. Many countries in sub-Saharan Africa have large numbers of women who farm in their own right, either because of a high incidence of female-headed households in rural areas (as in Botswana, Lesotho, Sierra Leone and Zambia, for instance), or because there is a traditional demarcation of crops into 'men's crops' and 'women's crops' (as in Cameroon, Ghana and Malawi) (FAO, 1986). Outside Africa, women are more likely to be managing post-harvest activities, such as processing, storage and marketing, than managing production of staple crops; or managing livestock-related activities or horticulture. But whatever the differences in the particular activities undertaken, the point remains that many women in agriculture do undertake management responsibilities.

The picture of farmers as men disadvantages women farmers and hinders attempts to improve agricultural productivity. When there is an implicit assumption that farmers are men, it is not surprising if new agricultural technology and inputs flow mainly to men – and there is a wealth of evidence that this is what has happened in developing countries over the last three decades. Despite the attempts of concerned researchers to make rural women 'visible' to policy-makers in the 1970s, most rural projects up to the early 1980s still addressed women through welfare and home economics programmes for farmers' wives. Governments still fail to collect comprehensive, reliable and unbiased statistics on the contribution women make to agricultural production (Safilios-Rothschild, 1987). But we know from village-level studies that when resources are redirected to women there are increases in agricultural productivity and efficiency (Jiggins, 1987; Staudt, 1987).

To see the male bias in analysis which is conducted in terms of abstract categories, we have to examine the implicit assumptions structuring the definition and use of the abstract categories. Is there a hidden assumption about the homogeneity of sectors of the economy regardless of gender? For instance, is it assumed that surplus labour can be withdrawn from agricultural production because those left behind will be able to make up any shortfall, without considering the division of agricultural tasks into 'men's tasks' and 'women's tasks'? This does seem to the implicit assumption in many dual-sector models of development – including Sen's (Sen, 1966). Ignoring the gender division of labour may result in failure to consider the overloading of women and reduction in agricultural productivity that male migration from the rural sector may induce. Is there a hidden assumption about the costs of reproduction of labour power and who bears them? Most models of the economy treat labour, like land, as an unproduced factor of production. In effect, there is an implicit assumption that the necessary inputs of time and effort required to ensure its continuing supply will be forthcoming even though these inputs are unpaid. These inputs are, of course, disproportionately supplied by women in their roles as wives, mother and daughters, ministering to the needs of other family members. Time budget studies from around the world show that these activities are undertaken by most women in addition to activities that are counted as 'economic' (Goldschmidt-Clermont, 1987). Women have far less leisure than men because of this 'double day'. Ignoring women's unpaid domestic work obscures both the burdens women bear and the constraints this work places upon women's capacity to respond to opportunities for paid work. There is no intrinsic reason why the work of caring for others should not be shared equally between women and men. But a reduction in this asymmetry is unlikely while male bias continues to deny the economic contribution such work makes.

Is there a hidden assumption about the benevolence of the ties that bind households together? The household is in some sense a social unit, but can we assume it is a unity? The models of the household constructed by neo-classical economists do assume unity, implying, for instance, that the welfare of household members can be judged on the basis of aggregate household income, and that extra income accruing to one household member will 'trickle down' to others. But a growing volume of case-study evidence supports alternative models, such as Sen's co-operative conflict model. They suggest a picture of a home divided over income and expenditure decisions (Dwyer and Bruce (eds), 1988). Households are in some sense pooling and sharing organisations but to imply that this pooling and sharing is unproblematic is to reveal male bias. There is considerable evidence to suggest that while women typically pool and share their income, especially with their children, men are more inclined to reserve part of their income for discretionary personal spending (*ibid.*). Uncritical theorisation of existing forms of family life is a barrier to securing real reciprocity. Overcoming male bias does not mean the disintegration of pooling and sharing of resources between women and men. Rather, it means more extensive pooling and sharing and a disintegration of unjust gender asymmetries in family relationships.

When supposed gender neutrality masks male bias, this serves to obscure the distribution of costs and benefits of development processes between men and women. It also serves to obscure the barriers that gender asymmetries constitute to the successful realisation of many development policy objectives. To overcome such bias, it is not enough to affix 'women' as an afterthought. For that tends to obscure the differences between different groups of women, and to perpetuate the gender blindness of analytical concepts. What is needed is a gender-aware conceptualisation in the first place. Otherwise, male bias will remain even though 'women' are present.

Male bias in development policy is encouraged by male bias in everyday attitudes and practices and by male bias in analysis, reinforced by male bias in politics. Until the late 1970s women were largely invisible to policy-makers, whose perspective might be summed up in the old Russian proverb: 'I thought I saw two people coming down the road, but it was only a man and his wife'. Women were treated merely as dependants of men. Development objectives were disaggregated on a household basis and it was assumed that resources targeted to men would equally benefit dependent women and children. For a variety of reasons, including the advocacy of 'women in development' experts, and the breakdown of family support systems leading to increases in the numbers of female-headed households in dire poverty, by the end of the 1970s women had become visible to policy-makers – but as recipients of welfare benefits rather than as producers and agents of development. There was a proliferation of special women's projects, many of which failed to become self-supporting because of lack of gender awareness in their design, perpetuating the idea of women as a drain on the public purse (Buvinic, 1986).

In the 1980s more attention has been paid to women as agents of development, but as agents of social development whose caring and nurturing could substitute for expenditure on heath, education and social services (Antrobus, 1988; Dwyer and Bruce (eds), 1988). Moreover, there remains the problem of male bias in the policy process itself. With few exceptions, women's interests are marginalised in the formulation and implementation of economic policy (Moser, 1989). Women's voices play little part. At the grass-roots level, there are frequently factors that inhibit women from speaking out in public meetings. A study of women and village government in ten villages in Tanzania in 1981 found that village officials gave the following reasons for women's low attendance at village meetings: women are still too shy to attend public meetings; women are too busy at home to come; it is not the woman's job to roam and survey the village and attend things like meetings – it is her job to watch the house; women can't speak Kiswahili; women are uneducated; women don't understand the discussions; it's difficult for a woman to get a pass from the house to come; only one person from a household need come, so it is always the man; women don't need to come because we ask the men to tell them what we have discussed; women are not used to sitting together with men; there are still some men here who don't like to see women in meetings (Wiley, 1985, p. 170).

In the corridors of power there are relatively few women. The experience of women's bureaux and ministries of women's affairs is particularly discouraging, as they tend to be underresourced, overstretched and cut off from the economic policy-making process (Gordon, 1984). Development objectives are defined in practice in ways that are more beneficial to men than to women. Thus in practice it is not more food output *per se* that tends to be sought, but more of the kind of food output which is produced under the direction and control of men (for examples, see Mblinyi, 1988); not more private trading *per se* but more of the kind of private trading undertaken by men – large scale and capital intensive – rather than the kind of trading undertaken by women – localised and with a quick turnover (for an example from Ghana, see Loxley, 1988).

That is not to claim that male policy-makers deliberately define objectives in terms that benefit men more than women, but rather that they tend to see as in the general interest policies that in practice are male-biased, and to perceive policies that reduce gender asymmetry as female-biased.

THE UNDERLYING SUPPORTS OF MALE BIAS

I have discussed the proximate causes of male-biased development outcomes in terms of male bias in everyday attitudes and decisions, in theoretical analysis, and in the process of defining and implementing public policy. Underlying these individual and collective acts are structural factors that circumscribe and shape them. The key structural factor is not the way in which getting a living is organised, nor the way that raising children is organised, but the way these two processes are interrelated.

The crucial question is how children, and those people engaged in raising children, get their living. Do they have an entitlement to the necessary resources in their own right? Or are they dependent on other family members to secure access to the required resources? It would be possible for children and those engaged in caring for them to have an adequate independent entitlement through independent access to a basic minimum income paid to all members of society regardless of the work they do;[2] or through independent access to adequate earning opportunities coupled with adequate child care facilities. But this would require the integration of getting a living and raising children not being confined to the family; it would require integration through social provision and the mediation of organisations in the public sphere; it would require access to a basic minimum income as of right, to be independent of family circumstances. Such an entitlement is extremely rare. In practice, most children and those caring for them are either dependent on other family members for access to income or resources, or, when there is no one to depend on, they suffer poverty and deprivation because of difficulties in combining child care and income generating activities. Relief may be available through charities or state welfare schemes but this relief is generally not an absolute entitlement and creates a new form of dependence. The opportunity to earn an income through selling labour or products in the market may seem to offer independence to women. But in practice this independence is open only to a small minority. For the market does not provide adequate and affordable child care facilities for most women, and does not guarantee them an adequate income. On the whole, markets tend to lead to the concentration of income in the hands of those who start off with most resources. Unless markets are socially regulated they offer only the semblance of independence for women, and not the reality (Elson, 1989). The lack of an independent and secure entitlement creates a bias operating against those people who have the task of child care and weakens their bargaining position in the co-operative conflicts of the family.

The desire for an independent entitlement is not confined to well-educated feminists. A study of ten villages in Tanzania in the early 1980s reports the following comments as typical of those made by village women about their situation:

> 'Now we are sitting in meetings, but my husband can still beat me if I complain. We are still dependent on men.'
>
> 'Women are still the same because the money belongs to the husband still.'
>
> 'A man can still refuse you anything because he owns all the things.'
>
> 'Our main problem here is that the men are drinking our money and we have no way to get more.'
>
> 'In my opinion it would be better if the Council gave every woman one acre for herself.'
>
> (Wiley, 1985, p. 171)

The lack of an independent entitlement for children and those who care for them tells against women. It gets women 'locked in' to child care. There are some phases of raising children which physically have to be undertaken by women – pregnancy, childbirth, breast-feeding – but the rest could be undertaken by men too. However, if lack of an independent entitlement forces women into dependence for these phases of child-rearing, phases which are particularly difficult to combine with income-earning, then women are likely to get locked in to all the other phases. Dependence and its associated lack of bargaining power at one stage are transmitted to later stages. The winners of a co-operative conflict in one round have enhanced bargaining power for the future. The transmission can also work intergenerationally, perpetuating asymmetry over time (Sen, 1987). It is biology that creates the initial link between women and children; but it is the socially determined lack of entitlement that turns this link into the underpinning of male bias.

Overcoming male bias is not simply a matter of persuasion, argument and changes in viewpoint in everyday attitudes, in theoretical reasoning and in the policy process. It also requires changes in the deep structures of economic and social life, and collective action not simply individual action. It requires profound changes in the way that raising children and getting a living are integrated, so as to make maternity economically autonomous. Marriage clearly cannot do this; and existing forms of market opportunity and state provision have not done it either. The question of how to make progress on this issue is among those addressed in the final chapter of this book.

[. . .]

NOTES

1 Sears, Roebuck & Co. argued that they were not guilty of discrimination against women. Rather, the gender structure of society meant that women did not offer themselves for certain types of job.
2 Such an entitlement differs from most existing welfare state provisions by being unconditional and paid to individuals. For more discussion of this issue, see the *Bulletin* of the Basic Income Research Group.

REFERENCES

Antrobus, P. (1988), 'Consequences and responses to social and economic deterioration: the experience of the English-speaking Caribbean', Workshop on Economic

Crisis, Household Strategies, and Women's Work, Cornell University.

Becker, G. (1981), *A Treatise on the Family*, Harvard University Press, Cambridge, Mass.

Block, N., and Dworkin, G. (eds) (1977), *The IQ Controversy*, Quartet Books, London.

Brydon, L., and Chant, S. (1989), *Women in the Third World*, Edward Elgar, Aldershot.

Buvinic, M. (1986), 'Projects for women in the Third World: explaining their misbehaviour', *World Development*, vol. 14, no. 5.

Dwyer, D., and Bruce, J. (eds) (1988), *A Home Divided: Women and Income in the Third World*, Stanford University Press, Stanford.

Elson, D. (1989), 'The impact of structural adjustment on women: concepts and issues', in B. Onimode (ed.), *The IMF, the World Bank and the African Debt*, vol. 2: *The Social and Political Impact*, Zed Books, London.

Elson, D., and Pearson, R. (1981), ' "Nimble fingers make cheap workers": an analysis of women's employment in Third World export manufacturing', *Feminist Review*, no. 7.

Folbre, N. (1986), 'Cleaning house: new perspectives on households and economic development', *Journal of Development Economics*, vol. 22.

FAO (1986), *Report of the Workshop on Improving Statistics on Women in Agriculture*, 21–23 October, Rome.

Goldschmidt-Clermont, L. (1987), *Economic Evaluations of Unpaid Household Work: Africa, Asia, Latin America and Oceania*, International Labour Office, Geneva.

Gordon, S. (1984), *Ladies in Limbo*, Commonwealth Secretariat, London.

ILO/INSTRAW (1985), *Women in Economic Activity: A Global Statistical Survey (1950–2000)*, International Labour Office, Geneva.

Jiggins, J. (1987), *Gender-Related Impacts and the Work of the International Agricultural Research Centers*, CGIAR Study Paper No. 17, World Bank, Washington DC.

—— (1988), 'Women and land in sub-Saharan Africa', Rural Employment Policies Branch, Employment and Development Department, International Labour Office, Geneva.

Joekes, S. (1987), *Women in the World Economy: An Instraw Study*, Oxford University Press, New York.

Johnson, H. and Bernstein, H. (eds) (1982), *Third World Lives of Struggle*, Heinemann, London.

Kessler-Harris, A. (1987), 'Equal Opportunity Commission v. Sears, Roebuck & Co.: a personal account', *Feminist Review*, no. 25.

Lipton, M. (1977), *Why Poor People Stay Poor – Urban Bias in World Development*, Temple Smith, London.

Loxley, J. (1988), *Ghana: Economic Crisis and the Long Road to Recovery*, North-South Institute, Ottawa.

Mbilinyi, M. (1988), 'Agribusiness and women peasants in Tanzania', *Development and Change*, vol. 19, pp. 549–83.

Moser, C. (1989), 'Gender planning in the Third World: meeting practical and strategic gender needs', *World Development*, vol. 19, no. 11.

Palmer, I. (1988), *Gender Issues in Structural Adjustment of Sub-Saharan African Agriculture and Some Demographic Implications*, Population and Labour Policies Programme, Working Paper No. 166, International Labour Organisation, Geneva.

Rosenzweig, M. (1986), 'Program interventions, intra-household distribution and the welfare of individuals: modelling household behaviour', *World Development*, vol. 14, no. 2.

Sen, A. K. (1966), 'Peasants and dualism with or without surplus labour', *Journal of Political Economy*, vol. 74, pp. 425–50.

—— (1984), *Resources, Values, and Development*, Blackwell, Oxford.

—— (1985), *Women, Technology, and Sexual Divisions*, UNCTAD and INSTRAW, Geneva.

—— (1987), 'Gender and co-operative conflicts' [mimeo.], World Institute of Development Economics Research, Helsinki.

Staudt, K. (1987), 'Uncaptured or unmotivated? Women and the food crisis in Africa', *Rural Sociology*, vol. 52, no. 1.

UN (1986), *World Survey on the Role of Women in Development*, Department of International Economic and Social Affairs, New York.

Wiley, L. (1985), 'Tanzania: the Arusha Planning and Village Development Project', in C. Overholt, M. Anderson, K. Cloud, and J. Austin (eds), *Gender Roles in Development Projects*, Kumarian Press, West Hartford.

SIX

'The Anti-Politics Machine: "Development" and Bureaucratic Power in Lesotho'

The Ecologist (1994)

James Ferguson (with Larry Lohmann)

Editors' Introduction

James Ferguson is an Anthropologist at Stanford University in California. He received his PhD from the Anthropology Department at Harvard University, and has gone on to author three important monographs, two of which are based on sustained historical and ethnographic research in Lesotho and Zambia. In *The Anti-Politics Machine: 'Development', Depoliticization, and Bureaucratic Power in Lesotho* (1990), summarised in the reading here, Ferguson investigates the mismatch between the workings of the development apparatus and local ground realities. Ferguson's next book, *Expectations of Modernity: Myths and Meanings of Urban Life on the Zambian Copperbelt* (1999) asks how Zambians think of 'modernisation' in the wake of economic decline. Ferguson revisits the classic anthropologies of the Rhodes Livingstone Institute and its defence of real and possible African urban modernities. He sets these against the lived understandings of former miners in the Copperbelt who are forced to 'return' to a countryside to which they have little actual connection. In two edited collections with his interlocutor Akhil Gupta, the author of the equally important *Postcolonial Developments* (1998), Gupta and Ferguson have sought to critique their own discipline of anthropology by rethinking 'culture' in a way that is more properly attentive to human geography, power and geopolitics. In a provocative and important short intervention, 'Seeing like an oil company', Ferguson sets out an agenda for a spatially sensitive analysis of contemporary Africa as islands of securitised, extractive accumulation in a sea of humanitarianism and war. Most recently, *Global Shadows: Africa in the Neoliberal World Order* (2006) collects Ferguson's writings on Africa's relations to the world. It does so in relation to crises of the state, NGO governmentality, corporate power, structural adjustment and the cultural politics of African modernity. Ferguson's current research concerns liberalism, neo-liberalism and poverty in South Africa.

In the reading that follows, Ferguson shows how the World Bank liked to construct Lesotho as an isolated locale, while in fact it had quite clearly become a migrant labour reserve for regional capital, particularly in apartheid South Africa. International institutions were funnelling vast amounts of aid and development interest into Lesotho and yet the development apparatus continued to 'fail' in alleviating poverty and inequality. Ferguson impresses upon us the importance of understanding what this 'failure' accomplishes. Within Lesotho, the machinery of development bureaucratises poverty alleviation, severing it from politics. While the development apparatus accomplishes this 'depoliticisation', it shifts attention from the socio-cultural dynamics through which people value cattle – what he calls 'the bovine mystique'. *The Anti-Politics Machine* has set a standard of excellence that few studies of actual development institutions and practices have matched. It helps us pose the question of development in the new millennium as one that *has* to be asked in multiple ways and at multiple scales: from the intentions and practices of global institutions to the workings of development policy and the social and cultural relations that are thereby elided. We caution the reader not to read Ferguson for easy prescriptions. Instead, the reader might use these intuitions as a guide to ask how institutions, ideologies and practical realities grate against each other, producing varied outcomes that must be analysed in their concreteness.

Key references

James Ferguson with Larry Lohmann (1994) 'The anti-politics machine: "development" and bureaucratic power in Lesotho', *The Ecologist* 24 (5): 176–81.

James Ferguson (1990) *The Anti-Politics Machine: 'Development', Depoliticization, and Bureaucratic Power in Lesotho*, Cambridge: Cambridge University Press.

— (1999) *Expectations of Modernity: Myths and Meanings of Urban Life on the Zambian Copperbelt*, Berkeley: University of California Press.

— (2005) 'Decomposing modernity: history and hierarchy after development', in Ania Loomba, Suvir Kaul, Matti Bunzl, Antoinette Burton and Jed Esty (eds) *Postcolonial Studies and Beyond*, Durham, NC: Duke University Press.

— (2006) *Global Shadows: Africa in the Neoliberal World Order*, Durham, NC: Duke University Press.

Akhil Gupta (1998) *Postcolonial Developments: Agriculture in the Making of Modern India*, Durham, NC: Duke University Press.

Akhil Gupta and James Ferguson (eds) (1997) *Culture, Power, Place: Exploration in Critical Anthropology*, Durham, NC: Duke University Press.

Akhil Gupta and James Ferguson (eds) (1997) *Anthropological Locations: Boundaries and Grounds of a Field Science*, Berkeley: University of California Press.

■ ■ ■ ■ ■ ■ ■

In the past two decades, Lesotho – a small land-locked nation of about 1.8 million people surrounded by South Africa, with a current Gross National Product (GNP) of US$816 million – has received "development" assistance from 26 different countries, ranging from Australia, Cyprus and Ireland to Switzerland and Taiwan. Seventy-two international agencies and non- and quasi-governmental organizations, including CARE, Ford Foundation, the African Development Bank, the European Economic Community, the Overseas Development Institute, the International Labour Organization and the United Nations Development Programme, have also been actively involved in promoting a range of "development" programmes. In 1979, the country received some $64 million in "official" development "assistance" – about $49 for every man, woman and child in the country. Expatriate consultants and "experts" swarm in the capital city of Maseru, churning out plans, programmes and, most of all, paper, at an astonishing rate.

As in most other countries, the history of "develop-ment" projects in Lesotho is one of "almost unremit-ting failure to achieve their objectives".[1] Nor does the country appear to be of especially great economic or strategic importance. What, then, is this massive and persistent internationalist intervention all about?

CONSTRUCTING A "DEVELOPER'S" LESOTHO

To "move the money" they have been charged with spending, "development" agencies prefer to opt for standardized "development" packages. It thus suits the agencies to portray developing countries in terms that make them suitable targets for such packages. It is not surprising, therefore, that the "country profiles" on which the agencies base their interventions frequently bear little or no relation to economic and social realities.

In 1975, for example, the World Bank issued a report on Lesotho that was subsequently used to justify a series of major Bank loans to the country.

One passage in the report – describing conditions in Lesotho at the time of its independence from Britain in 1966 – encapsulates an image of Lesotho that fits well with the institutional needs of "development" agencies:

> Virtually untouched by modern economic development . . . Lesotho was, and still is, basically, a traditional subsistence peasant society. But rapid population growth resulting in extreme pressure on the land, deteriorating soil and declining agricultural yields led to a situation in which the country was no longer able to produce enough food for its people. Many able-bodied men were forced from the land in search of means to support their families, but the only employment opportunities [were] in neighbouring South Africa. At present, an estimated 60 per cent of the male labour force is away as migrant workers in South Africa. . . . At independence, there was no economic infrastructure to speak of. Industries were virtually non-existent.[2]

THE INVENTION OF "ISOLATION"

To a scholar of Lesotho, these assertions appear not only incorrect but outlandish. For one thing, the country has not been a "subsistence" society since at least the mid-1800s, having entered the twentieth century as a producer of "wheat, mealies, Kaffir corn [sic], wool, mohair, horses and cattle" for the South African market.[3] Nor were the local Basotho people isolated from the market. When they have had surpluses of crops or livestock, the people have always known how to go about selling them in local or regional markets. According to *The Oxford History of South Africa*:

> In 1837 the Sotho of Basutoland . . . had grain stored for four to eight years: in 1844 white farmers "flocked" to them to buy grain. During 1872 (after the loss of their most fertile land west of the Caledon) the Sotho exported 100,000 *muids* [185-lb bags] of grain . . . and in 1877 when the demand for grain on the diamond fields had fallen, "large quantities" were held by producers and shopkeepers in Basutoland.[4]

Livestock auctions, meanwhile, have been held throughout the country since at least the 1950s, and animals from central Lesotho have been sold by the Basotho as far afield as South Africa for as long as anyone can remember. Far from being "untouched" by modern "development" at the time of independence, colonial rule had established a modern administration, airports, roads, schools, hospitals and markets for Western commodities.

The decline in agricultural surpluses, moreover, is neither recent nor, as the Bank suggests, due to "isolation" from the cash economy. More significant is the loss by the Basotho of most of their best agricultural land to encroaching Dutch settlers during a series of wars between 1840 and 1869. Nor is migration a recent response of a pristine and static "traditional" economy to "population pressure". As H. Ashton, the most eminent Western ethnographer of the Basuto, noted in 1952, "labour migration is . . . nearly as old as the Basuto's contact with Europeans"[5] – indeed, throughout the colonial period to the present, Lesotho has served as a labour reservoir exporting wage workers to South African mines, farms and industry.

Large-scale labour migration, moreover, preceded the decline in agriculture by many years and may even have contributed to it. Even in years of very good crop production, from the 1870s on intermittently into the 1920s, workers left the country by the thousand for work. In the early stages, it seems, migration was not related to a need to make up for poor food production but to buy guns, clothing, cattle and other goods, and, from 1869, to pay taxes.

LESOTHO REALITY

In fact, far from being the "traditional subsistence peasant society" described by the Bank, Lesotho comprises today what one writer describes as "a rural proletariat which scratches about on the land".[6]

Whilst the World Bank claims that "agriculture provides a livelihood for 85 per cent of the people",[7] the reality is that something in the order of 70 per cent of average rural household income is derived from wage labour in South Africa, while only six per cent comes from domestic crop production.[8] Similar myth-making pervades a joint FAO/World Bank report from 1975, which solemnly states that "about 70 per cent of [Lesotho's] GNP comes from the sale of pastoral products, mainly wool and mohair". A more conventional figure would be two or three per cent.[9]

Also false is the "development" literature's picture of Lesotho as a self-contained geographical entity whose relation with South Africa (its "rich neighbour") is one of accidental geographic juxtaposition rather than structural economic integration or political subordination, and whose poverty can be explained largely by the dearth of natural resources within its boundaries, together with the incompleteness with which they have been "developed". If the country is resource-poor, this is because most of the good Sotho land was taken by South Africa. Saying, as USAID does in a 1978 report, that "poverty in Lesotho is primarily resource-related" is like saying that the South Bronx of New York City is poor because of its lack of natural resources and the fact that it contains more people than its land base can support.

REARRANGING REALITY

A representation which acknowledged the extent of Lesotho's long-standing involvement in the "modern" capitalist economy of Southern Africa, however, would not provide a convincing justification for the "development" agencies to "introduce" roads, markets and credit. It would provide no grounds for believing that such "innovations" could bring about the "transformation" to a "developed", "modern" economy which would enable Lesotho's agricultural production to catch up with its burgeoning population and cut labour migration. Indeed, such a representation would tend to suggest that such measures for "opening up" the country and exposing it to the "cash economy" would have little impact, since Lesotho has not been isolated from the world economy for a very long time.

Acknowledging that Lesotho is a labour reserve for South African mining and industry rather than portraying it as an autonomous "national economy", moreover, would be to stress the importance of something which is inaccessible to a "development" planner in Lesotho. The World Bank mission to Lesotho is in no position to formulate programmes for changing or controlling the South African mining industry, and it has no disposition to involve itself in political challenges to the South African system of labour control. It is in an excellent position, however,

to devise agricultural improvement projects, extension, credit and technical inputs, for the agriculture of Lesotho lies neatly within its jurisdiction, waiting to be "developed". For this reason, agricultural concerns tend to move centre stage and Lesotho is portrayed as a nation of "farmers", not wage labourers. At the same time, issues such as structural unemployment, influx control, low wages, political subjugation by South Africa, parasitic bureaucratic elites, and so on, simply disappear.

TAKING POLITICS OUT OF "DEVELOPMENT"

One striking feature of the "development" discourse on Lesotho is the way in which the "development" agencies present the country's economy and society as lying within the control of a neutral, unitary and effective national government, and thus almost perfectly responsive to the blueprints of planners. The state is seen as an impartial instrument for implementing plans and the government as a machine for providing social services and engineering growth.

"Development" is, moreover, seen as something that only comes about through government action; and lack of "development", by definition, is the result of government neglect. Thus, in the World Bank's view, whether Lesotho's GNP goes up or down is a simple function of the current five-year "development" plan being well-implemented or badly-implemented: it has nothing to do with whether or not the mineworkers who work in South Africa get a raise in any particular year. Agricultural production, similarly, is held to be low because of the "absence of agricultural development schemes" and, thus, local ignorance that "worthwhile things could be achieved on their land". In this way, an extraordinarily important place is reserved for policy and "development" planning.[10]

Excluded from the Bank's analysis are the political character of the state and its class basis, the uses of official positions and state power by the bureaucratic elite and other individuals, cliques and factions, and the advantages to them of bureaucratic "inefficiency" and corruption. The state represents "the people", and mention of the undemocratic nature of the ruling government or of political opposition is studiously avoided. The state is taken to have no interests except "development": where "bureaucracy" is seen as a problem, it is not a political matter, but the unfortunate result of poor organization or lack of training.

Political parties almost never appear in the discourse of the Bank and other "development" institutions, and the explicitly political role played by "development" institutions such as Village Development Committees (VDCs), which often serve as channels for the ruling Basotho National Pary (BNP), is ignored or concealed. "The people" tend to appear as an undifferentiated mass, a collection of "individual farmers" and "decision makers", a concept which reduces political and structural causes of poverty to the level of individual "values", "attitudes" and "motivation". In this perspective, structural change is simply a matter of "educating" people, or even just convincing them to change their minds. When a project is sent out to "develop the farmers" and finds that "the farmers" are not much interested in farming, and, in fact, do not even consider themselves to be "farmers", it is thus easy for it to arrive at the conclusion that "the people" are mistaken, that they really are farmers and that they need only to be convinced that this is so for it to be so.

In fact, neither state bureaucracies nor the "development" projects associated with them are impartial, apolitical machines which exist only to provide social services and promote economic growth. In the case of the Canadian- and World Bank-supported Thaba-Tseka Development Project, an agricultural programme in Lesotho's central mountains, Sesotho-language documents distributed to villagers were found to have slogans of the ruling Basotho National Party (BNP) added at the end, although these did not appear in any of the English language versions. Public village meetings conducted by project staff were peppered with political speeches, and often included addresses by a high-ranking police officer on the "security threat" posed by the opposition Basutoland Congress Party. Any money remaining after project costs had been repaid went to the BNP's Village Development Committees – leading one villager to note caustically, "It seems that politics is nowadays nicknamed 'development'."

Tellingly, when I interviewed the Canadian Coordinator of the Thaba-Tseka Project in 1983, he expressed what appeared to be a genuine ignorance of the political role played by VDCs. The project hired labour through the committees, he stated, because the government had told them to. "We can't afford to get involved with politics," he said. "If they say 'hire through the Committees,' I do it."

It seems likely that such apparent political naivete is not a ruse, but simply a low-level manifestation of the refusal to face local politics which, for institutional reasons, characterizes the entire "development" apparatus.

INEVITABLE FAILURE

Because the picture of Lesotho constructed by the Bank and other "development" agencies bears so little resemblance to reality, it is hardly surprising that most "development" projects have "failed" even on their own terms. Thus after years of accusing local people of being "defeatist" or "not serious" about agriculture, and even implying that wage increases at South African mines were "a threat" to the determination of farmers to become "serious", Thaba-Tseka Project experts had to concede that local people were right that little beside maize for local consumption was going to come out of their tiny mountain fields, and that greater investment in agriculture was not going to pay handsome rewards.[11]

Casting themselves in the role of politically-neutral artisans using "development" projects as tools to grab hold of and transform a portion of the country according to a pre-determined plan, "development" officials assumed that the projects were givens and all they had to do was "implement" them.

In the case of the Thaba-Tseka Project, for example, planners assumed that it would be a relatively simple matter to devolve much of the decision-making to a newly constituted Thaba-Tseka district, in order to increase efficiency, enable the project to be in closer touch with the needs of "the people" and avoid its becoming entangled in government bureaucracy. But what the planners assumed would be a simple technical reform led – predictably – to a whole range of actors using the reforms for their own ends.

The project's Health Division, for example, was partly appropriated as a political resource for the ruling National Party. Power struggles broke out over the use of project vehicles. Government ministries refused to vote funds to the project and persisted in maintaining their own control over their field staff and making unilateral decisions on actions in the district. An attempt to hire a Mosotho to replace the project's expatriate Canadian director was rejected, since as long as the programme's image remained "Canadian", there could be no danger of bringing about a real "decentralization" of power away from Maseru, Lesotho's capital.

Instead of being a tool used by artisans to resculpt society, in short, the project was itself worked on: it became like a bread crumb thrown into an ant's nest. Plans for decentralization were thus abandoned in 1982. Yet Thaba-Tseka's planners continued to insist that the project's failure resulted somehow from the government's failure to understand the plan, or from the right organizational chart not having been found. Needing to construe their role as "apolitical", they continued to see government as a machine for delivering services, not as a political fact or a means by which certain classes and interests attempted to control the behaviour and choices of others.

A DIFFERENT KIND OF PROPERTY

Another example of "failure" stemming from the "development" discourse's false construction of Lesotho is that of livestock "development".

"Development" planners have long seen Lesotho's grasslands as one of the few potentially exploitable natural resources the country possesses,[12] and the country's herds of domestic grazing animals as an inertia-ridden "traditional" sector ripe for transformation by the dynamic "modern" cash economy. What is required, according to planners, is to develop "appropriate marketing outlets", control grassland use to optimize commercial productivity through destocking and grazing associations, introduce improved breeds, and convince "farmers to market their non-productive stock".[13]

Far from being the result of "traditional" inertia, however, the Basotho's reluctance to treat livestock commercially is deeply embedded in, and partly maintained by, a modern, capitalist labour reserve economy. In Lesotho's highly-monetized economy, an item such as a transistor radio or a bar of soap may be subject to the same market mechanisms of pricing, supply and demand as it is anywhere else. Cattle, goats and sheep, however, are subject to very different sorts of rules. Although cash can always be converted into livestock through purchase, there is a reluctance to convert grazing animals to cash through sale, except when there is an emergency need for food, clothes, or school fees.

This practice is rooted in, and reinforced by, a social system in which young working men are away in South Africa supporting their families for ten or eleven months of the year. (Mines hire only men, and it is very difficult for women from Lesotho to find work in South Africa.) If a man comes home from the mines with cash in his pocket, his wife may present him with a demand to buy her a new dress, furniture for the house or new blankets for the children. If, on the other hand, he comes home with an ox purchased with his wages, it is more difficult to make such demands.

One reason that men like to own large numbers of livestock is that they boost their prestige and personal networks in the community, partly since they can be farmed out to friends and relatives to help with their field work. They thus serve as a "placeholder" for the man in the household and the community, symbolically asserting his structural presence and prestigious social position, even in the face of his physical absence. After he has returned to the household because of injury, age or being laid off from the South African mines to "scratch about on the land", livestock begin to be sold in response to absolute shortages of minimum basic necessities. Grazing animals thus constitute a sort of special "retirement fund" for men which is effective precisely because, although it lies within the household, it cannot be accessed in the way cash can.

Hence a whole mystique has grown up glorifying cattle ownership – a mystique which, although largely contested by women, is constantly fought for by most men. Such conflict is not a sign of disintegration or crisis; it is part of the process of recreating a "tradition" which is never simply a residue of the past. If the cultural rules governing livestock in Lesotho persist, it is because they are made to persist; continuity as much as change has to be created and fought for.

Investment in livestock is thus not an alternative to migrant labour but a consequence of it. If livestock sellers surveyed by "development" experts report no source of income other than agriculture, this does not mean that they are "serious stock farmers" as opposed to "migrant labourers"; they may simply be "retired".

However useful and necessary they may be, moreover, livestock in Lesotho is less an "industry" or a "sector" than a type (however special) of consumer good bought with wages earned in South Africa when times are good and sold off only when times are bad. The sale of an animal is not "off-take" of a surplus, but part of a process which culminates in the destruction of the herd. A drop in livestock exports from Lesotho is thus not, as the "development" discourse would have it, a sign of a depressed "industry", but of a rise in incomes. For instance, when wages were increased in South African mines in the 1970s, Basotho miners seized the opportunity to invest in cattle in unprecedented numbers, leading to a surge in import figures from 4,067 in 1973 to 57,787 in 1978. Over the same period, meanwhile, cattle export figures dropped from 12,894 to 574. A boom in exports, on the other hand, would be the mark of a disaster.

Not surprisingly, attempts to "modernize" Lesotho's "livestock sector" have met with resistance. Within one year of the Thaba-Tseka Project attempting to fence off 15 square kilometres of rangeland for the exclusive use of "progressive", "commercially-minded" farmers, for example, the fence had been cut or knocked down in many places, the gates stolen, and the area was being freely grazed by all. The office of the association manager had been burned down, and the Canadian officer in charge of the programme was said to be fearing for his life.

This resistance was rooted in more than a general suspicion of the government and the "development" project. To join the official "grazing association" permitted to use the fenced-in land, stock owners were required to sell off many poor animals to buy improved ones, ending up with perhaps half as many. Such sales and restrictions in herd size were not appealing for most Basotho men. Joining the association not only meant accepting selection, culling and marketing of herds. It also meant acquiescing in the enclosure of both common grazing land and (insofar as any Mosotho's livestock are also a social, shared domain of wealth) animals. It thus signified a betrayal of fellow stock-owners who remained outside the organization, an act considered anti-social. Prospective association members also probably feared that their animals – which represent wealth in a visible, exposed, and highly vulnerable form – might be stolen or vandalized in retaliation.

THE SIDE EFFECTS OF "FAILURE"

Despite such disasters, it may be that what is most important about a "development" project is not so much what it fails to do but what it achieves through its "side effects". Rather than repeatedly asking the politically naive question "Can aid programmes ever be made really to help poor people?", perhaps we should investigate the more searching question, "What do aid programmes do *besides* fail to help poor people?"

Leftist political economists have often argued that the "real" purpose of "development" projects is to aid capitalist penetration into Third World countries. In Lesotho, however, such projects do not characteristically succeed in introducing new relations of production (capitalist or otherwise), nor do they bring about modernization or significant economic transformations. Nor are they set up in such a way that they ever could. For this reason, it seems a mistake to interpret them *simply* as "part of the historical expansion of capitalism" or as elements in a global strategy for controlling or capitalizing peasant production.

Capitalist interests, moreover, can only operate through a set of social and cultural structures so complex that the outcome may be only a baroque and unrecognizable transformation of the original intention. Although it is relevant to know, for instance, that the World Bank has an interest in boosting production and export of cash crops for the external market, and that industrialized states without historic links to an area may sponsor "development" projects as a way of breaking into otherwise inaccessible markets, it remains impossible simply to read off actual events from these known interests as if the one were a simple effect of the other. Merely knowing that the Canadian government has an interest in promoting rural "development" because it helps Canadian corporations to find export markets for farm machinery, for example, leaves many of the empirical details of the Canadian role in Thaba-Tseka absolutely mysterious.

Another look at the Thaba-Tseka Project, however, reveals that, although the project "failed" both at poverty alleviation and at extending the influence of international capital, it did have a powerful and far-reaching impact on its region. While the project did not transform livestock-keeping, it did build a road to link Thaba-Tseka more strongly with the capital.

While it did not bring about "decentralization" or "popular participation", it was instrumental in establishing a new district administration and giving the government a much stronger presence in the area than it had ever had before.

As a direct result of the construction of the project centre and the decision to make that centre the capital of a new district, there appeared a new post office, a police station, a prison and an immigration control office; there were health officials and nutrition officers and a new "food for work" administration run by the Ministry of Rural Development and the Ministry of Interior, which functioned politically to regulate the power of chiefs. The new district centre also provided a good base for the "Para-Military Unit", Lesotho's army, and near the project's end in 1983, substantial numbers of armed troops began to be garrisoned at Thaba-Tseka.

In this perspective, the "development" apparatus in Lesotho is not a machine for eliminating poverty that is incidentally involved with the state bureaucracy. Rather, it is a machine for reinforcing and expanding the exercise of bureaucratic state power, which incidentally takes "poverty" as its point of entry and justification – launching an intervention that may have no effect on the poverty but does have other concrete effects.

This does not mean that "the state", conceived as a unitary entity, "has" more power to extract surplus, implement programmes, or order around "the masses" more efficiently – indeed, the reverse may be true. It is, rather, that more power relations are referred through state channels and bureaucratic circuits – most immediately, that more people must stand in line and await rubber stamps to get what they want. "It is the same story over again," said one "development" worker. "When the Americans and the Danes and the Canadians leave, the villagers will continue their marginal farming practices and wait for the mine wages, knowing only that now the taxman lives down the valley rather than in Maseru."[14]

At the same time, a "development" project can effectively squash political challenges to the system not only through enhancing administrative power, but also by casting political questions of land, resources, jobs or wages as technical "problems" responsive to the technical "development" intervention. If the effects of a "development" project end up forming any kind of strategically coherent or intelligible whole, it is as a kind of "anti-politics" machine, which, on the model of the "anti-gravity" machine of science fiction stories, seems to suspend "politics" from even the most sensitive political operations at the flick of a switch.

Such a result may be no part of the planners' intentions. It is not necessarily the consequence of any kind of conspiracy to aid capitalist exploitation by incorporating new territories into the world system or working against radical social change, or bribing national elites, or mystifying the real international relationships. The result can be accomplished, as it were, behind the backs of the most sincere participants. It may just happen to be the way things work out. On this view, the planning apparatus is neither mere ornament nor the master key to understanding what happens. Rather than being the blueprint for a machine, it is a *part* of the machine.

WHAT IS TO BE DONE? BY WHOM?

If, then, "development" cannot be the answer to poverty and powerlessness in Lesotho, what is? What is to be done, if it is not "development"?

Any question of the form "What is to be done?" demands first of all an answer to the question "By whom?" The "development" discourse, and a great deal of policy science, tends to answer this question in a utopian way by saying "Given an all-powerful and benevolent policy-making apparatus, what should it do to advance the interests of its poor citizens?"

This question is worse than meaningless. In practice, it acts to disguise what are, in fact, highly partial and interested interventions as universal, disinterested and inherently benevolent. If the question "What is to be done?" has any sense, it is as a real-world tactic, not a utopian ethics.

The question is often put in the form "What should *they* do?", with the "they" being not very helpfully specified as "Lesotho" or "the Basotho". When "developers" speak of such a collectivity what they mean is usually the government. But the government of Lesotho is not identical with the people who live in Lesotho, nor is it in any of the established senses "representative" of that collectivity. As in most countries, the government is a relatively small clique with narrow interests. There is little point in asking what such entrenched and often extractive elites should do in order to empower the poor. Their own structural position makes it clear that they would be the last ones to undertake such a project.

Perhaps the "they" in "What should they do?" means "the people". But again, the people are not an undifferentiated mass. There is not one question – What is to be done? – but hundreds: What should the mineworkers do? What should the abandoned old women do? and so on. It seems presumptuous to offer prescriptions here. Toiling miners and abandoned old women know the tactics proper to their situations far better than any expert does. If there is advice to be given about what "they" should do, it will not be dictating general political strategy or giving a general answer to the question "what is to be done?" (which can only be determined by those doing the resisting) but answering specific, localized, tactical questions.

WHAT SHOULD WE DO?

If the question is, on the other hand, "What should *we* do?", it has to be specified, which "we"? If "we" means "development" agencies or governments of the West, the implied subject of the question falsely implies a collective project for bringing about the empowerment of the poor. Whatever good or ill may be accomplished by these agencies, nothing about their general mode of operation would justify a belief in such a collective "we" defined by a political programme of empowerment.

For some Westerners, there is, however, a more productive way of posing the question "What should we do?". That is, "What should we intellectuals working in or concerned about the Third World do?" To the extent that there are common political values and a real "we" group, this becomes a real question. The answer, however, is more difficult.

Should those with specialized knowledge provide advice to "development" agencies who seem hungry for it and ready to act on it? As I have tried to show, these agencies seek only the kind of advice they can take. One "developer" asked my advice on what his country could do "to help these people". When I suggested that his government might contemplate sanctions against apartheid, he replied, with predictable irritation, "No, no! I mean development!" The only advice accepted is about how to "do development" better. There is a ready ear for criticisms of "bad development projects", only so long as these are followed up with calls for "good development projects". Yet the agencies who plan and implement such projects – agencies like the World Bank, USAID, and the government of Lesotho – are not really the sort of social actors that are very likely to advance the empowerment of the poor.

Such an obvious conclusion makes many uncomfortable. It seems to them to imply hopelessness; as if to suggest that the answer to the question "What is to be done?" is: "Nothing." Yet this conclusion does not follow. The state is not the only game in town, and the choice is not between "getting one's hands dirty by participating in or trying to reform development projects" and "living in an ivory tower". Change comes when, as Michel Foucault says, "critique has been played out in the real, not when reformers have realized their ideas".[15]

For Westerners, one of the most important forms of engagement is simply the political participation in one's own society that is appropriate to any citizen. This is, perhaps, particularly true for citizens of a country like the US, where one of the most important jobs for "experts" is combating imperialist policies.

NOTES AND REFERENCES

1 Murray, C., *Families Divided: The Impact of Migrant Labour in Lesotho*, Cambridge University Press, New York, 1981, p. 19.

2 World Bank, *Lesotho: A Development Challenge*. World Bank, Washington, DC, 1975, p. 1.

3 "Basutoland", *Encylopedia Britannica*, 1910.

4 Wilson, M. and Thompson, L. (eds.) *The Oxford History of South Africa*, Vol. 1, Oxford University Press, New York, 1969.

5 Ashton, H., *The Basuto: A Social Study of Traditional and Modern Lesotho*, Oxford University Press, New York, 1967 (second edition), p. 162.

6 Murray, C., op. cit. 1.

7 FAO/World Bank, *Draft Report of the Lesotho First Phase Mountain Area Development Project Preparation Mission*, Vols. I and II), FAO, Rome, 1975, Annex 1, p. 7.

8 van der Wiel, A.C.A., *Migratory Wage Labour: Its Role in the Economy of Lesotho*, Mazenod Book Centre, Mazenod, 1977.

9 FAO/World Bank, op. cit. 7, Annex 1, p. 7.

10 World Bank, op. cit. 2, p. 9.

11 See "Appraisal of Project Progress During the Pilot Phase and Review of Plans to Expand Agricultural Programs in Phase II of Project Operations". CIDA, Ottawa, 1978, p. 39.

12 See, for example, FAO/World Bank, op. cit. 7, Annex 1, pp. 10–12. For a related South African history of government intervention into "traditional" livestock keeping, see Beinart, W. and Bundy, C., "State Intervention and Rural Resistance: The Transkei, 1900–1965", in Klein, M. (ed.) *Peasants in Africa*, Sage, Beverley Hills, 1981.

13 CIDA, op. cit. 11.

14 Quoted in Murphy, B., "Smothered in Kindness", *New Internationalist*, No. 82, 1979, p. 13.

15 Foucault, M., "Questions of Method: An Interview", *Ideology and Consciousness*, 8, 1981, p. 13.

S i X

Institutions, governance and participation

INTRODUCTION TO PART SEVEN

'If the model doesn't fit reality, change reality.' This well-known saying has become the mantra of some of the policy makers that delivered the counter-revolution in development studies in the 1970s and 1980s. Consider Anne Krueger, for example. An erstwhile research leader in the World Bank, who later moved to the International Monetary Fund, by far the more market fundamentalist of the two Bretton Woods institutions, Krueger has recently blamed the failures of the Washington Consensus model on the lack of effort invested by reforming countries. As Dani Rodrik reports in his reading here, for Anne Krueger (2004) it was a case of 'Meant well, tried little, failed much'. As he goes on to note, 'The policy implication that follows is simple: do more of the same [trade liberalization, privatization, fiscal balance or minimal deficits], and do it well'.

For Rodrik himself this advice will not do. (Joseph Stiglitz is equally scathing about the IMF's textbook economics when he discusses the fund's handling of the Asian financial crisis in the late 1990s: Stiglitz, 2002). Rodrik does not go all the way back to Polanyi – or even to Marx or Smith – to make an argument about the irredeemably social nature of pure or free markets. But he does commend a World Bank (2005) analysis of the 1990s 'reforms' for paying proper attention to 'the dynamic forces that lie behind the growth process', as opposed to the static efficiency gains that come from eliminating 'deadweight-loss triangles'.

This is an important observation, and it strikes at the heart of debates about economic development. For Rodrik and other institutionalists, one of the big problems of World Bank–IMF advice in the 1990s was its obsession with static resource allocation inefficiencies. In their view, this is akin to not seeing the wood for the trees. The wood, Rodrik insists, pointing to the remarkable post-1990 economic growth of China and India, two unconventional reformers, is the growth engine itself and the productivity gains that power it. And the second big problem with Krueger-style advice, Rodrik continues, is that it is like a laundry list. It provides a set menu of reforms that has to be followed by all countries regardless of the particular constraints on productivity growth that affect them. For Rodrik, this is bad public policy.

> Enhancing private investment incentives may require improving the security of property rights in one country, but enhancing the financial sector in another. Technological catch-up may call for better or worse patent protection, depending on the level of development. This explains why countries that are growing – the [World Bank] report cites Bangladesh, Botswana, Chile, China, Egypt, India, Lao PDR, Mauritius, Sri Lanka, Tunisia, and Vietnam – have such diverse policy configurations, and why attempts to copy successful policy reforms in another country often end up in failure.

Rodrik respects the power of markets but always with regard to their particularities, or differences. And these differences have to do both with global constraints and sequencing issues, and with the nature of local institutions: broadly, the rules of the game, including the distribution and enforcement of property rights. In this context, Rodrik takes care to cite an important paper by Acemoglu *et al.* (2001) which finds that 'security of property rights has been historically perhaps the single most important determinant of why some countries grew rich and others remained poor'.

Another paper he might have cited is the one by James E. Mahon, Jr, that is excerpted here. Mahon addresses himself to another key question in the development literature: Why is it that economic growth in Latin America has fallen behind that of the East Asian newly industrialised countries, given the substantial lead that Latin America enjoyed at the start of the twentieth century? His answer, briefly, is that Latin America paid the price of its earlier success. Many Latin American countries in the 1950s and 1960s enjoyed considerable success as primary commodity exporters. In Mahon's view, these same countries would have had to devalue their currencies very significantly in the 1970s to compete with East Asia by means of a new strategy of export-oriented industrialisation. Fearing the consequences of this at home, where real wages would also have had to be pushed down sharply in the short term, political elites in Latin America opted for the status quo. They continued with primary commodity dependence and expensive forms of import-substitution industrialisation.

Somewhat absent from Mahon's analysis is another key difference between East Asia and Latin America, which has to do with their agrarian policies. Bruce Johnston and Peter Kilby (1975) have argued that the land reforms imposed in South Korea and Taiwan after 1945/49 transferred land to those peasants who mainly tilled it. Although their surpluses were squeezed to pay for industrial accumulation, this was not done in draconian fashion. This was not the USSR in the 1930s. Peasants still had strong incentives to feed their families and produce marketed surpluses. They also found it rational to invest in their farms in a way that unprotected tenants and labourers in Latin America did not. There, for the most part, a bimodal agrarian structure worked to the advantage of a small number of capitalist farmers and estate owners. A majority of working families in the countryside continued to live in extreme poverty (as, for example, in the Brazilian north-east). Many more were pushed to urban slums and shanties to make a living. Balanced rural–urban or agricultural–industrial growth remained elusive in many countries in Latin America, even as it was being manufactured in East Asia.

Another example of virtuous institutional change is saluted in the next reading. In an important and much cited article, Jean C. Oi addresses herself to the peculiarities of reform in post-Mao China. How did China achieve such high rates of economic growth in the 1980s without privatisation? Did this run counter to some key claims of the Washington Consensus and World Bank orthodoxy? In Oi's view, the secret of China's post-1976 success is to be found in local state corporatism. Briefly, the post-Mao reforms transferred income rights over agricultural production from collectives to individual farming households. One vehicle for this transfer was the household responsibility system. The very success of this strategy, however, threatened local government with a significant loss of income, which could have led to political tension. The Chinese reformers, however, as Oi also makes clear, instituted a set of parallel fiscal reforms. These allowed local governments to keep a higher share of the tax revenues generated by economic growth. In other words, they were given a stake in the system. Local states now began to behave in entrepreneurial fashion. In the process they became more akin to developmental corporations than predatory instruments of governance.

Local governments can also be made more responsive to citizens, albeit under some fairly pressing conditions. These are discussed here by Patrick Heller (2002) in his comparative study of Porto Alegre (Brazil), Kerala (India) and South Africa. Heller draws an important distinction between two sets of political developments. On the one hand, he notes the emergence of formal and even decentralised democracies in the developing world since the 1980s. At the same time, he is less convinced there is a genuine commitment – in most cases – to the promotion of informed participation by ordinary people in the democratic process. For the latter to happen, Heller suggests, it is important that a process of formal institutional change is accompanied by 'a rich and dense tapestry of grassroots democratic organizations – the historical legacy of prolonged mass-based pro-democracy movements – capable of mobilizing constituencies traditionally excluded from policy-making arenas, and dislodging traditional clientelistic networks'.

Sadly, as Heller also proceeds to report, the development of well-informed, rights-bearing citizens has been slow to happen in most of the world. Again, this is a legacy of previous institutional arrangements, notably those produced within regimes of more or less direct colonial rule. Partha Chatterjee has argued that,

The story of citizenship in the modern West moves from the institution of civic rights in civil society to political rights in the fully developed nation-state. Only then does one enter the relatively recent phase where 'government from the social point of view' seems to take over. In countries of Asia and Africa, however, the chronological sequence is quite different. There the career of the modern state has been foreshortened. Technologies of governmentality often pre-date the nation-state, especially where there has been a long experience of European colonial rule.

(Chatterjee 2004: 36)

The subjects of colonial powers became formal citizens of new states at independence. In Chatterjee's view, however, the republican ideals that were put before them as members of the nationalist struggle were cast aside to make way for 'developmental state(s) which promised to end poverty and backwardness by adopting appropriate policies of economic growth and social reform' (ibid.: 37).

In India, the new 'citizen' was constituted as a supplicant or beneficiary of a ruling elite which sought the endorsement of the citizenry every few years in 'the great anonymous performance of citizenship [the vote]' (ibid.: 18). In many parts of Africa even this privilege was denied. Real power was conferred, as Mahmood Mamdani (1996) shows here, on the Native Authorities which had sat below the ruling white authorities, and which had been given licence to rule the peasantry with a degree of violence that was legitimised in the name of custom. For Mamdani, 'The core agenda that African states faced at independence was three-fold: deracializing civil society, detribalizing the Native Authority, and developing the economy in the context of unequal international relations.' He also insists that of these three objectives only the first – the substitution of African for white (or even 'Indian') elites – has been achieved with any degree of success, and then mainly in the highest reaches of the state. 'Without a reform of the local state, the peasantry locked up under the hold of a multiplicity of ethnically defined Native Authorities could not be brought into the mainstream of the historical process. In the absence of democratization, there could be no development of a home market.'

Mamdani concludes by pointing up the limits of a discourse of democratisation that is focused on civil society institutions and periodic multi-party elections. He also reflects on the politics of depoliticisation and the emergence of what he calls centralised despotism. For Mamdani, as for Chatterjee, the notion that the peasantry can be empowered directly by new technologies of rule informed by the good governance agenda is naive to the point of being 'unscrupulously charitable' (Chatterjee 2004: 38). (They have in mind such things as untutored participation, self-help groups, and perhaps even micro-finance.) This mistakes both the nature of the state in Africa, and the historical production of state–society relations characterised by vertical dependencies, violence and patrimonialism.

It also follows, David Mosse (2001) contends, in the final reading in this Part, that intended beneficiaries of participatory aid projects, in Africa as in the Indo-British Rainfed Farming Project (KRIBP) discussed here, often 'fail' to read these projects in the terms intended by their Western designers. They rather understand them from within a social map that highlights their dependence on local power brokers, on the one hand, and which encourages them to learn the language of the project to gain such advantages as are on offer. In the case of better-off villagers this means learning the languages of poverty and participation. As Mosse sharply demonstrates, villagers learn the language of formal institutional change – participatory development, decentralisation, the new public administration – in part to reproduce the informal structures of rule that govern their daily lives.

REFERENCES AND FURTHER READING

Acemoglu, D., Johnson, S. and Robinson, J. (2001) 'The colonial origins of comparative development: an empirical investigation', *American Economic Review* 91 (5): 1369–401.

Bayart, J-F., Ellis, S. and Hibou, B. (1999) *The Criminalization of the State in Africa*, Oxford: James Currey.

Chatterjee, P. (2004) *The Politics of the Governed: Reflections on Popular Politics in Most of the World*, New York: Columbia University Press.

Cooke, B. and Kothari, U. (eds) (2001) *Participation: The New Tyranny?* London: Zed.

Crook, R. and Manor, J. (1998) *Democracy and Decentralisation in South Asia and West Africa: Participation, Accountability and Performance*, Cambridge: Cambridge University Press.

Dirks, N. (2001) *Castes of Mind: Colonialism and the Making of Modern India*, Princeton, NJ: Princeton University Press.

Dixit, A. (2004) *Lawlessness and Economics: Alternative Modes of Governance*, Princeton, NJ: Princeton University Press.

Hyden, G. (2006) *African Politics in Comparative Perspective*, Cambridge: Cambridge University Press.

Johnston, B. and Kilby, P. (1975) *Agriculture and Structural Transformation*, Oxford: Oxford University Press.

Krueger, A. (2004) 'Meant well, tried little, failed much: policy reforms in emerging market economies', Remarks at the Roundtable Lecture at the Economic Honors Society, New York University, 23 March.

Kuczynski, P-P. and Williamson, J. (eds) (2003) *After the Washington Consensus: Restarting Growth and Reform in Latin America*, Washington, DC: Institute for International Economics.

Kumar, S. and Corbridge, S. (2002) 'Programmed to fail? Development projects and the politics of participation', *Journal of Development Studies* 39 (2): 79–103.

Posner, D. (2005) *Institutions and Ethnic Politics in Africa*, Cambridge: Cambridge University Press.

Rodrik, A., Subramanian, A. and Trebbi, F. (2004) 'Institutions rule: the primacy of institutions over geography and integration in economic development', *Journal of Economic Growth* 9 (2): 131–65.

Stiglitz, J. (2002) *Globalization and Its Discontents*, New York: W.W. Norton.

Tendler, J. (1997) *Good Government in the Tropics*, Baltimore, MD: Johns Hopkins University Press.

World Bank (2005) *Economic Growth in the 1990s: Learning from a Decade of Reform*, Washington, DC: World Bank.

'Goodbye Washington Consensus, Hello Washington Confusion?'

Journal of Economic Literature (2006)

Dani Rodrik

Editors' Introduction

Dani Rodrik was born in Istanbul, Turkey, in 1957 and is a Professor of International Political Economy at Harvard University's John F. Kennedy School of Government. He has worked for the National Bureau of Economic Research in the United States and is affiliated with several leading think-tanks concerned with economic performance.

Rodrik is often described as a heterodox or institutional economist, but it is perhaps more accurate to say that he is one of an influential group of political economists who are reworking the field of 'development economics'. Significantly, a book that Rodrik edited in 2003 – *In Search of Prosperity: Analytical Narratives on Economic Growth* – brought together a number of reasonably like-minded collaborators, including Daron Acemoglu, J. Bradford DeLong, Simon Johnson, James A. Robinson, Devesh Roy, Arvind Subramanian and Susan Wolcott. Earlier in his career, Rodrik worked alongside Robert Wade (also represented in this reader in Part 8). If anything unites this group of scholars it is a shared commitment to reworking standard economic growth models so that they more accurately capture the diversity of developing country institutions and growth paths. The intention, then, is not to throw out mainstream economic theory or to go back to the two-sector models that dominated a first generation of development economics. Nevertheless, there is in the work of Rodrik especially a healthy scepticism about open-economy models and about globalisation more generally. In a book from 1997 he asked *Has Globalization Gone Too Far?* And a 2001 report for UNDP was published as 'The global governance of trade as if development really mattered'. Rodrik's work is open to the suggestions that (a) the increased internationalisation of trade and capital flows might not benefit the poor directly by raising long-term growth rates, and (b) might even damage the living standards of the poor if (or where) 'globalisation' is used as an excuse for cutting back on welfare transfers to poorer people. Rodrik typically insists that these are open questions that must be investigated empirically and in particular countries. He also tends to insist, as the reading published here confirms, that development policies are at their worst when they derive from forms of economic theory that assume that 'one size [of model] can fit all'. This was the case with the Washington Consensus in the late 1980s and 1990s. None of this means, however, that Rodrik imagines that development policies must always be idiosyncratic, or that development agencies should embrace a 'nihilistic attitude where "everything goes" '. Far from it. In the second half of this article Rodrik commends an approach to thinking about growth strategies which combines three elements: a diagnostic analysis of constraints on growth, imaginative policy design to address these constraints effectively, and the institutionalisation of diagnosis and policy response in the longer term governance of the economy. As Rodrik explains, this is what *he* means by institutionalism.

Key references

Dani Rodrik (2006) 'Goodbye Washington consensus, hello Washington confusion?', *Journal of Economic Literature* XLIV (December): 973–87.
— (1997) *Has Globalization Gone Too Far?* Washington: Institute for International Economics.
— (2001) 'The global governance of trade as if development really mattered', prepared for UNPD Weatherhead Centre for International Affairs, Harvard University.
— (ed.) (2003) *In Search of Prosperity: Analytical Narratives on Economic Growth*, Princeton, NJ: Princeton University Press.
— (2007) *One Economics, Many Recipes: Globalization, Institutions and Economic Growth*, Princeton, NJ: Princeton University Press.
D. Rodrik, R. Hausmann and L. Pritchett (2005) 'Growth accelerations', *Journal of Economic Growth* 10: 303–29.
D. Rodrik and R. Waczaig (2005) 'Do democratic transitions produce bad economic outcomes?' *American Economic Review, Papers and Proceedings* 95 (2): 50–5.

1. INTRODUCTION

Life used to be relatively simple for the peddlers of policy advice in the tropics. Observing the endless list of policy follies to which poor nations had succumbed, any well-trained and well-intentioned economist could feel justified in uttering the obvious truths of the profession: get your macro balances in order, take the state out of business, give markets free rein. "Stabilize, privatize, and liberalize" became the mantra of a generation of technocrats who cut their teeth in the developing world and of the political leaders they counseled.

Codified in John Williamson's (1990) well-known Washington Consensus, this advice inspired a wave of reforms in Latin America and Sub-Saharan Africa that fundamentally transformed the policy landscape in these developing areas. With the fall of the Berlin Wall and the collapse of the Soviet Union, former socialist countries similarly made a bold leap toward markets. There was more privatization, deregulation, and trade liberalization in Latin America and Eastern Europe than probably anywhere else at any point in economic history. In Sub-Saharan Africa, governments moved with less conviction and speed, but there too a substantial portion of the new policy agenda was adopted: state marketing boards were dismantled, inflation reduced, trade opened up, and significant amounts of privatization undertaken.[1]

Such was the enthusiasm for reform in many of these countries that Williamson's original list of do's and don'ts came to look remarkably tame and innocuous by comparison. In particular, financial liberalization and opening up to international capital flows went much farther than what Williamson had anticipated (or thought prudent) from the vantage point of the late 1980s. Williamson's (2000) protestations notwithstanding, the reform agenda eventually came to be perceived, at least by its critics, as an overtly ideological effort to impose "neoliberalism" and "market fundamentalism" on developing nations.

The one thing that is generally agreed on about the consequences of these reforms is that things have not quite worked out the way they were intended. Even their most ardent supporters now concede that growth has been below expectations in Latin America (and the "transition crisis" deeper and more sustained than expected in former socialist economies). Not only were success stories in Sub-Saharan Africa few and far in between, but the market-oriented reforms of the 1990s proved ill-suited to deal with the growing public health emergency in which the continent became embroiled. The critics, meanwhile, feel that the disappointing outcomes have vindicated their concerns about the inappropriateness of the standard reform agenda. While the lessons drawn by proponents and skeptics differ, it is fair to say that nobody really believes in the Washington Consensus anymore.[2] The question now is not whether the Washington Consensus is dead or alive; it is what will replace it.

The World Bank's *Economic Growth in the 1990s: Learning from a Decade of Reform* (2005, henceforth *Learning from Reform*) is one of a spate of recent attempts at making sense of the facts of the last decade and a half, and probably the most intelligent. In fact, it is a rather extraordinary document insofar as it shows how far we have come from the original Washington Consensus. There are no confident assertions here of what works and what doesn't – and no blueprints for policymakers to adopt. The emphasis is on the need for humility, for policy diversity, for selective and modest reforms, and for experimentation. "The central message of this volume," Gobind Nankani, the World Bank vice-president who oversaw the effort, writes in the preface of the book, "is that there is no unique universal set of rules . . . [W]e need to get away from formulae and the search for elusive 'best practices' . . ." (p. xiii).[3] Occasionally, the reader has to remind himself that the book he is holding in his hands is not some radical manifesto, but a report prepared by the seat of orthodoxy in the universe of development policy.

2. THE RECORD

Here is how *Learning from Reform* summarizes the surprises of the 1990s. First, there was an unexpectedly deep and prolonged collapse in output in countries making the transition from communism to market economies. More than a decade into the transition, many countries had still not caught up to their 1990 levels of output. Second, Sub-Saharan Africa failed to take off, despite significant policy reform, improvements in the political and external environments, and continued foreign aid. The successes were few – with Uganda, Tanzania, and Mozambique the most commonly cited instances – and remained fragile more than a decade later. Third, there were frequent and painful financial crises in Latin America, East Asia, Russia, and Turkey. Most had remained unpredicted by financial markets and economists until capital flows started to reverse very suddenly. Fourth, the Latin American recovery in the first half of the 1990s proved short-lived. The 1990s as a whole saw less growth in Latin America in per capita GDP than in 1950–80, despite the dismantling of the state-led, populist, and protectionist policy regimes of the region. Finally, Argentina, the poster

boy of the Latin American economic revolution, came crashing down in 2002 as its currency board proved unsustainable in the wake of Brazil's devaluation in January 1999.

Significantly, the period since 1990 was *not* a disaster for economic development. Quite to the contrary. From the standpoint of global poverty, the last two decades have proved the most favorable that the world has ever experienced. Rapid economic growth in China, India, and a few other Asian countries has resulted in an *absolute* reduction in the number of people living in extreme poverty.[4] The paradox is that that was unexpected too! China and India increased their reliance on market forces, of course, but their policies remained highly unconventional. With high levels of trade protection, lack of privatization, extensive industrial policies, and lax fiscal and financial policies through the 1990s, these two economies hardly looked like exemplars of the Washington Consensus. Indeed, had they been dismal failures instead of the successes they turned out to be, they would have arguably presented stronger evidence in support of Washington Consensus policies.[5]

Along with this telling, if anecdotal, evidence has come a more skeptical reading of the cross-national relationship between policy reform and economic growth. Characteristically, it is the World Bank itself that has been prone to make grandiose claims on the impact of policy reform. In one particularly egregious instance cited by William Easterly (2005), Paul Collier and David Dollar (2001) argued that policy reform of the conventional type could cut world poverty by half. Work by Easterly (2005) and Francisco Rodríguez (2005) show that the data do not support such claims. The evidence that macroeconomic policies, price distortions, financial policies, and trade openness have predictable, robust, and systematic effects on national growth rates is quite weak – except possibly in the extremes. Humongous fiscal deficits or autarkic trade policies can stifle economic growth, but moderate amounts of each are associated with widely varying economic outcomes.[6]

The question is how to interpret this recent experience, and how to turn the interpretation into concrete policy advice. Here *Learning from Reform* makes some valuable progress. I summarize some of the main conclusions below, emphasizing those that depart most strongly from the earlier approach.

3. THE INTERPRETATION

One of the insights of *Learning from Reform* is that the conventional package of reforms was too obsessed with deadweight-loss triangles and reaping the efficiency gains from eliminating them, and did not pay enough attention to stimulating the dynamic forces that lie behind the growth process. Seeking efficiency gains does not amount to a growth strategy. Although the report does not quite put it in this way, what I think the authors have in mind is that market or government failures that affect accumulation or productivity change are much more costly, and hence more deserving of policy attention, than distortions that simply affect static resource allocation. They may also be harder to identify. Focusing on the latter instead of the former results in small benefits, and could even turn out to be counterproductive when policy makers face a political budget constraint (more reform in one area means less reform in another).

A second conclusion is that the broad *objectives* of economic reform – namely market-oriented incentives, macroeconomic stability, and outward orientation – do not translate into unique set of policy actions. In the words of the Report, "The principles of . . . 'macroeconomic stability, domestic liberalization, and openness' have been interpreted narrowly to mean 'minimize fiscal deficits, minimize inflation, minimize tariffs, maximize privatization, maximize liberalization of finance,' with the assumption that the more of these changes the better, at all times and in all places – overlooking the fact that these expedients are just *some* of the ways in which these principles can be implemented" (p. 11, emphasis in the original). The authors go on to point out that each of these ends can be achieved in a number of ways. For example, trade openness can be achieved through lower import tariffs, but also through duty drawbacks, export subsidies, special economic zones, export processing zones, and so on. This renunciation of standard "best practice" in World Bank policy advice is quite remarkable, and must not have come without a significant internal fight.

Third, different contexts require different solutions to solving common problems. Enhancing private investment incentives may require improving the security of property rights in one country but enhancing the financial sector in another. Technological catch-up may call for better or worse patent protection, depending on the level of development. This explains why countries that are growing – the report cites Bangladesh, Botswana, Chile, China, Egypt, India, Lao PDR, Mauritius, Sri Lanka, Tunisia, and Vietnam – have such diverse policy configurations, and why attempts to copy successful policy reforms in another country often end up in failure.

Fourth, *Learning from Reform* argues that there has been a tendency to exaggerate the advantages of rules over discretion in government behavior. Rules were meant to discipline the malfeasance of governments. But it turns out that "government discretion cannot be bypassed" (p. 14). Argentina's currency board, which removed monetary policy from the hands of the government, worked well when the binding constraint was lack of credibility, but led to disastrous outcomes when the binding constraint became an overvalued currency. There is no alternative to improving the processes of decisionmaking (better checks and balances, better guiding principles, better implementation) such that discretion leads to better outcomes.

Finally, reform efforts need to be selective and focus on the *binding constraints* on economic growth rather than take a laundry-list approach à la Washington Consensus. While there is no foolproof method of identifying these constraints, common sense and economic analysis can help (see below). When investment is constrained by poor property rights, improving financial intermediation will not help. When it is constrained by high cost of capital, improving institutional quality will hardly work. Experimentation and learning about the nature of the binding constraints, and the changes therein, are therefore an integral part of the reform process. Even though countries may face situations in which many constraints need to be addressed simultaneously, the report judges these situations to be rare: "In most cases, countries can deal with constraints sequentially, a few at a time" (p. 16).

Taking these conclusions at face value, what they entail is nothing less than a radical rethink of development strategies. Of course, it would be naïve to think that the World Bank's practice will therefore change overnight. There is little evidence that operational work at the Bank has internalized these lessons to any significant extent.[7] And, as I will discuss below, there are contending interpretations of what has gone wrong and how to move forward. But the mere fact that such views have been put forward in an official World Bank publication is indicative of the changing nature of the debate and of the space that is opening up within orthodox circles for alternative visions of development policy.

4. THE ALTERNATIVES I: INSTITUTIONS

Around the same time that the World Bank was grappling with the lessons of the 1990s, its sister institution across the street, the International Monetary Fund (IMF), put out a document that focused on much the same issues in the context of Latin America (Anoop Singh et al. 2005). This is an equally remarkable document which shows that in Washington there is anything but consensus these days. The IMF report starts from the same basic premise – growth has been disappointing – but its basic argument could not be more different. According to its authors, the problem was not with the approach taken to reform, but that it did not go deep and far enough. Using the report's own words, "reforms were uneven and remained incomplete" (p. xiv). "More progress was made," the IMF report claims, "with measures that had low up-front costs, such as privatization, relative to reforms that promised greater long-term benefits, such as improving macroeconomic and labor market institutions, and strengthening legal and judicial systems" (p. xiv). The same diagnosis is expressed succinctly in the title of one of Anne Krueger's speeches on policy reform: "Meant Well, Tried Little, Failed Much" (Krueger 2004). From this perspective, the failures have to be chalked up to too little reform of the kind that Washington has advocated all along and not to the nature of these reforms itself.[8] The policy implication that follows is simple: do more of the same, and do it well.

Several key ideas underpin this interpretation of the evidence. First, political leaders may have had the talk, but they didn't quite have the walk: their commitment to genuine reform was often "skin-deep" and there was "lack of follow-through" (Krueger 2004). Second, and more fundamentally, even committed reformers stopped well short of undertaking the full gamut of institutional changes needed to create well-functioning market economies. Regulatory and supervisory institutions in product and financial markets proved too weak. Poor governance and corruption remained a problem. Courts and the judiciary were ineffective. And labor market institutions were not sufficiently "flexible."

Of course this second point, about the lack of emphasis on institutional reform, is itself an implicit repudiation of the original version of the Washington Consensus, insofar as the latter did not feature institutional reform of the type that Krueger and the IMF have in mind in their interpretation of the 1990s. Most of the items in Williamson's original list were relatively simple policy changes (liberalize trade, eliminate currency overvaluation, reduce fiscal deficits, and so on) that did not require deep-seated institutional changes. Williamson did include "property rights" in his list, but that was the last item on the list and came almost as an afterthought.

What has become clearer to practitioners of the Washington Consensus over time is that the standard policy reforms did not produce lasting effects if the background institutional conditions were poor. Sound policies needed to be embedded in solid institutions. Moreover, there were significant complementarities across different areas of reform. Trade liberalization would not work if fiscal institutions were not in place to make up for lost trade revenue, capital markets did not allocate finance to expanding sectors, customs officials were not competent and honest enough, labor-market institutions did not work properly to reduce transitional unemployment, and so on. The upshot is that the original Washington Consensus has been augmented by a long list of so-called "second-generation" reforms that are heavily institutional in nature. The precise enumeration of these requisite institutional reforms depends on who is talking and when, and often the list seems to extend to whatever it is that the reformers may *not* have had a chance to do – which is one of the problems that I will discuss below. Nonetheless, one possible rendition is shown in Table 1, where I have listed ten second-generation reforms to maintain symmetry with the original Washington Consensus.

This focus on institutions has also received a strong boost from the (largely unrelated) rediscovery of institutions as a driver of long-term economic performance in the empirical literature on economic growth. In particular, Daron Acemoglu, Simon Johnson, and James A. Robinson's (2001) important work drove home the point that the security of property rights has been historically perhaps the single most important determinant of why some countries grew rich and others remained poor. Going

Original Washington Consensus	"Augmented" Washington Consensus the previous 10 items, plus:
1. Fiscal discipline	11. Corporate governance
2. Reorientation of public expenditures	12. Anti-corruption
3. Tax reform	13. Flexible labor markets
4. Financial liberalization	14. WTO agreements
5. Unified and competitive exchange rates	15. Financial codes and standards
6. Trade liberalization	16. "Prudent" capital-account opening
7. Openness to DFI	17. Non-intermediate exchange rate regimes
8. Privatization	18. Independent central banks/inflation targeting
9. Deregulation	19. Social safety nets
10. Secure Property Rights	20. Targeted poverty reduction

Table 1 The Augmented Washington Consensus

one step further, Easterly and Ross Levine (2003) showed that policies (i.e., trade openness, inflation, and exchange rate overvaluation) do not exert any independent effect on long-term economic performance once the quality of domestic institutions is included in the regression. Often, this work has taken a form that may be called "institutions fundamentalism" – to relate it to (and distinguish it from) the earlier wave of "market fundamentalism." Getting the institutions right is the mantra of the former, just as getting prices right was the mantra of the latter. The Augmented Washington Consensus derives its academic support largely from this work on the primacy of institutions.[9, 10]

Taken to its logical conclusion, the focus on institutions has potentially debilitating side effects for policy reformers. Institutions are by their very nature deeply embedded in society. If growth indeed requires major institutional transformation – in the areas of rule of law, property rights protection, governance, and so on – how can we not be pessimistic about the prospects for growth in poor countries? After all, such institutional changes typically happen very rarely – perhaps in the aftermath of war, civil wars, revolutions, and other major political upheavals. The cleanest cases that link institutional change to growth performance occur indeed at such historical junctures: consider for example the split between East and West Germany, or of North and South Korea. But what are poor countries that do not want to go through such upheavals to do?

Learning from Reform pays lip service to the importance of institutions, but to its credit it steers clear from too much institutions determinism. That is wise because the Augmented Washington Consensus' focus on institutional change proves to be largely a dead-end upon closer look. There are two major reasons for this, which I summarize here.

First, the cross-national literature has been unable to establish a strong causal link between any particular design feature of institutions and economic growth. We know that growth happens when investors feel secure, but we have no idea what specific institutional blueprints will make them feel more secure in a given context. The literature gives us no hint as to what the right levers are. Institutional function does not uniquely determine institutional form. If you think this is splitting hairs, just compare the experience of Russia and China in the mid-1990s. China was able to elicit inordinate amounts of private investment under a system of public ownership (township and village enterprises), something that Russia failed to do under Western-style private ownership. Presumably this was because investors felt more secure when they were allied with local governments with residual claims on the stream of profits than when they had to entrust their assets to private contracts that would have to be enforced by incompetent and corrupt courts. Whatever the underlying reason, China's experience demonstrates how common goals (protection of property rights) can sometimes be achieved under divergent rules. This is a theme that *Learning from Reform* loudly trumpets.

Second, we should not forget that Acemoglu, Johnson, and Robinson (2001) work and other related research focused on *long-term* economic performance. The typical dependent variable in this line of literature is the *level* of income in some recent year, not the rate of economic growth over a particular period. When institutional indicators are introduced in *growth* regressions, the results are much weaker and less robust. Empirical work focusing on transitions into and out of growth has found little evidence that large-scale institutional transformations play a role (Hausmann, Pritchett, and Rodrik 2005; Benjamin F. Jones and Benjamin A. Olken 2005). To take two important examples, China embarked on rapid growth in the late 1970s with changes in its system of incentives that were marginal in nature (and certainly with no ownership reform or significant change in its trade regime early on), and India's transition to high growth in the early 1980s was preceded (or accompanied) by no identifiable institutional changes. These and other experiences suggest that a policymaker interested in igniting economic growth may be better served by targeting the most binding constraints on economic growth – where the bang for the reform buck is greatest – than by investing scarce political and administrative capital on ambitious institutional reforms. Of course, institutional reform will be needed eventually to sustain economic growth. But it may be easier and more effective to do that when the economy is already growing and its costs can be spread over time.

In the limit, the obsession with comprehensive institutional reform leads to a policy agenda that is hopelessly ambitious and virtually impossible to fulfill. Telling poor countries in Africa or Latin America that they have to set their sights on the best-practice institutions of the United States or Sweden is like telling them that the only way to develop is to become developed – hardly useful policy advice! Furthermore, there is something inherently unfalsifiable about this advice. So open-ended is the agenda that even the most ambitious institutional reform efforts can be faulted ex post for having left something out. So you reformed institutions in trade, property rights, and macro but still did not grow? Well, it must be that you did not reform labor-market institutions. You did that too but still did not grow? Well, the problem must be with lack of safety nets and inadequate social insurance. You reformed those with little effect? Obviously the problem was that your political system was unable to generate sufficient credibility, lock-in, and legitimacy for the reforms. In the end, it is always the advisee who falls short, and never the advisor who is proved wrong.

5. THE ALTERNATIVES II: FOREIGN AID

Yet another vision of reform strategy is offered by the U.N. Millennium Project (2005), led by Jeffrey Sachs. This vision is no less holistic than that of the institutions fundamentalists, although the elements of the package and the weight placed on each differ. The U.N. Project calls for a comprehensive and simultaneous increase in "public investments, capacity building, domestic resource mobilization, and official development assistance," while providing "a framework for strengthening governance, promoting human rights, engaging civil society, and promoting the private sector" (p. xx). But it also abounds in concrete details of what can and should be done. Some of the "quick-win actions" it proposes include free distribution of bed nets against malaria, ending user fees for primary education and essential health services, expansion of school meals programs in hunger zones, and replenishment of soil nutrients on smallholder agriculture through subsidized or free distribution of chemical fertilizers.

The U.N. Millennium Project views current levels of foreign aid to be a significant constraint on the achievement of global poverty reduction. Hence it calls for a significant increase in aid – a doubling of annual official development assistance to $135 billion in 2006, rising to $195 billion by 2015 – to finance public investments in human capital and infrastructure and to develop the technologies needed to transform health and agriculture in poor societies. Sachs and his collaborators exhibit a certain impatience with those who argue that the real constraint is poor institutions and weak governance, and that large aid flows are more likely to disappear in the pockets of corrupt officialdom than to foster development. They argue that many of the poorest countries of the world (e.g., Benin, Mali, Senegal) have in fact made significant strides in improving their economic and political institutions, and that in any case the investments in human capital that they advocate would likely foster better institutions as well. In their view, the obsession with governance is often just an excuse for rich countries not doing more to help poor nations.

The theory underlying the U.N. Millenium Project's view of the world is that low-income countries in Africa (and possibly elsewhere) are stuck in a low-level equilibrium, a "poverty trap" (Sachs et al. 2004). The neoclassical production function assumes that the marginal product of capital is high at low levels of development (when the economy has low levels of capital). But if there are some increasing returns to scale (e.g., setting up a modern factory requires a minimum investment to be made), complementarities (e.g., running a modern factory needs an adequate supply of educated workers), or negative feedback effects (e.g., an increase in incomes raises population growth), the marginal return to capital is initially low rather than high. Small increments to capital yield very little fruit, and the economy can have multiple steady states, one of which involves a poverty trap. Since it does not pay to invest, households do not save and the economy remains poor. This very old idea (going back at least to Paul N. Rosenstein-Rodan (1943) and Richard R. Nelson (1956)) can be used to justify a "big push" – i.e., a large-scale, simultaneous effort to raise the capital stock (public, private, human) to levels where the neoclassical forces of convergence begin to operate and the economy breaks free of the poverty trap.

Several questions are raised by this take on African poverty. First, what do we make of the fact that historically few low income countries have embarked on high growth in this big-push fashion or through the infusion of large amounts of foreign aid? As Sachs's critics love to point out, there has not been a shortage of foreign aid in Africa, and some of the most rapidly growing countries of the past have done so without relying much on Western aid. Sachs and his collaborators counter that Africa is special because it suffers from high transport costs, low-productivity agriculture, a very heavy disease burden, adverse geopolitics, and slow diffusion of technology from abroad (Sachs et al. 2004, pp. 130–31), all of which make the region particularly prone to a poverty trap. But couldn't one have said much the same of Vietnam, a war-torn, impoverished country facing economic sanctions from the United States, which took off in the late 1980s even though it did not receive much aid from Western nations until the mid-1990s?

Or what do we make of the fact that economic growth is actually not uncommon among Sub-Saharan African nations themselves? The theory of poverty traps suggests that these countries are stuck in low-level equilibria from which they find it very hard to extricate themselves. The reality seems to be somewhat different. Most African countries have shown themselves capable of producing economic growth over nontrivial time horizons. A telling statistic produced by Jones and Olken (2005) is that three-quarters of Sub-Saharan African countries have grown fast enough to experience some convergence with U.S. income levels over at least one ten-year period since 1950. Similarly, in Hausmann, Pritchett, and Rodrik (2005), where we studied growth accelerations since the 1950s, we found such accelerations to be quite frequent in low-income countries, including among those in Africa. In fact, growth accelerations turned out to be more common in low-income countries than in middle- or high-income countries, in line with the neoclassical growth model. The trouble seems to be not that poor African countries are unable to grow, but that their growth spurts eventually fizzle out. This suggests a rather different remedy, one that focuses in the short run on selectively removing binding constraints on growth (which may well differ from country to country), and in the medium- to longer-run on enhancing resilience to external shocks.[11] I will elaborate on this remedy below.

Ultimately, where the U.N. Millennium Project differs most from *Learning from Reform* is in the extent of knowledge that it assumes we have and consequently in the degree of self-confidence exhibited by its authors. The U.N. Millennium Project is based on the view that we basically know enough to mount a bold, ambitious, and costly effort to eradicate world poverty. We have successfully identified all the margins that matter, and we better move on all of them simultaneously. *Learning from Reform*, by contrast, is an ode to humility. What we have learned, it says implicitly, is the folly of assuming that we know too much. We need to downplay grandiose claims, move cautiously, and concentrate our efforts where the payoffs seem the greatest.

6. A PRACTICAL AGENDA FOR FORMULATING GROWTH STRATEGIES

But what is the operational content of such a cautious, experimentalist approach? If we adopt the path recommended by *Learning from Reform*, can we say anything more than "different strokes for different folks" or avoid a nihilistic attitude where "everything goes"? *Learning from Reform* says little that is useful on this, but I think the answer is "yes" to both questions. Let me briefly outline here a way of thinking about growth strategies that avoids some of the obvious pitfalls.

This approach consists of three sequential elements. First, we need to undertake a *diagnostic analysis* to figure out where the most significant constraints on economic growth are in a given setting. Second, we need creative and imaginative *policy design* to target the identified constraints appropriately. Third, we need to *institutionalize* the process of diagnosis and policy response to ensure that the economy remains dynamic and growth does not fizzle out.

6.1 Step 1: Growth diagnostics

Policy reforms of the (Augmented) Washington Consensus type are ineffective because there is nothing that ensures that they are closely targeted on what may be the most important constraints blocking economic growth. The trick is to find those areas where reform will yield the greatest return. Other-wise, policymakers are condemned to a spray-gun approach: they shoot their reform gun on as many potential targets as possible, hoping that some will turn out to be the ones they are really after. A successful growth strategy, by contrast, begins by identifying the most binding constraints.

But can this be done? In Hausmann, Rodrik, and Velasco (2005), we develop a framework that we believe suggests a positive answer. We begin with a basic but powerful taxonomy (see Figure 1). In a low-income economy, economic activity must be constrained by at least one of the following two factors: *either* the cost of finance must be too high *or* the private return to investment must be low. If the problem is with low private returns, that in turn must be due *either* to low economic (social) returns *or* to a large gap between social and private returns (low private appropriability). The first step in the diagnostic analysis is to figure out which of these conditions more accurately characterizes the economy in question.

Fortunately, it is possible to make progress because each of these syndromes throws out different sets of diagnostic signals or generate different patterns of comovements in economic variables. For example, in an economy that is constrained by cost of finance we would expect real interest rates to be

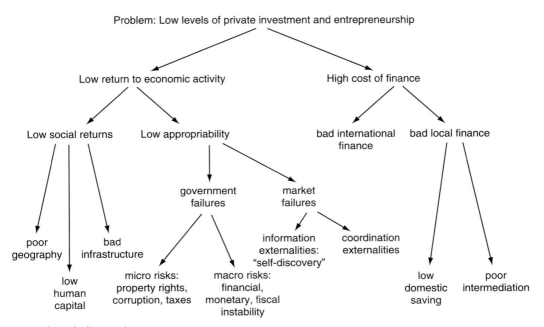

Figure 1 Growth diagnostics

high, borrowers to be chasing lenders, the current account deficit to be as large as the foreign borrowing constraint will allow, and entrepreneurs to be full of investment ideas. In such an economy, an exogenous increase in investible funds, such as foreign aid and remittances, will spur primarily investment and other productive economic activities rather than consumption or investment in real estate. This description comes pretty close to capturing the situation of countries such as Brazil or Turkey, for example. By contrast, in an economy where economic activity is constrained by low private returns, interest rates will be low, banks will be flush in liquidity, lenders will be chasing after borrowers, the current account will be near balance or in surplus, and entrepreneurs will be more interested in putting their money in Miami or Geneva than in investing it at home. An increase in foreign aid or remittances will finance consumption, housing, or capital flight. These in turn are the circumstances that characterize countries such as El Salvador and Ethiopia.

When we identify low private returns as the culprit, we will next want to know whether the source is low social returns or low private appropriability of those returns. Low social returns can be due to poor human capital, lousy infrastructure, bad geography, or other similar reasons. Once again, we need to be on the lookout for diagnostic signals. If human capital (either because of low levels of education or the disease environment) is a serious constraint, we would expect the returns to education or the skill premium to be comparatively high. If infrastructure is the problem, we would observe the bottlenecks in transport or energy, private firms stepping in to supply the needed services, and so on.

Appropriability problems – i.e., a large gap between private and social returns – can in turn arise under two sets of circumstances. One possibility has to do with the policy/institutional environment: taxes may be too high, property rights may be protected poorly, high inflation may generate macro risk, labor-capital conflicts may depress production incentives, and so on. Alternatively, the fault may lie with market failures such as technological spillovers, coordination failures, and problems of economic "self-discovery" (i.e., uncertainty about the underlying cost structure of the economy; see Hausmann and Rodrik 2003). As usual, we look for the tell-tale signs of each of these. Sometimes, the diagnostic analysis proceeds down a particular path not because of

direct evidence but because the other paths have been ruled out.[12]

It is possible to carry out this kind of analysis at a much finer level of disaggregation, and indeed any real-world application has to be considerably more detailed than the one I have sketched here. But I hope this summary conveys the value of an explicitly diagnostic framework. Even a rudimentary application of these principles can sometimes reveal important gaps or shortcomings in traditional reform packages. For example, when the cost of finance is an important binding constraint (as seems likely in Brazil), institutional improvements aimed at improving the "business climate" (i.e., reducing red tape, lowering taxes, and so on) will be not only ineffective (since the problem does not lie with investment demand), but it can also backfire (since an increase in investment demand will put further upwards pressure on interest rates).

6.2 Step 2: Policy design

Once the key problem(s) are identified, we need to think about the appropriate policy responses. The key in this step is to focus on the market failures and distortions associated with the constraint identified in the previous step. The principle of policy targeting offers a simple message: target the policy response as closely as possible on the source of the distortion. Hence if credit constraints are the main constraint, for example, and the problem is the result of lack of competition and large bank spreads, the appropriate response is to reduce impediments to competition in the banking sector.

Simple as it may be, this first-best logic often does not work, and indeed can be even counterproductive. The reason is that we are necessarily operating in a second-best environment, due to other distortions or administrative and political constraints. In designing policy, we have to be on the lookout for unforeseen complications and unexpected consequences. Let me return to an example from China. Formal ownership rights in China's township and village enterprises (TVEs) were vested not in private hands or in the central government, but in local governments (townships or villages). From the lens of first-best reform, these enterprises are problematic since, if our objective is to spur private investment and entrepreneurship, it would have been far preferable to institute private property rights (as Russia and other East European transition economies did). But the first-best logic is not helpful here because a private property system relies on an effective judiciary for the enforcement of property rights and contracts. In the absence of such a legal system, formal property rights are not worth much, as minority shareholders in Russia soon discovered to their chagrin. Until an effective judiciary is created, it may make more sense to make virtue out of necessity and force entrepreneurs into partnership with their most likely expropriators, the local state authorities. That is exactly what the TVEs did. Local governments were keen to ensure the prosperity of these enterprises as their equity stake generated revenues directly for them. In the environment characteristic of China, property rights were effectively more secure under direct local government ownership than they would likely have been under a private property-rights legal regime.

Such examples can be easily multiplied (Rodrik 2005a). As an additional illustration, consider the case of achieving integration with the world economy. Policymakers in countries such as South Korea and Taiwan in the early 1960s and China in the late 1970s had decided that enhancing their countries' participation in world markets was a key objective. For a western economist, the most direct route would have been to reduce or eliminate barriers to imports and foreign investment. Instead, these countries achieved the same ends (i.e., reduce the anti-trade bias of their economic policies) through unconventional means. South Korea and Taiwan employed export targets and export subsidies for their firms. China carved out special economic zones where foreign investors had access to a free-trade regime. Policymakers chose these unconventional solutions presumably because they created fewer adjustment costs and put less stress on established social bargains.

6.3 Step 3: Institutionalizing reform

The nature of the binding constraint will necessarily change over time. For example, schooling may not be a binding constraint initially, but as investment and entrepreneurship pick up, it will likely become one unless the quality and quantity of schools increase over time. In Hausmann, Rodrik, and Velasco (2005), we illustrate this issue using the example of the Dominican Republic. This country

was able to spur growth with a number of sector-specific reforms that stimulated investment in tourism and maquilas. But it neglected making the institutional investments required to lend resilience and robustness to economic growth – especially in the area of financial market regulation and supervision. When September 11 led to the drying of tourist inflows, the country paid a big price. A Ponzi scheme that had developed in the banking sector collapsed, and cleaning up the mess cost the government 20 percentage points of GDP and led the economy into a downward spiral. It turned out that the economy had outgrown its weak institutional underpinnings. The same can be said of Indonesia, where the financial crisis of 1997–98 led to total economic and political collapse. It may yet turn out to be case also of China unless this country manages to strengthen the rule of law and enhance democratic participation.

What is needed to sustain growth? Two types of institutional reform seem to become critical over time. First, there is the need to maintain productive dynamism. Natural resource discoveries, garment exports from maquilas, or a free-trade agreement may spur growth for a limited of time. Policy needs to ensure that this momentum is maintained with ongoing diversification into new areas of tradables. Otherwise, growth simply fizzles out. What stands out in the performance of East Asian countries is their continued focus on the needs of the real economy and the ongoing encouragement of technology adoption and diversification.

The second area that needs attention is the strengthening of domestic institutions of conflict management. The most frequent cause for the collapse in growth is the inability to deal with the consequences of external shocks – i.e., terms of trade declines or reversals in capital flows. Endowing the economy with resilience against such shocks requires strengthening the rule of law, solidifying (or putting in place) democratic institutions, establishing participatory mechanisms, and erecting social safety nets. When such institutions are in place, the macroeconomic and other adjustments needed to deal with adverse shocks can be undertaken relatively smoothly. When they are not, the result is distributive conflict and economic collapse (Rodrik 1999). The contrasting experiences of South Korea and Indonesia in the immediate aftermath of the Asian financial crisis in 1997–98 are quite instructive in this regard.

Institutional reforms in these areas are difficult to implement and they take time. Economic science typically provides very little guidance on how to proceed (Avinash K. Dixit 2004). But the point is that these difficulties do not need to stand in the way of formulating less ambitious, more selective, and more carefully targeted policy initiatives that can have very powerful effects on igniting economic growth in the short run. What is required to *sustain* growth should not be confused with what is required to *initiate* it.

7. CONCLUDING REMARKS

It is now time for a confession. As the preceding discussion ought to have made clear, I find *Learning from Reform* a useful and important document in no small part because its central themes parallel those that I have been advocating for some time along with a number of my colleagues at the Kennedy School (see in particular Rodrik 2005a; Hausmann, Rodrik, and Velasco 2005; and Hausmann, Pritchett, and Rodrik 2005). It is gratifying to see one's ideas being taken seriously, particularly by an institution that has frequently served as a target for one's criticisms. The report pays me compliments in other ways too: one of its two opening quotes is taken from my work (the other is from Al Harberger). And I return the compliment by acting as one of the endorsers on its back cover.[13] Had the editor of this *Journal* not insisted, I would not have found it proper to write this review essay.

But I would like to think that the laudatory note I have struck above has to do not just with an ego that is being stroked. Coming from the institution that is one of the chief architects of the reforms of the last twenty years, *Learning from Reform* is a genuinely interesting document: it represents a mea culpa as well as a way forward. It pushes us to think harder and deeper about the economics of reform than anything else out there. It warns us to be skeptical of top-down, comprehensive, universal solutions – no matter how well intentioned they may be. And it reminds us that the requisite economic analysis – hard as it is, in the absence of specific blueprints – has to be done case by case.

These should be music to any economist's ears. After all, what distinguishes professional economists from ideologues is that the former are trained to make *contingent* statements: policy A is to be recommended only if conditions x, y, and z obtain.[14] Sensible advice consists of a well-articulated mapping from observed conditions onto its policy implications. This simple but fundamental principle seems to have gotten lost in much of the thinking on economic reform in the developing world, which has often taken an a priori and mechanical form. Its rediscovery is therefore good news not just for poor nations, but for the economics profession as well.

NOTES

* Harvard University. I am grateful to Roger Gordon for his encouragement and comments; to Ricardo Hausmann, Lant Pritchett, and John Williamson for their reactions; and to Roberto Zagha for the many insights he has shared with me over the last few years. John Williamson reminded me that my title is far from original, having been used in almost identical form by Moises Naim (1999). In its present form, the title also makes allusion to the classic paper by Carlos Diaz-Alejandro (1985).

1 To cite just one example, fifty percent or more of the state-owned enterprises were divested during the 1990s in the Central African Republic, Cote d'Ivoire, Gambia, Ghana, Guinea-Bissau, Kenya, Mali, Tanzania, Togo, Uganda, and Zambia (John Nellis 2003). On the extent of trade reform in Africa, see Vinaye D. Ancharaz (2003).

2 In a book edited with Pedro-Pablo Kuczynski in 2003, John Williamson laid out an expanded reform agenda, emphasizing crisis-proofing of economies, "second-generation" reforms, and policies addressing inequality and social issues (Kuczynski and Williamson 2003).

3 Roberto Zagha led the team that prepared the report. Members of the team were J. Edgardo Campos, James Hanson, Ann Harrison, Philip Keefer, Ioannis Kessides, Sarwar Lateef, Peter Montiel, Lant Pritchett, S. Ramachandran, Luis Serven, Oleksiy Shvets, and Helena Tang.

4 According to World Bank estimates, there were roughly 400 million fewer people living below the $1 a day poverty line in 2001 compared to two decades earlier (Chen and Ravallion 2004).

5 See Dani Rodrik (2005a) for an interpretative survey of recent growth experience.

6 See also Rodrik (2005b) for a general methodological critique of growth regressions with policy variables on the right-hand side.

7 Along with Ricardo Hausmann and the lead author of the World Bank report, Roberto Zagha, I have been involved in an effort to bring some of these implications

to bear on the country operational work at the Bank. One thing we have discovered is how difficult it is to wean the Bank's country economists away from the Washington-Consensus, laundry-list, best-practice approach to reform.

8 But even within the IMF, there are divergent views. The IMF's Evaluation Office (nominally independent and headed until recently by a distinguished outsider, Montek Ahluwahlia, but staffed largely by IMF economists) has produced reports that often reach different conclusions.

9 A mea culpa here: My article on "Institutions Rule" (Rodrik, Arvind Subramanian, and Francesco Trebbi 2004) is frequently seen as being in the frontline of institutions fundamentalism (although there are important caveats in the second half of the paper).

10 The most serious challenge to institutions fundamentalism has been launched by Edward L. Glaeser, Rafael La Porta, Florencio Lopez-de-Silanes, and Andrei Shleifer (2004) who find the empirical approach in the institutions-cause-income literature flawed and think it is human capital (and dictators) that cause growth.

11 For an empirical analysis which emphasizes the role of external shocks (in interaction with weak institutions) as the culprit for growth collapses, see Rodrik (1999).

12 So in the case of El Salvador we concluded that lack of self-discovery was an important and binding constraint in part because there was little evidence in favor of the other traditional explanations (Hausmann and Rodrik 2005).

13 To add to the incestousness of the relationship, Lant Pritchett, my coauthor on Hausmann, Pritchett, and Rodrik (2005), served as the principal author of two of the chapters of *Learning from Reform*.

14 As a trite, but still useful illustration, consider trade liberalization, which is one of the most common policy reforms recommended to developing countries (typically unconditionally) (Rodrik 2005a). Economic theory says that trade liberalization is guaranteed to enhance welfare only under a long list of conditions: The liberalization must be complete or else the reduction in import restrictions must take into account the potentially quite complicated structure of substitutability and complementarity across restricted commodities. There must be no microeconomic market imperfections other than the trade restrictions in question, or if there are some, the second-best interactions that are entailed must not be adverse. The home economy must be "small" in world markets or else the liberalization must not put the economy on the wrong side of the "optimum tariff." The economy must be in reasonably full employment or, if not, the monetary and fiscal authorities must have effective tools of demand management at their disposal. The income redistributive effects of the liberalization should not be judged undesirable by society at large or, if they are, there must be compensatory tax-transfer schemes with low enough excess burden. There must be no adverse effects on the fiscal balance or, if there are, there must be alternative and expedient ways of making up for the lost fiscal revenues. The liberalization must be politically sustainable and hence credible so that economic agents do not fear or anticipate a reversal. And an even longer list of requirements would have to be present for trade liberalization to generate *economic growth*, i.e., go beyond static Harberger triangles. While the theory of the second-best should not paralyze us, neither should we hand-wave it away as easily as we seem to do in our role as policy advisors.

REFERENCES

Acemoglu, Daron, Simon Johnson, and James A. Robinson. 2001. "The Colonial Origins of Comparative Development: An Empirical Investigation." *American Economic Review*, 91(5): 1369–1401.

Ancharaz, Vinaye D. 2003. "Determinants of Trade Policy Reform in Sub-Saharan Africa." *Journal of African Economies*, 12(3): 417–43.

Chen, Shaohua, and Martin Ravallion. 2004. "How Have the World's Poorest Fared since the Early 1980s?" World Bank Policy Research Working Paper 3341.

Collier, Paul, and David Dollar. 2001. "Can the World Cut Poverty in Half? How Policy Reform and Effective Aid Can Meet International Development Goals." *World Development*, 29(11): 1787–1802.

Diaz-Alejandro, Carlos. 1985. "Good-Bye Financial Repression, Hello Financial Crash." *Journal of Development Economics*, 19(1–2): 1–24.

Dixit, Avinash K. 2004. *Lawlessness and Economics: Alternative Modes of Governance*. Princeton: Princeton University Press.

Easterly, William. 2005. "National Policies and Economic Growth: A Reappraisal." In *Handbook of Economic Growth, Vol. 1A*, ed. P. Aghion and S. Durlauf. Amsterdam: North-Holland, 1015–59.

Easterly, William, and Ross Levine. 2003. "Tropics, Germs, and Crops: How Endowments Influence Economic Development." *Journal of Monetary Economics*, 50(1): 3–39.

Glaeser, Edward L., Rafael La Porta, Florencio Lopez-de-Silanes, and Andrei Shleifer. 2004. "Do Institutions Cause Growth?" *Journal of Economic Growth*, 9(3): 271–303.

Hausmann, Ricardo, Lant Pritchett, and Dani Rodrik. 2005. "Growth Accelerations." *Journal of Economic Growth*, 10(4): 303–29.

Hausmann, Ricardo, and Dani Rodrik. 2003. "Economic Development as Self-Discovery." *Journal of Development Economics*, 72(2): 603–33.

Hausmann, Ricardo, and Dani Rodrik. 2005. "Discovering El Salvador's Production Potential." *Economia*, (6)1: 43–102.

Hausmann, Ricardo, Dani Rodrik, and Andrés Velasco. 2005. "Growth Diagnostics." John F. Kennedy School of Government, Harvard University. Mimeo.

Jones, Benjamin F., and Benjamin A. Olken. 2005. "The Anatomy of Start–Stop Growth." NBER Working Papers, no. 11528.

Krueger, Anne O. "Meant Well, Tried Little, Failed Much: Policy Reforms in Emerging Market Economies", remarks at the Roundtable Lecture at the Economic Honors Society, New York University, March 23, 2004.

Kuczynski, Pedro-Pablo, and John Williamson, eds. 2003. *After the Washington Consensus: Restarting Growth and Reform in Latin America*. Washington, D.C.: Institute for International Economics.

Naim, Moises. 1999. "Fads and Fashion in Economic Reforms: Washington Consensus or Washington Confusion?" Paper Prepared for the IMF Conference on Second Generation Reforms, Washington, D.C.

Nellis, John. 2003. "Privatization in Africa: What Has Happened? What Is To Be Done?" Center for Global Development Working Paper 25.

Nelson, Richard R. 1956. "A Theory of the Low-Level Equilibrium Trap in Underdeveloped Economies." *American Economic Review*, 46(5): 894–908.

Rodríguez, Francisco. 2005. "Cleaning Up the Kitchen Sink: On the Consequences of the Linearity Assumption for Cross-Country Growth Empirics." Unpublished.

Rodrik, Dani. 1999. "Where Did All the Growth Go? External Shocks, Social Conflict, and Growth Collapses." *Journal of Economic Growth*, 4(4): 385–412.

Rodrik, Dani. 2005a. "Growth Strategies." In *Handbook of Economic Growth, Vol. 1A*, ed. P. Aghion and S. Durlauf. Amsterdam: North-Holland, 967–1014.

Rodrik, Dani. 2005b. "Why We Learn Nothing from Regressing Economic Growth on Policies." Unpublished.

Rodrik, Dani, Arvind Subramanian, and Francesco Trebbi. 2004. "Institutions Rule: The Primacy of Institutions over Geography and Integration in Economic Development." *Journal of Economic Growth*, 9(2): 131–65.

Rosenstein-Rodan, Paul N. 1943. "Problems of Industrialization of Eastern and South-Eastern Europe." *Economic Journal*, 53: 202–11.

Sachs, Jeffrey D., John W. McArthur, Guido Schmidt-Traub, Margaret Kruk, Chandrika Bahadur, Michael Faye, and Gordon McCord. 2004. "Ending Africa's Poverty Trap." *Brookings Papers on Economic Activity*, (1): 117–216.

Singh, Anoop, Agnes Belaisch, Charles Collyns, Paula De Masi, Reva Krieger, Guy Meredith, and Robert Rennhack. 2005. "Stabilization and Reform in Latin America: A Macroeconomic Perspective of the Experience since the 1990s." IMF Occasional Paper 238.

U.N. Millennium Project. 2005. *Investing in Development: A Practical Plan to Achieve the Millennium Development Goals*. New York: United Nations.

Williamson, John, ed. 1990. *Latin American Adjustment: How Much Has It Happened?* Washington, D.C.: Institute for International Economics.

Williamson, John. 2000. "What Should the World Bank Think about the Washington Consensus?" *World Bank Research Observer*, 15(2): 251–64.

World Bank. 2005. *Economic Growth in the 1990s: Learning from a Decade of Reform*. Washington, D.C.: World Bank.

'Was Latin America Too Rich to Prosper? Structural and Political Obstacles to Export-Led Industrial Growth'

Journal of Development Studies (1992)

James E. Mahon, Jr

Editors' Introduction

James Mahon is a Professor in the Political Science Department at Williams College. His scholarship is on political economy, comparative politics, US–Latin America foreign relations and Latin American politics. He is working on fiscal politics and the reform of the state in Latin America, and he has also written on US foreign policy. Mahon's *Mobile Capital and Latin American Development* (1996) compares political economy and policy in Argentina, Brazil, Chile, Columbia, Mexico and Venezuela in order to explain why some countries had to contend with much higher rates of capital flight before the debt crisis of the 1980s. These countries typically had faced fewer trade problems and had not shifted their exchange policies to diversify exports and to deal with instability in exchange rates. In the turbulent period since the 1970s such countries suffered from tremendous capital flight. By the end of the twentieth century, these were also the most aggressively neo-liberal regimes in Latin America. Mahon shows how mobile foreign capital pressured neo-liberal reform in the interests of the elite and against those of the democratic majority. Mahon's work engages with the implicit debate between Smith and Marx in Part 2, and it explains how neo-liberalism actually works, *pace* Harvey (Part 9).

One of the key issues in twentieth-century development studies has been the great divergence between Latin America and East Asia. Why and how did the East Asian 'Tigers' – South Korea, Taiwan, Hong Kong and Singapore – grow rapidly through export-oriented industrialisation and banking, while the much older capitalist economies of Latin America grew so much more sluggishly in the second half of the twentieth century? Some of the most important arguments about this divergence have focused on the nature of the state, its 'embeddedness' and 'autonomy' from social classes, as Peter Evans (1995) framed the issues. Another important set of arguments has to do with colonial and agrarian legacies, which shaped state capacities and class forces in particular ways. Cristobal Kay (2002) has used this agrarian argument to make the case that the East Asian trajectories were built on radical land reform (or public housing, as Manuel Castells and his colleagues have argued) and on state power that effectively neutralised older landed elites. Latin American countries could not break from their semi-feudal past, or not decisively enough.

Mahon's argument in the reading here builds on these arguments but adds an important nuance. Latin American regimes could not shift from import-substitution industrialisation (ISI), or protected domestic industry, to export-oriented industrialisation (EOI) as rapidly as East Asian regimes because this would have required a much more substantial 'income sacrifice'. In other words, Latin American workers would have had to pay the price, and this was a political cost that parties in power could not easily take. They could not devalue currencies enough to shift industrial and trade policies in such a way as to make Latin American countries competitive with East Asian rates of EOI growth. Ironically, ISI in Latin America kept power in the hands of foreign investors and domestic oligarchs, and often supported their class alliances with military regimes. Both domestic industrial capital and the organised working class defended devaluation, and Latin American ISI regimes remained stuck with dependency on imports and primary product exports. Mahon shows how difficult the idea of imitating East Asian success can be for Latin American countries, particularly for the working classes. Those that have shifted to neo-liberal economic policies quickly have accentuated class inequalities between holders of foreign assets and the poor (see Harvey, Part 9).

Key references

James E. Mahon (1992) 'Was Latin America too rich to prosper? Structural and political obstacles to export-led economic growth', *Journal of Development Studies* 28: 241–63.
— (1996) *Mobile Capital and Latin American Development*, University Park, PA: Penn State University Press.
James E. Mahon with Javier Corrales (2002) 'Pegged for failure? Argentina's crisis', *Current History*, February.
Peter Evans (1995) *Embedded Autonomy: States and Industrial Transformation*, Princeton, NJ: Princeton University Press.
Cristobal Kay (2002) 'Why East Asia overtook Latin America: agrarian reform, industrialization and development', *Third World Quarterly* 23: 1073–102.

INTRODUCTION

Much of recent discussion about the political economy of development has turned upon a contrast between Latin America and East Asia. East Asian growth, seen largely as the result of dynamic export-oriented industrialisation (henceforth EOI), has been juxtaposed with a relatively more sluggish Latin American performance. This has raised the question: why did Latin America persist for so long with a more protectionist model emphasising import-substituting industrialisation (henceforth ISI), despite foreign exchange crises and IMF prodding? One common answer implicates political weaknesses that bloated the state, protected urban industrialists, and coddled labour unions. Alternatively, the divergence is explained with reference to the wishes of foreign actors, especially the US government. This essay emphasises another dimension of the contrast: the size of the structural and hence political obstacles faced by middle-income developing countries in the turn to exporting low-wage manufactured goods.

After recounting briefly the most important features of post-war industrialisation policies in each region, I review and classify some explanations for the inter-regional contrast. Looking closely at the late 1950s and the 1960s, I focus on the political and structural determinants of the foreign exchange and trade policies that formed the core of ISI and EOI. I present evidence for the hypothesis that Latin America's relatively productive primary exports posed an obstacle to a full reorientation of policy from ISI to EOI. In most of the major Latin American countries the embrace of EOI would have required a substantially greater income sacrifice to achieve world-competitive low wage levels via devaluation. This made the decision even riskier politically: it pushed the expected time at which pre-devaluation real wages might be reattained beyond the horizons of politicians. This helps explain how the very 'success' of Latin America's historical insertion into the world economy as a primary-product exporter served to discourage the policy reforms that would have been necessary to reinsert the region on more dynamic terms. The essay ends with reflections on the likelihood of successful reinsertion in the wake of the debt crisis.

As already noted, the contrast between East Asian EOI growth and Latin American ISI stagnation has been a prominent theme in the literature on development in the last five years.[1] My emphasis on the role of the primary export sector in discouraging EOI follows the analysis of Marcelo Diamand in his treatment of Argentina [*1973; 1986*]. It has been discussed in Keesing [*1981: 19–20*] and, within a comparative framework similar to mine, by Gereffi [*1987: 27–9*]. I am taking up Haggard's criticism of this line of argument, that it is 'too loose and too ahistorical to explain much' if it merely states that East Asia is resource poor and Latin America is resource rich [*1989: 148*]. I seek to show that during a crucial period when many Latin American countries were experiencing the kind of exchange crises that could provoke policy innovation, these countries also would have encountered large structural and political obstacles to low-wage, export-oriented industrial growth.

EXCHANGE POLICY AND LATIN AMERICAN DEVELOPMENT: THE RISE AND DECLINE OF ISI

The legacy of the Second World War to Latin American economies was a contradictory one. By 1945 the major countries of the region had accumulated large foreign exchange reserves, which though substantially inconvertible, were the largest they had ever held. There was postponed demand for investment goods, pent-up demand for foreign consumer goods (from newly-rich factory owners, among others), and a larger industrial work force. Wartime inflation had caused a decline in real wages in most places, but many governments repressed the demands of workers under laws ostensibly justified by the war emergency. Finally, in most major countries (except Argentina) price inflation had outrun similar trends in the United States during the 1940–46 period, while exchange rates stayed nearly fixed [*IMF, various issues*].[2] This made US goods seem relatively cheap in domestic currency terms. The stage was set for the appearance of several familiar problems of postwar Latin American economies: the rapid exhaustion of the foreign exchange reserves,

inflation-driven wage disputes, and the entrenchment of an import-substitution model whose chief instruments were trade protection and state subsidy, rather than a reliance on relative price effects.

The foreign-exchange constraint soon became a major driving force of economic policy in nearly every country. For various reasons, the policy responses were similar across the region.[3] Foreign exchange shortfalls were to be ended by attending to the import side-industrial import substitution behind administrative barriers [*Baer, 1972*]. Hence many large Latin American countries erected multiple exchange rate systems designed to subsidise industrial-sector imports at the expense of the traditional export sector [*König, 1968*]. But by the late 1950s it was clear that this elaborate method of industrial subsidy did not solve the foreign sector problem. Export revenue stagnated or declined in the second half of the 1950s and early 1960s, while import needs as a proportion of GDP did not fall quickly enough.[4] Recurrent balance-of-payments crises became a regular feature in many countries, notably Argentina, Brazil, Chile, and Colombia, from the mid-1950s to the late 1960s. But now it was politically much more difficult to undertake sharp and effective currency depreciation than it had been in the early 1930s.[5]

Two perverse and unexpected results of Latin American ISI were particularly disturbing. The first was that it made the economy more, not less dependent on its capacity to import [*United Nations, ECLA, 1964; Diaz Alejandro, 1965b; Felix, 1968*]. By eliminating consumer goods imports in favour of imports of raw, intermediate, or capital goods, it placed jobs and economic growth at the mercy of the balance of payments. Instead of affecting only the consumption styles of the upper classes who bought imported wares, now foreign exchange crises would send shock waves through an entire industrial economy that depended on imported inputs [*Baer, 1972: 106*]. The second unfortunate result was the persistence or worsening of dependence on a few primary-commodity exports. Exchange rates artificially supported by trade and exchange controls (intended to benefit domestically-oriented industry) were unfavourable to exports, especially new or marginal ones. Special rates established for minor exports were fiscally costly in proportion to their aggressiveness, and often varied too much to be effective. Over the long term, exports tended to remain restricted to those best established. Combined, the greater import dependence and the persistent concentration in primary exports meant that when the terms of trade went against the country or a bad harvest curtailed its exports, the loss of import capacity brought industrial recession.

Politically, this situation was disastrous. The very groups who incarnated the old 'colonial' style of development were the ones to whom the country had to turn in the event of exchange shortfalls. The hopes that ISI could liberate the country from the clutches of foreign investors or domestic oligarchs were dashed repeatedly.

REFORMS OF THE ISI MODEL IN LATIN AMERICA

A full transition out of ISI and into EOI never really took place in Latin America, but in the face of recurrent exchange crises several countries did reform their foreign exchange regimes[6] and conduct partial or discontinuous liberalisations of trade. (The least partial and most continuous was, of course, Chile after mid-1975.) The typical exchange policy reform included a crawling-peg rate system, designed to keep real rates stable under inflation for the benefit of exports and the control of capital flows, and some continued reliance on exchange control to reduce procyclical pressures in the exchange market. Where the reforms were durable enough to survive until the borrowing spree of the late 1970s, countries suffered less from problems of speculation and capital flight [*Mahon, 1989*].

In most of the region, however, periodic exchange shortfalls, heavy reliance on foreign finance, and the periodic resuscitation of ISI programmes were the norm. None of the foreign-sector policy reforms constituted a thoroughgoing reorientation toward manufactured exports, or even toward exports in general. The more successful were what Fishlow [*1987*] has called 'export-adequate' reforms: they retained the emphasis on a large internal market (as is implied by Brazil's debt-driven simultaneous pursuit of ISI and export objectives in the late 1970s), while diversifying and expanding exports in order to reduce commodity dependence and increase imports. They tended to benefit traditional rural elites who diversified agro-export production as much as they favoured export-minded industrial firms. Other countries, which had suffered few crises, left the funda-

mental structure of foreign-sector policy intact. In Mexico and Venezuela, for example, the regime of protected ISI was first supported but was then dramatically undermined by commodity prices and international capital flows.

FROM ISI TO EOI IN SOUTH KOREA AND TAIWAN

In many ways the experiences of South Korea and Taiwan with import substitution in the 1950s were similar to those of the larger Latin American countries. Analyses of economic development in these two countries place the stage of import substitution between 1953 and 1960 for South Korea and between 1953 and 1961 for Taiwan, with the latter boundary a gradual one, especially in Taiwan [*Mason, Kim, Perkins, et al., 1980: 126–7; Fei, Ranis and Kuo, 1979: 26–8;* also *Cole and Lyman, 1971; Jacoby, 1966; Gold, 1986*]. During these years policies were a familiar mix of subsidies, overvalued exchange rates, import protection, and credit rationing marked also by the discretionary use of political power to favour selected borrowers or importers. In South Korea especially, this system was associated with administrative complexity and corruption. One study of its foreign trade regime declared that 'the multiplicity of exchange rates and means of allocating foreign exchange from 1953 to 1960 are staggering to comprehend' [*Frank, Kim and Westphal, 1975: 36*]. Although both countries' trade policies also included substantial fiscal subsidies and other legal advantages for exporters, they ended the 1950s still immensely dependent on US aid to finance their imports.

Nevertheless, by 1965 both countries had embarked on concerted programmes to incorporate themselves into the international economy through the promotion of labour-intensive manufactured exports. Both had undertaken large devaluations (Korea twice in early 1961, then a 95 per cent devaluation in 1964; Taiwan in 1960) and had simplified their foreign exchange systems to allow the market more power in reducing imports and to increase incentives to export. Both had allowed domestic interest rates to rise, in part to help stabilise prices in the wake of the devaluations.

By the late 1960s foreign direct investment had risen greatly from its insubstantial previous level, especially in Taiwan, reinforcing these countries' positions as low-wage sites for export industry. The two countries were on a path of rapid employment growth and rising real wages; there were even labour shortages in Taiwan after the late 1960s and in Korea beginning somewhat later. In the 1970s and early 1980s backward linkages from export processing plants were strengthened and, along with new efforts at import substitution (especially in South Korea), each country began to export some higher value-added goods such as automobiles and personal computers. Although the problems of their EOI models became more evident in the late 1980s [*Bello and Rosenfeld, 1990*], the fact remains that these countries moved decisively from ISI to EOI and prospered as a result.

WHY DID LATIN AMERICA NOT FOLLOW THE EAST ASIAN PATH?

Despite the large foreign debt and recent labour protest in South Korea, by 1990 the contrast between advancing East Asia and debt-ridden Latin America was a stark one. As real wages rise in South Korea and Taiwan, Latin American per capita incomes have fallen to levels approximating those of the early 1970s. The inter-regional difference in real per capita GDP growth is especially strong for the 1986–89 period [*IBRD, 1989; OECD Development Centre, 1989; IMF, 1989*].

Apparently, the East Asian countries followed a development strategy more appropriate to the international environment than that pursued by most of Latin America. Hence the puzzle which preoccupies many recent observers: why the contrast between the relatively smooth, unproblematic, and generally sustained transition from ISI to EOI in Taiwan and South Korea, and the more partial and difficult policy reorientation in much of Latin America before the debt crisis?

Answering this question involves accounting for the differing capabilities of governments to respond to foreign exchange crises by effectively turning policy from ISI to EOI. I would like to suggest that there are four dimensions to the explanation: first, the differing alignments of interests in civil society (the relative power of rural vs. urban groups, the role of multinational corporations, etc.); second, the power of the state over these societal interests (institutional insulation, repressive capability, foreign support);

third, the severity and duration of the crises; and finally, the objective difficulty of an effective policy reorientation. That is, 'successful' countries may have had more societal allies and fewer enemies of policy change than did other countries; they also may have had stronger states, may have experienced more severe crises that provided stronger motivations for change, or may have faced a much more manageable task than did others.[7]

[. . .]

DEPENDENCY AND THE 'UNBALANCED PRODUCTIVE STRUCTURE'

Central to most accounts of twentieth-century Latin American political and economic development is an appreciation of the difficulty involved in breaking with an economic model of primary-product export that favoured a traditional, chiefly land-based oligarchy and its foreign allies. Industrialisation was delayed and difficult to achieve politically because the groups that commanded the majority of the productive base, the consumption share, and the political apparatus were positively against it. The Depression served as a catalyst for change because to varying degrees it weakened these established forces, often forcing them to rely on the state to support their income, while creating the conditions for industrial growth. But even after some amount of import substitution had been spurred on by the currency depreciations of the 1930s or by the trade disruption of the Second World War, the traditional oligarchy was still strong, especially if allied with the military in an anti-democratic conspiracy.

In short, it appeared that the crucial barrier to continued industrial development presented by the primary export sector (whether foreign or domestically controlled) was its political resistance. From this perspective, many of the roots of dependency could be traced to the persistence of an export sector which reflected the nineteenth-century conditions of its origin and often fought to re-establish them.

If the Diamand hypothesis of the 'unbalanced productive structure' has merit, it shows another side of Latin American dependency under ISI. A thoroughgoing insertion of a country into the world market on the basis of its primary exports (Argentina is the archetype) would have discouraged industrial exports for other reasons. The most fundamental would have been the effect of high primary-export productivity on the exchange rate.[8]

This conclusion puts the postwar history of Latin American development in an interesting light. The catastrophe of the 1930s pushed the major countries of Latin America to undertake an import substitution process that was driven mainly by real depreciations that made all imports expensive. Postwar ISI differed. It would hew more closely to an easier but less sustainable path of making selected imports cheap to selected buyers, while exploiting the favourable trends in Latin American terms of trade in the decade after 1945.

The new postwar policies responded more to an economically stronger and ideologically more assertive industrial bourgeoisie, and less to the interests of the traditional export sector. While the social base of the top industrialists was often not very distinct from that of the traditional rural elites (especially in the smaller countries), the economic base had changed, with commerce and primary commodities weakened and industry strengthened by a decade and a half of reduced or interrupted trade. Still, factory owners were not the only supporters of the new set of policies. The pro-ISI enthusiasm was shared by most sectors of the military, labour, and many technocrats.

Reasons for avoiding devaluation could be traced to both new and old elements of this political economy. The most important foes of devaluation were urban industrial capitalists, who tended to be favoured by the exchange control systems erected in lieu of devaluation, and urban workers, who would lose from any acceleration of inflation not compensated immediately by equivalent nominal wage gains. Organised labour was especially prominent in postponing devaluation. Policy-makers also found good 'technical' reasons for resisting devaluation – the crucial one being an 'elasticity pessimism' about the probable response in supply and (if the country was a dominant world producer) international price of the country's key commodity to the move [Diaz Alejandro, 1965a: Ch. 1].[9] Meanwhile, clamouring for devaluation and trade liberalisation were the traditional export sectors and international financial interests. All in all, many found it easy to see which side of the issue was the popular, 'national', and reasonable one.

These trends led to the familiar policy deadlock. A period of recurrent exchange crises accompanied industrial growth. By the time the crises put policy

reform on the agenda, the interests holding fast to ISI were even stronger than they had been in 1945.

Many policy-makers understood that the problem lay in the export sector. Accepting that the country could not expect permanently rising export revenue from primary products, it was clear that other (perhaps more income-elastic) exports had to be substituted. But the high productivity of the primary sector made this difficult. Beginning in the 1960s, one had to compete in the new sectors with exporters of similar productivity, whose exchange rates were not buoyed up by the material legacy of a 'successful' primary-export stage.

More: in Taiwan and Korea the material legacy of primary-export development was practically insignificant. Even within East Asia, Taiwan and South Korea are distinguished by the disastrous declines of their primary export sectors after 1945. The difference between Latin America and these two countries was especially striking in the first postwar decade. In 1954 Latin America was just coming off a fairly sustained rise in its terms of trade. The coffee boom was still on; the metals bonanzas had ended with the Korean War a year earlier. Still, relative prices were ahead of 1945 (and 1939). In contrast, South Korea's export sector was in shambles. Exports in 1956 had fallen to less than half of their real value of 1950 [Westphal and Kim, 1982: 212]. In Taiwan, the loss of the Japanese market for tropical foodstuff exports compounded the countrywide problems of hyperinflation and rapid real depreciation, attendant upon civil war in 1945–49. One result of these disasters was the abject dependence of both East Asian countries on US aid to finance imports. Another was a depreciated level of parallel exchange rates, despite the aid inflow.

The poverty of primary exports in the two Asian countries had other social consequences, too. Because this resource poverty permitted an ISI process only at a low level of income, the pie was too small for it to be used, Argentina-style, as a basis for a strong political alliance of newly-mobilised labour and domestically-oriented business. It was less tempting to continue with it, and its limitations meant that the interests defending it would be relatively weak.

For Latin American politicians to have duplicated the disaster that befell the primary export sectors of Taiwan and South Korea, they would have had to re-create the Depression experience through state policy. This, of course, was inconceivable to them. Yet the problem illustrates a bit of the complexity in the dependent relations of a primary-export country. It shows another side to a phenomenon which theorists pose in abstract terms, when they speak of the 'substantial contradictions to be overcome if a country enmeshed in the capitalist world system was to change its position in the international division of labour' [*Evans, 1987: 204*].

Ironically, under the typical structure of Latin American ISI a 'contradiction' was felt most directly by the politicians who were most sensitive to the interests of urban wage-earners. 'Populists' were the ones most loath to contemplate any drastic export reorientations – including programmes which would have started by redistributing rural property – because of the benefits to urban consumers that could be derived from a more aggressive exploitation of the existing rural export structure (or its accumulated surpluses). The experience of post-1959 Cuba may indicate that it is hard for leaders of any persuasion to refuse to support consumption through primary-product exports, when they have inherited the opportunity to do so.

Let me repeat that I am not trying to argue that only one feature distinguishes Taiwan and South Korea from Latin America. I have simply tried to provide one more reason why urban consumption interests were more powerful in Latin America. It was not just a matter of relative state power, societal interests, exchange crises, or international forces. Relative to Latin America, in East Asia the transition to EOI bore a much smaller income cost. This is a contrast that is often hidden when the successes of Taiwan and South Korea are compared to the less adept or permanent policy changes in Latin America.

Nor is this a policy brief in favour of low wages. The idea that all middle-income developing countries ought to be subjected to competition on the basis of wage rates would be immoral and absurd; it is nearly as contradictory as supposing that they could all become major exporters of simple manufactures at the same time. My purpose has been to counteract the somewhat voluntaristic tone that sometimes inhibits these inter-regional comparisons. The East Asian achievement, that rare example of a region rising quickly in the hierarchy of the capitalist world, was facilitated by other rare events, usually beyond the ability of politicians to duplicate.

CAN LATIN AMERICA NOW IMITATE EAST ASIAN SUCCESS?

The previous discussion treats explanations for Latin America's relative failure to adopt a development strategy oriented to the export of manufactures in the postwar period. It has given particular emphasis to the differences in primary export productivity which affected real wage levels under the ISI regime, and thus to the degree of sacrifice necessary to achieve export competitiveness. Latin America was 'disadvantaged' in this regard, for having been an important world producer of primary goods for centuries, and for not having suffered the kind of catastrophe which afflicted Taiwan or especially Korea in the first postwar decade.

But with the debt crisis, this catastrophe has come. Real wages have fallen steeply all across the region. Large real devaluations and, in Argentina, Brazil, and Peru, hyperinflation have been part of a trend pushing real wages down again. Now Latin American countries are among the world's low-wage industrial economies, as they struggle to boost exports and pay their debts. Does this mean that they will follow the East Asian example?

This is where the other elements of the Sachs argument may be instructive. According to him, Latin America also differs from East Asia in the *pressure* for 'populist' solutions, and this pressure is the result of unequal income distribution. The extreme poverty in Taiwan and Korea in the 1950s was broadly shared: more extreme in Korea, it was also shared more widely. Given the previous land reforms and the small (initial) economic size of rural export producers, policies set to benefit exports did not have the pro-oligarchic significance they had in Latin America.'

The same events responsible for making Latin America suddenly competitive in manufactured exports may have weakened the political basis of the low-wage model. Where capital flight played a prominent role in bringing on the debt crisis (Argentina, Mexico, Venezuela), the political problem is obvious: poor citizens forced to sacrifice so that the country can pay a debt that financed the purchase of foreign assets by the rich. But even if the foreign debts were forgiven tomorrow, there would be a problem. Among the prime beneficiaries from steep currency depreciation have been the holders of undepreciated foreign assets. Capital flight has favoured them over those without this kind of wealth.

Moreover, in all debt-strapped countries the shortage of capital has driven up real interest rates on domestic securities, enriching holders of government paper. In short, it is likely that the debt crisis has accentuated the regressive pattern of income distribution in Latin America in the same measure as it has made the debtors into internationally competitive manufacturers. If so, this is not a very favourable climate for a sustained low-wage export drive.

NOTES

This work is part of a larger project that was founded by the Center for Latin American Studies and the Institute for International Studies at the University of California, Berkeley, The Tinker Foundation, The S.W. Cowell Foundation, and the Institute for the Study of World Politics. The author thanks George T. Crane, Elizabeth Norville, Karl Fields, Wes Young, Mark Tilton, Paula Consolini, John Sheahan, and Henry Bruton for their comments on previous drafts.

1   A short bibliography would include Evans [*1987*]; Fishlow [*1987*]; Gereffi [*1987*]; Gereffi and Wyman [*1989*]; Gereffi and Wyman [*1989*]; Lin [*1989*]; special issues of *Economic Development and Cultural Change*, Vol.36, April 1988, *World Development*, Vol. 17, No. 9, 1989, and *The Annals of the American Academy of Political and Social Science*, Vol. 505, Sept. 1989.
2   Raymond Mikesell, in the US delegation at Bretton Woods, wrote in 1949 that the IMF allowed overvalued rates in many Latin American countries to persist because it 'evidently believed that stability of exchange rates was more significant than an appropriate rate pattern' [*1949: 397*].
3   Apart from imputing an influence to nationalist business, many analyses emphasise the roles of the Depression and the writings of ECLA economists on Latin American policy-makers. An exception to this line of argument is Maxfield and Nolt [*1990*], which traces the role of US government and business interests in promoting ISI.
4   In current dollar terms, Argentina did not surpass its export revenues of 1951 until 1963; Chile, favoured by the rise in copper prices in the 1960s, reached its 1955 level by 1963; Colombia's export earnings did not exceed their 1954 value until 1969; while worst of all, Brazil took until 1969 to better its export performance (even in current dollars) of 1951 [*IMF, International Financial Statistics, various issues*]. On the problematic trend of import elasticities, see the discussion and works cited below.

5   Although the argument has many elements familiar in this literature, several of these draw on observations made in Cooper [*1971*] and Hirschman [*1968*]. It is illustrated for Brazil in Arnold [*1972*].
6   Exchange policy reforms in the late 1950s and 1960s were more serious in proportion to the degree of crisis in the export sector, as Diaz-Alejandro [*1984*] observed about the 1930s. They were also more durable where the old export interests, despite the distress in the market for the traditional main commodity, were still powerful enough to provide significant political support for a sustained reorientation. For more detail on the determinants of exchange policy leading up to the capital flight of the early 1980s, see Mahon [*1989*].
7   In this scheme, international forces show up through their effects on the domestic political economy.
8   Another might have been the contribution of still-powerful traditional elites to the failure to make ISI-bred firms internationally competitive at this relatively appreciated exchange rate.
9   Diaz-Alejandro noted that policymakers tended in their 'elasticity pessimism' to discount the creation of *new* export activities. This tendency may be attributable to pro-ISI ideology.

REFERENCES

Arnold, S.H., 1972, 'The Politics of Export Promotion: Economic Problem-Solving in Brazil, 1956–1969', doctoral dissertation, Johns Hopkins University.

Baer, W., 1972, 'Import Substitution and Industrialization in Latin America: Experiences and Interpretations', *Latin American Research Review* Vol. 7, No. 1, Spring.

Bello, W. and S. Rosenfeld, 1990, 'Dragons in Distress: The Crisis of the NICs', *World Policy Journal*, Summer, pp. 431–68.

Cole, D.C. and P.N. Lyman, 1971, *Korean Development: The Interplay of Politics and Economics*, Cambridge, MA: Harvard University Press.

Cooper, R.N., 1971, *Currency Devaluation in Developing Countries* (Princeton Essays in International Finance No. 86), Princeton, NJ: Princeton University Press, June.

Diamand, M., 1973, *Doctrinas Económicas, Desarrollo e Independencia*, Buenos Aires: Paitlos.

Diamand, M., 1986, 'Overcoming Argentina's Stop-and-Go Economic Cycles', in J. Hartlyn and S. Morley (eds.), *Latin American Political Economy: Financial Crisis and Political Change*, Boulder, CO: Westview Press.

Diaz Alejandro, C.F., 1965a, *Exchange Rate Devaluation in a Semi-Industrialized Country: The Experience of Argentina 1955–1961*, Cambridge, MA: MIT Press.

Diaz Alejandro, C.F., 1965b, 'On the Import Intensity of Import Substitution', *Kyklos*, Vol.18, pp. 495–511.

Diaz Alejandro, C.F., 1984, 'Latin America in the 1930s', in R. Thorp (ed.), *Latin America in the 1930s: The Role of the Periphery in World Crisis*, London: Macmillan.

Evans, P., 1987, 'Class, State, and Dependence in East Asia: Lessons for Latin Americanists', in Frederic Deyo (ed.), *The Political Economy of the New Asian Industrialism*, Ithaca, NY: Cornell University Press.

Fei, J.C.H., Ranis, G. and S. Kuo, 1979, *Growth with Equity: The Taiwan Case*, Washington, DC: International Bank for Reconstruction and Development/Oxford University Press.

Felix, D., 1968, 'The Dilemma of Import Substitution – Argentina', in G.F. Papanek (ed.), *Development Policy: Theory and Practice*, Cambridge, MA: Harvard.

Fishlow, A., 1987, 'Some Comparative Reflections on Latin American Economic Performance and Policy', Working Paper 8754, Department of Economics, University of California, Berkeley, mimeo.

Frank, C.R., Kim, K.S. and L.E. Westphal, 1975, *Foreign Trade Regimes and Economic Development: South Korea*, Volume VII of NBER Special Conference Series, New York: National Bureau of Economic Research.

Gereffi, G., 1987, 'Industrial Restructuring and National Development Strategies: A Comparison of Taiwan, South Korea, Brazil, and Mexico', paper presented at the International Conference on 'Taiwan, R.O.C., A Newly Industrialized Society', National Taiwan University, Taipei.

Gereffi, G. and D. Wyman, 1989, 'Determinants of Development Strategies in Latin America and East Asia', in S. Haggard and C.I. Moon (eds.), *Pacific Dynamics: The International Politics of Industrial Change*, Boulder, CO: Westview Press.

Gereffi, G. and D. Wyman, 1989, *Manufactured Miracles: Paths of Industrialization in Latin America and East Asia*, Princeton, NJ: Princeton University Press.

Gold, T.B., 1986, *State and Society in the Taiwan Miracle*, Armonk, NY: M.E. Sharpe.

Haggard, S., 1989, 'The East Asian NICs in Comparative Perspective', *Annals of the American Academy of Political and Social Science*, Vol. 505, Sept.

Hirschman, A.O., 1968, 'The Political Economy of Import-Substituting Industrialization in Latin America', *Quarterly Journal of Economics*, Vol. 82, No. 1, Feb; also in 1971, *A Bias for Hope*, New Haven, CT: Yale.

IBRD (World Bank), 1989, *World Tables, 1988–89*, Washington, DC: World Bank.

IMF, various issues, *International Financial Statistics*, Washington, DC: International Monetary Fund.

IMF, 1989, *World Economic Outlook*, Washington, DC: International Monetary Fund, Oct.

Jacoby, N.H., 1966, *U.S. Aid to Taiwan: A Study of Foreign Aid, Self-Help, and Development*, New York: Praeger.

Keesing, D.B., 1981, 'Exports and Policy in Latin-American Countries: Prospects for the World Economy and for Latin-American Exports, 1980–90', in W. Baer and M. Gillis (eds.), *Export Diversification and the New Protectionism: The Experiences of Latin America*, Champaign, IL: National Bureau of Economic Research.

König, W., 1968, 'Multiple Exchange Rate Policies in Latin America', *Journal of Inter-American Studies*, Vol. 10, pp. 35–52.

Lin, C.Y., 1989, *Latin America versus East Asia: A Comparative Development Perspective*, Armonk, NY: M.E. Sharpe.

Mahon, J.E., 1989, 'Capital Flight and the Politics of Exchange Policy in Six Latin American Countries, 1930–1983', doctoral dissertation, University of California, Berkeley.

Mason, E.S., Kim, M.J., Perkins, D.H. *et al.*, 1980, *The Economic and Social Modernization of the Republic of Korea*, Cambridge, MA: Harvard.

Maxfield, S. and J.H. Nolt, 1990, 'Protectionism and the Internationalization of Capital: U.S. Sponsorship of Import Substitution Industrialization in the Philippines, Turkey and Argentina', *International Studies Quarterly*, Vol. 34, pp. 49–81, March.

Mikesell, R. 1949, 'The International Monetary Fund', *Journal of Political Economy*, Vol. 57, No. 5, Oct.

OECD Development Centre, 1989, *Latest Information on National Accounts of Developing Countries*, No. 21, Paris: Organisation for Economic Cooperation and Development.

United Nations, Economic Commission for Latin America (ECLA), 1964, *The Economic Development of Latin America in the Postwar Period*, New York: United Nations.

Westphal, L.E. and K.S. Kim, 1982, 'Korea', in Bela Balassa *et al.*, *Development Strategies in Semi-industrial Economies*, Baltimore, MD; Johns Hopkins University Press/IBRD.

SEVEN

'Fiscal Reform and the Economic Foundations of Local State Corporatism in China'

World Politics (1992)

Jean C. Oi

Editors' Introduction

Jean C. Oi is William Haas Professor of Chinese Politics at Stanford University, and Senior Fellow at the Freeman Spogli Institute for International Studies. Oi completed a BA in political science and Asian languages and literature from Indiana University. In 1972, Oi went to Taiwan for the language training that would prove crucial to her careful fieldwork to follow. Oi's PhD in political science from the University of Michigan, Ann Arbor, coincided with reforms under Deng Xiaopeng's Open Door Policy. Oi first visited China in this context, in 1980, and over repeated trips she studied the dramatic increases in farm house-hold earnings in the early 1980s under the 'household responsibility system'. What Oi also witnessed was the emergence of rural industrialisation in a variety of sectors – textiles, furniture, steel tubes, beer, cornstarch – through what appeared to be the initiative of entrepreneurial bureaucrats in the countryside. Oi continued a strong schedule of fieldwork while teaching at Lehigh and Harvard universities. Her work at this time, along with the research of her partner, sociologist Andrew Walder, would transform China studies in important, interdisciplinary ways. Oi maintains that it is necessary to treat Chinese development as part of the diversity of national development trajectories rather than as a strange case quarantined in Area Studies.

Oi and Walder were part of a new generation of scholarship on the dynamics of social institutions in China's transition. They tried to show that China's transition to the market offered quite important lessons vis-à-vis transitions from socialism to capitalism in the former Soviet Union and Eastern Europe. In the latter cases, Jeffrey Sach's 'shock therapy' of fast economic reform was proving to be disastrous. In contrast, post-Maoist China's market transition was seen by some economists as a *surprisingly* successful 'market socialism', which ought to be premised on culturally specific forms of cooperation. Confucionism and Daoism were dragged in to explain away the assumptions of the shock therapists. Oi turned instead to the actual transformations in power and practice at the village level in the reform period in *State and Peasant in Contemporary China* (1989).

The reading is important for trying to explain the social institutions that made possible China's rapid growth and industrialisation in the 1980s. This very important article would become part of a book, *Rural China Takes Off* (1999). The reader should take note as to how Oi explains the institutional dynamics of what she calls 'local state corporatism'. In particular, Oi argues that centralised control of the economy has been loosened in some ways, but that market transition has worked through existing social relations. Crucially, decollectivisation and fiscal decentralisation provided the broader conditions through which party officials also act as entrepreneurs. Decollectivisation meant that agriculture was no longer the main source of revenue. Local officials became more invested in the diversification of economic activity. Collective enterprises would prove to be the most efficacious form of organisation of rural industry in terms of absorbing surplus labour. Hard budget constraints from the centre increased local government incentives to help foster collective rural industry. Finally, local state corporatism allowed some reinvestment of the surplus in welfare – housing, education, pensions, health care – as a form of consent to a society with socialist expectations of redistribution. The argument is provocative, and the reader should ask how this evidence squares with evidence of rising labour militancy in the work of, among others, C. K. Lee (Part 9). Oi has continued to write on rural China, on state-owned enterprises, rural debt and development, and village politics. Oi and Walder have edited *Property Rights and Economic Reform in China* (1999), and Oi has co-edited *At the Crossroads of Empires: Middlemen, Social Networks and State-building in Republican Shanghai* (2007).

Key references

Jean C. Oi (1992) 'Fiscal reform and the economic foundations of local state corporatism in China', *World Politics* 45 (1): 99–126.

— (1989) *State and Peasant in Contemporary China: The Political Economy of Village Government*, Berkeley: University of California Press.

— (1999) *Rural China Takes Off: Institutional Foundations of Economic Reform*, Berkeley: University of California Press.

Jean Oi and Andrew Walder (eds) (1999) *Property Rights and Economic Reform in China*, Palo Alto, CA: Stanford University Press.

Nara Dillon and Jean Oi (eds) (2007) *At the Crossroads of Empires: Middlemen, Social Networks and State-building in Republican Shanghai*, Palo Alto, CA: Stanford University Press.

While clearly lagging behind the Eastern bloc countries and the former Soviet Union in political reform, China is arguably the most successful of the socialist states in implementing economic reform. Real GNP grew at an average annual rate of 10.4 percent from 1980 to 1988 compared with only 6.4 percent during 1965–80; total GNP grew more than twofold between 1978 and 1988. During 1978–88 per capita GNP doubled in real terms.[1] The most impressive strides were in the rural economy. Calculated on the basis of comparable prices, total agricultural output value increased by an average annual rate of 6.04 percent between 1978 and 1990. Output value of rural collective enterprises increased 26.7 percent for that same twelve-year period. Per capita net income among peasants increased from 133.57 to 629.79 yuan, with an average annual increase of 41.4 yuan, as compared with only a 3.2 yuan increase on average for the previous twenty-eight years.[2]

Far from standing in the way of change, local officials have spearheaded the reform drive. What explains their response to national reform initiatives? Bureaucratic change is a function of whether "net incentive effects upon key officials are positive or negative, and to what degree."[3] In China the necessary incentives were provided by fiscal reform. This article seeks to illuminate how this institutional change caused local officials in a socialist system to pursue rapid rural industrial growth.[4]

North and Weingast have pointed to the importance of secure property rights for economic development.[5] The discussion has been limited to a large extent to market economies, with a focus on individuals as entrepreneurs – those who own and operate the firms within a market economy. The importance of this principle underlies the felt need in formerly socialist countries such as Poland to privatize. There is, however, no inherent reason who only individuals, as distinct from governments, can be entrepreneurs. Similarly, there is no inherent reason why secure property rights will be an effective incentive only if they are assigned to individuals.

Growth will result as long as there are secure property rights for *some* organized unit and sufficient incentives for that unit to pursue growth. The impressive growth of collective rural industrial output between 1978 and 1988 is in large measure a result of local government entrepreneurship. Fiscal reform has assigned local governments property rights over increased income and has created strong incentives for local officials to pursue *local* economic development. In the process local governments have taken on many characteristics of a business corporation, with officials acting as the equivalent of a board of directors.[6] This merger of state and economy characterizes a new institutional development that I label *local state corporatism*.

I want to make clear that "corporatism" as used here differs from its use in previous studies. By local state corporatism I refer to the workings of a *local* government that coordinates economic enterprises in its territory as if it were a diversified business corporation. I am not concerned with the role of the central state in the vertical integration of interests within society as a whole.[7] Whereas the central state set the reform process in motion and provided localities with the incentives and the leeway to develop economically, it is *local* government that has determined the outcome of reform in China. Hence, my designation *local state* corporatism.

Local state corporatism suggests a loosening of central control in a Leninist system, but it does not refer to a process that some have labeled the re-emergence of civil society.[8] My argument also contradicts, or at least greatly qualifies, theories of market transition that posit a corrosive effect of markets upon the power of local officials.[9] Local state corporatism is similarly difficult to square with the view that market reform is destroying the cellular, or honeycomb, nature of Chinese rural society and allowing for greater penetration by the central state.[10] On the contrary, this analysis highlights the co-existence of a strong local officialdom and public enterprise with a thriving market economy and a weakened central state.

This paper shows how fiscal reforms have provided incentives for local government officials to pursue economic, especially industrial, development. I limit my discussion to an examination of local government at the county, township, and village levels – the lowest levels of government and those most directly involved in rural industrialization. First, I describe the ways that fiscal reforms changed the rules for the disposition of locally collected revenues. Second, I analyze the interests and incentives that these new rules have created. Third, I show why local governments chose to pursue rural industrialization. Finally, I examine how local governments plan and coordinate economic activity to maximize local interests. In the conclusion I consider some broader implications of local state corporatism for China's political future.

[. . .]

III FISCAL REFORM AND THE RISE OF RURAL INDUSTRY

The problem faced by local governments was how best to generate revenues. Decollectivization eliminated agriculture as a viable source of needed revenue. After the institution of the responsibility system that made the household the unit of production and accounting, local governments, especially at the village and to a lesser extent at the township levels, lost legitimate access to all income from agriculture, except for a percentage of the nominal agricultural tax. Income from the sale of agricultural products went to the household.

The combination of decollectivization and revenue sharing forced local officials to make careful calculations of the benefits and costs of the different sectors of the rural economy. The smaller return on funds invested and the limited access that local officials had to income from agriculture, especially grain production, made agriculture the least desirable investment option.[11] Once the diversification of the rural economy was encouraged, local officials diverted resources and energies from agriculture to more profitable enterprises. One possible source was to promote the development of the private sector and gain revenues through taxation.[12] That, as I noted above, can be quite lucrative for local coffers. However, because of the political taint still associated with private enterprise and the shifting political winds at the center toward private enterpreneurs, that route is taken cautiously.[13] The most lucrative and least problematic strategy politically was the development of rural collective enterprises (*xiangzhen qiye*).[14]

Not only did rural collective enterprises provide a solution (applauded by the national leadership) to the surplus labor problem that arose from

Proportion paid	Example	% of profits paid to government
1. Large	Beijing,	56.99
	Shanghai	71.67
2. Small	Jiangsu,	30.00
	Zhejiang,	30.00
	Hebei	30.00
3. Minimal	Zhejiang (Wenzhou)	management fee

Table 1 Regional differences in after-tax profits paid to township and village governments
Source: He Xian (fn. 21), 34.

decollectivization,[15] but they were also increasingly lucrative.[16] Beginning as a small part of the rural economy in 1980, by 1987 industry had already surpassed agriculture nationally as the dominant source of rural income.[17] For the twelve-year period from 1978 to 1990 the output value of these collective enterprises increased by 26.7 percent.[18] By the end of 1991 rural industry was producing 50 percent of total industrial output, equal to that of the state-owned industrial sector.

As the legal owner of these enterprises, local governments could legitimately extract nontax levies – the various fees and ad hoc surcharges described above. Revenue from rural collective enterprises has been the source of funds for public services, welfare, and subsidies to other, less profitable, sectors of the economy, such as agriculture, in addition to serving as the base for industrial expansion. Localities that have rural industry can provide services and are most likely to be strong effective governments; those that do not are most likely to be weak and ineffective governments.

Given the fiscal importance of rural industry, it is not surprising that it has mushroomed throughout rural China, in particular, rural industry owned by village governments. By 1988, 46.6 percent of all counties in China had rural industry with output value in excess of 100 million yuan.[19] Township and village enterprises total around 1.5 million. [A table] provides a twelve-year overview of the development of these enterprises. In absolute numbers they are only a minor portion of the more than eighteen million enterprises in the countryside, but they are the most important in terms of the amount of revenue and output value generated.[20]

The local variations and the ad hoc levies also make it difficult to determine precisely how much of enterprise revenues any township government and its economic commission take. A national study of township governments found a tremendous range in what they took from their enterprises: from the bulk of enterprise profits to a small portion to little more than the management fee (see Table 1).[21] The amount turned over was correlated with the extent to which government was involved in investment and development of production. Unfortunately, the study does not tell us which method is the most common. It is important to note that the "small" amount is still 30 percent of after-tax profits.[22] Aggregate statistics suggest, however, that the percentage of such profits going to government may be decreasing. Comparing 1984, 1986, and 1987, the amount of after-tax profits paid to either township or village governments has decreased from 52.17 percent to only 39.1 percent (see Table 2).[23]

Regardless of the precise amounts, overall revenues from enterprises at the township, and particularly at the village, level are controlled either directly or indirectly by local officials. Profits that are

Year	Income tax as % of total profits	% of after-tax profits taken by government
1984	24.8	52.17
1986	–	40.78
1987	21.7	39.1

Table 2 Division of township and village enterprise profits, 1984–87

Source: *Xiangzhen qiye nianjian 1978–1987* (fn. 23), 270.

not extracted are strictly regulated. Rules stipulate *how* retained profits must be used – this holds for township as well as village enterprises, and there is a formula that sets strict limits on the amount that can be used for workers' benefits and bonuses; the bulk of profits must be used for reinvestment.

IV THE SINEWS OF LOCAL STATE CORPORATISM

Some might read the above statistics on high rates of extraction as evidence that local officials are corrupt and bleed the collective enterprises under their control – behavior that could be seen as confirmation of the predatory nature of states. But one should go further, to examine how this income is used rather than simply accepting this behavior as rent seeking. A framework that would account for both the heavy local extractions and the significant growth of China's rural economy is to analyze local government as a corporation: it takes profits from its factories to pay for expenditures and for reinvestment.

The extraction of profits from enterprises is one of the most important mechanisms for allowing local governments to operate as a corporation. As the holders of rights over income flows, local governments can decide as a corporation how to use profits from its various enterprises and how to redistribute income.[24] But a successful corporation also needs other means to control the operation of its enterprises. Among the most important of these are the selection of management personnel, control over the allocation of scarce production inputs, provision of services, and control of investment and credit decisions. Much of the infrastructure for exercising government direction remains from the Maoist period.

[. . .]

V IMPLICATIONS

The successful role of local governments in China's development points to the need to specify further the existing theories about the relationship between economic growth, property rights, and government intervention. In this study I have shown that individuals need not have property rights over enterprise profits for economic growth to occur. Given the large number of formerly socialist systems attempting market reform and economic development, the Chinese model may stand as a less radical but viable alternative to privatization. Furthermore, the Chinese case raises questions about the adequacy of rent-seeking theories of the state to capture the role of extracted revenues when the local state is the entrepreneur and redistributive orientations hold sway.

The economic outcome of local state corporatism in China is a thriving rural economy with an industrial sector that now rivals state-owned industry in production value and factor productivity. Some economists have questioned the economic efficiency and value of rural industrialization. Critics stress that because growth in China's reform economy has been driven by the fiscal needs of individual local governments rather than by a nationally coordinated effort, rural enterprise may be profitable locally, but they may be redundant if viewed nationally.[25] Nonetheless, the rise of rural collective industry has created market competition and reduced the role of the central state in dictating patterns of economic growth. In the short run localities might subsidize enterprises simply to generate sales and product taxes, which are unrelated to profits, and to employ increasing numbers of surplus laborers from agriculture.[26] Over the long run, however, severe budget constraints will cause local officials to change product lines or close inefficient rural industries. Certainly, rural enterprises have shortcomings, but some economists have begun to consider them as the necessary competition to make state enterprises more efficient.[27] In the short run the problem for the central state is that state industry can no longer show a high rate of profits, which in turn reduces revenues to the central government.[28]

The Chinese case also demonstrates that the introduction of markets has not resulted in a proportionate decrease in the role of local government. Authors who originally stressed the transformative effects of markets now emphasize the ways in which rural governments are like local corporations.[29] A distinguishing characteristic of local state corporatism is that markets are a key part of the local economy but government coordination and intervention continue. Most of the materials used by the rural enterprises are secured through the market; but ability to pay is necessary but not necessarily sufficient for access. Local officials have assumed new roles as entrepreneurs, selectively allocating scarce resources to shape patterns of local economic growth.

Moreover, the combination of the fiscal reforms and market competition has reinforced rather than broken down the cellular nature of rural society in a number of respects. Although there certainly is more mobility, opportunity, and trade across "cells," there is strong parochialism. Rural industrial development has made possible substantial redistribution of income and the provision of greater services and benefits than was ever possible under the Maoist system, but provision is limited to registered members of the community. The benefits given to outside workers in village-run collective enterprises are less than those given to workers from the village. The growing protectionist sentiment is evident in the unwillingness of local grain stores to sell low-priced, subsidized grain to any urban residents holding valid grain-ration coupons other than to those registered in the locality. In the past localities routinely complied with national regulations stipulating that anyone holding valid coupons should be sold subsidized grain. These regulations still hold, but budget constraints have made it too costly for localities to spend their valuable subsidies on outsiders.

Up to this point, I have focused on the economic importance of local state corporatism, but its importance goes beyond the economy. Local state corporatism may provide China with an alternative to the pains of privatization and regime collapse experienced by the Eastern bloc countries and the Soviet Union. Local state corporatism has provided China with a relatively nonthreatening alternative economic system that allows for strong local state intervention. Communist party officials can play a leading economic as well as political role. In most of the localities where I have done fieldwork, it is the local party bosses – the first party secretaries of the county, township, or village – who are at the helm of the drive for economic development. Party secretaries are involved most intimately with the development of rural collective industry, leaving other matters to their subordinates or to their counterparts in the administrative portion of government. This is especially so in villages, where the village party secretary oversees industry while the village head is responsible for agriculture. Many a village party secretary also serves as the chairman of the board of directors of the village industrial corporation, the body established in some villages to oversee village industry. It is perhaps this aspect of local state corporatism that best explains why local

officials in China, unlike those in the Soviet Union, have embraced rather than resisted economic reforms. For Chinese local officials, both party and nonparty, the reforms could be achieved without any real threat to their power.

No simple, direct correlation exists between decollectivization, market reform, and the power of local officials in China. The Chinese experience has shown that some local governments have been weakened by the reforms, whereas others have been strengthened. The key variable seems to be the sources of income in a village, specifically the degree of local industrialization after the adoption of the household-responsibility system. Local governments that control only an agricultural, particularly a grain-based, economy are left with few options other than to levy ad hoc surcharges and various other fees and penalties, which the Chinese press often refers to as "increasing the peasant burden." Local governments that have continued to be strong and effective have developed nonagricultural sources of income that accrue directly to local government. The higher the level of industrialization, the more likely that local government will act in a corporate manner to intervene, extract, and redistribute income.[30] Carrying out economic reform and pursuing growth hold the potential of strengthening, not weakening, local official power.

Local state corporatism is also a path of least resistance because it allows China, a country socialized in the doctrines of Marxism-Leninism Mao Zedong Thought, to maintain its old values. As I have elaborated on elsewhere,[31] the income from rural industry has allowed local governments to redistribute substantial amounts of money to its members in the form of various subsidies, including grain subsidies and subsidies for education, health care, old-age pensions, and housing. In addition, local state corporatism allows local officials to be not just businessmen but also overseers of community well-being. And furthermore, it allows local officials to take social need into consideration when making economic decisions; the profitability of an enterprise may not be the most important factor in determining whether it will remain in operation. The need to employ the surplus labor released by decollectivization may make it economically rational for villages to keep open an unprofitable factory.[32] One can wonder how efficient this growth will be over the long term, but at least in the medium term the emergence of local state corporatism allows the redistribution of income to remain an established and viable part of economic life in China's rural communities.

Local state corporatism holds the most threatening consequences for central–local relations.[33] The issue is how the pursuit of local interests conflicts with larger national concerns. Withholding of income and manipulation of tax revenues are only two examples of how the fiscal reforms have colored the way that central directives are implemented at the local level. More generally, the fiscal reforms have given local authorities access to and the ability to increase their resource endowments, to be used however they wish, regardless of constraints imposed by the center. The Chinese reforms have invigorated the economy, but the success of local state corporatism may in the long run force the emergence of something akin to a federal system that more clearly recognizes the rights and power of localities.

NOTES

An earlier draft of the present paper was presented at the Program in Agrarian Studies, Colloquium Series for the Spring Term 1992, at Yale University, New Haven, February 28, 1992, and at the Meeting of the Public Choice Society, New Orleans, March 20–22, 1992. For their comments on earlier drafts of this paper, I would like to thank the members of these seminars and Houchang Chehabi, Jorge Dominguez, Frances Hagopian, Roderick Mac-Farquhar, Charles Sabel, James Scott, Kamal Sheel, Kenneth Shepsle, Ashutosh Varshney, Andrew Walder, Barry Weingast, and Jennifer Widner. Research was supported by the Committee on Scholarly Communication with the People's Republic of China and by Harvard University. The National Fellows Program of The Hoover Institution provided support for writing an earlier draft of this paper.

1 Kang Chen, Gary Jefferson, and Inderjit Singh, "Lessons from China's Economic Reform," *Journal of Comparative Economics* 16 (June 1992).

2 Zhang Hongyu, "China's Land System Transformation and Adjustment of Agricultural Structure: Reviewing China's Rural Reform and Development since 1978," *Liaowang Overseas Edition* 47 (November 25, 1991), 16–17, translated in *FBIS-CHI-91–241*, December 16, 1991, pp. 52–54. The most impressive growth in agriculture was prior to 1985. For a useful economic overview of the impact of the reforms on agriculture, see Terry Sicular, "China's Agricultural Policy during the Reform Period," in Joint Economic Committee, Congress of the United States, *China's Economic Dilemmas in the 1990s: The Problems of Reforms, Modernization, and Interdependence* (Washington, D.C.: Government Printing Office, 1991), 1:340–64. After 1985 the most dynamic growth was in rural industry.

3 Anthony Downs, *Inside Bureaucracy* (Boston: Little, Brown, 1967), 201.

4 On institutions and the incentives within them that shape economic and political behavior, see James G. March and Johan P. Olsen, *Rediscovering Institutions: The Organizational Basis of Politics* (New York: Free Press, 1989).

5 Douglass C. North and Barry, R. Weingast, "Constitutions and Commitment: The Evolution of Institutions Governing Public Choice in Seventeenth-Century England," *Journal of Economic History* 49 (December 1989).

6 I present a preliminary version of this argument in Oi, "The Chinese Village, Inc.," in Bruce Reynolds, ed., *Chinese Economic Policy* (New York: Paragon House Press, 1988). Others have drawn a similar analogy between local government and business entities. See, e.g., articles in William Byrd and Lin Qingsong, eds., *China's Rural Industry: Structure, Development, and Reform* (New York: Oxford University Press, 1990).

7 Schmitter, for example, defines corporatism "as a system of interest representation in which the constituent units are organized into a limited number of singular, compulsory, noncompetitive, hierarchically ordered and functionally differentiated categories, recognized or licensed (if not created) by the state and granted a deliberate representational monopoly within their respective categories in exchange for observing certain controls on their selections of leaders and articulation of demands and supports" (pp. 94–95). See Philippe C. Schmitter, "Still the Century of Corporatism?" reprinted in Fredrick B. Pike and Thomas Strich, eds., *The New Corporatism: Social-Political Structures in the Iberian World* (Notre Dame, Ind.: University of Notre Dame Press, 1974). Joseph Fewsmith uses a version of Schmitter's understanding of corporatism to describe China during the Republican period. See Fewsmith, *Party, State, and Local Elites in Republican China: Merchant Organizations and Politics in Shanghai, 1890–1930* (Honolulu: University of Hawaii Press, 1985). Alfred Stepan provides one of the most useful general treatments of corporatism. See Stepan, *The State and Society: Peru in Comparative Perspective* (Princeton: Princeton University Press, 1978), esp. pt. 1. Stepan uses corporatism to refer to "a particular set of policies and institutional arrangements for structuring interest representation. . . . In return for such prerogatives and monopolies the state claims the right to monitor representational groups by a variety of mechanisms so as to discourage the expression of 'narrow' class-based, conflictual demands. Many . . .

have used such corporatist policies for structuring interest representation" (p. 46).

8 This link is explicit in David Ost's treatment of Solidarity
as an example of societal corporatism. See Ost,
"Towards a Corporatist Solution in Eastern Europe: The
Case of Poland," *Eastern European Politics and Societies 3*
(Winter 1989). A similar link to civil society is pursued
in the writings on China, although with many more
reservations. See, e.g., Anita Chan, "Revolution or
Corporatism? Chinese Workers in Search of a Solution"
(Paper presented at the conference Toward the Year
2000: Socioeconomic Trends and Consequences in
China, Asia Research Centre, Murdoch University,
Western Australia, January 29–31, 1992); Mayfair Yang,
"Between State and Society: The Construction of Cor-
porateness in a Chinese Socialist Factory," *Australian
Journal of Chinese Affairs* 22 (July 1989); Peter Lee, "The
Chinese Industrial State in Historical Perspective: From
Totalitarianism to Corporatism," in Brantly Womack,
ed., *Contemporary Chinese Politics in Historical Perspective*
(Cambridge: Cambridge University Press, 1991); and
Margaret Pearson, "The Janus Face of Business Associ-
ations in China: Socialist Corporatism in Foreign
Enterprises" (Manuscript, Dartmouth College, 1992).

9 Victor Nee, "A Theory of Market Transition: From
Redistribution to Markets in State Socialism," *American
Sociological Review* 54 (October 1989).

10 Vivienne Shue, *The Reach of the State: Sketches of the
Chinese Body Politic* (Stanford, Calif.: Stanford University
Press, 1988).

11 For a more detailed discussion of the diverse effects
of the fiscal reforms on agriculture, see Oi, "Chinese
Agriculture: Modernization, But at What Costs?" in
Thomas Robinson and Zhiling Lin, eds., *The Chinese and
Their Future: Beijing, Taipei, and Hong Kong* (Lanham,
Md.: University Press of America, 1992).

12 I examine this option in Oi, "Private and Local State
Entrepreneurship: The Shandong Case" (Paper pre-
sented at the annual meeting of the Association for
Asian Studies, Washington, D.C., April 2–5, 1992).

13 The exception is in places like Wenzhou, which is
famous for its private businesses.

14 The term *rural enterprises* includes various types of
ventures, from factories to food stalls, that range from
collective to private owned. The largest and economically most significant of the enterprises in terms of output value, number of people employed, and taxes paid
are those owned by local government, either the township or village. Unless otherwise noted, the discussion in
this article is limited to collectively owned enterprises.

15 The percentage of the rural labor force engaged in all
types of rural enterprises rose from 9.5% in 1978 to
22.1% in 1990. The proportion is even higher in areas
with developed rural industries. Zhang Hongyu (fn. 2).

16 See Byrd and Lin (fn. 6) for detailed studies of rural
enterprises based on survey data. See also Christine P.
W. Wong, "Interpreting Rural Industrial Growth in the
Post-Mao Period," *Modern China* 14 (January 1988);
and Oi, "The Fate of the Collective after the Commune,"
in Deborah Davis and Ezra Vogel, eds., *Chinese Society
on the Eve of Tiananmen: The Impact of Reform*, Contemporary China Series, no. 7 (Cambridge: Council on
East Asian Studies, Harvard University, 1990).

17 *Zhongguo tongji nianjian 1988* (Beijing: Zhongguo tongji
chubanshe, 1988), 214.

18 Zhang Hongyu (fn. 2).

19 *Zhongguo nongye nianjian 1989* (Beijing: Nongye
chubanshe, 1989), 20.

20 The others are private or joint-stock enterprises.

21 He Xian, "Woguo xiangzhen qiye shouru fenpei wenti,"
Zhongguo nongcun jingji 3 (March 21, 1988), 34.

22 Ibid.

23 *Xiangzhen qiye nianjian 1978–1987* (Beijing: Nongye
chubanshe, 1989), 270.

24 See Oi (fn. 16) for a more detailed discussion of how
this income keeps the village together as a "collective"
by providing the necessary funds to pay for welfare and
public services and to support agriculture, as well as for
reinvestment in industry.

25 See Christine Wong, "Fiscal Reform and Local Industrialization: The Problematic Sequencing of Reform in
Post-Mao China," *Modern China* 18 (April 1992); and
idem, "Central–Local Relations in an Era of Fiscal
Decline: The Paradox of Fiscal Decentralization in
Post-Mao China," *China Quarterly* 128 (December
1991).

26 Walder, "County Finance and Local Industry: A View
from Zouping County, Shandong" (Paper presented at
the Jackson School of International Studies, University
of Washington, Seattle, February 15, 1990) has found
the same phenomena operating with regard to the state-
owned factories at the county level.

27 Chen, Jefferson, and Singh (fn. 1).

28 See Barry Naughton, "Implications of the State
Monopoly over Industry and Its Relaxation," *Modern
China* 18 (January 1992).

29 Compare Nee (fn. 9) with Victor Nee, "Organizational
Dynamics of Market Transition: Hybrid Forms,
Property Rights, and Mixed Economy in China,"
Administrative Science Quarterly 37 (March 1992).

30 Writing shortly after the reforms started, Kathleen
Hartford points to the general need for cadres to
increase their control over finances if they are to
keep the collective alive and their power strong. See
Hartford, "Socialist Agricultural Is Dead: Long Live
Socialist Agriculture! Organizational Transformations
in Rural China," in Elizabeth J. Perry and Christine
Wong, eds., *The Political Economy of Reform in Post-Mao*

China, Contemporary China Series, no. 2 (Cambridge: Council on East Asian Studies, Harvard University, 1985).

31 Oi (fn. 16).

32 China Interview 8691.

33 See Oi, "Fiscal Reform, Central Directives, and Local Autonomy in Rural China" (Paper presented at the annual meeting of the American Political Science Association, Washington, D.C., August 29–September 1, 1991).

'Moving the State: The Politics of Democratic Decentralization in Kerala, South Africa and Porto Alegre'

Politics and Society (2001)

Patrick Heller

Editors' Introduction

Patrick Heller is Associate Professor of Sociology and Director of the Political Economy of Development Program at the Watson Institute, at Brown University. Heller is a Swiss citizen, and was strongly influenced by his family's commitments to European social democracy. Teenage years in India drew him into engagement with South Asian development and labour politics. He completed a BA (Hons) in Sociology and South Asian Studies at the University of California at Santa Cruz, and went on to postgraduate study in Sociology at the University of California at Berkeley, where he finished a PhD in 1994. Heller taught at the School of International and Public Affairs at Columbia University from 1995 to 2000. In 2000-2001, Heller was a visiting researcher at the Institute for Social and Economic Research at the former University of Durban, Westville; a consultant for the European Union and Unicity Committee for Durban's local government; and a visiting senior researcher for the Centre for Policy Studies, Johannesburg.

Heller's first substantial research project explored the foundations of development with strong redistribution, particularly for women and workers, and under the aegis of communist parties, in the state of Kerala in South India. Heller shows how the mobilisation of poor tenants and landless labourers lies behind the institutionalisation of class politics in a kind of social democratic transition. The book emerging from this research, *The Labor of Development* (2000), is important for bringing worker mobilisation to the centre of the much vaunted 'Kerala model' of social development.

Heller's subsequent work was in South Africa, on local government politics in Durban and on civic associations in Johannesburg. Heller has also been part of a network of comparative scholarship on democratic decentralisation in India, Brazil, South Africa and the United States. In the reading that follows, he brilliantly synthesises insights from his and others' work, through a comparison of the Indian state of Kerala, the Brazilian city of Porto Alegre and post-apartheid South Africa. In all three cases, organised working-class action enabled a process of democratisation under the aegis of highly organised, cohesive, left-of-centre political parties. However, this has not led to similar kinds of transformative capacities. Heller explains this divergence in terms of the state's rationality as well as mobilisation dynamics. The ruling African National Congress (ANC) has retained an instrumentalist, centralised, 'Leninist' conception of its role in post-apartheid South Africa, while left political parties have become less vanguardist and bureaucratic in Porto Alegre and Kerala. Moreover, popular mobilisation has continued to be a force for the state to reckon with in the content of democratic decentralisation in Porto Alegre and Kerala, while civil society has been effectively controlled or marginalised by the ANC's hegemony in South Africa. The reader should pay close attention to the way in which Heller uses these grounded insights to correct a polarised debate between 'technocrats' and 'anarcho-communitarians' on what he calls the 'delicate equilibrium between representation and participation, public goods and local preferences, and between technocracy and democracy'. Heller continues to write on the possibilities for social democracy in the South; for instance, he was one of the authors of the comparative study, *Social Democracy in the Global Periphery: Origins, Challenges, Prospects* (2007). His recent work has also explored avenues for middle-class mobility in two important democratic experiments in development in the South, contemporary India and South Africa.

Key references

Patrick Heller (2001) 'Moving the state: the politics of democratic decentralization in Kerala, South Africa and Porto Alegre', *Politics and Society* 29(1): 131–63.
— (2000) *The Labor of Development: Workers and the Transformation of Capitalism in Kerala*, Ithaca, NY: Cornell University Press.
Erik Olin Wright and Archon Fung (2003) *Deepening Democracy: Institutional Innovations in Empowered Participatory Governance (Real Utopias Project)*, London: Verso.
Patrick Heller and Leela Fernandes (2006) 'Hegemonic aspirations: new middle class politics and India's democracy in comparative perspective', *Critical Asian Studies* 38(4): 495–522.
Richard Sandbrook, Marc Edelman, Patrick Heller and Judith Teichman (2007) *Social Democracy in the Global Periphery: Origins, Challenges, Prospects*, Cambridge: Cambridge University Press.

DECENTRALIZATION AND DEMOCRACY

Over the past decade, a large number of developing countries have made the transition from authoritarian rule to democracy. The rebirth of civil societies, the achievement of new freedoms and liberties have all been celebrated with due enthusiasm. But now that the euphoria of these transitions has passed, we are beginning to pose the sobering question of what difference democracy makes to development, or to be more precise, whether democracy can help redress the severe social and economic inequalities that characterize developing countries.

Two separate problems are involved here. The first parallels the western European literature on the rise of the welfare state and is centrally concerned with patterns of interest aggregation, and specifically the dynamics and effects of lower class formation. This literature has convincingly argued that political rights can be translated into social rights, and procedural democracy becomes substantive democracy, only to the extent that lower class demands are organized and find effective representation in the state.

In the developing world however, uneven capitalist development, resilient social cleavages and various forms of bureaucratic authoritarianism have blunted lower class collective action. The three cases examined here, however, break with this pattern. In South Africa, Brazil, and the Indian state of Kerala, working-class mobilization has driven political transformations and democratization has increased the overall political clout of popular sectors.[1] Working-class politics in all three cases exhibit a high degree of organizational capacity and cohesiveness marked by the presence of strong labor federations and influential left-of-center political parties (although this is less true of Brazil). But working-class political power does not necessarily cumulate into transformative capacity, especially in an era when globalization has weakened the ability of nation-states to deploy the regulatory and redistributive instruments through which European states expanded labor's share of the social surplus. Equity-enhancing reforms in both South Africa and Brazil have been frustrated.[2] And even in Kerala, where working-class mobilization has a longer history and has wielded significant redistributive results, disappointing economic growth, the pressures of liberalization, and the declining service efficiency of the state have all combined to threaten earlier gains in social development.

This leads us to the second problematic of democratization, namely the institutional character of democratic states. Even where formal democracy has been consolidated, the question arises as to just how responsive these democracies are. Developing states have become politically answerable through periodic elections, but have the bureaucratic institutions they inherited from authoritarian or colonial rule become more open to participation by subordinate groups? Have they really changed their modes of governance, the social partners they engage with and the developmental goals they prioritize? Is the reach and robustness of public legality sufficient to guarantee the uniform application of rights of citizenship? The state has certainly been transformed, but has it, in the language that now dominates the posttransition discourse on development, become closer to the people? There are of course many dimensions to this particular problematic, but none that is more central, and that has garnered more attention, than the challenge of democratic decentralization.

Across the political spectrum, the disenchantment with centralized and bureaucratic states has made the call for decentralization an article of faith. Strengthening and empowering local government has been justified not only on the grounds of making government more efficient but also on the grounds of increasing accountability and participation. But to govern is to exercise power, and there are no a priori reasons why more localized forms of governance are more democratic.

Indeed, the history of colonial rule was largely a history of decentralized authority in which order was secured and revenues extracted through local despots. And in its contemporary incarnation, decentralization in the developing world, especially when driven by international development agencies, has more often than not been associated with the rolling back of the state, the extension of bureaucratic control, and the marketization of social services.[3]

The purpose of this article is to explore the conditions under which a distinctly democratic variant of decentralization – defined by an increase in the scope and depth of subordinate group participation in authoritative resource allocation – can be initiated and sustained. Because such a project is tantamount to fundamentally transforming the exercise of state

power, it requires an exceptional, and in most of the developing world improbable, set of political and institutional opportunities. In South Africa, the Indian state of Kerala, and the Brazilian city of Porto Alegre, new political configurations and underlying social conditions have converged to create just such a set of opportunities. Most visibly, left-of-center political parties that were born of popular struggles have come to power and inherited significant transformative capacities. The ascendancy of the African National Congress (ANC), the Communist Party of India–Marxist (CPM), and the Partido dos Trabalhadores (PT) have all been associated with the formulation of clear and cohesive transformative projects in which the democratization of local government was given pride of place. Although the parties in question have captured power at different levels of the state – the national, provincial, and municipal, respectively – they have all enjoyed, and indeed used, their authoritative powers to initiate fundamental reforms in the character of local government.[4]

If a committed political agent is a necessary ingredient for administrative and fiscal devolution, the *democratic empowerment* of local government is critically dependent on the associational dynamics and capacities of local actors. Again, the cases examined here are quite exceptional. All three boast a rich and dense tapestry of grassroots democratic organizations – the historical legacy of prolonged mass-based prodemocracy movements – capable of mobilizing constituencies traditionally excluded from policy-making arenas, and dislodging traditional clientalistic networks.

But the building of local democratic government, even under the most favorable of conditions, is anything but linear. It requires not only that a favorable political alignment be maintained but that a delicate and workable balance between the requirements of institution building and grassroots participation be struck. Subtle differences in political configurations and relational dynamics can thus produce divergent trajectories. In the cases of Kerala and Porto Alegre, initial reforms that increased the scope of local participation have been sustained, and have seen a dramatic strengthening of local democratic institutions and planning capacity. In contrast, in South Africa a negotiated democratic transition that has been rightfully celebrated as one of the most inclusive of its kind, and foundational constitutional and programmatic commitments to building "demo-cratic developmental local government" have given way to concerted political centralization, the expansion of technocratic and managerial authority, and a shift from democratic to market modes of accountability. If democratic decentralization in Kerala and Porto Alegre has been conceived as a means of resurrecting socially transformative planning in an era of liberalization, local government in South Africa has become the frontline in the marketization of public authority. Given the similarity of favorable preconditions – capable states and democratically mobilized societies – we are confronted with an intriguing divergence in outcomes.

Taken in isolation, South Africa's failure to deepen democracy might readily be ascribed to the pressures of globalization, and specifically structural pressures to submit to a neoliberal strategy of economic development and its attendant managerial vision of local government. While it is indeed the case that the ANC has, since 1996, embraced a surprisingly orthodox strategy of growth-led development, a capital-logic argument carries little weight. Unlike most African countries, South Africa has not been subjected to a formal structural adjustment program, and its relatively low level of external debt, high levels of domestic investment capital, significant foreign currency reserves, diversified manufacturing base, and natural resource endowments have made it much less dependent on global financial and commodity markets than most developing economies. If anything, India and Brazil have come under far more pressure to liberalize their economies.

An explanation for the sustainability of democratic decentralization in Kerala and Porto Alegre and its unraveling in South Africa must instead be located in domestic configurations, and in particular in the relational dynamics that have governed state/civil society and party/social movement engagements. The comparison developed in this article highlights two analytical clusters. The first concerns the nature of the political project of local government reform and specifically addresses how critical differences in the broader political context, as well as key internal political party dynamics, have led what are otherwise quite similar parties to develop dramatically different visions of the role of the state in deepening democracy. In the case of South Africa, an electorally hegemonic party, that for historical reasons has developed an instrumentalist understanding of state

power, has succumbed to insulationist and oligarchical tendencies. In Kerala and Porto Alegre, subnational parties operating in more competitive environments have broken with their vanguardist traditions, become critical of bureaucratic state-led development, and have committed themselves to building democracy from the bottom up. The second set of variables begins with the recognition that civil society and social movements are critical to any sustainable process of democratic decentralization. In both Kerala and Porto Alegre, social movements that have retained their autonomy from the state have provided much of the ideological and institutional repertoire of democratic decentralization, and party–social movements relations have generated functional synergies between institution building and mobilization. In South Africa, a once strong social-movement sector has been incorporated and/or marginalized by the ANC's political hegemony, with the result that organized participation has atrophied and given way to a bureaucratic and commandist logic of local government reform.

STRATEGIES OF DECENTRALIZATION

If making a democratic state has been difficult, making the state responsive has been even more so. Although decentralization has become the centerpiece of the modernizing-democratic discourse of developing world state elites and the international development community, shifting decision-making, allocative, and implementation functions of the central state or provincial states to local governments has proven to be difficult for three reasons. The first and most obvious is that most states – and those who control them – have little interest in decentralization. To move the locus of public authority is to shake up existing patterns of political control and patronage. The second is that there is much institutional inertia to overcome. The postcolonial state was born in the heyday of developmentalism. Central and commandist states were the anointed agents of development. The political imperatives of unifying ethnically fractious nations, whose boundaries were the shaky legacies of colonial rule, only heightened the imperative of political centralization. Vast organizational resources have thus been invested in highly centralized and often insulated modes of governance. Though top-down planning has lost much of its luster in the past decade, it remains a powerful organiza-tional reflex. The third is that even when state elites commit themselves to decentralization, the task at hand is Herculean: new laws and regulations have to be passed, personnel have to be redeployed, resources have to be rechannelled, and local adminis-trative capacities have to be built up. In other words, much institution building and training must take place before local government can work effectively.

The challenge of decentralization is thus formid-able and has given rise to two diametrically opposed transformative visions. The first and the most influen-tial is the technocratic vision. Here, decentralization is equated with the task of designing appropriate institutions, the structure of which can be derived from an accumulated corpus of (mostly Western) knowledge of public administration, finances, and planning. The technocratic view is informed by an unbounded faith in the ability of experts to appre-hend and transform the world. This utopian rational-ism has deeply depoliticizing and autocratic impulses. The agent of transformation, as Centeno has written of Mexico's technocratic neoliberal elites, becomes a "state elite committed to the imposition of a single, exclusive policy paradigm based on the application of instrumental rational techniques."[5] This requires that decision makers be insulated from the noisy world of politics. Democracy, to borrow Centeno's words, must be kept "within reason," and the trans-formative thrust must of necessity come from above.

At the other end of the spectrum one finds what Bardhan has called the anarcho-communitarians (hereafter ACs).[6] Here the problem is not too much, but too little democracy. The ACs argue persuasively that the advent of liberal democracy does little to change the overly centralized, and elite-controlled character of postcolonial states. Democracy, they argue, can only achieve its full potential when its for-mal representative institutions are supplemented by a vibrant and participatory civil society. The anarchical element comes in the rejection of traditional vehicles of popular mobilization, namely parties and unions, as oligarchic and too-beholden-to-state-centered models of development. The communitarian element comes from their faith in the capacity of local actors to know and express their interests. Decentralization for the ACs must be driven by the prefigurative actions of social movements – building up local capacity, grassroots institutions, and extraparlia-mentary arenas of participation.

Both the technocratic and AC view suffer from utopianism. As a form of what James Scott has called high modernism,[7] the technocratic view has unbounded faith in science, and argues that its only a matter of getting the institutions right. In this telos, the answers moreover are there for all to behold with the good governance and best practices of the west to be emulated. With their postmodernist impulses, the ACs of course reject the means–end rationality and utilitarianism of the technocratic vision, and place their faith instead in the emergent qualities of civil society. Implicit in this view is that all forms of association contribute to democracy and that the resurgence of civil society is self-sustaining. If the technocratic vision understands but reifies institu-tions at the expense of mobilization, the ACs under-stand and reify mobilization at the expense of institutions.

The technocratic view is certainly correct when it points to the need to develop the necessary manager-ial, organizational, and technical capacities to make local government work. Three technical/organiza-tional problems must be addressed. First, any success-ful planning process requires certain technical inputs. Data must be gathered, plans drawn up, and budgets made. Second is the problem of coordination between levels and the provision of public goods. Not all governance functions can be decentralized. The provision of public goods requires central authority and capacity. Successful planning must be integrated. This calls for effective and constant coordination between levels of government. Decentralization thus requires building a delicate balance of division of authority and competence between levels while creating structures and avenues for coordination. Third, as Weber always insisted, democracy requires bureaucracy. Because partici-pation can never be comprehensive or continuous (capacities for participation are uneven and cannot be sustained throughout the planning and implementa-tion process), there is a need to routinize and for-malize the process through which participatory inputs are translated into outputs; hence, the technical requirement for rules of transparency, accountability, representation, and decisional authority.

Where the technocratic vision is lacking is in its impulse to sanitize decentralization of everything political. For starters, any effort to move the state requires redistributing political power. Democratic decentralization is a political project. The idea that governance is largely a technical/organizational proposition also suffers from two critical flaws. On one hand, no technical or organizational process is ever immune from power and politics. Speaking of what he calls the pragmatic (vs. political) school of thought on decentralization, Schönwälder writes that

> This preoccupation with the practical aspects of decentralization reflects a fundamental belief on the part of the pragmatic school that it is flaws in the planning and execution of decentralization programs, and not the social, economic, cultural or political environment in which these programs are set, which ultimately determine their success or failure.[8]

There is of course an institutional logic to such discourses. Technocratic rationality produces universally applicable blueprints, and decentralization-by-design promises smooth implementation. This appeals to donors and international development organizations because a legible world, cleansed of messy political and social issues, is a world ripe for expert interventions.

But successful decentralization has been the exception to the rule. Technocratic visions have failed because they suffer from an exaggerated sense of the rational mutability of the world (the fiction of induced development) and because their apolitical and frictionless visions of the world are invariably frustrated by politics and friction. Blueprints developed in the West are hardly appropriate to Third World contexts of uneven economic development, pervasive social inequalities, cultural heterogeneity, large-scale social exclusion, the resilience of pre-democratic forms of authority, and weak state capacity. The technocratic view is symptomatic of what Foucault called governmentality, "the idea that societies, economies, and government bureaucracies respond in a more or less reflexive, straight-forward way to policies and plans."[9] This hubris reaches its apogee in the technocratic belief that increased participation can be engineered through appropriate policy design.

An important corrective to the technocratic view is the AC idea that democracy can only be nurtured from below. Social movements can help democratize society by mobilizing previously marginalized actors and by promoting horizontal solidarities. Social movements can moreover nurture grassroots democratic institutions such as civics, shop-floor democracy, and women's groups. But social movements and community initiatives confront two problems. To begin with, most "communities" are inflected with power relations and fragmented by social divisions. In those rare moments that communities have solidarity and are mobilized, the balance of power nonetheless remains weighted against them. And even when historical conjunctures create unique opportunities for transformation from below – such as South Africa's democratic transition – the mobilizational momentum required to effect institutional transformation is notoriously difficult to sustain. The life cycle of associational efforts often depend on the success with which they can scale-up and institutionalize their demands.[10] A critical question, then, is how civil associations can coordinate their activities with state agencies without compromising their associational autonomy.

If the sustainability of popular participation is problematic, so are its effects. Increased associationalism can promote narrow and parochial interests, resulting in the "mischief of factions" and demands for state patronage (rents), just as much as it can result in the promotion of the public interest or broad-based reforms that benefit the majority.[11] Here the issue of creating linkages with more aggregated forms of interest representation – political parties or states – that can check parochial or narrow interests, becomes critical. Moreover, because the capacity of collective action and resource mobilization is unevenly distributed, decentralization runs the risk of aggravating spatial and social inequalities. Contrary to the AC view that emphasizes the need for entirely decentralized units, there is both the need on one hand for "coordinated decentralization" in which articulation between levels allows for resource coordination, the diffusion of innovation, and information feedback[12] and on the other hand for the maintenance of a bounded aggregated authority – the state – to provide nonlocal public goods (including regulatory frameworks) and to aggressively redress regional inequalities. And given the race-to-the-bottom regional competitive pressures unleashed by globalization, central authority remains critical to defining and enforcing basic minimum standards of social citizenship.

Between these bookends of the debate, it is possible to carve out a more balanced analytical position that for want of a better term can be labeled the *optimist-conflict model*. Conflict, because it recognizes both the tension between institutions and demand making that the technocratic vision emphasizes, and the tension between mobilization and institutionalization that the AC framework highlights. It is optimist however, in that it rejects the notion that these conflicts are of a zero-sum nature. High levels of demand making need not necessarily result in institutional overload and ungovernability, much as some routinization of movement dynamics need not result in demobilization. Instead, the optimist-conflict model views the dynamic tensions of development in a more relational and contingent perspective, a view that specifically recognizes the transformative potential of politics. It also recognizes the potential, through such politics, of forging recombinant institutions that can creatively manage a delicate equilibrium between representation and participation, public goods and local preferences, and between technocracy and democracy.

[. . .]

NOTES

I thank Archon Fung, Erik Olin Wright, and William Freund for their extensive comments. I am also grateful to Gianpaolo Baiocchi, Steven Friedman, T.M. Thomas Isaac, David Hemson, Doug Hindson, K.P. Kannan, and Xolela Mangcu for their insights. This paper was first presented at the International Conference on Democratic Decentralisation, Thiruvananthapuram, May 23–27, 2000.

1 For a comparative discussion of the role of labor movements in the democratization process in South Africa and Brazil, see Gay W. Seidman, *Manufacturing Militance: Workers' Movements in Brazil and South Africa, 1970–1985* (Berkeley: University of California Press, 1994). For a discussion of the Kerala case, see Patrick Heller, *The Labor of Development: Workers in the Transformation of Capitalism in Kerala, India* (Ithaca, NY: Cornell University Press, 1999).

2 For Brazil, see Kurt Weyland, *Democracy without Equity: Failures of Reform in Brazil* (Pittsburgh, PA: University of Pittsburgh Press, 1996). For South Africa, see Hein Marais, *South Africa: Limits to Change* (London: Zed Books, 1998).

3 For a particularly illuminating case study, see James Ferguson, *The Anti-Politics Machine: "Development," Depoliticization and Bureaucratic Power in Lesotho* (New York: Cambridge University Press, 1994).

4 Even if the regimes in question are not commensurate as political units, the fact that each is, within its own context, empowered to effect the reform of local government, and in particular to promote greater participation in the process of allocating resources, serves as the basis for comparison. In South Africa, the recent transformation of the nation-state and the political imperative of undoing the spatial and social segregation of apartheid has of course made national government and legislation the critical agent of local government reform. In India, as E.M.S. Namboodiripad famously remarked, since Independence, relations between the center and the states have been governed by democracy, but the relations between the states and local government have been governed by bureaucracy. This however changed with the passage in 1993 of the Seventy-third and Seventy-fourth Amendments to the Constitution that granted local governments new democratic powers but left most of its developmental functions to the discretion of state governments. In Brazil, federalism and a long history of party fragmentation and oligarchical politics have made municipalities critical sites of interest aggregation and patronage. In the postauthoritarian period, the Partido dos Trabalhadores's (PT) efforts to build participatory democracy have thus focused on cities. Porto Alegre represents only one, but by far the most carefully documented, of successful PT-led initiatives to institute "popular budgeting."

5 Miguel Centeno, *Democracy within Reason: Technocratic Revolution in Mexico* (Philadelphia: Pennsylvania University Press, 1997).

6 Pranab Bardhan, "The State against Society: The Great Divide in Indian Social Science Discourse," in Sugata Bose and Ayesha Jalal, eds., *Nationalism, Democracy and Development* (New Delhi: Oxford University Press, 1999), 184–95.

7 James Scott, *Seeing Like a State: How Certain Schemes to Improve the Human Condition Have Failed* (New Haven, CT: Yale University Press, 1998).

8 Gerd Schonwalder, "New Democratic Spaces at the Grassroots? Popular Participation in Latin American Local Governments," *Development and Change* 28, (1997): 757.

9 James Ferguson, *The Anti-Politics Machine*, 194.

10 Jonathan Fox, "How Does Political Society Thicken?: The Political Construction of Social Capital in Mexico," in Peter Evans, ed., *State-Society Synergy: Government and Social Capital in Development* (Berkeley, CA: International and Area Studies, 1996), 119–49.

11 Joshua Cohen and Joel Rogers, "Secondary Associations and Democratic Governance," in Erik O. Wright, ed., *Association and Democracy* (London: Verso, 1995).

12 Archon Fung and Erik Olin Wright, "Deepening Democracy: Innovations in Empowered Participatory Governance," this issue.

'Conclusion: Linking the Urban and the Rural'

from *Citizen and Subject: Contemporary Africa and the Legacy of Late Colonialism* (1996)

Mahmood Mamdani

Editors' Introduction

Mahmood Mamdani is Herbert Lehman Professor of Government in the Department of Anthropology and Political Science and the School of International and Public Affairs at Columbia University. Mamdani is a third-generation East African from Kampala, Uganda. He received a US government scholarship to attend the University of Pittsburgh, where he earned a BA in 1967. He went on to receive an MA in Law and Diplomacy at Tufts University's Fletcher School in 1969, and a PhD from Harvard University in 1974. Mamdani taught at the University of Dar es Salaam from 1973 to 1979, at Makerere University from 1980 to 1993, and was the founding Director of the Centre for Basic Research in Kampala, from 1987 to 1996. He then took up the position of A. C. Jordan Professor of African Studies and Director for the Center for African Studies at the University of Cape Town, South Africa, from 1996 to 1999, before becoming President of the Council for the Development of Social Research in Africa (CODESRIA) in Dakar, Senegal, from 1999 to 2002. In 2004, Mamdani was one of the members of the Global Meeting of Intellectuals from Africa and the African Diaspora organised by the African Union in Dakar.

Mamdani's important early work was a critique of the neo-Malthusianism of population policy. His *Myth of Population Control* (1973) offered a close study of the reasons why rural families in Punjab, India, might choose to have more children for entirely rational reasons. Mamdani then turned to a series of important articles on the agrarian question, imperialism and development in Uganda, culminating in *Politics and Class Formation in Uganda* (1976) and *Imperialism and Fascism in Uganda* (1983). Mamdani's abiding interests have included the politics and remit of 'African Studies' in relation to social movements, academic freedom, and the politics of 'rights' and 'culture', as explored in his edited collection *Beyond Rights Talk and Culture Talk: Comparative Essays on Rights and Culture* (2000). Mamdani has held to a pan-African view of development, and has written on the Democratic Republic of Congo while at CODESRIA.

The reading here is excerpted from his *Citizen and Subject: Contemporary Africa and the Legacy of Late Colonialism* (1996). *Citizen and Subject* poses a provocative argument about the colonial ante-cedents of postcolonial African political crises. Mamdani deliberately generalises across tropical Africa and South Africa as a response to South African exceptionalism. His main argument is that colonial policy in the 1920s and 1930s propped up indigenous authorities and customary law in a system of 'decentral-ised despotism' in the countryside, while forging the elements of modern statecraft in cities. After colonial-ism, or apartheid in the case of South Africa, postcolonial governments have taken one of two paths. On the conservative path, states have maintained decentralisation through existing chiefs or new forms of clientelism. On the radical path, states have centralised authority in one-party structures that exacerbate rural–urban conflict. Postcolonial states have jostled between authoritaianism and ethnic conflict as a consequence of the specific legacies of late colonialism. This is a provocative, even polemical argument, but it does cast the question of decentralisation, widespread in development literature, in new light.

Mamdani extends this argument in his explanation of the colonial origins of racial conflict in the Rwandan genocide, in *When Victims Become Killers: Colonialism, Nativism, and Genocide in Rwanda* (2001). Finally, Mamdani has been writing on US imperialism and the fetishism of the Muslim in numerous articles, and in *Good Muslim, Bad Muslim: America, The Cold War, and the Roots of Terror* (2004).

Key references

Mahmood Mamdani (1996) *Citizen and Subject: Contemporary Africa and the Legacy of Late Colonialism*, Princeton, NJ: Princeton University Press.
— (1973) *Myth of Population Control: Family, Caste and Class in an Indian Village*, New York: Monthly Review Press.
— (1976) *Politics and Class Formation in Uganda*, New York: Monthly Review Press.
— (1983) *Imperialism and Fascism in Uganda*, Trenton, NJ: Africa World Press.
— (ed.) (2000) *Beyond Rights Talk and Culture Talk: Comparative Essays on Rights and Culture*, New York: St Martin's Press.
— (2001) *When Victims Become Killers: Colonialism, Nativism, and Genocide in Rwanda*, Princeton, NJ: Princeton University Press.
— (2004) *Good Muslim, Bad Muslim: America, the Cold War, and the Roots of Terror*, New York: Pantheon/Random House.
Frederick Cooper (1996) *Decolonization and African Society: The Labor Question in French and British Africa*, Cambridge: Cambridge University Press.

As the dawn of independence broke on a horizon of internal conflict, reconsideration of the African colonial experience began. Could it be that the African problem was not colonialism but an incomplete penetration of traditional society by a weak colonial state or deference to it by prudent but shortsighted colonizers? Could it be that Europe's mission in Africa was left half finished? If the rule of law took centuries to root in the land of its original habitation, is it surprising that the two sides of the European mission – market and civil society, the law of value and the rule of law – were neither fully nor successfully transplanted in less than a century of colonialism? And that this fragile transplant succumbed to caprice and terror on the morrow of independence?

With the end of the cold war, this point of view has crystallized into a tendency with a name, Afropessimism, and a claim highly skeptical of the continent's ability to rejuvenate itself from within. Whether seen as a problem of incomplete conquest or as one of unwise deference to traditional authorities, both sides of the Afro-pessimist point of view lead to the same conclusion: a case for the recolonization of Africa, for finishing a task left unfinished. Part of the argument of this book is that Afro-pessimism is unable to come to grips with the nature of the colonial experience in Africa precisely because it ignores *the mode* of colonial penetration into Africa.

Yet another set of questions coheres around a perspective that is not evolutionist but particularistic, whose impetus is not toward highlighting African "backwardness" but underlining its difference. That difference is said to be the tendency to fragmentation and particularism, hitherto held in check and obscured by a shared dilemma, colonial racism. Was not racism the general aspect of the African experience – its colonial and external aspect – and tribalism its particular, indigenous and internal, aspect? Generally emancipated from racism with the end of colonialism, did not Africa once again come to be in the grip of a specifically African particularism: tribalism, ethnic' conflict, and primordial combat? Another part of the argument in this book is that it is too naive to think or racism and tribalism as simple opposites, for alien (racial) domination was actually grounded in and mitigated through ethnically organized local power. In the colonial period, ethnic identity and separation were politically enforced. Although forged through colonial experience, this form of the state survived alien domination. Reformed after independence, purged of its racial underpinnings, it emerged as a specifically African form of the state.

THE FORM OF THE STATE

Colonial genesis

I have argued that to grasp the specificity of colonial domination in Africa, one needs to place it within the context of Europe's larger colonizing experience. The trajectory of the wider experience, particularly as it tried to come to grips with the fact of resistance, explains its midstream shift in perspective: from the zeal of a civilizing mission to a calculated preoccupation with holding power, from rejuvenating to conserving society, from being the torchbearers of individual freedom to being custodians protecting the customary integrity of dominated tribes. This shift took place in older colonies, mainly India and Indochina, but its lessons were fully implemented in Africa, Europe's last colonial possession. Central to that lesson was an expanded notion of the customary.

Britain was the first to marshal authoritarian possibilities in indigenous culture. It was the first to realize that key to an alien power's achieving a hegemonic domination was a cultural project: one of harnessing the moral, historical, and community impetus behind local custom to a larger colonial project. There were three distinctive features about the customary as colonial power came to define it. First, the customary was considered synonymous with the tribal; each tribe was defined as a cultural group with its own customary law. Second, the world of the customary came to be all-encompassing; more so than in any other colonial experience, it came to include a customary access to land. Third, custom was defined and enforced by customary Native Authorities in the local state – backed up by the armed might of the central state.

To appreciate the significance of this, we need to recall only one fact. Although the use of force was outlawed in every British colony in the aftermath of the First World War (and in French colonies after the Second), this applied to the central state and usually to European officials supervising Native Authorities in the local state, but not to the Native Authorities. For this, there was one reason. So long as the use of force could be passed off as customary it was considered legitimate, and – to complete the tautology – force decreed by a customary authority was naturally regarded as customary. No wonder that when force was needed to implement development measures on reluctant peasants, its use was restricted to Native Authorities as much as possible. In the language of power, custom came to be the name of force. It was the halo around the regime of decentralized despotism.

The customary was never singular, but plural. As far as possible, every tribe was governed by its customary law. Europe did not bring to Africa a tropical version of the late-nineteenth-century European nation-state. Instead it created a multicultural and multiethnic state.[1] The colonial state was a two-tiered structure: peasants were governed by a constellation of ethnically defined Native Authorities in the local state, and these authorities were in turn supervised by white officials deployed from a racial pinnacle at the center.

Another peculiarity of this form of the state was that the relation between force and market was not antithetical. It was not simply that force framed market institutions. It was more that force and market came to be two alternative ways of regulating the process of production and exchange. To the extent that the scope of the customary included land and labor, that of the market was limited. To flush either labor or its products out of the realm of the customary required the use of force. Clearly, there was and is no particular and fixed balance between force and market. Its degree remains variable: the customary was never a Chinese wall keeping the tide of market relations at bay; nor was it of nominal significance. The customary was porous. Within its parameters, market relations were enmeshed with extra-economic coercion. Free peasants were differentiated, and those better off were shielded from the regime of force.

Postcolonial reform and variations

Characteristic of Afro-pessimism, whether in its left-wing or right-wing version, is a "roots of the crisis" literature that reduces the past to a one-dimensional reality. The result is a reconstruction of the past as if the only thing that happened was laying the foundations of a present crisis. The result is not an analysis that appropriates the past as a contradictory mix, but one that tends to debunk it.

The core agenda that African states faced at independence was threefold: deracializing civil society, detribalizing the Native Authority, and developing the economy in the context of unequal international relations. In a state form marked by bifurcated power, deracialization and detribalization were two aspects that would form the starting point of an overall process of democratization. By themselves, even if joined together, they could not be tantamount to democratization. Together, this amalgam of internal and external imperatives signified the limits and possibilities of the moment of state independence.

Of this threefold agenda, the task undertaken with the greatest success was deracialization. Whether formulated as a program of "indigenization" by mainstream nationalist regimes – conservative or moderate – from Nigeria to Zaire to Idi Amin's Uganda, or as one of nationalization by radical ones, from Ghana to Guinea to Tanzania, the tendency everywhere was to erode racially accumulated privilege in erstwhile colonies. Whether they sought to Africanize or to nationalize, the historical legitimacy of postindependence nationalist governments lay mainly in the program of deracialization they followed. The difference between them, however, was an effect of the strategy of distribution each one employed. Whether the tendency was privatization or etatism, both strategies opened opportunities for nepotism and corruption, for clientelism.

In contrast to deracialization, the task undertaken with the least success was democratization. Key to democratization was the Native Authority in the local state: its detribalization would have to be the starting point in reorganizing the bifurcated power forged under colonialism. The failure to democratize explains why deracialization was not sustainable and why development ultimately failed. Without a reform in the local state, the peasantry locked up under the hold of a multiplicity of ethnically defined Native Authorities could not be brought into the mainstream of the historical process. In the absence of democratization, development became a top-down agenda enforced on the peasantry. Without thoroughgoing democratization, there could be no development of a home market. This latter failure opened wide what

was a crevice at independence. With every downturn in the international economy, the crevice turned into an opportunity for an externally defined structural adjustment that combined a narrowly defined program of privatization with a broadly defined program of globalization. The result was both an internal privatization that recalled the racial imbalance that was civil society in the colonial period and an externally managed capital inflow that towed alongside a phalanx of expatriates – according to UN estimates, more now than in the colonial period!

But if the limits of the postindependence period were reflected in a deracialization without democratization, I will argue that the Achilles' heel of the contemporary "second independence movement" lies in its political failure to grasp the specificity of the mode of rule that needs to be democratized. Theoretically, this is reflected in an infatuation with the notion of civil society, a preoccupation that conceals the actual form of power through which rural populations are ruled. Without a reform of the local state, as I will soon show, democratization will remain not only superficial but also explosive.

MAINSTREAM NATIONALISM

The mainstream nationalists who inherited the central state at independence understood colonial oppression as first and foremost an exclusion from civil society, and more generally as alien rule. They aimed to redress these wrongs through deracialization internally and anti-imperialism externally. The new state power sought to indigenize civil society institutions and to restructure relations between the independent state and the international economy and polity.

In the absence of the detribalization of rural power, however, deracialization could not be joined to democratization. In an urban-centered reform, the rural contaminated the urban. The tribal logic of Native Authorities easily overwhelmed the democratic logic of civil society. An electoral reform that does not affect the appointment of the Native Authority and its chiefs – which leaves rural areas out of consideration as so many protectorates – is precisely about the reemergence of a decentralized despotism! In such a context, electoral politics turned out to be about more than just who represents citizens in civil society, because victors in that contest would also have a right to rule over subjects through Native Authorities, for the winner would appoint chiefs, the Native Authority, everywhere. More than the rule of law, the issue in a civil society-centered contest comes to be who will be master of all tribes. As a Kenyan political scientist once remarked to me, the ethnicity of the president is the surest clue to the ethnic tinge of the government of the day. This is why civil society politics where the rural is governed through customary authority is necessarily patrimonial: urban politicians harness rural constituencies through patron–client relations. Where despotism is presumed, clientelism is the only noncoercive way of linking the rural and the urban.

Confined to civil society, democratization is both superficial and explosive: superficial because it is interpreted in a narrowly formal way that does not address the specificity of customary power – democratization equals free and fair multiparty elections – and explosive because, with the local state intact as the locus of a decentralized despotism, the stakes in any multiparty election are high. The winner would not only represent citizens in civil society, but also dominate over subjects through the appointment of chiefs in the Native Authority. The winner in such an election is simultaneously the representative power in civil society and the despotic power over Native Authorities.

Tribalism is more one-sidedly corrosive in an urban context than in the rural one. Stripped of the rural context, where it is also a civil war, tribalism in urban areas has no democratic impetus. It becomes inter-ethnic only. This practice is not confined to propertied strata. We have seen that migrants who became involved in the inter-ethnic politics of civil society did so partly to protect customary rural rights. In the absence of the democratization of Native Authorities and the custom they enforced, the more civil society was deracialized, the more it came to be tribalized. Urban tribalism appeared as a post-independence problem in states that reproduced customary forms of power precisely because deracialization was a postindependence achievement of these states.

RADICAL NATIONALISM

The accent of mainstream nationalism was on deracializing civil society, but it is the radical regimes that sought to detribalize Native Authority. The institutional basis of that effort was the single party, the inheritor of militant anticolonial nationalism, which symbolized a successful linkup between urban militants and rural insurrectionary movements against Native Authorities. Militant urban nationalism was the social and ideological glue that cemented otherwise heterogeneous peasant-based struggles. From that experience arose the single party as yet another noncoercive link between the rural and the urban.

The single party was simultaneously a way to contain social and political fragmentation reinforced by ethnically organized Native Authorities and a solution imposed from above in lieu of democratization from below, for the militants of the single party came to distrust democracy, by which they understood a civil society-centered electoral reform. A democratic link between the urban and the rural was in their eyes synonymous with a civil society-based clientelism. Seen as the outcome of an urban multiparty project, clientelism appeared as the other side of a deepening fragmentation along ethnic lines.

Whereas multiparty regimes tended toward a superficial and explosive democratization of civil society, their single-party counterparts tended to depoliticize civil society. The more they succeeded, the more the single party came to be bureaucratized. As the center of gravity in the party-state shifted from the party to the state, the method of work came to rely more on coercion than on persuasion. Whether heralding development or waging revolution, the single party came to enforce it from above on a reluctant peasantry. Although depoliticization contained inter-ethnic tensions within civil society – and as a consequence within the whole polity – the result of a forced developmental march was to exacerbate tensions between the rural and the urban. The single party turned from a mobilizing organ into a coercive apparatus; in the words of Fanon, militants of yesterday turned into informers of today. True, there was a significant break with the formal institutions of indirect rule, but there was no such break with the form of its power. An institution such as chiefship may be abolished, only to be replaced by another with similar powers. The ideological text may change from the customary to the revolutionary – and so may political practice – but, in spite of real differences, there remains a continuity in administrative power and technique: radical experiences have not only reproduced, but also reinforced fused power, administrative justice and extra-economic coercion, all in the name of development.

The reform of decentralized despotism turned out to be a centralized despotism. So we come to the seesaw of African politics that characterizes its present impasse. On one hand, decentralized despotism exacerbates ethnic divisions, and so the solution appears as a centralization. On the other hand, centralized despotism exacerbates the urban–rural division, and the solution appears as a decentralization. But as variants both continue to revolve around a shared axis – despotism.

[. . .]

NOTE

1 Notwithstanding Basil Davidson's claim to the contrary in *The Black Man's Burden: Africa and the Curse of the Nation-State* (New York: Three Rivers Press, 1993).

' "People's Knowledge", Participation and Patronage: Operations and Representations in Rural Development'

from Bill Cooke and Uma Kothari (eds)
Participation: The New Tyranny? (2001)

David Mosse

Editors' Introduction

David Mosse teaches in the Department of Anthropology and Sociology at London University's School of Oriental and African Studies. He gained his PhD in Anthropology from Oxford University for a study of caste, Christianity and Hinduism that was carried out in a drought-prone village in southern Tamil Nadu, India, in the early 1980s. Mosse has returned to that field site on a regular basis, including in the late 1980s when he was working on drought relief plans for Oxfam. Mosse has also worked extensively for the British development agency, the Department for International Development (DFID). Mosse is unusual among Western anthropologists in working for so long on the boundaries between development practice and rigorously intellectual enquiry. His first book, *The Rule of Water: Statecraft, Ecology and Collective Action in South India* (2003), along with several important articles on tank irrigation and water governance in Tamil Nadu, might easily be categorised as 'academic', although they draw on important practical (or policy) work as well. Mosse's second book, however, *Cultivating Development* (2005), is an outstanding ethnography that is directly focused on aid policy and practice. That book, and the reading which we reproduce here, draw on Mosse's long work for, and as a critical observer of, the Indo-British Rainfed Farming Project (KRIBP), a joint-venture of the governments of India and the UK that was inaugurated in 1992 and which became a flagship development project for both countries.

The KRIBP aimed to improve the livelihoods of mainly poorer farmers, many of them *adivasis* ('tribals'), in the western Indian states of Gujarat, Maharashtra and Madhya Pradesh. A sister project, the Eastern India Rainfed Farming Project (EIRFP), would later be set up in Bihar, Orissa and West Bengal, again mainly to work with *adivasi* clients. What was notable about both projects, as Mosse explains, was their commitment to mainstreaming both improved agricultural technologies and practices of participation among beneficiary communities. Simply put, the KRIBP was a model project of and for the 1990s. It drew heavily on the work of Robert Chambers and sought with real integrity to solicit the views of farmers about their 'problems', and to work with and through these farmers to plan solutions. As a development project the KRIBP was avowedly not top-down. Indeed, the funding agencies of the KRIBP and the EIRFP came to define the success of both projects not so much in terms of yields or other outputs, important though these are, as in terms of being farmer-led and participatory. What Mosse provides is an inside account of what happened next. The reading that follows provides a gripping account of the two-way processes that always define development interventions in the project mode. As Mosse shows, it is not simply the project officials who 'read' the environments, villages and villagers they work with; the 'objects of development' (in Ferguson's terms, see Part 1) are also busy reading the project. In the process they learn to bend or reinvent the project for purposes that can be very much at odds with the fine sentiments and participatory methods proposed by project managers.

Key references

David Mosse (2001) ' "People's Knowledge", participation and patronage: operations and representations in rural development', in Bill Cooke and Uma Kothari (eds) *Participation: The New Tyranny*? London: Zed Books.
— (2003) *The Rule of Water: Statecraft, Ecology and Collective Action in South India*, New Delhi: Oxford University Press.
— (2005) *Cultivating Development: An Ethnography of Aid Policy and Practice*, London: Pluto Press.
David Mosse and D. Lewis (eds) (2006) *Development Brokers and Translations*, Bloomfield, CT: Kumarian Press.
Li, T. M. (2007) *The Will to Improve*, Durham, NC: Duke University Press.
T. Mitchell (2002) *Rule of Experts: Egypt, Techno-politics, Modernity*, Berkeley: University of California Press.

INTRODUCTION

An important principle of participatory development is the incorporation of local people's knowledge into programme planning. In some circles this is now the dominant understanding of participation, particularly where techniques of participatory learning and planning (PRA/PLA) are taken as defining features of 'participation' in development. Clearly the meaning of 'participation' is not confined to 'people's knowledge' and planning, but it is an important element.

Firmly embedded in the literature on PRA and participation is the supposition that the articulation of people's knowledge can transform top-down bureaucratic planning systems. Chambers, for instance, posits PRA as a key instrument in challenging the institutionally produced ignorance of development professional 'uppers', which not only denies the realities of 'lowers' but imposes its own uniform, simplified (and wrong) realities on them.

> The essence of PRA is changes and reversals – of role, behaviour, relationship and learning. Outsiders do not dominate and lecture; they facilitate, sit down, listen and learn . . . they do not transfer technology; they share methods which local

people can use for their own appraisal, analysis, planning action, monitoring and evaluation.
>
> (Chambers 1997: 103)

Deferring to local knowledge provides the key to the reversal of hierarchies of power in development planning; PRA reduces dominance and is empowering to the poorest. It stands, in Chambers' view, as a major counterbalance to the power of dominant development discourses. PRA, Chambers (1997) argues, 'draws on, resonates with, and contributes to' a wider new paradigm in which positivist, reductionist, mechanistic, standardized-package, top-down models and development blueprints are rejected, and in which 'multiple, local and individual realities are recognized, accepted, enhanced and celebrated' (ibid.: 188).

With a few project-based illustrations, this chapter will question the potential that a PRA-based focus on 'people's knowledge' has to provide a radical challenge to existing power structures, professional positions and knowledge systems. It will indicate ways in which, on the contrary, participatory approaches have proved compatible with top-down planning systems, and have not necessarily heralded changes in prevailing institutional practices of development. The possibility of transformation is not denied, but the need to examine the social practices of 'local knowledge' production is emphasized, especially given the growing popularity of PRA-based planning and its spread from NGOs into public sector development bureaucracies. The critical point is that what is taken as 'people's knowledge' is itself constructed in the context of planning and reflects the social relationships that planning systems entail. As Long and Villareal point out, knowledge must be looked at relationally, that is, as a product of social relationships and not as a fixed commodity (1994).

PARTICIPATION AND BUREAUCRATIC PLANNING

'Participation' no longer has the radical connotations it once had (e.g. in the radical popular movements of the 1960s). More prominent in present-day discourse are such pragmatic policy interests as 'greater productivity at lower cost', efficient mechanisms for service delivery, or reduced recurrent and maintenance costs (Rahnema 1992: 117). Under the influence of both international donors and domestic policy shifts towards local resource management and cost recovery, participatory planning techniques are now incorporated into the routines of public sector implementation agencies. Here, however, they place new demands on resources, imply a significant departure from normal procedures and decision-making systems, and/or are implemented in the field by people who may as yet have little to gain from the new accountabilities they signify. In short, there are often strong disincentives to adopting participatory approaches. Indeed, in India participatory approaches are still mostly pursued where external agency funding is available to cover the perceived additional risks. It is the optimistic belief of 'agrarian populists' (Bebbington 1994: 205) (and the intention of several development programmes) that exposure to participatory experiences and 'rural people's knowledge' will effect change in values, attitudes and behaviours in authoritarian bureaucracies (ibid.: 207). In this chapter, I look at the accommodation that 'participatory planning' makes with organizational practices, and the way in which organizations seek to secure the benefits (financial, political and symbolic) but avoid the costs of 'participation'.

Any useful discussion on the meaning of 'partici-pation' requires a context, and here this is provided by the experience of the Kribhco Indo-British Farming Project (KRIBP), a donor-funded programme of a large public sector organization in India, which is managed by a special unit combining features of both NGO and government systems. This project is not selected because it, in particular, illustrates problems and constraints in participatory planning. On the contrary, the project is one of very few that has given explicit attention to the processes and dilemmas involved in implementing participatory approaches, and has sought constantly to engage in critical reflection on practice and to modify its planning approach and strategy in the light of experience.[1] Nevertheless, many of these dilemmas come from the need to weld a donor policy interest in 'partici-pation' onto existing organizational interests and processes.

KRIBP (described in detail elsewhere, see Jones et al. 1994; Mosse 1994) is a participatory farming systems development project situated in the Bhil tribal region of western India (the border areas of Gujarat, Rajasthan and Madhya Pradesh states). The project strategy, oriented towards the goal of improving the livelihoods of poor farming families, involves generating location-specific natural resources development plans through a PRA-based process,[2] organized locally by a team of field-based Community Organizers. In principle, local problems are identified and prioritized by villagers, workable solutions found (a joint process) and implementation regimes agreed and negotiated between project staff and members of communities.

Programme activities cover a range of farming system areas: crop trials and community seed multi-plication, agro-forestry and 'wasteland' development, horticulture, soil and water conservation, minor irrigation, livestock development, and credit management for input supply. As far as possible these interventions are low-cost, involve minimal subsidies and/or encourage cost-recovery. Planning such activities requires a high degree of commitment on the part of women and men from villages, and the sustainability of benefits beyond the life of the project depends upon continued management of local resources, and access to external capital and state development programmes through village-based groups (e.g. irrigation groups, credit management groups). In terms of most 'scales of participation', the project has a fairly high (or deep) level of participation: it aims for an intense relationship with farmers at early stages in decision-making (Biggs 1989; Farrington and Bebbington 1993).

The project, benefiting from uncommon clarity in its participatory planning methodology (Mosse et al. forthcoming), nevertheless faced some characteristic problems in putting this into practice. These throw light on the way in which what is read or presented as 'local knowledge' (such as community needs, interests, priorities and plans) is a construct of the planning context, behind which is concealed a complex micro-politics of knowledge production and use. I will briefly comment on four aspects of this: first, the shaping of knowledge by local relations of power; second, the expression of outsider agendas as 'local knowledge'; third, local collusion in the planning consensus; and finally the direct manipulation of 'people's planning' by project agents.

Knowledge and local relations of power

'Local knowledge' reflects local power. PRA events have become a crucial medium through which local perspectives are identified and expressed. But as I have argued elsewhere (Mosse 1994), in KRIBP (as in other contexts) these events can be seen as producing a rather peculiar type of knowledge, strongly shaped by local relations of power, authority and gender (e.g. women being constrained in their expression of opinions). While 'local knowledge' is highly differentiated in terms of who produces it and in terms of different ways of knowing (see also Hobart 1993), it is precisely these relevant differences that are concealed in planning PRAs. What make PRAs especially subject to the effects of dominance and muting is their character as *public* events – events taking place in the presence of local authority or outsiders and directed towards community action. But it is not only the 'public' nature of participatory planning that makes it political, but also its 'open-endedness' When definitions of need, programme activity and 'target group' are open, much is at stake in controlling these. As Christoplos, writing on rural development in Vietnam, says: '[B]y leaving open the definition of the poor farmer, the most significant variable in the planning process, participatory projects become tools for various actors (even the poor themselves) in the political arena' (1995: 2).

Outsider agendas as 'local knowledge'

Project actors are not passive facilitators of local knowledge production and planning. They shape and direct these processes. At the most basic level, project staff 'own' the research tools, choose the topics, record the information, and abstract and summarize according to project criteria of relevance. Given project–villager power relations, it is not greatly surprising that what was recorded in village PRAs reflected (and endorsed) a broad project analysis, for example, that the long-term loss of soil fertility (along with deforestation) was a major cause of declining agricultural productivity in the area. More generally, PRAs 'did not reveal an alternative to the official view of poverty ... but served to further legitimize (the official) discourse with farmer testimonies' (Christoplos 1995: 17–18, commenting on PRA analyses in the Mekoing Delta, Vietnam).[3] In fact, farmers' practical interest in soil erosion arose from the more urgent need for off-season wage labour, which project works offered. Subsequent research on rural livelihoods in the project gave a more central place to wage labour, as well as indebtedness, relationships of dependence and the advanced sale of migrant labour (Mosse et al. forthcoming), while more detailed ethnographic research indicates considerable farmer investment in soil and water conservation (SWC) in the region, leading to an increase in fertility and intensification of cultivation with population increases (Sjoblom 1999). Had other issues such as credit or wage labour been given a more central place in the early analysis of livelihoods (unclothed in 'farming system' concerns), some rather different interventions might have been conceived. The prior emphasis on declining soil fertility is, of course, unsurprising in a farming project integrated into the national goal of increasing the productivity of hitherto neglected rainfed areas (Jones et al. 1995).

Projects clearly influence the way in which people construct their 'needs'. Not all the information recorded in PRAs will register as legitimate 'needs' and so influence technology preferences or programme decisions. In KRIBP villages, for example, the matrix ranking of tree species was used in initial PRAs to identify a wide range of species and multiple uses for them.[4] The focus of discussion was on the actual *uses* of trees. When, however (in 1993), village-level nurseries were being established and farmers (women and men) were asked about their needs, and which species should be raised in the nurseries, a far more limited range of options was considered. Indeed, there was an overwhelming preference (reflected in the nurseries raised) for one particular species – eucalyptus.[5] There was a significant gap between patterns of usage (reflected in PRAs) and the expressed needs (or desires) that ultimately influenced decisions. Actual uses were even reinterpreted in terms of needs expressed in the light of project deliverables. Some villagers, for example, expressed a strong preference for eucalyptus as timber for housing when, in fact, they had little or no experience of using the species for this purpose. It happened that the village nursery programme was sponsored by the State Forest Department, which was perceived as strongly favouring eucalyptus (which was indeed the most commonly planted tree under 'social forestry' locally). Villager 'needs' were significantly shaped by perceptions of what the agency was able to deliver.[6] The expressed need for eucalyptus was, like the desire for soil and water conservation, in effect a low-risk community strategy for securing known benefits in the short term (trees or wages) that might have been jeopardized by some more complex and differentiated statement of preferences.[7] Farmers used a wide range of trees and were well aware of soil erosion effects, but these ideas *about* livelihood constraints would not be the same as those employed *for* action involving change with external agents. The latter take account of technology *availability*, and perceptions about which demands are likely to be considered *legitimate* (i.e. compatible with given project objectives). Following a KRIBP-organized visit to the local Krishi Vigyan Kendra (agriculture science centre), for example, some village women prioritized the planting of subabul and lemon, species to which they were exposed during the visit. Important and valuable though these innovations are, they may not be those that arise first from the more descriptive understanding of women's livelihoods.

Local collusion in the planning consensus

Clearly needs are socially constructed (Pottier 1992) and 'local knowledge' shaped both by locally dominant groups and by project interests. 'Insider' and 'outsider' are inseparable in what would more correctly be referred to as 'planning knowledge'

rather than 'people's knowledge'. Arguably, through participatory learning, it is farmers who acquire new 'planning knowledge' and learn how to manipulate it, rather than professionals who acquire local perspectives. This, then, is the third point, namely that people themselves actively concur in the process of problem definition and planning, manipulating authorized interpretations to serve their own interests. It was farmers, after all, who were able to use a consensus on the loss of soil fertility as the means to address the far more urgent requirement for paid employment and to secure delivery of more certain benefits of wage labour in the short term. Local power hierarchies intersect with project priorities as a multitude of local perspectives and interests struggle to find a place within the authorizing framework of the project. While expression of 'illegitimate' interests gets suppressed, some individuals or groups have the skill or authority to present personal interests in more generally valid terms, others do not.

Over time and through negotiation, project staff *and* villagers (in the first instance only men of influence, but later a wider cross-section of people) collude in translating idiosyncratic local interests (such as in wage labour, wells, pumps, housing support and loans) into demands that can be read as legitimate. Both benefit. Villagers gain sanction for activities in their neighbourhoods, and field staff, by delivering desirable goods and schemes and wage labour, win support from locals who agree to 'participate', attend meetings, train as volunteers, host visitors, save and make contributions, do things for the poor, and in other ways validate both the wider project and staff performance within it. This planning process, driven by a shared interest in producing a plan for concrete action, invariably suppresses difference in favour of consensus, and prioritizes action over detailed design. Staff who try to be *too participatory*, spend too much time investigating 'real needs' or women's needs rather than delivering schemes, are soon seen as under-performing by both project and community.

Of course, these planning negotiations are not between equals. Whatever the rhetoric, the reality is that people participate in agency programmes and not the other way round. In relation to its tribal villagers, the KRIBP project was clearly the most powerful player. Moreover, it is project outsiders who need and use 'local knowledge' about livelihoods, often to bargain with villagers, to challenge claims on the project, to reject as well as accept villager proposals, to negotiate subsidy levels, savings, cost recovery, and resource-sharing arrangements, and to allocate labour benefits or gender roles on project works. In this sense, 'local knowledge' is part of the project's exercise of power in *constraining* as well as enabling 'self-determined change'. The polarity set up between extractive and participatory modes of learning obscures the fact that, once produced, information will be used in various ways in a project system, including to privilege certain subordinate perspectives within communities. People's knowledge is also used to advance and legitimize the project's own development agenda, or even to negotiate its participatory approach with other stakeholders such as funders, technical consultants, senior management.[8] The fact that 'PRA-type' information has been set as a new scientific standard by donor and other agencies does not, in itself, democratize power in programme decision-making. Participatory approaches and methods also serve to represent external interests *as* local needs, dominant interests *as* community concerns, and so forth.

Manipulation of 'people's planning'

'Rural people's knowledge' (including, for example, analysis of problems, needs and plans) is collaboratively produced in the context of planning. As 'planning knowledge' it is a rather unusual type of knowledge with some specific characteristics: it is strongly shaped by dominant interests and agency objectives/analyses; it is conditioned by perceptions of project deliverables and the desire for concrete benefits in the short term;[9] it is consensual and obscures diverging interests both within villages and between the village and project (simplifying and rationalizing local livelihood needs to ensure consistency with project-defined models); it closely matches and supports programme priorities; and it involves bargaining and negotiation between agency staff and villagers but ultimately is a collaborative product, concealing both villager and project manoeuvres. 'People's knowledge' is undoubtedly a powerful normative construct that serves to conceal the complex nature of information production in 'participatory' planning, especially the role of outsiders.

The final point here is that, not infrequently, programme decisions take place with little reference to

locally produced knowledge at all. PRA charts and diagrams provide attractive wall decorations, making public statements about participatory intentions, legitimizing decisions already made – in other words symbolizing good decision-making without influencing it. Even where considerable effort is directed towards the involvement of local people and their knowledge in planning, there are compelling reasons why locally generated plans do not provide the basis for programme choices. Quite apart from problems surrounding the use of 'people's knowledge' in planning systems, programme decisions are usually influenced by other interests altogether. The simplistic assumption that better access to local perspectives (even supposing this is unproblematic) will ensure that programme decisions are more participatory is, perhaps, only too obviously blind to the institutional realities of rural development.

For one thing, a new project such as KRIBP has its own needs. First, the project had to work out an acceptable compromise with villagers (in practice, key village leaders), a compromise between their hopes and expectations and project objectives, as a basis for continuing to work in the area. At a local level KRIBP project field staff (like others; see Arce and Long 1992) initially found that the acceptability of their presence in villages was largely based upon benefits they could, or promised to, deliver. They therefore felt constrained to initiate activities and programmes as a way of meeting new social obligations, demonstrating their influence and retaining their status as educated experts. Indeed, early programme choices were often shaped by the pragmatic need to manage villager petitioning while securing a social position locally. This may have been exaggerated by competition with peers working in other villages or by a perception that concrete actions would be rewarded over knowledge-building.[10]

Second, priorities are influenced by a project's wider institutional setting and its need to maintain relationships with local government, senior management, research institutions or donor advisers, with distinct development agendas that require the introduction of a stream of frequently flawed or inappropriate schemes such as the promotion of new winter crops, grain banks, farm machinery, mushroom cultivation, women's handicraft and drip irrigation, all presented as 'local needs'. More generally programme action is shaped by the project's engagement in wider coalitions contending for influence within national or international policy arenas (cf. Biggs 1995a, b). A project such as KRIBP may, in fact, participate in several coalitions pursuing different objectives – for example, agricultural production, environmental protection, poverty reduction and gender equality.

Third, choices and programme delivery are constrained by organizational systems and procedures (for example, budgeting time-frames, procedures for approval, sanctioning, fund disbursement and procurement). New concepts of 'process' have not obviated these institutionally grounded needs. Project managers still face other pressures to get things done, and other measures of efficiency than those provided by measures of participation. There are pressures for a local planning system to be sensitive to organizational realities as well as to villagers' livelihood constraints, and fieldworkers working under pressure to 'keep up momentum', to meet expenditure targets and to maximize quantifiable achievements may find themselves giving priority to familiar, conventional programmes over innovative initiatives where approval may be uncertain or delayed. There is therefore often a tendency for project works to cluster around a fixed set of standard interventions, limiting the potential creativity of participatory problem-solving.

Fieldworkers develop their own operational interpretation of both villager needs and project goals, and their own strategies of intervention, which are sensitive to the managerial and institutional environment as well as the village contexts in which they work.[11] Moreover, as villagers shape their needs and priorities to match the project's schemes and administrative realities, validating imposed schemes with local knowledge and requesting only what is most easily delivered, the project's institutional interests become built into community perspectives and project decisions become perfectly 'participatory'. So, if projects end up ventriloquizing villagers' needs it is not only, or primarily, because artful and risk-averse villagers ask for what they think they will get. It is also because development agencies are able to project their own various institutional needs onto rural communities. In short, through project systems of participatory planning, 'local knowledge', far from modifying project models, is articulated and structured by them.

In certain institutional contexts, 'participatory' processes can produce not diverse and locally varied

development programmes, but strong convergence into a fixed set – crop varieties, soil and water conservation measures, agro-forestry and a range of *ad hoc* welfare programmes, in the case of KRIBP. Indeed there are inherent disincentives (among staff and villagers) to exploring or developing complexity and a preference for deep and reinforcing grooves. Far from being continually challenged, prevailing preconceptions are confirmed, options narrowed, information flows into a project restricted, in a system that is increasingly controllable and closed. The danger, of course, is that a participatory project that has diverged away from analysing problems will have limited impact, miss opportunities, or, at worst, by mis-specifying the problem, contribute to an aggravation of poverty or environmental decline (see Starkloff 1996).

This shift from an open, exploratory system towards a closed one is not to be understood as intentional. It is the side-effect of institutional factors that are unlikely to be perceived by project actors themselves, by their supporting bureaucracies, or even by external observers. Ironically, it is often when the expressed needs of client villagers most perfectly 'mirror' organizational systems, when programmes have become most impervious to variable local needs and new perspectives, that they begin to be acclaimed for their participatory processes or the sophistication of their methods. Indeed, through the mirroring of project assumptions in local plans, a project can advertise its participatory achievements while retaining control over an increasingly standard set of project activities, reproduced through conservatism, convenience, and risk aversion on the part of both villagers and staff.

Over time, and through a Weberian process of routinization, the operational demands of a project such as KRIBP (i.e. timely implementation of high-quality programmes) can become divorced from its participatory methods and goals. These latter have invariably been established and promoted by donors, rather than implementing bureaucracies, and place enormous demands on existing procedures, or conflict with existing accountabilities (e.g. setting targets of quantitative achievements). Tensions between established and donor demands for 'participation' produce a characteristic 'dual logic' in projects. One logic, set out in donor project documents and 'log-frames', gives emphasis to local-level integrated planning and local capacity-building. The emphasis is on participation and sustainability. The other is the operational logic of the project agency, which emphasizes upward accountability, proper use of funds, and the planning and delivery of quality programmes. The emphasis is on *delivery*.

[. . .]

CONCLUSIONS

The first part of my argument can be summarized by saying that the popular 'PRA' assumption that learning and 'local knowledge' defines, or redefines, the relationship between local communities and development institutions needs to be reversed. It is often the case that the 'local knowledge' and 'village plans' produced through participatory planning are themselves shaped by pre-existing relationships – in the present case, by patronage-type relationships between a project organization and tribal villagers. Rather than project plans being shaped by 'indigenous knowledge', it is farmers who acquire and learn to manipulate new forms of 'planning knowledge'. In this way local knowledge becomes compatible with bureaucratic planning.

The second part of my argument looks at what ideas of participation *do* rather than at what they fail to do. 'Participation' can be seen primarily as a representation (or a theory) oriented towards concerns that are external to the project location. Such representations do not speak directly to local practice and provide little guide of implementation, but are important in negotiating relationships with donors, and part of wider development policy arguments. Participation as a set of development ideas or interpretative frameworks is sustained locally through its links to the wider policy process.

Models of participation and normative schemes are prevalent and powerful. In various forms, they provide the lenses through which we (project workers, consultants, academics) see and judge projects such as KRIBP as successes or failures. Laudatory or critical commentaries both draw attention to and endorse the same models of participatory development. In doing so, however, they privilege the model over project practices, first, obscuring the agency of project organizations and their staff, and second, producing ignorance of project impacts.

As the long-term 'social development' consultant to KRIBP, I myself held a strong external monitoring concern with participation, local autonomy, sustainability and project withdrawal, which challenged the project's operational emphasis on patronage and programme delivery, largess and local dependency. The problem is that such a normatively driven perspective obscured the agency of the development organization itself as a politically conscious, strategically operating marketing organization, and the particular institutional context and constraints that this provided for participatory development. In practice, it proved rather difficult to get the donor/consultant interpretation of participation in terms of local control, autonomy and villager capacity-building procedurally internalized in the project (in terms, for example, of targets, monitoring and staff rewards), even though it remained a key part of the project as represented. There are many reasons for this difficulty that cannot be gone into here, but some of them at least may be less surprising when we consider that the particular understanding of participation advocated was, in some quite fundamental ways, at odds with what was driving the project. After all, the project's reputation, the validation of its participatory approach, the performance of local field staff, indeed the core rationale of the project from the organization's point of view, were all based on its network of patronage and the delivery of an expanding range of programmes, increasingly through village 'volunteers' (*jankars*) who operated as the lower orders of the project delivery mechanism. This structure was not (easily) going to be replaced with a striving to independent capacities, local autonomy and the withdrawal of the project. Why would the organization want to rid itself of its best customers, and villagers take leave of a serviceable patron?

Furthermore, the privileging of participation models over project practices overlooks some significant impacts. For instance, a more pragmatic appraisal of project activities would be able to regard the project's achievements – such as new input lines for improved technology, new marketing possibilities, new avenues of patronage and the matching of local needs and the organization's own marketing strategy – as advantageous in a remote tribal area, rather than as a failure to meet the objectives of participation. The benefits of high-quality soil and water conservation, improved seed inputs, assisted seed distribution and storage, mediated links to national and international agricultural research agendas are highly significant. They may depend upon the permanent and expanding presence of the organization as a parastatal extension service – offering better technology and more affordable inputs rather than autonomy and independence to remote tribal villages. This may be a subversion of the currently dominant international development idea of 'farmer-managed development', but it is undoubtedly an operationally successful and institutionally supported version of participation.

NOTES

1 It should be clearly understood that this is in no way a descriptive account of the KRIBP project. Experience from the earlier years of the project is selectively cited only to illustrate some more generally applicable points. The perspective I offer comes from my own experience working as a social development consultant to the project over several years. It remains a personal perspective, and the KRIBP project, its consultants and donor bear no responsibility for the views and opinions expressed here.

2 See an earlier paper (Mosse 1994) for discussion of the project's early experience with PRA, and the lessons learned.

3 The worry that rapid participatory research methods would, in practice, often perform a legitimizing function for decisions already taken was raised well over a decade ago in the early debate on RRA (Wood 1981).

4 In one women's group 37 species were ranked in relation to eight different uses.

5 Of course the earlier PRAs were not free of omission or selectivity. Fruit trees, for example, emerged as an important priority but were not mentioned during the initial tree matrix exercises (Bezkorowajnyj et al. 1994).

6 This relationship has changed over time, and correspondingly there has been a change in the proportion of eucalyptus seedlings raised. This has fallen over the years of the project.

7 There is of course no uniformity of needs. A strong bias towards eucalyptus might meet some people's needs, but not those of others. Some people have the power and authority to influence the collective decision in favour of options that better meet their particular needs (cf. Mosse 1994). Women, typically, lack this power, and in this case their unarticulated experience of burdensome labour and time devoted to the collection of fuel and fodder (for which eucalyptus is not a first choice) or the economic and nutritional importance of forest species and the collection of non-timber forest produce

did not overly shape programme choices, even though they clearly featured in separate informal PRAs.

8 PRA information on the customary role of tribal women in decision-making about household finance, in livestock management, manuring and seed selection and management, for example, is necessary in arguing a case for their central role (as policy) in the project activities of credit, input supply or crop development that would otherwise by default come to be controlled by men.

9 In another example of project-related 'short-termism' Starkloff (1996) shows how participatory mechanisms generate support for 'coping mechanisms' rather than addressing underlying environmental problems.

10 The way in which gender relations ensure that women fieldworkers are placed at a distinct disadvantage in generating information from women and presenting this in terms of readily implementable programmes is properly the subject of a separate discussion.

11 Arce and Long (1992) examine in some detail the way in which a Mexican field-worker (a *tecnico* or technical agronomist) devises his own strategies of intervention in both the village and official administrative arenas, which enable him to retain legitimacy in the eyes of both villagers and bureaucrats.

REFERENCES

Arce, A. and N. Long (1992) 'The Dynamics of Knowledge: Interfaces Between Bureaucrats and Peasants', in N. Long and A. Long (eds), *Battlefields of Knowledge: the Interlocking of Theory and Practice in Social Research and Development*, Routledge, London and New York.

Bastian, S. and N. Bastian (eds) (1996) *Assessing Participation: A Debate from South Asia*, Konark, Delhi.

Bebbington, A. (1994) 'Theory and Relevance in Indigenous Agriculture: Knowledge, Agency and Organisation', in D. Booth (ed.), *Rethinking Social Development: Theory, Research and Practice*, Longman Scientific and Technical, Harlow.

Bezkorowajnyj, P. G., S. Jones, J. N. Khare and P. S. Sodhi (1994) 'A Participatory Approach to Developing Village Tree Programmes: The KRIBP Experience', *KRIBP Working Papers*, Centre for Development Studies, Swansea.

Biggs, S. (1989) 'Resource-poor Farmer Participation in Research: a Synthesis of Experience from Nine National Agricultural Research Systems', *OFCOR Project Study*, No. 3, ISNAR, The Hague.

——(1995a) *Participatory Technology Development: A Critique of the New Orthodoxy*, Olive Information Service, AVOCADO series 06/95, Durban.

—— (1995b) 'Participatory Technology Development: Reflections on Current Advocacy and Past Technology Development', paper prepared for the workshop 'Participatory Technology Development (PTD)', Institute of Education, London, March.

Chambers, R. (1997) *Whose Reality Counts? Putting the First Last*, IT Publications, London.

Christoplos, I. (1995) 'Representation, Poverty and PRA in the Mekong Delta', EPOS Environment Policy and Society, *Research Report*, No. 6, Linköping University, Sweden.

Farrington, J. and A. Bebbington with K. Wellard and D. J. Lewis (1993) *Reluctant Partners? Non-Governmental Organisations, the State and Sustainable Agricultural Development*, Routledge, London.

Hobart, M. (ed.) (1993) *An Anthropological Critique of Development: The Growth of Ignorance*, Routledge, London and New York.

Jones, S., J. Witcombe and D. Mosse (1995) 'The Development Choice between High and Low Potential Areas', *KRIBP Working Paper*, No. 7, Centre for Development Studies, University of Wales, Swansea.

Jones, S., J. N. Khare, D. Mosse, P. Smith, P. S. Sodhi and J. Witcombe (1994) 'The Kribhco Indo-British Rainfed Farming Project: Issues in the Planning and Implementation of Participatory Natural Resource Development', *KRIBP Working Paper*, No. 1, Centre for Development Studies, University of Wales, Swansea.

Long, N. and M. Villareal (1994) 'The Interweaving of Knowledge and Power in Development Interfaces', in I. Scoones and J. Thompson (eds), *Beyond Farmer First: Rural People's Knowledge, Agricultural Research and Extension Practice*, IT Publications, London.

Mosse, D. (1994) 'Authority, Gender and Knowledge: Theoretical Reflections on the Practice of Participatory Rural Appraisal', *Development and Change*, Vol. 25, pp. 497–526.

Mosse, D., S. Gupta, M. Mehta, V. Shah, J. Rees and the KRIBP Project Team [2002] 'Brokered Livelihoods: Debt, Labour Migration and Development in Tribal Western India', in A. de Haan and B. Rogaly (eds), *Migration and Sustainable Livelihoods*.

Pottier, J. (ed.) (1992) *Practising Development: Social Science Perspectives*, Routledge, London and New York.

Rahnema, M. (1992) 'Participation', in W. Sachs (ed.), *The Development Dictionary: A Guide to Knowledge as Power*, Zed Books, London, pp. 116–31.

Sjoblom, D. K. (1999) 'Land Matters: Social Relations and Livelihoods in a Bhil Community in Rajasthan, India', PhD Thesis, School of Development Studies, University of East Anglia.

Starkloff, R. (1996) 'Participatory Discourse and Practice in a Water Resource Crisis in Sri Lanka', in S. Bastian and N. Bastian (eds), *Assessing Participation*.

Wood, G. D. (1981) 'The Social and Scientific Context of Rapid Rural Appraisal', *IDS Bulletin*, Vol. 12, No. 4, pp. 3–7.

PART EIGHT

Globalisation, security and well-being

Questions about globalisation and its effects have been centre-stage in discussions of development since about 1990. Very often they have run in parallel with debates on good governance and institutions, although they have their own curiosities. To begin with there are important matters of chronology and definition. Only a fool would argue that globalisation begins in 1990, or that it is coterminous with the Internet age. Globalisation is often defined as a process of time–space compression. We have a sense that the world is speeding up and that our lives are now radically entwined with those of distant strangers. People all over the world can consume certain events at the same time – the World Cup, for example, or coverage of the US-led invasion of Iraq. People with high-end mobile phones can speak to each other more or less wherever they are.

At the same time, we know that many people living in 1800 or 1900 thought that they too were living in a period of tremendous technological change. Recall Adam Smith, from Part 2. He suggested that the 'discovery of America and that of a passage to the East Indies by the Cape of Good Hope' amounted to two of 'the most important events recorded in the history of mankind'. Recall, too, the sense of wonder that people expressed when the telephone and motor car were invented, or when a trans-Atlantic cable was first laid. Arguably, it was only in the middle part of the twentieth century that a sense of deepening global integration was pared back in the West. Decolonisation, the Cold War and the Bretton Woods system (roughly 1944–1971/3) created the conditions for systems of geographical separation and national economic planning. Countries through this period were linked mainly by traded goods and services, rather than by significant flows of industrial or financial capital.

Globalisation, then, is an ongoing set of processes. It has been deepening, unevenly and inconsistently, since at least the dawn of capitalism. To examine its effects on such important matters as growth, poverty or security we need to define our terms precisely. And even then we are likely to find that one-to-one correspondence effects are rare. This is very much what we see in the first two readings here, those by Martin Wolf and by one of his long-term correspondents, Robert Hunter Wade. Wolf is a booster for globalisation. He takes the Smithian view that increased global integration can work to the advantage of all participants. Integration is defined here in terms of traded goods and services, capital flows and flows of people. At the same time, Wolf recognises that there are political obstacles to the free flow of people in today's world economy, the negative effects of which are considerable. Dani Rodrik has suggested that 'liberalizing cross-border labour movements can be expected to yield benefits that are roughly 25 times larger than would accrue from the traditional agenda focusing on goods and capital flows' (2002: 19–20). (It is worth noting that global GDP increased by over 700 per cent from 1973 to 2003, during which time exports of goods and services increased by over 1300 per cent. Over the same period, the international flow of people increased by less than 250 per cent.)

Robert Wade doesn't pick up this argument so much as one of Wolf's related claims: namely, that greater densities of international economic integration can be expected to reduce extreme poverty and income inequality levels at the global scale, and to a considerable degree is already doing so. He is taking aim, that is to say, at what has become another World Bank orthodoxy. In the words of former Bank President, James Wolfensohn, 'Over the past twenty years the number of people living on less than $1 a day has fallen by 200 million, after rising steadily for 200 years.' For Wade, such comforting notions will not do. Wade suggests that World Bank poverty numbers are subject to large margins of error. In addition, any general argument about globalisation and its effects since c.1980 is heavily affected by the inclusion/ exclusion of China, or China and India. Further, even if poverty headcounts can be shown to be declining at a more rapid rate since c.1980, this is not a priori evidence of a positive globalisation effect. In India, for example, other factors might have spurred a reduction in headcount poverty totals and ratios. One obvious candidate is the spread of a state-sponsored Green Revolution. Another is significant government invest-ments in health care and targeted anti-poverty schemes. In Wade's view, the jury is still out. We need to be cautious about making any sort of causal argument about 'globalization (or greater trade integration) and its effects' beyond the regional level. We also need to ask serious questions about why it is 'at the level of the whole, the increasing returns of the Matthew effect – "To him who hath shall be given" – continues to dominate decreasing returns in the third wave of globalization'.

If the Wolf–Wade debate dominates one area of contemporary political economy, another set of debates has begun to explore the environmental and gender security implications of global economic growth (see also Dyson on climate change in Part 9). The third reading in this section comes from Partha Dasgupta, an economist at Cambridge University. Dasgupta notes that measures of human well-being have generally taken shape without reference to what he calls the natural environment. In the 1950s, 1960s and 1970s, development was measured largely in terms of growth of GDP, or of GDP per capita. In the 1980s and 1990s this index was supplemented by the Human Development Index (HDI), particularly in the annual reports of the United Nations Development Programme. Different versions of the HDI took account of such things as infant mortality and literacy rates, or gender inequality scores. Dasgupta and some other environmental economists are now trying to devise an accounting framework that will move beyond both measures. They want to take account of changes (build-ups/depletion) in stocks of natural capital, as well as in stocks of material, human and social capital. According to Dasgupta, preliminary findings suggest that 'real development' in China since c.1980 has been only marginally positive once natural capital scores are factored in. In Pakistan and India it has been negative.

More rapid economic growth in China and India has also been strongly gendered in its effects, as it has been around the world. Economic reforms in both countries have helped pull many households out of absolute poverty. At the same time, a rising tide of wealth has not always empowered women. Sometimes it has, sometimes it hasn't. Much depends on context: for example, on patterns of female participation in the paid workforce and the public sphere. In India, we hear of young women who find work in a call-centre in Mumbai or Bangalore. But many more are condemned to low paying and low productivity jobs in the countryside, even as local men take advantage of new urban labour market opportunities. Within India's richer northern and western states (including Punjab, Himachal Pradesh, Haryana and Gujarat) there is mounting evidence that more girls are going to school. Yet one key aim of their parents is to secure better terms for their daughters in overheated dowry markets. Female foeticide, and the relatively poor treatment of infant girls, are also at their worst in these parts of India.

In the reading reproduced here, Amartya Sen reminds us that there are about 100 million missing women in the world. This appalling holocaust is centred in South Asia and China, and there is little reason to suppose that it is getting better under a regime of globalisation. More might be expected of the expanding possibilities for collective action among women in this region. This has certainly been a rallying cry for women's organisations in India, where it has been stimulated by the growth of village forest committees and the reservation of seats for women in elected *panchayats* (units of local government). As Bina Agarwal explains here, however, when women lack meaningful command over land or other assets, the prospects for effective political participation remain poor. Women in South Asia still very often engage the political process through male relatives or brokers. The so-called second democratic upsurge in India, which has led to the political enfranchisement of many of the country's Other Backward Castes (those just 'above' the *adivasis* (tribals) and ex-untouchables), has touched the lives of male family members much more than it has female family members.

The growing spread of HIV-AIDS is also a threat to female (and male) security in India. Total levels of infection there were almost as high in 2006 as in southern Africa, although the rate of infection was far lower. Again, the growing integration of the country's economy is one possible source of the rise of AIDS. Another is the social obstacles women face in their inability to insist on contraception during heterosexual intercourse, whether in the domain of marriage or sex work. In many countries in sub-Saharan Africa, however, though by no means in all, women's biological vulnerability to HIV-AIDS has been hugely increased by the economic crises that took hold in the 1980s and which still continue.

Brooke Grundfest Schoepf explores these multiple intersections in the reading reproduced here. Her work is largely concerned with practical issues of AIDS prevention among Africa's women and youth. As Helen Epstein has recently argued, these must address some of the reasons why African men and women are more likely to have concurrent long-term sexual relationships (but no higher number of lifetime partners) than men and women elsewhere. In southern Africa, long-distance migration for work might be key (see also Wolpe, in Part 5, for discussion of the roots of segregation and work-based migration in this region). But Schoepf also points to the dark underbelly of contemporary globalisation. Proponents of global capitalism sometimes exhibit 'Africa' as a case study of failed integration. The median age at death remains horribly low in sub-Saharan Africa – in the early 1990s it was less than 5 years (Sen, 2003: xi) – and some blame this on the region's reluctance to abandon *dirigiste* practices of economic mismanagement. Africa has failed to join India, China, South-East Asia and even Latin America in seeking a place at the table of the world economy.

There is clearly an element of truth in this claim (though see also Collier in Part 9). But what is missing is any sense that Africa continues to be haunted by its colonial pasts, and by a contemporary resource curse. One reason for the lack of good governance in the region today is the dominance of what are now called international bads: the unregulated arms trade, for example, money laundering by elites through offshore banks, and the willingness of the West and China to pay such high prices for the region's natural resources. The formation of effective and capable states is difficult when governments can avoid raising taxes to pay for public spending. No taxation too often means no representation or accountability. Resource-rich countries can then be cursed with endemic corruption and mass poverty. Poor people search for patrons within political society rather than seeking the blessings of rational law and justice in civil society. A sense of citizenship remains weakly developed, at best.

In such situations, too, where the state is either de facto privatised by ruling elites, or is simply collapsed (as in parts of Sudan, Somalia or Afghanistan), it is natural that people will seek non-state protectors. These might include warlords, militant armies and mafia groups. The reading by Charles Hirschkind and Saba Mahmood gives a sense of these developments in the course of an important critique of feminism, the Taliban, and the politics of counter-insurgency. Hirschkind and Mahmood also reveal the historical complicity of the United States (and the British and the Russians, it should be said) in 'creating the miserable conditions under which Afghan women were living' at the time of the US-led invasion in 2001. They begin to reveal the social double helix that binds the export of capital (infrastructure and reconstruction contracts, more recently) to the coercive and disciplinary powers of the state and American empire. One of the objectives of the globalisation literature, it might be contended, is to disguise these interweavings. An abstracted globalisation then floats freely in intellectual and political space. Instead of recognising its service on behalf of capital (rather than for undocumented labour, say, or vast pools of potential migrants to the West), globalisation functions as a metric by or under which different countries can be disciplined against a liberal, market-access norm.

Equally disturbing, suggest Hirschkind and Mahmood, is the representation of those movements, including the Taliban, which stand outside this privileged space and which allegedly threaten Islamic women with such horrors that the West is again forced to come to their rescue. For these authors, this optic amounts to little more than Modernisation Redux. Hirschkind and Mahmood insist upon the importance of a broader field of vision. They aim to parochialise the desires of Western liberals and progressives alike. They ask us to see the lives of Muslim women as anything but one-dimensional. Theirs is an important call for more careful and grounded research on inequality and difference within today's global – or imperial – geographies.

REFERENCES AND FURTHER READING

Auty, R. (1997) 'Natural resources, the state and development strategy', *Journal of International Development* 9: 651–63.

Bhagwati, J. (2004) *In Defence of Globalization*, Oxford: Oxford University Press.

Birdsall, N. and Subramanian, A. (2004) 'Saving Iraq from its oil', *Foreign Affairs* July/August.

Chatterjee, P. (2004) *The Politics of the Governed: Reflections on Popular Politics in Most of the World,* New York: Columbia University Press.

Corbridge, S., Williams, G., Srivastava, M. and Véron, R. (2005) *Seeing the State: Governance and Governmentality in India*, Cambridge: Cambridge University Press.

Dollar, D. and Kray, A. (2002) 'Growth is good for the poor', *Journal of Economic Growth* 7: 195–225.

Epstein, H. (2007) *The Invisible Cure: Africa, the West and the Fight Against Aids*, New York: Farrar, Straus and Giroux.

Firebaugh, G. (2003) *The New Geography of Global Income Inequality*, Cambridge, MA: Harvard University Press.

Government of the United Kingdom (2004) *Migration and Development: How to Make Migration Work for Poverty Reduction* (House of Commons, International Development Committee), London: The Stationery Office.

Gregory, D. (2004) *The Colonial Present: Afghanistan, Palestine, Iraq*, Oxford: Blackwell.

Harriss-White, B. (2003) *India Working: Essays on Society and Economy*, Cambridge: Cambridge University Press.

Mitchell, T. (2002) *Rule of Experts: Egypt, Techno-politics, Modernity*, Berkeley: University of California Press.

Nakkeeran, N. (2003) 'Women's work, status and fertility: land, caste and gender in a South Indian village', *Economic and Political Weekly* 38: 3931–9.

Norton, A. (2004) *Leo Strauss and the Politics of American Empire*, New Haven, CT: Yale University Press.

Rodrik, D. (2002) *Feasible Globalizations*, Harvard University: John F. Kennedy School of Government, Working Paper Series RWP02–029.

Sen, A.K. (2003) 'Foreword', in Paul Farmer, *Pathologies of Power: Health, Human Rights and the New War on the Poor*, Berkeley: University of California Press.

Stiglitz, J. (2002) *Globalization and Its Discontents*, New York: W. W. Norton.

'The Market Crosses Borders'

from *Why Globalization Works* (2004)

Martin Wolf

Editors' Introduction

Martin Wolf read Philosophy, Politics and Economics at Oxford University for his first degree, before earning a Master's degree in Economics from the same university in 1971. Wolf then joined the young professionals programme of the World Bank, where he later stayed as a senior economist until 1981. After six years at the Trade Policy Research Centre in London, Wolf joined the *Financial Times* newspaper in 1987, where he established himself not only as its chief economics commentator but also as a persuasive and passionate champion of what was soon being called 'globalisation'. Wolf continues to write regular columns for the *FT*, on the British economy and on the world economy. He debates regularly with other economists, both in the *FT* and in international bodies including the World Economic Forum at Davos, where he has been a Forum Fellow since 1999. An expert on international trade policy, Wolf came to broader public notice – even beyond *FT* readers – with his 2004 book, *Why Globalization Works*.

One reason *Why Globalization Works* itself works so well is because Wolf combines the analytical skills of an economist and the writing skills of a journalist. Wolf has the great virtue, as did Keynes and Friedman before him, of presenting economic ideas in a way that can be understood by a broad public. What Wolf takes to be the eternal virtues of classical economics and political economy, notably those beginning with Adam Smith, are presented straightforwardly and without jargon. This is certainly true in the excerpt from *Why Globalization Works* that is presented here: Wolf's account of the benefits that accrue to ordinary people when markets cross borders. Wolf is a strong proponent of more liberal trade regimes and of direct foreign investment. While bullish about the benefits of liberalisation, however, he does enter two caveats: the first in regard to unstable flows of debt-creating capital, and the second in regard to political obstacles to greater movements of workers. Globalisation might be working, in Wolf's view, but it is a long way from working perfectly.

Key references

Martin Wolf (2004) *Why Globalization Works*, New Haven, CT: Yale University Press.
— (2001) 'Will the nation state survive globalization', *Foreign Affairs* 80 (Jan.–Feb.): 178–90.
— (2003) 'The morality of the market', *Foreign Policy* (Sept.–Oct.): 46–50.
— (various) columns for the *Financial Times*: www.ft.com/comment/columnists/martinwolf.
R. Wade and Martin Wolf (2002) 'Are global poverty and inequality getting worse?' *Prospect* (March): 16–21.

What is prudence in the conduct of every family can scarce be folly in that of a great kingdom. If a foreign country can supply us with a commodity cheaper than we ourselves can make it, better buy it of them with some part of the produce of our own industry, employed in a way in which we have some advantage.

Adam Smith[1]

I perform a specialized function – commentary on the world economy – within the global division of labour. I work for a publication that sells more than three-fifths of its copies outside what was once its home market. These copies are published on the same day in some twenty different places around the world. It is also possible to subscribe to the Internet version of the newspaper. Moreover, my personal transactions do not stop at the seas surrounding Britain. I am not limited to British cameras, computers and vegetables or to sightseeing and investing only in Britain. I, as is true of the *FT* itself, am part of an internationally integrated economy. Here then are two realities of the contemporary marketplace: it crosses borders, ably assisted by modern technology; and it allows people to perform specialized functions.

Behind these realities is a more important one: I make my transactions because, given my resources and opportunities, I expect to benefit from them. As Adam Smith said more than two centuries ago, what is beneficial *within* a country is also beneficial *for* a country. People buy and sell with residents of their country because they expect to be made better off. They buy and sell with non-residents for the same reason.

When the statisticians add up transactions with non-residents, they call these a country's external transactions. But these are not a country's transactions, except statistically. Other than where a government is directly involved, a country's transactions are the aggregate of individual transactions by its residents. Moreover, because the motivation for such transactions is the same as for transactions with fellow residents, they are just as likely to contribute to the welfare of those who undertake them. This, in a nutshell, is the logic of global economic integration.

Yet there are also some important differences between transactions within jurisdictions and transactions across them. These differences fall under three categories: economic, jurisdictional and values.

The *economic* difference is that there are special obstacles – legal and non-legal – to activities that cross frontiers. Borders matter. Countries do not normally transact, but governments regulate transactions. Without exception, they regulate transactions between residents and non-residents differently from how they regulate transactions among residents. But policy and practice vary all the way from total prohibition (North Korea) to almost total freedom (Hong Kong).

The *political* difference is that more than one legal jurisdiction is unavoidably affected when transactions cross borders. Consequently, such transactions create an inescapable challenge – that of multiple jurisdictions.

The *value* difference is that economic analysis and political discourse generally proceed as if the welfare of foreigners or non-residents counts for nothing. It is one of the ironies of current debate that critics of globalization, who tend to present themselves as cosmopolitan, include many who take this assumption to extremes. But the assumption is not implausible: actual political institutions behave as if the welfare of foreigners counted for far less than those of nationals and residents.

Behind these differences between domestic and international transactions is one of the most obvious facts about the world: markets want to be cosmopolitan; states do not.[2] Technology permitting, a market for mutually enriching exchange will span the globe, because people will want to buy at the cheapest price (for any given quality) and sell at the highest. Jurisdictions, however, are territorial. Yet markets, as we have seen in Chapters 3, 4 and 5 above, are dependent on states, since states alone provide the legal order and regulatory structure they require. Some people, it is true, suggest that there are alternative ways of providing these conditions for productive commerce – through the active involvement of non-governmental organizations or through transnational institutions. But, significant though these new actors may become, they cannot replace states as repositories of power and authority.[3] This conflict between the natural tendency of markets to cross borders and the need for the states that define those borders to support markets is at the heart of all the challenges created by a global economy.

The world is not merely divided into states. It is divided into unequal ones. The richest country (the United States) has a real income per head seventy times that of the poorest (Sierra Leone). The biggest country (China) has a population of 1.26 billion, while the World Bank's latest World Development Indicators lists nineteen independent and semi-independent economies with populations of fewer than 100,000. The dollar purchasing power of the biggest economy (the United States) is 36,000 times bigger than that of St Kitts and Nevis. The reality of the world is one of states with very different standards of living, technological sophistication, populations and economic size. Only the United States is a heavyweight on all these scales.

Yet the difference in scale is not that fundamental. The contrast between Finland and the United States is trivial compared to the gap between both and, say, Nigeria or Pakistan. The first two are, by historical standards, effective and powerful states: the latter are defective in almost all significant respects. Alas, many of the world's states are incapable of providing the basic requirements of a civilized existence, let alone of a dynamic market economy. Inevitably, given the role of the state in supporting the market, theirs are also the countries that neither develop successfully nor play a significant role in the world economy. Outsiders will not contract with people who live in countries that lack the rule of law and at least a basic infrastructure. Ultimately, the inequality in the capacities of states makes for black holes in the world economy, countries from which little but desperate people and capital flight emerge.

ECONOMICS OF INTEGRATION

Economic transactions involve goods and services and what economists call 'factors of production' – labour, capital and land. By assumption, land cannot move. It is bound in space and, accordingly, provides a definition of a territorial state. But goods, services, labour and capital can and do move. So do people, as purchasers of goods and services in other countries, usually as tourists. What makes the notion of a country real economically is restrictions on the movement of one or more of these across frontiers. Today, virtually all countries have some restrictions on cross-border movement of goods and services. Restrictions on movements of people as tourists vary, with controls imposed by both sending and receiving countries. Restrictions on capital movement also vary. The advanced countries generally have no restrictions on movement of portfolio capital, but restrictions on foreign direct investment – ownership regulations, for example – as well as subsidies to inward investment remain. Restrictions on movement of labour are pervasive.

This makes the obvious point that liberalization is not an all-or-nothing proposition. Conceptually, it is possible to combine restrictions on trade and movement of capital and labour in eight different ways. At least five of these combinations have existed. The United Kingdom of the nineteenth century was close to freedom on all three dimensions. Within the European Union of today, all three are again free, in principle. Most developing countries were, until recently, highly restrictive on all three. Many still are. The Hong Kong of today has free trade and free movement of capital, but tight restrictions on movement of people, especially from China. The United States of the nineteenth century had restrictive trade policies, but movement of capital and labour was almost completely free. There are costs to any restrictions, in terms of freedom of choice, the emergence of black markets and corruption, and their effectiveness. But various combinations are certainly possible.

Trade [4]

Any analysis of trade starts from the assumption that movement of capital and labour is prevented. A country is a jurisdiction with a circumscribed pool of labour, capital and land. The argument for trade is that it increases opportunities for owners of these factors of production to engage in mutually beneficial transactions. It is an extension across frontiers of the division of labour. We need merely ask what our standard of living would be if we had to grow our own food, make our own clothes and shoes, build our own houses, make our own furniture, write our own books and newspapers, build our own vehicles or be our own doctors and dentists. But opportunities for the division of labour do not cease within a single national jurisdiction.

As Douglas Irwin of Dartmouth University argues, John Stuart Mill, one of the intellectual giants of the nineteenth century, divided the gains from trade into three categories. These were direct advantages, indirect advantages and intellectual and moral advantages.[5]

In the first category come the standard static gains from trade – exploitation of economies of scale and comparative advantage. David Ricardo propounded the latter idea, perhaps the cleverest in economics. The Nobel-laureate Paul Samuelson was once asked to name 'one proposition in all of the social sciences which is both true and non-trivial'.[6] His answer was comparative advantage. It is true and cannot be trivial, because author after author fails to understand the theory's most powerful implication.[7] Specialization makes sense, argued Ricardo, even if one country is more efficient at everything than its trading partners. Countries should specialize at what they are *relatively* most efficient at doing. Countries do not compete in trade, as companies do.[8] Rather, industries compete inside countries for the services of factors of production. Opening a country to trade moves output in the direction of activities that offer domestic factors of production the highest returns. The shrinking import-competing industry is not competing with imports from foreigners, but with what its own domestic export industry can pay.

Trade in accordance with comparative advantage is similar to a productivity increase. Instead of making a particular good, an economy can obtain more of it, indirectly, by exporting something else. These gains can be large. A classic example was the opening of Japan in 1858, under American pressure. Before opening, the prices of silk and tea were much higher in the world than in Japan, while the prices of cotton and woollen goods were far lower. After opening, Japan exported silk and tea and imported cotton and woollen goods. This is estimated to have increased Japan's real income by 65 per cent without considering the long-run productivity and growth impact of its joining the world economy.[9]

In the second indirect category come the dynamic gains from trade. Trade promotes competition and productivity growth. Companies innovate in response to competitive pressure. Widening the market to include more competitors increases this pressure.

Trade is also a conduit for foreign technology, via imports of capital and intermediate goods that embody significant innovations. Professor Irwin observes that even in the United States between a quarter and a half of growth in so-called 'total factor productivity', the part of productivity growth not explained by capital accumulation and improved skills, is attributable to new technology embodied in capital equipment. No developing country would have access to the world's advanced technologies without trade. This, as I discovered when I worked on India in the 1970s, was one of the reasons why its productivity growth was so low, even though it was operating far below the level of the world's best practice. Its imports were too restricted by the government to allow producers access to the world's best machinery.

It is virtually impossible to prove the correlation between trade liberalization and growth, beyond doubt. But as Professors Peter Lindert of the University of California, Davis and Jeffrey Williamson of Harvard state, there are *no* examples of countries that have risen in the ranks of global living standards while being less open to trade and capital in the 1990s than in the 1960s. 'There are', they continue, 'no anti-global victories to report for the postwar' developing world.[10] As one Indian observer has remarked of his own country's policies, 'by suppressing economic liberty for forty years, we destroyed growth and the future of two generations'.[11]

The third set of benefits are, claimed Mill, intellectual and moral. To the extent that trade facilitates growth, for example, as it has done in the most successful post-war developing countries, it has made a powerful contribution to the arrival of democracy. One of the most encouraging developments of the past two decades is that South Korea, Taiwan and Chile, all of which began their rapid outward-looking development under dictatorships, have now become stable and vibrant democracies. Even in China, market reforms, including trade liberalization, have brought an enormous reduction in the repressiveness of the political regime, compared to its totalitarian apogee under a Maoist dictatorship much admired by western leftists.

The bottom line then is that liberal trade is beneficial. The obstacles to it, largely created by governments, need to be reduced.

Capital

Recently, people have tended to argue for capital controls on the view that capital mobility has proved problematic for national economic management.[12] Against this should be set the strong arguments in favour of capital mobility. The most important of these is personal freedom. Controlling the ability of people to export their capital has been among the first steps of despotic or economically destructive regimes. It also imposes a constraint on the malfeasance of governments, particularly on the overt or covert theft of their people's savings or the all-too-frequent abuse of the financial system. From this point of view, freedom of movement of capital is even more important than that to buy and sell goods.

There are also efficiency reasons for being in favour of capital mobility, which allows shifting of consumption between periods, higher returns and risk-spreading. In particular, countries with surplus savings (such as today's Japan) or companies with a large stock of valuable knowledge (such as many of today's leading multinational companies) should be able to deploy what they own abroad, to mutual advantage. Placing a part of one's capital abroad is also a way to diversify risks. Since one is always heavily exposed to the country one lives and works in, it makes sense to diversify that risk away. Thus the suppliers and recipients of finance should both gain.[13]

Capital mobility also changes the analysis of trade, which is usually based on the assumption that capital and labour are immobile. But, as Ronald Jones of the University of Rochester has written:

> the idea of comparative advantage is linked to the notion that inputs are trapped by national boundaries, so that the only decision that needs to be made concerns the allocation *within* the country of these inputs. Ricardian theory stressed that a comparison of absolute productivities of such inputs between countries had no bearing on the allocation issue or on subsequent patterns of international trade. Instead, it was the *comparative* advantage of these inputs among sectors that matters. . . . However, once any input has the choice of country location, . . . [t]he doctrine of comparative advantage, with its emphasis on the question of what a factor *does* within the country, needs to share pride of place with the doctrine of absolute advantage guiding the question of where an internationally mobile factor *goes*.[14]

This is an important point. If capital moved freely to equalize expected returns across the world, only the relative abundance and productivity of human and natural resources would determine the pattern of trade. This is indeed one of the fears of many critics of globalization.

At present, however, this qualification to theory seems far less important than one might expect (and, with the welfare of poor countries at heart, also hope). As will be discussed further below (Chapter 8), the investible resources of the world are locked into already rich countries, with much the largest cross-border flow to the United States from other rich countries (especially Japan). Movement of capital from rich countries to poorer ones has been not only modest, but crisis-prone.[15] In 2000, for example, the gross dollar savings of the high-income developed countries were $5,600 billion. If a mere 10 per cent of these gross savings were to flow to the developing world, this would amount to $560 billion a year. But the highest net flow of long-term capital to developing countries in the 1990s was only $341 billion in 1997, just before the Asian crisis.[16] More than a decade ago, the Nobel-laureate Robert Lucas noted this low level of capital flow to developing countries, which is contrary to standard assumptions about where returns would be highest.[17]

Again, the multinational companies possess vast knowledge and experience in virtually every area of economic activity. In particular, as was asserted in Chapter 4, it is inside these institutions that much of the economically useful knowledge of the advanced countries is developed, retained and applied. It is in the interests of countries around the world, but especially of developing countries, to gain access to that knowledge via foreign direct investment. Now, many developing countries recognize that foreign direct investment brings benefits that foreign borrowing does not. The direct investor is locked in and cannot flee the country whenever trouble strikes. If investors make reasonable profits, they will consider themselves long-term participants in the host country's economy and often bring in more capital. Last and most important, direct investment brings substantial additional benefits that spill over into the domestic economy, including transfer of technology and managerial skills.[18] In many developing countries, multinational corporations have been the most important way to train nationals in modern management and technology. Some of these benefits, notably the last, are not restricted to developing country recipients. Ireland and the United Kingdom are two relatively advanced economies to have gained hugely from inward direct investment.[19]

One explanation for the modest flows of capital to developing countries is the shortage of complementary human skills. Another explanation would be external economies in the use of that capital: people are more productive if they work with other productive people. But it is also important to remember the discussion of financial fragility in Chapter 4. Difficulties exist within any financial system, because of asymmetric information, problems in foretelling the future, self-fulfilling panics and herd behaviour. But when funds cross borders, these vulnerabilities become still greater.[20]

In emerging market economies, difficulties over the design of the exchange rate and monetary regimes interact with the fragility of financial arrangements to create a host of obstacles to stable and sizeable capital flows. First, ignorance of financial and economic conditions in foreign countries is greater than at home. This applies particularly to emerging-market economies. Second, confidence in the probity of the governments and legal systems of countries abroad is low. Third, important elements of the legal and regulatory systems malfunction or do nor exist. Fourth, banking institutions often have comprehensive guarantees, while being used by governments or owners for their own purposes, which makes them fundamentally unsound.[21] Fifth, financial and accounting information is often lacking altogether, or is totally unreliable. Sixth, foreigners may expect to be discriminated against during a financial crisis, especially when the state or well-connected domestic interest is insolvent. Seventh, the government may have a long history of financial profligacy and default and so a poor reputation. Finally, when there is cross-border lending in a currency other than the borrower's, which is normal in lending to emerging markets, there is the additional foreign currency risk. All these problems are relatively small if, say, American banks lend in the United Kingdom. They are large if they lend to, say, Argentina. Their net effect is to make finance expensive, small in size and, worst of all, unstable.

The low level of foreign direct investment in many emerging market economies is explained by not dissimilar factors. Confidence in the probity and effectiveness of the governmental and legal systems is often very low. The economy is likely to be unstable, as may be the politics. The risk of nationalization or some other form of expropriation may well loom. The market may be modest in size, while the inputs needed to use the country as an export platform could well be lacking. It is not surprising, for all these reasons, that foreign direct investment in

Africa is so low, to take just one example. It is also unsurprising that the profit requirements on investments there are so high.

Yet none of these difficulties, real though they are, is sufficient reason for giving up on the aim of greater net capital flows to poorer countries. These should be treated as constraints to be lifted, not ones to be accepted. For if they are treated as binding and reinforced with capital controls, the world's poorest people will find it still more difficult to escape their plight.

People and geography

Traditional economic theory suggests that if trade is free and, still more, if capital flows freely, people should not need to move to gain higher incomes. Trade and capital mobility should equalize returns to labour across the world. Unfortunately, the assumptions underlying these orthodox models have turned out to be very far from true. We see the surplus capital of some rich countries pouring into the richest country in the world. In terms of purchasing power, a bus driver in Germany receives thirteen times one in Kenya, though his skills are much the same.[22] Workers with given skills earn vastly more in rich countries than in poor ones. Yet the rich countries that pay these higher wages for the same skills do not offer lower returns on capital than poor ones. If anything, the reverse is the case. Apparently capital-scarce countries have dismally low returns on capital, while apparently capital-abundant ones offer high returns. Similarly, free trade does not equalize wages, because productive efficiency diverges immensely across countries.

This paradoxical situation has two powerful – and disturbing – implications. The first is that the simplest thing we can do to alleviate mass human poverty is to allow people to move freely or their labour services to be traded freely, though perhaps temporarily. This is not a cause critics of globalization have embraced. That is not surprising, since it would kill support for their cause in high-income countries. Yet the US could fit in another billion people. There is even much empty space in Europe, too. This is not a recommendation. It is an observation of human hypocrisy.

The second implication is that there is no end of geography. Geography combined with jurisdiction matters more today than ever before in history. The quality of one's life depends even more on where one is born than on the class into which one is born. This means, in turn, that free trade and capital movement, albeit beneficial, will not, on their own, equalize global incomes. The big question then is why the productive possibilities of different jurisdictions differ to such an extent. Beyond the financial failings already mentioned above, there are three additional reasons.

The first is that richer countries can invest far more in the skills of their people. A large supply of human skills not only raises returns to capital and unskilled labour, but allows all productive processes to be run more efficiently, for it is the skilled who understand how things need to be done.

The second reason is that there are increasing returns to activities in specific locations.[23] Agglomerations of skill raise returns on all those skills. But skill also begets knowledge, which begets more knowledge, which then begets more skill. These increasing returns are location specific. We are indeed social animals: our incomes depend on the skills of those around us.

The last reason is that different jurisdictions differ in their ability to offer the requirements of productive market activity – property-rights protection, an honest and effective bureaucracy, judicial independence, good-quality infrastructure, decent health and education services, and so forth.[24] This is the hand of history at work: some places may have started from a poor resource endowment or geographical isolation; they may then have suffered predatory forms of colonization; they may be afflicted by cultural handicaps; and they may have made serious policy arrors.[25] Quite possibly they may have experienced all these things. This unhappy past may then have cumulated over time into a vicious circle of poor government, low growth, low skill formation and so back to poor growth.

The notion of such poverty traps can be overdone. Few imagined four decades ago that South Korea would be among the most successful developing countries. But history matters and so, therefore, does geography.

This is not the end of the story of human mobility Assume the increasing-returns story is right. Then skilled people in poor countries will earn less than they could in rich ones. They will want to move and, given their skills, may well be allowed to do so. Assume also that skilled people raise the wages of the unskilled where they live. Then free movement of skilled people from poor countries benefits rich countries and the skilled migrants, but harms the poor they leave behind. This is a reverse-aid programme. It is also one in which the rich countries are deeply engaged at present. The British National Health Service, for example, could not function without imports of skilled staff from developing countries.

Getting policy right

The ability of a country to take advantage of opportunities offered by international economic integration depends on the quality of the state and the policies it follows. What is necessary for a well-functioning domestic economy is just as necessary, if not more so, for a smooth engagement with international commerce. Indeed, they must be the same, since commerce is commerce, regardless of whom it is with. If anything, these conditions are more important for transactions with foreigners, since they always have more alternatives.

The second important policy conclusion is that, in keeping with the principles of good policy outlined in the previous chapter, the best interventions are direct ones. This is why modern trade theory insists that the case for free trade is stronger than that for *laissez-faire*. The reason for this is that market failures are nearly, though not always, domestic in origin. The best policy is one targeted directly at that failure.[26]

Consider, for example, one of the best-known arguments for protection – the infant-industry argument. The argument for infant-industry protection is that there is learning-by-doing or other network benefits that are not captured by a company, but benefit an entire industry. If these benefits are large, a potentially profitable industry may never even be started. This, it is argued, is a justification for protecting the nascent industry from foreign competition. But protection is an indirect and ineffective policy for promoting infants. Apart from the cost it imposes on consumers, it has two other seriously negative side-effects: first, it limits the new industry to the domestic market, since protection, by definition, raises returns only on domestic sales; and, second, it provides protection from the world's most potent competitors. The first limitation may not matter much for countries with relatively big and rapidly growing domestic markets (such as the United States in the nineteenth century), but it is significant for most developing countries, which have tiny markets: Nigeria's dollar purchasing power in 2000 was less than a tenth of London's. The second limitation means that, protected from effective competition, the infants almost always fail to grow up.[27]

Is an intervention at the border ever the most direct one? The answer is yes. National security is a reason for such protection. A country also has the right to decide who lives within its borders. One might describe this as a country's defining right. That is also why controls on the movement of people across borders are universal. There are also arguments for imposing either tariffs on imports or taxes on exports where a country has monopoly power in trade. In this way, a country can drive down the prices of its imports or raise the prices of its exports. The strategic trade policy literature of the 1980s and 1990s, which examined trade policy for oligopolistic industries, was a more sophisticated application of the same basic notion. But the underlying idea of exploiting foreigners is disturbing. There are also big problems with applying strategic trade policy: knowledge of the degree and sustainability of monopoly power is quite limited; results of strategic trade policy models are also extraordinarily unstable; and governments can be 'gamed' by the companies to obtain undeserved and costly protection.

The conclusion then is that the principle of directness rules out almost all use of trade policy as a way of dealing with defects in the domestic economy. Similarly, the same principle makes controls on financial transactions with foreigners generally undesirable, since most (though not quite all) of the problems that arise in financial markets reflect failures that are just as relevant for the domestic economy. But directness does justify control on the movement of people at the border.

The third important policy objective is to be wary of costly interactions among policies. Among the most important is the danger of attracting inward direct investment by offering protection. Think of a simple example of a car company that is enticed by protection against imports into setting up an assembly plant that uses completely knocked down kits. It is perfectly possible for the foreign currency cost of each kit to be the same as a complete car. Then the profit that is exported comes from a tax on domestic consumers, as does the bill for wages and other raw materials. The foreign currency balance of the country is now worse than before, since the kits cost the same as cars used to do, but it expatriates profits as well. This is not an imaginary example. It is an often repeated case of stupidity, justified on infant-industry grounds.

Conclusion

Trade in goods and services is helpful for the performance of the economy. More trade in capital would help as well, particularly greater direct investment in poor countries. But there is too little of it, while flows of debt-creating capital are unstable. Movement of people and trade in labour services are, in the present circumstances, probably the best thing one could do for the poor of the world. But the political obstacles are mountainous.

[. . .]

NOTES

1 Adam Smith, *An Inquiry into the Nature and Causes of the Wealth of Nations* (Oxford: Clarendon Press, 1976), p. 457.

2 The tension between the forces creating global markets and political fragmentation were the central theme of a classic book by Richard Cooper of Harvard University: *The Economics of Interdependence: Economic Policy in the Atlantic Community* (New York: McGraw-Hill, 1968). Since Professor Cooper wrote this seminal work, the problem of the Atlantic community has become a global one.

3 An interesting analysis of alternative views of the foundations for the governance of the world economy is contained in Robert Gilpin, *Global Political Economy: Understanding the International Economic Order* (Princeton: Princeton University Press, 2001), chapter 15. Professor Gilpin discusses 'the new medievalism' and 'transgovernmentalism', but concludes that states remain the foundation of any global economic order, particularly the major states.

4 Excellent recent discussions of the underlying arguments for liberal trade are contained in Jagdish N. Bhagwati, *Free Trade Today* (Princeton: Princeton University Press, 2002). Professor Bhagwati is the world's leading trade economist. Also excellent are Douglas A. Irwin, *Free Trade under Fire* (Princeton: Princeton University Press, 2002) and *Against the Tide: An Intellectual History of Free Trade* (Princeton: Princeton University Press, 1996). Much of the discussion

below draws on these sources. Also helpful is W. Max Corden, *Trade Policy and Economic Welfare* (Oxford: Clarendon Press, 1974, second edition 1997).

5 Irwin, *Free Trade under Fire*, pp. 29–48.

6 Retold in Irwin, *Free Trade under Fire*, p. 25, note 5.

7 As will be seen in Part IV below, among the distinguished authors whose analysis of the consequences of globalization is vitiated by this error is John Gray in his *False Dawn: The Delusions of Global Capitalism* (London: Granta, 1998).

8 The economist who has argued most powerfully that countries do not compete like companies is Paul Krugman. See 'Competitiveness: A Dangerous Obsession', *Foreign Affairs 73* (March-April 1994), pp. 28–44 and 'Ricardo's Difficult Idea: Why Intellectuals Don't Understand Comparative Advantage', in Gary Cook (ed.), *The Economics and Politics of International Trade*, Volume 2 of *Freedom and Trade* (London: Routledge, 1998).

9 Irwin, *Free Trade under Fire*, p. 30.

10 Peter H. Lindert and Jeffrey G. Williamson, 'Does Globalization Make the World More Unequal?', paper presented at the National Bureau of Economic Research conference on Globalization in Historical Perspective Santa Barbara, California, 3–6 May 2001, p. 25.

11 Gurchuran Das, *India Unbound* (New York: Alfred A. Knopf, 2001), p. 175.

12 Even Jagdish Bhagwati, a staunch free-trader, has taken this position. See Jagdish N. Bhagwati, 'The Capital Myth', *Foreign Affairs*, Vol. 77 (May–June, 1998), pp. 7–12.

13 A discussion of many of the issues is contained in Forest Capie, *Capital Controls: A 'Cure' Worse than the Problem* (London: Institute of Economic Affairs, 2002).

14 Ronald W. Jones, *Globalization and the Theory of Input Trade* (Cambridge, Massachusetts: MIT Press, 2000), pp. 135–6.

15 On crises, see *Finance for Growth: Policy Choices in a Volatile World* (Washington DC: World Bank, 2001), chapter 2.

16 Data are from *Global Development Finance: Financing the Poorest Countries, Analysis and Summary Tables 2002* (Washington DC: World Bank, 2002).

17 Robert E. Lucas, 'Why doesn't Capital Flow from Rich to Poor Countries?' *American Economic Review*, 80 (May 1990), pp. 92–6.

18 See Edward M. Graham, *Fighting the Wrong Enemy: Antiglobal Activists and Multinational Enterprises* (Washington DC: Institute for International Economics, 2000), p. 172.

19 The positive impact of inward foreign direct investment on British manufacturing is one of the themes of Geoffrey Owen's book, *From Empire to Europe: The Decline and Revival of British Industry since the Second World War* (London: HarperCollins, 1999).

20 See Lucas, 'Why doesn't Capital Flow?', and Wendy Dobson and Gary Clyde Hufbauer, *World Capital Markets: Challenge to the G-10* (Washington DC: Institute for International Economics, 2001), chapter 1.

21 An interesting analysis of the Asian financial crisis of 1997–8 that stresses moral hazard created by government guarantees is contained in Giancarlo Corsetti, Paolo Pesenti and Nouriel Roubini, 'Paper Tigers? A Model of the Asian Crisis', National Bureau of Economic Research, Working Paper 6783, www.nber.org, November 1998.

22 World Bank, *World Development Report 1995: Workers in an Integrating World* (Oxford: Oxford University Press, for the World Bank, 1995), pp. 10–14.

23 The increasing-return story is well told by William Easterly in *The Elusive Quest for Growth: Economists' Adventures and Misadventures in the Tropics* (Cambridge, Massachusetts: MIT Press, 2001), chapter 8.

24 This is the core of the late Mancur Olson's theory of why economies diverge to such a large extent. See his article 'Big Bills Left on the Sidewalk: Why Some Countries are Rich and Others Poor', *Journal of Economic Perspectives*, Vol. 10 (Spring 1996), pp. 3–24.

25 Daron Acemoglu, Simon Johnson and James A. Robinson, in 'Reversal of Fortune: Geography and Institutions in the Making of the Modern World Income Distribution', National Bureau of Economic Research Working Paper 8460, 2001, and *Quarterly Journal of Economics*, Vol. 117, argue that the rich places of today were poor in 1500 and vice versa, because European colonizers imposed extractive institutions on the rich places they seized (such as India and Mexico) and wealth-generating institutions on the poor ones (such as North America). The theory is neat. But many of the rich countries of 1500 already had efficient wealth-extracting institutions imposed by already established élites. All the colonizers needed to do was take them over. That was certainly true in India. The difference may be that in rich agrarian societies extractive systems were well in place and have survived in the hands of local politicians to this day. But in sparsely settled places, institutions that generated wealth had to be created. It is probably no accident that all those institutions were introduced by the British.

26 See Bhagwati, *Free Trade Today*, especially p. 29.

27 The examples of failed infants are without end. My personal favourite is of the Morris Oxford, a not particularly successful car designed in the 1950s, which was still being manufactured and sold as the Ambassador in India in the 1990s. The Indian car industry finally started to grow up only with investment by foreign multinational businesses. Producers of failed infants are, alas, an important obstacle to trade liberalization.

EIGHT

'Is Globalization Reducing Poverty and Inequality?'

World Development (2004)

Robert Hunter Wade

Editors' Introduction

Robert Hunter Wade, a New Zealander, was born in Sydney, Australia, in 1944 and is currently a Professor of Political Economy and Development at the London School of Economics and Political Science. Wade has broad-ranging expertise in development studies. Trained in economics and anthropology, he made an early impact in development and South Asian studies with a penetrating analysis of 'the system of administrative and political corruption' that organised, in a very real sense, a major unit of canal irrigation in South India. Drawing both on ethnography and critical political economy, Wade was able to show how a canal irrigation system worked (or did not properly work) in agricultural and water delivery terms, while also moving outwards from the canals to explore the workings of the local political society. In a widely admired article in the *Journal of Development Studies* (1982), and in his classic book *Village Republics* (1988), Wade was able to show how highly organised systems of corruption and kickbacks within government line departments, and between government officers and their political superiors or counterparts, worked to ensure the production of irrigation systems that were characterised by pervasive uncertainty and sub-optimality. It was mainly by creating insecurity of water supply for farmers that executive and assistant engineers created opportunities for corruption or rent-seeking. The engineers in turn were required to raise considerable rental income in part to pay off the politicians who could threaten their occupation of remunerative posts.

Neo-liberal analysis of corruption often recommend privatisation as one way to deal with the 'misuse of public office for private gain'. Wade, however, refused to embrace the suggestion that a pure market in water could solve South India's irrigation and farming problems. He noted that India's canal systems are generally many times bigger than is the case with established water markets, and that this would create huge enforcement problems. Wade also noted that farmers would not come to such a market on an equal basis, nor could they be expected to have anything like perfect information on the likely value of present or future water supplies. His preferred solution to institutionalised corruption was thus institutional: stronger lines of accountability within the irrigation department and the strengthening of irrigator user groups. A similar regard for the importance of institutions, or the rules of the game, was also to the fore in his next and perhaps best-known book, *Governing the Market: Economic Theory and the Role of Government in East Asian Industrialization* – a book published, remarkably, in 1990, just a couple of years after *Village Republics*. Here Wade displays the interest in political economy that continues to distinguish his work. Eschewing standard World Bank analysis of the East Asian 'miracle' – the view, crudely put, that the Asian Tigers developed by embracing the global market on an open-economy basis – Wade joined with Alice Amsden and others in pointing up, here with particular reference to Taiwan, how governments in East Asia had pursued systems of growth-enhancing governance that actively managed 'the market' to produce rapid industrial development. Wade was also careful to acknowledge, however, that even Taiwan and South Korea had graduated from 'governed market' to more 'open market' strategies over time. He also insisted that it was a mistake to suppose that all governments in the developing world had the administrative or political capacity to 'do a Taiwan'; another mistake was to suppose that countries in the 1980s or 1990s would face the same geopolitical or international economic circumstances as South Korea and Taiwan had faced in the 1950s or 1960s. One of the key strengths of an institutionalist approach is precisely this close regard for geographical difference and historical sequencing.

More recently, Wade has drawn on these insights to present an important challenge to the Washington Consensus and various boosters of globalisation. It is perhaps not too fanciful to argue that the reading reproduced here shows Wade's accumulated skills as an ethnographer (well to the fore in his close reading of how data sets are produced on poverty and inequality) and as a political economist attentive to the logical problems that must surround any attempt to read off the effects of something so loosely defined as 'globalisation'. It is a combination of skills that is becoming less common among students of development.

Key references

Robert Hunter Wade (2004) 'Is globalization reducing poverty and inequality?', *World Development* 32 (4): 567–89.
— (1982) 'The system of administrative and political corruption: canal irrigation in South India', *Journal of Development Studies* 18: 287–328.
— (1988) *Village Republics: The Economic Conditions of Collective Action in India*, Cambridge: Cambridge University Press.
— (1990) *Governing the Market: Economic Theory and the Role of Government in East Asia's Industrialization*, Princeton, NJ: Princeton University Press.
— (2003) 'The invisible hand of American empire', *Ethics and International Affairs* 17 (2): 17–88.
— (2003) 'What strategies are viable for developing countries today?', *Review of International Political Economy* 10 (4): 621–44.
— (2007) 'A new global financial architecture?', *New Left Review* NS 46 (July–August).

Over the past 20 years the number of people living on less than $1 a day has fallen by 200 million, after rising steadily for 200 years.
(James Wolfensohn, president of the World Bank, World Bank, 2002b)

The *best evidence available* shows . . . the current wave of globalization, which started around 1980, has actually promoted economic equality and reduced poverty.
(Dollar & Kraay, 2002; emphasis added)

Evidence suggests the 1980s and 1990s were decades of declining global inequality and reductions in the proportion of the world's population in extreme poverty.
(Martin Wolf, *The Financial Times*, 2002)

[G]lobalization has dramatically increased inequality between and within nations.
(Jay Mazur, US union leader, 2000)

1 INTRODUCTION

The neoliberal argument says that the distribution of income between all the world's people has become more equal over the past two decades and the number of people living in extreme poverty has fallen for the first time in more than a century and a half. It says that these progressive trends are due in large part to the rising density of economic integration between countries, which has made for rising efficiency of resource use worldwide as countries and regions specialize in line with their comparative advantage. Hence the combination of the "dollar-Wall Street" economic regime[1] in place since the breakdown of the Bretton Woods regime in the early 1970s, and the globalizing direction of change in the world economy since then, serves the great majority of the world's people well. The core solution for lagging regions, Africa above all, is freer domestic and international trade and more open financial markets, leading to deeper integration into the world economy.

Evidence from the current long wave of globalization thus confirms neoliberal economic theory – more open economies are more prosperous, economies that liberalize more experience a faster rate of progress, and people who resist further economic liberalization must be acting out of vested or "rent-seeking" interests. The world economy is an open system in the sense that country mobility up the income/wealth hierarchy is unconstrained by the structure. The hierarchy is in the process of being flattened; the North–South, core–periphery, rich country–poor country divide is being eroded away as globalization proceeds. The same evidence also validates the rationale of the World Trade Organization (WTO), the World Bank, the International Monetary Fund (IMF) and other multilateral economic organizations as agents for creating a global "level playing" field undistorted by state-imposed restrictions on markets. This line of argument is championed by the more powerful of the centers of "thinking for the world" that influence international policy making, including the intergovernmental organizations such as the World Bank, the IMF and the WTO, also the US and UK Treasuries, and opinion-shaping media such as *The Financial Times* and *The Economist*.

The standard Left assumption, in contrast, is that the rich and powerful countries and classes have little interest in greater equity. Consistent with this view, the "anti-globalization" (more accurately, "anti-neoliberal") argument asserts that world poverty and inequality have been rising, not falling, due to forces unleashed by the same globalization (for example, union leader Jay Mazur's quote above).[2] The line of solution is some degree of tightening of public policy limits on the operation of market forces; though the "anti-neoliberal" camp embraces a much wider range of solutions than the liberal camp.

The debate tends to be conducted by each side as if its case was overwhelming, and only an intellectually deficient or dishonest person could see merit in other's case. For example, Martin Wolf of *The Financial Times* claims that the "anti-globalization" argument is "the big lie."[3] If translated into public policy it would cause more poverty and inequality while pretending to do the opposite.

This paper questions the empirical basis of the neoliberal argument. In addition, it goes beyond the questions to suggest different conclusions about levels and trends, stated in terms not of certainties but stronger or weaker probabilities. Finally it explains why we should be concerned about probably-rising world inequality, and how we might think about the neglected subject of the political economy of statistics.

2 THE REGIONAL COLLAGE

The growth rate of world GDP, measured in US dollars and at current exchange rates, fell sharply from around 5.5% in 1970–80 to 2.3% in 1980–90 to 1.1% in 1990–2000.[4] This is bad news, environmental considerations aside. But it still grew a little faster than world population over the past two decades; and the (population-weighted) GDP of developing countries as a group grew a little faster than that of the high-income countries. On the other hand, regional variation within the global South is large. Table 1 shows the trends of regional per capita GNP to the per capita GNP of the "core" regions (with incomes converted to US$ at current exchange rates as a measure of *international* purchasing power). During 1960–99 the per capita incomes of sub-Saharan Africa, Latin America, and West Asia and North Africa fell as a fraction of the core's; South Asia's remained more or less constant; East Asia's (minus China) rose sharply; China's also rose sharply but from a very low base. The most striking feature is not the trends but the size of the gaps, testimony to the failure of "catch-up." Even success-story East Asia has an average income only about 13% of the core's.[5] It is a safe bet that most development experts in 1960 would have predicted much higher percentages by 2000.

The variation can also be shown in terms of the distribution of world income by regions and income percentiles. Figure 1 shows the regional distribution of people at each income percentile for two years, 1990 and 1999. Here incomes are expressed in "purchasing power parity" dollars (PPP$),[6] in order to

Region	1960	1980	1999
Sub-Saharan Africa	5	4	2
Latin America	20	18	12
West Asia and North Africa	9	9	7
South Asia	2	1	2
East Asia (w/o China and Japan)	6	8	13
China	1	1	3
South	5	4	5
North America	124	100	101
Western Europe	111	104	98
Southern Europe	52	60	60
Australia and NZ	95	75	73
Japan	79	134	145
North (= core)	**100**	**100**	**100**

Table 1 GNP per capita for region as % of core's GNP per capita[a]

Source: Arrighi, Silver, and Brewer (2003).

[a] Based on World Bank data. GNP at current exchange rates.

measure, notionally at least, *domestic* purchasing power. One sees the African collapse in the increased share of the African population in the bottom quintile; also the falling back of the Eastern and Central European populations from the second to the third quintile; and the rising share of the East Asian population in the second quintile.

Figure 2 shows, in the top half, the world's population plotted against the log of PPP$ income, taking account of both between-country and within-country income distribution; and the breakdown by region. The bottom half shows the world's income plotted against income level, hence the share of income accruing to people at different income levels and in different regions. Residents of South Asia and East Asia predominate at income levels below the median, and residents of the OECD countries predominate at the top.

Finally, Figure 3 shows the movement in the bimodal shape of the overall PPP$ income-to-population distribution during 1970–99. The 1999 distribution has shifted forward compared to the 1970 one, especially the lower of the two income humps, reflecting the arrival of large numbers of South and East Asians into the middle deciles of the world income distribution.

How does the collage – positive world per capita growth and wide divergence of economic performance between developing regions – net out in terms of global trends in poverty and inequality?

3 POVERTY

Figure 2 shows the two standard international poverty lines, $1 per day and $2 per day; and also the line corresponding to an income of 50% of the world's median income. Notice that even the higher $2 per day absolute poverty line is below the conventional "minimum" *relative* poverty line of half of the median. Notice too how small a share of world income goes to those on less than $1 per day, and how small a share of the income of the richest earners would be needed to double the income of the poorest.

Figures 1–3 are based on a data set on income inequality compiled by the United Nation's World Institute for Development Economics Research (WIDER).[7] But the standard poverty numbers – the ones normally used in discussions about the state of the world – come from the World Bank's data set. This is the source of the claims that, in the words of President James Wolfensohn, "Over the past 20 years the number of people living on less than $1 a day has fallen by 200 million, after rising steadily for 200 years"[8] and "the proportion of people worldwide living in absolute poverty has dropped steadily in recent decades, from 29% in 1990 to a record low of 23% in 1998."[9] The opening sentence of the Bank's *World Development Indicators 2001* says, "Of the world's 6 billion people 1.2 billion live on less than $1 a day," the same number in 1987 and 1998.[10]

No ifs or buts. I now show that the Bank's figures contain a large margin of error, and the errors *probably* flatter the result in one direction.[11]

To get the world extreme poverty headcount the Bank first defines an international poverty line for a given base year by using purchasing power parity conversion factors (PPPs) to convert the purchasing power of an average of the official national poverty lines of a set of low-income countries into the US dollar amount needed to have the same notional purchasing power in the United States in the same year. In its first global poverty estimation this procedure yielded a conveniently understandable US$1 per day for the base year of 1985.[12] Then the Bank

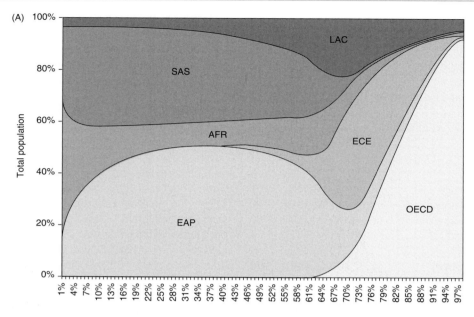

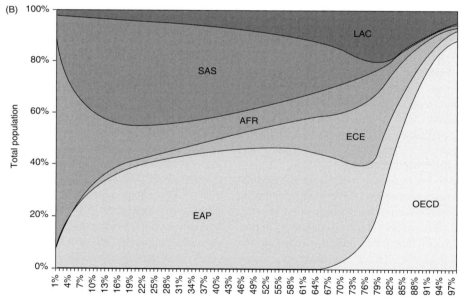

Figure 1 World income distribution, by region, at each percentile of global income distribution: (A) 1990 and (B) 1999 (population at any particular income = 100)
Source: Dikhanov & Ward (2003).

uses PPP conversion factors to estimate the amount of local currency, country by country, needed to have the same purchasing power in the same year as in the US base case. This gives an international extreme poverty line equivalent to US$1 per day, expressed in domestic currency. By way of illustration, Rs. 10 may have the same purchasing power in India in 1985 as US$1 in the United States in the same year, in which case India's international extreme poverty line is Rs. 10 per day. From household surveys the Bank then

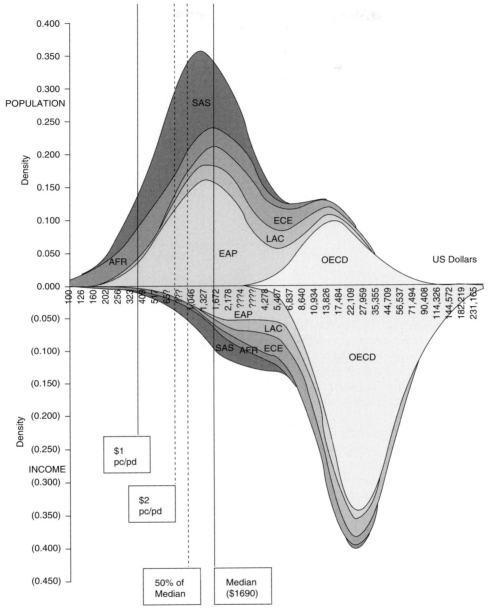

Figure 2 World income distribution, by region: top half, distribution of world population against income; bottom half, distribution of world income against income, 1999
Source: Dikhanov & Ward (2003).

estimates the number of people in the country living on less than this figure. It sums the country totals to get the world total. It uses national consumer price indices to keep real purchasing power constant across time, and adjusts the international poverty line for each country upwards with inflation.

(a) Large margin of error

There are several reasons to expect a large margin of error, regardless of direction. First, the poverty head-count is very sensitive to the precise level of the international poverty lines. This is because the shape

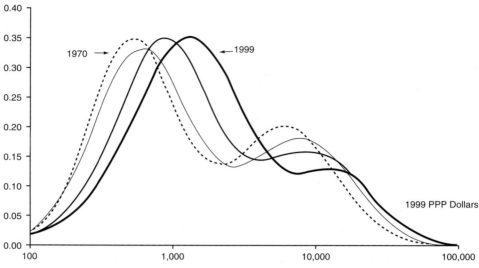

Figure 3 World income distribution, 1970, 1980, 1990, 1999 (Source: Dikhanov & Ward (2003).

of income distribution near the poverty line is such that, in most developing countries, a given percentage change in the line brings a similar or larger percentage change in the number of people below it. Recent research on China suggests that a 10% increase in the line brings a roughly 20% increase in the poverty headcount.

Second, the poverty headcount is very sensitive to the reliability of household surveys of income and expenditure. The available surveys are of widely varying quality, and many do not follow a standard template. Some sources of error are well known, such as the exclusion of most of the benefits that people receive from publicly provided goods and services. Others are less well known, such as the sensitivity of the poverty headcount to the survey design. For example, the length of the recall period makes a big difference to the rate of reported expenditure – the shorter the recall period the higher the expenditure. A recent study in India suggests that a switch from the standard 30-day reporting period to a seven-day reporting period lifts 175 million people from poverty, a nearly 50% drop. This is using the Indian official poverty line. Using the higher $1/day international line the drop would be even greater.[13] The point here is not that household surveys are less reliable than other possible sources (for example, national income accounts); simply that they do contain large amounts of error.

Third, China and India, the two most important

countries for the overall trend, have PPP-adjusted income figures that contain an even bigger component of guess work than for most other significant countries. The main sources of PPP income figures (the Penn World Tables and the International Comparison Project) are based on two large-scale international price benchmarking exercises for calculating purchasing power parity exchange rates, one in 1985 in 60 countries, the other in 1993 in 110 countries. The government of China declined to participate in both. The purchasing power parity exchange rate for China is based on guestimates from small, *ad hoc* price surveys in a few cities, adjusted by rules of thumb to take account of the huge price differences between urban and rural areas and between eastern and western regions. The government of India declined to participate in the 1993 exercise. The price comparisons for India are extrapolations from 1985 qualified by later *ad hoc* price surveys. The lack of reliable price comparisons for China and India – hence the lack of reliable evidence on the purchasing power of incomes across their distributions – compromises any statement about levels and trends in world poverty.[14]

Fourth, the often-cited comparison between 1980 and 1998 – 1.4 billion in extreme poverty in 1980, 1.2 billion in 1998 – is not valid. The Bank introduced a new methodology in the late 1990s which makes the figures noncomparable. The Bank has recalculated the poverty numbers with the new method only back to 1987.[15]

The change of method amounts to: (i) a change in the way the international poverty line was calculated from the official poverty lines of a sample of low- and middle-income countries (and a change in the sample countries), which resulted in, (ii) a change in the international poverty line from $PPP 1 per day to $PPP 1.08 per day, and (iii) a change in the procedure for aggregating, country by country, the relative price changes over 1985–93 for a standard bundle of goods and services.

We do not know what the 1980 figure would be with the new method. We do know however that the new method caused a huge change in the poverty count even for the same country in the same year using the same survey data.[16] Table 2 shows the method-induced changes by regions for 1993. Angus Deaton, an expert on these statistics, comments that "Changes of this size risk swamping real changes . . . it seems impossible to make statements about changes in world poverty when the ground underneath one's feet is changing in this way."[17]

(b) Downward bias

Further sources of error bias the results downward, making the number of people in poverty seem lower than it really is; and the bias probably increases over time, making the trend look rosier than it is. There are at least three reasons.

First, the Bank's international poverty line underestimates the income or expenditure needed for an individual (or household) to avoid periods of food-clothing-shelter consumption too low to maintain health and well-being. (Moreover, it avoids altogether the problem that basic needs include unpriced public goods such as clean water and access to basic healthcare.) The Bank's line refers to an "average consumption" bundle, not to a basket of goods and services that makes sense for measuring poverty (though "$1 per day" does have intuitive appeal to a Western audience being asked to support aid). Suppose it costs Rs. 30 to buy an equivalent bundle of food in India (defined in terms of calories and micronutrients) as can be bought in the United States with $1; and that it costs Rs. 3 to buy an equivalent bundle of services (haircuts, massages) as $1 in the United States, such services being relatively very cheap in developing countries.[18] Current methods of calculating purchasing power parity, based on an

	Old poverty rate (%)	New poverty rate (%)
Sub-Saharan Africa	39.1	49.7
Latin America	23.5	15.3
Middle East/N Africa	4.1	1.9

Table 2 1993 poverty rate, using old and new World Bank methodology[a]

Source: Deaton (2001).

[a] The poverty rate is the proportion of the population living on less than $1 a day.

average consumption bundle of food, services and other things, may yield a PPP exchange rate of $PPP 1 = Rs. 10, meaning that Rs. 10 in India buys the equivalent average consumption bundle as $1 in the United States. But this is misleading because the poor person, spending most income on food, can buy with Rs. 10 only one-third of the food purchasable with $1 in the United States. To take the international poverty line for India as Rs. 10 therefore biases the number of poor downward.

We have no way of knowing what proportion of food-clothing-shelter needs the Bank's international poverty line captures. But we can be fairly sure that if the Bank used a basic needs poverty line rather than its present artificial one the number of absolute poor would rise, because the national poverty lines equivalent to a global basic needs poverty line would probably rise (perhaps by 30–40%).[19] A 30–40% increase in a basic-needs-based international poverty line would increase the world total of people in extreme poverty by at least 30–40%. Indeed a recent study for Latin America shows that national extreme poverty rates, using poverty lines based on calorific and demographic characteristics, may be more than *twice* as high as those based on the World Bank's $1/day line. For example, the World Bank estimates Brazil's extreme poverty rate (using its international poverty line) at 5%, while the Economic Commission for Latin America, using a calories-and-demography poverty line, estimates the rate at 14%.[20]

In short, we can be reasonably confident that switching from the Bank's rather arbitrarily derived international extreme poverty line to one reflecting the purchasing power necessary to achieve elementary human capabilities would substantially raise the number of people in extreme poverty.

The second reason is that the Bank's new international poverty line of $1.08/day probably increases the downward bias, leading the Bank to exaggerate the decline in the poverty headcount between the years covered by the old methodology and those covered by the new one. The new international poverty line of $PPP 1.08 *lowers* the equivalent national poverty lines in most countries compared to the earlier $PPP 1 line. It lowers them in 77% of the 94 countries for which data are available, containing 82% of their population. It lowers the old international poverty line for China by 14%, for India, by 9%, for the whole sample by an average of 13%.[21] As noted, even a small downward shift in the poverty line removes a large number of people out of poverty.

Third, future "updating" of the international poverty line will continue artificially to lower the true numbers, because average consumption patterns (on which the international poverty line is based) are shifting toward services whose prices relative to food and shelter are lower in poor than in rich countries, giving the false impression that the cost of the basic consumption goods required by the poor is falling.[22]

All these problems have to be resolved in one way or another in any estimate of world poverty, whoever makes it. But the fact that the World Bank is the near-monopoly provider introduces a further complication. The number of poor people is politically sensitive. The Bank's many critics like to use the poverty numbers as one of many pointers to the conclusion that it has accomplished "precious little," in the worlds of US Treasury Secretary O'Neill; which then provides a rationale for tighter US control of the Bank, as in the statement by the head of the US Agency for International Development, "Whether the US way of doing things drives some multilateral institutions, I think it should, because, frankly, a lot of the multilateral institutions don't have a good track record."[23]

A comparison of two recent Bank publications suggests how the Bank's statements about poverty are affected by its tactics and the ideological predispositions of those in the ideas-controlling positions. *The World Development Report 2000/2001: Attacking Poverty* says that the number of people living on less than $1 a day *increased* by 20 million from 1.18 billion in 1987 to 1.20 billion in 1998. When it was being written in the late 1990s the key ideas-controlling positions in the Bank were held by Joe Stiglitz and Ravi Kanbur (respectively, chief

economist and director of the *World Development Report 2000/2001*), not noted champions of neoliberal economics.[24] At that time the Bank was trying to mobilize support for making the Comprehensive Development Framework the new template for all its work, for which purpose *lack* of progress in development helped. Then came the majority report of the Meltzer Commission, for the US Congress, which said the Bank was failing at its central task of poverty reduction and therefore should be sharply cut back – as shown by the fact that the number of people in absolute poverty remained constant at 1.2 billion during 1987–98.[25] Now the Bank needed to emphasize progress. The next major Bank publication, *Globalization, Growth, and Poverty: Building an Inclusive World Economy*, claimed that the number of people living in poverty *decreased* by 200 million in the 18 years over 1980–98.[26] By this time Stiglitz and Kanbur were gone and David Dollar, a prominent Bank economist, was ascendant. He was chief author of *Globalization, Growth and Poverty*.[27]

(c) Conclusions about poverty

We can be fairly sure that the Bank's poverty headcount has a large margin of error in *all* years, in the sense that it may be significantly different from the headcount that would result from the use of PPP conversion factors based more closely on the real costs of living of the poor (defined in terms of income needed to buy enough calories, micronutrients and other necessities in order not to be poor). By the same token we should question the Bank's confidence that the trend is downward.

We do not know for sure how the late 1990s revision of the method and the PPP numbers alters the poverty headcount in any one year and the trend. But it is likely that the Bank's numbers substantially underestimate the true numbers of the world's population living in extreme poverty, and make the trend look brighter.

On the other hand, it is quite plausible that the *proportion* of the world's population living in extreme poverty has fallen over the past 20 years or so. For all the problems with Chinese and Indian income figures we know enough about trends in other variables – including life expectancy, heights, and other non-income measures – to be confident that their poverty headcounts have indeed dropped dramatically over the past 20 years. If it is the case (as some experts claim) that household surveys are more likely to miss the rich than the poor, their results may *overstate* the proportion of the population in poverty. The magnitude of world population increase over the past 20 years is so large that the Bank's poverty numbers would have to be *huge* underestimates for the world poverty rate not to have fallen. Any more precise statement about the absolute number of the world's people living in extreme poverty and the change over time currently rests on quicksand.

[. . .]

5 GLOBALIZATION

I have raised doubts about the liberal argument's claim that (a) the number of people living in extreme poverty worldwide is currently about 1.2 billion, (b) it has fallen substantially since 1980, by about 200 million, and (c) that world income inequality has fallen over the same period, having risen for many decades before then. Let us consider the other end of the argument – that the allegedly positive trends in poverty and inequality have been driven by rising integration of poorer countries into the world economy, as seen in rising trade/GDP, foreign direct investment/GDP, and the like.

Clearly the proposition is not well supported at the world level if we agree that globalization has been rising while poverty and income inequality have not been falling. Indeed, it is striking that the pronounced convergence of economic policy toward "openness" worldwide over the past 20 years has gone with divergence of economic performance. But it might still be possible to argue that globalization explains differences between countries: that more open economies or ones that open faster have a better record than less open ones than open more slowly.

This is what World Bank studies claim. The best known, *Globalization, Growth and Poverty*,[28] distinguishes "newly globalizing" countries, also called "more globalized" countries, from "nonglobalizing" countries or "less globalized" countries. It measures globalizing by *changes* in the ratio of trade to GDP over 1977–97. Ranking developing countries by the amount of change, it calls the top third the more globalized countries, the bottom two-thirds, the less globalized countries. It finds that the former have had faster economic growth, no increase in inequality, and

faster reduction of poverty than the latter. "Thus globalization clearly can be a force for poverty reduction," it concludes.

The conclusion does not follow. First, using "change in the trade/GDP ratio" as the measure of globalization skews the results.[29] The globalizers then include China and India, as well as countries such as Nepal, Côte d'Ivoire, Rwanda, Haiti, and Argentina. It is quite possible that "more globalized" countries are *less* open than many "less globalized" countries, both in terms of trade/GDP and in terms of the magnitude of tariffs and nontariff barriers. A country with high trade/GDP and very free trade policy would still be categorized as "less globalized" if its *increase* in trade/GDP over 1977–97 put it in the bottom two-thirds of the sample. Many of the globalizing countries initially had very *low* trade/GDP in 1977 and still had relatively low trade/GDP at the *end* of the period in 1997 (reflecting more than just the fact that larger economies tend to have lower ratios of trade/GDP). To call relatively closed economies "more globalized" or "globalizers" and to call countries with much higher ratios of trade/GDP and much freer trade regimes "less globalized" or even "nonglobalizers" is an audacious use of language.

Excluding countries with high but not rising levels of trade to GDP from the category of more globalized eliminates many poor countries dependent on a few natural resource commodity exports, which have had poor economic performance. The structure of their economy and the low skill endowment of the population make them dependent on trade. If they were included as globalized their poor economic performance would question the proposition that the more globalized countries do better. On the other hand, including China and India as globalizers – despite relatively low trade/GDP and relatively protective trade regimes – guarantees that the globalizers, weighted by population, show better performance than the nonglobalizers. Table 3 provides an illustration.

The second problem is that the argument fudges almost to vanishing point the distinction between trade quantities and trade policy, and implies, wrongly, that rising trade quantities – and the developmental benefits thereof – are the consequence of trade liberalization.

Third, the argument assumes that fast trade growth is the major cause of good economic performance. It does not examine the reverse causation,

	Exports/GDP			GNPRG 1988–99 (%)
	1990	1999	% Change	
Nonglobalizers				
Honduras	36	42	17	−1.2
Kenya	26	25	−0.04	0.5
Globalizers				
India	7	11	57	6.9
B'desh	6	14	133	3.3

Table 3 Trade-dependent nonglobalizers and less-trade-dependent globalizers

Source: World Bank, *World Development Report 2000/01*, Tables 1 and 13.

from fast economic growth to fast trade growth. Nor does it consider that other variables correlated with trade growth may be important causes of economic performance: quality of government, for example. One reexamination of the Bank's study finds that the globalizer countries do indeed have higher quality of government indicators than the nonglobalizer countries, on average.[30] Finally, trade does not capture important kinds of "openness," including people flows and ideas flows. Imagine an economy with no foreign trade but high levels of inward and outward migration and a well-developed diaspora network. In a real sense this would be an open or globalized economy, though not classified as such.

Certainly many countries – including China and India – have benefited from their more intensive engagement in international trade and investment over the past one or two decades. But this is not to say that their improved performance is largely due to their more intensive external integration. They began to open their own markets *after* building up industrial capacity and fast growth behind high barriers.[31] In addition, throughout their period of so-called openness they have maintained protection and other market restrictions that would earn them a bad report card from the World Bank and IMF were they not growing fast. China began its fast growth with a high degree of equality of assets and income, brought about in distinctly nonglobalized conditions and unlikely to have been achieved in an open economy and democratic polity.[32]

Their experience – and that of Japan, South Korea and Taiwan earlier – shows that countries do not have to adopt liberal trade policies in order to reap large benefits from trade.[33] They all experienced relatively fast growth behind protective barriers; a significant part of their growth came from replacing imports of consumption goods with domestic production; and more and more of their rapidly growing imports consisted of capital goods and intermediate goods. As they became richer they tended to liberalize their trade – providing the basis for the misunderstanding that trade liberalization drove their growth. For all the Bank study's qualifications (such as "We label the top third 'more globalized' without in any sense implying that they adopted pro-trade policies. The rise in trade may have been due to other policies or even to pure chance"), it concludes that trade liberalization has been the driving force of the increase in developing countries' trade. "The result of this trade liberalization in the developing world has been a large increase in both imports and exports," it says. On this shaky basis the Bank rests its case that developing countries must push hard toward near-free trade as a core ingredient of their development strategy, the better to enhance competition in efficient, rent-free markets. Even when the Bank or other development agencies articulate the softer principle – trade liberalization is the necessary direction of change but countries may do it at different speeds – all the attention remains focused on the liberalization part, none on how to make protective regimes more effective.

In short, the Bank's argument about the benign effects of globalization on growth, poverty and income distribution does not survive scrutiny at either end. And a recent cross-country study of the relationship between openness and income distribution strikes another blow. It finds that among the subset of countries with low and middle levels of average income (below $5,000 per capita in PPP terms, that of Chile and the Czech Republic), higher levels of trade openness are associated with *more* inequality, while among higher-income countries more openness goes with less inequality.[34]

6 CONCLUSION

It is plausible, and important, that the proportion of the world's population living in extreme poverty has probably fallen over the past two decades or so, having been rising for decades before then. Beyond this we cannot be confident, because the World Bank's poverty numbers are subject to a large margin of error, are probably biased downward, and probably make the trend look rosier than it really is. On income distribution, several studies suggest that world income inequality has been rising during the past two to three decades, and a study of manufacturing pay dispersions buttresses the same conclusion from another angle. The trend is sharpest when incomes are measured at market-exchange-rate incomes. This is less relevant to relative well-being than PPP-adjusted incomes, in principle; but it is highly relevant to state capacity, interstate power, and the dynamics of capitalism. One combination of inequality measures does yield the conclusion that income inequality has been falling – PPP-income per capita weighted by population, measured by an averaging coefficient such as the Gini. But take out China and even this measure shows widening inequality. Falling inequality is thus not a *generalized* feature of the world economy even by the most favorable measure. Finally, whatever we conclude about income inequality, absolute income gaps are widening and will continue to do so for decades.

If the number of people in extreme poverty is not falling and if global inequality is widening, we cannot conclude that globalization in the context of the dollar-Wall Street regime is moving the world in the right direction, with Africa's poverty as a special case in need of international attention. The balance of probability is that – like global warming – the world is moving in the wrong direction.

The failure of the predicted effects aside, the studies that claim globalization as the driver are weakened by (a) the use of *changes* in the trade/GDP ratio or FDI/GDP ratio as the index of globalization or openness, irrespective of level (though using the level on its own is also problematic, the level of trade/GDP being determined mainly by country size); (b) the assumption that trade liberalization drives increases in trade/GDP; and (c) the assumption that increases in trade/GDP drive improved economic performance. The problems come together in the case of China and India, whose treatment dominates the overall results. They are classed as "globalizers," their relatively good economic performance is attributed mainly to their "openness," and the deviation between their economic policies – substantial trade protection and capital controls, for example – and the core economic policy package of the World Bank and the other multilateral economic organizations is glossed.

At the least, analysts have to separate out the effect of country size on trade/GDP levels from other factors determining trade/GDP, including trade policies, because the single best predictor of trade/GDP is country size (population and area). They must make a clear distinction between statements about (i) levels of trade, (ii) changes in levels, (iii) restrictiveness or openness of trade policy, (iv) changes in restrictiveness of policy, and (v) the content of trade – whether a narrow range of commodity exports in return for a broad range of consumption imports, or a diverse range of exports (some of them replaced imports) in return for a diverse range of imports (some of them producer goods to assist further import replacement).

(a) Should we worry about rising inequality?

The neoliberal argument says that inequality provides incentives for effort and risk-taking, and thereby raises efficiency. As Margaret Thatcher put it, "It is our job to glory in inequality and see that talents and abilities are given vent and expression for the benefit of us all."[35] We should worry about rising inequality only if it somehow makes the poor worse off than otherwise.

The counterargument is that this productive incentive effect applies only at moderate, Scandinavian, levels of inequality. At higher levels, such as in the United States over the past 20 years, it is likely to be swamped by social costs. Aside from the moral case against it, inequality above a moderate level creates a kind of society that even crusty conservatives hate to live in, unsafe and unpleasant.

Higher income inequality within countries goes with: (i) higher poverty (using World Bank data and the number of people below the Bank's international poverty line);[36] (ii) slower economic growth, especially in large countries such as China, because it constrains the growth of mass demand; (iii) higher unemployment; and (iv) higher crime.[37] The link to higher crime comes through the inability of unskilled men in high inequality societies to play traditional male economic and social roles, including a plausible contribution to family income. But higher crime and violence is only the tip of a distribution of social relationships skewed toward the aggressive end of the spectrum, with low average levels of trust and social capital. In short, inequality at the national level should certainly be a target of public policy, even if just for the sake of the prosperous.

The liberal argument is even less concerned about widening inequality between countries than it is about inequality within countries, because we cannot do much to lessen international inequality directly. But on the face of it, the more globalized the world becomes, the more that the reasons why we should be concerned about within-country inequalities also apply between countries. If globalization within the current framework actually increases inequality within and between countries, as some evidence suggests, increases in world inequality above moderate levels may cut world aggregate demand and thereby world economic growth, making a vicious circle of rising world inequality and slower world growth.

Rising inequality between countries impacts directly the national political economy in the poorer states, as rich people who earlier compared themselves to others in their neighborhood now compare themselves to others in the United States or Western Europe, and feel deprived and perhaps angry. Inequality above moderate levels may, for example, predispose the elites to become more corrupt as they compare themselves to elites in rich countries. They may squeeze their own populations in order to sustain a comparable living standard, enfeebling whatever norms of citizenship have emerged and preventing the transition from an "oligarchic" elite, concerned to maximize redistribution upward and contain protests by repression, to an "establishment" elite, concerned to protect its position by being seen to operate fairly. Likewise, rapidly widening between-country inequality in current exchange rate terms feeds back into stress in public services, as the increasing foreign exchange cost of imports, debt repayment and the like has to be offset by cuts in budgets for health, education, and industrial policy.

Migration is a function of inequality, since the fastest way for a poor person to get richer is to move from a poor country to a rich country. Widening inequality may raise the incentive on the educated people of poor countries to migrate to the rich countries, and raise the incentive of unskilled people to seek illegal entry. Yet migration/refugees/asylum is the single most emotional, most atavistic issue in Western politics. Polls show that more than two-thirds of respondents agree that there should be fewer "foreigners" living in their countries.[38]

Rising inequality may generate conflict between states, and – because the *market-exchange-rate* income gap is so big – make it cheap for rich states to intervene to support one side or the other in civil strife. Rising inequality in market-exchange-rate terms – helped by a high US dollar, a low (long-run) oil price, and the WTO agreements on intellectual property rights, investment, and trade in services – allows the United States to finance the military sinews of its postimperial empire more cheaply.[39]

The effects of inequality within and between countries depend on prevailing norms. Where power hierarchy and income inequality are thought to be the natural condition of man the negative effects can be expected to be lighter than where prevailing norms affirm equality. Norms of equality and democracy are being energetically internationalized by the Atlantic states, at the same time as the lived experience in much of the rest of the world is from another planet.

In the end, the interests of the rich and powerful should, objectively, line up in favor of greater equity in the world at large, because some of the effects of widening inequality may contaminate their lives and those of their children. This fits the neoliberal argument. But the route to greater equity goes not only through the dismantling of market rules rigged in favor of the rich – also consistent with the neoliberal argument – but through more political (nonmarket) influence on resource allocation in order to counter the tendency of free markets to concentrate incomes and power. This requires international public policy well beyond the boundaries of neoliberalism.

The need for deliberate international redistribution is underlined by the evidence that world poverty may be higher in absolute numbers than is generally thought, and quite possibly rising rather than falling; and that world income inequality is probably rising too. This evidence suggests that the income and prosperity gap between a small proportion of the world's population living mainly in the North and a large proportion living entirely in the South is a structural divide, not just a matter of a lag in the South's catch-up. Sustained preferences for the South may be necessary if the world is to move to a single-humped and more narrowly dispersed distribution over the next century.

(b) The political economy of statistics

Concerns about global warming gave rise to a coordinated worldwide project to get better climatological data; the same is needed to get better data on poverty and inequality. The World Bank is one of the key actors. It has moved from major to minor source of foreign finance for most developing countries outside of Africa. But it remains an important global organization because it wields a disproportionate influence in setting the development agenda, in offering an imprimatur of "sound finance" that crowds in other resources, and in providing finance at times when other finance is not available. Its statistics and development research are crucial to its legitimacy.[40] Other regional development banks and aid agencies have largely given up on statistics and research, ceding the ground to the World Bank. Alternative views come only from a few "urban guerrillas" in pockets of academia and the UN system.[41] Keynes' dictum on practical men and long-dead economists suggests that such intellectual monopolization can have a hugely negative impact.

Think of two models of a statistical organization that is part of a larger organization working on politically sensitive themes. The "exogenous" model says that the statistics are produced by professionals exercising their best judgment in the face of difficulties that have no optimal solutions, who are managerially insulated from the overall tactical goals of the organization. The "endogenous" model says that the statistics are produced by staff who act as agents of the senior managers (the principals), the senior managers expect them to help advance the tactical goals of the organization just like other staff, and the statistics staff therefore have to massage the data beyond the limits of professional integrity, or quit.

Certainly the simple endogenous model does not fit the Bank; but nor does the other. The Bank is committed to an Official View of how countries should seek poverty reduction, rooted in the neo-liberal agenda of trade opening, financial opening, privatization, deregulation, with some good governance, civil society and environmental protection thrown in; it is exposed to arm-twisting by the G7 member states and international nongovernmental organizations (NGOs); it must secure their support and defend itself against criticism.[42] It seeks to advance its broad market opening agenda not through coercion but mainly by establishing a sense that the agenda is right and fitting. Without this it would lose the support of the G7 states, Wall Street, and fractions of developing country elites. The units of the Bank that produce the statistics are partly insulated from the resulting pressures, especially by their membership in "epistemic communities" of professionals inside and outside the Bank; but not wholly insulated. To say otherwise is to deny that the Bank is subject to the Chinese proverb, "Officials make the figures, and the figures make the officials;" or to Goodhart's law, which states that an indicator's measurement will be distorted if it is used as a target. (Charles Goodhart was thinking of monetary policy, but the point also applies to variables used to make overall evaluations of the performance of multilateral economic organizations.) To say otherwise is equally to deny that the Bank is affected by the same pressures as the Fund, about which a former Fund official said, "The managing director makes the big decisions, and the staff then puts together the numbers to justify them."[43] But little is known about the balance between autonomy and compliance in the two organizations, or the latitude of their statisticians to adjust the country numbers provided by colleagues elsewhere in the organization which they believe to be fiddled (as in the China case, above).[44]

Some of the Bank's statistics are also provided by independent sources, which provide a check. Others, including the poverty numbers, are produced only by the Bank, and these are more subject to Goodhart's law. The Bank should appoint an independent auditor to verify its main development statistics or cede the work to an independent agency, perhaps under UN auspices (but if done by, say, UNCTAD, the opposite bias might be introduced). And it would help if the Bank's figures on poverty and inequality made clearer than they do the possible biases and the likely margins of error.

All this, of course, only takes us to the starting point of an enquiry into the causes of the probable poverty and inequality trends,[45] their likely consequences, and public policy responses; but at least we are now ready to ask the right questions. Above all, we have to go back to a distinction that has all but dropped out of development studies, between increasing returns and decreasing returns or, more generally, between positive and negative feedback mechanisms. The central question is why, at the level of the whole, the increasing returns of the Matthew effect – "To him who hath shall be given" – continues to dominate decreasing returns in the third wave of globalization.

NOTES

I thank without implicating Sanjay Reddy, Michael Ward, Branko Milanovic, Ron Dore, David Ellerman, Martin Wolf, Timothy Besley, and James Galbraith; and the Institute for Advanced Study, Berlin, and the Crisis States Program, DESTIN, LSE, for financial support. Final revision accepted: 30 October 2003.

1 Gowan (1999).
2 Mazur (2000).
3 Wolf (2000).
4 International Monetary Fund (2003). The trend is, however, highly sensitive to the dollar's strong depreciation in the 1970s and appreciation in the 1990s. When this is allowed for, the world growth rate may be closer to trendless.
5 In more concrete terms the number of hours of work it took for an entry-level adult male employee of McDonalds to earn the equivalent of one BigMac around 2000 ranged from: Holland/Australia/NZ/UK/US, 0.26–0.53 h; Hong Kong, 0.68 h; Malaysia/South Korea, 1.43–1.46 h; Philippines/Thailand, 2.32–2.2.66 h; China, 3.96 h; India, 8 h.
6 Purchasing power parity is a method of adjusting relative incomes in different countries to take account of the fact that market exchange rates do not accurately reflect purchasing power – as in the common observation that poor Americans feel rich in India and rich Indians feel poor in the United States.
7 The WIDER data set marries consumption from household surveys with consumption from national income accounts, and makes an allowance for (nonpublic sector) nonpriced goods and services.
8 World Bank (2002a) and World Bank (2002b, p. 30).
9 Wolfensohn (2001).
10 World Bank (2001b, p. 3). The $1 a day is measured in purchasing power parity. See also World Bank (2002c).
11 I am indebted to Sanjay Reddy for discussions about the Bank's poverty numbers (Reddy & Pogge, 2003a). See also Ravallion (2003), and Reddy and Pogge (2003b). In this paper I do not consider the additional problems that arise when estimating the impact of economic growth on poverty. See Deaton (2003).
12 The Bank also calculates a poverty headcount with $2/day, which suffers from the same limitations as the $1/day line.
13 Reported in Deaton (2001).
14 See Reddy and Pogge (2003a).
15 Also "[Since 1980] the most rapid growth has occurred in poor locations. Consequently the number of poor has declined by 200 million since 1980" (Dollar & Kraay, 2002, p. 125).
16 The new results were published in World Development Report 2000/2001 (World Bank, 2001a).
17 Deaton (2001, p. 128).
18 I take this example from Pogge and Reddy (2003).
19 The 25–40% figure is Reddy and Pogge's estimate, the range reflecting calculations based on PPP conversion factors for 1985 and 1993, and for "all-food" and "bread-and-cereals" indices.
20 Also, Bolivia's extreme poverty rate according to the World Bank line was 11%, according to the ECLA line, 23%; Chile, 4%, 8%; Colombia, 11%, 24%; Mexico, 18%, 21% (ECLA, 2001, p. 51).
21 Reddy and Pogge (2003a).
22 This effect is amplified by the widespread removal of price controls on "necessities" and the lowering of tariffs on luxuries.
23 Gopinath (2002).
24 See Wade (2002a). It uses Stiglitz's firing and Kanbur's resignation to illuminate the US role in the Bank's generation of knowledge.
25 Meltzer Commission (2000). Meltzer later described the drop in the proportion of the world's population in poverty from 28% in 1987 to 24% in 1998 as a "modest" decline, the better to hammer the Bank (Meltzer, 2001).
26 World Bank (2002c). See Deaton (2002).
27 Dollar was ascendant not in terms of bureaucratic position but in terms of epistemic influence, as seen in the Human Resource department's use of him as a "metric" for judging the stature of other economists. When reporters started contacting the Bank to ask why

it was saying different things about the poverty numbers – specifically why two papers on the Development Research Complex's web site gave different pictures of the trends – the response was not, "We are a research complex, we let 100 flowers bloom," but rather an assertion of central control. Chief economist Nick Stern gave one manager "special responsibility" for making sure the Bank's poverty numbers were all "coherent" (Stern to research managers, email, April 4, 2002).

28 World Bank (2002c).

29 In this Section 1 draw on the arguments of Rodrik (1999, 2001).

30 Besley (2002). Besley uses indicators such as press freedom, democratic accountability, corruption, civil rights.

31 Cf. "As they reformed and integrated with the world market, the 'more globalized' developing countries started to growth rapidly, accelerating steadily from 2.9% in the 1970s to 5% through the 1990s" (World Bank, 2002c, p. 36, emphasis added).

32 Rodrik (1999).

33 Wade (2003a [1990]).

34 Milanovic (2002). Milanovic finds that in countries below the average income of about $PPP 5,000, higher levels of openness (imports plus exports/GDP) are associated with lower income shares of the bottom 80% of the population.

35 Quoted in George (1997).

36 Besley and Burgess (2003).

37 Lee and Bankston (1999), Hsieh and Pugh (1993), Fajnzylber, Lederman, and Loayza (1998) and Freeman (1996).

38 Demeny (2003).

39 Wade (2003b, 2003c).

40 Kapur (2002).

41 For a good example of a heterodox book from a corner of the UN system, see UNDP (2003). The WTO lobbied to prevent its publication.

42 Wade (2003d).

43 Gopinath (1999).

44 Key experts in the relevant statistical unit thought that colleagues had fiddled the China income numbers reported in Table 4, but their boss ignored their objections.

45 For discussion of causes see Wade (2002b).

REFERENCES

Arrighi, G., Silver, B., & Brewer, B. (2003). Industrial convergence, globalization and the persistence of the North–South divide. *Studies in Comparative International Development, 38*(1), 3–31.

Besley, T. (2002). Globalization and the quality of government. Manuscript, Economics Department, London School of Economics, March.

Besley, T., & Burgess, R. (2003). Halving world poverty. *Journal of Economic Perspectives, 17*(3), 3–22.

Castles, I. (2001). *Letter to The Economist*, May 26.

Deaton, A. (2001). Counting the world's poor: problems and possible solutions. *The World Bank Research Observer, 16*(2), 125–147.

Deaton, A. (2002). Is world poverty falling? *Finance and Development, 39*(2), 34. Available: http://www.imf.org/external/pubs/ft/fandd/2002/06/deaton.htm.

Deaton, A. (2003). Measuring poverty in a growing world (or measuring growth in a poor world). NBER Working Paper 9822. http://d.repec.org/n?u=Re-PEc:nbr:nberwo:9822&r=dev.

Demeny, P. (2003). Population policy dilemmas in Europe at the dawn of the twenty-first century. *Population and Development Review, 29*(1), 1–28.

Dikhanov, Y., & Ward, M. (2003). Evolution of the global distribution of income in 1970–99. In *Proceedings of the Global Poverty Workshop*, Initiative for Policy Dialogue, Columbia University. Available: http://www-l.gsb.Columbia.edu/ipd/povertywk.html.

Dollar, D., & Kraay, A. (2002). Spreading the wealth. *Foreign Affairs* (January/February), 120–133.

ECLA (2001). *Panorama social de America Latina 2000–01*, ECLA (CEPAL), Santiago.

Fajnzylber, P., Lederman, D., & Loayza, N. (1998). What causes violent crime? Typescript, Office of the Chief Economist, Latin America and the Caribbean Region. World Bank, Washington, DC.

Freeman, R. (1996). Why do so many young American men commit crimes and what might we do about it? *Journal of Economic Perspectives, 10*(1), 25–42.

George, S. (1997). How to win the war of ideas: lessons from the Gramscian right. *Dissent, 44*(3), 47–53.

Gopinath, D. (1999). Slouching toward a new consensus. *Institutional Investor, 1*(September), 79–87.

Gopinath, D. (2002). Poor choices. *Institutional Investor*, September 1, 41–50.

Gowan, P. (1999). *The global gamble*. London: Verso.

Hsieh, C. C., & Pugh, M. (1993). Poverty, income inequality, and violent crime: a meta-analysis of recent aggregate data studies. *Criminal Justice Review, 18*, 182–202.

International Monetary Fund (2003). *World economic outlook*. Database IMF, Washington, DC, April.

Kapur, D. (2002). The changing anatomy of governance of the World Bank. In J. Pincus & J. Winters (Eds.), *Reinventing the World Bank* (pp. 54–75). Ithaca, NY: Cornell University Press.

Lee, M. R., & Bankston, W. (1999). Political structure, economic inequality, and homicide: a cross-sectional analysis. *Deviant Behavior: an Interdisciplinary Journal, 19*, 27–55.

Mazur, J. (2000). Labor's new internationalism. *Foreign Affairs, 79*(January/February), 79–93.

Meltzer, A. (2001). The World Bank one year after the Commission's report to Congress. In *Hearings before the Joint Economic Committee*, US Congress, March 8.

Meltzer Commission (United States Congressional Advisory Commission on International Financial Institutions) (2000). Report to the US Congress on the International Financial Institutions. Available: http://www.house/gov/jec/imf/ifiac.

Milanovic, B. (2002). Can we discern the effect of globalization on income distribution? Evidence from household budget surveys. World Bank Policy Research Working Papers 2876. Available: http://econ.worldbank.org.

Pogge, T., & Reddy, S. (2003). Unknown: the extent, distribution, and trend of global income poverty. Available: http://www.socialanalysis.org.

Ravallion, M. (2003). Reply to Reddy and Pogge. Available: http://www.socialanalysis.org.

Reddy, S., & Pogge, T. (2003a). How not to count the poor. Available: http://www.socialanalysis.org.

Reddy, S., & Pogge, T. (2003b). Reply to Ravallion. Available: http://www.socialanalysis.org.

Rodrik, D. (1999). The new global economy and developing countries: making openness work. In *Policy Essay 24, Overseas Development Council*. Baltimore: Johns Hopkins University Press.

Rodrik, D. (2001). Trading in illusions. *Foreign Policy, 123*(March/April), 55–62.

UNDP (2003). *Making global trade work for people*. London: Earthscan.

Wade, R. H. (2002a). US hegemony and the World Bank: The fight over people and ideas. *Review of International Political Economy, 9*(2), 201–229.

Wade, R. H. (2002b). Globalisation, poverty and income distribution: does the liberal argument hold. In *Globalisation, Living Standards and Inequality: Recent Progress and Continuing Challenges* (pp. 37–65). Sydney: Reserve Bank of Australia.

Wade, R. H. (2003a [1990]). *Governing the market*. Princeton, NJ: Princeton University Press.

Wade, R. H. (2003b). The invisible hand of the American empire. *Ethics and International Affairs, 17*(2), 77–88.

Wade, R. H. (2003c). What strategies are viable for developing countries today. The WTO and the shrinking of development space. *Review of International Political Economy, 10*(4), 621–644.

Wade, R. H. (2003d). The World Bank and its critics: the dynamics of hypocrisy. *Studies in Comparative International Development*, in press.

Wolf, M. (2000). The big lie of global inequality. *Financial Times*, February 8.

Wolf, M. (2002b). Doing more harm than good. *Financial Times*, May 8.

Wolfensohn, J. (2001). Responding to the challenges of globalization: Remarks to the G-20 finance ministers and central governors. Ottawa, November 17.

World Bank (2001a). *World development report 2000/2001, Attacking Poverty*. New York: Oxford University Press.

World Bank (2001b). *World development indicators 2001*. Washington, DC: The World Bank.

World Bank (2002a). *World development indicators 2002*. Washington, DC. The World Bank.

World Bank (2002b). *Global economic prospects and the developing countries 2002: making trade work for the world's poor*. Washington, DC: The World Bank.

World Bank (2002c). *Globalization, growth, and poverty: building an inclusive world economy*. New York: Oxford University Press.

'Wealth and Well-Being'

from *Human Well-Being and the Natural Environment* (2001)

Partha Dasgupta

Editors' Introduction

Partha Dasgupta was born in Dhaka, then in British India now in Bangladesh, in 1942, and read both Physics and Mathematics for his first degrees, at Delhi and Cambridge universities respectively. After gaining a PhD in Economics from Cambridge in 1968, Dasgupta worked for three years as a Research Fellow at the university before joining the London School of Economics in 1971. Dasgupta stayed at the LSE until the end of 1984, when he moved back to Cambridge as a Professor of Economics. Dasgupta is now the Frank Ramsey Professor of Economics at Cambridge. Knighted in 2002 for his 'service to economics', Sir Partha Dasgupta has also held visiting positions in Stanford and Stockholm, worked with Stephen Marglin and Amartya Sen for the United Nations (on guidelines for project evaluation, 1972), received many honours from learned societies, and in 2002 and 2004, respectively, shared the Volvo Environment Prize and the Kenneth E. Boulding Memorial Award of the International Society for Ecological Economics.

Dasgupta has long worked in the intersecting fields of environmental and development economics, and it is here that he has published his best-known books: the monumental *An Inquiry into Well-Being and Destitution* (1993) and *Human Well-Being and the Natural Environment* (2001). The reading that is reproduced here comes from the later book. Although only a short piece, it is pregnant with implications for public policy. Dasgupta has become increasingly active as a critic of mainstream conceptions, or measurements, of 'development' that are focused either on per capita income growth or human development (incomes, literacy, life expectancy, etc.) more broadly. Neither of these traditions, Dasgupta suggests, pays sufficient attention to questions of inter-generational equity or sustainability. Specifically, they have nothing to say about levels of natural resource or biodiversity depletion that should give ordinary persons pause for thought before announcing the success of present-day development. What marks out Dasgupta's work, however, and that of World Bank economists including Hamilton and Clemens, is an insistent concern to theorise 'natural capital loss' in a rigorous and yet also empirically measurable fashion. Dasgupta accepts that such work is in its infancy and that very often we have to use proxy measures for environmental damage. He has been in the vanguard, nonetheless, of attempts to 'account for the environment' in development circles (especially), and thus to rethink the boosterism that has surrounded (in particular) China's seeming success as a developing country since c.1980. And China is by no means the 'worst case' country: the data that Dasgupta tables here suggest that changes in per capita wealth have been marginally positive in China but actually negative across large parts of South Asia and sub-Saharan Africa. So much for 'development'.

Key references

Partha Dasgupta (2001) *Human Well-Being and the Natural Environment*, Oxford: Oxford University Press.
— (1993) *An Inquiry into Well-Being and Destitution*, Oxford: Clarendon Press.
— (2004) 'Are we consuming too much?', *Journal of Economic Perspectives* 18 (3): 147–72.
— (2007) 'Nature and the economy', *Journal of Applied Ecology* 44: 475–87.
K. Hamilton and M. Clemens (1999) 'Genuine savings rates in developing countries', *World Bank Economic Review* 13: 333–56.

▪ ▪ ▪ ▪ ▪ ▪

[. . .]

GENUINE INVESTMENT: APPLICATIONS[1]

Reporting studies undertaken at the World Bank, Hamilton and Clemens (1999) have provided estimates of genuine investment in a number of countries. They call it 'genuine saving'. The authors included in the list of a country's assets its manufactured, human, and natural capital. There is a certain awkwardness in many of the steps they took to estimate genuine investment. For example, investment in human capital in a given year was taken to be public expenditure on education. It is an overestimate, because each year people die and take their human capital with them. This is depreciation, and should have been deducted.

Among the resources making up natural capital, only commercial forests, oil and minerals, and the atmosphere as a sink for carbon dioxide were included. (Not included were water resources, forests as agents of carbon sequestration, fisheries, air and water pollutants, soil, and biodiversity.) So there is an undercount, possibly a serious one. Nevertheless, one has to start somewhere, and theirs is a useful compendium of estimates.

Being a global commons, Earth's atmosphere poses an intriguing problem. When a country adds to the atmosphere's carbon content, it damages the commons. In calculating the value of the change in the country's capital assets, how much of the reduction should we include?

Two possibilities suggest themselves.[2] One is to attribute to each country the fraction of Earth's atmosphere that reflects the country's size relative to the world as a whole, using population as a means of comparison (or GNP, or whatever). The other is to regard the global commons as every country's asset. In that case, the entire cost of global warming inflicted by a country would be regarded as that country's loss. Hamilton and Clemens (1999) follow the latter route.[3] They use 20 US dollars as the figure for damages caused by a tonne of emitted carbon dioxide.[4]

The accounting value of forest depletion is taken to be the stumpage value (price minus logging costs) of the quantity of commercial timber and fuelwood harvested in excess of natural regeneration rates. This is an awkward move, since Hamilton and Clemens don't say what is intended to happen to the land that is being deforested. For example, if the deforested land is converted into an urban sprawl, the new investment in the sprawl would be accounted for in conventional accounting statistics; but if it is intended to be transformed into farmland, matters are different: the social worth of the land as a farm should be included as an addition to the economy's productive base.

Despite these limitations in the data, it is instructive to put the theory we have developed to work in the context of the world's poorest regions. Table 1 does just that. The account that follows covers sub-Saharan Africa, the Indian subcontinent, and China. Taken together, they contain nearly half the world's population. They also comprise pretty much all the world's poorest countries.

The first column of figures in Table 1 contains estimates of genuine investment, as a proportion of GNP, in Bangladesh, India, Nepal, Pakistan, China, and sub-Saharan Africa, over the period 1970–93. Notice that Bangladesh, Nepal, and sub-Saharan Africa have disinvested: their capital assets shrank during the period in question. In contrast, genuine investment was positive in China, India, and Pakistan. This could suggest that the latter countries were

	I/Y	g(L)	g(W/L)	g(Y/L)	g(HDI)
Bangladesh	−0.013	2.3	−2.60	1.0	3.3
India	0.080	2.1	−0.10	2.3	2.2
Nepal	−0.024	2.4	−3.00	1.0	5.3
Pakistan	0.040	2.9	−1.90	2.7	1.8
Sub-Saharan Africa	−0.028	2.7	−3.40	−0.2	0.9
China	0.100	1.7	0.80	6.7	−0.2

Table 1 Genuine investment and capital deepening in selected regions, 1970–1993

Notes

I/Y: genuine investment as proportion of GNP.
Source: Hamilton and Clemens (1999: table 3).

g(L): average annual percentage rate of growth of population, 1965–96.
Source: World Bank (1998: table 1.4).

g(W/L): average annual percentage rate of change in *per capita* wealth.
g(Y/L): average annual percentage rate of change in *per capita* GNP, 1965–96.
Source: World Bank (1998: table 1.4).

g(HDI): average annual percentage rate of change in UNDP's Human Development Index, 1987–97.
Source: UNDP (1990, 1999).
Assumed output-wealth ratio: 0.25.

wealthier at the end of the period than at the beginning. But when population growth is taken into account, the picture changes.

The second column contains the annual percentage rate of growth of population in the various places over the period 1965–96. All but China have experienced rates of growth in excess of 2 per cent per year, sub-Saharan Africa and Pakistan having grown in numbers at nearly 3 per cent per year. We now want to estimate the average annual change in wealth per capita during 1970–93. To do this, we should multiply genuine investment as a proportion of GNP by the average output–wealth ratio of an economy to arrive at the (genuine) investment–wealth ratio, and then compare changes in the latter ratio to changes in population size.

Since a wide variety of natural assets (human capital and various forms of natural capital) are unaccounted for in national accounts, there is an upward bias in published estimates of output–wealth ratios, which traditionally have been taken to be something like 0.30. In what follows I use 0.25 as a check against the upward bias in traditional estimates. This is almost certainly still a conservatively high figure.

The third column in Table 1 contains estimates of the annual percentage rate of change in wealth per head. The procedure for this has been to multiply genuine investment as a proportion of GNP by 0.25, and to substract the annual percentage rate of growth of population from the product. This is a crude way to adjust for population change, but more accurate adjustments would involve greater computation.

The striking message of the third column is that all but China have *decumulated* their capital assets on a per capita basis during the past thirty years or so. Earlier we noted that total factor productivity has declined in sub-Saharan Africa over the past four decades. Taken together, the evidence is that, relative to its population, the region's productive base has shrunk considerably. This may not cause surprise, since sub-Saharan Africa is widely known to have regressed in terms of most socio-economic indicators. But the figures in Table 1 for Bangladesh, India, Nepal, and Pakistan should cause surprise. They have all decumulated their capital assets on a per capita basis. Earlier we noted that there has probably not been any growth in total factor productivity in South Asia as a whole. Taken together, the evidence is that, relative to its population, the region's productive base has shrunk.

How do changes in wealth per head compare with changes in conventional measures of the quality of life? The fourth column of the table contains estimates of the annual percentage rate of change in per capita GNP during 1965–96; and in the fifth column I have compiled figures for the average growth rate in UNDP's Human Development Index (HDI) over the period 1987–97.

Notice how misleading our retrospective assessment of long-term economic development in the Indian subcontinent would be if we were to look at growth rates in GNP per head. Pakistan, for example, would be seen as a country where GNP per head grew at a healthy 2.7 per cent per year, implying that the index doubled in value between 1965 and 1993. In fact, the average Pakistani became poorer by a factor of nearly 2 during that same period.

Bangladesh too has disinvested in her capital assets. The country is recorded as having grown in terms of GNP per head at a rate of 1 per cent per year during 1965–96. In fact, at the end of the period the average Bangladeshi was less than half as wealthy as she was at the beginning.

The case of sub-Saharan Africa is, of course, especially sad. At an annual rate of decline of per capita wealth of 3.4 per cent, the average person in the region becomes poorer by a factor of 2 every twenty years. And this doesn't include the decline in total factor productivity. The ills of sub-Saharan Africa are routine reading in today's newspapers and magazines. But they aren't depicted in terms of a decline in wealth. Table 1 shows that sub-Saharan Africa has experienced an enormous decline in its capital base over the past three decades. As total factor productivity in the region has fallen too, its entire productive base has shrunk. It will take years for the region to recover, if it is able to do so at all.

India can be said to have avoided a steep decline in its capital base. But it has been at the thin edge of economic development, having managed not quite to maintain its capital assets relative to population size. Total factor productivity in India has risen during the past decade, but, as I argued earlier, the estimates have an upward bias. If the figures in Table 1 are taken literally, the average Indian was slightly poorer in 1993 than in 1970.

Even China, so greatly vaunted for its progressive economic policies, has just managed to accumulate wealth in advance of population increase. For a poor country, a growth rate of 0.8 per cent per year in wealth per head isn't something about which one gets too excited. In any event, a more accurate figure for the output-wealth ratio would almost surely be considerably lower than 0.25. Moreover, the estimates of genuine investment don't include soil erosion or urban pollution, both of which are thought to be especially problematic in China.[5]

What of the HDI?[6] In fact, it misleads even more than GNP per head. As the third and fifth columns show, HDI offers precisely the opposite picture of the one we should obtain when judging the performance of countries. It has, however, been suggested to me by a colleague that, as HDI has been so constructed that it can assume values only between 0 and 1, it should not be used for intertemporal comparisons. He reminded me also that the United Nations Development Programme has used HDI mostly for cross-country comparisons. But an index that purports to reflect average well-being in a group cannot arbitrarily be limited to group comparisons at a point in time. If the index is deemed suitable for the purposes of such comparisons, it simply has to be made available for the purpose of well-being comparisons among groups separated by time; otherwise, the index isn't a measure of well-being. Furthermore, the fact that HDI lies in the interval [0,1] is no reason why it must not be used for intertemporal comparisons. That well-known measure of income inequality, the Gini coefficient, also lies in the interval [0,1]; but it is routinely asked whether the Gini coefficient of income (or wealth) distribution in a country has worsened or improved over time. I don't have an intuitive feel for what may or may not be a healthy growth rate for HDI in today's poor countries, but for sub-Saharan Africa the index improved during the 1990s, at 0.9 per cent per year.[7] Bangladesh and Nepal have been exemplary in terms of HDI. However, both countries have decumulated their assets at a high rate.

These are all rough and ready figures, but they show how accounting for human and natural capital can make for substantial differences in our conception of the development process. The implication should be heart-breaking: over the past three decades, the Indian subcontinent and sub-Saharan Africa, two of the poorest regions of the world, comprising something like a third of the world's population, have become even poorer. In fact, some of the countries in these regions have become a good deal poorer.

NOTES

1   That an economy needs to raise the worth of its productive base if social well-being is to increase was the
topic of discussion, starting in 1993, among members
of an Advisory Council created by Ismail Serageldin,
then vice president for environmentally sustainable
development at the World Bank. Serageldin (1995)
provides an outline of empirical work on genuine
investment that was initiated in his vice presidency. For
estimates of the depreciation of natural capital on a
regional basis, see Pearce, Hamilton, and Atkinson
(1996), and World Bank (1997).

2   Dasgupta, Kriström, and Mäler (1995).

3   Under optimum global management, the country
would be required to pay the total cost in the form of an
international tax. The tax is Pigovian (Ch. 8).

4   The estimate is due to Fankhauser (1994).

5   Hussain, Stern, and Stiglitz (2000) contains an analysis
of why China has been the economic success it is
widely judged to have been in recent years. However,
there is no mention of what may have been happening
to China's natural resource base in the process of the
country's economic development.

6   I am grateful to Sriya Iyer for suggesting that I should
check to see what HDI says about the progress of the
poorest nations.

7   Using the cross-country statistics for 1987 in UNDP
(1990), my calculations lead me to conclude that HDI
(1987) for sub-Saharan Africa was 0.111. In UNDP
(1999) the figure for HDI (1997) in sub-Saharan Africa
was reported to be 0.463. If the figures are to be
believed, HDI increased over the decade 1987–97 by a
factor of 4.17, yielding an average annual rate of growth
of about 14%, a statistic I find difficult to believe! In any
event, I was unable to reproduce the figure from the
statistics offered in the publication. My calculations,
based on the statistics given in UNDP (1999), led me to
conclude that HDI (1997) for sub-Saharan Africa was
0.122. This yields an annual growth rate of 0.9% per
year for HDI during 1987–97.

REFERENCES

Dasgupta, P., B. Kriström, and K.-G. Mäler (1995), 'Current
Issues in Resource Accounting', in P. O. Johansson,
B. Kriström, and K. G. Mäler, eds., *Current Issues in
Environmental Economics* (Manchester: Manchester
University Press).

—— Fankhauser, S. (1994), 'Evaluating the Social Costs of
Greenhouse Gas Emissions', *Energy Journal*, 15: 157–84.

—— Hamilton, K., and M. Clemens (1999), 'Genuine
Savings Rates in Developing Countries', *World Bank
Economic Review*, 13: 333–56.

—— Hussain, A., N. Stern, and J. Stiglitz (2000),
'Chinese Reforms from a Comparative Perspective', in
P. J. Hammond and G. D. Myles, eds., *Incentives, Organiza-
tion, and Public Economics* (Oxford: Oxford University
Press).

—— Pearce, D., K. Hamilton, and G. Atkinson (1996),
'Measuring Sustainable Development: Progress on
Indicators', *Environment and Development Economics*,
1: 85–101.

—— Serageldin, I. (1995), 'Are We Saving Enough for the
Future?' in *Monitoring Environmental Progress: Report on
Work in Progress, Environmentally Sustainable Develop-
ment*, World Bank, Washington.

UNDP (United Nations Development Programme) (1990,
1994, 1998, 1999), *Human Development Report* (New
York: Oxford University Press).

World Bank (1997), *Expanding the Measure of Wealth:
Indicators of Environmentally Sustainable Development*
(Washington: World Bank).

—— (1998), *World Development Indicators* (Washington:
World Bank).

'More Than 100 Million Women are Missing'

New York Review of Books (1990)[*]

Amartya Sen[*]

Editors' Introduction

Amartya Kumar Sen was born in Dhaka, then in British India now in Bangladesh, in 1933. After a teaching career that began in Jadavpur University at age 23, Sen worked successively at the Delhi and London Schools of Economics, before moving in turn to Oxford, Harvard and Cambridge universities, and then back to Harvard again in 2005 as Lamont University Professor. Sen was awarded the Nobel Prize for Economics in 1998, partly in respect of his work in the field of social choice theory and partly to recognise his extraordinary work on famines, gender issues and inequality. No one individual has contributed more, or more widely, to the field of development studies since *c.*1950 than Amartya Sen. His work in this area has covered such topics as surplus labour in peasant economies and the relationship between farm size holdings and productivity in Indian agriculture, as well as key contributions on the measurement of human development, on collective action problems and the isolation paradox, and on the intrinsic meaning of development (as the expansion of people's capabilities and thus freedom: see his *Development as Freedom*, 2000). Beyond mainstream development studies, Sen has also written extensively on Rabindranath Tagore and the Bengal renaissance, as well as on secularism, tolerance and the rise of religious nationalisms (especially Hindutva in India). His reach is extraordinary.

Sen's work also has clear policy relevance, even if Sen himself has generally refused invitations to work directly for the major development agencies (certainly the World Bank and the IMF, less so the UNDP where his work in the 1970s informed the construction of that agency's Human Development Index). This is especially true of Sen's work on the causes and consequences of famine. Sen has forcibly reminded us that famines rarely happen because of a regional collapse in the food supply (the Food Availability Decline thesis); rather, famines emerge where vulnerable people see their entitlements to food collapse sharply for one reason or another (including, most notably, when demand for the labour of food-purchasing house-holds shrinks or dries up). Sen has also insisted, rightly, that the effects of famines are gendered, with famines generally being more dangerous for women than for men. This insight is at one with his truly frightening contention, reproduced here, that more than 100 million women are missing from the world today. The exact figure cannot be properly known, but there is little doubt, as Sen suggests, that about 50 million women were missing from China alone in 1990, notwithstanding bloated talk about 'equality' in that country. Large numbers of women are also missing in India, where there are about 1,070 men for every 1,000 women today, and from Bangladesh, Pakistan and North Africa. In this reading, Sen tries to explain why this geography of missing women exists and persists. He does so, typically, in a manner that avoids both economic and cultural essentialism. Sen's focus is on differential female education and employment rights, but throughout he pays close attention to local constructions and understandings of these practices. As so often, too, Sen suggests how theoretical and empirical analysis might be linked to proposals for politics and public action.

Key references

Amartya Sen (1990) 'More than 100 million women are missing', *New York Review of Books* 37 (20).
— (1981) *Poverty and Famines: An Essay on Entitlement and Deprivation*, Oxford: Clarendon Press.
— (1982) *Choice, Welfare and Measurement*, Oxford: Blackwell.
— (1989) 'Food and freedom', *World Development* 17: 769–81.
— (2000) *Development as Freedom*, New York: Anchor Books.
Corbridge, S. (2002) 'Development as freedom: the spaces of Amartya Sen', *Progress in Development Studies* 2: 183–217.

1.

At birth, boys outnumber girls everywhere in the world, by much the same proportion – there are around 105 or 106 male children for every 100 female children. Just why the biology of reproduction leads to this result remains a subject of debate. But after conception, biology seems on the whole to favor women. Considerable research has shown that if men and women receive similar nutritional and medical attention and general health care, women tend to live noticeably longer than men. Women seem to be, on the whole, more resistant to disease and in general hardier than men, an advantage they enjoy not only after they are forty years old but also at the beginning of life, especially during the months immediately following birth, and even in the womb. When given the same care as males, females tend to have better survival rates than males.[1]

Women outnumber men substantially in Europe, the US, and Japan, where, despite the persistence of various types of bias against women (men having distinct advantages in higher education, job specialization, and promotion to senior executive positions, for example), women suffer little discrimination in basic nutrition and health care. The greater number of women in these countries is partly the result of social and environmental differences that increase mortality among men, such as a higher likelihood that men will die from violence, for example, and from diseases related to smoking. But even after these are taken into account, the longer lifetimes enjoyed by women given similar care appear to relate to the biological advantages that women have over men in resisting disease. Whether the higher frequency of male births over female births has evolutionary links to this potentially greater survival rate among women is a question of some interest in itself. Women seem to have lower death rates than men at most ages whenever they get roughly similar treatment in matters of life and death.

The fate of women is quite different in most of Asia and North Africa. In these places the failure to give women medical care similar to what men get and to provide them with comparable food and social services results in fewer women surviving than would be the case if they had equal care. In India, for example, except in the period immediately following birth, the death rate is higher for women than for men fairly consistently in all age groups until the late thirties. This relates to higher rates of disease from which women suffer, and ultimately to the relative neglect of females, especially in health care and medical attention.[2] Similar neglect of women vis-à-vis men can be seen also in many other parts of the world. The result is a lower proportion of women than would be the case if they had equal care – in most of Asia and North Africa, and to a lesser extent Latin America.

This pattern is not uniform in all parts of the third world, however. Sub-Saharan Africa, for example, ravaged as it is by extreme poverty, hunger, and famine, has a substantial excess rather than deficit of women, the ratio of women to men being around 1.02. The "third world" in this matter is not a useful category, because it is so diverse. Even within Asia, which has the lowest proportion of women in the world, Southeast Asia and East Asia (apart from China) have a ratio of women to men that is slightly higher than one to one (around 1.01). Indeed, sharp diversities also exist within particular regions – sometimes even within a particular country. For example, the ratio of women to men in the Indian states of Punjab and Haryana, which happen to be among the country's richest, is a remarkably low 0.86, while the state of Kerala in south-western India has a ratio higher than 1.03, similar to that in Europe, North America, and Japan.

To get an idea of the numbers of people involved in the different ratios of women to men, we can estimate the number of "missing women" in a country, say, China or India, by calculating the number of extra women who would have been in China or India if these countries had the same ratio of women to men as obtain in areas of the world in which they receive similar care. If we could expect equal populations of the two sexes, the low ratio of 0.94 women to men in South Asia, West Asia, and China would indicate a 6 percent deficit of women; but since, in countries where men and women receive similar care, the ratio is about 1.05, the real shortfall is about 11 percent. In China alone this amounts to 50 million "missing women," taking 1.05 as the benchmark ratio. When that number is added to those in South Asia, West Asia, and North Africa, a great many more than 100 million women are "missing." These numbers tell us, quietly, a terrible story of inequality and neglect leading to the excess mortality of women.

2.

To account for the neglect of women, two simplistic explanations have often been presented or, more often, implicitly assumed. One view emphasizes the cultural contrasts between East and West (or between the Occident and the Orient), claiming that Western civilization is less sexist than Eastern. That women outnumber men in Western countries may appear to lend support to this Kipling-like generalization. (Kipling himself was not, of course, much bothered by concerns about sexism, and even made "the twain" meet in romantically masculine circumstances: "But there is neither East nor West, Border, nor Breed, nor Birth,/When two strong men stand

face to face, tho' they come from the ends of the earth!") The other simple argument looks instead at stages of economic development, seeing the unequal nutrition and health care provided for women as a feature of underdevelopment, a characteristic of poor economies awaiting economic advancement.

There may be elements of truth in each of these explanations, but neither is very convincing as a general thesis. To some extent, the two simple explanations, in terms of "economic development" and "East–West" divisions, also tend to undermine each other. A combined cultural and economic analysis would seem to be necessary, and, I will argue, it would have to take note of many other social conditions in addition to the features identified in the simple aggregative theses.

To take the cultural view first, the East–West explanation is obviously flawed because experiences within the East and West diverge so sharply. Japan, for example, unlike most of Asia, has a ratio of women to men that is not very different from that in Europe or North America. This might suggest, at least superficially, that real income and economic development do more to explain the bias against providing women with the conditions for survival than whether the society is Western or Oriental. In the censuses of 1899 and 1908 Japan had a clear and substantial deficit of women, but by 1940 the numbers of men and women were nearly equal, and in the postwar decades, as Japan became a rich and highly industrialized country, it moved firmly in the direction of a large surplus, rather than a deficit, of women. Some countries in East Asia and Southeast Asia also provide exceptions to the deficit of women; in Thailand and Indonesia, for example, women substantially outnumber men.

In its rudimentary, undiscriminating form, the East–West explanation also fails to take into account other characteristics of these societies. For example, the ratios of women to men in South Asia are among the lowest in the world (around 0.94 in India and Bangladesh, and 0.90 in Pakistan – the lowest ratio for any large country), but that region has been among the pioneers in electing women as top political leaders. Indeed, each of the four large South Asian countries – India, Pakistan, Bangladesh, and Sri Lanka – either has had a woman as the elected head of government (Sri Lanka, India, and Pakistan), or has had women leading the main opposition parties (as in Bangladesh).

It is, of course, true that these successes in South Asia have been achieved only by upper-class women, and that having a woman head of government has not, by itself, done much for women in general in these countries. However, the point here is only to question the tendency to see the contrast between East and West as simply based on more sexism or less. The large electoral successes of women in achieving high positions in government in South Asia indicate that the analysis has to be more complex.

It is, of course, also true that these women leaders reached their powerful positions with the help of dynastic connections – Indira Gandhi was the daughter of Jawaharlal Nehru, Benazir Bhutto the daughter of Zulfikar Bhutto, and so on. But it would be absurd to overlook – just on that ground – the significance of their rise to power through popular mandate. Dynastic connections are not new in politics and are pervasive features of political succession in many countries. That Indira Gandhi derived her political strength partly from her father's position is not in itself more significant than the fact that Rajiv Gandhi's political credibility derived largely from his mother's political eminence, or the fact (perhaps less well known) that Indira Gandhi's father – the great Jawaharlal Nehru – initially rose to prominence as the son of Motilal Nehru, who had been president of the Congress party. The dynastic aspects of South Asian politics have certainly helped women to come to power through electoral support, but it is still true that so far as winning elections is concerned, South Asia would seem to be some distance ahead of the United States and most European countries when it comes to discrimination according to gender.

In this context it is useful also to compare the ratios of women in American and Indian legislatures. In the US House of Representatives the proportion of women is 6.4 percent, while in the present and the last lower houses of the Indian Parliament, women's proportions have been respectively 5.3 and 7.9 percent. Only two of the 100 US Senators are women, and this 2 percent ratio contrasts with more than 9 and 10 percent women respectively in the last and present "upper house," Rajya Sabha, in India. (In a different, but not altogether unrelated, sphere, I had a much higher proportion of tenured women colleagues when I was teaching at Delhi University than I now have at Harvard.) The cultural climate in different societies must have a clear relevance to differences between men and women – both in survival and in other ways as well – but it would be hopeless to see the divergences simply as a contrast between the sexist East and the unbiased West.

How good is the other (i.e., the purely economic) explanation for women's inequality? Certainly all the countries with large deficits of women are more or less poor, if we measure poverty by real incomes, and no sizable country with a high gross national product per head has such a deficit. There are reasons to expect a reduction of differential female mortality with economic progress. For example, the rate of maternal mortality at childbirth can be expected to decrease both with better hospital facilities and the reduction in birth rate that usually accompanies economic development.

However, in this simple form, an economic analysis does not explain very much, since many poor countries do not, in fact, have deficits of women. As was noted earlier, sub-Saharan Africa, poor and underdeveloped as it is, has a substantial excess of women. Southeast and East Asia (but not China) also differ from many other relatively poor countries in this respect, although to a lesser degree. Within India, as was noted earlier, Punjab and Haryana – among the richest and most economically advanced Indian states – have very low ratios of women to men (around 0.86), in contrast to the much poorer state of Kerala, where the ratio is greater than 1.03.

Indeed, economic development is quite often accompanied by a relative worsening in the rate of survival of women (even as life expectancy improves in absolute terms for both men and women). For example, in India the gap between the life expectancy of men and women has narrowed recently, but only after many decades when women's relative position deteriorated. There has been a steady decline in the ratio of women to men in the population, from more than 97 women to 100 men at the turn of the century (in 1901), to 93 women in 1971, and the ratio is only a little higher now. The deterioration in women's position results largely from their unequal sharing in the advantages of medical and social progress. Economic development does not invariably reduce women's disadvantages in mortality.

A significant proportional decline in the population of women occurred in China after the economic and social reforms introduced there in 1979. The Chinese Statistical Yearbooks show a steady decline in the already very low ratio of women to men in the population, from 94.32 in 1979 to 93.42 in 1985 and

1986. (It has risen since then, to 93.98 in 1989 – still lower than what it was in 1979.) Life expectancy was significantly higher for females than for males until the economic reforms, but seems to have fallen behind since then.[3] Of course, the years following the reforms were also years of great economic growth and, in many ways, of social progress, yet women's relative prospects for survival deteriorated. These and other cases show that rapid economic development may go hand in hand with worsening relative mortality of women.

3.

Despite their superficial plausibility, neither the alleged contrast between "East" and "West," nor the simple hypothesis of female deprivation as a characteristic of economic "underdevelopment" gives us anything like an adequate understanding of the geography of female deprivation in social well-being and survival. We have to examine the complex ways in which economic, social, and cultural factors can influence the regional differences.

It is certainly true that, for example, the status and power of women in the family differ greatly from one region to another, and there are good reasons to expect that these social features would be related to the economic role and independence of women. For example, employment outside the home and owning assets can both be important for women's economic independence and power; and these factors may have far-reaching effects on the divisions of benefits and chores within the family and can greatly influence what are implicitly accepted as women's "entitlements."

Indeed, men and women have both interests in common and conflicting interests that affect family decisions; and it is possible to see decision making in the family taking the form of the pursuit of co-operation in which solutions for the conflicting aspects of family life are implicitly agreed on. Such "cooperative conflicts" are a general feature of many group relations, and an analysis of cooperative conflicts can provide a useful way of understanding the influences that affect the "deal" that women get in the division of benefits within the family. There are gains to be made by men and women through following implicitly agreed-on patterns of behavior; but there are many possible agreements – some more favorable to one party than others. The choice of one such cooperative arrangement from among the range of possibilities leads to a particular distribution of joint benefits. (Elsewhere, I have tried to analyze the general nature of "cooperative conflicts" and the application of the analysis of such conflicts to family economics.[4])

Conflicts in family life are typically resolved through implicitly agreed-on patterns of behavior that may or may not be particularly egalitarian. The very nature of family living – sharing a home and experiences – requires that the elements of conflict must not be explicitly emphasized (giving persistent attention to conflicts will usually be seen as aberrant behavior); and sometimes the deprived woman would not even have a clear idea of the extent of her relative deprivation. Similarly, the perception of who is doing "productive" work, who is "contributing" how much to the family's prosperity, can be very influential, even though the underlying principles regarding how "contributions" or "productivity" are to be assessed may be rarely discussed explicitly. These issues of social perception are, I believe, of pervasive importance in gender inequality, even in the richer countries, but they can have a particularly powerful influence in sustaining female deprivation in many of the poorer countries.[5]

The division of a family's joint benefits is likely to be less unfavorable to women if (1) they can earn an outside income; (2) their work is recognized as productive (this is easier to achieve with work done outside the home); (3) they own some economic resources and have some rights to fall back on; and (4) there is a clear-headed understanding of the ways in which women are deprived and a recognition of the possibilities of changing this situation. This last category can be much influenced by education for women and by participatory political action.

Considerable empirical evidence, mostly studies of particular localities, suggests that what is usually defined as "gainful" employment (i.e., working outside the home for a wage, or in such "productive" occupations as farming), as opposed to unpaid and unhonored housework – no matter how demanding – can substantially enhance the deal that women get.[6] Indeed, "gainful" employment of women can make the solution of "cooperative conflicts" less unfavorable to women in many ways. First, outside employment for wages can provide women with an income to which they have easier access, and it can also serve as a means of making a living on which women can rely, making them less vulnerable. Second, the social respect that is associated with being a "bread winner" (and a "productive" contributor to the family's joint prosperity) can improve women's status and standing in the family, and may influence the prevailing cultural traditions regarding who gets what in the division of joint benefits. Third, when outside employment takes the form of jobs with some security and legal protection, the corresponding rights that women get can make their economic position much less vulnerable and precarious. Fourth, working outside the home also provides experience of the outside world, and this can be socially important in improving women's position within the family. In this respect outside work may be "educational" as well.

These factors may not only improve the "deal" women get in the family, they can also counter the relative neglect of girls as they grow up. Boys are preferred in many countries because they are expected to provide more economic security for their parents in old age; but the force of this bias can be weakened if women as well as men can regularly work at paid jobs. Moreover, if the status of women does in general rise and women's contributions become more recognized, female children may receive more attention. Similarly, the exposure of women to the world through work outside the home can weaken, through its educational effect, the hold of traditional beliefs and behavior.

In comparing different regions of Asia and Africa, if we try to relate the relative survival prospects of women to the "gainful employment" of both sexes – i.e., work outside the home, possibly for a wage – we do find a strong association. If the different regions of Asia and Africa (with the exception of China) are ranked according to the proportion of women in so-called gainful employment relative to the proportion of men in such employment, we get the following ranking, in descending order:[7]

1 Sub-Saharan Africa
2 Southeast and Eastern Asia
3 Western Asia
4 Southern Asia
5 Northern Africa

Ranking the ratios of life expectancy of females

to those of males produces a remarkably similar ordering:

1 Sub-Saharan Africa
2 Southeast and Eastern Asia
3 Western Asia
4 Northern Africa
5 Southern Asia

That the two rankings are much the same, except for a switch between the two lowest-ranking regions (lowest in terms of both indicators), suggests a link between employment and survival prospects. In addition to the overall correspondence between the two rankings, the particular contrasts between sub-Saharan Africa and North Africa, and that between Southern (and Western) Asia and Southeast (and Eastern) Asia are suggestive distinctions *within* Africa and Asia respectively, linking women's gainful employment and survival prospects.

It is, of course, possible that what we are seeing here is not a demonstration that gainful employment causes better survival prospects but the influence of some other factor correlated with each. In fact, on the basis of such broad relations, it is very hard to draw any firm conclusion; but evidence of similar relations can be found also in other comparisons.[8] For example, Punjab, the richest Indian state, has the lowest ratio of women to men (0.86) in India; it also has the lowest ratio of women in "gainful" employment compared to men. The influence of outside employment on women's well-being has also been documented in a number of studies of specific communities in different parts of the world.[9]

4.

The case of China deserves particular attention. It is a country with a traditional bias against women, but after the revolution the Chinese leaders did pay considerable attention to reducing inequality between men and women.[10] This was helped both by a general expansion of basic health and medical services accessible to all and by the increase in women's gainful employment, along with greater social recognition of the importance of women in the economy and the society.

There has been a remarkable general expansion of longevity, and despite the temporary setback during the terrible famines of 1958–1961 (following the disastrous failure of the so-called Great Leap Forward), the Chinese life expectancy at birth increased from the low forties around 1950 to the high sixties by the time the economic reforms were introduced in 1979. The sharp reduction in general mortality (including female mortality) is all the more remarkable in view of the fact that it took place despite deep economic problems in the form of widespread industrial inefficiency, a rather stagnant agriculture, and relatively little increase in output per head. Female death rates declined sharply – both as a part of a general mortality reduction and also relatively, vis-à-vis male mortality. Women's life expectancy at birth overtook that of men – itself much enhanced – and was significantly ahead at the time the economic and social reforms were introduced in 1979.

Those reforms immediately increased the rate of economic growth and broke the agricultural stagnation. The official figures suggest a doubling of agricultural output between 1979 and 1986 – a remarkable achievement even if some elements of exaggeration are eliminated from these figures. But at the same time, the official figures also record an *increase* in the general mortality rates after the reforms, with a consistently higher death rate than what China had achieved by 1979. There seems to be also a worsening of the relative survival of women, including a decline, discussed earlier, of the ratio of women to men in the population, which went down from 94.3 in 1979 to 93.4 in 1985 and 1986. There are problems in interpreting the available data and difficulties in arriving at firm conclusions, but the view that women's life expectancy has again become lower than that of men has gained support. For example, the World Bank's most recent *World Development Report* suggests a life expectancy of sixty-nine years for men and sixty-six years for women (even though the confounded nature of the subject is well reflected by the fact that the same *Report* also suggests an average life expectancy of seventy years for men and women put together).[11]

Why have women's survival prospects in China deteriorated, especially in relative terms, since 1979? Several experts have noted that recently Chinese leaders have tended, on the whole, to reduce the emphasis on equality for women; it is no longer much discussed, and indeed, as the sociologist Margery Wolf puts it, it is a case of a "revolution postponed."[12]

But this fact, while important, does not explain why the relative survival prospects of women would have so deteriorated during the early years of the reforms, just at the time when there was a rapid expansion of overall economic prosperity.

The compulsory measures to control the size of families which were introduced in 1979 may have been an important factor. In some parts of the country the authorities insisted on the "one-child family." This restriction, given the strong preference for boys in China, led to a neglect of girls that was often severe. Some evidence exists of female infanticide. In the early years after the reforms, infant mortality for girls appeared to increase considerably. Some estimates had suggested that the rate of female infant mortality rose from 37.7 per thousand in 1978 to 67.2 per thousand in 1984.[13] Even if this seems exaggerated in the light of later data, the survival prospects of female children clearly have been unfavorably affected by restrictions on the size of the family. Later legal concessions (including the permission to have a second child if the first one is a girl) reflect some official recognition of these problems.

A second factor relevant to the survival problems of Chinese women is the general crisis in health services since the economic reforms. As the agricultural production brigades and collectives, which had traditionally provided much of the funding for China's extensive rural health programs, were dismantled, they were replaced by the so-called "responsibility system," in which agriculture was centered in the family. Agricultural production improved, but cutbacks in communal facilities placed severe financial restrictions on China's extensive rural medical services. Communal agriculture may not have done much for agricultural production as such, but it had been a main source of support for China's innovative and extensive rural medical services. So far as gender is concerned, the effects of the reduced scope of these services are officially neutral, but in view of the pro-male bias in Chinese rural society, the cutback in medical services would have had a particularly severe impact on women and female children. (It is also the pro-male bias in the general culture that made the one-child policy, which too is neutral in form, unfavorable to female children in terms of its actual impact.)

Third, the "responsibility system" arguably has reduced women's involvement in recognized gainful employment in agriculture. In the new system's more traditional arrangement of work responsibilities, women's work in the household economy may again suffer from the lack of recognition that typically affects household work throughout the world.[14] The impact of this change on the status of women within the household may be negative, for the reasons previously described. Expanded employment opportunities for women outside agriculture in some regions may at least partially balance this effect. But the weakening of social security arrangements since the reforms would also have made old age more precarious, and since such insecurity is one of the persistent motives for families' preferring boys over girls, this change too can be contributing to the worsening of care for female children.[15]

5.

Analyses based on simple conflicts between East and West or on "under-development" clearly do not take us very far. The variables that appear important – for example, female employment or female literacy – combine both economic and cultural effects. To ascribe importance to the influence of gainful employment on women's prospects for survival may superficially look like another attempt at a simple economic explanation, but it would be a mistake to see it this way. The deeper question is why such outside employment is more prevalent in, say, sub-Saharan Africa than in North Africa, or in Southeast and Eastern Asia than in Western and Southern Asia. Here the cultural, including religious, backgrounds of the respective regions are surely important. Economic causes for women's deprivation have to be integrated with other – social and cultural – factors to give depth to the explanation.

Of course, gainful employment is not the only factor affecting women's chances of survival. Women's education and their economic rights – including property rights – may be crucial variables as well.[16] Consider the state of Kerala in India, which I mentioned earlier. It does not have a deficit of women – its ratio of women to men of more than 1.03 is closer to that of Europe (1.05) than those of China, West Asia, and India as a whole (0.94). The life expectancy of women at birth in Kerala, which had already reached sixty-eight years by the time of the last census in 1981 (and is estimated to be seventy-two years now), is considerably higher than men's

sixty-four years at that time (and sixty-seven now). While women are generally able to find "gainful employment" in Kerala – certainly much more so than in Punjab – the state is not exceptional in this regard. What is exceptional is Kerala's remarkably high literacy rate; not only is it much higher than elsewhere in India, it is also substantially higher than in China, especially for women.

Kerala's experience of state-funded expansion of basic education, which has been consolidated by left-wing state governments in recent decades, began, in fact, nearly two centuries ago, led by the rulers of the kingdoms of Travancore and Cochin. (These two native states were not part of British India; they were joined together with a small part of the old Madras presidency to form the new state of Kerala after independence.) Indeed, as early as 1817, Rani Gouri Parvathi Bai, the young queen of Travancore, issued clear instructions for public support of education:

> The state should defray the entire cost of education of its people in order that there might be no backwardness in the spread of enlightenment among them, that by diffusion of education they might be better subjects and public servants and that the reputation of the State might be advanced thereby.[17]

Moreover, in parts of Kerala, property is usually inherited through the family's female line. These factors, as well as the generally high level of communal medicine, help to explain why women in Kerala do not suffer disadvantages in obtaining the means for survival. While it would be difficult to "split up" the respective contributions made by each of these different influences, it would be a mistake not to include all these factors among the potentially interesting variables that deserve examination.

In view of the enormity of the problems of women's survival in large parts of Asia and Africa, it is surprising that these disadvantages have received such inadequate attention. The numbers of "missing women" in relation to the numbers that could be expected if men and women received similar care in health, medicine, and nutrition, are remarkably large. A great many more than a hundred million women are simply not there because women are neglected compared with men. If this situation is to be corrected by political action and public policy, the reasons why there are so many "missing" women must first be better understood. We confront here what is clearly one of the more momentous, and neglected, problems facing the world today.

NOTES

* This essay was published in 1990. Subsequent work by Dr Sen is summarized in "Missing Women," in *British Medical Journal*, March (1992) and "Missing Women Revisited," in *British Medical Journal*, December (2003).

1 An assessment of the available evidence can be found in Ingrid Waldron's "The Role of Genetic and Biological Factors in Sex Differences in Mortality," in A.D. Lopez and L.T. Ruzicka, eds., *Sex Differences in Mortality* (Canberra: Department of Demography, Australian National University, 1983). On the pervasive cultural influences on mortality and the difficulties in forming a biological view of survival advantages, see Sheila Ryan Johansson, "Mortality, Welfare and Gender: Continuity and Change in Explanations for Male/Female Mortality Differences over Three Centuries," in *Continuity and Change*, forthcoming.

2 These and related data are presented and assessed in my joint paper with Jocelyn Kynch, "Indian Women: Wellbeing and Survival," *Cambridge Journal of Economics*, Vol. 7 (1983), and in my *Commodities and Capabilities* (Amsterdam: North-Holland, 1985), Appendix B. See also Lincoln Chen et al., "Sex Bias in the Family Allocation of Food and Health Care in Rural Bangladesh," in *Population and Development Review*, Vol. 7 (1981); Barbara Miller, *The Endangered Sex: Neglect of Female Children in Rural North India* (Cornell University Press, 1981); Pranab Bardhan, *Land, Labor, and Rural Poverty* (Columbia University Press, 1984); Devaki Jain and Nirmala Benerji, eds., *Tyranny of the Household* (New Delhi: Vikas, 1985); Barbara Harriss and Elizabeth Watson, "The Sex Ratio in South Asia," in J.H. Momsen and J.G. Townsend, eds., *Geography of Gender in the Third World* (State University of New York Press, 1987); Monica Das Gupta, "Selective Discrimination against Female Children in Rural Punjab, India," in *Population and Development Review*, Vol. 13 (1987).

3 See the World Bank's *World Development Report 1990* (Oxford University Press, 1990), Table 32. See also Judith Banister, *China's Changing Population* (Stanford University Press, 1987), Chapter 4, though the change in life expectancy may not have been as large as these early estimates had suggested, as Banister herself has later noted.

4 "Gender and Cooperative Conflicts," Working Paper of the World Institute of Development Economics Research (1986), in Irene Tinker, ed., *Persistent Inequalities: Women and World Development* (Oxford University Press, 1990). In the same volume see also the papers of Ester Boserup, Hanna Papanek, and Irene Tinker on closely related subjects.

5 The recent literature on the modeling of family relations as "bargaining problems," despite being usefully suggestive and insightful, has suffered a little from giving an inadequate role to the importance of perceptions (as opposed to objectively identified interests) of the parties involved. On the relevance of perception, including perceptual distortions (a variant of what Marx had called "false perception"), in family relations, see my "Gender and Cooperative Conflicts." See also my *Resources, Values and Development* (Harvard University Press, 1984), Chapters 15 and 16; Gail Wilson, *Money in the Family* (Avebury/Gower, 1987).

6 See the case studies and the literature cited in my "Gender and Cooperative Conflicts." A pioneering study of some of these issues was provided by Ester Boserup, *Women's Role in Economic Development* (St. Martin's, 1970). See also Bina Agarwal, "Social Security and the Family," in E. Ahmad, et al., *Social Security in Developing Countries*, to be published by Oxford University Press in 1991.

7 Details can be found in my "Gender and Cooperative Conflicts."

8 For example, see Pranab Bardhan, *Land, Labor, and Rural Poverty* on different states in India and the literature cited there.

9 See the literature cited in my "Gender and Cooperative Conflicts."

10 See Elisabeth Croll, *Chinese Women Since Mao* (M.E. Sharpe, 1984).

11 See *World Development Report 1990*, Tables 1 and 32. See also Banister, *China's Changing Population*, Chapter 4, and Athar Hussain and Nicholas Stern, *On the recent increase in death rate in China*, China Paper #8 (London: STICERD/London School of Economics, 1990).

12 See Margery Wolf, *Revolution Postponed: Women in Contemporary China* (Stanford University Press, 1984).

13 See Banister, *China's Changing Population*, Table 4.12.

14 On this and related matters, see Nahid Aslanbeigui and Gale Summerfield, "The Impact of the Responsibility System on Women in Rural China: A Theoretical Application of Sen's Theory of Entitlement," in *World Development*, Vol. 17 (1989).

15 These and other aspects of the problem are discussed more extensively in my joint book with Jean Drèze, *Hunger and Public Action* (Oxford University Press, 1989).

16 For interesting investigations of the role of education, broadly defined, in influencing women's well-being in Bangladesh and India, see Martha Chen, *A Quiet Revolution: Women in Transition in Rural Bangladesh* (Schenkman Books, 1983); and Alaka Basu, *Culture, the*

Status of Women and Demographic Behavior (New Delhi: National Council of Applied Economic Research, 1988).

17 Kerala has also had considerable missionary activity in schooling (a fifth of the population is, in fact, Christian), has had international trading and political contacts (both with east and west Asia) for a very long time, and it was from Kerala that the great Hindu philosopher and educator Sankaraharya, who lived during AD788–820, had launched his big movement of setting up centers of study and worship across India.

'Conceptualising Environmental Collective Action: Why Gender Matters'

Cambridge Journal of Economics (2000)

Bina Agarwal*

Editors' Introduction

Bina Agarwal is a Professor of Economics at the Institute of Economic Growth, Delhi University. She has also taught at Michigan, Harvard and Minnesota universities in the United States. Agarwal is a previous vice president of the International Economic Association, serves on the board of the Global Development Network, and was a recipient in 2002 of the Malcolm Adhiseshiah award for distinguished contributions to development studies. In addition to being a published poet – her collection, *Monsoon Poems*, came out in 1976 – Agarwal is the author of two notable books, *Cold Hearth and Barren Slopes: The Woodfuel Crisis in the Third World* (1986) and *A Field of One's Own: Gender and Land Rights in South Asia* (1994). The second of these books is a classic of scholarship and quiet advocacy. Agarwal demonstrates how the single most important factor affecting the lives of women in South Asia is their lack of command over property, particularly arable land but also common property resources. Her work also highlights the gap that typically opens up in South Asia between the law as written, which in India grants substantial inheritance rights to women, and the operation of various social conventions which prevent women from exercising effective control over land. Low female literacy rates, male dominance in legal and administrative bodies, female seclusion practices, and patrilocal post-marital residence practices make it difficult for women to enter the public sphere on anything like an equal basis with men. As the reading published here also shows, poor women in South Asia often find it hard to mobilise collectively to manage the broader environments (including forest resources) on which so many of them depend.

  None of this means, however, that women should be painted simply as the victims of 'development' or of patriarchal structures of rule. Agarwal has worked hard in India to use her academic expertise to engage in struggles that seek the greater legal empowerment of women (especially in regard to land). In 1988 she chaired a government of India committee to make devolution rules in land tenure laws gender equal, and she was later at the forefront of a campaign that led to the passing of the Hindu Succession (Amendment) Act, 2005. As Agarwal explained in an interview in the *Indian Express* newspaper (13 September 2005), the Amendment Act proposes significant improvements in the legal rights of Hindu women in six north Indian states – Delhi, Haryana, Himachel Pradesh, Jammu and Kashmir, Punjab and Uttar Pradesh – where tenurial laws have been strongly geared to the interests of men. It now remains for the grassroots movement which spearheaded this legal challenge to mobilise women more effectively on a long-term basis to better exploit an improving legal landscape. Empowerment and development are not simply handed down from on high; they have to be struggled for over long periods and often in the teeth of well-organised resistance.

Key references

Bina Agarwal (2000) 'Conceptualising environmental collective action: why gender matters', *Cambridge Journal of Economics* 24: 283–310.
— (1986) *Cold Hearth and Barren Slopes: The Woodfuel Crisis in the Third World*, London: Zed Books.
— (1996) *A Field of One's Own: Gender and Land Rights in South Asia*, Cambridge: Cambridge University Press.
— (1994) 'Gender, resistance and land: interlinked struggles over resources and meanings in South Asia', *Journal of Peasant Studies* 22: 81–125.
— (1997) 'Environmental action, gender equity and women's participation', *Development and Change* 28 (1): 1–44.
C. Jackson (1993) 'Doing what comes naturally? Women and environment in development', *World Development* 23: 1007–22.

[. . .]

1. ASSESSING GROUP FUNCTIONING: PARTICIPATION, EQUITY AND EFFICIENCY[1]

Three important criteria for judging the performance of community institutions for environmental management would be: the extent of community participation in decision-making, equity in the distribution of costs and benefits, and efficiency in protecting and regenerating the resource. On all these counts, institutions which look successful may be found lacking from a gender perspective, as is illustrated by South Asia's experience in community forest management.

A range of community forestry groups (henceforth called CFGs) have emerged here in recent years. In India, some groups have been state-initiated, taking the form of various co-management arrangements, such as the Joint Forest Management (JFM) programme launched in 1990, which so far covers 19 states. Under it, village communities and the government share the responsibility and benefits of regenerating degraded local forests. Non-governmental organisations (NGOs) sometimes act as intermediaries and catalysts. Other groups have been initiated autonomously by a village council, youth club or village elder, and are found mainly in the eastern states of Orissa and Bihar. Yet others have a mixed history, such as the *van panchayats* or forest councils in the Uttar Pradesh (UP) hills of north-west India, created by the British in the 1930s to manage certain categories of forest. Many of the councils have

survived or been revived in recent years by NGOs or villagers. Of the groups of various origins, those initiated under the JFM programme are the most widespread, both geographically and in area covered: there are today an estimated 21,000 such groups, covering about 2.5 million hectares (or 4%) of largely degraded forest land (SPWD, 1998, p. ix). The programme is ultimately expected to include all states of India. Self-initiated autonomous groups and *van panchayats* are more regionally concentrated.

Similarly, in Nepal, under the Community forestry programme launched in 1993, the users of a given forest are constituted into forest user groups entrusted with managing and drawing benefits from that tract of State forest. Unlike most JFM groups in India, Nepal's CFGs can receive even good forest land and so far manage 15% of the country's forest area, the target being 61%. Micro-level forest management groups have also emerged elsewhere in Asia (Poffenberger *et al.*, 1997).

CFG management is through a two-tier structure: a general body of members (which can include all village households) and a smaller executive committee. The CFGs perform a range of functions: framing rules on forest use, deciding on penalties if rules are broken, resolving conflicts, organising cleaning and cutback operations, distributing forest produce or benefits thereof, and organising patrol groups or hiring watchmen.[2] Who has a voice in these bodies thus has a critical bearing on how well they function, and who gains or loses from their interventions.

In terms of immediate regeneration, many of these initiatives have done well. Sometimes replanting is undertaken, but where the rootstock is intact, restrictions on entry and protection efforts in themselves can lead to rapid natural revival. For instance, several degraded forest lands that I visited in the semi-arid zones of western India, which in the early 1990s provided little other than dry twigs and monsoon grass, have been covered with young trees within five to seven years of CFG protection. Apart from an increase in tree density, incomes are reported to have risen and biodiversity to have been enhanced.[3] Some regions also report an improvement in the land's carrying capacity, reflected in a notable rise in milch cattle numbers since protection began (Arul and Poffenberger, 1990). Several other parts of the country show an increase in earnings from the sale of items made from forest raw materials (Kant et al., 1991), and a fall in seasonal outmigration (Viegas and Menon, 1991; Chopra and Gulati, 1997).[4] A number of villages have even received awards for conservation.

Viewed from a gender perspective, however, these results look less impressive in terms of participation, the distribution of costs and benefits, and efficient functioning.

1.1 Participation

Women usually constitute less than 10% of the CFG general body membership in both India and Nepal.[5] In India's JFM programme, for instance, membership at the household level is 70–80% in many villages, and in some cases it is 100%. But eight out of 19 JFM states allow only one member per household – this is inevitably the male household head. In some states, both spouses are members, but this still excludes other household adults. Only two states allow membership to all village adults.[6] In the autonomous groups, the customary exclusion of women from village decision-making bodies has been replicated in the CFGs. But even where membership is open to women, their presence is sparse.

Women's presence on executive committees is also typically low; or there is an incongruity created by the mandatory inclusion of one or two women on the executive committee with very few women in the general body. The women so included usually constitute a nominal rather than an effective presence, since they are seldom selected or elected by village women as their representatives or for their leadership qualities. Membership apart, when women

do attend meetings, they seldom speak up, and when they do speak, their opinions are given little weight. Nepal's CFGs present a similar picture (Moffatt, 1998).

In effective terms, therefore, most CFGs in South Asia are 'men's groups' with, at best, a marginal female presence. Mixed groups with significant female presence are proportionately few and there is a small percentage of all-women's groups – an estimated 3% of all groups in Nepal (Moffatt, 1998, p. 37), and probably even less in India. These all-women CFGs are usually found in areas of high male outmigration, or where they have been especially promoted by a local NGO or donor agency. They typically control very small plots of mostly barren land, while male-controlled CFGs receive the larger and better forest areas.

Despite their virtual absence from male-controlled CFGs, women often play an active role in the protection efforts, keeping an informal lookout or forming patrol groups parallel to men's, because they feel men's patrolling is ineffective. In almost all the villages I visited, many women recounted cases of apprehending intruders, persuading any women they saw breaking rules to desist, fighting forest fires alongside men (or even in men's absence), and so on. Women's limited participation in decision-making, however, means that they have little say in the framing of rules on forest use, monitoring, benefit distribution, etc., with implications for both distributional equity and efficiency.

1.2 Distributional equity

Gender inequities characterise CFGs in the sharing of both costs and benefits. While costs associated with membership fees, patrolling time or the forest guard's pay are usually borne by men, the costs of forgoing forest use are largely borne by women. This includes time spent in searching for alternative sites for firewood and fodder, using inferior substitute fuels, stallfeeding animals, losing income earlier obtained from selling forest products, and so on.

Of the 87 CFGs I interviewed on my 1998–99 field visits, for instance, 52% have banned firewood collection. About half of these do not open the forest at all, and the rest allow restricted collection for a few days a year. Where previously women could fulfil at least a part of their needs from the protected area, they are now forced to travel to neighbouring sites, involving additional time, energy and the risk of being treated as intruders.[7]

In some sites in the Indian states of Gujarat (West India) and West Bengal (east India), when protection started, women's collection time increased from 1–2 hours to 4–5 hours for a headload of firewood, and journeys of half a kilometre lengthened to 8–9 kilometres (Sarin, 1995; my fieldwork in 1993, 1995). Sometimes, mothers seek help from school-going daughters, with negative effects on the daughters' education.

Where possible, women shift to substitute fuels: twigs, dung cakes, agricultural waste, etc. These require extra time to ignite or to keep alight, and constant tending. Some economise on fuel by heating bath water in winter only for their husbands and not for themselves, eating cold leftovers, and so on. Many in the poorest households are compelled to steal from their own or neighbouring tracts of protected forest, and risk being caught and fined. As some poor, low-caste women in the UP hills told me in 1998: 'We don't know in the morning how we will cook at night.'

Over time, these hardships have at best been alleviated in some areas; rarely have they been eliminated. Firewood shortages continue to be reported even 8 or 9 years after protection began in many of the villages I visited across several states, including in 18 of the 19 Gujarat sites. Some existing estimates suggest that several times more can be extracted sustainably than is currently being allowed (Shah, 1997). The persistent shortages women face in these contexts thus appear to have more to do with their lack of voice and bargaining power in the CFGs, than from a lack of aggregate availability.

Inequities also stem from the distribution of benefits from protection. In some cases the benefits are not distributed at all but put into a collective fund and used by the groups as they see fit. A number of the autonomous groups in Orissa (east India) managed by all-male youth clubs, for instance, have been selling forest products, including the wood obtained from thinning operations, and using the proceeds for religious festivities, a club house, or club functions. In many poor households that cannot afford to buy firewood and other products (which they had earlier collected free), the burden of finding alternative collection sites, or doing without, again falls mostly on women.

Where the CFGs distribute the benefits, women of non-member households receive none, since entitlements are linked to membership. But even in member households it is men who usually receive the benefits directly, either because they alone are members, or because distribution is on a household basis, so that despite both spouses being members they get only one share, which the man receives. Women might gain indirectly if the benefits are in kind (say as firewood), but if they are in cash, money distributed to male members is seldom shared equitably within the family. In many cases, the men have spent the money on gambling, liquor or personal items (Guhathakurta and Bhatia, 1992). This is in keeping with the pattern also noted among poor households outside the context of forest management, where women are found to spend most of the income they control on the family's (especially the children's) basic needs and men are found to spend a significant part of their income on personal consumption.[8] In the absence of direct claims to CFG benefits, both women's and children's welfare can thus be affected adversely. Not surprisingly, in a meeting of three JFM villages of West Bengal, women, when asked about benefit-sharing, all wanted equal and separate shares for husbands and wives (Sarin, 1995). Similarly, in a number of Gujarat villages I found that attempts to enlist more women members into CFGs were proving unsuccessful, since the women were demanding their own share in the benefits as a condition for joining, while existing CFG rules allowed only one share per household.

Thus, many cases, that an ungendered evaluation would deem success stories of participative community involvement in resource regeneration, are found to be largely non-participative and inequitable from a gender perspective.

Of course, a lack of participation in CFG decision-making is not the only cause of gender-unequal sharing in costs and benefits. A number of other factors would also impinge on this, including the pre-existing gender division of labour and the initial resource endowments (such as land and assets) that women or their households possess. For instance, since the main responsibility for firewood and fodder collection, animal care, cooking, etc. falls on women, they also end up shouldering the burden of finding other fuel and fodder sources when the forest is closed. Again, women who neither themselves own land or trees, nor belong to households that do, bear the biggest costs of forest closure. Such inequalities are often sharp. Briscoe's (1979) village study in Bangladesh is indicative: he found that 89% of all fruit and fodder trees were owned by 16% of households, which also owned 55% of the cropped area and 46% of the cattle. Over and above these considerations, however, women's absence from the decision-making forums of the CFGs makes a critical difference to gender distribution since that is where the rules on cost and benefit sharing are made.

Moreover, women's absence from the CFGs can indirectly affect intrahousehold benefit sharing in so far as relative contributions affect *perceptions* about claims (Agarwal, 1997B; A. K. Sen, 1990). Women and girls who were seen to be contributing to such activity would be better placed to claim benefits.

1.3 Efficiency

Women's lack of participation in CFG functioning can also have adverse implications for efficiency and sustainability. At least three types of inefficiency could arise. One, some initiatives may fail to take off. Two, those that do take off (such as the cases of successful regeneration cited above) may show efficiency gains in the short run, but be unsustainable in the long run. Three, and relatedly, there may be a significant gap between the efficiency gains realised and those potentially realisable (in terms of resource productivity and diversity, satisfying household needs, enhancing incomes, stemming outmigration, etc.). These inefficiencies could arise from one or more of the following problems, some of which have already surfaced, and others may be anticipated.

First, there are *rule enforcement problems*. Since (as noted) it is women who regularly have to collect firewood, grasses, and non-timber forest products, their lack of involvement in framing workable rules for protection and use creates tendencies to circumvent the rules. In almost all the villages I visited, there were at least a few cases of violation. Violations by men are usually for timber (for self-use or sale) but violations by women are typically for firewood, especially if they are poor and landless. In Agrawal's (1999) study of a *van panchayat*, 70–80% of the reported violations were by women (either from the same or nearby villages), most of whom appear to have been poor and low caste. In some Orissa villages that I visited, women found the forest closure rules formulated by the all-male committee so strict (the forest was not opened even for a few days per year) that they finally took up a separate patch for protection. In most regions, however, women lack this alternative. Many express deep resentment at the unfairness of existing rules.[9]

If consulted, women usually suggest less stringent and more egalitarian rules (my fieldwork in 1998–99). A women's group in the UP hills recognising that 'the male members of the forest committee have difficulties implementing the rules', persuasively argued that if the men were to discuss the problems with the women, more 'mid-way' rules could be devised which would prove more effective and viable in the long run (Britt, 1993, p. 148).

A second source of inefficiency lies in *information flow imperfections* along gender lines, both within and outside households. Information about the rules framed, or changes in rules, such as in membership eligibility conditions or on other aspects of forest management, do not always filter down to the women, nor is there any inbuilt mechanism for their feedback. In a study of two West Bengal villages, only a small percentage of women in one village and none in the other had been consulted before CFG formation, or were aware of the role of members within the CFG (Sarin, 1998, p. 40). Similarly, male officials seldom consult the women when preparing village micro-plans for forest development, or at best do so at the very early stages and without a follow-up. Some women hear about the plans through their husbands, others not at all.[10] These communication problems can prove particularly acute in regions of high male outmigration.

Thirdly, efficiency issues can arise from *inaccurate assessments of resource depletion*. For example, there can be gender differences in abilities to identify the state of the local resource base. During my field visit to Gujarat in 1995, a woman's informal forest patrol group took me to their patrol site, and pointing out the illegal cuttings that the men had missed, noted: 'Men don't check carefully for illegal cuttings. Women keep a more careful look-out.' Part of this gender difference arises because women, as the most frequent collectors of forest products, are more familiar with the forest than men who use the forest sporadically. Culling information on the frequency of fuelwood collection from 13 regions in six states in India, I found that in nine regions women collected daily, and in four others once every two to four days (Agarwal, 1997A, p. 12).

Fourthly, and relatedly, inefficiencies arise due to *problems in catching transgressors*. Where protection is informal, women, given their greater contact with the forest, are more likely than men to spot transgressors. But even where formal patrol groups exist, all-male patrols or male guards face cultural constraints in physically catching women intruders. Sometimes, usually where the intruders are from another village, these women's families even threaten to register police cases against members of the patrol.

Where women voluntarily take up patrolling by forming informal groups to supplement men's efforts, it can significantly improve protection. In Sharma and Sinha's (1993) study of 12 *van panchayats*, all the four that they deem 'robust' and successful have active women's associations. They note (1993, p. 173): 'If the condition of the forests has improved in recent years, much of the credit goes to these women's associations.' Even though these associations have no formal authority for forest protection, they spread awareness among women of the need to conserve forests, monitor forest use, and exert social pressure on women who violate usage rules. Pandey (1990, p. 30) similarly observes from her Nepal study: 'Without [women's] genuine support in this venture, an unfenced forest, located in the middle of four villages and containing such favoured species could not have existed for nearly a decade without a watchman.'

However, women's informal groups lack the authority to penalise offenders, who must be reported to the formal (typically all-male) committees. This bifurcation of authority and responsibility along gender lines systematically disadvantages women, while increasing their work burden, and is likely to prove less efficient than where responsibility and authority coincide. In a number of cases, I found that women had abandoned their informal efforts because men's committees or male forest officials had time and again failed to take any action against those whom the women apprehended.

Fifthly, and relatedly, the effective *conflict resolution* that is necessary for efficient functioning is made problematic with women's virtual exclusion from the formal committees. For instance, when women catch intruders, they are seldom party to discussions or decisions on appropriate sanctions. Women often get excluded from conflict resolution meetings even when the dispute directly involves them.[11] Where they

are called, and the conflict involves men, they often feel the settlements are male-biased (Roy *et al.*, 1993).

A sixth form of inefficiency stems from the *non-incorporation of women's specific knowledge of species-varieties*. While there is little to support the romanticised view (e.g., Shiva, 1988) that women are the main repositories of environmental knowledge, there is evidence that women and men are often privy to different types of knowledge. This difference arises from the gender division of labour, and gender differences in spatial mobility and age. Where women are the main seed selectors and preservers, they are substantially better informed about seed varieties than are the men (Burling, 1963; Acharya and Bennett, 1981). Similarly, women as the main fuel and fodder collectors can often explain the attributes of trees (growth rates, quality of fuelwood, medicinal and other uses, etc.) better than the men (Pandey, 1990), or can identify a large number of trees, shrubs and grasses in the vicinity of fields and pastures (Chen, 1993). Knowledge of medicinal herbs is similarly use-related and gender-specific.[12] Gender-differentiated knowledge can also result from differences in male-female spatial domains: men are often better informed about species found in distant areas, and women about the local environment where they collect (Jewitt, 1996; Gaul, 1994).

The systematic exclusion of one gender from consultation, decision-making, and management of new planting programmes is thus likely to have negative efficiency implications, by failing to tap either women's knowledge of diverse species for enhancing biodiversity, or their understanding of traditional silvicultural practices when planting species they are better informed about. Some NGOs have recognised the potential of women's specific knowledge and tapped it for promoting medicinal herbs in the protected areas. There are also examples of women's groups resisting male pressure for planting the commercially profitable Eucalyptus and instead selecting diverse species, using their substantial knowledge of local trees and shrubs (Sarin and Khanna, 1993). But such examples are rare.

A seventh form of inefficiency can arise from ignoring possible *gender differences in preferences for trees and plants*. Women often prefer trees which have more domestic use value (as for fuel and fodder), or which provide shade for children grazing the animals, while men more typically opt for trees that bring in cash.[13] The exceptions are cases where existing forests provide adequate fuel and fodder and where women too may choose commercial species for new planting (Chen, 1993). Women's greater involvement in forest development would ensure that forest microplans provide for a larger portion of household needs, thus enhancing their commitment to the initiative.

Basically, assessments of environmental initiatives in terms of participation, distributional equity as well as efficiency can all prove inaccurate if gender differences are ignored. Ignoring gender also violates several of the conditions deemed by many scholars as necessary for building successful and enduring institutions for managing common pool resources (see discussion in Baland and Platteau, 1996). These include conditions such as:

'Most individuals affected by the operational rules . . . participate in modifying [them]' (Ostrom, 1990, p. 90); and the rules are kept simple and fair (McKean, 1992).

'Monitors, who actively audit [common pool resource] conditions and appropriator behaviour, are accountable to the appropriators or are the appropriators' (Ostrom, 1990, p. 90).

'Appropriators who violate operational rules . . . [are] assessed graduated sanctions . . . by other appropriators, by officials accountable to these appropriators, or by both' (Ostrom, 1990, p. 90).

There are effective mechanisms for resolving conflicts between parties, and arrangements for discussing problems (Ostrom, 1990; Wade, 1988).

The first condition is violated by excluding women from the process of framing and modifying rules, and by framing rules that are unfair to women and resented by them. The second and third conditions are violated in that (i) men who monitor formally are usually not accountable to female appropriators, (ii) women who monitor informally are often not accountable to the formal committees, and (iii) women as appropriators or monitors are excluded from decisions on sanctions imposed by the formal committees. The fourth condition is violated where women are excluded from conflict resolution discussions.

[. . .]

NOTES

* University of Delhi. I am most grateful to Paul Seabright, Janet Seiz and Nancy Folbre for their comments on an earlier draft.

1 See also Agarwal (1997A, 2000).

2 In the case of JFM, most of these functions are undertaken jointly with a forest department official.

3 Raju *et al.* (1993), Arul and Poffenberger (1990). My field visit to Gujarat in 1995 also confirmed this.

4 See also Raju *et al.* (1993) and SPWD (1994) for documentation on returns from CFG protection, in various regions.

5 See Roy *et al.* (*c.* 1992), Guhathakurta and Bhatia (1992), and Narain (1994) on JFM; Kant *et al.* (1991), Singh and Kumar (1993) on India's autonomous groups; Ballabh and Singh (1988), Sharma and Sinha (1993) on *van panchayats*; and Moffatt (1998) for Nepal.

6 See Agarwal (2000) for details of JFM membership conditions for the general bodies and executive committees in different states.

7 See Sarin (1995) and Agarwal (1997A).

8 See Mencher (1988) and Noponen (1991) for India. See also Blumberg (1991) for some other countries.

9 See also Shah and Shah (1995), Singh and Kumar (1993) and Agarwal (1997A).

10 See, for example, Guhathakurta and Bhatia (1992), Singh (1997) and Correa (1995).

11 My field visits, 1998–99; see also Sarin (1995), Nightingale (1998).

12 My fieldwork, 1993; see also Gaul (1994), Jewitt (1996), Kelkar and Nathan (1991).

13 See Agarwal (1992), Brara (1987) and Sarin and Khanna (1993).

BIBLIOGRAPHY

Acharya, M. and Bennett, L. 1981. *An Aggregate Analysis and Summary of Village Studies, The Status of Women in Nepal*, II, Part 9, Kathmandu, CEDA, Tribhuvan University

Agarwal, B. 1992. The gender and environment debate: lessons from India, *Feminist Studies*, vol. 18, no. 1, 119–58.

Agarwal, B. 1997A. Environmental action, gender equity and women's participation, *Development and Change*, vol. 28, no. 1, 1–44

Agarwal, B. 1997B. 'Bargaining' and gender relations: within and beyond the household, *Feminist Economics*, vol 1, no. 5, 1–51

Agarwal, B. 2000. 'Group Functioning and Community Forestry in South Asia: A Gender Analysis and Conceptual Framework, WIDER Working Paper, World Institute for Development Economics Research, Helsinki

Agrawal, A. 1999. State formation in community spaces: control over forests in the Kumaon Himalaya, India, paper prepared for presentation at the University of California, Berkeley, Workshop on Environmental Politics, 30 April

Arul, N.J. and Poffenberger, M. 1990. FPC case studies, pp. 13–25 in Pathan, R. S., Arul, N. J. and Poffenberger, M. (eds), *Forest Protection Committees in Gujarat: Joint Management Initiative*, Working Paper no. 7, Ford Foundation, New Delhi

Baland, J. M. and Platteau, J. P. 1996. *Halting Degradation of Natural Resources: Is there a Role for Rural Communities?* Oxford, Clarendon Press

Ballabh, V. and Singh, K. 1988. 'Van (Forest) Panchayats in Uttar Pradesh Hills: A Critical Analysis', Research Paper, Institute for Rural Management, Anand

Blumberg, R. L. 1991. Income under female vs. male control: hypotheses from a theory of gender stratification and data from the Third World, pp. 97–127 in Blumberg, R. L. (ed.), *Gender, Family and Economy: The Triple Overlap*, Newbury Park, Sage

Brara, R. 1987. 'Shifting Sands: A Study of Right in Common Pastures', Report, Institute of Development Studies, Jaipur

Briscoe, J. 1979. Energy use and social structure in a Bangladeshi Village, *Population and Development Review*, vol. 5, no. 4, 615–41

Britt, C. 1993. 'Out of the Wood? Local Institutions and Community Forest Management in two Central Himalayan Villages', draft monograph, Cornell University, Ithaca

Burling, R. 1963. *Rengsanggri: Family and Kinship in a Garo Village*, Philadelphia, University of Pennsylvania Press

Chen, M. 1993. Women and wasteland development in India: an issue paper, pp. 21–90 in Singh, A. and Burra, N. (eds), *Women and Wasteland Development in India*, Delhi, Sage

Chopra, K. and Gulati, S. C. 1997. Environmental degradation and population movements: the role of property rights, *Environment and Resource Economics*, vol. 9, 383–408

Correa, M. 1995. *Gender and Joint Forest Planning and Management: A Research Study in Uttara Kannada District, Karnataka*, Dharwad, Karnataka, Indian Development Service

Gaul, K. K. 1994. 'Negotiated Positions and Shifting Terrains: Apprehension of Forest Resources in the Western Himalaya', Doctoral Dissertation, Department of Anthropology, University of Massachusetts, Amherst

Guhathakurta, P. and Bhatia, K. S. 1992. *A Case Study on Gender and Forest Resources in West Bengal*, Delhi, World Bank

Jewitt, S. 1996. 'Agro-ecological Knowledges and Forest Management in the Jharkhand, India: Tribal Develop-ment or Populist Impasse', PhD Dissertation, Department of Geography, University of Cambridge

Kant, S., Singh, N. M. and Singh, K. K. 1991. *Community-based Forest Management Systems (Case Studies from Orissa)*, SIDA, New Delhi; Indian Institute of Forest Management, Bhopal; and ISO/Swedforest, New Delhi

Kelkar, G. and Nathan, D. 1991. *Gender and Tribe: Women, Land and Forests in Jharkhand*, London, Zed Books.

McKean, M. A., 1992. Management of traditional common lands (*Iriaichi*) in Japan, pp. 63–98 in Bromley, D. W. (ed.), *Making the Commons Work: Theory, Practice and Policy*, San Francisco, Institute for Contemporary Studies Press

Mencher, J. 1988. Women's work and poverty: women's contribution to household maintenance in two regions of South India, pp. 99–119 in Dwyer, D. and Bruce, J. (eds), *A Home Divided: Women and Income Control in the Third World*, Stanford, Stanford University Press

Moffatt, M. 1998. 'A Gender Analysis of Community Forestry and Community Leasehold Forestry in Nepal with a Macro-Meso-Micro Framework', MA Dissertation in Development Policy Analysis, Department of Economics and Social Studies, University of Manchester

Narain, U. 1994. 'Women's Involvement in Joint Forest Management: Analyzing the Issues', draft paper, 6 May

Nightingale, A. 1998. 'Inequalities in the Commons: Gender, Class and Caste in Common Property Regimes: A Case from Nepal', MacArthur Consortium Working Paper Series, Institute of International Studies, University of Minnesota

Noponen, H. 1991. The dynamics of work and survival for the urban poor: a gender analysis of panel data from Madras, *Development and Change*, vol. 22, no. 2, 233–60

Ostrom, E. 1990. *Governing the Commons*, Cambridge, Cambridge University Press

Pandey, S. 1990. 'Women in Hattidunde Forest Management in Dhading District, Nepal', MPE Series no. 9, Kathmandu, Nepal, International Center for Integrated Mountain Development (ICIMOD),

Poffenberger, M., Walpole, P., D'Silva, E., Lawrence, K. and Khare, A. 1997. 'Linking Government Policies and Programs with Community Resource Management Systems: What is Working and What is Not', Research Network Report no. 9, A Synthesis Report of the Fifth Asia Forest Network Meeting held at Surajkund, India, December 1996

Raju, G., Vaghela, R. and Raju, M. S. 1993. *Development of People's Institutions for Management of Forests*, Ahemdabad, VIKSAT

Roy, S. B., Mukherjee, R., Roy, D. S., Bhattacharya, P. and Bhadra, R. K. 1993. 'Profile of Forest Protection Committees at Sarugarh Range, North Bengal', Working Paper no. 16, IBRAD, Calcutta

Roy, S. B., Mukerjee, R. and Chatterjee, M. *c*.1992. 'Endogenous Development, Gender Role in Participatory Forest Management', IBRAD, Calcutta

Sarin, M. 1995. Regenerating India's forest: reconciling gender equity and joint forest management, *IDS Bulletin*, vol. 26, no. 1, 83–91

Sarin, M. 1998. *Who is Gaining? Who is Losing? Gender and Equality Concerns in Joint Forest Management*, New Delhi, Society for Promotion of Wasteland Development

Sarin, M. and Khanna, R. 1993. Women organize for wasteland development: a case study of SARTHI in Gujarat, pp. 91–127 in Singh, A. and Burra, N. (eds), *Women and Wasteland Development in India*, New Delhi, Sage

Sen, A. K. 1990. Gender and cooperative conflicts, pp. 123–49 in Tinker, I. (ed.), *Persistent Inequalities: Women and World Development*, New York, Oxford University Press

Shah, A. 1997. Jurisdiction versus equity: tale of two villages, *Wastelands News*, February–April, 58–63

Shah, M. K. and Shah, P. 1995. Gender, environment and livelihood security: an alternative viewpoint from India, *IDS Bulletin*, vol. 26, no. 1, 75–82.

Sharma, A. and Sinha, A. 1993. 'A Study of the Common Property Resources in the Project Area of the Central Himalaya Rural Action Group', mimeo, Indian Institute of Forest Management, Bhopal, Madhya Pradesh

Shiva, V. 1988. *Staying Alive: Women, Ecology and Survival*, London, Zed Books

Singh, M. 1997. 'Lumping and Levelling: Gender Stereotypes and Joint Forest Management', paper presented at the seminar on the Social Construction of Community Participation in Joint Forest Management, organised by the University of Edinburgh and Indian Council of Forestry Research and Education, India International Centre, 9–11 April 1997

Singh, N. and Kumar, K. 1993. 'Community Initiatives to Protect and Manage Forests in Balangir and Sambalpur Districts', SIDA, New Delhi

SPWD 1994. *Joint Forest Management Update, 1993*, New Delhi, Society for Promotion of Wastelands Development

SPWD 1998. *Joint Forest Management Update, 1997*, New Delhi, Society for the Promotion of Wastelands Development

Viegas, P. and Menon, G. 1991. 'Forest Protection Committees of West Bengal: Role and Participation of Women', paper prepared for the ILO Workshop on Women and Wasteland Development, International Labour Organization, New Delhi, 9–11 January

Wade, R. 1988. *Village Republics: Economic Conditions for Collective Action in South India*, Cambridge, Cambridge University Press

'AIDS, Gender, and Sexuality during Africa's Economic Crisis'

from Gwendolyn Mikell (ed.), *African Feminism: The Politics of Survival in Sub-Saharan Africa* (1997)

Brooke Grundfest Schoepf

Editors' Introduction

Brooke Grundfest Schoepf is an economic and medical anthropologist, currently Senior Fellow at the Institute for Health and Social Justice and lecturer in the Department of Social Medicine at Harvard University Medical School. Schoepf obtained a doctorate at Columbia University in 1969. She spent the next several years researching in the United States, France and England, teaching at the University of Connecticut Medical School. After spending four years at the Tuskegee Institute, Schoepf went on a Fulbright Senior Scholar's Research Award to Zimbabwe in 1983. She has since had many years of experience across eleven African countries, engaged in teaching, research and training. Schoepf's engagement with Zaire or DRC began in 1974, when she taught economic and medical anthropology at the National University in Lubumbashi. Her research was then in the area of agricultural and ecological change. Between 1985 and 1990, Schoepf led the CONNAISSIDA Project, a collaborative research group conducting ethnographic and action research on AIDS prevention, in Zaire. Following this, she served as an AIDS policy and planning consultant to various universities, NGOs, UNICEF, UNESCO and other international organisations. In 1994 Schoepf commenced research on the Rwandan genocide and its aftermath in Rwanda, and on the Kivu frontier. She has been a Visiting Professor at the National University of Rwanda. Schoepf has also been involved in research on the interface of human rights and humanitarian assistance in the protection of refugee children in Guinea. This is part of a larger project on forced migration in West Africa.

In the reading that follows, Schoepf offers clear insight into the dynamics of sexuality, economic crisis and the spread and prevention of HIV/AIDS, based on the CONNAISSIDA study. CONNAISSIDA (meaning *connaître*, 'to know' + *SIDA*, or AIDS) takes an ethnographic action-research approach, locating people's understandings in their structured vulnerabilities as they are shaped by wider political economic and ideological pressures. Schoepf's argument is similar to Elson's insight about women's vulnerability to structural adjustment. However, by focusing on one context in some depth, Schoepf is able to attend to the historical trajectory of Zaire's crisis and its gendered implications, and also to representations of sex, gender and AIDS as they structure risks for women.

Key references

Brooke Grundfest Schoepf (1997) 'AIDS, gender and sexuality during Africa's economic crisis', in G. Mikell (ed.)
    African Feminism: The Politics of Survival in Sub-Saharan Africa, Philadelphia: University of Pennsylvania Press.
— (2001) 'International AIDS research in anthropology: taking a critical perspective on the crisis', *Annual Review of Anthropology*, 30: 335–61.
— (2002) ' "Mobutu's disease": a social history of AIDS in Kinshasa', *Review of African Political Economy*, 29 (93–4): 561–73.
— (1986) 'Primary health care in Zaïre', *Review of African Political Economy*, Summer: 54–8.
Paul Farmer (1993) *AIDS and Accusation: Haiti and the Geography of Blame*, Berkeley: University of California Press.

INTRODUCTION

AIDS has spread rapidly across the globe, with cases reported in 162 countries, including 47 in Africa. Twelve million Africans are estimated to have been infected since the start of the pandemic through 1994.[1] That number may double by the year 2000 as infection continues to spread. The human immuno-deficiency virus (HIV) that causes AIDS is transmitted from infected persons by sexual intercourse, blood, and from mother to infant during pregnancy, birth, and lactation. In Africa, where heterosexual transmission accounts for more than 80 percent of infections, women outnumber men among both HIV-infected (seropositive) persons and identified AIDS cases.[2] Ten to 30 percent of sexually active adults in major cities are HIV-infected. Prevalence is high in some rural areas as well, as the virus reaches new populations through trade, tourism, migration, and war. Because years can elapse between infection and onset of disease, many people who live with the virus look healthy and are unaware that they may transmit the virus to others.

Most of those infected are expected to progress to fatal disease eventually. AIDS is the leading cause of adult deaths in high-prevalence areas, exceeding even pregnancy-related morbidity in women. The future health and survival of many millions already are compromised. AIDS not only strains inadequate health resources and adds to the hardships of families. Its economic, psychological, and socio-political impacts will be felt increasingly throughout African societies.

Since development of a cure or vaccine will take many years, even decades, reduction of sexual risk, especially through regular condom use, is needed to limit the epidemic. Information can raise awareness but seldom leads to widespread change in complexly motivated social behaviors. Propelled by erotic desire, culturally constructed, freighted with moral values, and often silenced, sexual relations are among the most complex. Therefore, mass-media campaigns need to do more than transmit information. They need to use imaginative dramatic scenarios based on people's lived experience, and to model behavior changes that lead to successful prevention. In addition, interactive, socially empowering, community-based risk-reduction interventions are needed to enable people to decide upon changes, support one another, and, in effect, change their culture.

Links to sex, reproduction, and death endow AIDS with extraordinary symbolic power. Deep, contextualized knowledge about beliefs and meanings of AIDS, and about the motivations, social pressures, and economic circumstances surrounding sexuality and health, is needed for effective prevention campaigns. The burgeoning epidemic challenges social scientists to link theory and basic research with the search for ways to enable people to avoid HIV infection.

This chapter draws on findings of the trans-disciplinary CONNAISSIDA Project in Zaire, which investigated popular representations and responses to AIDS. The name CONNAISSIDA, formed from the French word *connaître* (to know) and the acronym *SIDA* (AIDS), stands for "meaning of AIDS." We used it to encompass our own understandings and those of our informants, linked in a dialogic methodology (see Schoepf 1993a). Grounded in medical and economic anthropology, the project incorporated understandings from several other fields, including social psychology, public health, and development studies.

From February 1985 through June 1990, CON-NAISSIDA researchers conducted more than 1,800 open-ended interviews, mainly in Kinshasa and Lubumbashi. Interviews with many individuals and groups were repeated over time and supplemented by participant-observation in several popular neighborhoods and elite networks and also by collection of life-history narratives.[3] Topics ranged widely as people variously situated socially were asked what they knew about AIDS, what problems they saw in ensuring their own protection and that of persons close to them, and, given their understanding of their culture, how obstacles might be overcome. Results were used to design community-based education using participatory-empowerment methods based on group dynamics.[4] Linking macrolevel political economy to microlevel ethnography illuminates women's risk. It shows how poverty, inequality, and gendered perceptions of AIDS hamper prevention. Many findings have been replicated by research undertaken elsewhere in sub-Saharan Africa.[5]

POLITICAL ECONOMY IN CRISIS

Disease epidemics often appear in conjunction with economic and political crisis. AIDS has emerged and spread in Africa during two decades of deepening crisis that has roots in distorted political economies and policies inherited from the colonial period. Following independence, few countries invested substantially in peasant farming. Most agricultural investment benefited the owners of large plantations, including local elites and multinational corporations. Low productivity of agricultural labor, relatively low producer prices, extensive privatization, and population pressure on remaining arable land contributed to a decline in peasant farming systems and led to an exodus to cities already crowded with unemployed. At the international level, oil price increases, neo-colonial investment policies, and declining terms of trade for commodities exported to world markets were accompanied by the appropriation of public resources by African ruling classes. This continuing capital drain to the developed countries was accompanied by increasing internal disparities in wealth and power.

In some areas, male labor migration in search of wages intensified, separating families and delaying marriage for many youth. In other areas prolonged low-intensity wars, civil disturbance, droughts, and insect plagues uprooted populations and caused hunger, disease, death, and despair. In the 1980s, debt-service payments and structural-adjustment measures imposed by Western creditors as a condition for further borrowing brought still more intense hardships to the poor and middle classes. Results included soaring food prices, collapse of social infrastructure (particularly health services and education), family disruption, and increased malnutrition and sickness in both cities and rural areas.

Zaire's political economy, closely linked to mineral exports and to the strategic concerns of the Cold War, was one of the first to be rocked by the shock waves of the world economic crisis that began in 1973. The Mobutu regime looted the economy, channeling public resources into private capital funds. Despite rich resource endowment, per capita incomes are among the world's lowest, and many of the poor live in absolute misery, while a small group has grown wealthy from control of the state. This "political-commercial bourgeoisie" maintained political control through corruption, violence, and massive foreign support. Dwindling external support for the regime, riots and looting by the soldiery, and politically orchestrated ethnic violence have resulted in successive waves of economic decline. These conditions underlie the current crisis in state–society relations. Zaire arguably constitutes a worst-case scenario. Nevertheless, the crisis and its underlying causes are found across the continent.

CRISIS AND GENDER

Macrolevel crisis creates conditions for microlevel dislocation. For example, men whose incomes are low and uncertain, or who become unemployed and hopeless, are frequently unwilling to assume responsibility for children. In families suffering from material want and psychological stress, alcoholism and violence increase. Marriage ties, already tenuous for many, become more fragile. The feminization of poverty, observed throughout the world, is not well documented for Zaire. Nevertheless, several small-scale studies have found that economic crisis is experienced most severely by poor women and their children. Many youth growing up without education, skills, or job prospects face bleak futures and can only live for the moment. They become available for hire in illegal enterprises and armed gangs sponsored by repressive regimes. Poverty and violence propel the epidemic spread of HIV and other sexually transmitted diseases (STDs), tuberculosis, cholera, and most recently, Ebola fever.

Most women still shoulder traditional responsibilities for providing food and other household necessities. However, in both rural and urban areas, many now do so without the traditional role-complementarity provided by husbands and lineage members. Male dominance in family and community is not simply inherited from traditional social organization. Culturally constructed gender relations varied widely in the region. In many precolonial societies, women held important religious and political offices, including village headships and chiefships. In other societies, women's membership in corporate kin groups protected their access to resources, while collective retaliation sanctioned men who abused their power. Colonial institutions – the "trinity" of state, church, and employers – altered the balance of forces. Elder men acquired new power over women and youth, which in many areas was far in excess of their former status. For example, cash cropping provided incentives and opportunities for elder men to take numerous wives who were excluded from the proceeds of their own and children's labor. Discourses of "tradition" and, later, of "authenticity" were invented to aid efforts to control women's labor and sexuality.

In the wake of structural adjustment, the state placed new demands for revenue on local communities. These, in turn, have responded by increased taxation and users' fees. In families hard pressed to sell more produce, women's labor is most easily harnessed, for they are seldom able to refuse patriarchal authority. Increasing numbers of young women have sought escape from rural drudgery by migrating to the city. Sex ratios in most large cities of Zaire are virtually equal, but waged jobs are segregated by gender and educational attainment, which is also gender stratified. Since families perceive that men are more likely to obtain employment, sons are given preference. Although sex discrimination in employment is illegal, males, whose labor is abundant and cheap, are preferred by employers. Women make up about 4 percent of the formal-sector workforce. Outside of agriculture, where women find seasonal employment on large farms and plantations at extremely low pay, few waged jobs are available to women without secondary-school diplomas. Nevertheless, because they must provide their own support and that of dependents, women without capital resort to casual employment. In the cities they work chiefly as low-paid labor – as housemaids, nannies, traders, seamstresses, cooks, and hairdressers (often for women already established in commerce or the professions) – or as barmaids and "sex workers."

Most prosperous entrepreneurs are men who control capital and other scarce resources, generally through privileged links to the state. Less likely than men to benefit from such relations, many women embark upon microenterprises such as food processing, petty trade, sewing, and market gardening. Easy entry to the sector for women with little capital carries with it ease of exit. Competition among the self-employed is intense. While some manage to succeed, most income-generating occupations provide only bare subsistence, and many women remain impoverished despite long hours of arduous labor. A 1987–89 study of household budgets in Kinshasa found the situation of poor and middle-class women had deteriorated in recent years, despite their trading activities. When their income-generating efforts fail, women may supplement inadequate incomes by trading sexual services. Some seek "spare tires" to help meet immediate cash needs, such as health care for a sick child or contributions to funerals of friends and relatives. Women who lose their trading capital need to meet daily expenses for food and rent. Although the actual monetary value of such exchanges may be extremely low, they are needed to support poor households.[6] The proliferation of multiple-partner strategies is a direct consequence of deepening economic crisis.

The large-scale, capital-intensive import-substitution industry has failed to provide a basis for mass employment, sustainable development, and capital formation. The "informal" or small-business sector has attracted international attention since the 1970s as the most dynamic aspect of African economies. A woman shopkeeper in Lubumbashi commented on the usefulness of the informal sector in maintaining the status quo:

Women "break stones" to make ends meet. Their struggle to survive and support dependants relieves individual men of responsibility for ensuring family welfare. Men don't protest, so employers are spared the expense of paying wages that families can live on, and the state isn't threatened by political contestation.[7]

Even before human rights became a subject for official discourse, she and several colleagues ranked gender inequality and extreme poverty as important violations. They viewed lack of marital property rights as a deterrent to marriage for some women, and they pointed to the vulnerability of windows and divorcées. Informants argued that poverty robs women of the ability to fulfill their socially designated responsibilities and thus debases them, often forcing them into prostitution. Although they recognized this necessity and did not condemn the women who used it, they disagreed with scholars who view sex work with equanimity and pointed to serious reproductive health risks. By the 1980s, AIDS had transformed what was once a survival strategy into a route to early, painful death.

There are limits to women's patience. In 1989 and again in 1990, women market traders in Kinshasa demonstrated against government policies that had led to hyperinflation and frustrated their efforts to feed their families. Their protests were harshly repressed, and the leaders jailed. Since prisoners are reported to be frequently subjected to gang rape, and since seroprevalence is thought to be particularly high among the military, the threat of AIDS acts as a deterrent to women's militancy.

The next section considers some of the special cultural and biological risk factors for women in the region.

WOMEN'S BIOLOGICAL VULNERABILITY

Elite men are highly visible among the AIDS-patient population of Kinshasa and other cities of the region. Their names are known; their deaths cause comment. Some men believe that women are more resistant to HIV and AIDS. This perception notwithstanding, women are more easily infected than men; they constitute the majority of AIDS cases, of seropositives, and of those at risk.

Regardless of the type of partner relationship, specific conditions and sexual practices place women at special risk if their partners are infected. "Classic" STDs, which often led to reproductive health problems in the past, have become epidemic.[8] Their presence substantially increases the risk of acquiring HIV from an infected sex partner. Because signs are often subtle and many men do not notify their partners, women may not know when they are infected. Even when they suspect an infection, shame may prevent women from seeking treatment. Often the care available is inadequate, prohibitively expensive, or undignified and lacking in confidentiality.

Trauma causing tears in the vaginal skin (mucosa) allows the HIV to enter. This may occur at first intercourse, particularly in the case of girls and young adolescents penetrated by mature men. Before menstruation begins, the lower reproductive tract is anatomically and physiologically immature. The multiple cell layers and secretions that provide adult women with some protection develop gradually. The condition of the vagina is also a factor in adult trauma. For example, in many cultures, men who prefer intercourse in a tight, dry vagina may omit erotic foreplay. Women explain that without lubrication, "Men feel as though they are penetrating a virgin." In these cultures, copious secretions may cause a woman to be mocked or shamed for "liking sex too much." A variety of astringent herbal preparations and baths are employed to induce vaginal constriction. In the cultures where these practices are "traditional," and among others to which they are spreading, women say that they are ashamed of what they perceive as a vagina widened by successive births. In the presence of high levels of background infection, these conditions, taken together with the likelihood of acquiring other STDs from sexually experienced male partners, help to explain the high susceptibility of young females, many of whom become infected at first coitus.

Following menopause, the female genital mucosa again become thin and fragile due to lack of estrogen, and secretions are often limited. Although some African cultures deem it unseemly for a woman and her daughter to give birth at the same time, this is not the case everywhere. There is a saying that "Good soup is made in old pots." Men may return to their first wives, even as they seek sex with younger women. Few men protect their wives by using condoms at home. It is evident that biological risk is amplified by sociocultural forces, and the next section examines some related popular representations.

POPULAR REPRESENTATIONS OF AIDS

Cultural politics make the issues discussed in this chapter a sensitive subject. Racist constructions of African sexuality have been elaborated in Western discourse about AIDS in Africa. Zairians, especially, have been represented in the Western press as too fond of sex, too poorly educated, too "primitive," and too irrational to protect themselves from AIDS. In reply, Kinshasa university students, playing upon the French acronym SIDA, coined the dismissive phrase, "Syndrome Imaginaire pour Decourager les Amoureux" – an imaginary syndrome to discourage lovers (Schoepf 1991a).

There is no evidence that Africans are more "promiscuous" than other peoples, nor can behavior found today be considered "traditional." Not everyone is at risk. Some couples have followed Christian tenets to the letter, married without prior sexual experience, and remained faithful to one another. Some men are polygynous but do not seek women other than their wives. Moreover, even among the most sexually active people, access to formally and informally transmitted information can lead to rational reflection and risk reduction. Nevertheless, numerous constraints related to sex, gender, and power impede HIV prevention.

AIDS was first identified among Zairians in 1983, and international bio-medical research began in Kinshasa soon after. Still, AIDS remained a politically tabooed subject. Public discussion was muted, and little information appeared in national news media. Prodded by international donors, however, the government campaign began in 1987. As health officials, the mass media, and voluntary organizations cautiously began to provide information, people started talking more about AIDS. Ideas regarding transmission and prevention, disease origins and etiology, varied widely and changed over time. Urban elites, who had access to television, international publications, and friends in the health professions, were most informed.

Most people's knowledge was sketchy, however, and misinformation common. For example, the media told of insect transmission, and despite later disclaimers, people continued to cite it. On the other hand, few were aware of the risk of mother–infant transmission. Advice to "avoid prostitutes" was heard, but just who is a prostitute? Advice to "stay faithful to one partner" was impractical for many and misleading for those whose partners were already infected. Advice about safer sex was extremely limited and seldom cited by the public. As predicted, messages of the mass campaign created considerable awareness of AIDS, but relatively few people changed their sexual behavior sufficiently to reduce their own risk of infection or to protect partners. In mid-1987 the most common reaction to AIDS in Kinshasa was denial. Mass-media campaigns did not adequately inform the public about the slow action of the virus. People found it difficult to grasp that a healthy looking person could harbor a fatal HIV infection, could infect others, and would be likely to die in a few years. Failure to comprehend the lengthy and variable incubation period contributed to confusion and blame casting.

Numerous popular misconceptions bolstered peoples' avoidance of threatening personal risk assessments. Because AIDS was first discovered among Africans treated in Europe, it was said to affect mainly the wealthy and prostitutes. Since some rich and powerful men widely reputed for their sexual exploits apparently were unaffected, people joked that AIDS could not be too serious in Kinshasa. Some working-class men believed themselves to be free of risk even while they engaged in risky behavior. For example, two garage mechanics in their twenties said that AIDS is not a danger for them because: "We are too poor to travel to all those foreign places. Anyway, our girlfriends are young and healthy schoolgirls."

Although the government's advice to "avoid prostitutes" was heard, risk was redefined. Fear of AIDS propelled some men to seek very young girls who they believed were likely to be free of infection. The cars of businessmen and government officials could be seen parked at school-yard gates, waiting for girls to emerge. Male school teachers claimed sex as a fringe benefit of their poorly paid profession.[9] Boulevard hookers (*londoniennes*) donned school uniforms in an effort to allay the fears of prospective clients. Some men reported that they sought plump women, since they knew weight loss to be a sign of AIDS. Others sought women from the peripheral neighborhoods since they believed AIDS to be an urban disease (Schoepf 1991b, c).

Some who believed that they already were infected, however, said they saw no point in taking precautions:

A plumber said that he believes that there is nothing he can to do help himself live a long life.

"If you are going to get AIDS you'll get it, regardless." However, he was a minority of one in a discussion with friends, two other artisans and a sales manager. These three men had already eliminated extramarital adventures. But, "since anyone can have occasional relapses," they stated that they intended to use condoms.

These men were unusual, for at that time, most treated AIDS as just another disease, one misfortune among the many with which they had to contend. Interviewed over a three-year period, one highly educated official's attitudes evolved from skepticism to blame:

In 1985 the informant considered AIDS to be an invention of Western propagandists seeking to discredit Africans. The official believed that this "imaginary syndrome" was intended not only to discourage African lovers, but also to discourage European and Japanese tourists and investors whose money is needed to redress Africa's economic crisis. Why else, he reasoned, would scientists engage in irresponsible speculation about an African origin for AIDS? In 1986, when the death of some prominent people made it difficult to deny the existence of AIDS, it was widely attributed to women's sexual congress with Westerners. In 1987 the informant said that since he became aware of the danger, he has limited his sexual relations to three current wives. If he should find himself infected, he "knows" that it would be due to their infidelity. He does not believe that he might have been infected by previous partners. "Women are the major transmitters of AIDS, because they are more promiscuous than men, who if they desire a woman, marry her." By 1988 his fear of AIDS had increased; he said that seropositive women should be quarantined. The prospect of interning thousands of women for many years did not give him pause. Nor did he recognize that infected men would continue to spread the virus.

The identification of AIDS as an STD has made it easy to blame "promiscuous" women for its spread. The wife of a former cabinet minister told of several neighbors who were said to have died of AIDS. They included a doctor's first wife and her last child, as well as his second wife.

"He is still well and though he might be a healthy carrier, people suspect the women of infidelity. A professor down the street also died of AIDS. Since his wife is a long-distance trader people are sure that she gave him the disease."

Women traders who appear to be wealthy are said to have traded sex and sometimes even to have sacrificed children to the mermaid river spirit, Mamy Wata, in order to attain success in business.

Constructions of risk and attribution of responsibility follow existing patterns of power and control. Condom use, too, is linked to control issues. As with other forms of contraceptive technology, men say that wives who have access to condoms will no longer fear pregnancies and thus will feel free to conduct extramarital affairs. Faithfulness means women being faithful to men. Religious leaders have been heard to tell women that if they are "innocent" – that is, faithful to their husbands – they are not at risk for AIDS.

The prospect of thousands of adult deaths from AIDS occurring in the years to come was difficult to imagine in the midst of day-to-day hardships. Numerous cognitive blockages bolstered denial. Even though on one level they stated that neither biomedicine nor folk healers could cure AIDS, many people reacted casually, as though antibiotic injections would provide a cure for this as for many other sexually transmitted diseases.

Diseases that available biomedical services cannot cure are categorized by many as "African diseases," believed to be caused by spirits, cursing, or sorcery. Diseases believed to be sexually transmitted are surrounded by special moral stigma in both traditional and Christian religions. Thus people have incentives to push the fatal outcome of AIDS from their minds. Because the ultimate causes of AIDS are "obscure," however, does not mean that people must reject biological causality. Nevertheless, in the context of medical pluralism, when biomedicine fails, many seek to know why a specific person was (or was not) attacked by the deadly virus.

Some intellectuals rejected condoms along with other forms of contraception as an imperialist design to limit African populations. By 1989 their objections had been stilled by the evidence of mounting deaths. A new phrase was coined: AIDS (SIDA) became the Acquired Income Deficiency Syndrome, caused by *Salaries Insuffisants Depuis des Années*. Popular representations linked AIDS to poverty, especially among women.

WOMEN AT RISK

Although infection is not confined to special "risk groups," rates are highest among people with multiple sexual partners; the more partners, the greater the risk. Women sex workers are highest at risk for HIV. In Kinshasa 27 percent of a sample of women soliciting in Matonge bars were seropositive in 1985; in 1988 35 percent of a larger cohort were infected. Rates are higher still in other cities: 88 percent of poor sex workers in Kigali in 1985; more than 90 percent in Nairobi in 1988; 70 percent in Abidjan in 1990; in Dar-es-Salaam, 50 percent of female bar workers were seropositive in 1990. Prevention campaigns most often target people of marginal social status, such as prostitutes, migrant workers, and truck drivers, rather than high-status officials, military officers, and businessmen. Since sex workers are frequently unable to refuse clients who reject condoms, behavior-change interventions must target men – wealthy, powerful men as well as working-class men.

Associating AIDS with morally stigmatized prostitution is a hindrance to prevention. Women, particularly those who seek to escape male control, are especially likely to be blamed for spreading STDs. Deepening economic crisis appears to have increased gender conflict, as men seek to maintain control of scarce resources. There is danger that moral panic will lead to roundups, witch hunting, and increased violence against women.[10] Such campaigns to eradicate prostitution by rounding up women are counterproductive, however. They generally drive sex workers underground and alienate them from disease-prevention campaigns. Conducted in the name of "morality," roundups violate women's human rights to work, to walk unaccompanied, to choose their associates, and to go about their daily lives.

The focus on prostitutes and "promiscuity" also obscures the risk to other partners of many men and women who have had multiple sex partners over the past decade. Polygyny remains a customary form of marriage in most cultures and is found in all social classes. Men also enjoy other types of multiple-partner sexual relations with varying degrees of social recognition. Thus many wives are at risk even if they have obeyed normative proscriptions regarding extramarital sex, which are imposed in some, but not all, central African cultures. The focus on prostitutes as a risk for men obscures the fact that most women can neither refuse risky sex nor negotiate condom use with men they suspect have HIV.

Among the many Kinshasa women who in 1987 earned their livelihood by supplying sexual services to multiple partners, information levels and risk-reduction responses varied according to education and access to information in their social networks. The stylish professionals who solicit in the gambling casinos, nightclubs, and high-priced hotels of the city center were best informed. This was partly a consequence of an AIDS education meeting organized on their behalf by an enterprising vocational school director. In June 1987 some said they insisted that clients use condoms:

> Two well-dressed young women hitched a lift to a downtown gambling casino popular with Europeans. Asked if they feared AIDS, one exclaimed: "Not at all! We have our protection! Any man who doesn't use protection can stay with his money. We want to stay alive!" Both extracted packets of condoms from their handbags.

Other sex workers, their earning power less secure, said that they proposed condoms but would risk infection with noncompliant customers. Londoniennes, the peripatetic hookers who work the principal downtown boulevard, were less informed and less likely to be protected by condoms.

The poorest sex traders (*mingando*), who work from rented rooms in the popular quarters, were least likely to use protection and reported the greatest frequency of sexual encounters, averaging from five hundred to more than a thousand annually. They often suffered from untreated STDs. Most at risk, they also were the least informed about safer sex. Although several knew women who had died from AIDS, they used denial to avoid feelings of despondency. Others expressed their existential dilemma as fatalism: "We all have to die of something. What does it matter if we die of hunger now or AIDS later?" Nevertheless, small-group empowerment training can develop a sense of efficacy:

> In October 1987, one network of Kinshasa sex workers who had practiced negotiating condom

protection in a CONNAISSIDA workshop decided to try their new skills with clients. In their case, avoidance and fatalism had been fueled by the belief that they could do nothing to prevent AIDS. All but the eldest were successful. Following the empowerment workshops, their status rose in the neighborhood; they became defined as possessors of expert knowledge. They demanded: "Teach us to do what you do," and conducted AIDS education in the community. Action, spurred by a sense of empowerment, took the place of their former denial.

Not all women with multiple sex partners consider themselves prostitutes. Some are mistresses (*deuxieme bureaux*) regularly supported by married men of means; other are students or single working women who seek extra income when they visit bars or dancing clubs for an evening's entertainment. Although they expect to receive gratuities, these women rejected the idea of using condoms, which were stigmatized by their association with STDs and prostitution.

Two young women, aged eighteen and twenty, were interviewed at a nightclub in Matonge. One described herself as the mistress of a successful married man who paid her rent and other basic expenses. On nights when she expected him to be otherwise engaged, she sought other paying partners. Asked if she would use a condom to prevent AIDS infection, she replied vehemently: "I am not sick! If I were my Bwana would know it and he would tell me. He wouldn't come back." Nevertheless, when pressed, she said that she would accept a partner who insisted on using condoms if he provided them. The younger woman studied at a secretarial college. She believed the AIDS danger to be greatly exaggerated: "All those people who are dying now, are they really dying of AIDS?"

This skepticism was expressed by many people to bolster their denial of HIV risk.

Categorizing all "free women" as prostitutes, and AIDS as a disease spread by them, is counter-productive. This stigmatizing practice is not limited to laymen; it is also found in biomedical literature on AIDS. The AIDS epidemic will diminish as men come to terms with their own risks, accept responsibility

to protect others, and change their sexual behavior. Some who had begun to use condoms were turned from their course when, in November 1987, the government announced that a drug to cure AIDS had been discovered. Although the Zairian researcher later scaled down his claims at a public meeting in February 1988, the disclaimer was not broadcast. Wealthy people traveled from neighboring countries in search of treatment, and many months elapsed before skepticism won out over hope. Public health advice which focuses on social categories ("risk groups" or "core transmitters") rather than on actual risky behaviors feeds this stigmatization and denial. As we saw above, some who are at risk develop a false sense of security and fail to use protection.

Other informants – mainly women – recognized that neither monogamous wives nor women and men who had reduced their numbers of sex partners over the past few years were exempt from HIV risk. Both elite and working-class wives frequently expressed powerlessness in the face of what they knew, assumed, or suspected to be their husbands' multiple-partner relations. The wife of a high-government official with five grown children said that she is faithful to her husband. However, she believes herself to be at risk because she assumes that he takes advantage of sexual opportunities that go with his position.

I can't ask him to stop, but I wish he would use condoms. Use condoms with the other women, with me, whatever. But I just don't feel I can introduce the subject. It wouldn't do any good. My husband would get angry and tell me to mind my own business.

A restaurant owner in her thirties, married, with a formal-sector job and four young children, elaborated on the difficulties faced by women with AIDS:

Society rejects you. When you die you will not even be missed because you have died of a shameful disease. They will say that this woman has strayed. They will not see that maybe she has remained faithful while her husband has strayed. Given the status of women in most of our African societies, AIDS is doubly stigmatizing for women.

I asked her: What might be done to protect women and children?

I pray to God that he sets us on the road to good conduct . . . not make love so much and especially that couples can remain faithful. . . . Using condoms will have to be initiated by men. Husbands give the orders and wives obey. If a wife were to suggest – and I emphasize the word suggest because a married woman cannot insist in such matters! – her husband immediately would react unfavorably. He would think: "She is accusing me of infidelity!" . . . In our African societies, when a husband and a wife have a dispute, it doesn't stop with them. The entire family mixes in. And if couples begin to use condoms, they will not produce children. Children are the goal of marriage. . . . A woman without children is an insignificant woman!

A childless woman is not only diminished in the eyes of others, but also in her own. Because she has no descendants to name their children after her, she drops from the genealogical immortality conferred by positional succession, which for many individuals is tantamount to reincarnation.

This informant suggested that once they understood that many children (about one-third) born to infected mothers may develop AIDS, families might open channels of communication between spouses to help them assess their level of risk and take protective measures. Other informants agreed that "since all families want to have descendants, they might contribute toward finding a solution." Strong social pressures to produce children, particularly where bridewealth has been paid, militate against HIV-positive women following medical advice to abstain from procreation.

By the end of 1987, AIDS was more than just another disease; it was very frightening to see young adults waste away. Poor women's vulnerability and feelings of powerlessness increased with escalating hardships. A market trader in her twenties with two years of post-primary education explained the interconnected circumstances that render many women powerless to prevent HIV and other infections. In 1987 she reported that her small business had collapsed. "Prices are so high that people can't buy as much now and there are too many sellers of everything!" She knew that AIDS is real: "You only have to go to Pavillon 5 at Mama Yemo (Hospital) if you are not convinced!" She said that she stopped having extramarital sex when she became aware of

the danger about a year ago (1986). However, her husband is a riverboat captain and she doubted that he remains faithful during his frequent journeys. She did not know about condoms and, when informed, said she did not believe her husband would agree to use them.

She also worried about her sisters who are not married and are without salaried employment. Since her own business had failed, she could no longer help them. She said that their way of surviving the crisis is to seek several partners, since "Most men cannot support all a woman's needs now. They have to give more to their wives."[11]

Indeed, some wives reported that they used both the crisis and fear of AIDS to persuade husbands to remain at home. Casual recreational sex appeared to have diminished in response to insecurity. Following rampages by the military in some neighborhoods, bars were virtually empty at night in 1990–1991 for periods of several months before picking up again in 1992. The deepening crisis also lessened the material assistance provided to poor women from their extended families. Abandoned or neglected wives, mistresses, and widows are at high risk for AIDS when they use multiple-partner strategies to make up funds that they formerly got from one man.

Women who are informed and economically independent are more able to control their sexual interactions. For example, some market women whose partners depend upon their earnings are able to prevail upon them to use condoms. Similarly, well-paid professional women may choose between celibacy and partners who use condoms. However, given the strong socialization of girls for obedience to men, many women remain psychologically bound: Some women who "should" be aware of their risky situations may use denial to avoid feelings of powerlessness.

The wife of an army colonel said that she could not get AIDS because she is faithful to her husband. She mentioned that he travels frequently to distant garrisons, but avoided speculating on his activities when away from home. Then she told of a friend who committed suicide upon discovering that her husband had a second family in another part of town. She failed to see a parallel between her friend's lack of suspicion and her own.

Another dimension of differences in power between spouses appeared when informants were asked to consider how they would respond if a spouse were to develop AIDS. Men generally replied that they would divorce their wives; women replied that they would feel sad and cease sexual relations with their husbands. Condoms were not mentioned by either sex. The difference in responses is attributable to women's financial dependency and social powerlessness. In cultures where a married woman's refusal of sexual services to her husband is grounds for divorce, abstinence may be an option only for those desiring to end their marriages. The divorced state now is more precarious than ever, as is widowhood for women who do not have stable waged employment or successful enterprises. Thus not only poor or unmarried women are vulnerable. Wives of prosperous men cannot always act in their best interests because few control independent resources.

> The parents of an official with AIDS prevailed upon his wife to continue sexual relations unprotected by condoms as a demonstration of her devotion. The wife became fatally ill following her husband's death.
>
> A physician who knew he had AIDS kept the knowledge from his wife, with whom he continued to have unprotected intercourse until he became very sick. She only learned the nature of his disease at his death. Friends and acquaintances were sure that she must have been infected.

These were extreme cases for the men were already sick. Because AIDS is stigmatized and wives tend to be accused by their husbands' relatives as the source of HIV infection, many women are afraid to confront husbands as they might have done in the past regarding STDs.

Women's relative powerlessness in relationships with men increases not only their own vulnerability to HIV infection, but their risk of bearing infected infants. AIDS has created tens of thousands of widows and orphans in areas of high prevalence, straining the coping capacity of families and communities. Survivors are especially vulnerable economically since property accumulated during marriage is often seized by the deceased husband's relatives.

[. . .]

CONCLUSION

The rapid spread of AIDS in Africa results from the deep, multistranded crisis in political economy and health. Transmitted via sex and blood, AIDS is surrounded by dense meanings to which cultural constructions of gender roles are central. Women are especially at risk because of their poverty, their relative powerlessness in the overall organization of African societies, and their subordinate position with respect to men. These conditions circumscribe their options so that few are able to practice safer sex. Those who have reduced their risk are women with decision-making autonomy based on their capacity to support themselves without resorting to sex within or outside of marriage. Although the youngest and poorest are most at risk, the experience of married women dependent on wealthy husbands confirms the thesis that women do not automatically share their husbands' class position.

Official advice about fidelity does not address questions of risk within marital relations, which is where many women become infected. Early in the epidemic, women in Kinshasa pointed out that narrowly targeted campaigns were inadequate and, in the context of gender-biased stigma, would be counterproductive. Women's groups recommended that action research with men on HIV prevention, gender relations, and sexual health receive the highest priority.

Ethnographic action research offers an opportunity to explore prevention issues with people, variously situated in society, whose intimate lives are affected. CONNAISSIDA's "political economy and culture" approach is related to methodological advances made in the study of African societies over the past quarter century. One is understanding how macro-level political economies affect sociocultural dynamics at the microlevel – including the political ecology of disease and social response to epidemics. The rapid spread of HIV among women is patent evidence that the informal sector cannot be relied upon to create development imperiled by stagnation in the formal sector. The fact that AIDS is propelled by class, age, and gender inequality underscores the need for sustainable development to reduce the rapid global spread. The epidemic is emblematic of the process of capital accumulation, which drains resources away from the villages, upward to national ruling classes, and outward to world markets.

AIDS prevention also stimulates countercurrents of resistance to dominant cultural norms. Gender relations need not be static, and recent scholarship highlights many examples of women's struggles to change their condition. However, these struggles take place in circumstances not of women's making; without external solidarity they may be overwhelmed. Community-based empowerment methods can foster realistic risk assessment and begin to overcome existing cultural constraints. In the final analysis, however, effective AIDS prevention requires changing the status of women and youth to increase their material independence, their psychological autonomy, and their social power. The multiplex crisis of the state has pushed "women's issues" to the rear, yet, as this chapter argues, they are basic issues of African social and cultural survival. Because the impact of AIDS will be so devastating, prevention might be used to initiate far-ranging dialogues about the consequences of persistent inequality.

NOTES

This chapter is a revised version of an article published in 1988 in the *Canadian Journal of African Studies* 22(3): 625–44. Grateful acknowledgement is made to CONNAISSIDA colleagues Mme. Veronique Engundu Walu, Dr. Alphonse wa Nkera Rukarangira, Prof. Pascale Ntsomo Payanzo, and Mr. Claude Schoepf. The contents of this chapter are solely my responsibility.

1 WHO/GPA 1995:5. This includes eleven million adults and more than a million children.
2 Many references are omitted to save space. Biomedical references can be found in Schoepf, Rukarangira et al. 1988a, b; and Schoepf 1993a. Social science and interview citations appear in Schoepf 1988; 1992a, b, c; and 1993b; a historical literature review is in Schoepf 1991a.
3 These include ethnographic studies using participant-observation of household economics and family life that Veronique Walu and I conducted. In 1987 Walu recorded eighteen month-long household budgets (Walu 1991; Schoepf and Walu 1991; Schoepf, Walu, Schoepf, and Russell 1991). Dr. Rukarangira studied the informal economy and cross-border trade in southeast Shaba between 1983 and 1987 (Rukarangira and Schoepf 1991). I also collected life histories of women in Lubumbashi and Kinshasa between 1975 and 1990 with the aid of Mme. Walu and the late Mmes. Beatrice Hateyana Makyla and Bernadette Nsengimana.
4 For descriptions of CONAISSIDA's action research, see Schoepf, Walu, Rukarangira et al. 1991; and Schoepf 1993a, 1995.

5 Studies are reviewed in De Bruyn 1992 and Schoepf 1992b, 1993b.
6 A rapid sexual encounter in a poor neighborhood of Kinshasa cost sixty zaires in July 1987, equivalent to fifty cents. This sum could buy a large bowl of cassava meal, one large or two small onions, three eggs, a bottle of palm oil, or a beer.
7 Author's fieldnotes from a study of women in the informal economy of Lubumbashi, Zaire, 1977–79.
8 Untreated STDs can lead to pelvic inflammatory disease (and sterility), cancer, miscarriage, and congenital blindness.
9 Secondary-school teachers earned 2,500–4,000 zaires per month in mid-1987, equivalent to between twenty and thirty-three U.S. dollars.
10 A section on scapegoating women has been omitted, see Schoepf 1988.
11 Interview by Walu, April 1987.

REFERENCES

De Bruyn, Maria
    1992 "Women and AIDS in Developing Countries." *Social Science and Medicine* 34(3): 249–62.
Rukarangira, wa Nkera, and Brooke G. Schoepf
    1989 "Social Marketing of Condoms in Zaire." *WHO/AIDS Health Promotion Exchange* 3: 2–4.
    1991 "Unrecorded Trade in Shaba and Across Zaire's Southern Borders." In *The Real Economy of Zaire*, ed. Janet MacGaffey, pp. 72–96. London: James Currey; Philadelphia: University of Pennsylvania Press.
Schoepf, Brooke G.
    1988 "Women, AIDS, and Economic Crisis in Zaire." *Canadian Journal of African Studies* 22(3): 625–44.
    1991a "Ethical, Methodological, and Political Issues of AIDS Research in Central Africa." *Social Science and Medicine* 33(7): 749–63.
    1991b "Political Economy, Sex, and Cultural Logics: A View from Zaire." *African Urban Quarterly* 6(1–2): 96–106. Special issue on AIDS, STDs, and urbanization in Africa.
    1991c Représentations du SIDA et pratiques populaires à Kinshasa. *Anthropologie et Sociétés* 15(2–3): 149–66.
    1992a "AIDS, Sex, and Condoms: African Healers and the Reinvention of Tradition in Zaire." *Medical Anthropology* 14: 225–42.
    1992b "Gender Relations and Development: Political Economy and Culture." In *Twenty-First Century Africa: Towards a New Vision of Self-Sustainable Development*, ed. A. Seidman and F. Anang, pp. 203–41. Trenton, N.J.: Africa World Press.

1992c "Women at Risk: Case Studies from Zaire." In *Social Analysis in the Time of AIDS*, ed. G. Herdt and S. Lindenbaum, pp. 259–86. Newbury Park, Calif.: Sage Publications.

1993a "AIDS Action Research with Women in Kinshasa." *Social Science and Medicine* 37(11): 1401–13.

1993b "Gender, Development, and AIDS." In *The Women and International Development Annual*, vol. 3, ed. R. Gallin, A. Ferguson, and J. Harper. Boulder, Colo.: Westview Press.

1995 "Culture, Sex Research and AIDS Prevention in Africa." In *Culture and Sexual Risk: Anthropological Perspectives on AIDS*, eds. H. T. Brummelhuis and G. Herdt, pp. 29–51. New York: Gordon and Breach.

Schoepf, Brooke G., wa Nkera Rukarangira, Ntsomo Payanzo, Engundu Walu, and Claude Schoepf

1988 "AIDS, Women, and Society in Central Africa." In *AIDS, 1988: AAAS Symposium Papers*, ed. R. Kulstad, pp. 175–81. Washington, D.C.: American Association for the Advancement of Science.

Schoepf, Brooke G., wa Nkera Rukarangira, Claude Schoepf, Ntsomo Payanzo, and Engundu Walu

1988 "AIDS and Society in Central Africa: The Case of Zaire." In *AIDS in Africa: Social and Policy Impact*, ed. N. Miller and R. Rockwell, pp. 211–35. Lewiston, N.Y.: Mellen Press.

Schoepf, Brooke G., and Engundu Walu

1991 "Women's Trade and Contribution to Household Budgets in Kinshasa." In *The Real Economy in Zaire*, ed. J. MacGaffey, pp. 124–51. London: James Currey; Philadelphia: University of Pennsylvania Press.

Schoepf, Brooke G., Engundu Walu, wa Nkera Rukarangira, Ntsomo Payanzo, and Claude Schoepf

1991 "Gender, Power, and Risk of AIDS in Central Africa." In *Women and Health in Africa*, ed. M. Turshen, pp. 187–203. Trenton, N.J.: Africa World Press.

Schoepf, Brooke G., Engundu Walu, Diane Russell, and Claude Schoepf

1991 "Women and Structural Adjustment in Zaire." In *Structural Adjustment and African Women Farmers*, ed. C. Gladwin, pp. 151–68. Gainesville: University of Florida Press.

Walu, Veronique Engundu

1991 "Women's Survival Strategies in Kinshasa." MA thesis, Institute for Social Studies, Women and Development Program, The Hague, Netherlands.

World Health Organization Global Programme on AIDS (WHO/GPA)

1995 "Cumulative Infections Approach 20 Million." *Global AIDS News* 1: 5.

'Feminism, the Taliban, and Politics of Counter-Insurgency'

Anthropological Quarterly (2002)

Charles Hirschkind and Saba Mahmood[1]

Editors' Introduction

Saba Mahmood is Associate Professor in the Department of Anthropology, University of California at Berkeley. Mahmood is originally from Pakistan. Her first degrees were in Architecture: a BA from the University of Washington, Seattle, in 1985, an MA in Architecture and an MA in Urban Planning from the University of Michigan. Mahmood practised Architecture, and had a short foray into Political Science before turning to Anthropology, completing a PhD from Stanford University in 1998. Mahmood then held a Chancellor's Post-doctoral Fellowship at the University of California, Berkeley, during 1998–99, and was a Harvard Academy Scholar for two years between 2000 and 2003, during which time she also became Assistant Professor at the Divinity School at the University of Chicago. Mahmood was Visiting Professor at the International Institute for Islam at Leiden University in the Netherlands in 2003. Mahmood is the author of *Politics of Piety* (2005), a book which quite fundamentally questions the way we think of liberalism, feminism, Islamism and embodied practice. Through a careful ethnography of women in Cairo's 'Mosque movement', Mahmood calls attention to the ways in which ethics are embodied in daily practices rather than in passive submission to ideology. Arguing against liberal feminist notions of female subjection, Mahmood looks at how specific subjects and agencies are produced through engagements in Islamism. Mahmood's current work extends these important theoretical and methodological insights to the question of 'liberal Islam', or what Mamdani sarcastically calls 'the Good Muslim'. Mahmood's ethnographic research on secularism in Lebanon and Egypt is bound to tell us things we cannot already know, despite our best (Western, liberal, conservative or radical) expectations.

Charles Hirschkind is Associate Professor in the Department of Anthropology at the University of California at Berkeley. Hirschkind completed his PhD in Anthropology at the New School for Social Research. He is the author of *The Ethical Soundscape* (2006), and co-editor of *Powers of the Secular Modern: Talal Asad and His Interlocutors* (2006). *The Ethical Soundscape* explores religious practice as it is mediated by media technologies and specific phonic/aural traditions. Hirschkind looks closely at how cassette-recorded sermons are used by a variety of people in contemporary Egypt, drawing on Islamic traditions of aural discipline and recitation. Hirschkind's current project focuses on the politics of memory and Islamic identity in Andalusia, Spain.

The reading that follows is excerpted from Mahmood and Hirschkind's incisive response to the prose of US militarism in Afghanistan, and its misrepresentations on three key issues. First, they show how US counter-insurgency reproduces a wilful blindness to Northern (US, British, Russian) complicity in the production of regional political and economic crisis, and the way in which the Taliban's methods in certain respects have been shaped by groups supported in the Cold War 'proxy wars' of the 1980s. Second, they show how a 'humanitarian' sensibility in the US public sphere has relied on stereotypes of rescuing the hapless Afghan woman. This blatantly neo-colonial form of sympathy, garbed in liberal feminism, ignores the historical and social complexities of gender and inequality. Third, support for imperial militarism has been built through an inability to understand the public and political practices of Islam, as is clear in various debates on veiling. Even when the veil is not a symbol of anti-colonialism, *pace* Fanon (see Part 3), Muslim women often chose to veil, or to express their piety, for reasons that have little to do with freedom or bondage. They ask sarcastically, 'Can our bras, ties, pants, miniskirts, underwear and bathing suits all be so easily arrayed on one or the other side of this divide?' Thinking about 'political Islam', Muslim women, imperialism and counter-insurgency requires more careful reflection, and action. The reader might consider how this analysis recasts questions of security and well-being in relation to the fetishism of the veiled Muslim woman, *pace* McClintock (Part 1), and the power of Occidentalist discourse, *pace* Coronil (Part 9).

Key references

Charles Hirschkind and Saba Mahmood (2002) 'Feminism, the Taliban, and Politics of Counter-Insurgency', *Anthropological Quarterly* 75 (2): 339–54.

Saba Mahmood (2005) *Politics of Piety: The Islamic Revival and the Feminist Subject*, Princeton, NJ: Princeton University Press.

Charles Hirschkind (2006) *The Ethical Soundscape: Cassette Sermons and Islamic Counterpublics*, New York: Columbia University Press.

Charles Hirschkind and David Scott (eds) (2006) *Powers of the Secular Modern: Talal Asad and His Interlocutors*, Palo Alto, CA: Stanford University Press.

Joan Wallach Scott (2007) *The Politics of the Veil*, Princeton, NJ: Princeton University Press.

On a cool breezy evening in March 1999, Hollywood celebrities turned out in large numbers to show their support for the Feminist Majority's campaign against the Taliban's brutal treatment of Afghan women. Jay and Mavis Leno hosted the event, and the audience included celebrities like Kathy Bates, Geena Davis, Sidney Potier, and Lily Tomlin. Jay Leno had tears in his eyes as he spoke to an audience that filled the cavernous Directors Guild of American Theater to capacity. It is doubtful that most people in this crowd had heard of the suffering of Afghan women before. But by the time Mellissa Etheridge, Wynonna Judd, and Sarah McLachlan took to the stage, following the Afghan chant meaning "We are with you," tears were streaming down many cheeks.[2]

The person spearheading this campaign was Mavis Leno, Jay Leno's wife, who had been catapulted into political activism upon hearing about the plight of Afghan women living under the brutal regime of the Taliban. This form of Third World solidarity was new for Mavis Leno. Prior to embarking on this project, reports *George* magazine, "Leno restricted her activism to the Freddy the Pig Club, the not-so radical group devoted to a rare series of out-of-print children's books."[3] She was recruited by her Beverly Hills neighbor to join the Feminist Majority, an organization formed by Eleanor Smeal, a former president of NOW. Little did members of the Feminist Majority know that Leno would make the plight of Afghan women living under the Taliban rule a cause celebre: not only did the Hollywood celebrities join the ranks of what came to be called the "Stop Gender Apartheid in Afghanistan" campaign, but a large number of popular women's magazines (like *Glamour, Jane, Teen*, etc.), in addition to feminist journals like *Sojourner, Off our Backs* and *Ms.*, carried articles on the plight of Afghan women living under the Taliban. The Lenos personally gave a contribution of $100,000 to help kick off a public awareness campaign. Mavis Leno testified before the Senate Foreign Relations Committee, spoke to Unocal shareholders to dissuade them from investing in Afghanistan, and met with President Bill Clinton to convince him to change his wavering policy toward the Taliban. In addition, the Feminist Majority carried out a broad letter writing campaign targeted at the White House. The Feminist Majority claims that it was their work that eventually dissuaded Unocal officials to abandon their plans to develop a natural gas pipeline in Afghanistan, and convinced the Hollywood-friendly Bill Clinton to condemn the Taliban regime.

Even skeptics who are normally leery of Western feminists' paternalistic desire to "save Third World women" were sympathetic to the Feminist Majority's campaign. This was in part because the restrictions that the Taliban had imposed on women in Afghanistan seemed atrocious by any standard: they forbade women from all positions of employment, eliminated schools for girls and university education for women in cities, outlawed women from leaving their homes unless accompanied by a close male relative, and forced women to wear the *burqa* (a head to toe covering with a mesh opening to see through). Women were reportedly beaten and flogged for violating Taliban edicts. There seemed to be little doubt in the minds of many that the United States, with its impressive political and economic leverage in the region, could help alleviate this sad state of affairs. As one friend put it, "Finally our government can do something good for women's rights out there, rather than working for corporate profits." Rallying against the Taliban to protest their policies against Afghan women provided a point of unity for groups from a range of political perspectives: from conservatives to liberals and radicals, from Republicans to Democrats, and from Hollywood glitterati to grass roots activists. By the time the war started, feminists like Smeal could be found cozily chatting with the generals about their shared enthusiasm for Operation Enduring Freedom and the possibility of women pilots commanding F-16s.[4]

Among the key factors that facilitated this remarkable consensus, there are two in particular that we wish to explore here: the studied silence about the crucial role the United States had played in creating the miserable conditions under which Afghan women were living; and secondly, a whole set of questionable assumptions, anxieties, and prejudices embedded in the notion of Islamic fundamentalism. It was striking how a number of commentators, in discussions that preceded the war, regularly failed to connect the predicament of women in Afghanistan with the massive military and economic support that the US provided,

as part of its Cold War strategy, to the most extreme of Afghan religious militant groups. This silence, a concomitant of the recharged enthusiasm for the US military both within academia and among the American public more generally, also characterized much of the response both to reports of mounting civilian casualties resulting from the bombing campaign, and to the widespread famine that the campaign threatened to aggravate. For example, as late as early December, the Feminist Majority website remained stubbornly focused on the ills of Taliban rule, with no mention of the 2.2 million victims of three years of drought who were put at greater risk of starvation because US bombing severely restricted the delivery of food aid. Indeed, the Feminist Majority made no attempts to join the calls issued by a number of humanitarian organizations – including the Afghan Women's Mission – to halt the bombing so that food might have been transported to the Afghans before winter set in.[5] In the crusade to liberate Afghan women from the tyranny of Taliban rule, there seemed to be no limit of the violence to which Americans were willing to subject the Afghans, women and men alike. Afghanistan, so it appeared, had to bear another devastating war so that, as the *New York Times* triumphantly noted at the exodus of the Taliban from Kabul, women can now wear *burqas* "out of choice" rather than compulsion.

The twin figures of the Islamic fundamentalist and his female victim helped consolidate and popularize the view that such hardship and sacrifice were for Afghanistan's own good. Following the September 11th attacks, the *burqa*-clad body of the Afghan woman became the visible sign of an invisible enemy that threatens not only "us," citizens of the West, but our entire civilization. This image, one foregrounded initially by the Feminist Majority campaign though later seized on by the Bush administration and the mainstream media, served as a key element in the construction of the Taliban as an enemy particularly deserving of our wrath because of their harsh treatment of women. As Laura Bush put it in her November 17th radio address to the nation: "Civilized people throughout the world are speaking out in horror – not only because our hearts break for the women and children of Afghanistan, but also because in Afghanistan, we see the world the terrorists would like to impose on the rest of us." Not surprisingly, the military success of Operation Enduring Freedom was celebrated first and foremost as the liberation of Afghan women from Taliban control.

Our main concern here is not simply to dwell on the inadequacies of the campaign to rescue Afghan women by the Feminist Majority or other groups, but to address the larger set of assumptions and attitudes undergirding this campaign and that are reflected widely in American public opinion: attitudes about the proper place of public religious morality in modern Islamic societies, and in particular how such morality is seen to shape and constrain women's behavior. The Taliban in many ways have become a potent symbol of all that liberal public opinion regards as grievously wrong with islamic societies these days, proof of the intense misogyny long ascribed to islam, and most emphatically to those movements within Islam referred to as fundamentalist. That from the rubble left behind by the game of super power politics played out on Afghan bodies and communities, we can only identify the misogynist machinations of the islamic fundamentalist testifies to the power this image bears, and the force it exerts on our political imagination.

COUNTER-INSURGENCY

It is striking that even among many of those who came to acknowledge the US involvement in the civil war in Afghanistan, the neat circuit of women's oppression, Taliban evil, and Islamic fundamentalism remained largely unchallenged. It is worthwhile here to briefly recall some of the stunning history of the conflict in Afghanistan. US concern for what was until then a neglected part of South West Asia was greatly heightened when the Soviet Union invaded Afghanistan in 1979. President Jimmy Carter signed a directive to begin covert operations in Afghanistan in order to harass the Soviet occupying forces by supplying funds, weapons, and other forms of support to the Afghan fighters known as the *mujahedeen*. By 1986, under the Reagan administration, this project had mushroomed into the largest covert operation in US history since WW II. Overall, the US funneled more than $3 billion to the mujahedeen, with an equal if not greater amount coming from Saudi Arabia, one of the staunchest US allies. The Saudi monarchy had historically been lavish funders of anti-leftist forces around the globe. The aims of the Saudi monarchy to root out any communist

influence from the Muslim world dovetailed with the Reagan Doctrine which had increased US support for anticommunist insurgencies against Soviet-backed regimes in various parts of the Third World.

Pakistan was the ground from which this covert operation was staged. The then military dictator of Pakistan, General Zia ul-Haq, who had just overthrown the democratically elected prime minister Zulfikar Ali Bhutto, was more than eager to oblige the Americans, not only as a means to obtain US economic aid but also to bolster the legitimacy of his military rule. Pakistan's Inter-Services Intelligence agency, or ISI, was a key player in both channeling US arms to the Afghan *mujahedeen*, as well as training them. The strategy that the CIA pursued in this covert operation was quite different from the one pursued in Nicaragua and Angola insofar as no Americans trained the *mujahedeen* directly – instead the CIA trained Pakistani instructors and members of the ISI.[6]

Throughout the Afghan war, critics of the CIA's covert operation voiced two major complaints: first, that the bulk of US aid was being funneled to the most extreme and conservative Islamic groups from the Afghan opposition; second, that as an indirect consequence of the CIA operation, the Afghanistan-Pakistan region was now the largest producer of heroin as well as a sizeable marketplace for illicit arms. Let us consider each of these. When Moscow first intervened militarily in Afghanistan in 1979, there were a variety of both Islamic and secular-nationalist Afghan groups opposed to the Moscow-backed Communists, some of them espousing political and religious positions we would label "moderate." Yet the majority of the US aid (as much as 75%) was channeled to the most extremist of these opposition groups, an important consequence of which was the marginalization of moderate and secular voices. It is widely understood that the Pakistani agency ISI was instrumental in choosing these groups. But as the *World Policy Journal* noted, "There is no evidence to indicate that CIA officials or other US policymakers strenuously objected to the channeling of aid to the most extreme authoritarian elements of the Afghan resistance".[7]

One of the most favored of these groups was headed by Gulbuddin Hekmatyar, a man known for throwing acid in the faces of women who refused to wear the veil, and whose group received as much as 50% of US aid. When questioned about the US support of Hekmatyar, a CIA official in Pakistan explained, "Fanatics fight better."[8] This policy of promoting extremist Islamic groups in the region, and equipping them with the most sophisticated military and intelligence equipment, had gradually, over a period of ten years, created the political climate in which the emergence of the Taliban was a predictable outcome. Even though the Taliban did not come into power until 1995, well after both the US and Soviet Union had withdrawn from the region, their methods were not much different from groups that the US and its allies had supported. Neither, for that matter, are the practices of the United States' more recent allies, the Northern Alliance, a fact that is becoming evident since their seizure of power in Kabul. After the exodus of the Taliban, as the Northern Alliance were being legitimized in Germany, the widely respected Afghan women's organization, Revolutionary Association of the Women of Afghanistan, put out a statement saying "The people of the world need to know that in terms of widespread raping of girls and women from seven to 70, the track record of the Taliban can no way stand up against that of these very same Northern Alliance associates."[9]

The arms pipeline established between the US-ISI-*Mujahedeen* was notoriously corrupt, and many of the arms that the CIA supplied ended up being sold in the open market as well as being channeled to groups of fighters already known for their excessively violent tactics against non-combatant peoples living within the area of conflict. The CIA turned a blind eye to this arms leak, chalking it up to the necessary cost of a covert operation, and in so doing, turned the region into one of the most heavily armed areas in the world.[10] In addition, as the Afghan *mujahedeen* gained control over liberated zones in Afghanistan, they required that their supporters grow opium to support the resistance. Under CIA and Pakistani protection, Pakistan military and Afghan resistance fighters opened heroin labs on the border between the two countries. By 1981 this region was supplying 60% of the US demand for heroin. In Pakistan the results were particularly horrendous: the number of heroin addicts rose from a handful in 1979 to one million two hundred thousand by 1995.[11]

In its literature, the Feminist Majority claims that "Afghanistan, under the Taliban rule, [had] become the number one producer of illicit opium and heroin in the world."[12] Insomuch as the Taliban did not come to power until 1995 and Afghanistan was already the major supplier of world heroin by 1985, this was a misrepresentation of facts. On the contrary, according to the United Nations, the Taliban all but eliminated heroin production in the first year from the areas under their control.[13] Where heroin production did continue to flourish was in areas controlled by the Northern Alliance. Its cultivation has remained an important source of revenue for them, and indeed, since their rise to power, poppy cultivation has been revived in many of the areas from which the Taliban had managed to eliminate it. The Feminist Majority's misrepresentation of the Taliban drug policy was consistent with the overall picture that the group sought to present, one that held the Taliban solely responsible for the catastrophic situation that the Afghans, in particular women, faced.

Feminist Majority statements consistently ignored the devastation wrought by two decades of warfare in which women and children had suffered most heavily, and instead suggested a relatively benign picture of women's lives prior to Taliban rule. For example, in 1998 when the Lenos announced their $100,000 contribution to the Feminist Majority campaign, Mavis Leno said, "Two years ago women in Afghanistan could work, be educated, and move about freely. Then the Taliban seized power. Today women are prohibited from leaving their homes unless accompanied by a close male relative and are forced to wear the burqa. Girls and women are banned from schooling. . . . No healthcare . . . no education . . . no freedom of movement. This nightmare is reality for 11.5 million women and girls in Afghanistan."[14] It has been common knowledge for anyone interested in the region that Afghan men and women have long suffered from many of the ills that the Feminist Majority attributed to the Taliban. For example, in addition to being one of the poorest nations of the world, Afghanistan had, for a number of years, one of the highest infant and maternal mortality rates. These conditions were only exacerbated by twenty years of war during which the delicate balance of tribal power was radically destabilized by the influx of weapons, making ordinary people subject to violence on an unprecedented scale. As is often the case, the increased militarization of Afghan society made women more subject to violence than at any time before. During this period of civil war, perhaps two million Afghans were killed, and six million made refugees – 75% of whom are women and children. Afghanistan today remains one of the most heavily landmined countries in the world, with people being maimed and killed on a daily basis. And if those weapons are inadequate, among the many types of collateral that the US has put into its recent deal with the country is a new stratum of unexploded munitions. Given these conditions, the narrow focus on Taliban rule by the Feminist Majority and other groups, and their silence on the channeling of US aid to the most brutal and violent Afghan groups (of which the Taliban were only one), must be seen as a dangerous simplification of a vastly more complicated problem. Why were conditions of war, militarization, and starvation considered to be less injurious to women than the lack of education, employment, and, most notably, in the media campaign, Western dress styles?

The silence among scholars and women's advocacy groups around these issues was coupled with a highly selective and limited representation of Afghan life under Taliban rule, one that filtered out all information that might contribute to a more nuanced understanding of Afghan women's situation. For example, the Taliban decree to ban girls and women from schools affected only a tiny minority of urban dwellers since the majority of the population reside in the rural areas where there are few schools: approximately 90% of women and 60% of men in Afghanistan are illiterate. Likewise, rarely was it mentioned that the Taliban policy of disarming the population, and the strict surveillance of all major areas under their control had made it possible for the first time in years for women to move outside their homes without fear of being raped (of course, being beaten for a variety of moral transgressions remained a distinct possibility). According to recent reports, this security is rapidly disintergrating. As the Agence France-Presse recently reported, "Just 10 weeks after the Taliban fled Kabul city, Afghans are already starting to say they felt safer under the now-defeated hardline militia than under the power-sharing interim administration that has replaced it. Murders, robberies and hijackings in the capital, factional clashes in the north and south of the country, instability in Kandahar and banditry on roads linking main centres are beginning to erode the optimism that greeted the inauguration of the interim administration on December 22."[15]

Equally relevant here is the fact that even though Taliban policies had made conditions much worse for urban women, they did not substantially affect the lives of the vast majority of rural women either because many of the Taliban edicts already mirrored facts of rural life, or because those edicts were never enforced. Sensitive writers documenting the catastrophy unfolding in Afghanistan have occasionally pointed this out. For example, an article published in the *New Yorker* noted that just outside of the urban centers, "one sees raised paths sub-dividing wheat fields . . . in which men and women work together and the women rarely wear the burka; indeed, since they are sweating and stooping so much, their heads often remain uncovered. The Taliban has scarcely altered the lives of uneducated women, except to make them almost entirely safe from rape."[16] As the article suggested, one consequence of the admittedly oppressive regulations put into place by the Taliban was that life for the majority of Afghans had become considerably safer.[17] Despite the availability of this kind of data, the Feminist Majority and other advocacy groups carefully kept any ambiguities out of their case against the Taliban as the sole perpetrators of the ills committed against Afghan women.

Taking these realities into account demands a more nuanced strategy on the part of anyone who wishes to help the women of Afghanistan in the long run. Already before the bombing began, one consequence of the campaign to rescue Afghan women was the dramatic reduction of humanitarian aid to Afghanistan, the brunt of which was borne by women and children as the most destitute members of the population.[18] When some of those concerned protested this outcome, they were chided for being soft on the Taliban.[19] It seemed like any attempt to widen the discussion beyond the admittedly brutal practices of the Taliban was doomed to be labeled as antithetical to women's interests.

[. . .]

PUBLIC RELIGION

One reason why Islamic movements make many liberals and progressives uncomfortable is the Islamists' introduction of religious concerns into what are considered to be properly political issues. The argument is often made that if the Muslim world is to become modern and civilized, it must assign Islam to the space of the private and personal. When religion is allowed to enter into public debate and make political claims, we are told, it results in rigid and intolerant policies that are particularly injurious to women and minorities. Once again we quote from Salman Rushdie who reiterates this admonishment to the Muslim world: "The restoration of religion to the sphere of the personal, its depoliticization, is the nettle that all Muslim societies must grasp in order to become modern. . . . If terrorism is to be defeated, the world of Islam must take on board the secular-humanist principles on which the modern is based, and without which Muslim countries' freedom will remain a distant dream."[20]

One of the many problems with such a formulation is that it ignores the multiple ways in which the public and private are linked in contemporary society. As many scholars have argued for some time now, the division between the public and the private is quite porous; the two are ineluctably intertwined in myriad ways. The most striking example of this linkage is the reaction that the adoption of the veil provoked in some European and Middle Eastern countries. In France, for example, the decision on the part of Muslim schoolgirls to wear the headscarf was denounced as injurious to French public life and in 1994 the French government banned the headscarves from public schools. Similarly, between 1998–2000, more than 25,000 women were barred from Turkey's college campuses because they refused to remove the headscarves, and hundreds of government employees were fired, demoted or transferred for the same reason.[21] In all of these instances, the pleas of the young women that their adoption of the veil was an expression of their personal faith, and not an endorsement of state-censured Islamist politics, went unheeded.

Both of these examples demonstrate not simply that the private and the public are inter-twined, but more importantly that only certain expressions of "personal faith" – and not others – are to be tolerated even in modern liberal societies. That is, what gets relegated to the sphere of the personal is still a *public* decision. Thus we need to put to question the idea suggested by Rushdie, among others, that were Muslims simply to privatize their faith, their behavior would become acceptable to secular sensibilities.

One of the reasons why the veil provoked such a passionate response even among feminists in France is the assumption that it potently symbolizes women's subordinate status within Islam. A number of French feminists supported the ban on the headscraf because, as a leading French feminist intellectual, Elizabeth Badinter, put it, "The veil . . . is the symbol of the oppression of a sex. Putting on torn jeans, wearing yellow, green, or blue hair, this is an act of freedom with regard to social conventions. Putting a veil on the head, this is an act of submission. It burdens a woman's whole life."[22] While the veil's symbolic meaning has been frequently discussed, particularly by those opposed to it, the question is far more complicated than suggested here. The veil has been freighted with so many meanings in contemporary social and political conflicts that any ascription of a singular meaning to it – such as 'symbol of women's oppression' – is unconvincing. Think of the very different contexts within which the practice of veiling is undertaken, for example, in Afghanistan, France, Turkey, or for that matter the US. Whereas the veil was forced on urban women in Afghanistan by the Taliban under the threat of physical violence, in France its adoption has, in many instances, come in the context of young women going against their parents' more assimilated lifestyles. In Turkey, on the other hand, the coercive powers of the law were marshaled, back in the 1920s, to force women to *unveil*. More recently, the practice of veiling has gained ascendancy as part of an opposition movement protesting the rigid policies of a state that insists on dictating the ways in which personal practices of religious piety should appear in public. Note that this is not to say the veil never works to signify women's oppression. The point is that to speak about the meaning of the veil in any of these contexts requires a lot more analytical work than that undertaken by those who oppose its adoption.

It is interesting that Badinter opposes the decision to veil by young Muslims girls on the grounds that, as an act in accord with (and therefore not in contest with) Islamic norms of female modesty, it does not rise to the status of "an act of freedom in regards to social conventions." This points out the degree to which the normative subject of feminism remains a liberatory one: one who contests social norms (by wearing torn jeans and dying her hair blue), but not one who finds purpose, value, and pride in the struggle to live in accord with certain tradition sanctioned virtues. Women's voluntary adoption of what are considered to be patriarchal practices are often explained by feminists in terms of false consciousness, or an internalization of patriarchal social values by those who live within the asphyxiating confines of traditional societies. Even those analyses that demonstrate the workings of women's subversive agency in the enactment of social conventions remain circumscribed within the singular logic of subordination and insubordination. A Muslim woman can only be one of two things, either uncovered, and therefore liberated, or veiled, and thus still, to some degree, subordinate. Can our bras, ties, pants, mini-skirts, underwear, and bathing suits all be so easily arrayed on one or the other side of this divide? Can our daily activities and life decisions really be captured and understood within this logic of freedom or captivity?

We need a way to think about the lives of Muslim women outside this simple opposition. This is especially so in those moments of crisis, such as today, when we tend to forget that the particular set of desires, needs, hopes, and pleasures that liberals and progressives embrace do not *necessarily* exhaust the possibilities of human flourishing. We need to recognize that, whatever effect it has had on the women who wear it, the veil has also had a radical impact on our own field of vision, on our capacity to recognize Muslim societies for something other than misogyny and patriarchal violence. Our ability to respond, morally and politically, in a responsible way to these forms of violence will depend on extending these powers of sight.

NOTES

1 We would like to thank Noah Solomon and Scott Richard for their research assistance in gathering the pertinent data for this article.

2 Stacie Stukin, "Warrior Princes," *George*, July 1999.

3 Stukin, p. 45.

4 Sharon Lerner, "Feminists Agonize Over War in Afghanistan: What Women Want," *The Village Voice*, October 31, 2000.

5 The Afghan Women's Mission reported that according to UNICEF, "Two million people do not have enough food to last the winter, and 500,000 of them will be unreachable after snow begins to fall in mid-November." From the press release issued by Afghan Women's Mission, October 2001, http://www.afghanwomensmission.org.

6 Steve Coll, "Anatomy of a Victory: CIA's Covert Afghan War," *The Washington Post*, July 19, 1992.

7 *World Policy Journal*, "The Unintended Consequences of Afghanistan," Spring 1994, p. 81.

8 Diego Cordovez and Selig Harrison, *Out of Afghanistan: The Inside Story of Soviet Withdrawal* (New York: Oxford University Press), pp. 62–63.

9 *Reuters Wire Service*, "Afghan Women's Group Gloomy of the Post-Taliban Era," December 10, 2001.

10 *Electronic Telegraph* reported that the CIA had spent "more than L70 million in a belated and often bungled operation to buy back the missiles" it had provided to the Afghan resistance (Daniel Mcgrory, "CIA Stung by Its Stingers," November 3, 1996).

11 Alfred McCoy, *The Politics of Heroin* (New York: Lawrence Hill Books), pp. 445–460.

12 From "Campaign to Stop Gender Apartheid in Afghanistan Organizational Co-Sponsor Resolution," http://www.feminist.org/afghan/afghanresolution/html (July 4, 2000).

13 *New York Times* reported that the Taliban's "ban on opium-poppy cultivation appears to have wiped out the world's largest crop in less than a year" (Barbara Crossette, "Taliban's Ban on Poppy A Success, U.S. Aides Say, May 20, 2001). The *Times* acknowledged that poor farmers were most adversely affected by this ban since they could not grow any other crop that would fetch them the same kind of income.

14 "Mavis Leno to Chair Feminist Majority Foundation's Campaign to Stop Gender Apartheid," http://www.feminist.org/news/pr/pr102198.html, July 4, 2001.

15 Agence France-Presse, Friday January 25, 2002, http://sg.news.yahoo.com/020125/1/2crv0.html.

16 William Vollmann, "Letters from Afghanistan: Across the Divide," *The New Yorker*, May 15, 2000, p. 67.

17 Vollmann, pp. 64–65.

18 See Megan Reif, "Beyond the Veil – Bigger Issues," *Christian Science Monitor*, May 3, 2000. Also see "Afghanistan – UN Denies Aid," *Off Our Backs*, 28, no. 5 1998.

19 See, for example, the response to Megan Reif's article (May 3, 2000) by Mavis Leno in the *Christian Science Monitor*, May 18, 2000, p. 8.

20 Salman Rushdie, "Yes, This is About Islam," *New York Times*, November 2, 2001.

21 Molly Moore, "The Problems of Turkey Rest on Women's Heads; Islamic Scarves Seen as Threat to Secular State," *The Washington Post*, October 9, 2000.

22 Quoted in Norma Claire Moruzzi, "A Problem with Headscarves: Contemporary Complexities of Political and Social Identity," *Political Theory*, vol. 22, no. 4, November 1994, p. 662.

Development in the twenty-first century

INTRODUCTION TO PART NINE

Trying to conclude *The Development Reader* is no easy matter. The selection of readings that follows is only one of many possible selections. Other editors would have chosen differently, as doubtless they would have done across all parts of this book. That said and accepted, however, our aim here is to present a group of readings that speaks to two linked concerns. On the one hand, we want to go back to our core themes of markets, empire, nature and difference. In our view these will remain defining issues for development in the twenty-first century, although of course they will be taken up and examined in new ways and from different perspectives. Linked to this, we want to suggest that development issues in the next century will increasingly be phrased at the global scale, with more attention being paid both intellectually and politically to what are being called global public goods and bads (international migration, climate change, disease transmission, money-laundering, trade negotiations, the arms trade, terrorism, planets of slums, and so on). Almost all of these issues, incidentally, were being discussed at the end of the nineteenth century.

Given pressures on space, we cannot cover all of these issues here, but we do want to flag up the continued rise of Asia within the world economy and the problems that some African countries might face in terms of integrating into that system. We also want to call attention to what some are calling the 'new imperialism' (with China and India not being alone in joining the United States in an expanded hunt for natural resources the world over), and the problems that weakened states will face in dealing with these new imperial pressures. And lastly we want to deal with global warming – arguably the most critical issue in the new development agenda – and the possibility that globalisation is intimately linked to the production of new geographies of anger aimed at minority groups. As we have tried to suggest throughout this reader, development is far too important an issue to be left to development policy alone. We must also remain alert to the ways in which the object of development (Part 1) is theorised and brought to public attention by competing social groups with very different endowments of power and resources.

We can start on an upbeat note. More than one hundred and fifty years after Marx was writing about the effects of British rule in India, the question of Asia is back at the heart of development studies. The recent rapid growth of China and India, not to mention of the South-East Asian newly industrialised countries, is profoundly reshaping the map of world economic activity. According to Stephen Radelet and Jeffrey Sachs, in just over forty years, from 1950 to 1992, Asia's share of world income rose from 19 per cent to 32 per cent while its share of world population remained more or less constant at two-thirds. Since that time, China's GDP has continued to grow at around 10 per cent per annum, and now India is recording sustained rates of growth of 7–9 per cent. By 2025 it is likely that 55–60 per cent of the world's income will be earned in Asia, with future relative gains continuing at least up to 2050. 'Standards of living will still be much higher in the West, but average per capita income in Asia will probably increase to around one-third of the U.S. level [by 2025], compared with a meager 13 percent today.' Rates of absolute poverty are also expected to fall sharply over the same period. It is predicted that less than 15 per cent of Indians will have to survive on less than a dollar a day by 2015 – a particularly brutal definition of poverty, it has to be said – as compared to about 30 per cent (300 million people) in the year 2000.

These changes in relative economic power can be expected over the coming decades to be linked to profound shifts in the distribution of political power in the inter-state system. They will also lead to changes in the architectures of international governance, where the United States can be expected to lose its veto powers in the Bretton Woods institutions. The World Bank, too, will be forced to define itself more as a knowledge-providing institution than as a loan dispenser. This will certainly be true of its relations with middle-income countries, where easy access to private capital markets will continue to reduce the attractiveness of loans from the IBRD.

Just why these global shifts are happening, however, remains the subject of considerable debate. Radelet and Sachs advance a version of the pro-globalisation thesis that we have met before (see Wolf, Part 8). They discount Paul Krugman's argument that East Asia's success has been based on heavy investment spending rather than on high productivity growth. Instead, they favour a version of the flying geese model wherein different regions of East Asia have successively acted as export platforms, in the process pulling entire countries more squarely into the global marketplace. Export promotion rather than import-substitution industrialisation has been key, they maintain.

Many scholars disagree with this explanation. A number of institutional economists and sociologists, led by Alice Amsden, Peter Evans and Robert Wade, and ably followed by Ha-Joon Chang, Mushtaq Khan and others, have argued that East Asia's recent successful integration into the global economy followed a long period when states governed domestic markets and tried hard to 'get relative prices wrong'. The key instruments used to support such growth-enhancing governance (as opposed to market-enhancing governance: see Khan, 2005) varied from one country to another, but they included property rights violations (land reform), protectionism and favourable credits for infant industries. Ha-Joon Chang maintains that no country has yet industrialised on the basis of the good governance agendas promoted by the Bretton Woods institutions. No country in the West, Chang argues, secured the structural transformation of their economy as a free democracy. (Strong states can solve certain collective action problems better than popular democracies, no matter that Amartya Sen's account of *Development as Freedom* (2000) argues to the contrary.) And no country in the West took off into self-sustaining growth without active government intervention amid political systems riddled with corruption, defined here as the abuse of public office for private gain. It is thus more than galling, Chang concludes, that the West is now using the agendas of participation, free trade and good governance to 'kick away the ladder' (a phrase first used by Friedrich List: see Introduction to Part 2) that it once used to climb out of mass poverty and low levels of economic development.

Chang and Khan, like James Mahon earlier (see Part 7), express particular concern about the ability of Latin American countries to chart a route to sustainable economic development by means of the neo-liberalism that Radelet and Sachs commend, and which has been commended to Latin America by the Bretton Woods institutions since the early 1980s. The emergence in recent years of political leaders including Hugo Chavez in Venezuela and Eva Morales in Bolivia has encouraged some critics to suggest that Latin America will turn its back on an economic model that has brought little prosperity to poorer or even middle-class households. This seems unlikely, but there are signs that neo-liberalism in Latin America is mutating into something softer, or with safety nets – what Gwynne and Kay (2000) call neostructuralism – even as indigenous voices are being raised more loudly in favour of forms of decision making that will be less dominated by ruling (white) elites.

Meanwhile, in sub-Saharan Africa, the prospect that countries might climb out of poverty by means of neo-liberal globalisation has been challenged more sternly still. The economist Paul Collier is no fan of dirigiste development. He nonetheless takes issue with the suggestion that sub-Saharan Africa can develop simply by exploiting its cheap wages, or by making its economies more attractive to foreign capital inflows. He also doesn't buy into the suggestion that Africa will benefit hugely by allowing its citizens to find work abroad, perhaps in the European Union. Collier argues that the world economy's bottom billion may have missed the boat when it comes to globalisation. He maintains that Africans, especially, will have to wait a long time before they can open up a wage gap with Asia that will be sufficient to attract capital to a region that is chronically short of skills, security and basic infrastructure. As things stand, investment there is simply too risky, as it is in some parts of India and Pakistan. Many African countries remain trapped by poor geography and poor institutions. And worse, to the extent that some Africans can make it to Europe, they are generally the 'best and brightest', the very people the continent can least afford to lose.

Collier accepts that all this 'adds up to a depressing picture of what globalization is doing for the bottom billion', and certainly it is at odds with much conventional wisdom. Even so, it is possible that Collier is underestimating the problems facing some African countries. Security, after all, and infra-structure, generally have to be provided by the state. And yet in large parts of sub-Saharan Africa (and not only in Africa) the talk today is of pervasive state failure, or of states in crisis or under stress. Some states are unable to secure control over their territory. Many are the prisoners of post-imperial partitions that failed to match national boundaries with key ethnic identities. And many more find it hard to raise the revenues that state-building and legitimacy requires, save perhaps from the rents provided by natural resources or foreign aid. In such circumstances it is difficult to build the institutions needed for concerted economic growth and development. (Nor are matters helped by easy access to cheap arms or the ability of ruling elites to place funds in offshore accounts.) It is perhaps rather the case that weak African countries will be unable to protect their national interests and resources against forms of accumulation driven by what David Harvey calls 'the new imperialism'.

Harvey explores the tensions that are played out between the molecular and the territorial moments of capital accumulation on a global stage. In his view, capital is constantly on the lookout for new investment opportunities in the global periphery and violence is always a handmaiden of this molecular search. Land is seized from indigenous people or working-class communities, and state assets are stripped by private monopolists. Labouring people are intimidated by hired thugs or private armies, or, in the case of China, by party elites and the local state. They are paid minimum wages or worse. This is how high rates of growth are secured, albeit growth that is immiserating for the most powerless. At the same time, apparently rootless capitals always flee 'home' in a crisis, and look for support from 'their' governments, whether this be in the United States, China or India. The new imperialism is much like the old imperialism in this respect. It is strongly territorialised and wars are likely to break out for control of space and scarce natural resources. This is how Harvey interprets recent events in Iraq and the broader Middle East. It is all about the oil, he says.

Harvey contends that the rise of China poses a particular threat to current maps of world order and disorder. In part, although he does not elaborate this point, this is because the repression of labour in China is occurring alongside the production of massive unemployment and peasant labour migration to the cities. Ching Kwan Lee develops this argument in the reading reproduced here, and it has become a staple of new wave Chinese cinema. Lee argues strongly that the instability occasioned by the transition from state socialism to market socialism in China is producing high levels of worker activism. In the first years of the twenty-first century this instability was being contained, more or less, by the Communist government. Lee is concerned, nonetheless, that the political settlement is precarious. In the longer run, she says, 'The political consequences of reform may well depend on whether class conflicts, among other types of social conflicts, can . . . be contained within the fledgling system of "socialist legality", in a polity with a widening rift between central and local state power, and in a society rife with contradictions between entrenched socialist rhetoric and emergent capitalist realities.'

To some degree, too, the political settlement within China, and perhaps also in India, will come to depend on how those countries strike bargains over natural resources in Africa, West and South-East Asia, and Latin America. The new imperialism differs in this specific respect from the imperialism of the nineteenth century: the two major Asian powers see themselves as potential superpowers and are begin-ning to conduct their foreign policies accordingly. In Africa already the major Western donors, including the UK's Department for International Development (DFID), are noting with some concern – and perhaps self-righteously – that some African elites are disinclined to listen to agendas of democratisation and good governance when the Chinese are prepared to buy natural resources – for example, oil from Sudan – no questions asked. (It should be noted, too, that the rise of China and, less so, India, as manufacturing powers will lead to a relative lowering of the prices of manufactured goods relative to those of primary commodities. This might fuel more rapid growth in parts of Africa, while also deepening the region's dependence on natural resources.)

The rapid movement of China and India into Africa is also likely to change the nature of 'the gatekeeper state' that is described by Frederick Cooper in his reading here (and which is hinted at by Paul Collier). Cooper notes that 'one of the origins of instability in Africa is the inability of gatekeepers to keep the gate. Traders bypass customs collection; oppositional networks establish connections with outside powers and get arms and support; and people often try to live their lives within and across territorial boundaries as if the state has limited power over them'. In large part, the porosity of these states, together with their hollowed-out nature, is the result of the peculiar, and peculiarly damaging, nature of the incorporation of African territories first into European colonial systems of rule, and second – as Jeffrey Herbst has argued – into a world of aid-fed development centred on the alleged primacy of the nation state.

Cooper is quick to point out that there is no generic gatekeeper state in Africa – he contrasts the recent histories of Ghana, Congo–Zaire, Senegal, Nigeria, Kenya and Tanzania – but he is in agreement with Herbst in this last respect. There is a tendency for African 'big men' to court what Achille Mbembe calls a culture of vulgarity, sometimes using aid funds to 'eat their countries' even as their personal wealth and sexual appetites are brazenly flaunted. As Cooper also notes, however, the power that derives from rents or looted money is not read in unambiguous fashion by ordinary men and women. 'Popular culture is at one level angry . . . but at another level desirous. The vulgarity itself marks both an evil – dangerously close to witchcraft, for supernatural forces are also seen to "eat" their victims – and a sign of power, something deeply desirable as well as immoral.'

For all this, Cooper ends his account on an ambivalent note that is some way removed from Afro-pessimism. He suggests that African big men may or may not choose to believe they can gain from 'fostering economic growth in which many people share'. Whether they go for more inclusive growth or the opposite strategy of denying 'potential rivals access to resources' and offices will depend on how they read these strategies in terms of their own political careers. Fernando Coronil, in the next reading, also dares to imagine a better future, one that – following Walter Benjamin – seeks to 'understand the past in order to find within the present the seeds of a desirable future'. The focus of Coronil's work, however, is more discursive than Cooper's, and he helps point us back to some big-picture issues relating to the object and purpose of development. We have included Coronil here because he asks us to destabilise 'the old imperial maps shaded in black and white' – those maps which later sought to establish sharp boundaries between the West and the Rest; the developed and the underdeveloped; the First, Second and Third Worlds – in favour of what he calls 'nonimperial geohistorical categories'.

Coronil recognises this is no easy task, and he warns against embracing forms of utopian thought which would, as Terry Eagleton once put it, 'simply make us ill' (Eagleton *et al.*, 1990: 25). Presumably, this would mean thinking critically about forms of anti-developmentalism, and especially those which presume to wish away the social forces that produce any recognisable account of the modern condition: markets and divisions of labour, systems of international relations that are more or less imperial, and dynamic interrelationships with nature (or produced nature). Rather like Gandhi, some contemporary proponents of anti-development (see the essays in Rahnema and Bawtree, 1997) opt for a politics of blunt refusal and cultural essentialism in which the West, Development, Science and Technology are capitalised and set in stone. They are all read as social practices that are one-sidedly disempowering, rather than as agencies of empowerment and disempowerment that are keenly contested and which have social effects that are always context specific and often highly gendered. (Consider the washing machine. Hugely wasteful of water, no doubt, but a very considerable boon to women where patriarchal societies make the prior judgement that women will have the sole or primary responsibility for washing clothes and other items.)

But there is a lot more to post-developmental thought than hardline anti-developmentalism, and it would be wrong to conclude this reader without taking seriously two arguments that are being raised against development as unconsidered modernisation, or globalisation as unrestrained economic growth. On the one hand there is the matter of what Arjun Appadurai calls the 'fear of small numbers'. Much like the war on terrorism launched by George Bush after the 9/11 attacks, development is (still) being used by some governments as an excuse to silence dissident voices or to erase social practices of which they disapprove. Nor is this only a feature of Western governments or Western imperialism. The commitment of the Chinese Communist Party to a bricks-and-mortar conception of development as modernisation has encouraged neither sensitive urban upgrading nor great respect for the rights of minority populations to defend their cultural differences.

In addition, and this is more Appadurai's concern here, while globalisation needn't lead to cultural genocide – it more often promotes cultural intermingling and hybridisation – it can produce 'new incentives for cultural purification as more nations lose the illusion of national economic sovereignty or well-being' (2006: 7). Globalisation might then be linked directly with the rise of ethnic cleansing post-1990, in democratic states like India as well as in Rwanda and the former Yugoslavia. In these changed circumstances, Appadurai concludes, minorities become 'the major site for displacing the anxieties of many states about their own minority or marginality (real or imagined) in a world of few megastates, of unruly economic flows and compromised sovereignties'. In much the same way, strong advertisements of claims to cultural difference and self-determination, as, for example, in Tibet, are likely to bring forth concerted efforts at developmentalism. In this case, the aim is to furnish precisely those dull landscapes of compulsion and homogenisation – the abstract spaces of mobile capital – that post-developmentalism and rights-based development advocacy is concerned to problematise and speak against.

(Thankfully, as Appadurai points out elsewhere in his book, there is some evidence that the production of flat worlds and abstract space will be countered by transnational advocacy groups and non-governmental organisations (NGOs), or by what he calls, more broadly, 'grassroots globalization'. Appadurai gives as one example the Shack/Slumdwellers International, an organisation that is 'focused on building the capacity of poor people in cities to explore and practice specific means of urban governance with an eye to building their own capacity to set goals, achieve expertise, share knowledge, and generate commitment' (see also Heller on Porto Alegre in Part 7).)

And lastly, looming over everything, there is the spectre of human-induced climate changes on such a scale that we are faced with the end of the world as we know it, as Tim Dyson puts it in our final reading. Among the many virtues of this reading is that Dyson shows us just why we should be worried about climate change, and why – as with many disease epidemics – it is unlikely to call forth the global public action that is required to deal with it either soon enough or on a grand enough scale. Dyson takes us soberly through the evidence that is already available to us. He shows that its impacts will bite particularly hard in the developing world, where it might be linked among other things to crises of food production, water scarcity, distress migration and the still greater urbanisation of poverty, all of which could spark political conflict. Dyson doesn't hector and he doesn't use scare tactics. But he does enough, we think, to suggest that few ideas were more dangerous in the middle part of the twentieth century than that man had achieved mastery over nature – whether armed with DDT in Sri Lanka, irrigation in California or the thoughts of Chairman Mao in China. Rhoads Murphey, Rachel Carson and James Scott have issued warnings before in this reader about the arrogance of such assumptions, and about the sting in the tail they might yet reveal. Here Dyson does this same. As much as any reading in this book, he demands that we look again at that most contentious of words – development – and the world of social and economic practices that are sponsored in its name.

REFERENCES AND FURTHER READING

Amsden, A. (1989) *Asia's Next Giant: South Korea and Late Industrialisation*, Oxford: Oxford University Press.

Appadurai, A. (2006) *Fear of Small Numbers: An Essay on the Geography of Anger*, Durham, NC: Duke University Press.

Benjamin, W. (1969) *Illuminations*, New York: Schocken Books.

Chang, H-J. (2002) *Kicking Away the Ladder: Development Strategies in Historical Perspective*, London: Anthem Press.

Davis, M. (2006) *Planet of Slums*, London: Verso.

Devji, F. (2005) *Landscapes of the Jihad: Militancy, Morality and Modernity*, Ithaca, NY: Cornell University Press.

Eagleton, T., Jameson, F. and Said, E. (1990) *Nationalism, Colonialism and Literature*, Minneapolis: University of Minnesota Press.

Evans, P. (1995) *Embedded Autonomy: States and Industrial Transformation*, Princeton, NJ: Princeton University Press.

Government of the United Kingdom, Department for International Development (2006) *Eliminating Poverty: Making Governance Work for Poor People*, Command 6876, London: The Stationery Office.

Gwynne, R. and Kay, C. (2000) 'Views from the periphery: futures of neoliberalism in Latin America', *Third World Quarterly* 21: 141–56.

Harvey, D. (2003) *The New Imperialism*, Oxford: Clarendon.

Herbst, J. (2000) *States and Power in Africa: Comparative Lessons in Authority and Control*, Princeton, NJ: Princeton University Press.

Khan, M. (2005) 'Governance, economic growth and development since the 1960s', Mimeo: SOAS (Background paper for World Economic and Social Survey, 2006).

Krugman, P. (1994) 'The myth of Asia's miracle', *Foreign Affairs*, November–December.

Mbembe, A. (2001) *On the Postcolony*, Berkeley: University of California Press.

Payne, A. (2005) *The Global Politics of Unequal Development*, London: Palgrave.

Rahnema, M. and Bawtree, V. (eds) (1997) *The Post-Development Reader*, London: Zed Books.

Sen, A. K. (2000) *Development as Freedom*, New York: Anchor.

Stern, N. (2007) *The Economics of Climate Change: The Stern Report*, Cambridge: Cambridge University Press.

Wade, R. (1990) *Governing the Market: Economic Theory and the Role of Government in East Asian Industrialization*, Princeton, NJ: Princeton University Press.

'Asia's Reemergence'

Foreign Affairs (1997)

S. Radelet and J. Sachs

Editors' Introduction

Stephen Radelet was born in 1957 and is a Senior Fellow at the Center for Global Development in the United States. In the 1990s, Radelet worked alongside Jeffrey Sachs at the Harvard Institute for International Development. From January 2000 to June 2002 he was Deputy Assistant Secretary of the US Treasury for Africa, the Middle East and Asia, and in that capacity he worked on the US response to Turkey's financial crisis, Pakistan's debt restructuring, and India's changing economic relationships with the United States. An expert on trade, aid and debt, Radelet has lived and worked in Indonesia and The Gambia, and has worked closely with the president and minister of finance of Liberia. He is in this sense a classic development professional, mixing theoretical and empirical work and active policy advising in a manner that probably no one has perfected more over the past twenty or so years than his co-author here, Jeffrey Sachs. Born in Detroit, Michigan, in 1954, Sachs's website at Columbia University's Earth Institute, where he has been the director since leaving Harvard in 2002, notes that he was listed as one of the 100 most influential people in the world by *Time* magazine in 2004 and 2005. The website also notes that Sachs 'is internationally renowned for his work as economic advisor to governments in Latin America [his first big break came during the Bolivian hyperinflation of the mid-1980s], eastern Europe, the former Soviet Union, Asia and Africa, and his work with international agencies on problems of poverty reduction, debt cancellation for the poorest countries, and disease control'. The pro-poor relevance of Sachs's work in the ex-Soviet Union has not been universally admired – although he rejects the phrase, Sachs has been charged by some fellow economists (and many others) with foisting 'shock therapy' on Yeltsin's Russia – but there is considerable regard for Sachs's important work with ex-UN Secretary General, Kofi Annan, around the Millennium Development Goals, and for his activism on behalf of various public health and debt cancellation initiatives, particularly in Africa. In 2005, Sachs published his best-selling book, *An End to Poverty?*, a book that at once posed as an economist's biography and as a vital action plan for making poverty history in Africa. Though forcefully challenged by William Easterly, among others, *An End to Poverty?*, with its foreword by the rock singer Bono, remains required reading in the development field.

In the reading published here, Radelet and Sachs combine to present an optimistic account of Asia's changing position in the world economy since the 1950s. In strong, clear language they reject Paul Krugman's rather dismissive take on the East Asian miracle (plenty of hard work, not much in the way of innovation or productivity gains), and they insist that Asia as a whole, notwithstanding the financial crisis that hit much of the continent in the late 1990s (when they were writing), is set fair to secure another fifty years of relative and absolute growth within the world economic system. They also note, very properly, that the speed with which Asia's share of the world economy can be made to match or better its population share, will depend above all on what happens over the next few decades in India and China. Bad policies can still upset a future that looks rosy.

Key references

S. Radelet and J. Sachs (1997) 'Asia's Reemergence', *Foreign Affairs* 76 (6): 44–59.

S. Radelet and J. Sachs (1998) 'The East Asian financial crisis: diagnosis, remedies, prospects', *Brookings Papers on Economic Activity* 1: 1–74.

S. Radelet and J. Sachs (2000) 'The onset of the East Asian currency crisis', in Paul Krugman (ed.) *Currency Crises*, Chicago, IL: University of Chicago Press. Also available as National Bureau of Economic Research Working Paper #6680.

S. Radelet, J. Sachs and J-W. Lee (2001) 'The determinants of and prospects for economic growth in Asia', *International Economic Journal* 15 (3).

D. Perkins, S. Radelet and D. Lindauer (2006) *Economics of Development*, sixth edition, New York: W. W. Norton.

CAPITALISM LEAVES ITS WESTERN ENCLAVE

Beginning in the early 1500s, for more than four centuries now, the West has been ascendant in the world economy. With but 14 percent of the world's population in 1820, Western Europe and four colonial offshoots of Great Britain (Australia, Canada, New Zealand, and the United States) had already achieved around 25 percent of world income. By 1950, after a century and a half of Western industrialization, their income share had soared to 56 percent, while their population share hovered around 17 percent. Asia, with 66 percent of the world's population, had a meager 19 percent of world income, compared with 58 percent in 1820. In 1950, however, one of the great changes of modern history began, with the rapid growth of many Asian economies. By 1992, fueled by high growth rates, Asia's share of world income had risen to 33 percent.[1] This tidal shift is likely to continue, with Asia reemerging by the early 21st century as the world's center of economic activity.

Asia's sudden ascent has become something of a Rorschach test for the economics profession and the foreign policy community. For some, Asia's rapid growth is an economic miracle that calls for a re-evaluation of Western economics strategies. For others, such as the MIT economist Paul Krugman, writing in the November/December 1994 *Foreign Affairs*, the rapid growth has looked hollow. Not only has there been no miracle, but there was reason to believe that Asian growth might display weaknesses similar to those of the period of rapid Soviet growth in the 1950s and 1960s. These doubts seemed to find support in the sudden, sharp currency crises

that gripped several high-flying Southeast Asian economies (especially Indonesia, Malaysia, the Philippines, and Thailand) in mid-1997. Even money managers formerly enamored of the region decried underlying institutional weaknesses, including corruption, nepotism, populist policies, and insufficient banking regulation.

The Southeast Asian currency crises of 1997 are not a sign of the end of Asian growth but rather a recurring – if difficult to predict – pattern of financial instability that often accompanies rapid economic growth. Just as Indonesia, Malaysia, and Korea rapidly recovered from financial crises in the 1970s and 1980s, so the Asian economies are likely to resume rapid growth within two to three years. In the long term, growth will continue because most of Asia has adopted capitalism as the organizing basis of economic life and become deeply integrated into the global economy. This has been true for more than a century in Japan, since the Meiji Restoration of 1868. Korea and Taiwan adopted essentially capitalist development strategies in the 1960s, while most of Southeast Asia made similar choices in the 1970s. Even China in recent years can be considered to have adopted an essentially capitalist development model, despite continued Communist Party rule and a state sector that still employs around 18 percent of the labor force. India began turning away from a milder version of socialism in the early 1990s, though Indian domestic politics still contains strong doses of anticapitalist rhetoric.

If there is anything to the "Asian miracle," it is that several governments, benefiting from Japan's early experience and from each other's experiences since the 1960s, have been able to create an economic environment for profitable, private investment – almost always with important foreign partners – despite serious shortcomings in overall political and economic conditions. They did so, in most cases, by creating in the midst of weaker economic institutions a capitalist enclave that has gradually spread throughout the economy. Put another way, Asia's challenge, so far accomplished, has been to create a virtuous circle, in which a modern economic sector originally confined to an enclave has not only expanded through new investments but has fueled a much broader modernization of political and economic institutions. Addressing governmental weaknesses is the largest hurdle facing Asian countries, particularly the region's two colossi that constitute around two-thirds of Asia's population and around 38 percent of the world's population – China and India. Corruption is rife, judicial systems are weak, and local governments often lack authority and adequate finances. But global capitalism stirs powerful forces for economic growth even in the face of serious limitations in law, economic structure, and politics.

The West currently has a disproportionate share of world income, but its share will diminish as capitalism pervades Asia. By 2025, Asia will likely reassume its place at the center of the world economy. Asia may account for 55 to 60 percent of world income in the year 2025, with the West's share falling from around 45 percent today to between 20 and 30 percent. Standards of living will still be much higher in the West, but average per capita income in Asia will probably increase to around one-third of the U.S. level, compared with a meager 13 percent today.[2]

CURRENCY CRISES

These long-term projections might seem heedlessly optimistic in the face of the economic shocks buffeting Asia. Recently the darlings of the international investment community, Southeast Asian economies have taken a beating at the hands of money managers inside and outside their countries. Equity markets fell by around 50 percent (in dollar terms) in Thailand between January and September 1997, while declines in other Asian markets have been in the range of 40 percent. Has the bubble of East Asian growth burst, the years of rapid growth already a thing of the past?

The currency upheavals probably reflect short-run financial considerations rather than a long-term crisis of regional growth. The simplest part of the explanation is that the Southeast Asian countries increasingly pegged their currencies to the U.S. dollar during the 1990s, even though the region's trade depends not just on the United States but on the European and Asian markets (with Japan playing an important role). After mid-1995 the U.S. dollar began to appreciate sharply vis-à-vis the yen and the major continental European currencies such as the deutsche mark, French franc, and Spanish peseta. As a result, the Southeast Asian currencies also appreciated sharply against yen, European currencies, and other national currencies, such as the Chinese yuan, that also depreciated vis-à-vis the dollar. It took 3.5 yen to buy one Thai baht in May 1995; by May 1997, on the eve of the currency crisis, it took 4.6 yen, a rise of 31 percent. In effect, the Southeast Asian exporters were pricing themselves out of the European and Japanese markets. They were also facing stiff competition from China, which had devalued the yuan in January 1994. Naturally, exchange rates came under pressure.

There is a second, related aspect of the financial crisis. The Southeast Asian countries all attracted considerable foreign capital investments in the 1990s. Many of these investments enhanced export potential and thereby contributed to the main engine of long-term growth. In the mid-1990s, however, a rising share of foreign flows appears to have headed for speculative investments in the real estate markets. Following financial market deregulation in many countries, commercial banks got into the act by borrowing dollars from abroad and lending the funds domestically to real estate developers. When they made these loans in local currency, the banks exposed themselves to the risk of currency depreciation, since the value of such loans would fall relative to the value of their dollar borrowing. Even when the domestic real estate loans were in dollars, however, the banks were at risk, since domestic property developers would be unable to repay the dollar-denominated loans in the event of a weakening of the domestic currency. The pegged exchange rate regimes gave (misplaced) confidence to the financiers to accept such risks. Thus when property markets weakened in 1996 and the currencies depreciated in 1997, the banks were hit by a double whammy of nonperforming loans, and many faced insolvency. The banking crisis was exacerbated, especially in

Thailand, by inadequate levels of bank capital and poor supervision.

The currency crises will likely have continuing effects for two or three years on the overall health of the banking system and on the construction sector, so that growth during 1997 and 1998 will be slower than in previous years. The crises also underscore better regulation and supervision of the financial markets as a condition for more stable growth in the future. These are serious challenges. Nonetheless, looking forward more than a couple of years, the currency crises (appropriately dealt with) do not call into question the underlying export-growth strategy of the region or the medium-term growth prospects.

THE KRUGMAN CRITIQUE

Paul Krugman's critique of East Asia's rapid growth was that it was unimpressive and even suspect because it was based largely on heavy investment spending rather than productivity growth. He riled Asian policymakers by noting that rapid Soviet growth had been similarly based on heavy investment spending. The implication was that East Asian growth was fragile, and, indeed, likely to founder. Krugman was right to dispel the notion that Asia's "miraculous" growth could continue at very high rates forever, but he was wrong about the solidity of Asia's economic development, and he gave a misleading impression of Asia's prospects for the future.

First, Krugman's empirical claims about investment versus productivity as factors of growth were much too stark. Most researchers have found that rapid growth in Asia has been due to both productivity growth and capital investment, though it is probably true that investment spending, not pure productivity gains, has been the major source of overall GDP growth. Second, and perhaps more important, growth based on rapid capital accumulation (that is, physical investment spending) can be highly desirable, as long as the investment spending itself meets the market test, in the sense of rates of return that exceed the cost of capital. Here the contrast between East Asia and the former Soviet Union could not be greater. Soviet capital was allocated by bureaucratic fiat, not market forces. Rates of return in the Soviet Union were low and falling rapidly as early as the late 1950s. Rates of return in East Asia, in contrast, have been high and have declined only

gradually over time. In Korea, for example, after 30 years of rapid growth, the marginal productivity of capital is estimated at around 20 percent, far higher than estimates for the United States of around 11 percent.

Krugman's critique is correct on a general point. To the extent that growth is driven by capital accumulation rather than pure productivity gains, the marginal productivity of capital is likely to decline as the capital stock deepens — that is, as capital per worker in the Asian economies rises to the level of Western economies. But this point does not really answer the question of how *fast* the decline in growth is likely to be. All other things being equal, growth rates tend to fall gradually (over decades) as developing countries close the income gap with the United States (at around $27,000 per capita). A country at one-fourth the U.S. income level experiences a growth rate roughly 2.8 percentage points above the U.S. rate. If the United States manages per capita growth of about 2.0 percent per year, a country at $7,000 per capita (such as Thailand) could have per capita growth around 4.8 percent per year, equivalent to aggregate GDP growth of 6.5 percent to 7.0 percent per year. When the income gap narrows to around half the U.S. level, growth diminishes by about 1.4 percent per year, so a country at around $13,500 might be expected to have a per capita growth rate near 3.4 percent per year. Interestingly, Japan itself — once viewed naively as a country that would quickly soar ahead of U.S. income levels — demonstrated this kind of tapering off of growth as its economic success led to a narrowing (and virtual elimination) of the income gap with the United States.

Of course growth also depends on policy choices, geography, and demography. For example, demographic changes supported high savings and rapid growth in the past, as falling fertility rates led to an increasing share of workers in the population. In the future, however, Asia's aging populations — most immediately in Japan — will put pressure on pension and public health insurance systems and slow economic growth.

The main conclusion, shown in Table 1, is that while all of Asia has continued room for significant growth, the high-income East Asian economies are likely to grow more slowly in the next 30 years than in the past 30 because of both capital deepening and demographic changes. Southeast Asia and China, where incomes are low compared with the advanced

Economy	GDP per capita relative to U.S.			Projected per capita GDP growth rate, 1996–2025
	1965	1995	Projected in 2025	
Four Tigers	17.3%	72.2%	98.5%	2.8
Hong Kong	30.1	98.4	116.5	2.1
Singapore	15.9	85.2	107.0	2.5
South Korea	9.0	48.8	82.6	3.5
Taiwan	14.2	56.2	88.0	3.1
China	3.2	10.8	38.2	6.0
Southeast Asia	10.0	21.2	45.7	4.5
Indonesia	5.2	13.1	35.8	5.0
Malaysia	14.3	36.8	71.2	3.9
Philippines	10.7	9.4	28.5	5.3
Thailand	9.7	25.6	47.4	3.8
South Asia	8.5	9.2	21.3	4.4
Bangladesh	9.9	8.5	17.2	3.9
India	6.5	7.8	24.4	5.5
Pakistan	7.7	7.7	18.1	4.4
Sri Lanka	10.1	12.6	25.3	3.9

Table 1 Growth prospects for Asian economies, 1995–2025

Source: *Emerging Asia: Changes and Challenges.*

economies, should be able to grow at about the same rate as in the past 30 years, as the impact of capital deepening is offset by continued policy reform and institutional upgrading. In these countries, the slowing effect from the aging of society is still several decades away. South Asia is likely to accelerate its growth rate, as policy reforms take root and demographic shifts work in favor of high savings and faster growth.

EAST ASIA'S GROWTH STRATEGY

Table 2 reminds us of the extraordinary achievements of Asian economies. From 1965 to 1995, per capita income rose more than sevenfold in the four "tigers" and about fourfold in Southeast Asia and China. As is well known, all these high-flying economies exhibited certain similarities. They achieved rapid export growth, followed prudent fiscal policies, recorded high rates of saving, pursued a public policy in support of rising literacy and basic education, did not undermine agriculture, and

achieved a rapid transition to low rates of population growth. While these basic features of high growth seem straightforward – and indeed are the workhorse components of typical development programs of the World Bank – something went very right in East Asia that did not click in other parts of the developing world.

Developing countries typically lag many years behind the advanced countries in the adoption of new technologies in manufacturing and services. Infrastructure is poor. Research and development is generally far behind the world standard and is useful mainly to support the adoption of proven technologies from the advanced economies. How can the lagging countries hope to catch up with the world leaders? In development thinking over the past half-century, three types of answers have been given to this question. The first has been the doctrine of the "big push," according to which a government should contrive to put all the supporting pieces in place at nearly the same time through large-scale physical investments in infrastructure, basic industry, and research and development, as well as legal and

Four Tigers	6.6%
Hong Kong	5.6
Korea	7.2
Singapore	7.2
Taiwan	6.2
China	5.6
Southeast Asia	3.9
Indonesia	4.7
Malaysia	4.8
Philippines	1.2
Thailand	4.8
South Asia	1.9
Bangladesh	1.6
India	2.2
Pakistan	1.6
Sri Lanka	2.3
OECD	2.1
Latin America	0.9
Sub-Saharan Africa	0.2

Table 2 Asian annual per capita GDP growth, 1965–1995
Source: *Emerging Asia: Changes and Challenges.*

institutional changes. The Stalinist drive toward rapid industrialization in the 1930s and China's Great Leap Forward of 1958–61 were the most destructive manifestations of this thinking, but numerous failed lesser big pushes litter the development scene.

The second idea has been the doctrine of "import substitution" or "infant industry protection," which holds that national industry requires breathing space to catch up with foreign competitors. This venerable doctrine, extending from Alexander Hamilton in 1790 to Friedrich List in the 1840s to Raul Prebisch in the 1950s, has achieved some successes but a much larger number of failures. Infant industry protection often becomes senile industry protection: domestic firms in small markets never attain the scale at which they could overcome foreign cost advantages, and protection leaves enterprises lazy, dependent on state handouts, and behind in adopting technology. Infant industry protection works best in large markets. Its track record in the United States, Germany, Japan, and more recently in Brazil, China, and Korea shows some modest successes (balanced by many high costs). Its record in much smaller economies in Latin

America, South Asia, Central Europe, and elsewhere is one of almost unremitting failure.

The third doctrine, which best exemplifies the Asian paradigm, was aptly named by the Japanese economist Kaname Akamatsu in the 1930s: the "flying geese" model, according to which countries gradually move up in technological development by following in the pattern of countries just ahead of them in the development process. In this vision, Korea and Taiwan take over leadership in textiles and apparel from Japan as Japan moves into the higher-technology sectors of electronics, transport, and other capital goods. A decade or so later, Korea and Taiwan are able to upgrade to electronics and auto components, while the textile and apparel industries move to Indonesia, Thailand, and Vietnam.

To some extent, the flying geese pattern can be seen as the natural outcome of market forces: labor-abundant, capital-scarce economies will be internationally competitive in labor-intensive sectors, such as apparel, and will graduate to more capital- or skill-intensive sectors as savings and education deepen the pool of capital and skilled workers. And yet, as the Asian economies demonstrated, more than markets are required. Even the simplest labor-intensive products (apparel, footwear, electronics assembly) are part of a sophisticated international division of labor, one increasingly determined by multinational enterprises and technological designs created in the advanced economies. The trick is to bring multinational production enterprises and their technologies into the poorer economies to link them to the engines of growth of the advanced economies.

If the paradigmatic institution of the big push was state ownership of industry, and for import substitution was private ownership backed by protectionism, for flying-geese development it is the export platform. The idea behind an export platform is to create an enclave economy hospitable to foreign investors and integrated into the global economy, without the problems of infrastructure, security, rule of law, and trade policies that plague the rest of the economy. Asian governments introduced several variations of the export platform, including export processing zones (EPZs), bonded warehouses, special economic zones, and duty drawback systems. Governments supported these institutions with macroeconomic policies that strengthened the incentives for labor-intensive exports, especially via appropriate exchange rates.

The export platform strategy began with textiles and apparel but really took off with electronics. With the emergence of the semi-conductor industry and the early realization by Hewlett-Packard, Texas Instruments, National Semiconductor, and others that even in this very high-tech sector there were several very low-tech processes such as chip assembly, the new industry leaders began a search for low-wage production sites. Advances in information technology greatly enhanced the applicability of export platform production. Computer-assisted design and manufacturing allows digitized instructions for design, fabric-cutting, or other technical specifications to be sent from engineering headquarters to local production sites via phone lines. Reductions in transport costs, for example through the computerization and faster turnaround time of container ports, also facilitated the outsourcing of production.

Some early candidates for low-wage production sites were in America's back yard. Texas Instruments began production in El Salvador, only to leave in the mid-1980s when political instability and *yanqui*-baiting heated up. Hong Kong was a more likely candidate: stable, with low wages, and under British law and political control. Hong Kong's early success in the 1950s was not lost on its neighbors such as Taiwan and Korea, and then Singapore in 1965 and Malaysia in the early 1970s. The Southeast Asian countries began to compete furiously for footloose electronics firms that were neglected or shunned in the Caribbean, Mexico, Central America, or other potential sites. By 1975 East Asia employed roughly 95 percent of worldwide offshore electronics assembly workers. By the mid-1970s, the die was cast: poor East Asian countries were swept up in the worldwide electronics revolution, while Latin America, Central Europe, and sub-Saharan Africa were bypassed. Starting in 1980, China created several special economic zones up and down the coast, but especially in Shenzhen City, on the border of Hong Kong. Within a few years, one of the world's greatest export booms was under way, with millions of new jobs in China directed toward labor-intensive export production.

EPZs, bonded warehouses, and duty drawbacks accounted for the bulk of the manufacturing exports of the East Asian tiger economies in the early years, but they also served as models. In Korea, for example, there was more stress on joint ventures and technology licensing than on foreign investment in EPZs.

But the underlying model was the same: domestic production linked with worldwide technology through the direct involvement of foreign firms. Joint ventures, original equipment manufacturing, outsourcing under license, and similar arrangements all facilitate export-oriented manufacturing by poorer economies.

The flying geese of East Asia caught the updraft of global electronics production, which has helped carry them through more than 15 years of rapid economic growth. In this one sector lies much of the manufacturing export "miracle" of Malaysia, Singapore, Korea, Taiwan, and to a lesser extent Hong Kong (which instead became the service center for export platform production in southern China). Table 3 highlights the dramatic shift in exports from primary commodities to labor-intensive sectors, including apparel and textiles, but especially electronics components and machinery.

Table 3 also highlights the fallacy of the argument that import substitution rather than export promotion was the key to East Asia's success. Korea and to a lesser extent Taiwan, Indonesia, and Malaysia, have combined the export-promotion strategy with an import-substitution strategy. Korea, for example, spent considerable effort fostering heavy industry, such as steel and chemicals, behind protectionist barriers. But the protected sectors, by and large, have played a small direct role in export success and have not become the export champions. They have apparently also not even played a large role as inputs into the export sectors. Korea's electronics boom did not rely on its steel and chemicals. The connection between protectionism and export success is equally tenuous in Malaysia and Indonesia.

The underlying assumption of the flying geese approach is that the sophistication of domestic production will move forward one position at a time. That is, a country assembling shoes is not likely to get stuck at that stage; experience, education, and further physical investment will lead from footwear to simple electronics assembly and from there to more sophisticated consumer goods and then to automotive components, heavy machinery, and perhaps on to high-technology goods. Critics of the labor-intensive exports strategy charge that it is a dead end. They are probably correct that EPZ production alone does not guarantee a foothold on the next rung on the ladder. But all the early East Asian export-platform graduates – Hong Kong, Korea, Singapore, Taiwan – were able to develop higher levels of local technology

Economy	Year	Primary products	Textiles and other labour-intensive items	Electronics and machinery	Other manufactures
Indonesia	1970	98.6%	0.2%	0.0%	1.2%
	1994	60.7	22.8	4.8	11.7
Malaysia	1970	94.7	0.9	1.0	3.4
	1994	29.4	8.3	44.4	17.9
Singapore	1970	71.2	6.8	8.0	14.0
	1994	17.1	4.9	57.7	20.3
South Korea	1970	35.2	39.6	6.3	18.9
	1994	8.1	22.2	34.5	35.2
Thailand	1970	94.0	1.4	0.1	4.5
	1994	31.1	23.8	27.7	17.4

Table 3 Composition of Asian exports, 1970 and 1994

Source: *Emerging Asia: Changes and Challenges.*

and sophistication, typically continuing to rely on joint ventures and strategic alliances with more sophisticated multinational firms. Acer cut its teeth on computer production under license for U.S. brands; Samsung went from chip assembly to global leadership in 64K random access memory chip production allied with IBM and other electronics leaders.

THE CHALLENGES OF GOVERNANCE

The emphasis here is on the industrial component of the overall growth nexus, since it is crucial and has been poorly understood by outside observers. Consider the debate over industrial policy. All of the successful East Asian countries shared a common industrial policy: promotion and support of labor-intensive exports. It involved picking winners, at least in the narrow sense of recognizing early on that electronics assembly operations could provide a strong impetus to growth and were therefore worth attracting through special zones, tax holidays, and other investment incentives. (It is truer to say that the electronics "winners," like Texas Instruments, Hewlett-Packard, and Intel, picked Asia, rather than the other way around.) Traditional industrial policies based on import substitution to promote heavy industry were also carried out, but not in all the successful countries. To the extent these policies brought any net benefits – and it is far from clear that they did –

their success was limited to Japan, Taiwan, and Korea.

Asia has achieved rapid growth *despite* severe limitations in its institutions. The most general challenge facing the region is the creation of systems of governance and law beyond the export platforms that are consistent with the needs of sophisticated, high-income economies. In much of Asia, the rule of law remains weak. Strong central governments control powerful and politicized bureaucracies that can override local interests, the judiciary, and even private property rights. Unsurprisingly, much of Asia, including some of its fastest growing countries, ranks very poorly on international opinion surveys regarding the extent of corruption and bribery. Far from greasing the wheels of commerce, corruption was a factor in the weak financial market regulation that contributed to this year's currency crises. A recent empirical study by Shang-Jin Wei of Harvard University found that Asian corruption greatly discourages foreign direct investment, equivalent to a tax on multinational firms of 20 percent or more.[3] Local governments are also weak, unable to address urgent infrastructural and regulatory challenges. The *Emerging Asia* study documented that tens of billions of dollars in environmental degradation takes place every year in Asia's megacities due to extreme but remediable levels of pollution and congestion.

The challenge of governance will be most acute in the two mega-states. China faces severe and growing strains on a centralized political system that for more than two millennia has been predicated on a largely sedentary peasant population. The very underpinning of Chinese statecraft is called into question by the social dynamism and geographic mobility of modern Chinese society. The share of the Chinese population engaged in agriculture has declined rapidly from around 70 percent in 1980 to below 55 percent today. On the other hand, the proportion of the labor force on the move within China has burgeoned to almost unimaginable (and still undocumented) proportions. As estimated 100 to 150 million are migrating within the country, mainly from countryside to urban areas and from the interior to the coasts, which offers profound advantages for export based activity. Local governments and transport infrastructure have been overwhelmed by these population movements. Similarly, social systems are breaking down, since they too were predicated on an immobile population locked into villages or into centrally planned state enterprises that expected and guaranteed lifetime employment.

China will have to struggle with privatization, banking reform, and legal modernization, as well as the daunting challenges of environmental degradation, wide regional inequalities, pervasive corruption, and profoundly inadequate infrastructure. There is a path ahead, based as in the rest of Asia on the proposition that institutional upgrading must proceed in step with economic development. In the final analysis, however, sustained economic development will depend on the ability of China's political system to move beyond traditional models of statecraft.

India came late to the process of global integration and market reform. While its economy was never subjected to the extreme versions of socialism practiced in the Soviet Union and China, the much milder form of Indian socialism held economic growth in check for nearly four decades. Until 1991, economic life was tied up in a mind-numbing and corrupt system of licenses and government approvals for investments, imports, exports, employment, land purchases, hiring and dismissals, and virtually every other aspect of economic activity – the so-called License Raj. When a balance-of-payments crisis threatened India in mid-1991, the system finally cracked, with Nehruvian socialism succumbing to history in the same year as the Soviet Union. Since then, a considerable portion of the License Raj has been dismantled, and India's eco-

nomic growth has increased to more than six percent per year.

Like China, India is in a complex and gradual transition from socialism to markets, from traditional political structures to those needed for a rapidly growing economy. India has the vast advantage over China of a constitutional order predicated on federalism, democratic legitimacy, and the rule of law. The problems in India lie in specific practices. While India is under the rule of law, the laws are often terrible and antiquated. Federalist arrangements purportedly give the states broad authority to carry out regional policies, but in almost every policy area, ranging from land use to labor rights to trade to infrastructure development, a morass of relations between the central government and the states can stall necessary actions for years.

The remarkable period since mid-1996, when the once-dominant Congress Party lost its governing majority, suggests reasonable prospects for further reform. A short-lived government led by the Hindu-nationalist Indian People's Party (BJP) committed itself to continued economic liberalization and market reforms, albeit with a nationalist rhetorical stance. The BJP government was followed by a United Front coalition government, in which a remarkably disparate group of centrists and leftists committed to a "minimum program" of continued market reform. As a result of these twists and turns, nearly every major political group in India has committed itself programmatically to market reforms. Prospects for sustained economic growth of six percent or more per year (roughly four percent or more per capita) in the next few years seem a reasonable bet.

AN ASCENDANT ASIA AND THE WEST

In a fundamental sense, the system of market capitalism, which first appeared in Western Europe, has finally become a global – and in particular, Asian – instrument of economic development. Asia has demonstrated that it can mold capitalist institutions into a vehicle for rapid economic catch-up. The implications of a more globally balanced economic prosperity will be profound and require sustained analysis. The West today represents around 45 percent of world GDP, although it has just 13 percent of the world's population. According to the baseline estimates that underlie the growth projections in Table 1, continued economic success in Asia, in the context of an open, market-based world economy, is likely to reduce the West's share of world income to around 30 percent, while Asia's share could well rise above half. Much of the force behind this shift is the reasonable expectation that three of the four largest countries – China, India, and Indonesia – together comprising 40 percent of the world's population, could achieve per capita growth rates of 5 percent or more for the next three decades. In this forecast, Asia's share of world income happens to rise to 58 percent, just about the same share that Asia had in 1820, at the outset of the Industrial Revolution. Over two centuries of economic development, Asia's share of world income slid until 1950, but it will probably continue to rise through 2025.

Perhaps the wisest observations about this possibility were voiced 222 years ago by Adam Smith, who noted that the discovery of the sea passage between Western Europe and Asia came at a time of unprecedented European military advantage over the Asians, so that the Europeans "were enabled to commit with impunity every sort of injustice in those remote countries." But increased trade itself would be the vehicle for raising the incomes and thereby the political defenses of the Asian powers: "Hereafter . . . the inhabitants of all the different quarters of the world may arrive at that equality of courage and force which, by inspiring mutual fear, can alone overawe the injustice of independent nations into some sort of respect for the rights of one another. But nothing seems more likely to establish this equality of force than the mutual communication of knowledge and all sorts of improvements which an extensive commerce . . . carries along with it." The reemergence of Asia in the world economy will be an opportunity for mutual gain and a more balanced international system. This is the time for the Western nations to work for long-term interests by encouraging Asia to rejoin a world economic system based on commitments to the international rule of law, political and economic freedom, and open opportunities for trade and development by all countries that subscribe to shared international values.

NOTES

1 The 1820–1950 data are from Angus Maddison, *Monitoring the World Economy, 1820–1992*, Paris: Organization for Economic Cooperation and Development, 1995, pp. 226–27. The more recent data are from Alan Heston and Robert Summers, "The Penn World Tables 5.6," on-line, University of Pennsylvania, 1994.
2 This analysis draws heavily on the Asian Development Bank's study, *Emerging Asia: Changes and Challenges*, Manila: 1997, prepared in conjunction with a team of academic advisers drawn mainly from the Harvard Institute for International Development. See also the technical background paper by Radelet, Sachs, and Jong-Wha Lee, "Economic Growth in Asia," HIID, 1997.
3 Shang-Jin Wei, "How Taxing Is Corruption on International Investors?" NBER Working Paper No. 6030, 1997.

'On Missing the Boat: The Marginalization of the Bottom Billion in the World Economy'

from The Bottom Billion: Why the Poorest Countries are Failing and What Can be Done About It (2007)

Paul Collier

Editors' Introduction

Paul Collier studied Philosophy, Politics and Economics at Oxford University in the 1960s before carrying out doctoral research there. His academic career has largely been based at the same university, where he is now a Professor of Economics, Fellow of St Anthony's College and Director of the Centre for the Study of African Economies. Collier, however, is very far from being an ivory tower academic. He was senior adviser to UK Prime Minister Tony Blair's Commission on Africa, worked for many years at the World Bank, and has addressed the General Assembly of the United Nations. Collier's prominence in this field rose hugely in 2007 following the publication of his book, *The Bottom Billion*, from which we extract here. The book became a best-seller in the United States, helped no doubt by its no-nonsense style and its sharp underlying message.

Collier's first contention is that we now live in a 1–4–1 world. That is to say, development is working for most people. In addition to the one billion people at the top of the world economy who lead affluent lifestyles, a further four billion people are getting their feet on the ladder of economic growth and are rapidly being pulled away from poverty. Collier's own work, however, is mainly devoted to the people in the bottom billion. These are people living on less than a dollar a day, usually in states that are failing or crisis-ridden in large degree. Many of these states are in sub-Saharan Africa, although just as many poor people live in South Asia (in Pakistan, Nepal, Bangladesh and the eastern parts of India). According to Collier, the bottom billion are beset by four traps: the conflict trap, the natural resource trap, the 'landlocked with bad neighbours' trap and the 'bad governance in a small country' trap. Each of these traps needs to be dealt with by different sets of policy instruments. For example, in a resource-rich country, the primary need is probably not to downsize the state; rather, it is to improve accountability mechanisms to link oil or mineral revenues, say, to government spending policies. In landlocked countries, spending on infrastructure might be key, as Jeffrey Sachs and others have proposed more widely. Collier's work is distinguished by its attention to the details of policy making: how much aid to give, to whom, when, in what form, for how long, on what terms, and so on. In 2007, *The Bottom Billion* was welcomed as a significant response to an ongoing debate on aid and development that recently had been contributed to by Jeffrey Sachs and William Easterly, among others.

In the chapter here, Collier takes aim at those who propose globalisation as a solution to the problems of the bottom billion. Collier disputes the idea that capital will flow to cheap wage countries if they are trapped in civil wars or scarred by poor infrastructure. This is to put the cart before the horse, he suggests. In any case, to mix a metaphor, the horse may already have bolted. It will take some time before African countries can hope to compete in global markets with Asian countries that have low wages, high productivity and well-provided public goods (including security). Unhappily, the short-term future for the bottom billion is not a bright one.

Key references

Paul Collier (2007) *The Bottom Billion: Why the Poorest Countries are Failing and What Can be Done About It*, Oxford: Oxford University Press.

(2006) 'African growth: why a "big push"?', *Journal of African Economies* 15: 188–211.

P. Collier and A. Hoeffler (2004) 'Greed and grievance in civil war', *Oxford Economic Papers* 56 (4): 563–95.

Commission for Africa (2005) *Our Common Interest: Report of the Commission for Africa*, London: Penguin.

J. DiJohn (2007) 'Oil abundance and violent political conflict: a critical assessment', *Journal of Development Studies* 43: 961–86.

All the people living in the countries of the bottom billion have been in one or another of the traps that I have described in the preceding four chapters. Seventy-three percent of them have been through civil war, 29 percent of them are in countries dominated by the politics of natural resource revenues, 30 percent are landlocked, resource-scarce, and in a bad neighborhood, and 76 percent have been through a prolonged period of bad governance and poor economic policies. Adding up these percentages, you will realize that some countries have been in more than one trap, either simultaneously or sequentially.

But when I speak of traps, I am speaking figuratively. These traps are probabilistic; unlike black holes, it is not impossible to escape from them, just difficult. Take as an example the trap of bad governance and poor policies, and remember that the mathematical expectation of being stuck with bad policies is nearly sixty years. That expectation is built up from the very small chance, less than 2 percent, of escaping from the trap in a single year. But of course that small chance implies that periodically countries do escape. This is true of all the traps: a peace holds (as is currently the case in Angola), natural resources get depleted (as is looming in Cameroon, which has nearly exhausted its oil reserves), reformers succeed in transforming governance and policies (as is now under way in Nigeria). And such transformations have implications for the landlocked: as Nigeria turns itself around, Niger, though still landlocked, is now in

a better neighborhood. The focus of this chapter is to ask what happens next.

You might think that if a country escapes from a trap, it can then start to catch up – it will begin to grow, and grow pretty fast. The professional term for catch-up is "convergence." The best-studied example of convergence is the European Union. The countries that were initially the poorest members, such as Portugal, Ireland, and Spain, have grown the fastest, whereas the country that was initially richest, Germany, has grown slowly, and so the states that make up the European Union have converged. That is partly why relatively poor countries such as Poland and the other countries of Eastern Europe have been keen to join, whereas the countries that are richer than the European Union, Norway and Switzerland, have decided not to do so. Convergence is also working on a global scale: the lower-income countries are, on the whole, growing faster than the developed countries. People in the developed world are starting to get worried that China is converging on us so fast. The fact that the countries of the bottom billion have bucked this trend to convergence is the puzzle with which I started. And so far my explanation has been that they have been stuck in one or another of the four traps.

Will the countries that emerge from the traps follow the path blazed by the successful majority of developing countries? Will they join the rush to convergence? Globalization arouses passions: it is considered either wonderful or terrible. I think the sad reality is that although globalization has powered the majority of developing countries toward prosperity, it is now making things harder for these latecomers. The purpose of this chapter is to explain why the countries of the bottom billion have missed the boat.

What is globalization? Its effects on the economies of developing countries come from three distinct processes. One is trade in goods, the second is flows of capital, and the third is the migration of people. The three aspects of globalization are so distinct that even the idea that economies have become more globalized depends upon which dimensions you take. In terms of both capital movements and migration, the developing countries were more globalized a century ago than they are now. It is only trade in goods that has grown to unprecedented levels. And even that has not been a continuous process. Between 1914 and 1945 world trade collapsed because of wars and protectionism. It is often said that globalization is inevitable, but those interwar years cast doubt on this assertion: for those who hate globalization, the retreat of trade, capital flows, and migration during the period 1914–45 should be interesting because they are a kind of a natural experiment. Unfortunately, they were a ghastly experiment: the reversal of globalization, though feasible, looks massively undesirable based on the one occasion when we did it.

But the consequences of globalization for the bottom billion are different. Let's take the three aspects of globalization in turn, and see how they affect the bottom billion.

TRADE AND THE BOTTOM BILLION

International trade has taken place for several thousand years. However, the most dramatic transformation of the size and composition of trade has been during the past twenty-five years. For the first time in history, developing countries have broken into global markets for goods and services other than just primary commodities. Until around 1980 developing countries' role was to export raw materials. Now, 80 percent of developing countries' exports are manufactures, and service exports are also mushrooming. The production of primary commodities is basically land-using, and exporting them is most likely to benefit the people who own the land. Some-

times the land is owned by peasant farmers, but often the key beneficiaries are mining companies and big landowners. So trade based on primary commodity exporting is likely to generate quite a lot of income inequality. And its scope is inherently limited by the size of the market: as exports grow, prices turn against exporters. By contrast, manufactures and services offer much better prospects of equitable and rapid development. They use labor rather than land. The opportunity to export raises the demand for labor. Since the defining characteristic of developing countries is that they have a lot of unproductive labor, these exports are likely to spread the benefits of development more widely. And because the world market in manufactures and services is huge and was initially dominated by the rich countries, the scope for expansion by developing countries is massive.

However, before getting starry-eyed about this transformation in developing countries' trade, let us ask why it took so long. In the 1960s and 1970s the rich world dominated global manufacturing despite having wages that were around forty times as high as those in the developing world. Why did this massive wage gap not make developing countries competitive? Part of the answer is that the rich world imposed trade restrictions on the poor world. Another part of the answer is that the poor world shot itself in the foot with its own trade restrictions, which made exporting into a competitive world market unprofitable. But trade restrictions are only part of the explanation for the persistence of the wage gap for so long. The more important explanation is that the rich world could get away with a big wage gap because there are spatial economies of scale in manufacturing. That is, if other firms are producing manufactures in the same location, that tends to lower the costs for your firm. For example, with lots of firms doing the same thing, there will be a pool of workers with the skills that your firm needs. And there will be plenty of firms producing the services and inputs that you need to function efficiently. Try moving to someplace where there are no other firms, and these costs are going to be much higher even if raw labor is much cheaper.

The professional term for this is "economies of agglomeration." It was the key building block for the big insight of Paul Krugman and Tony Venables. They asked what would happen if the wage gap widened until it became big enough to offset this advantage from scale economies. Imagine yourself as the first firm successfully to jump the wage gap – that is, you relocate from the high-wage world to the low-wage world. At first you do not make a fortune. You just about break even – if by moving it was possible to instantly make a fortune, someone else would already have done so. You are the first to move and not go bankrupt, and you just get by. It is lonely being the first firm; there are no other firms around to generate those agglomeration economies, but you just hang on. And now here comes the important step. How do things look to a second firm that is thinking of relocating? Well, for the second firm it all looks a bit better than it did for the first firm because there is already another firm there. So the second firm relocates. And that also helps the first firm. They both start to do better than just getting by. And the third firm? Better still. What happens is an explosive shift of manufacturing to the new location. Does this sound familiar – like the shift of manufacturing from the United States and Europe to Asia? The change has been explosive because once activity started to relocate, agglomerations grew in low-wage Asia. In the process, wages are being driven up in Asia, but the gap was initially enormous and there is a huge amount of cheap labor in Asia, and so this process of convergence is going to run for many more years. I have described it as firms relocating. Sometimes this is precisely what happens – outsourcing, or "delocalization." But it need not be, and you do not stop it by banning firms from moving. It could equally well be that new firms set up in the low-wage locations and outcompete the existing firms in the high-wage locations. Firms do not have to move in order for industrial activity to shift location, since births of firms in one place and deaths of firms in another come to the same thing.

In effect, in order to break into global markets for manufactures it is necessary to get over a threshold of cost-competitiveness. If only a country can get over the threshold, it enjoys virtually infinite possibilities of expansion: if the first firm is profitable, so are its imitators. This expansion creates jobs, especially for youth. Admittedly, the jobs are far from wonderful, but they are an improvement on the drudgery and boredom of a small farm, or of hanging around on a street corner trying to sell cigarettes. As jobs become plentiful they provide a degree of economic security not just for the people who get them but for the families behind the workers. And gradually, as jobs expand, the labor market tightens and wages start to rise. This started to happen in

Madagascar in the late 1990s. The government established an export processing zone and created policies good enough that firms were sufficiently cost-competitive to take advantage of an American trade arrangement called the Africa Growth and Opportunity Act. Almost overnight the zone grew from very few jobs to 300,000 jobs. That is a lot of jobs in a country with only 15 million people. The jobs would probably have kept on growing, but politics got in the way. When the president, Admiral Didier Ratsiraka, lost the election he refused to step down, and he got his cronies to blockade the port, a city his supporters controlled. For eight months the worthy admiral attempted to get his job back through economic strangulation of the wayward electorate. Unsurprisingly, by then the export processing zone had been decimated. By the time it restarted there were only 40,000 jobs and firms were wary of returning. I remember a manager of an American garment company telling me in disbelief that the former president had chosen to wreck his own country. He said, "If it's like that, then count us out. We'll stick to Asia."

Madagascar is a country of the bottom billion that in the 1990s almost broke into world markets. How about the bottom billion more generally? In this initial shift out of Europe and America the bottom billion are those low-income countries that for one reason or another did not get chosen by firms as a good place to relocate. How has this affected their chances of convergence? It suggests to me that there was a moment – roughly the decade of the 1980s – when the wage gap was sufficiently wide that any low-wage developing country could break into global markets as long as it was not stuck in one of the traps. During the 1990s this opportunity receded because Asia was building agglomerations of manufactures and services. These agglomerations became fabulously competitive: low wages combined with scale economies. Neither the rich countries nor the bottom billion could compete. The rich countries did not have low wages, and the bottom billion, which surely had low wages, did not have the agglomerations. They had missed the boat.

I decided to try to test this out empirically. This time my co-researcher was Steve O'Connell, who had already worked with me on the problems of the landlocked. Our question was whether the bottom billion had shot themselves in the foot during the 1980s, closing off their opportunities for export diversification.

So far, we have only looked at Africa itself, not "Africa+." That is because Steve and I did this work in the context of an African research network: the African Economic Research Consortium. I expect that what is true of Africa will turn out to be true of the rest of the bottom billion. Generally, I find that there is no "Africa effect": Africa often looks distinctive because it is dominated by the characteristics of the bottom billion. However, it is an empirical matter, and I might turn out to be wrong.

First of all, recall that Africa is disproportionately either landlocked or resource-rich. For different reasons, these two categories are very likely to be out of the game as far as export diversification is concerned. In the rest of the developing world the two groups combined account for only 12 percent of the population. In Africa they account for two-thirds of the population. Therefore, even if all of Africa's coastal, resource-scarce societies had been ready to break into global markets in the 1980s, two-thirds of the population of the region would have been left out. But were the coastal, resource-scarce economies ready to diversify into global markets? This was the group for which the issue of shooting themselves in the foot arose. Steve and I applied a classification that a group of us had developed to describe debilitating configurations of governance and policy: we considered them failing states, as defined in Chapter 5. During the 1980s only 4 percent of the population of Africa's coastal, resource-scarce countries were in countries that were free of these debilitating configurations. In fact, it comes down to Mauritius and not much else. So if you were a firm looking to relocate to a cheap labor country in the 1980s you might have chosen Mauritius, and indeed many firms did, but you were unlikely to have chosen anywhere else in Africa.

But would firms have chosen Africa even if governance and policies had been better? This sort of counterfactual question is difficult to tackle. Steve and I approached it by investigating whether those of Africa's coastal, resource-scarce countries that had subsequently escaped from being failing states had been able to diversify their exports. We found that each year of being free of the gross failures of governance and policy added significantly to the success of export diversification. The countries that stopped shooting themselves in the foot were able to break into new export markets. This is encouraging. It suggests that although Africa's coastal countries did

indeed shoot themselves in the foot during the 1980s, they might still be able to break into global markets. It seems likely, however, that the process of breaking in is now harder than before Asia managed to establish itself on the scene.

If there really has been a process of missing the boat, it is pretty depressing. For one thing, it implies that the incentive for governments in the bottom-billion countries to reform, make peace, or do whatever else is needed to break free of the traps is greatly reduced. Courageous people face down the powerful interests lined up against them and implement reform only to find that little happens. The reactions to reforms that do not deliver economic success can be ugly. All the old vested interests have their knives out to kill off reform attempts. Another type of reaction is the quack remedy: people are liable to become victims of populism. The most depressing reaction is for people to see the society as intrinsically flawed. Their prolonged period of economic failure in Africa and the other countries of the bottom billion has deeply eroded the self-confidence of their societies. The expectation of continued failure reinforces the pressures for the brightest people to leave.

In Part 4, I will be arguing that this bleak prospect is not inevitable. There is something that can be done about it: we need to get serious about supporting the heroes in the struggle that is already being waged within the societies of the bottom billion. But for the moment stick with the world as it is, and let's see how it is likely to play out. When will the boat come around again? That is, when will the bottom billion actually be able to break into global markets? The automatic processes of the global economy will eventually bring the boat back around. But the bottom billion will have to wait a long time until development in Asia creates a wage gap with the bottom billion similar to the massive gap that prevailed between Asia and the rich world around 1980. This does not mean that development in the bottom billion is impossible, but it does make it much harder. The same automatic processes that drove Asian development will impede the development of the bottom billion.

So the growth of agglomerations in Asia has made the export diversification route more difficult for the bottom billion. Another effect of this growth is that Asians are increasingly desperate to secure supplies of natural resources. The Chinese are all over the countries of the bottom billion, securing natural resource deals. Superficially this is good news: it is certainly raising prices, most obviously of oil, which some countries of the bottom billion export. But you saw in Chapter 5, on the trap of poor policy, that high prices for resource exports are likely to chill the impetus for reform. In Chapter 2, on the conflict trap, you saw that the spread of high natural resource prices increased the risk of conflict. And you saw in Chapter 3, on the natural resource trap, that natural resources are not the royal road to growth unless governance is unusually good. In the bottom billion it is already unusually bad, and the Chinese are making it worse, for they are none too sensitive when it comes to matters of governance. When Zimbabwe's Robert Mugabe was looking for money to bail himself out of the ruinous consequences of his political choices, he came up with the "look east" strategy. East did not mean Russia, it meant China. And China has welcomed his overtures with open arms. The same goes for Angola. After the defeat of Jonas Savimbi's UNITA, the developed countries finally decided to put the squeeze on the government of Angola, trying to clean up grotesque misuse of the oil money. China came in with over $4 billion in loans, and the Angolan government was off the hook. So the bottom billion are locked into natural resource exports twice over: by the threshold effects of Asian export agglomerations and by Asia's desperate need for natural resources.

The growth of global trade has been wonderful for Asia. But don't count on trade to help the bottom billion. Based on present trends, it seems more likely to lock yet more of the bottom-billion countries into the natural resource trap than to save them through export diversification.

CAPITAL FLOWS AND THE BOTTOM BILLION

The economies of the bottom billion are short of capital. Traditionally, aid has been supposed to supply the capital that the bottom billion lack, but even where this works it supplies only public capital, not private capital. Public capital can supply much of the infrastructure that these societies need, but it cannot begin to supply the equipment that workers need in order to be productive; that can be supplied only by private investors. As part of the work I describe below, we have measured the capital stock available

for each member of the workforce, country by country. Africa is the most capital-scarce region, but this becomes dramatically more pronounced when capital is separated into its private and public components. In a successful region such as East Asia there is more than twice as much private capital as public capital. By contrast, Africa has twice as much public capital as private capital. What it and the other economies of the bottom billion really lack is private investment. This translates into a lack of equipment for the labor force to work with, and this in turn condemns workers to being unproductive and so to having low incomes. The labor force of the bottom billion needs private capital, and in principle globalization can provide it. Basic economic theory would suggest that in the societies that are short of capital, the returns on capital would be high, and this would attract an inflow of private capital.

Private capital inflows

Global capitalism does often work like this. China, for example, is attracting huge private capital inflows. Of course, the East Asian crisis of 1998, during which foreign money panicked and fled the region, showed that short-term financial inflows can be a mixed blessing, exposing countries to financial shocks. But longer-term investment is likely to be beneficial all around. Workers in developing countries get jobs and increased wages, and the firms that move capital to developing countries get higher returns on it. Such capital movements, like trade, normally generate mutual gains. Since political contests are usually presented as zero-sum games – your gain is my loss – the people who are most politically engaged have the hardest time believing in mutual gains. Hence, perhaps, the exaggerated suspicions of globalization.

But what about the bottom billion? Again, I think that the effect of globalization – this time through capital flows – is different. The biggest capital flows are not going to the countries that are most short of capital; they are bypassing the bottom billion. The top of the league for investment inflows has been Malaysia, a highly successful middle-income country. The only substantial inflows of private investment to the bottom billion have been to finance the extraction of natural resources – the top of the league among the bottom billion has been Angola, due to the opportunities for offshore oil.

Why are the most capital-scarce countries not attracting a larger capital inflow? Historically, part of the answer has been poor governance and policy. Obviously, this does not impede capital inflows for resource extraction – hence Angola – but it has curtailed the footloose investment in manufacturing, services, and agribusiness. Since the 1990s quite a few of the societies of the bottom billion have implemented significant reforms of governance and policies. The problem is that even these reforming countries are not attracting significant inflows of private capital. The key question is why not. To try to answer it I teamed up with Cathy Pattillo, an African American now working at the IMF.

The answer is that the perceived risk of invest-ment in the economies of the bottom billion remains high. Investor perceptions of risk can be measured – one useful indicator is a survey, done by the magazine *Institutional Investor*, that scores the perceived risk for each country on a scale of 1 to 100. A score of 100 implies the sort of maximum safety appropriate for your grandmother's nest egg, and a score of 1 is only for kamikaze investors. Risk ratings such as this one show up as significant in statistical explanations of private investment; unsurprisingly, high risk dis-courages investment.

The problem for the reforming countries of the bottom billion is that the risk ratings take a long time to reflect turnarounds. I first came across this problem when I was advising the reforming govern-ment of Uganda in the early 1990s. The government had taken some remarkably brave decisions, and the economy was starting on what was to prove a prolonged period of rapid growth. At that time the *Institutional Investor* rating gave Uganda 5 out of 100, the worst rating in Africa. This was so far out of line with what the government was doing that it was worth mounting an image-building campaign with investors. Gradually, the ratings improved. I remem-ber bumping into the Ugandan economic team at a meeting in Hong Kong in 1997. The latest issue of *Institutional Investor* had just come out, and they rushed up to me in excitement, saying, "Have you seen it?" They had achieved one of the largest improvements in the world, with their score rising from 18 to 23 – but it was still well below the level at which serious investment inflows were likely, which is about 30 to 40. Why does it take so long for investors to revise their views of the bottom billion? There are three reasons for the problem.

Paradoxically, the countries with the strongest reforms are those that started from the worst govern-ance and policies. Often things have to get really bad to provoke incisive change. And so the reforms start from a truly terrible rating, much as happened in Uganda. If you start from 5, it is going to take a while before you get to the range at which investment flows set in.

The second problem is that the typical economy of the bottom billion is very small. A corollary is that the community of private investors knows virtually nothing about it – absorbing information is costly, if only in time, and these places are simply not sufficiently important enough to bother with. This became evident when the government of Uganda was trying to change the country's image. The last time Uganda had been in the news had been because of Idi Amin, the publicity-obsessed coup leader who, not content with being styled president, had also made himself a field marshal (or to give him his fuller title, His Excellency, President for Life, Field Marshal Al Hadji Dr. Idi Amin, VC, DSO, MC, Lord of All the Beasts of the Earth and Fishes of the Sea, and Con-queror of the British Empire in Africa in General and Uganda in Particular). By the early 1990s Amin had been gone for over a decade, but most potential investors still thought he was president. There are fifty-eight countries in the bottom billion, and inves-tors do not track them individually but think of them collectively as "Africa" and dismiss them. Contrast this with China: every major international company knows that it has to keep abreast of developments in China. This even shows up statistically: one team of researchers has shown that the investor ratings sys-tematically exaggerate the problems of the countries of the bottom billion.

The third reason is that policy improvements are often genuinely fairly fragile: many incipient turnarounds subsequently abort. Reform is always politically difficult and, as we will see in the next chapter, it has not been helped by donor policy con-ditionality. Even the governments that genuinely want to reform are usually pushed into the role of opposing some of the reforms urged on them by the donors, because the donors want everything to happen at once. And governments that do not want to reform periodically take the money, embark on a few reforms, and then abandon them. So the genuine reformers have not been able to distinguish them-selves from the bogus reformers. Because they cannot distinguish themselves, investors lump them all together and say, "Don't call us, we'll call you." They go to China instead.

Fundamentally, the problem is one of credibility. Reforms induced by donor money are not credible with investors, and even without donor money they are high-risk. What can a government that is genuinely committed to reform do about it? Economic theory does give us the right answer, but it is not very attractive. The government needs to create a convincing signal of its intentions, and to do this it has to adopt reforms that are so painful that a bogus reformer is simply not prepared to adopt them. It thereby reveals its true type, to use the language of economics. The Ugandan government actually did this. It restored property to its rightful owners, the Asians who had been expelled by Amin. In the run-up to a presidential election the Ugandan government also slashed the size of the civil service, throwing thousands out of work. Such decisions raised its risk rating so sharply. Though necessary to change investor perceptions, this signaling strategy both is harsh and runs the risk of creating a political backlash. In Part 4, I am going to discuss ways in which credibility might be achieved less painfully.

Private capital outflows

The lack of capital inflows is only half the story of why global capital markets are not working for the bottom billion. The other half is that their own capital flows out of them. Much of this is illegal, and so it is hidden. It is called capital flight. To find out whether capital is flowing out of the bottom billion you need to get under the skin of the official numbers. This was a big task, and it took three of us to crack it: I joined forces with both Cathy Pattillo and Anke Hoeffler, and it took us a very long time.

Suppose you live in a bottom-billion country and want to get your money out. You have to get hold of foreign currency – dollars. It's often illegal; in many cases all foreign currency has to be sold to the central bank at the official exchange rate, so what can you do? There are various tricks, one of which is to falsify the documentation on exports. You find someone who is exporting $1,000 worth of coffee to the United States. That individual bribes a few people in the customs office so that the documentation says $500. This way, the exporter only has to hand over $500 to

the central bank. He can then sell the other $500 to you, and you can deposit it in a foreign bank. To find evidence of such schemes, we looked for discrepancies in the numbers – the coffee exporter bribes the local customs officers but not the American customs officials, so the documentation at the U.S. end of the transaction correctly records that $1,000 worth of coffee has been imported into the United States. By comparing export figures with import figures and using other discrepancies, it is possible to tease out capital flight year by year for each country. This allows you to discover, for example, that by the end of military rule in Nigeria in 1998 Nigerians were holding around $100 billion of capital outside the country. It became a newspaper sensation when I reported it to the annual conference of the Central Bank of Nigeria in Abuja.

We then estimated the value of private wealth held in each country, year by year. This may sound difficult, but you can work it out from data on private investment using something called the "perpetual inventory method." Finally, we added the private wealth held as capital flight abroad to the private wealth held within the country, to see what proportion of total private wealth was held abroad. This yielded what rapidly became one of the famous numbers about Africa: By 1990, 38 percent of its private wealth was held abroad. This was a greater proportion than in any other region. It was even higher than the Middle East, where oil wealth and deserts, unsurprisingly, tend to encourage investment abroad. Africa integrated into the global financial economy, but in the wrong direction: the most capital-scarce region in the world exported its capital. (As can be surmised from my description of how we arrived at that 38 percent figure, the technique is not precise. We can reliably say that capital flight has been substantial, but quite how big we do not really know.)

So Africans were voting with their wallets, taking their money out of the region. What was driving this massive capital flight? If you ask Africans, they tell you it is corruption. Those in power loot public money and get it safely abroad. This is surely part of the story, but it is not at the heart of what is going on. For example, Indonesia had corruption on a world-class scale. President Suharto took what we might politely term "Asian family values" to extraordinary heights of paternalistic generosity. But most of the money stayed in the country. Africans took their money, whether corruptly acquired or honestly acquired, out of Africa because the opportunities for investment were so poor. One reason why the investment opportunities were so poor was because the countries were stuck in one or another of the traps. Capital flight was a response to the traps. In the sophisticated language of professional economics, capital flight was a "portfolio choice": people were holding their assets where they would yield a reasonable and a safe return. How do we know? We tried to explain the portfolio choices statistically. Why, for example, did Indonesians in 1980 hold nearly all their wealth domestically, whereas Ugandans in 1986 held two-thirds of their wealth abroad? We tried a whole range of explanations, such as measures of corruption and measures of the returns on capital. We found that in addition to the problem that the traps depressed the returns on capital, investment opportunities were judged poor because of the perceived high level of risk, as measured by means of indices such as the *Institutional Investor* ratings described above. The credibility problem was not just scaring off foreigners – it was scaring off domestic investors as well.

So, despite being chronically short of private capital, the bottom billion are integrating into the global economy through capital flight rather than capital inflows. They are losing capital partly because the traps involve conditions such as political instability and poor policies, which make countries unsuited for investment. But even when countries succeed in shedding these characteristics they are still perceived as risky, and fears of retrogression keep capital out. So don't count on global capital mobility to develop the bottom billion, capital-scarce as they are. It is more likely to reinforce the traps.

MIGRATION AND THE BOTTOM BILLION

The bottom billion have not only integrated into the world economy through capital flight. They are increasingly integrating through migration. People vote with their feet as well as with their wallets. Historically, migration has been the great equalizer. In the nineteenth century the vast movement of people from Europe to North America did more to raise and equalize incomes than trade or capital movements. And more recently for some developing countries, migration has been a very good thing. For example, the Indian diaspora in the United States was probably

critical in India's breakthrough into the world market for e-services. For those bottom-billion countries with the least favorable prospects, migration offers a safety valve; as I discussed in Chapter 4, it is one strategy for countries such as Niger. But how does it look more generally for the bottom billion?

Having studied capital flight, Cathy, Anke, and I decided to try a similar approach with migration. We distinguished between the educated and the uneducated. With a bit of imagination you can think of education as a form of wealth: in one of the ugliest phrases in economics, educated people are "human capital," so labeled because their skills are valuable. We wondered whether the migration behavior of the educated from developing countries looked more like that of uneducated people or more like the portfolio choices of capital. I have to say I rather hoped that the educated would look more like people than like portfolios, that all the myriad features that humanity holds in common would swamp the value that the educated and portfolios have in common. But it was not to be. The migration decisions of educated people looked very like the portfolio decisions that determine where wealth is held, and not much at all like the migration decisions of the uneducated, who are more likely to migrate the wider the differential between their earnings at home and what they can earn abroad.

What does this imply for the bottom billion? It suggests that these countries will hemorrhage their educated people to a far greater extent than their uneducated people. Migration takes time to build up, but it accelerates. There is a simple reason for this: migration becomes easier if other family members have already moved. Our analysis predicts that the exodus of capital from the bottom billion was only phase one of the global integration of the bottom billion. Phase two will be an exodus of educated people. As Somalia continues to fail and other places continue to develop, more Somalis will leave, as there will be more places for them to go. But emigration will be selective: the brightest and the best will have most to gain from moving. They are also the ones most likely to be welcomed in host countries. Ordinary Somalis will have less incentive to leave because they lack the skills to gain employment, and indeed, they will become increasingly unwelcome and so will find it harder to leave Somalia. Those who do get out will not return, and their remittances will dwindle after a generation of separation. Emigration helps those who leave, but it can have perverse effects on those left behind, especially if it selectively removes the educated. Yet this is precisely what we predict: having already hemorrhaged capital, the countries at the bottom will increasingly hemorrhage educated labor – people like my friend Lemma Sembet, an Ethiopian who is one of America's leading professors of finance. Meanwhile, back in the countries of the bottom billion, the financial sectors are run by people whose understanding of financial economics does not equip them to manage much more than a piggy bank.

Remember from Chapter 5 that to achieve a turnaround from being a failing state, a country is helped by having a critical mass of educated people. The countries of the bottom billion are already desperately short of qualified people, and the situation is likely to get worse. The flight of the skilled is at its most rapid in precisely those bottom-billion environments where there is most scope for change: postconflict societies. So, whereas migration has generally been helpful as part of the development process, I am skeptical of it as a force for transforming the bottom billion. I think that by draining these countries of their talent, migration is more likely to make it harder for these nations to decisively escape the trap of bad policy and governance.

LIFE IN LIMBO: OUT OF THE FRYING PAN . . .

This all adds up to a depressing picture of what globalization is doing for the bottom billion. To get a chance to play in the global economy, you need to break free of the traps, and that is not easy. Remember, in order to turn a country around it helps to have a pool of educated people, but the global labor market is draining the bottom billion of their limited pool of such people. Even once they reform, many of these economies find it difficult to attract private investment inflows, and may continue to hemorrhage their own modest private wealth. And they face a high hurdle in trying to break into diversified markets for exports because China, India, and the other successful developing countries have already done so. Even once free of the traps, countries are liable to be stuck in a kind of limbo – no longer falling

apart, but not able to replicate the rapid growth of Asia, and so failing to converge.

This indeed seems to describe a lot of bottom-billion countries that have recently come out of the traps. Remember that in the past four years the average country of the bottom billion has at last started to grow. I have interpreted that as a temporary phenomenon linked to the global boom in commodities. But suppose you were to put the most favorable gloss on it – that they have broken free of the traps. Well, although they are growing, it is at a very sedate pace – much more slowly than the other developing countries even during the slow decade of the 1970s. Even if their present growth rate is sustained, they will continue to diverge rapidly. It will take them many decades to reach what we now consider to be the threshold of middle income, and by that time the rest of the world will have moved on.

There is also a yet more depressing variant of the future for these limbo countries: the traps still await them. As long as they have low incomes and slow growth they continue to play Russian roulette. Côte d'Ivoire survived low income and slow growth for a couple of decades but then fell into conflict as the result of a coup. Zimbabwe survived the same and then fell into bad governance. Tanzania, currently among the most hopeful low-income countries, is about to become resource-rich due to new discoveries of gas and gold. Malawi grew remarkably well for the first decade of its independence, considering that it is landlocked and resource-scarce, but then its neighbors fell into the conflict trap and, being dependent upon them, it too began to decline. And so a miserable but possible scenario is that countries in the bottom billion oscillate between the traps and limbo, perhaps switching in the process from one trap to another.

In the next part of the book we will at last turn from the depressing scenarios of traps and limbos to what we can do about them. Let me be clear: *we* cannot rescue *them*. The societies of the bottom billion can only be rescued from within. In every society of the bottom billion there are people working for change, but usually they are defeated by the powerful internal forces stacked against them. We should be helping the heroes. So far, our efforts have been paltry: through inertia, ignorance, and incompetence, we have stood by and watched them lose.

'Consent to Coercion'

from *The New Imperialism* (2003)

David Harvey

Editors' Introduction

David Harvey is Distinguished Professor of Anthropology at the City University of New York, and probably the most important human geographer in the world today. Harvey is originally from Gillingham in Kent, England. He completed his BA (Hons) and PhD in Geography at Cambridge University. Following this, he has worked at the University of Uppsala, the University of Bristol, Johns Hopkins University, Oxford University, and the London School of Economics. Harvey's early work was on the historical geography of hops production in nineteenth-century Kent. Harvey also wrote on questions of philosophy and method, culminating in his call for more rigorous theorization in *Explanation in Geography* (1969). Harvey's shift to Baltimore in the United States paralleled his turn to radical geography and Marxism in order to explain urban social inequality. The spatial expression of racism in the problem of ghetto formation is a key theme in *Social Justice and the City* (1973). Following Henri Lefebvre's theory of social space, Harvey's magisterial *The Limits to Capital* (1982) rethinks the way capitalism works through time *and* space. While capital searches for a 'spatial fix', to invest surpluses of capital and labour, Harvey argues that this search for a 'fix' is in vain, and all spatial formations have to be devalued or destroyed for capital accumulation to proceed. Through this insight, spatial formations like cities, infrastructure, built environments, boom towns, rust belts, decaying ghettos, regions, and countries appear to be spatial constructions tied to the contradictions of capitalism. Like Marx, Harvey thinks these contradictions lead to wider spaces of destruction, tending to global catastrophe in the nuclear age, and one might add in relation to global warming. Like Marx, Harvey thinks we ignore social crises at our collective peril.

In parallel to this spatial extension of Marxist theory, Harvey wrote equally important historical geographic essays on Second Empire Paris, some of which were later republished in the monograph *Paris, Capital of Modernity* (2003). In the late 1980s, Harvey turned to debates about modernity, postmodernity and transition to a new mode of capitalist regulation he called flexible accumulation in *The Condition of Postmodernity* (1989). Harvey then wrote on environmental questions in *Justice, Nature and the Geography of Difference* (1996), and on urbanisation, embodiment and utopia in *Spaces of Hope* (2000). Responding to the expansion of neo-liberalism and US militarism since 2001, Harvey returns to questions of imperialism explored at the end of *The Limits to Capital* in *The New Imperialism* (2003), excerpted in the reading that follows.

The reading turns specifically to the intertwined dynamics of capitalist expansion and political–military expansion. Both are spatial phenomena, Harvey argues, but they do not exactly coincide. The global economy has become more volatile with the ascendancy of finance capital after 1973 (the capitalist dynamic), and states have been led by the Wall Street–IMF–US Treasury alliance to restructure economic policy in line with neo-liberal ideology. On the one hand, a transnational capitalist class alliance of financiers, rentiers and CEOs have found it in their interests for a military–political dynamic to secure the conditions of securitised profitability, a role that US neo-conservatives had planned to provide for a while. On the other hand, neo-liberalism has brought chronic insecurity to working-class citizens, not least in the United States. Various forms of defence against internal enemies – immigrants, single mothers, moral outcasts, and now Muslims – fuel fascist-like tendencies within capitalist societies, particularly in the United States and Europe. The Bush administration's key neo-conservatives had planned since at least 1997 to take advantage of the opportunity that nationalist fervour provided after the 9-11 attacks on the World Trade Center and the Pentagon, as detailed in the *Project for a New American Century*.

While we may now question neo-conservative expectations to 'reconstruct' Iraq in the manner of post-war Japan or Germany, Harvey's argument about the intentions of the US imperial establishment are provocative and important. We suggest the reader pays attention not just to the scenario Harvey that lays out, but to the way he interrogates US imperial ambitions for Pax Americana, or for control of the oil spigot in the Middle East region, against the dynamics of capitalist accumulation, and particularly against the growing centrality of China as a centre of global accumulation. Harvey's subsequent book, *A Brief History of Neo-liberalism* (2005), continues to explain the expansion of neo-liberalism through elite class alliances, and his current research extends this to the most important case, China.

Key references

David Harvey (2003) *The New Imperialism*, Oxford: Oxford University Press.
— (1969) *Explanation in Geography*, New York: St Martin's Press.
— (1973) *Social Justice and the City*, Baltimore, MD: Johns Hopkins University Press.
— (1982) *The Limits to Capital*, Chicago, IL: University of Chicago Press.
— (1989) *The Condition of Postmodernity*, Oxford: Blackwell.
— (1996) *Justice, Nature and the Geography of Difference*, Oxford: Blackwell.
— (2000) *Spaces of Hope*, Berkeley: University of California Press.
— (2003) *Paris, Capital of Modernity*, London: Routledge.
— (2005) *A Brief History of Neo-liberalism*, Oxford: Oxford University Press.

Project for a New American Century (www.newamericancentury.org).

■ ■ ■ ■ ■ ■

Imperialism of the capitalist sort arises out of a dialectical relation between territorial and capitalistic logics of power. The two logics are distinctive and in no way reducible to each other, but they are tightly interwoven. They may be construed as internal relations of each other. But outcomes can very substantially over space and time. Each logic throws up contradictions that have to be contained by the other. The endless accumulation of capital, for example, produces periodic crises within the territorial logic because of the need to create a parallel accumulation of political/military power. When political control shifts within the territorial logic, flows of capital must likewise shift to accommodate. States regulate their affairs according to their own distinctive rules and traditions and so produce distinctive styles of governance. A basis is here created for uneven geographical developments, geopolitical struggles, and

different forms of imperialist politics. Imperialism cannot be understood, therefore, without first grappling with the theory of the capitalist state in all its diversity. Different states produce different imperialisms, as was obviously so with the British, French, Dutch, Belgian, etc. imperialisms from 1870 to 1945. Imperialisms, like empires, come in many different shapes and forms. While there may be much that is contingent and accidental – indeed it could not be any other way given the political struggles contained within the territorial logic of power – I believe we can go a long way to establishing a solid interpretative framework for the distinctively capitalistic forms of imperialism by invoking a double dialectic of, first, the territorial and capitalist logics of power and, secondly, the inner and outer relations of the capitalist state.

Consider, in this light, the case of the recent shift in

form from neo-liberal to neo-conservative imperialism in the United States. The global economy of capitalism underwent a radical reconfiguration in response to the overaccumulation crisis of 1973–5. Financial flows became the primary means of articulating the capitalistic logic of power. But once the Pandora's box of finance capital had been opened, the pressure for adaptive transformations in state apparatuses also increased. Step by step many states, led by the United States and Britain, moved to adopt neo-liberal policies. Other states either sought to emulate the leading capitalist powers or were forced to do so through structural adjustment policies imposed by the IMF. The neo-liberal state typically sought to enclose the commons, privatize, and build a framework of open commodity and capital markets. It had to maintain labour discipline and foster 'a good business climate'. If a particular state failed or refused to do so it risked classification as a 'failed' or 'rogue' state. The result was the rise of distinctively neo-liberal forms of imperialism. Accumulation by dispossession re-emerged from the shadowy position it had held prior to 1970 to become a major feature within the capitalist logic. In this it did a double duty. On the one hand the release of low-cost assets provided vast fields for the absorption of surplus capitals. On the other, it provided a means to visit the costs of devaluation of surplus capitals upon the weakest and most vulnerable territories and populations. If volatility and innumerable credit and liquidity crises were to be a feature of the global economy, then imperialism had to be about orchestrating these, through institutions like the IMF, to protect the main centres of capital accumulation against devaluation. And this is exactly what the Wall Street–Treasury–IMF complex successfully engaged upon, in alliance with the European and Japanese authorities, for more than two decades.

But the turn to financialization had many internal costs, such as deindustrialization, phases of rapid inflation followed by credit crunches, and chronic structural unemployment. The US for one lost its dominance in production, with the exception of sectors such as defence, energy, and agribusiness. The opening up of global markets in both commodities and capital created openings for other states to insert themselves into the global economy, first as absorbers but then as producers of surplus capitals. They then became competitors on the world stage. What might be called 'sub-imperialisms' arose, not

only in Europe but also in East and South-East Asia as each developing centre of capital accumulation sought out systematic spatio-temporal fixes for its own surplus capital by defining territorial spheres of influence. But these spheres of influence were overlapping and interpenetrating rather than exclusive, reflecting the ease and fluidity of capital mobility over space and the networks of spatial interdependency that increasingly ignored state borders.

The benefits of this system were, however, highly concentrated among a restricted class of multinational CEOs, financiers, and rentiers. Some sort of transnational capitalist class emerged that nevertheless focused on Wall Street and other centres such as London and Frankfurt as secure sites for placements of capital. This class looked, as always, to the United States to protect its asset values and the rights of property and ownership across the globe. While economic power seemed to be highly concentrated within the United States, other territorial concentrations of financial power could and did arise. Capital concentrated in European and Japanese markets could take its cut, as could almost any rentier class that positioned itself correctly within the matrix of capitalistic institutions. Debt crises might rock Brazil and Mexico, liquidity crises might destroy the economies of Thailand and Indonesia, but rentier elements within all those countries could not only preserve their capital but actually enhance their own internal class position. Privileged classes could seal themselves off in gilded ghettos in Bombay, São Paulo, and Kuwait while enjoying the fruits of their investments on Wall Street. Just because Wall Street was awash with money did not mean, therefore, that Americans owned that money. Wall Street's problem was to find profitable uses for all the surplus money it commanded, no matter whether it was held by Americans or foreigners.

This geographical dispersal of capitalistic class power did not only apply to rentiers and financial interests; production capital took advantage of the spatial volatility and the shifting territorial logics. The large multinationals in electronics, shoes, and shirts gained remarkably through geographical mobility. But then so did certain other social groups. The Chinese business diaspora, for example, improved its position precisely because it had both the means and the inclination to extract profits out of mobility. Taiwanese and South Korean sub-contractors moved into Latin America and Southern Africa and did extraordinarily well, while those they employed suffered appallingly.[1]

But it was a peculiar feature of this world that an increasingly transnational capitalist class of financiers, CEOs, and rentiers, should look to the territorial hegemon to protect their interests and to build the kind of institutional architecture within which they could gather the wealth of the world unto themselves. This class paid very little heed to place-bound or national loyalties or traditions. It could be multi-racial, multi-ethnic, multicultural, and cosmopolitan. If financial exigencies and the quest for profit required plant closures and the diminution of manufacturing capacity in their own backyard, then so be it. US financial interests were perfectly content to undermine US hegemony in production, for example. This system reached its apogee during the Clinton years, when the Rubin–Summers Treasury Department orchestrated international affairs greatly to the advantage of rentier interests on Wall Street, though they often took very high risks in doing so. The culmination was the disciplining of competition from East and South-East Asia in 1997–8 in such a way as to allow the financial centres of Japan and Europe, but above all the United States, to snap up assets for almost nothing and thereby augment their own profit lines at the cost of massive devaluations and the destruction of livelihoods elsewhere. This was, however, only one example of the innumerable debt and financial crises that afflicted many parts of the developing world after 1980 or so.

Neo-liberal imperialism abroad tended to produce chronic insecurity at home. Many elements in the middle classes took to the defence of territory, nation, and tradition as a way to arm themselves against a predatory neo-liberal capitalism. They sought to mobilize the territorial logic of power to shield them from the effects of predatory capital. The racism and nationalism that had once bound nation-state and empire together re-emerged at the petty bourgeois and working-class level as a weapon to organize against the cosmopolitanism of finance capital. Since blaming the problems on immigrants was a convenient diversion for elite interests, exclusionary politics based on race, ethnicity, and religion flourished, particularly in Europe where neo-fascist movements began to garner considerable popular support. The corporate and financial elites gathered at Davos in 1996 then worried that a 'mounting backlash' against globalization within industrial

democracies might have a 'disruptive impact on economic activity and social stability in many countries'. The prevailing mood of 'helplessness and anxiety' was conducive to 'the rise of a new brand of populist politician' and this could 'easily turn into revolt'.[2]

But by then the anti-globalization movement was beginning to emerge, attacking the powers of finance capital and its primary institutions (the IMF and the World Bank), seeking to reclaim the commons, and demanding a space within which national, regional, and local differences could flourish. With the state so clearly siding with the financiers and in any case performing as a prime agent in the politics of accumulation by dispossession, this movement looked to the institutions of civil society to transform the territorial logics of power on a variety of scales, from intensely local to global (as in the case of the environmental movement). The prevalence of fraud, rapine, and violence provoked many violent responses. The surface civilities that supposedly attach to properly functioning markets were little in evidence. The protest movements that surfaced throughout the world were, for the most part, ruthlessly put down by state powers. Low-level warfare raged across the world, often with US covert involvement and military assistance.

Eschewing traditional forms of labour organization, such as unions, political parties, and even the pursuit of state power (now seen as hopelessly compromised), these oppositional movements looked to their own autonomous forms of social organization, even setting up their own unofficial territorial logics of power (as did the Zapatistas), oriented to improving their lot or defending themselves against a predatory capitalism. A burgeoning movement of non-governmental organizations (some of them sponsored by governments) sought to control these social movements and orient them towards particular channels, some of which were revolutionary but others of which were about accommodation to the neo-liberal regime of power. But the result was a ferment of local, dispersed, and highly differentiated social movements battling either to confront or to hold off the neo-liberal practices of imperialism orchestrated by finance capital and neo-liberal states.

The volatility inherent in neo-liberalism ultimately returned to haunt the heartland of the United States itself. The economic collapse that began in the high-tech dot.com economy in 1999 soon spread to reveal that much of what passed for finance capital was in fact unredeemable fictitious capital supported by scandalous accounting practices and totally empty assets. Even before the events of 9/11, it was clear that neo-liberal imperialism was weakening on the inside, that even the asset values on Wall Street could not be protected, and that the days of neo-liberalism and its specific forms of imperialism were numbered. The big issue was what kind of relation between the territorial and capitalistic logics of power would now emerge and what kind of imperialism it would produce.

The fortuitous election of George W. Bush, a born-again Christian, to the US presidency brought a neo-conservative group of thinkers close to power. The neo-conservatives, well funded and organized in numerous 'think-tanks' like the neo-liberals before them, had long sought to impose their agenda on government. And it is a different agenda from that of neo-liberalism. Its primary objective is the establishment of and respect for order, both internally and upon the world stage. This implies strong leadership at the top and unwavering loyalty at the base, coupled with the construction of a hierarchy of power that is both secure and clear. To the neo-conservative movement, adherence to moral principle is also crucial. In this it finds its backbone and electoral base with fundamentalist Christians who hold to beliefs of a very special kind. In the wake of 9/11, for example, Jerry Falwell and Pat Robertson (two major leaders within the movement) expressed the view that the event was a sign of God's anger at the permissiveness of a society that tolerated abortion and homosexuality. Later, on one of the most watched current affairs programmes on American television, Falwell declared that Muhammad was the first great terrorist, while others expressed support for Zionism and for Sharon's violence towards the Palestinians since this would lead to Armageddon and the Second Coming. Belief in the book of Revelation and Armageddon is very widespread (Reagan espoused it, for example). It is hard for Europeans in particular to understand that around a third of the US population holds firmly to such beliefs (including creationism rather than evolution), which imply acceptance of the horrors of war (particularly in the Middle East) as a prelude to the achievement of God's will on earth. Much of the US military is now recruited from the south, where these views are prevalent.

While the neo-conservatives know they cannot stay in power holding to such a platform, the influ-ence of the Christian right cannot be underestimated. The failure to place any constraints on Sharon's violent repression of the Palestinians (interpreted by fundamentalists as a positive step towards Armageddon) is a case in point. And in the conflict with the Arab world it is hard not to let these attitudes slip into the rhetoric of a Christian crusade versus an Islamic jihad, thus converting Huntington's unconvincing thesis of an imminent clash of civilizations into a geopolitical fact.[3]

The neo-conservative charter for foreign policy was laid out in *The Project for the New American Century* that got under way in 1997.[4] The title speaks, as did Luce back in 1941, of a century rather than of territorial control. It deliberately repeats, therefore, all the evasions that Smith exposes in Luce's presentation.[5] The Project is 'dedicated to a few fundamental propositions: that American leadership is good both for America and for the world; that such leadership requires military strength, diplomatic energy, and commitment to moral principle; and that too few political leaders today are making the case for global leadership'. The principles involved were clearly laid out in Bush's statement on the anniversary of 9/11 (cited in Chapter 1 above). Though recognized as distinctive American values, these principles are presented as universals, with terms like freedom and democracy and respect for private property, the individual, and the law bundled together as a code of conduct for the whole world. The Project also seeks to 'rally support for a vigorous and principled policy of American international involvement'. This means exporting and if necessary imposing appropriate codes of conduct upon the rest of the world. Most of the core members of the Project came, however, from the defence establishment of the former Reagan and Bush administrations. They are key representatives of that 'military-industrial complex' against whose power Eisenhower had long ago so clearly warned and which had grown so much more powerful in the Reagan years. Most of them joined the new Bush administration. Whereas the key positions in the Clinton administration were in the Treasury (where Rubin and Summers ruled supreme), the new Bush administration looks to its defence experts – Cheney, Rumsfeld, Wolfowitz, and Powell – to shape international policy, and relies upon a Christian conservative – Ashcroft – as Attorney General to enforce order at home. The Bush administration is, therefore, dominated by neo-conservatives, deeply indebted to the military-industrial complex (and a few other major sectors of American industry, such as energy and agribusiness), and supported in its moral judgements by fundamentalist Christians. Their task was to consolidate power behind a minority-led political agenda within the territorial logic of power. In this they well understood the connection between internal and external order. They intuitively accepted Arendt's view that empire abroad entails tyranny at home, but state it differently. Military activity abroad requires military-like discipline at home.

Iraq had long been a central concern for the neo-conservatives, but the difficulty was that public support for military intervention was unlikely to materialize without some catastrophic event 'on the scale of Pearl Harbor', as they put it. 9/11 provided the golden opportunity, and a moment of social solidarity and patriotism was seized upon to construct an American nationalism that could provide the basis for a different form of imperialist endeavour and internal control. Most liberals, even those who had formerly been critical of US imperialist practices, backed the administration in launching its war against terror and were prepared to sacrifice something of civil liberties in the cause of national security. The accusation of being unpatriotic was used to suppress critical engagement or meaningful dissent. The media and the political parties fell into line. This enabled the political leadership to enact repressive legislation with scarcely any opposition – most notably the Patriot and Homeland Security Acts. Draconian curbs on civil rights were instituted. Prisoners were held illegally and without representation in Guantanamo Bay, indiscriminate round-ups of 'suspects' occurred, and many were held for months without access to legal advice, let alone a trial. Police could arbitrarily detain anyone suspected of 'terrorism', which could include, it soon became clear, even those in the anti-globalization movement. Draconian surveillance techniques were introduced (the FBI was to have access to records of book-borrowing from libraries, book purchases, internet connections, records of student enrolment, membership of scuba-diving clubs, etc.). The administration also seized the opportunity to cut all kinds of programmes for the poor (in the name of sacrifice for a national cause). It imposed a tax-cut programme that grossly favoured the wealthiest 1 per cent of the population (in the name of stimulating the economy)

and even proposed the elimination of taxes on dividends in the vain hope that this might bolster asset values on Wall Street. But such policies, coupled with flagrant violations of the Bill of Rights and of American constitutionality, could only be sustained, as Washington, Madison, and many others had long ago recognized and feared, through foreign entanglements of an imperialist sort. Given the threats implied in the events of 9/11, and the climate of suppression of dissent, even liberal opinion swung behind the idea of the invasion of Afghanistan, the routing of the Taliban, and the global hunt for al Qaeda.

To sustain the momentum and realize their ambitions, the paranoid style of American politics had to be put to work. The neo-conservatives had long dwelt on the threats posed by Iraq, Iran, and North Korea, and several other so-called 'rogue states', to the global order. Behind this, however, there always lurked the figure of China, long feared as both unpredictable and potentially a powerful competitor on the world stage. The alliance between the neo-conservatives and the military-industrial complex had pressured Clinton during the 1990s to increase military expenditures and be prepared to fight two regional wars – against, for example, 'rogue states' such as Iraq and North Korea – simultaneously. Iraq was central, in part because of its geopolitical position and dictatorial regime, which was immune to financial disciplining because of its oil wealth, but also because it threatened to lead a secular pan-Arab movement that might dominate the whole of the Middle Eastern region and be able to hold the global economy hostage to its powers over the flow of oil. President Carter, recall, had insisted that any attempt to use oil in this way would not be tolerated, and direct US military commitment to the region dates back to at least 1980. The first Gulf War did not produce regime change in Baghdad, in part because there was no UN mandate for it. The settlement imposed on Iraq was unsatisfactory to both sides. The Iraqis baulked and sanctions were imposed, weapons inspectors were sent in and then expelled, the Kurds were protected in an autonomous zone in the north by military threats, and a low-level war continued in the skies above Iraq as the US and Britain jointly patrolled no-fly zones in both the north and the south. Clinton designated Iraq a 'rogue state' and adopted a policy of regime change in Baghdad but restricted the means to covert action and overt economic sanctions which, the neo-conservatives vociferously argued, would not work.

After 9/11, the neo-conservatives had their 'Pearl Harbor'. The difficulty was that Iraq plainly had no connection with al Qaeda and the fight against terrorism had to take preference. In the invasion of Afghanistan the military tested out much of its new weaponry in the field, almost as a dress rehearsal for what they might do in Iraq and elsewhere. In the process, the US secured a military presence in Uzbekistan and Kyrgyzstan, within striking distance of the Caspian Basin oilfields (where the extent of reserves is still a mystery and where China is battling fiercely to gain a foothold in order to ensure its own supplies to satisfy its rapidly increasing internal demands). Within six months, and with the defeat of the Taliban in Afghanistan behind it, the US administration began to switch its attention to Iraq. By the summer of 2002 it was clear that the US was committed to force regime change on Baghdad militarily no matter what. The only interesting question was how this would be justified to the American public and internationally. From this point on, the administration resorted to all manner of smokescreens, shifting rhetoric daily, putting out undocumented assertions as if they were proven facts (of the sort described in Chapter 1). It sought to construct a coalition of the willing in which Britain, since it was already heavily involved in daily military action in Iraq (and from which it would have been very difficult to extricate itself), was to take a leading role. At first the US denied any role to the UN and even asserted it had no need for Congressional approval, but on these points it had to concede somewhat to political pressures both domestically and internationally. But it assiduously cultivated the new-found nationalism that was created after 9/11 and harnessed it to the imperial project of regime change in Iraq as essential for domestic security, at the same time as it used the imperial project to put in place ever tighter internal controls (fuelled by terror alerts and other security fears on the domestic front). Unfortunately, as Arendt again so astutely remarks, the coupling of nationalism with imperialism cannot be accomplished without resort to racism, and the degraded popular image of Arabs and Islam and official policies towards visitors and immigrants from Arab countries are all too indicative of a rising tide of racism in the US that may do untold future damage both internally and internationally.

While the situation is now one of rapid flux, accompanied by the usual smoke and mirrors of official pronouncement, it is nevertheless possible to discern roughly where the neo-conservative imperial project wants to go. I therefore conclude with a synopsis of that direction and an assessment of the forces ranged against it.

The neo-conservatives look to the reconstruction of Iraq along the lines pioneered in Japan and Germany after the Second World War. Iraq will be liberalized for open capitalistic development with the aim of ultimately creating a wealthy consumerist society along Western lines as a model for the rest of the Middle East. The necessary social, institutional, and political infrastructures will be put in place under US administration, but gradually give way to a clientelist Iraqi political administration (preferably as weak as the Japanese liberal party). Iraq will remain demilitarized but be protected by US forces that will remain in the Gulf region.[6] Iraqi oil will be used to finance the reconstruction and pay for some of the cost of the war, and, it is hoped, will be delivered to the markets of the world (conveniently denominated in dollars rather than euros) at a sufficiently low price to spark some kind of recovery in the global economy.

This is not, however, the limit to neo-conservatives' imperial ambition. They have already begun to speak of Iran (which after the occupation of Iraq will be totally surrounded by the US military and clearly threatened) and have launched accusations against Syria that speak of 'consequences'. So obvious have these remarks become that the British Foreign Secretary thought it important to state categorically that Britain would absolutely refuse to participate in any military action against either Syria or Iran. But the neo-conservative position, as articulated by Secretary of Defense Rumsfeld all along, is that the US does not need Britain to accomplish its objectives and that it will go it alone if necessary. Pressure on both Syria and Iran is mounting, while the US also looks to internal reform in Saudi Arabia both to forestall any attempt at a takeover by Islamicists (this was, after all, bin Laden's primary objective) and to deal with the fact that much of the fundamentalist teaching that has fuelled opposition to the US is supported by the Saudis. Meanwhile, the US has now honed, and experimented with in Iraq, a military capacity named 'shock and awe' which would have the power to simultaneously destroy the hundreds of long-range guns that the North Koreans have targeted on Seoul. When it cares to, it can destroy all of North Korea's military power and nuclear capacity in one twelve-hour strike.

Lurking behind all of this appears to be a certain geopolitical vision. With the occupation of Iraq and the possible reform of Saudi Arabia and some sort of submission on the part of Syria and Iran to superior American military power and presence, the US will have secured a vital strategic bridgehead, as was pointed out in Chapter 2, on the Eurasian land mass that just happens to be the centre of production of the oil that currently fuels (and will continue to fuel for at least the next fifty years) not only the global economy but also every large military machine that dares to oppose that of the United States. This should ensure the continued global dominance of the US for the next fifty years. If the US can consolidate its alliances with east European countries such as Poland and Bulgaria, and (very problematically) with Turkey, down to Iraq and into a pacified Middle East, then it will have an effective presence that slashes a line through the Eurasian land mass, separating western Europe from Russia and China. The US would then be in a military and geostrategic position to control the whole globe militarily and, through oil, economically. This would appear particularly important with respect to any potential challenge from the European Union or, even more important, China, whose resurgence as an economic and military power and potentiality for leadership in Asia appears as a serious threat to the neo-conservatives. The neo-

conservatives are, it seems, committed to nothing short of a plan for total domination of the globe.[7] In that ordered world of a Pax Americana, it is hoped that all segments may flourish under the umbrella of free-market capitalism. In the neo-conservative view, the rest of the world (or at least all property-owing classes) should and will be grateful for the space allowed for economic development under free-market capitalism everywhere.

[. . .]

NOTES

1 G. Hart, *Disabling Globalization: Places of Power in Post-Apartheid South Africa* (Berkeley: University of California Press, 2002).

2 Klaus Schwab and Claude Smadja, cited in D. Harvey, *Spaces of Hope* (Edinburgh: Edinburgh University Press, 2000), 70.

3 S. Huntington, *The Clash of Civilizations and the Remaking of the World Order* (New York: Simon & Schuster, 1997).

4 The website is ⟨www.newamericancentury.org⟩.

5 See N. Smith, *American Empire: Roosevelt's Geographer and the Prelude to Globalization* (Berkeley: University of California Press, 2003).

6 This formula is well described in C. Johnson, *Blowback: The Costs and Consequences of American Empire* (New York: Henry Holt, 2000).

7 D. Armstrong, 'Dick Cheney's Song of America: Drafting a Plan for Global Dominance', *Harper's Magazine*, 305 (Oct. 2002), 76–83.

'From the Specter of Mao to the Spirit of the Law: Labor Insurgency in China' [1]

Theory and Society (2002)

Ching Kwan Lee

Editors' Introduction

Ching Kwan Lee is Associate Professor of Sociology at the University of Michigan. Lee completed a BA (Hons) in Sociology from Hong Kong University in 1987 before receiving a scholarship to pursue postgraduate study in Sociology at the University of California at Berkeley, where she completed her PhD in 1994. Her award-winning *Gender and the South China Miracle* (1998) is a comparative ethnography of women workers in Hong Kong and China's Guangdong province, and the ways in which they participate in very different gendered labour regimes. Lee's work builds on and extends Michael Burawoy's now classic thesis on *The Politics of Production* (1985), on the ways in which workers participate in the labour process under broader conditions set by the state. Lee shows how workplace regimes, in her cases and in general, are profoundly gendered. In her subsequent work, Lee has contrasted the experiences of workers in the old industrial north with those of the dynamic south. This work has also led Lee to make sense of the proliferation of forms of worker (and one should add peasant) protest that has demonstrated how China has had to contend with some of the spatially uneven developmental dynamics that David Harvey has written of as intrinsic to capitalism. Lee's most recent book on this topic is *Against the Law: Labor Protests in China's Rustbelt and Sunbelt* (2007). Lee has also edited two volumes, *Working in China* and *Re-envisioning the Chinese Revolution*.

The reading that follows is excerpted from an article linked to *Against the Law*. Lee's argument here is vital to our understanding of contemporary China. One of Lee's larger arguments is to critique what she identifies in a variety of accounts of Chinese development, a ' "metaphysical pathos" celebrating the Chinese experience as a historical success story' (p. 221). Turning to the experiences of subaltern classes, Lee finds class conflict, confusion, despair, and also new forms of contestation using 'bourgeois' instruments of the law. Provocatively, Lee calls this the emergence of the 'insurgent worker of post-socialism', of militant class consciousness that gives socialist language new content through the experience of new forms of inequality. Importantly, Chinese workers continue to make claims to a socialist state for the general welfare of citizens sacrificed in the process of market transition. Chinese subalterns, in her view, provide a different view on 'market socialism', drawing on socialist legalism to protect workers in a market society. Lee's analysis is rare for its ethnographic texture and for its perspective beyond the local state, *pace* Oi (Part 7), and beyond elite class alliances, *pace* Harvey (this Part).

Key references

Ching Kwan Lee (2002) 'From the specter of Mao to the spirit of the law: labor insurgency in China', *Theory and Society* 31 (2): 189–228.

— (1998) *Gender and the South China Miracle: Two Worlds of Factory Women*, Berkeley: University of California Press.

— (2007) *Against the Law: Labor Protests in China's Rustbelt and Sunbelt,* Berkeley: University of California Press.

— (ed.) (1998) *Working in China: Ethnographies of Labor and Workplace Transformation*, New York: Routledge.

Ching Kwan Lee and Guobin Yang (eds) (2007) *Re-envisioning the Chinese Revolution: The Politics and Poetics of Collective Memory in Reform China*, Palo Alto, CA: Stanford University Press.

Michael Burawoy (1985) *The Politics of Production*, London: Verso.

> Hopes for the future are slim. Rather than pushing forward with confidence that we know the way, we see the growing failure of the most visible alternatives, which all seem profoundly flawed. . . .
>
> Alvin Gouldner, *The Dark Side of the Dialectic*[2]

China now confronts the most massive scale of unemployment and peasant labor migration in the history of the People's Republic. This is a potentially volatile time, marked by soaring numbers of labor disputes, petitions, and protests, prompting the regime to warn unambiguously of "new internal contradictions among the people."[3] Yet, so far, Communist rule has endured without effective popular challenge. A most intriguing paradox of the Chinese reform is thus the regime's capacity to maintain overall social stability, while market socialism has also intensified labor discontent and radicalized labor activism. This article unravels this paradox by proposing a thesis of "postsocialist labor insurgency." My overall argument is that the transition from state socialism to market socialism occasions a simultaneous radicalization of worker politics and the state's attempt to bolster its regulatory capacity by institutionalizing a "rule by law." Both tendencies, as I explicate below, have to do with the multi-faceted consequences of introducing a market economy. The precarious balance between labor activism and its partial incorporation[4] by this regulatory regime accounts for current political stability. So far, only some workers are involved in such activism, while many are demoralized and atomized in the wake of massive unemployment. However statistically unrepresentative, labor insurgency in specific locales remains theoretically significant in that it constitutes a "critical case," capable of fleshing out the dynamics, the limits, and hence also the potentials of worker politics in a period of structural transformation. History has shown that reform in China has always been the harbinger of revolutionary changes. The political consequences of reform may well depend on whether class conflicts, among other types of social conflicts, can in the long run be contained within the fledgling system of "socialist legality," in a polity with a widening rift between central and local state power, and in a society rife with contradictions between entrenched socialist rhetoric and emergent capitalist realities.

REVOLUTIONARY POTENTIAL OF SOCIALIST LABOR

One of the most provocative theoretical postulations on socialist working-class radicalism and its fate under reform is found in Michael Burawoy and Janos Lukacs's *The Radiant Past.*[5] State socialism, rather than capitalism, they argue, is more conducive to the making of a revolutionary working class in favor of workers' socialism. The institutional foundation nurturing such potentials is the state socialist regime of production, predicated on the shortage economy and the state apparatus inside the enterprise. The shortage economy creates constant need for worker autonomy to overcome anarchy in production, while the party-state apparatus at the point of production renders transparent the extraction of surplus and therefore a common class exploiter. These conditions forge labor's critical consciousness, encompassing, on the one hand, a negative immanent critique of state socialism for its failure to live up to its claims of superior efficiency and equality, and, on the other hand, a positive vision for an alternative order based on worker self-management. Market reform, Burawoy and Lukacs further contend, has the tendency to pacify working-class radicalism by providing alternative channels of mobility based on individual rather than group effort. The contrast in working-class movements in Poland and Hungary highlights how in Poland the lack of market opportunity and the presence of civil society organizations combined to make for Solidarity's revolt against the state. In Hungary, on the other hand, workers were politically demobilized after the failed 1956 revolt and instead invested their energy in exploiting entrepreneurial opportunities in the second economy. If the above argument can be summarized as the "revolutionary socialist working-class" thesis, the Chinese experience analyzed in this article reveals anomalies that compel its reconstruction into what can be called an "insurgent postsocialist labor" thesis. Both theses are concerned with the effect of the lived experience of state socialism and market reform on worker radicalization in terms of class consciousness and class capacity.

THE INSURGENT WORKER OF POSTSOCIALISM

The "insurgent postsocialist labor" thesis argues that the passage of state socialism is prone to trigger labor insurgency because market reform seriously compromises working-class interests and throws into sharp relief the lost potentials of socialism. Postsocialism, not state socialism, is the moment of labor radicalization. There are four aspects of change involved. First, against the original postulation that the lived experience of socialism is generative of oppositional class consciousness, I argue that workers' socialist experience is a necessary but not sufficient condition for its formation. *Critical class consciousness* emerges only under, and not prior to, market reform. The Chinese experience seems to suggest that there are variants of "socialist regime of production" and that socialist production per se is not necessarily conducive to the kind of inherent critique, even antagonism, of socialism found in Eastern Europe or Russia. The Chinese regime of production was configured differently. Rather than

anarchic production requiring worker autonomy and improvisation, the emphasis was always on political activism and mobilizational production campaigns under party control in the workplace, which pre-empted worker control or self-management.[6] Only when economic reform ushers in the exploitative forces of the market, reconstitutes shop-floor production relations, and incites class conflicts do workers activate the cognitive resources buried in received socialist ideology and rhetoric. Then, and only then, do these cultural repertoires make experiential sense and lead to a collective conceptual achievement of "class." The reconstitution of the material, political, and ideological moments of production occurs in the crucible of a new regime of production. In China, what I call "disorganized despotism" has emerged under reform to replace the former "neo-traditional" system.

Second, the impact of market reform on *mobilization*, rather than always negative as the original thesis suggests, is actually more varied. It is true that market opportunity dissipates collective mobilization – labor protests and demonstrations in China are more concentrated in the northeast and the interior provinces than in the more prosperous south. Nevertheless, market reform can be conductive to mobilization, even without civil society or social movement organizations, if its uneven development produces collective losers and frees them from state or enterprise control. Where the political and organizational resources necessary for class-specific social movements, cross-class alliances, or civil society associations are not available, insurgent mobilization can still be built on the organizational ruins of state socialism. Chinese workers in bankrupt enterprises or in the rust belt regions, who were previously organized by state socialism into work units, now engage in work-unit activism, the predominant mode of mobilization.

The third argument concerns workers' vision of an alternative social order and labor *subjectivity*. Instead of workers' socialism, Chinese workers' alternative society is not a total repudiation of state socialism. Rather, it is a social order where the state assumes responsibility for guaranteeing a minimum level of general welfare and plays a pivotal role in instituting and enforcing the rules of the market economy, which is deemed necessary to provide the freedom, opportunity, and dynamism lacking in a planned economy. Again, this positive vision of the state and its role has to do with workers' past experience with state paternalism, the interwined history of state making and class making after the 1949 Revolution, and the sustained political dominance of the party-state in the reform period. I show here that the subject of Chinese labor insurgency has a strong statist orientation, while also being "interpellated" by the state's ideological categories of class, comrade, and citizen.

Finally, *the state* undergoes self-transformation in the reform process. Instead of focusing on the role of the state in production and redistribution. I find a reconfiguration of state capacity in the realm of regulation. "Rule by law" is a new statecraft under one party rule and market liberalization. Although it is still a work in progress, the rudimentary ideology and apparatus of "socialist legality" do have the effect of inciting new forms of popular resistance riding on the call for legality and rights. But "rule *by* law" is an instrument of a repressive state, not tantamount to a "rule *of* law" capable of restraining the state. Workers' seizure of the rights rhetoric ironically means that their activism is at least partially channeled into, and restrained by, the state's new regulatory machinery and its discourse of legality.

A word about what I mean by "insurgency" and "postsocialism" is in order. The "revolutionary" potentials Burawoy and Lukacs find under state socialism refer to workers' opposition to state socialism and their espousal of a Marxian utopia, i.e., classless communism or workers' socialism. But workers' political aspirations and sensibilities are more wide-ranging and the notion of "insurgency" is used here because it is more open and less teleological in terms of workers' goals of struggle. It allows for politics outside the problematic of emancipation. It also better captures the "ambivalent"[7] terrain of politics rooted in entrenched dependence of the subaltern classes on a repressive state, a condition generating orientations of both entrapment and solidarity with the dominant power. Thus, following Guha, "insurgency" is "the site where two mutually contradictory tendencies . . . – a conservative tendency made up of the inherited and uncritically absorbed material of the ruling culture and a radical one oriented towards a practical transformation of the rebel's conditions of existence – met for a decisive trial of strength."[8] Finally, "postsocialism" refers to a historical condition in which (1) the centrally planned economy no longer plays a

predominant role in the production and redistribution of resources;[9] (2) and when the socialist state perceives a need to articulate "actually existing socialism" to capitalism, conditioned both by the structure of the former and an avowed attempt to overcome the deficiency of the latter.[10]

[. . .]

This article specifies the concrete social processes and institutional mechanisms out of which labor insurgency emerges under market reform. The transformation of labor system and worker life-worlds entails the mutually determining interplay among the state, the shop floor, and workers' political agency, each of which will be taken up in the following discussion. The first part looks at the new institutional apparatus of state domination over labor after the decline of the planned economy and the weakening of the organizational capacity of the party-state. I examine the Chinese state's attempts to forge a new basis for legitimacy based on a so-called "socialist legality" and the staunch resistance at the local level this has triggered. Local state agents and enterprise management share common interests in evading labor law and regulations, resulting in a regime of "disorganized despotism," as discussed in the second part. It is characterized on the shop floor by the ascendance of "scientific management," despotic disciplinary practices, and the demise of the party and the union inside the enterprise. Despotism constitutes worker interests in opposition to management and incites critical class consciousness. This brings me to the third part of the analysis, which turns to labor insurgency and subjectivity as these emerge out of the gap between the state project of socialist legality and the reality of disorganized despotism. Labor insurgency takes the form of "work-unit mobilization," facilitated by the organizational and identity resources bequeathed by state socialism. Labor subjectivity is characterized by a double consciousness of class and citizenship. The former is fostered by workers' community of memories, focusing on the rhetoric and practices of Maoist socialism, and the latter is an unintended consequence of the new state discourse of legality and citizen rights.

[. . .]

The strategic deployment of a language of legality and rights does not imply that workers attain the status and consciousness of "citizenship" à la western liberal democracy. It means, however, that they struggle to perform and thus realize a legal status, bestowed by the central government's law and regulations but denied them by local state agents. Riding on the ideology of legality, they bolster their bargaining power and legitimate their public activism in defense of their material interests. This is a site of insurgency, not revolution, where one finds a conservative tendency to accept as legitimate the existing order (central state power remains unchallenged) but also a radical one of practically transforming one's conditions of existence given new institutional resources. Finally, although the labor subject of such insurgency (at least the one by state workers) fuses identities of class, comradeship, and citizenship, all in the context of a strong statist orientation, that subjectivity is no less political or potent.

CONCLUSION

In this article, I have tried to specify the concrete social processes and institutional mechanisms leading to labor insurgency in postsocialist China. To explore the theoretical payoff of an analysis of the Chinese experience, I frame it as a series of anomalies vis-à-vis the thesis of the "revolutionary socialist worker" grounded in production politics in other state socialist transitions. My overall argument is that the passage of state socialism, not state socialism per se, offers the potential for labor radicalization in terms of critical consciousness and mobilizational capacity. First, the state attempts to rebuild legitimacy through a "rule by law," but this remains abortive due to collusion between local officials and enterprise owners. Second, disorganized reform fosters the emergency of despotism at the point of production, which seriously compromises workers' material interests and their standard of justice. Worker mobilization is thus fueled by survival needs and moral outrage, their demands incited and channeled by the state's discourse of legality and citizen rights. Interestingly, worker solidarity is also partially founded on collective memories of Maoist socialism and its class rhetoric. Among veteran workers in declining industries in rust-belt regions bereft of the much touted but ever elusive market opportunities, the propensity to stage mobilization targeting the state is particularly prominent. I have found that even in the absence of civil society associations or broad-based social movement unionism, market reform creates collec-

tive losers forging locality-based, work-unit activism. The Chinese material also indicates that postsocialist labor insurgency does not imply the total rejection of state socialism or passive embrace of the market. State workers' political aspirations challenge the state to retain its self-proclaimed role as the guardian of universal interest and to safeguard the socialist ideal of securing a basic livelihood for all, even under a market-driven economy. This statist orientation has roots in institutional and moral relations between the state and the working class under state socialism. But it also has to do with the quite unique capacity of the Chinese party-state to remain intact under postsocialism, when it also promotes accumulation and incubates market mechanisms.[11] Therefore, one can expect that as long as the state monopolizes political power, worker struggles will also tend to enlist state involvement, thus generating a precarious mode of political stability.[12]

Through tracing some of the dynamics of labor radicalization under postsocialism, I also want to underline its limits and openness. One reading of the above analysis casts doubt on the political potential of such patterns of working class insurgency. After all, the most restive workers are those in declining industries and bankrupt state enterprises, and their radicalism occurs at the moment of exit from the working class, i.e., when they become unemployed, thus permanently losing their status as "worker." In other words, the Chinese working class is made only at the moment when it is unmade by reform! With time, and without sustained organization, such localized, enterprise-based labor insurgency will lose steam as workers become more demoralized and unemployment is normalized. In addition, unrelenting state repression against any form of organized dissent, coupled with the entrenched exclusionary elitism of the Chinese intellectuals, makes cross-class social movement à la Polish Solidarity an unlikely outcome of Chinese reform.[13]

But there is potential for an alternative scenario. Because postsocialist labor politics fuses class solidarity with claims for workers as citizens with legal rights, it can galvanize broader political resonance. There are already clear signs that peasants are increasingly using the rhetoric of law and legal rights in their protests, rallying around policy slogans promulgated by the central government in their struggles against local level taxation and corruption.[14] The convergence of unorganized popular activist groups that share similar cognitive repertoires, and perhaps institutional targets (local state officials) as well, can be politically unsettling. Sociologists have noted the powerful political impact in state socialist systems of the "large numbers" phenomenon: the centralization of power and uniformity of social institutions creates large number of individuals with similar behavioral patterns, interests, and demands targeting the state. In China, widespread but uncoordinated collective action (or inaction) by peasants and rural cadres, for instance, has even vetoed or shaped state policies as fundamental as the de-collectivization of agriculture.[15] The same political potential applies to the postsocialist period and to the mode of locality-based labor radicalization analyzed in this article, not the least because such agitation dovetails with the central government interest in establishing a robust regulatory state apparatus based on rule by law. As one looks ahead at China's further integration into global capitalism, formally marked by its accession to the World Trade Organization, worsening urban and rural unemployment and intensified labor insurgency loom large in the horizon. Can the government manage to keep at bay the eruption of a large number of local protests by doing out emergency subsistence funds and launching anti-corruption campaigns? Or will the politics of legal activism initiate a process of nurturing society's self-organizing capacity so much that more sustainable efforts emerge as part of society's demands for citizen's legal rights? In any case, the irony is that rather than the entrepreneurs or the rising middle classes, whose interests reside in evading the law rather than promoting it, it is the popular classes, viz., workers and peasants, who champion the cause of "bourgeois" legal rights![16] In short, the demise of state socialism is compatible with a wide range of political outcomes.[17] Scholars of Chinese politics who focus on the new middle classes or the local state elite have respectively postulated the rise of "civil society,"[18] "corporatism,"[19] "symbiotic clientelism,"[20] and "local state corporatism"[21] as the possible directions of Chinese political development. Taking an alternative perspective and looking at one massive group of losers produced by market reform, this article suggests an alternative path of political change. Legal activism, and the concomitant and simultaneous rise in class and citizen consciousness, whether collective or individual, may chart a significant course of political development under market reform.

Finally, I have tried in this analysis to restore a critical perspective on the limits of reforms by excavating labor's standpoint and experience in the reform process. Many of the most influential studies of Chinese reform have offered us insightful analyses of institutional and organizational changes prompting high growth rates in the Chinese economy. At the same time, a "metaphysical pathos" celebrating the Chinese experience as a historical success story underlies many of these studies. But it is an optimism that flies in the face of working-class realities. The latter reveal a world of deepening class conflicts, moral confusion, economic dislocation, and decay. I do not suggest that the standpoint of labor is a privileged one, offering a more comprehensive or a "truer" view of reality than that of the bourgeoisie or the state. Yet, it is an opportunity for critical knowledge, based on workers' objective structural location in society and their practical engagement with those local realities. Living through two historical alternatives, i.e., state socialism and postsocialism, generates a sense of historicity of the current social order, and the contradictions between appearance and reality. Workers' collective nostalgia of certain aspects of the Maoist past (e.g., moral incentives, more egalitarian shop floor, camaraderie, revolutionary idealism, mass participation in national development) coexists with critique of the more tyrannical and dehumanizing aspects of that past history. This present analysis selects only those elements of their mentality that focus on what they consider lost achievements of socialism. Workers' lived experience under reform also points to the gap between the appearance and reality of a market driven society. In contrast to the pervasive neo-liberal ideology celebrating the rationality of the market, workers have found that opportunity, prosperity, freedom and legality are limited, elusive, and uneven in their availability. This embedded critical perspective can be a departure for understanding postsocialist society and for constructing postsocialist theory. At least, it sensitizes us to an alternative agenda of inquiry to that of "neoclassical sociology," characterized by an emphasis on strategic action in the deployment of capital, property forms, elite circulation, and optimistic assessments of the potentialities of capitalism as the end of history.[22] What this latter research program, including the most prominent works on the Chinese transition, has downplayed and what this article only begins to address instead is the *dialectical* interplay of elite and subaltern transformation, market formation and class formation, institution building and popular resistance, economics and politics.

NOTES

1 I thank Marc Blecher, Howard Kimeldorf, Mark Selden, Dorothy Solinger, Vivienne Shue, Victor Nee, and three reviewers of *Theory and Society* for stimulating comments and suggestions on an earlier version of this article. I am most indebted to Michael Burawoy and Jeff Paige for their inspiration, encouragement, and criticisms.

2 Alvin Gouldner, *The Dark Side of the Dialectic* (Dublin: Economic and Social Research Institute, 1974), 25.

3 Organization Department Project Group of the Chinese Communist Party, *China Survey Report, 2000–2001: A Study on Internal Contradictions Among the People Under New Circumstances* (Beijing: Central Compilation and Translation Press, 2001) in Chinese.

4 I borrow this notion from Ruth Berins Collier and David Collier, *Shaping the Political Arena* (Princeton: Princeton University Press, 1991). In their study of Latin American state–labor relations, "incorporation" refers to the attempt by the state to create channels for resolving labor conflicts, superseding the ad hoc use of repression. The state comes to assume a major role in institutionalizing a new system of class bargaining, and in shaping an institutionalized labor movement. In China, the state has from the beginning controlled the labor movement by recognizing only the official union. In the reform period, although repression is still applied to unofficial unionizing, labor conflicts are explicitly recognized and new regulatory mechanisms are put in place, in addition to reforming the role of the official union.

5 Michael Burawoy and Janos Lukacs, *The Radiant Past: Ideology and Reality in Hungary's Road to Capitalism* (Chicago: University of Chicago Press, 1992).

6 Andrew Walder, *Communist Neo-traditionalism: Work and Authority in Chinese Industry* (Berkeley: University of California Press, 1986). Tak-chuen Luk, "The Bureaucratization and Informalization of Working Class Lives: A Case Study of South Wind in State Socialist China, 1949–1996," Ph.D. Thesis, University of Chicago, 1998. Katharyne Mitchell, "Work Authority: the Happy Demise of the Ideal Type," *Comparative Study of Society and History* 34 (1992): 679–694.

7 The theoretical importance of "ambivalence" is elaborated in Neil Smelser, "The Rational and the Ambivalent in the Social Sciences," *American Sociological Review* 63/1 (1998): 1–16. Conditions of dependence, he argues, breed ambivalence. Ambivalence entails entrapment and solidarity, opposing and unstable affective orientations, all of which go against the usual assumption of rationality and choice under conditions of freedom in sociological theorizing.

8 Ranajit Guha, *Elementary Aspects of Peasant Insurgency in Colonial India* (Durham: Duke University Press, 1999), 11.

9 In China, the state sector now accounts only for about one-third of total industrial output. The workplace-based welfare system is almost totally dismantled and replaced by a contribution-based system.

10 Arif Dirlik, "Postsocialism? Reflections on 'Socialism With Chinese Characteristics' " in *Marxism and the Chinese Experience*, ed. Arif Dirlik and Maurice Meisner (Armonk: M.E. Sharpe, 1989), 364.

11 Michael Burawoy, "The State and Economic Involution: Russia Through a China Lens," *World Development* 24/6 (1996): 1105–1117.

12 Elizabeth J. Perry, "Challenging the Mandate of Heaven: Popular Protest in Modern China," *Critical Asian Studies* 33/2 (2001): 163–180.

13 See analyses by Elizabeth J. Perry on these issues in "From Paris to the Paris of the East and Back: Workers as Citizens in Modern Shanghai," *Comparative Study of Society and History* (April 1999): 348–373; "Casting a Chinese 'Democracy' Movement: the Roles of Students, Workers and Entrepreneurs," in *Popular Protest and Political Culture in Modern China*, ed. Jeffrey N. Wasserstrom and Elizabeth J. Perry (Boulder: Westview, 1994).

14 Thomas Bernstein, "Farmer Discontent and Regime Responses," in *The Paradox of China's Post-Mao Reforms*, ed. Merle Goldman and Roderick Macfarquhar (Harvard University Press, 1999), 197–219. For more recent reports, see Erik Eckholm, "Spreading Protests by China's Farmers Meet With Violence," *The New York Times*, February 1, 1999; "Farmers Seek Lawyer's Release," *South China Morning Post*, August 12, 2000. On peasants' nostalgia for political campaigns directed against cadre corruption, see Kevin J. O'Brien and Lianjiang Li, "Campaign Nostalgia in the Chinese Countryside," *Asian Survey* 39/3 (1999): 375–393.

15 Zhou Xueguang, "Unorganized Interests and Collective Action in Communist China," *American Sociological Review* 58/1: 54–73; Daniel Kelliher, *Peasant Power in China* (New Haven: Yale University Press, 1992); Dali Yang, *Calamity and Reform in China* (Stanford: Stanford University Press, 1996).

16 Thanks to Jeff Paige who shared this observation with me.

17 Andrew G. Walder, "The Decline of Communist Power: Elements of a Theory of Institutional Change," *Theory and Society* 23/2 (1994): 297–323.

18 Thomas Gold, "The Resurgence of Civil Society in China," *Journal of Democracy* 1/1 (1990): 18–31; David

Strand, "Protest in Beijing: Civil Society and Public Sphere in China," *Problems of Communism* 39 (1990): 1–19.

19 Gordon White et al., *In Search of Civil Society* (Oxford: Clarendon Press, 1996); Anita Chan, "Revolution or Corporatism?" *Australian Journal of Chinese Affairs* 29 (January 1993): 31–61.

20 David Wank, *Commodifying Communism* (Cambridge: Cambridge University Press, 1999).

21 Jean Oi, *Rural China Takes Off* (Berkeley: University of California Press, 1998); Andrew Walder, editor, *Zouping in Transition* (Harvard University Press, 1999); Nan Lin, "Local Market Socialism: Local Corporatism in Action in Rural China," *Theory and Society* 24/3 (1995): 301–354.

22 Michael Burawoy, "Neoclassical Sociology: From the End of Communism to the End of Classes," *American Journal of Sociology* 106/4 (January 2001): 1099–1120.

'The Recurrent Crises of the Gatekeeper State'

from *Africa Since 1940: The Past of the Present* (2002)

Frederick Cooper

Editors' Introduction

Frederick Cooper is Professor of History at New York University, and a leading historian of Africa. Cooper finished a BA at Stanford University in 1969, followed by a PhD at Yale in 1974. He has taught in the History departments at Harvard, Michigan and New York universities. Cooper's work has proceeded through a series of careful studies engaged with broader debates in the social sciences. His first book is on the particularities of slavery in the nineteenth-century plantation economy of the East African coast, using a combination of documentary evidence and interviews with descendants (*Plantation Slavery on the East Coast of Africa*, 1977). His subsequent book traces the transition from slavery to what comes next, between the interests of slaves, British officials, and Arab and Swahili landlords. He shows how the transition to wage labour is contested, incomplete and linked to ex-slave struggles for land through squatting (*From Slaves to Squatters: Plantation Labor and Agriculture in Zanzibar and Coastal Kenya, 1890-1925*, 1980).

Cooper's next book shifts to the period between 1930 and 1950, and focuses on transformations in work and urban politics. His focus is on the dynamics of casual labour and the colonial imperative to settle a working class and stable work relations in the city, for political and economic reasons consequent on heightened labour unrest (*On the African Waterfront: Urban Disorder and the Transformation of Work in Colonial Mombasa*, 1987). While working on this project, Cooper explored the broader argument in an important edited collection on migrant and casual labour in Africa's cities, called *Struggle for the City* (1983). The collection shows how workers acted and protested in the context of a colonial urbanism in which neither space nor time were ever fully commodified.

Cooper's monumental comparative history of British and French Africa between the mid-1930s and late 1950s assesses the range of imperial strategies towards labour recruitment, stabilisation and the welfare of colonial subjects. Not unlike Fanon, but through close historical research, Cooper shows how colonial instruments are used by labour unions and anti-colonial leadership for other ends (*Decolonization and African Society: The Labor Question in French and British Africa*, 1996). A shorter version of Cooper's argument about decolonisation and the challenges of independence across Africa appears in his *Africa Since 1940: The Past of the Present* (2002), from which the reading that follows is excerpted. A more recent collection of essays addresses questions of nationalism, imperialism, postcoloniality, along with a critique of the troika of 'identity', 'globalisation' and 'modernity', in *Colonialism in Question: Theory, Knowledge, History* (2005). Cooper has also co-authored two important sets of essays, one (with Allen Isaacman, Florencia Mallon, William Roseberry and Steve Stern) in response to the new historiography of subalternity and capitalism (*Confronting Historical Paradigms: Peasants, Labor, and the Capitalist World*

System in Africa and Latin America, 1993) and the second (with Rebecca Scott and Thomas Holt), on the comparative dynamics of empancipation (in *Beyond Slavery: Explorations of Race, Labor, and Citizenship in Postemancipation Societies*, 2000). Finally, Cooper has edited important books including *Tensions of Empire* with Ann Stoler (1997), and, particularly important for readers of this volume, *International Development and the Social Sciences: Essays on the History and Politics of Knowledge* (with Randall Packard, 1997), as well as, more recently, *Lessons of Empire* (with Craig Calhoun and Kevin Moore, 2005).

The reading makes an important synthetic argument about the state and development in Africa at the millennium. Cooper reminds us that a variety of constituencies saw new opportunities in the period after 1945. These groups came together in decolonisation movements around the failure of reformist colonialism and the importance of national 'self-determination'. Yet, they had inherited regimes with sovereignties recognised externally, but not entirely internally. They became 'gatekeeper states', sitting at the interface of African communities and the world, deriving rents from their gatekeeper status, and regulating access to resources, visas, permits, money, and so on. Outcomes have varied through the specific interactions of 'European' and 'African' histories of colonialism and its aftermath. Some have relied on authoritarian 'big men', others have been more flexible. The most crisis-prone gatekeeper states were propped up by Cold War imperialist rivalries. Indeed, gatekeeper states have been welcome to Western superpowers eager to retain influence and access to resources.

The gatekeeper state is not a static model. A conjuncture of economic and political forces has called it into question since the 1990s, again to varied effect. The reader should note the ways in which this concept shifts from some of the standard tropes used to discuss crises in Africa, whether in terms of 'corruption', 'good governance', or 'humanitarianism'. Moreover, Africa provides sobering evidence for critics of the state as a key site of conflict resolution and provision of public goods. Finally, notions of federalism and Pan-Africanism have kept open other possible kinds of states that might emerge. Cooper notes that popular conceptions of 'Other Africas' abound in the arts. More responsive forms of sovereignty, and broader social citizenships might be built, Cooper challenges, when African, European and North American (and, one might add, Chinese) states recognise their past, present and future complicity with the gatekeeper state.

Key references

Frederick Cooper (2002) *Africa Since 1940: The Past of the Present*, Cambridge: Cambridge University Press.

— (1977) *Plantation Slavery on the East Coast of Africa*, New Haven, CT: Yale University Press.

— (1980) *From Slaves to Squatters: Plantation Labor and Agriculture in Zanzibar and Coastal Kenya, 1890–1925*, New Haven, CT: Yale University Press.

— (1983) *Struggle for the City: Migrant Labor, Capital, and the State in Urban Africa*, London: Sage Publications.

— (1987) *On the African Waterfront: Urban Disorder and the Transformation of Work in Colonial Mombasa*, New Haven, CT: Yale University Press.

— (1996) *Decolonization and African Society: The Labor Question in French and British Africa*, Cambridge: Cambridge University Press.

— (2005) *Colonialism in Question: Theory, Knowledge, History*, Berkeley: University of California Press.

Frederick Cooper and Ann Stoler (eds) (1997) *Tensions of Empire*, Berkeley: University of California Press.

Frederick Cooper, Allen Isaacman, Florencia Mallon, William Roseberry and Steve Stern (1993) *Confronting Historical Paradigms: Peasants, Labor, and the Capitalist World System in Africa and Latin America*, Madison: University of Wisconsin Press.

Frederick Cooper and Randall Packard (eds) (1997) *International Development and the Social Sciences: Essays on the History and Politics of Knowledge*, Berkeley: University of California Press.

Frederick Cooper, Rebecca Scott and Thomas Holt (2000) *Beyond Slavery: Explorations of Race, Labor, and Citizenship in Postemancipation Societies*, Chapel Hill: University of North Carolina Press.

Frederick Cooper, Craig Calhoun and Kevin Moore (eds) (2005) *Lessons of Empire*, New York: New Press.

African states were successors in a double sense. First, they were built on a set of institutions – bureaucracies, militaries, post offices, and (initially) legislatures – set up by colonial regimes, as well as on a principle of state sovereignty sanctified by a community of already existing states. In this sense, African states have proven highly durable: borders have remained largely unchanged, and virtually every piece of Africa is recognized from outside as a territorial entity, regardless of the effective power of the actual government within that space. Even failed states – those unable to provide order and services for their citizens – are still states and derive resources from outside for that reason.

Second, African states took up a particular, and more recent, form of the state project of colonialism: development. African political parties in the 1950s and 1960s generally insisted that only an African government could ensure that development would serve the interests of "their" people. Here, continuity is less striking. By the 1970s in most African states, the development slogan had become either tragedy or farce, and people now viewed such claims with either bitterness at the politicians who developed their own wealth at the people's expense, or a continued yearning for development in the form of schools, hospitals, marketing facilities, and a chance to earn money and respect.

The early governments thus aspired both to define their authority over territory which, however arbitrary its borders, was now theirs, and to build something on that territory. But the dual project was born into the limitations of the old one, the colonial version of development. African states, like their predecessors, had great difficulty getting beyond the limitations of a gatekeeper state. Their survival depended precisely on the fact that formal sovereignty was recognized *from outside*, and that resources, such as foreign aid and military assistance, came to governments for that reason. Like colonial regimes, they had trouble extending their power and their command of people's respect, if not support, inward. They had trouble collecting taxes, except on imports and exports; they had trouble setting economic priorities and policies, except for the distribution of resources like oil revenues and customs receipts; they had trouble making the nation-state into a symbol that

inspired loyalty. What they could do was to sit astride the interface between a territory and the rest of the world, collecting and distributing resources that derived from the gate itself: customs revenue and foreign aid; permits to do business in the territory; entry and exit visas; and permission to move currency in and out.

Colonial states had ultimately derived their authority from the movement of military force from outside; their coercive power was more effective at staging raids and terrorizing resistors than at routinizing authority throughout a territory; they had built garrison towns and railways to extend their capacities for such deployments; and they had come to rely on localized systems of "traditional" legitimacy and obedience. Their successors faced similar limitations and often hesitated between alliances with the same decentralized authority structures, with the risk of reinforcing ethnic power brokers or breaking them via another deployment of military power likely to provoke regionalist opposition. Colonial states could turn to the armed forces of the distant metropole to ensure control of the gate. Their successors could not call in such support. Keeping the gate was more ambition than actuality, and struggles for the gate – and efforts of some groups to get around it – bedeviled African states from the start.

What independence added was the possibility of weaving patron–client relationships within the state, something colonial officials did too, but not so well. One also has to avoid the temptation of being overly institutional in one's analysis – to look for a "state" (bureaucracy, executive, military, legislature) attempting to rule over and interact with "civil society" (interconnections and collectivities formed among the people of a territory). At least as important were the vertical ties formed by a political elite, the connections between a "big man" and his supporters, who in turn had linkages to their own supporters. A person might hold an office – for example as a commissioner in charge of a region – with certain duties and powers, but his daily reality might be to distribute rewards and cultivate support, shaping vertical ties that did not quite coincide with his official role. It is too simple to call that corruption, although many Africans as well as outsiders would do so.

NINE

Not all manage vertical ties successfully, for one of the origins of instability in Africa is the inability of gatekeepers to keep the gate. Traders bypass customs collection; oppositional networks establish connections with outside powers and get arms and support; and people often try to live their lives within and across territorial borders as if the state had limited power over them. Some states, Senegal for instance, have managed patron–client systems with relative stability, whereas others, like Zaire, turned them into crude machines by which a single leader and his shifting entourage extracted resources arbitrarily from poor peasants and rich merchants to such a degree that the formal dimension of state rule collapsed. In Sierra Leone or Angola, rivals developed cross-border networks that smuggled out valuable resources, notably diamonds, and smuggled in arms and luxury goods for the gate-evading leadership.

During the struggles for independence, leaders of parties, trade unions, farmers' organizations, merchants' groups, students, and intellectuals aspired to a view of state-building with a strong "civic" dimension: the state would act in the interest of citizens as a body, through institutions accessible to all. Once in power, African regimes proved distrustful of the very social linkages and the vision of citizenship which they had ridden to power. Nkrumah set a dangerous and revealing precedent when he attacked trade unions and cocoa farmers' organizations in his first years in office. New presidents and prime ministers used their control of access to import-export markets and revenues to reward followers and exclude rivals.

The gatekeeper state was vulnerable: it made the stakes of control at a single point too high. Politics was an either/or phenomenon at the national level; local government was almost everywhere given little autonomy. Leaders often saw opposition as "tribal," which could encourage regional leaders to accuse the "national" elite of being tribalist itself. Gatekeeper states' insistence on the unity of the people and the need for national discipline revealed the fragility of their all-or-nothing control; they left little room for seeing opposition as legitimate. All states, undoubtedly, function via a mixture of personal ties and formal structures. That African history has encouraged people to form rich webs of connections, based on kinship, trading diasporas, religious networks, and so on is not necessarily a political liability; it was most often a strength. A combination of predictable state

institutions, a civic-minded political culture that emphasizes the accountability of leaders to the citizens, and a measure of personal politics and networking may produce tensions, but also balance civic virtue and personal connections. What is problematic about gatekeeper states is the focus of patronage systems on a single point and the undermining, in the midst of intense rivalry for that point, of alternative mechanisms for influencing decisions and demanding accountability.

Western governments and international institutions were only too willing to work with gatekeeper states, which were permeable to their influence and useful in cold war rivalries. Even a modest amount of economic or military aid could be a major patronage resource to a leader, or to an insurgency trying to evade the gatekeeping state, when other forms of access to the world economy were weak. From the moment of independence, notably in the Congo crisis of 1960 (see below), western states signalled that they might intervene against a radical *national* policy because they judged that its international implications were unfavorable to their interests. Some of Africa's most conflict-ridden states in the 1980s and 1990s had been among the largest recipients of economic and military aid intended to foster Cold War alliances: Angola, Zaire, Somalia, and Ethiopia. Both sides sometimes left in place former proxy combatants, either ruling cliques or insurgent movements, which continued to reap havoc – in Angola and Somalia for instance – after the Cold War ended.

Even more important, the huge disparities in the world economy accentuated the importance of gatekeeping to state elites: they could not bring their domestic economies to the level of Europe, North America, or East Asia, but they could try to police their citizens' access to the wealth that lay outside and find a profitable niche for themselves. Even a small quantity of resources – a marketing deal, remittances from migrant labor, foreign aid, automatic weapons – could make a decisive difference to whoever could control the asset. Hence the importance of guarding the gate to those who possess it or of building networks to get around state-regulated export-import institutions to those who do not.

When African states proved unable to cope with the difficulties of the post-1973 world economy, western development institutions were quick to blame the corruption and ill-considered policies of

African governments. The remedies proposed, under the rubric of structural adjustment, do not necessarily address the structural or historical conditions that gave rise to gatekeeper states. Both African rulers' fear of the kinds of mobilized citizenries through which they challenged colonial rulers and the diminished resources which African states have to offer their citizens in the age of structural adjustment make gatekeeping and patronage more attractive strategies of rule than the democratic bargain by which governments ask for votes and taxes and provide needed services. Gatekeeper states are thus not "African" institutions, nor are they "European" impositions; they emerged out of a peculiar Euro-African history.

[. . .]

Can the gatekeeper state become democratic?

Fifteen years ago, some observers thought that pro-trade, agriculturally oriented countries like Kenya and Côte d'Ivoire were exceptions to the rule of African economic stagnation; they were "miracles of the market." Since then, they have proven to be less than miraculous, constrained by their inability to do more than produce larger quantities of primary commodities which world markets want less of and by their leaders' perceived need to put their own control of patronage ahead of economic growth and diversification.

At the other end of the policy spectrum, African states like Nkrumah's Ghana tried to restructure a colonial economy radically, but had little other than mounting debts to show for the effort. The disasters of African political economy range from the self-proclaimed market-oriented (Zaire, Nigeria) to the self-proclaimed socialist (Tanzania, Guinea). The grossest violations of human rights have occurred in states whose formal ideologies have covered a wide spectrum (Zaire on the right and Guinea on the left), and in others where ideology was less important than the *folies de grandeur* of a leader (Idi Amin in Uganda or Jean-Bedel Bokassa in the so-called Central African Empire). Some states have broken down nearly completely, Somalia notably, where not only basic institutions – bureaucracy, judiciary, army – have failed but where rival warlords construct distinct patron–client systems.

Ideology is not the theme around which the most important variations occur. But gatekeeper states can be managed so tightly that they destroy all sources of social cohesion and economic dynamism other than that which can be controlled from the top, as in Mobutu's Zaire, or they can be more flexible, allowing somewhat diverse economic and social relationships to develop, as in Senegal. As I have argued, gatekeeper states are distinguished not by effective control of the gate, but by the intensity of struggle over it, which has had varying outcomes. While it is clear that it is a mistake to turn the worst failures of gatekeeper states – Liberia, Sierra Leone, Zaire, Guinea – into stand-ins for all of Africa, it is harder to look at the other end of the spectrum, to find clear paths for a way out.

In the classificatory scheme of political scientists Michael Bratton and Nicolas van de Walle, there were in 1989 only five multiparty regimes in Africa, compared to eleven military oligarchies, sixteen one-party regimes that gave people no choice but ritual ratifications of pre-selected candidates, and thirteen one-party systems that allowed some choice of candidates within the party. In the 1990s, there seemed to be an opening, and some commentators spoke of a "wave" of democracy pushing aside dictatorships and military governments across the world, including Africa. Was this a conjuncture in which the exercise of national citizenship was reinvigorated, perhaps comparable to the post-1945 conjuncture in which the notion of imperial citizenship opened up and then exploded the very idea of reformist colonialism? Some elements were coming together that shook up the gatekeeper state:

1 Over a decade of economic contraction eroded the patronage resources of ruling elites and left them searching for new ways to obtain support.
2 Key categories of African societies, notably professionals and students, found their paths to the future blocked and little outlet for their talents.
3 Workers and the urban poor may have remained desperate enough to keep trying to use whatever personal connections they could, but were doing so badly that many were *available* (as with youth in the 1940s) to be mobilized in strikes, demonstrations, and sometimes anti-regime violence.
4 Donor agencies like the World Bank were frustrated at the dissipation of aid via corruption and were demanding "good governance" as a

condition for the assistance which rulers desperately needed.

5 An expanding range of nongovernmental organizations, concerned with human rights, legal reform, women's empowerment, and ecology, were involved in Africa, and they were interacting more closely than before with African activists.

Just as the revolution which overthrew colonial rule came about not as the bottom overthrew the top but as interaction became more intense and the path of change less controllable by rulers, the uncertainties within African regimes seemed to promise an unfolding dynamic of political restructuring. Movements for multiparty elections and increased freedom of speech can be infectious; a contested election in one country suggests a possibility in another.

But the results were mixed and short of the dynamic of the 1950s and 1960s. A wave of national conventions, of rewriting of rules of electoral competition, of legalizing multiple parties, and of elections with at least a measure of fairness took place after 1989, starting with Bénin's calling of a national convention to write a more democratic constitution. In the 1990s, there have been more elections – some of them offering genuine choice – than in any time period since the early days of independence. In Bénin, Zambia, and Senegal, old-guard leaders were voted out of office; second rounds of more or less open elections have occurred in several countries. Bratton and van de Walle claim that between 1990 and 1994 progress toward democratic elections occurred in sixteen African countries (out of forty-seven), and limited transitions in others, and civil liberties improved at least a little in thirty-two countries.

The degree to which electoral choice and freedom of expression and organization have progressed varies, but there are only a few clear models – Botswana, Mali, and newly "free" South Africa – of the institutionalization of political reform and new expectations of ordinary people that their voices will be heard. Leaders remain wedded to winner-take-all models of government, fearful of allowing rivals to have access to provincial or local patronage and other resources. If they concede multiple party elections, they still use patronage and coercion to maintain control of the process. Opposition parties, meanwhile, have tended to be built around the same kind of patron–client ties as the ruling party. Moi's Kenya remains a clear example of a regime facing all the pressures cited above, but manipulating a multiparty system as it had a single party system, as a divided and clientelistic democracy movement fails to bring together professional and popular opponents of the status quo. In Bénin and Zambia victorious oppositions soon replicated the regime they had replaced. Nigerian political scientist Julius Ihonvbere fears that "elections and more elections" have "recycled" old leaders, old ideologies, old political styles, and old suspicions. They have neither ended the politics of patronage and suppression of dissidence nor put in place governments capable of addressing fundamental problems.

At first glance, pressures to "liberalize" politics – to reduce rulers' arbitrary power, foster electoral competition, and encourage debate – may seem congruent with pressures to "liberalize" economies – to reduce the size and power of government, privatize public enterprises, and stimulate market competition. But the effects of such double liberalization can be contradictory. Decreased government services, under pressure from the IMF, are not the best selling points for democracy. Thandika Mkandawire fears that without more resources, African states will become at best "choiceless democracies." Privatization of state enterprises may not release assets into an open "market" but put them into the hands of leading politicians and their clients, who in their "private" capacity as businessmen do what they did as "public" managers.

Indeed, the private–public distinction is a misleading one, where politics depends on the "big man" who controls a range of resources, from money to kinsmen to clients to state office. Downsizing the state might do less to reduce clientelism than to reduce the effectiveness of those institutions, including the civil service, which provide services to the population as a whole, making the search for a patron all the more necessary. At an extreme, public service can collapse altogether and the "state" loses control of even the gate, leaving in place a series of power brokers who acquire followers and weapons and transform themselves into warlords. This has happened most clearly in Sierra Leone and Somalia and was resisted where state institutions retain at least some respect and effectiveness, as in Senegal or Kenya. At the positive end of the spectrum is Botswana, where a relatively coherent leadership, with traditional legitimacy, education, and business acumen, has maintained a strong civil service, governed through recognized institutions rather than personal deals, allowed room for debate and decision-making at the district level, bargained firmly with foreign corporations to insure adequate state compensation and control, aided private enterprise, and insisted that state enterprises be commercially viable. Electoral democracy has remained intact, and Botswana has maintained one of Africa's highest rates of growth – 9.9 percent annually in 1960–80 and 5.6 percent in 1980–91 – although the distribution of income and wealth is wide and growing.

Despite the variations, the widespread sense of frustration after the era of "democratic experiments" in the early 1990s should be no surprise, for the historical patterns and global conditions which gave rise to gatekeeper states in the first place have not fundamentally been altered, and the political economy of an African state cannot necessarily be remade by an act of will of even the most enlightened leadership. But the fact that economic and political reform have been debated and attempted in different African countries is itself of great importance. Such debate has, perhaps in a way unseen since the 1950s or early 1960s, enhanced among ordinary citizens a sense that citizenship entails multiple possibilities: that officials can be held accountable for their actions; that constituencies for reform can be built; that associational life and public discussion can be enriched; and that trade unions, women's associations, and political parties can be organized.

OTHER AFRICAS: CONNECTIONS BEYOND THE NATION-STATE

As emphasized in previous chapters, the national focus of African elites in the mid-1950s represented a shrinking of spatial perspectives. Senghor, for one, regretted the narrow territorial focus of politics in the late 1950s, but his broad defense of equality within the French Union and even within smaller federations in West Africa proved untenable. By hosting the All-Africa Peoples' Congress in 1958, Nkrumah tried to revive Pan-Africanism, which had lost ground to territorial politics since the Manchester Congress of 1945. Pan-Africanism, however, was becoming a relationship of states, not of people. Nkrumah's hope for a United States of Africa achieved little support

from African leaders intent on protecting the sovereignty they had so strenuously fought for.

State-centered Pan-Africanism became institutionalized as the Organization of African Unity (OAU). It met regularly, discussed common action on various fronts – above all seeking global economic policies more favorable to developing countries – but its possibilities were constrained by what it was: an assembly of African heads of state, many of whom were part of the anti-democratic trends of their own countries. The OAU failed its most obvious moral test, failing to act in the name of values shared across Africa against those leaders who were the most egregious violators of them.

Economic cooperation, as described in chapter 5, had its failures, such as the East African Federation, and also its moderate successes in regard to tariff reduction, shared banking institutions, and other interstate arrangements, notably the Economic Community of West African States (ECOWAS) and SADCC. ECOWAS also has a political and military component, which performed more credibly than previous efforts in its interventions in Sierra Leone, where it contained a rebellion against an elected government. But after 1994, African institutions failed to contain the worst border-crossing conflict, when Uganda, Rwanda, Angola, and Zimbabwe fanned the flames of conflict in Central Africa and became involved in the trafficking of Congo's resources of gold and diamonds. Charles Taylor, once a rebellious warlord in Liberia, laid successful claim to recognized sovereignty over that country, while intriguing with shadowy movements in Sierra Leone, Guinea Bissau, and other neighbors in pursuit of the diamond trade. In short, intra-African cooperation has sometimes taken the form of covering up the vices of gatekeeper states and participating in the vices of cross-border networks that bypass the gate. Constructive interaction of African states remains an elusive goal.

African states have been more likely to be the targets of intervention from outside than active participants in international regulatory processes. Their need for periodic financial bail outs has made it hard for them to pose alternatives to IMF policies. Africa has been the focus of operations by refugee and famine relief organizations, whose services have been desperately needed, as in the aftermath of the Rwandan genocide of 1994 or the Sahelian famine of 1974. International humanitarian organizations have given Africans a means to find allies in their struggles against abusive regimes, in obtaining support for sustainable agriculture, or in combating AIDS, but some have defined the problem in terms of African helplessness – its need of outside help to feed starving babies – rather than that of cooperating with Africans to address issues of injustice and impoverishment.

At the same time, Africans have actively participated in international organizations which address the continent's problems. A Ghanaian, Kofi Annan, heads the UN. The United Nations Development Program (UNDP), where "Third World" economists and other specialists play important roles, has been instrumental in keeping alive the idea that growth and structural change in Africa require participation of its citizens in the process. The United Nations Educational, Scientific, and Cultural Organization (UNESCO), which for many years was directed by Mochtar M'Bow of Senegal, has called attention to the diversity of historical and cultural contributions of peoples throughout the world, and while opponents have criticized it for wastefulness, intellectual and cultural innovators have made use of its resources to promote the idea that not all literary, artistic, and musical contributions fit within canons derived from European experience. The UNESCO-sponsored history of Africa has its virtues and its flaws, but the fact that a substantial portion of its editorial board have spent time in detention or in exile reveals clearly that African intellectuals can both argue for specifically African perspectives and criticize the nation-states which claim to embody Africa's uniqueness.

Cross-border relationships take more forms than this. Every border gives rise to specialists who figure out how to cross it; they constitute a "community" as much as any people located firmly within a given set of boundaries. Islamic networks link much of Africa to other parts of the world, through annual pilgrimages, the training of religious leaders in Egypt or Saudi Arabia, and financial support from oil-producing states in the Middle East. Islam, as in Senegal, may mean participation in a brotherhood deeply rooted in local social and political relations, and it can mean participation in a universalistic faith with adherents around the world. Christian churches, as noted in other contexts, participated in a variety of transnational linkages. The numerous Africans who have obtained higher education in Europe and North America, who participate in international professional organizations, and who work for NGOs, participate in world-wide networks. The extent of labor migration and the fact that African communities are now well established in European countries shape other sorts of connections.

All this suggests that political imaginations of Africans, from poor migrants to sophisticated professionals, have multiple roots and may provide alternative models for changes. Some forms of political organization have been – for good, one hopes – excluded from the realm of possibility: the empire and the white supremacist state, for instance. The 1950s and 1960s witnessed an opening of opportunities for Africans to participate in politics, but the very possibility of attaining power focused imaginations on the units in which power was available.

Some argue that the problem is the nation-state itself. Basil Davidson, a stalwart supporter of African nationalism in its heyday, now sees the nation-state as the "black man's burden," an imposed institution inappropriate to the conditions of Africa. We have, however, seen something worse than the nation-state: its absence. The collapse of state institutions in Somalia, and their weakness in the face of warlordism in Sierra Leone, Liberia, and Zaire, reveal how much, in today's world, we depend on state institutions to regulate conflict and provide basic infrastructure. Anthropologists can celebrate the ability of the Somali clan structure to balance the relative strengths of different kinships, foster equality among males, and settle conflicts within the kinship system itself, but once – following the cynical manipulations of both the United States and the Soviet Union – Somalis got access to AK-47s and truck-mounted artillery, clan conflict took on an altogether different aspect. Seeing the possibility of actually dominating others, clan leaders carried warfare to a more devastating level, tearing down state structures altogether. The effects of vacuums in state power in Sierra Leone and Zaire are no less devastating.

It may also be too simple to lament the "western" character of the state, for states as they exist throughout the world reflect a history shaped by the struggles of once-colonized people, and state institutions take new forms when they are used in different contexts. But the history of decolonization did foreclose alternatives that were once the object of attention – supranational federations and Pan-Africanism – and it put in place a particular kind of state as well as a ruling class conscious of its own fragility. The devolution of sovereignty by European states was accompanied by a denial of responsibility for the historical process which had put in place this institutional nexus. This leaves the question of whether in the future the range of possible forms of political action can be expanded as well as constricted. Here, the question is not whether Africa should maintain or abolish the nation-state, but rather what kind of state can be constructed, what kinds of relationships can be forged across state lines, and what kinds of recognition within states can be given to the variety of forms of affinity to which citizens subscribe.

The tragedy of Africa's decolonization lies in the foreclosure of possibilities that once seemed genuine: of a citizenry choosing its leader; of a socially-conscious citizenship, in which education, health services, and other services are seen as a duty of government; and of cooperation across borders against the injustices of imperialism and for recognition of Africa's contributions to global culture. That states are units of electoral politics and territorial administration does not preclude a politics of larger and smaller units, of activism across borders. If some see sovereignty as a barrier against any questioning of the good or evil done within borders, others argue that sovereignty is a part of an effort to share the planet and that the abuse of sovereignty concerns people everywhere. African states will probably remain stuck within the limitations of the gatekeeper state – with its brittle and heavily guarded sovereignty – unless nation-states in Europe and North America as well as Africa acknowledge a shared responsibility for the past which shaped them and the future to which they aspire.

OTHER AFRICAS: POPULAR CULTURE AND POLITICAL CRITIQUE

One of the mistakes of western critics of African governments and their policies, coming from journalists and international financial "experts" alike, is to think that they were the first to diagnose problems. The fact is that African intellectuals became aware of fundamental difficulties within their countries while most of their western counterparts were still celebrating the arrival of Africans into the "modern" world. In city after city and village

after village, even under oppressive conditions, popular discontent has found powerful forms of expression.

A key moment in the emergence of a critical intelligentsia was the publication in 1968 of Ayi Kwei Armah's novel *The Beautyful Ones Are Not Yet Born*. Armah evoked the disillusionment of young Ghanaians who had been captivated by the idealism and dynamism of the new state, only to see their classmates and families caught up in the greed and corruption of the new order. The book ends with the chaos of a coup, in which the hero helps his corrupted former classmate escape; an ending which evokes the shared humanity of people whose political and ethical decisions differ and which opens the possibility that a new generation will have another chance. Other novelists and playwrights of that era – Mongo Beti, Chinua Achebe, Wole Soyinka, Ngugi wa Thiong'o for example – developed a politicized fiction written for Africans literate in French or English and for foreigners as well, and which brought out the moral dilemmas of people who had struggled against colonialism and were facing new struggles. Other writers refused to be typecast into the mold of writing political fiction and saw themselves opening up a wide variety of imagined worlds, as writers in many contexts have done. But most important were the efforts of intellectuals familiar with a variety of oral and written literary forms to perform in ways accessible to a wider populace. The emergence of a Yoruba popular theater in western Nigerian cities is a case in point, as was the performance of plays in Kikuyu by Ngugi wa Thiong'o, an effort which, unlike his equally critical writing in English, landed this distinguished writer in prison.

Critical artistic creation took more forms than the literary. The Nigerian singer Fela Ransome Kuti, the scion of a family active in the Nigerian nationalist movement, developed a politicized and highly popular musical style in Lagos, for which he was harassed and detained by the police. In Kinshasa, renowned for decades for its musical creativity, a subtle form of criticism of Zairian society under Mobutu crept into what at another level were love songs and dance music. The café/club/bar scene was entertainment, but it was also part of forming new kinds of sociability, which as in the 1940s and 1950s was not without political content. Zairian musicians – including Papa Wemba and Franco – were influenced by African-American jazz and Ghanian high life, and they in turn influenced music throughout Africa, Europe, and North America. So, too, with visual art. Kinshasa was the home of a vigorous community of painters, Cheri Samba and Kanda Matulu Tshibumba the best-known of them, who were influenced by comic book art as much as by formal schooling in painting to create a style that was in part moral lesson, in part political critique, and in part a colorful, visually appealing form of self-expression.

African writers, artists, singers, and film-makers have long debated what it means to "decolonize" African culture as well as to distance the artist from the rigidities of the post-colonial successor states. Does it mean rejecting European ideas of what constitutes a novel, a painting, or a film as much as refusing Eurocentric content? Does it mean searching for some sort of African "authenticity"? Or does it mean that the artist is open to whatever influence he or she wishes to turn to, to whatever themes he or she wishes to engage?

It would be a mistake to see African urban culture, in Kinshasa, Lagos, or elsewhere, as simply a popular, critical form standing in stark opposition to a detached form of "power." Power is much more ambiguous than that. Achille Mbembe finds a culture of "vulgarity" in African cities. The "big men" of the regime are seen literally to "eat" the country, to have huge sexual appetites as well, to flaunt their wealth and power. Popular culture is at one level angry at this, but at another level desirous. The vulgarity itself marks both an evil – dangerously close to witchcraft, for supernatural forces are also seen to "eat" their victims – and a sign of power, something deeply desirable as well as immoral.

The insight is especially crucial when one is not limited to seeing society divided into categories – elite versus popular classes or ethnic, racial, or gender division – but rather stresses *relationships*, and in particular vertical relationships. Rich and poor see each other not just in terms of antagonism, and not just via the desire of the latter to become the former, but by interaction and mutual expectations between the two. The poor seek access not just to wealth, but to the wealthy. The rich have access to many poor – who can provide political support, muscular action, and cheap services – while poor people have limited access to the rich. The jockeying for access gives rise to both connection and antagonism and gives African political culture, particularly in cities, a high degree of volatility. Outbursts of popular rage, whether riots against price increases in Zambia or strike action against the Abacha regime in Nigeria, alternate with periods when patron–client networks offer at least a possibility for vulnerable members of society to get something – even if most will end up disillusioned.

The importance of the quest for a rich relative, a marabout, a former classmate, or a well-placed fellow-villager is thus incorporated into popular culture, even as patron–client relationships are also crucial to the very summit of power. Patron–client relationships are part of any political system, but a particularly big part of African polities. And the reason is partly that of the structure of the world economy, above all the enormous gap between the resources available within African countries and those in Europe, North America, East Asia, and parts of the Middle East.

Africans as much as outsiders have expected state institutions to perform certain functions: to organize schools and clinics; to deliver mail; to guarantee security of persons and of transactions; and also to promote a common sense of affinity and shared responsibility among the nationally-defined community. States can and should be judged on whether they can perform such basic tasks for the benefit of the large majority of citizens. But states consist of people and people build networks. The institutional strength of states and the networks of its most important leaders may reinforce each other or stand antagonistically to each other. A political leader may benefit from fostering economic growth in which many people share, adding to his political capital, or he may also try to deny potential rivals access to resources and insure control by eliminating alternatives. One process fosters institution-building and resource generation; the other weakens institutions and consumes resources. Both are rational strategies, and both are used by leaders of African states in various combinations. Nothing guarantees that the virtuous circle will triumph over the vicious one. It helps little to point out the failings of governments without understanding the constraints and their causes, but historical reflections should deepen discussions of accountability, not foreclose them. Governments, oppositions, and concerned citizens have choices, and their consequences are enormous.

SUGGESTED READING

A full bibliography for this book may be found on the website of Cambridge University Press at http://uk.cambridge.org/resources/0521776007. It will be updated periodically.

Armah, Ayi Kwei. *The Beautyful Ones Are Not Yet Born.* Boston: Houghton Mifflin, 1968.

Bayart, Jean-François. *The State in Africa: The Politics of the Belly.* London: Longman, 1993. Trans. by Mary Harper. (Originally published as *L'état en Afrique: La politique du ventre.* Paris: Fayard, 1989.)

Bratton, Michael, and Nicolas van de Walle. *Democratic Experiments in Africa: Regime Transitions in Comparative Perspective.* Cambridge: Cambridge University Press, 1997.

Callaghy, Thomas. *The State–Society Struggle: Zaire in Comparative Perspective.* New York: Columbia University Press, 1984.

Hutchinson, Sharon. *Nuer Dilemmas: Coping with Money, War and the State.* Berkeley: University of California Press, 1996.

Joseph, Richard, ed. *State, Conflict and Democracy in Africa.* Boulder: Lynne Rienner, 1999.

Mbembe, Achille. *On the Postcolony.* Berkeley: University of California Press, 2001.

Reno, William. *Corruption and State Politics in Sierra Leone.* Cambridge: Cambridge University Press, 1995.

'Beyond Occidentalism: Toward Nonimperial Geohistorical Categories'

Cultural Anthropology (1996)

Fernando Coronil

Editors' Introduction

Fernando Coronil is a Venezuelan citizen, and Associate Professor in Anthropology and History, and Director of the Latin American and Caribbean Studies Program at the University of Michigan. Coronil grew up in Caracas, where he attended the famous school, Liceo Andres Bello. He became student body president in the highly politicised 1960s, a moment which would shape his enduring interest in the state and power. He went on to take an Honours in Social Thought and Institutions at Stanford University, while the United States was shaken by counter-cultural, anti-Vietnam and civil rights movements. He then pursued a PhD in Anthropology at the University of Chicago. Fieldwork in Cuba was interrupted by changes in Cuban policy towards foreigners, which led Coronil to decide against writing up this project. When legal troubles barred Coronil's re-entry to the United States, he and the anthropologist Julie Skurski moved to Venezuela to teach at the Andres Bello Catholic University and to research at Centro de Estudio del Desarollo (Centre for the Study of Development) at the Universidad Central de Venezuela. Here, Coronil and Skurski conducted important research on the automobile industry as the 'motor of national development', on political violence, and on nationalism in Venezuela.

Coronil's *The Magical State* (1997) rethinks the relationship between nature, labour and money in a society in which 'development' is fundamentally shaped by state control of wealth derived from oil. This 'devil's excrement' lends a spectral quality to the state, as if it was insulated from society, as governments shift between claims of 'dictatorship' and 'democracy' in a series of coups. Coronil manages to bring together our four central themes – markets, empire, nature and difference – in a powerful critique of Venezuelan modernity. Some of Coronil's ideas about postcoloniality and modernity/postmodernity appear in the extended introduction to the republication of Fernando Ortiz's classic *Cuban Counterpoint* and in important articles, one of which is excerpted as the reading here. Coronil and Skurski have also edited *States of Violence* (2006) on the violent foundations of modern nations. Coronil is currently writing a book on aesthetics and politics, focusing on images surrounding revolution in Cuba, and another book on the 2002 coup against Chavez in Venezulea.

The reading that follows returns us to a seemingly obvious question which could be asked of the many histories of 'development' we have sought to map in this reader: 'How to represent the contemporary world?'. Maps, Coronil suggests, have been both the means of representation of the world, as well as tools for critique. While some of the key metaphors of the modern world have centred on time – history, progress, improvement, growth and, in large part, 'development' – imperial dynamics have always been spatial. Imperialism past and present relies on spatial imaginations, maps being one kind. In an important work, Edward Said argued in *Orientalism* (1976) that representations of the 'Orient' were intrinsic to the extension of Europe's empires. Coronil extends this argument by asking how a kind of spatial myopia becomes complicit in hiding actual processes of inequality, as well as of actual, conflictual connections between human collectivities. By 'Occidentalism', Coronil means all the ways in which a world of difference is partitioned into bounded spaces that are then hierarchised. By treating these hierarchies as natural, Occidentalism helps reproduce the inequalities in power that keep the world grossly unequal. Spatial representation is key to this process of keeping the world unequal. The reader might think about how spatial metaphors in development thought are used in specific ways: by partitioning space into developed/underdeveloped, advanced/backward, North/South, East/West, communist/capitalist, or 'free'/'terrorist'. In a section not included in the present excerpt, Coronil carefully traces various ways in which writers, often quite critical of imperialism, have tried to take on these tendencies of Occidentalism, usually to limited effect.

Development thought has always wrestled with the challenges posed by various persisting inequalities. Development theory and practice have witnessed spectacular failures in the ability to intentionally transform these inequalities, whether through forced collectivisation in the USSR or structural adjustment imposed by the World Bank and IMF. In rethinking possible futures beyond the gross inequalities of the present, Coronil calls for reflection on our toolboxes so as to represent a world of difference and transcultural exchange and its many actual and ongoing forms of critique. Indeed, if 'development' continues to make sense for a world 'beyond Occidentalism', it is decidedly not just in remaking partitioned maps of the West vs. the dangerous, terrorist rest, or in the bland language of development consultants continuing to provide tutelage to the hapless 'Global South'. Development thought beyond Occidentalism lies in many forms of ongoing refusal, and ongoing attempts to fight with 'development', to consign it, everywhere, Coronil suggests, to the past.

Key references

Fernando Coronil (1996) 'Beyond Occidentalism: toward nonimperial geohistorical categories', *Cultural Anthropology* 11: 51–87.
— (1997) *The Magical State: Nature, Money and Modernity in Venezuela*, Chicago, IL: University of Chicago Press.
— (1995) 'Transculturation and the politics of theory: countering the center, Cuban Counterpoint', Introduction to Fernando Ortiz *Cuban Counterpoint: Tobacco and Sugar*, Durham, NC: Duke University Press.
Fernando Coronil and Julie Skurski (eds) (2006) *States of Violence*, Ann Arbor: University of Michigan Press.
Edward Said (1976) *Orientalism,* New York: Vintage Books.

■ ■ ■ ■ ■ ■

Are you sure it is my name?
Have you got all my particulars?
Do you already know my navigable blood,
my geography full of dark mountains,
of deep and bitter valleys
that are not on the map?
　　　　　　Nicolás Guillén, "My Last Name"

A place on the map is also a place in history.
　　　　　　Adrienne Rich, "Notes toward
　　　　　　　　a Politics of Location"

Frantz Fanon begins the conclusion of *Black Skins, White Masks* with the following epigraph taken from Marx's *The Eighteenth Brumaire of Louis Bonaparte*:

The social revolution ... cannot draw its poetry from the past, but only from the future. It cannot begin with itself before it has stripped itself of all its superstitions concerning the past. Earlier revolutions relied on memories out of world history in order to drug themselves against their own content. In order to find their own content, the revolutions of the nineteenth century have to let the dead bury the dead. Before, the expression exceeded the content; now the content exceeds the expression.

[1967:223]

Imagining a future that builds on the past but is not imprisoned by its horror, Fanon visualized the making of a magnificent monument: "On the field of battle, its four corners marked by scores of Negroes hanged by their testicles, a monument is slowly built that promises to be majestic. And, at the top of this monument, I can already see a white man and a black man *hand in hand*" (1967:222).[1] Drawing his poetry from the future, Fanon sought to counter the deforming burden of racialist categories and to unsettle the desire to root identity in tradition in order to liberate both colonizer and colonized from the nightmare of their violent history.

In a shared utopian spirit, here I explore representational practices that portray non-Western peoples as the Other of a Western Self. By examining how these practices shape works of cultural criticism produced in metropolitan centers and subtly bind them to the object of their critique, I seek room for a decentered poetics that may help us imagine geohistorical categories for a nonimperial world.[2]

IMPERIAL MAPS

How to represent the contemporary world? Maps have often served as a medium for representing the world as well as for problematizing its representation. From Jorge Luis Borges's many mind-twisting stories involving maps, I remember the image of a map, produced under imperial command, that replicates the empire it represents. The map is of the same scale as the empire and coincides with it point for point. In this exact double of the empire's domain, each mountain, each castle, each person, and each grain of sand finds its precise copy. The map itself is thus included in the representation of the empire, leading to an infinite series of maps within maps. The unwieldy map is eventually abandoned and is worn away by the corrosive force of time even before the decline of the empire itself. Thus, history makes the map no longer accurate, or perhaps turns it into a hyperreal representation that prefigures the empire's dissolution.

Unlike cartographers' maps produced under imperial orders, the representations I wish to examine are discursive, not graphic, and seem to be the product of invisible hands laboring independently according to standards of scholarly practice and common sense. Yet they involve the use of a shared spatial imagery and have the strange effect of producing a remarkably consistent mental picture or map of the world. In everyday speech as much as in scholarly works, terms such as the "West," the "Occident," the "center," the "first world," the "East," the "Orient," the "periphery," and the "third world" are commonly used to classify and identify areas of the world. Although it is not always clear to what these terms refer, they are used as if there existed a distinct external reality to which they corresponded, or at least they have the effect of creating such an illusion.

This effect is achieved in part by the association they conjure up as a group of terms. Often combined into binary sets, these sets forge links in a paradigmatic chain of conceptions of geography, history, and personhood that reinforces each link and produces an almost tangible and inescapable image of the world. For instance, the West is often identified with Europe, the United States, us, or with that enigmatic entity, the modern Self. In practice, these paradigmatic elements are frequently interchangeable or synonymous, so that such terms as "We" or "Self" are often employed to mean Europe, the United States, or the West – and vice versa. The term "third world," used since its creation during World War II to define the "underdeveloped" areas caught between the first (capitalist) and second (socialist) worlds, has remained the preferred home for the Other.[3] Although many of these categories are of only recent origin, they have gained such widespread acceptance that they seem almost unavoidable. Drawing on the naturalizing imagery of geography, they have become second nature.

Despite the apparent fixity of their geographic referents, these categories have historically possessed remarkable fluidity. With postmodern élan, they have taken on various identities and have come to identify places and peoples far removed from their original territorial homes. Japan, until recently an emblem of the East, has increasingly been accepted as a member of the West in international organizations as well as in popular culture. Raymond Williams, in a discussion tracing the origins of the West-East distinction to the Roman Empire and to the separation between the Christian and Muslim worlds, argues that the West "has so far lost its geographical reference as to allow description of, for example, Japan as a Western or Western-type society" (1983:333). Noam Chomsky, in turn, explains, "I'm using the phrase 'Europe,' of course, as a metaphor. Europe includes and in fact is led by the former European colonies in the Western Hemisphere and Asia. And of course Europe now includes Japan, which we may regard as honorary European" (1991:13). Historians of Europe are still of many minds about the birth of "Europe" as a meaningful category, and warn against the habit of reading history backward, extending the existence of present-day Europe into the past beyond a time when one could reasonably recognize its presence. The "third world," for years firmly anchored in the "periphery" – that is, in Asia, Africa, and Latin America – seems now to be moving toward the United States, where the term is being applied not just to areas populated by migrants from the original "third world" but to spaces inhabited by old domestic "minorities" such as "women of color," and to "underprivileged" ethnic and social groups. Los Angeles is increasingly referred to as "the capital of the third world," a designation that also serves as the title of a recent book (Rieff 1991).

While one may wish to question the imperial conceit that lies behind this move to elect as the capital of the "third world" a metropolitan city located within the territorial boundaries of the old first world, this ironic twist raises even more basic questions about the stability and meaning of these categories. If, like Chomsky's "Europe," these terms are used as metaphors, what are their original referents? Were they ever *not* metaphors? Yet, aren't these terms unavoidable precisely because they seem to designate tangible entities in the world, because they appear to be as natural as nature itself? In the face of their slippery fluidity, should our task be, as in the case of Borges's imperial map, to construct a perfect map by finding words that faithfully match reality "out there" point for point? And if we managed to freeze history and replicate geography in a map, wouldn't this representation be ephemeral? Since space too is located in time and is changing constantly, how could a map represent geography without apprehending its movement? But perhaps this shows that maps do not mirror reality, but depict it from partial perspectives, figuring it in accordance with particular standpoints and specific aims.

Within academia, the growing awareness of the limitations and ideological bias of the three worlds schema as a "primitive system of classification" (Pletsch 1981) has not stopped or significantly altered its almost inescapable use. The common practice among some scholars of indicating discomfort with the categories of this classificatory scheme by means of quotes or explicit caveats only confirms its stability and the lack of an alternative taxonomy. If we were to choose not to employ the term "third world," would we be better served by such categories as "the underdeveloped world," "backward areas," or the euphemism "developing nations"? As soon as new conceptions are constructed, as in the case of the call by the South Commission presided over by Nyerere to promote a "new world order," they seem to be resituated within the semantic field defined by the old binary structure, as was the case when George Bush appropriated this phrase months after it was formulated to create his own version of a "new world order" during the rhetorical war that preceded the Gulf War (Chomsky 1991:13). The shrinking of the second world has not dissolved the three world scheme, only realigned its terms. Thus, a noted journalist can say straightforwardly that the "Evil Empire turned out to be a collection of third-world countries" (Quindlen 1994).

With the consolidation of U.S. hegemony as a world power after 1945, the "West" shifted its center of gravity from Europe to "America," and the United States became the dominant referent for the "West." Because of this recentering of Western powers, "America," ironically, is at times a metaphor for "Europe." Perhaps one day Japan, today's "honorary European," will become the center of the West. In this string of historical turns, it is another historic irony, as well as a pun, that what began as an accident – the discovery of America as the "Eastern Indies" – gave birth to the Occident. Columbus, sailing from the west to reach the east, ended up founding the West. Perhaps if one day Japan becomes the West, and today's West recedes to the East, it will turn out that Columbus indeed reached, as he insisted, the East.

Given the intimate association between Europe and Empire, it is significant that in colonial and post-colonial studies Europe is primarily equated with the nations of its north western region. This exclusion of southern Europe is accompanied by the analytical neglect of Spain and Portugal as pioneering colonial powers that profoundly transformed practices of rule and established modular forms of empire that influenced the imperial expansion of Holland, England, and France. So ingrained has the association between European colonialism and northern Europe become that some analysts identify colonialism with its northern European expression (Klor de Alva 1992), thus excluding the first centuries of Spanish and Portuguese control in the Americas.

THE POLITICS OF EPISTEMOLOGY: FROM ORIENTALISM TO OCCIDENTALISM

The problem of evaluating the categories with which the world is represented was compellingly faced by Edward Said in *Orientalism* (1979), a pathbreaking work that raised to a higher level the discussion of colonial discourse in the United States. I propose to advance a related argument concerning Western representations of cultural difference that focuses on the politics of geohistorical categories.

In *Orientalism*, Said defines Orientalism as taking three interdependent forms: the study of the Orient; a "style of thought based upon an epistemological and ontological distinction made between the 'Orient' and (most of the time) the 'Occident' "; and a corporate institution dealing with the Orient (1979:2–3). While Said's discussion of each of these forms relates Orientalism to the exercise of power, his major concern is the connection between modern Orientalism and colonialism. Yet at times Said's discussion ambiguously moves between an abstract conception of the inevitable partiality of any representation and a historically situated critique of the limits of specific representations as the effect of unequal power relations. This unresolved tension may create the impulse to approach the gap between Western representations of the Orient and the "real" Orient by searching for more complete maps without inquiring into the sources of partiality of Orientalist representations.

Said confronted the ambiguity of his formulation in "Orientalism Reconsidered" (1986), written in response to the persistence of Orientalist representations in works produced by critics of imperialism. He called for an inclusion of "Orientalists" as part of the study of Orientalism: "because the social world includes the person or subject doing the studying as well as the object or realm being studied, it is imperative to include them both in any consideration of Orientalism" (1986:211).

For Said, the inclusion of the Orientalists entails a fundamental critique of the forms of Western knowledge informing their works in the following terms:

What, in other words, has never taken place is an epistemological critique at the most fundamental level of the connection between the development of a historicism which has expanded and developed enough to include antithetical attitudes such as ideologies of western imperialism and critiques of imperialism on the one hand and, on the other, the actual practice of imperialism by which the accumulation of territories and population, the control of economies, and the incorporation and homogenization of histories are maintained. If we keep this in mind we will remark, for example, that in the methodological assumptions and practice of world history – which is ideologically anti-imperialist – little or no attention is given to those cultural practices like Orientalism or ethnography affiliated with imperialism, which in genealogical fact fathered world history itself; hence the emphasis in world history as a discipline has been on economic and political practices, defined by the processes of world historical writing, as in a sense separate and different from, as well as unaffected by, the knowledge of them which world history produces. The curious result is that the theories of accumulation on a world scale, or the capitalist world state, or lineages of absolutism depend (a) on the same displaced percipient and historicist observer who had been an Orientalist or colonial traveller three generations ago; (b) they depend also on a homogenizing and incorporating world historical scheme that assimilated non-synchronous developments, histories, cultures and peoples to it; and (c) they block and keep down latent epistemological critiques of the institutional, cultural and disciplinary instruments linking the incorporative practice of world history with partial knowledges like Orientalism on the one hand and, on the other, with continued western hegemony of the non-European, peripheral world.

[1986:223–224]

This provocative challenge invites multiple responses. Here I propose to move beyond a predominantly epistemological critique of Western knowledge cast in its own terms toward a political understanding of the constitution of the "West" that encompasses an examination of its categorical system. To the extent that "the West" remains assumed in Said's work, I believe that Said's challenge, and the ambiguity in his discussion of Orientalism, may be creatively approached by problematizing and linking the two entities that lie at the center of his analysis: the West's Orientalist representations and the West itself.

I wish to take a step in this direction by relating Western representations of "Otherness" to the implicit constructions of "Selfhood" that underwrite them. This move entails reorienting our attention from the problematic of "Orientalism," which focuses on the deficiencies of the West's representations of the Orient, to that of "Occidentalism," which refers to the conceptions of the West animating these representations. It entails relating the observed to the observers, products to production, knowledge to its sites of formation. I would then welcome Said's call to include "Orientalists" in our examination, but I will refer to them as "Occidentalists" in order to emphasize that I am primarily interested in the concerns and images of the Occident that underwrite their representations of non-Western societies, whether in the Orient or elsewhere. This perspective does not involve a reversal of focus from Orient to Occident, from Other to Self. Rather, by guiding our understanding toward the relational nature of representations of human collectivities, it brings out into the open their genesis in asymmetrical relations of power, including the power to obscure their genesis in inequality, to sever their historical connections, and thus to present as the internal and separate attributes of bounded entities what are in fact historical outcomes of connected peoples.

Occidentalism, as I define it here, is thus not the reverse of Orientalism but its condition of possibility, its dark side (as in a mirror). A simple reversal would be possible only in the context of symmetrical relations between "Self" and "Other" – but then who would be the "Other"? In the context of equal relations, difference would not be cast as Otherness. The study of how "Others" represent the "Occident" is an interesting enterprise in itself that may help counter the West's dominance of publicly circulating images of difference. Calling these representations "Occidentalist" serves to restore some balance and

has relativizing effects.[4] Given Western hegemony, however, opposing this notion of "Occidentalism" to "Orientalism" runs the risk of creating the illusion that the terms can be equalized and reversed, as if the complicity of power and knowledge entailed in Orientalism could be countered by an inversion.

What is unique about Occidentalism, as I define it here, is not that it mobilizes stereotypical representations of non-Western societies, for the ethnocentric hierarchization of cultural difference is certainly not a Western privilege, but that this privilege is intimately connected to the deployment of global power. In a broad-ranging discussion of constructions of cultural difference, John Comaroff defines ethnicity, in contrast to totemism, as a classificatory system founded on asymmetrical relations among unequal groups, and reminds us that "classification, the meaningful construction of the world, is a necessary condition of social existence," yet the "marking of identities" is always the product of history and expresses particular modes of establishing cultural and economic difference (1987:303–305). As a system of classification that expresses forms of cultural and economic difference in the modern world, Occidentalism is inseparably tied to the constitution of international asymmetries underwritten by global capitalism. Linking Eurocentrism to capitalism, Samir Amin argues that "Eurocentrism is thus not a banal ethnocentrism testifying simply to the limited horizons beyond which no people on this planet has truly been able to go. Eurocentrism is a specifically modern phenomenon" (1989:vii).[5]

While classificatory systems may construct the relations among their terms as unidirectional, in effect they always entail different forms of mutuality. Noting that Said has not analyzed the impact of Orientalist images upon the people who use them, Nancy Armstrong has shown how Occidentalism involves the formation of specific forms of racialized and gendered Western Selves as the effect of Orientalist representations of non-Western Others.[6] In my view, Occidentalism is inseparable from Western hegemony not only because as a form of knowledge it expresses Western power, but because it establishes a specific bond between knowledge and power in the West. Occidentalism is thus the expression of a constitutive relationship between Western representations of cultural difference and worldwide Western dominance.

Challenging Orientalism, I believe, requires that Occidentalism be unsettled as a style of representation that produces polarized and hierarchical conceptions of the West and its Others and makes them central figures in accounts of global and local histories. In other words, by "Occidentalism" I refer to the ensemble of representational practices that participate in the production of conceptions of the world, which (1) separate the world's components into bounded units; (2) disaggregate their relational histories; (3) turn difference into hierarchy; (4) naturalize these representations; and thus (5) intervene, however unwittingly, in the reproduction of existing asymmetrical power relations.

[. . .]

HISTORY AND THE FETISHIZATION OF GEOGRAPHY

Borges's cartographers produced maps for the emperor. Here I have discussed the often implicit maps of empire produced by invisible hands and reproduced, with varying degrees of critical distance, by critics of colonialism for the metropolitan academic community and the public at large. I have focused on how certain representational practices assume a privileged center – the Occident, the first world, the West, the Self – from which difference continues to be defined as Otherness. Whether Otherness is dissolved in the service of the Self, subsumed within the Self, or celebrated in opposition to the Self, as in the three modalities discussed here, is in this respect less significant than its ongoing definition as a counterimage to a Self in need of confirmation, critique, or destabilization.

If in this discussion I have called attention to the way these maps reinscribe certain imperial boundaries, it is because, as Nicolás Guillén's poem suggests, they do not sufficiently educate us to see forms of humanity "that are not on the map." If Occidentalism is an imperial malady, one of its major symptoms is the ongoing reproduction of a colonial Self–Other polarity that mystifies the present as much as the past and obscures its potential for transformation.

In his last book, *State, Power, Socialism* (1978), Nicos Poulantzas argued that states establish a "peculiar relationship between history and territory, between the spatial and the temporal matrix"

(1978:114). Taking the nation as his fundamental unit, he characterized the unity of modernity in terms of the intersection of temporal and spatial dimensions: "national unity or modern unity becomes a historicity of a territory and territorialization of a history" (1978:114). Before his death, Poulantzas was building on Lefebvre's pathbreaking work *La production de l' espace* (1974), which attempts to integrate the study of geography with that of history and has inspired an important body of work by contemporary thinkers who have also reacted against the historicist conception of space as the static stage where time dynamically unfolds.[7] I wish now to bring this literature to bear on Occidentalism through a brief commentary on Poulantzas's insight.

Poulantzas's notion that modernity entails the territorialization of a history and the historicization of a territory does not indicate how this interaction works, but his wording gives the impression of a symmetrical exchange. Yet, the prevailing understanding of history as fluid, intangible, and dynamic and of geography as fixed, tangible, and static suggests that modernity is constituted by an asymmetrical integration of space and time. A telling example is Laclau's argument, in *New Reflections on the Revolution of Our Time* (1990), that space is fundamentally static while time is dynamic.[8] Paradoxically, therefore, the historicization of territories takes place through the obscuring of their history; territories are largely assumed as the fixed, natural ground of local histories. The territorialization of histories, in turn, occurs through their fixation in nonhistorical, naturalized territories. As a consequence, the histories of interrelated peoples become territorialized into bounded spaces. Since these spaces appear as being produced naturally, not historically, they serve to root the histories of connected peoples in separate territories and to sever the links between them. Thus, the illusion is created that their identities are the result of independent histories rather than the outcome of historical relations. There is a dual obscuring. The histories of various spaces are hidden,[9] and the historical relations among social actors or units are severed.[10]

In other words, history and geography are fetishized. As with commodities, the results of social–historical relations among peoples appear as intrinsic attributes of naturalized, spatialized, bounded units. Although Poulantzas focused on nations, we could consider these units to be groups of nations or supranational entities: the West, the Occident, the third world, the East, the South, as well as localized intranational subunits, such as peasants, ethnic "minorities," "slum dwellers," the "homeless," forms of "communalism," and so forth. With the generalization of commodity relations, modes of reification involved in commodity fetishism radiate from the realm of the production of things to the production of social identities. Typical markers of collective identities, such as "territory," "culture," "history," or "religion," appear as autonomous entities. Identified by these markers, interconnected peoples come to lead separate lives whose defining properties appear to emerge from the intrinsic attributes of their "histories," "cultures," or "motherlands." As commodity fetishism becomes deeply rooted in society, it works as a cultural schema that permeates other sociocultural domains. As with commodities, the material, thinglike, tangible form of geographical entities becomes a privileged medium to represent the less tangible historical relations among peoples. Through geographic fetishism, space is naturalized and history is territorialized. Thus, the West is constituted as an imperial fetish, the imagined home of history's victors, the embodiment of their power.

Every society represents other societies as part of the process of constructing its own collective identity, but each does so in ways that reflect its unique historical trajectory and cultural traditions. What distinguishes Occidentalism as an ethnocentric style of representation is that it is linked to the West's effective global dominance. While this linkage raises a number of questions concerning the relationship between Western knowledge about the world and power over it, it must be noted that this dominance is always partial and that it takes place through processes of transculturation which also transform the West. Westernization entails not the homogenization of the world's societies under the force of capitalism but their reciprocal transformation under diverse historical conditions. In this light, capitalism appears not as a self-identical system that emanates from the West and expands to the periphery but as a changing ensemble of worldwide relations that assumes different forms in specific regional and national contexts.

MODERNITY AND OCCIDENTALISM

The 19th-century thinkers who insightfully examined the making of the modern world before its categories became second nature initiated a polemical discussion of the relationship between modernity and capitalism. Yet it is striking that even divergent ideological positions often coincide in their assumption that the West is the source and locus of modernity. If we expand our focus so as to bring the West and the non-West within a unified field of vision that encompasses the historical terrain of their mutual formation (for example, Cooper and Stoler 1989), the modern world appears larger and more complex, formed by universalizing and innovating impulses that continuously redefine geographical and cultural boundaries and set new against old, Self against Other. If the West is involved in the creation of its obverse and the modern is unimaginable without the traditional, the West's preoccupation with alterity can be seen as being constitutive of modernity itself rather than as an incidental by-product of Western expansionism. The examination of Western representations of Otherness, from the perspective of a critique of Occidentalism, could then be encompassed within an interrogation of why Otherness has become such a peculiarly modern concern.

Bourgeois modernity is torn by contradictory tendencies. Its universalizing force is inseparably linked to expansive and yet exclusionary movements of capital that polarize nations across the globe as well as people within societies. Spurred by the pursuit of profit, capital's continuous transformation of economic relations dissolves established customs and makes obsolete the new, yet its innovative force is constrained by the structures of privilege within which novelty itself is produced. Commodities come to occupy a central place in the formation of individual and collective life projects, generating forms of power that rely on the possession and consumption of things. Through the medium of things, modernity promises abundance and endless progress. This promise is fulfilled within conditions of inequality that redefine its meaning and is constrained by powerful interests that confine and condition its fulfillment. "Progress" is thus constituted through a contradictory movement that erodes and establishes boundaries, that releases and contains energies. The future, as a modern construct, is rent by these tensions. The expansion of capital across space and time entails the dissolution of barriers to "development" but also the construction of walls against "disorder." While capital's expansion is the condition of its stability, stability is the condition of its expansion. In the modern world, as Marx and Engels observed, "all that is solid melts into air," but air itself is rendered solid, turned into another object.

Premised on a teleology of progress, capitalist development is embodied in reified institutions and categories. Cultural constructs such as the West and the third world come to acquire, like a commercial brand, an independent objective existence as well as the semblance of a subjective life. As part of their social intercourse, these forms feed the collective imagination and participate in the making of desires and needs, circulating as objects of libidinal attraction (Bhabha 1986) and as subjects of political action that define the terms of political intercourse. As fetishes of modernity, these cultural formations stand for social powers by alienating them; parts replace wholes. The West comes to be identified with leading capitalist nations, the economy with the market, democracy with universal elections, difference with Otherness. Embodying the contradictions of capitalist society, these formations help shape the landscape within which, with mesmerizing allure despite its disruptive social consequences, capitalist arrested development parades as modern progress.

This map of modernity is being redrawn by global changes in culture, aesthetics, and exchange that are commonly associated with the emergence of postmodernity. These transformations have multiple determinants and expressions, of which I can register only a few: the simultaneous integration and fragmentation of social space through new forms of communication; the globalization of market relations and financial networks; the shift from Fordism to flexible accumulation; the increasing tension between the national basis of states and the global connections of national economies; and the growing polarization of social classes both domestically and internationally.[11] As a result of these changes, familiar spatial categories are uprooted from their original sites and attached to new locations. As space becomes fluid, history can no longer be easily anchored in fixed territories. While deterritorialization entails reterritorialization, this process only makes more visible the social constructedness of space, for this "melting" of space is met partly with the "freezing" of history. With the generalization of the commodity form, as Lukács noted, "time sheds its qualitative, variable, flowing nature; it freezes into an exactly delimited, quantifiable continuum filled with quantifiable

'things' . . . in short, it becomes space" (1971:90). This spatialization of time serves as the location of new social movements, as well as of new targets of imperial control; it expands the realm of imperial subjection but also of political contestation.

As a result of these transformations, contemporary empires must now confront subaltern subjects within reconfigured spaces at home and abroad, as the Other, once maintained on distant continents or confined to bounded locations at home, simultaneously multiplies and dissolves. Collective identities are being defined in fragmented places that cannot be mapped with antiquated categories. The emergence of a new relationship between history and geography may permit us to develop a critical cartography and to abandon worn imperial maps shaded in black and white. Perhaps one day "their tattered fragments will be found in the western Deserts, sheltering an occasional Beast or beggar" (Borges 1970:90) or, in a world without beggars, an archaeologist of modernity.

TOWARD NONIMPERIAL GEOHISTORICAL CATEGORIES

Daughter: Mom, Why did all those people lose their jobs? Will we be poor too?
Mother: Because the factories where they worked were moved to places where it's cheaper to make cars, as often happens when capitalists compete to make more money. If we worked for GM we would have a hard time now.
Daughter: Why can't we just say no to capitalism? Do you think in a few years human beings are going to be extinct? Is the world going to be so polluted that if there is a God, God will say, "I'm tired of all this"? But if that happens, there won't be any Santa Claus. I just can't imagine there never being any more people in the world, never ever.
Dialogue between Andrea Coronil, 10, and Julie Skurski, following the televised announcement that 74,000 GM employees will lose their jobs. Ann Arbor, December 18, 1991.

How can we articulate the future historically? In seeking to prefigure an emancipatory future, we may track down its marks in the tensions of the present. As Terry Eagleton argues, "a utopian thought that does not risk simply making us ill is one able to trace within the present that secret lack of identity with itself which is the spot where a feasible future might germinate – the place where the future overshadows and hollows out the present's spurious repleteness" (1990:25). Walter Benjamin, who sought to understand the past in order to find within the present the seeds of a desirable future, asserted that

> to articulate the past historically does not mean to recognize it "the way it really was" (Ranke). It means to seize hold of a memory as it flashes up at a moment of danger. . . . Only that historian will have the gift of fanning the spark of hope in the past who is firmly convinced that even the dead will not be safe from the enemy if he wins. And this enemy has not ceased to be victorious.
>
> [1969:255]

It may be that only that historian who is convinced that the living cannot be safe as long as the dead remain unburied will have the gift of fanning the spark of hope in the future. "If you can write this," said a relative of peasants massacred in the town of Amparo, Venezuela, on the pretext that they were Colombian guerrillas, "tell them that despite all the lies they [the powerful] will tell, they won't be able to hide the truth. Sooner or later, the truth will be known. . . . Even though those people may not believe it, the dead also speak" (personal communication, July 17, 1989). The dead speak in many ways. In February 1989 another massacre took place in Venezuela in which several hundred people were killed following rioting against an IMF austerity program. The effort to exhume the secret mass graves of the army's victims became the focus of popular struggle around the massacre, as the government sought to prevent the bodies of the victims from speaking of how they had met death. When the stakes of history are high, the safety of the living rests on the voices of the dead who speak through the actions of the living. Establishing this link across time, the Maya rebels of the contemporary Zapatista movement in Mexico define their opposition through a collective history, proclaiming; "Zapata lives, the struggle continues!" while their spokesperson Subcommander Marcos underlines that the people who now speak "are the dead people of always, those who have to die in order to live" (quoted in Poniatowska 1994).

The interaction between geography and history thus involves an exchange not only between past and present but between present and future. Fanon, like Marx, drew on the poetry of the future to imagine a world in which the dead may bury the dead so that the living may be freed from the nightmare of the past. Reflecting on his position as an African American, Henry Louis Gates expresses the tension energizing an aspiration to identity informed by history and yet unconstrained by the past: "So I'm divided. I want to be black, to know black, to luxuriate in whatever I might be calling blackness at any particular time – but to do so in order to come out the other side, to experience a humanity that is neither colorless, nor reducible to color" (1994:xv). It is also in the spirit of freeing the living into the future that Carolyn Steedman concludes her powerful analysis of working-class longing, in which after illuminating everyday formations of desire within working-class culture she calls "for a structure of political thought that will take all of this, all these secret and impossible stories, recognize what has been made out on the margins; and then, recognizing it, refuse to celebrate it; a politics that will, watching this past, say 'So What?' and consign it to the dark" (1987:144).

As the future flashes up to a child in the form of a disenchanted, inhospitable, and depopulated world, the safety of those who follow us comes to depend as well on the poetry of the present.

NOTES

Acknowledgments. This essay is the product of a seminar I taught on Occidentalism in the fall of 1991 and has benefited from discussions in many other contexts: the Power Conference, University of Michigan (1992); the Department of Anthropology, University of Chicago (1993); American Anthropological Association (1993); and above all in two other seminars I taught at the University of Michigan, one in conjunction with Walter Mignolo. In these contexts I have been offered more helpful suggestions than I could include in the paper or acknowledge here. I would like to thank in particular my students at the University of Michigan, and Arjun Appadurai, Lauren Berlant, John Comaroff, Paul Eiss, Raymond Grew, David Hollinger, Brink Messick, Walter Mignolo, Colleen O'Neal, Sherry Ortner, Seteney Shami, Carolyn Steedman, and Gary Wilder for their comments, and Julie Skurski, who shared in the production of this article.

1 The gender bias of this utopian image shows that utopian visions, however universal in their intent, are

necessarily saturated by the history they seek to overcome and limited by the local position from which they are enunciated.

2   Like Mignolo's "pluritopical" and Shohat's and Stam's "polycentric," I use here "decentered" as a sign of relationality and differentiation among human communities (Mignolo 1995; Shohat and Stam 1994).

3   Pletsch insightfully discusses the genesis of the three worlds taxonomy and its ideological character as a primitive system of classification (1981).

4   After I presented this paper at the Power Conference, Michigan (January 1992), I read an article by Carrier where he makes various useful distinctions: "ethno-Orientalism," by which he means "essentialist renderings of alien societies by members of those societies themselves"; "ethno-Occidentalism," which refers to "essentialist renderings of the West by members of alien societies"; and "Occidentalism," by which he means "the essentialistic rendering of the West by Westerners" (1992:198–199). Carrier's classification helps us recognize various approaches to this general topic, such as Chen 1992; Keesing 1982; and Nader 1989. Carrier's attempt to analyze the process of producing of Orientalist representations, and to relate dialectically representations of Otherness to representations of the West, parallels my own aims in this article.

5   Amin defines Eurocentrism as "an essential dimension of the ideology of capitalism" (1989:ix) and explains his choice of this term over others, including "occidentalocentrism" (1989:xii–xiii).

6   Armstrong (1990) uses the term "Occidentalism" to refer to the "effects" of Orientalism on Western selves. I see these effects as one dimension of Occidentalism, as I define it here.

7   For example, works produced by political geographers (Entrikin 1991; Harvey 1989; Smith 1990; and Soja 1989), literary critics (Jameson 1984), and social philosophers (de Certeau 1988; Foucault 1980).

8   Massey offers a persuasive critique of Laclau's conservative understanding of space and develops an important argument concerning the relationship between time and space (1992).

9   For this discussion, I find useful de Certeau's conception of "space" as a "practiced place" (1988:117).

10 This point is supported by the pioneering work of African and African American scholars who have discussed the erasure of links between Greece and Africa in dominant historiography (for example, Diop 1974) as well as by Martin Bernal's forceful argument in *Black Athena* (1987).

11 For a lucid discussion of central issues in the study of globalization and transnationalism, see Rouse 1995.

REFERENCES CITED

Amin, Samir
1989   Eurocentrism. New York: Monthly Review Press.
Armstrong, Nancy
1990   Occidental Alice. Differences: A Journal of Feminist Cultural Studies 2(2):3–40.
Benjamin, Walter
1969   Illuminations. New York: Schocken Books.
Bernal, Martin
1987   Black Athena. New Brunswick: Rutgers University Press.
Bhabha, Homi
1986   The Other Question: Difference, Discrimination and the Discourse of Colonialism. *In* Literature, Politics and Theory. Francis Barker, Peter Hulme, Margaret Iversen, and Diana Loxley, eds. Pp. 148–172. London: Methuen.
Borges, Jorge Luis
1970   Dreamtigers. Austin: University of Texas Press.
Carrier, James G.
1992   Occidentalism: The World Upside Down. American Ethnologist 19(2):195–212.
Chen, Xiaomei
1992   Occidentalism as Counterdiscourse: *He shang* in Post-Mao China. Critical Inquiry 18:686–712.
Chomsky, Noam
1991   The New World Order. Agenda 62:13–15.
Cooper, Fred, and Ann Stoler
1989   Introduction. Tensions of Empire: Colonial Control and Visions of Rule. Special issue of American Ethnologist 16:609–621.
Comaroff, John
1987   Of Totemism and Ethnicity: Consciousness, Practice, and the Signs of Inequality. Ethos 52:301–323.
Coronil, Fernando, and Julie Skurski
1991   Dismembering and Remembering the Nation: The Semantics of Political Violence in Venezuela. Comparative Studies in Society and History 33(2):288–337.
de Certeau, Michel
1988   The Practice of Everyday Life. Berkeley: University of California Press.
Diop, C. A.
1974   The African Origin of Civilization: Myth or Reality? Westport, CT: L. Hill.
Eagleton, Terry, Fredric Jameson, and Edward Said
1990   Nationalism, Colonialism, and Literature. Minneapolis: University of Minnesota Press.

Entrikin, Nicholas
  1991 The Betweenness of Place: Towards a Geography of Modernity. Baltimore: Johns Hopkins Press.
Fanon, Frantz
  1967 Black Skins, White Masks. New York: Grove Press.
Foucault, Michel
  1980 Questions of Geography. In Power/Knowledge: Selected Interviews and Other Writings. Colin Gordon, ed. New York: Pantheon.
Gates, Henry Louis, Jr.
  1994 Colored People: A Memoir. New York: Alfred A. Knopf.
Harvey, David
  1989 The Condition of Postmodernity. Cambridge: Basil Blackwell.
Jameson, Fredric
  1984 Postmodernism, or the Cultural Logic of Late Capitalism. New Left Review 146:53–92.
Keesing, Roger
  1982 Kastom in Melanesia: An Overview. Mankind 13:297–301.
Klor de Alva, Jorge
  1992 Colonialism and Postcolonialism as (Latin) American Mirages. Colonial Latin American Review 1(1–2):3–23.
Laclau, Ernesto
  1990 New Reflections on the Revolution of Our Time. London: Verso.
Lefebvre, Henry
  1974 La production de l'espace. Paris: Anthropos.
Lukács, George
  1971 History and Class Consciousness. Cambridge, MA: MIT Press.
Marx, Karl
  1981 Capital, Vol. 3. New York: Vintage Books.
Massey, Doreen
  1992 Politics and Space/Time. New Left Review. 196:65–84.
Mignolo, Walter
  1995 The Darker Side of the Renaissance. Ann Arbor: University of Michigan Press.
Nader, Laura
  1989 Orientalism, Occidentalism, and the Control of Women. Cultural Dynamics 2:233–355.

New York Times
  1988 Moscow Summit: Excerpts from the President's Talks to Artists and Students. June 1: A12.
Pletsch, Carl
  1981 The Three Worlds, or the Division of Social Scientific Labor, circa 1950–1975. Comparative Studies in Society and History 23(4):565–590.
Poniatowska, Elena
  1994 El País. La Jornada, August 16: 19.
Poulantzas, Nicos
  1978 State, Power, Socialism. London: New Left Books.
Quindlen, Anna
  1994 Public and Private: Game Time. New York Times, June 25: A23.
Rieff, David
  1991 Los Angeles: Capital of the Third World. New York: Simon & Schuster.
Rouse, Roger
  1995 Thinking Through Transnationalism. Public Culture 7(2):353–402.
Said, Edward
  1979 Orientalism. New York: Vintage Books.
  1986 Orientalism Reconsidered. In Literature, Politics, and Theory. Francis Barker, Peter Hulme, Margaret Iversen, and Diana Loxley, eds. London: Methuen.
  1993 Culture and Imperialism. New York: Knopf.
Shohat, Ella, and Robert Stam
  1994 Unthinking Eurocentrism. New York: Routledge.
Smith, Neil
  1990 Uneven Development. Cambridge: Basil Blackwell.
Soja, Edward
  1989 Postmodern Geographies. London: Verso.
Steedman, Carolyn Kay
  1987 Landscape for a Good Woman: A Story of Two Lives. New Brunswick: Rutgers University Press.
Williams, Raymond
  1983 Keywords: A Vocabulary of Culture and Society. New York: Oxford University Press.

'Globalization and Violence'

from *Fear of Small Numbers: An Essay on the Geography of Anger* (2006)

Arjun Appadurai

Editors' Introduction

Arjun Appadurai is John Dewey Professor in the Social Sciences at the New School University in New York City. Born in Bombay in 1949, Appadurai completed his MA (1973) and PhD (1976) from the University of Chicago. His first work, *Worship and Conflict Under Colonial Rule* (1981), was a historical anthropology of religion, caste and class in the workings of an important temple in South India. Following this, Appadurai wrote important articles on Indian development, expanding on the entitlements approach of Amartya Sen by bringing power and politicisation into Sen's model. Appadurai's next major work was *The Social Life of Things* (1986), a novel edited collection of ethnographic studies of commodities, now considered a foundation for the field of material culture studies. Appadurai and Carol Breckenridge of the University of Chicago also formed the journal *Public Culture*, a venue for innovative work in anthropology and cultural studies. In more recent years, Appadurai's *Modernity at Large* (1996) poses the imagination as a social force in a globalising world, a provocation that has led to considerable debate. His research has shifted to a massive collective project on post-industrial Mumbai, in relation to other mega-cities.

In *Fear of Small Numbers* (2006), excerpted in the reading here, Appadurai turns to the interplay of majorities and minorities in the current moment of globalisation and 'war on terror'. He now shows how fears of the eradication of cultural difference – one of our four key themes going back to Gandhi (Part 2) – returns in new forms in our time. Unlike Frederick Cooper (this Part), Appadurai continues to defend the novelty of something called 'globalisation', and he pins this novelty on the whirl of finance capital, information technology and growing inequality. He also claims that this volatile world creates new (and more) friction and conflict, and that the geography of violence is no longer about violence between states but a 'worldwide genocidal impulse towards minorities'. Minorities everywhere are under threat. Minorities are not pre-given; they are made by what the philosopher Michel Foucault calls 'biopolitical' processes – state instruments to classify and control the life and death of populations and individuals. In a way, Appadurai is updating Anne McClintock's argument about racial degeneracy in late Victorian colonialism, in which populations were marked down as irredeemable. Now, populations, particularly in liberal democracies, are made into various kinds of majorities and minorities, rather than into a level playing field of citizens. Specific minorities are at risk in specific periods. They come to embody economic and political crisis, as fetish objects (Ann McClintock, Part 1), or as objects of humanitarian intervention through war (Hirschkind and Mahmood, Part 8). Appadurai's chapter is broad and suggestive. He points to old themes of cultural difference, but through an attention to new processes. Other scholars have delved into specific localities to ask how populations become minoritised in specific ways, for instance, as tribal authorities, youth gangs or NGOs in the turbulent Niger Delta (Watts, 2007). What is clear is that McClintock's insistence that we think of fetishism as inscribed in particular kinds of bodies – Nigerian youth gangs, Afghani women, South Africans living with AIDS – but also attend to the technologies through which minorities are constructed, is a priority also for Appadurai.

Key references

Arjun Appadurai (2006) 'Globalization and violence', in *Fear of Small Numbers: An Essay on the Geography of Anger*, Durham, NC: Duke University Press.
— (1981) *Worship and Conflict Under Colonial Rule: A South Indian Case*, Cambridge: Cambridge University Press.
— (1984) 'How Moral is South Asia's Economy: A Review Article', *Journal of Asian Studies* 43 (3): 481–97.
— (ed.) (1986) *The Social Life of Things: Commodities in Cultural Perspective*, Cambridge: Cambridge University Press.
— (1996) *Modernity at Large: Cultural Dimensions of Globalisation*, Minneapolis: University of Minnesota Press.
Michael Watts (2004) 'Resource curse?: governmentality, oil and power in the Niger Delta, Nigeria', *Geopolitics* 9(1): 50–80.

Globalization is a source of debate almost everywhere. It is the name of a new industrial revolution (driven by powerful information and communication technologies) which has barely begun. Because of its newness, it taxes our linguistic resources for understanding it and our political resources for managing it. In the United States and in the ten or so most wealthy countries of the world, globalization is certainly a positive buzzword for corporate elites and their political allies. But for migrants, people of color, and other marginals (the so-called South in the North), it is a source of worry about inclusion, jobs, and deeper marginalization. And the worry of the marginals, as always in human history, is a worry to the elites. In the remaining countries of the world, the underdeveloped and the truly destitute ones, there is a double anxiety: fear of inclusion, on draconian terms, and fear of exclusion, for this seems like exclusion from history itself.

Whether we are in the North or the South, globalization also challenges our strongest tool for making newness manageable, and that is the recourse to history. We can do our best to see globalization as just a new phase (and face) of capitalism, or imperialism, or neocolonialism, or modernization or developmentalism. And there is some force to this hunt for the analogy that will let us tame the beast of globalization in the prison house (or zoo) of language. But this historicizing move (for all of its technical legitimacy) is doomed to fail precisely in accounting for the part of globalization that is unsettling in its newness. Recourse to the archives of prior world systems, old empires, and known forms of power and capital can indeed soothe us, but only up to a point. Beyond that point lurks the intuition of many poor people (and their supporters in the world) that globalization poses some new challenges which cannot be addressed with the comforts of history, even those of the history of bad people and nasty world conquerors. This hazy intuition is at the heart of the uncertain coalitions and uneasy dialogues that surround globalization, even in the streets of Seattle, Prague, Washington, and many other less dramatized locations.

Where exactly does this newness lie and why do many critical intellectuals fail to understand it better? In my opinion, there are three interrelated factors which make globalization difficult to understand in terms of earlier histories of state and market. The first is the role of finance capital (especially in its speculative forms) in the world economy today: it is faster, more multiplicative, more abstract, and more invasive of national economies than ever in its previous history. And because of its loosened links to manufacture and other forms of productive wealth, it is a horse with no apparent structural rider. The second reason has to do with the peculiar power of the information revolution in its electronic forms. Electronic information technologies are part and parcel of the new financial instruments, many of which have technical powers which are clearly ahead of the protocols for their regulation. Thus, whether or not the nation-state is fading out, no one can argue that the idea of a national economy (in the sense first articulated by Friedrich List) is any more an easily sustainable project. Thus, by extension, national sovereignty is now an unsettled project for specific technical reasons of a new sort and scale. Third, the new, mysterious, and almost magical forms of wealth generated by electronic finance markets appear directly responsible for the growing gaps between rich and poor, even in the richest countries in the world.

More importantly, the mysterious roamings of finance capital are matched by new kinds of migration, both elite and proletarian, which create unprecedented tensions between identities of origin, identities of residence, and identities of aspiration for many migrants in the world labor market. Leaky financial frontiers, mobile identities, and fast-moving technologies of communication and transaction together produce debates, both within and across national boundaries, that hold new potentials for violence.

There are many ways that we can approach the problems of globalization and violence. One could take the United States and ask whether the growth in the prison industry (and what is sometimes called the carceral state) is tied to the dynamics of regional economies which are being pushed out of other more humane forms of employment and wealth creation. One could consider Indonesia and ask why there is a deadly increase in intrastate violence between indigenous populations and state-sponsored migrants. One could study Sri Lanka and ask whether there are real links between the incessant civil war there and the global diaspora of Tamils, with such results as eelam.com, an example of cyber-secession (Jeganathan 1998). One could worry about conventional secessionist movements from Chechnya and Kashmir to the Basque Country and many parts of Africa and ask whether their violence is strictly endogenous. One could look at Palestine and ask whether the intimate violence of internal colonialism is now so deeply tied to mass media and global intervention that it is doomed to permanent institutionalization. One could position oneself in Kosovo or Iraq and ask whether the violent humanitarianism of NATO air strikes is the newest from of biblical retribution by the armed gods of our times. Or one could identify with the perspective of terrified minorities in many national spaces, such as Palestine, Timor, or

Sierra Leone, often living in detention camps parading as neighborhoods or refugee camps, and ask about the violence of displacement and relocation.

Cutting across all these locations and forms of violence is the presence of some major global factors. The growing and organized violence against women, famously in the Taliban regime, is also clearly evident in many other societies that seek to cast the first stone, such as the United States, where domestic violence remains prevalent. The mobilization of youth armies, notably in Africa but also in many other sites of intrastate warfare, is producing war veterans who have hardly seen adulthood, much less peace. Child labor is sufficiently troubling as a globalized form of violence against children, but the labor of fighting in civilian militias and military gangs is a particularly deadly form of induction into violence at an early age. And then there are the more insidious forms of violence experienced by large numbers of the world's poor as they undergo displacements by huge dam projects or by projects of slum clearance. Here they experience the effects of the global politics of security states as victims of economic embargos, police violence, ethnic mobilization, and job losses. The shutdown of small-scale industries in Delhi in the past decade is a vivid example of the collusion of high-minded environmental discourses, corrupt city politics, and the desperate scramble for jobs and livelihood. This is part of the reason that the poor sometimes subject themselves to the intimate violence of selling their body parts in global organ markets, selling their whole bodies to domestic labor in unsafe countries, and offering their daughters and sons into sex work and other permanently scarring occupations.

Let us pull back for a moment and consider some objections to this line of thought. What does this catalogue have to do with globalization as such? Is it not just one more chapter in the story of power, greed, corruption, and exclusion that we can find as far back in human history as we please? I would argue otherwise. Many of the examples I have cited above are tied in specific ways to transformations in the world economy since 1970, to specific battles over indigenism and national sovereignty produced by the battle between competing universalisms such as freedom, market, democracy, and rights, which simply did not operate in the same way in earlier periods. Above all, the many examples I have given fit with the major empirical fact of macroviolence in the past two decades, which is the relative and marked growth in intrastate versus interstate violence. Thus, the maps of states and the maps of warfare no longer fit an older, realist geography. And when we add to this the global circulation of arms, drugs, mercenaries, mafias, and other paraphernalia of violence, it is difficult to keep local instances local in their significance.

Of all these contexts for violence, ranging from the most intimate (such as rape, bodily mutilation, and dismemberment) to the most abstract (such as forced migration and legal minoritization), the most difficult one is the worldwide assault against minorities of all kinds. In this matter, every state (like every family) is unhappy in its own way. But why are we seeing a virtually worldwide genocidal impulse toward minorities, whether they are numerical, cultural, or political minorities and whether they are minorities through lack of the proper ethnicity or proper documentation or by being visible embodiments of some history of mutual violence or abuse? This global pattern requires something of a global answer, and that is the aim of this book.

The existing answers do not take us very far. Is this a clash of civilizations? Not likely, since many of these forms of violence are intracivilizational. Is it a failure of states to fulfill the Weberian norm of monopolizing violence? Partly, but this failure itself requires further explanation, along with the concomitant worldwide growth in "private" armies, security zones, consultants, and bodyguards. Is it a general worldwide numbing of our humanitarian impulses, as someone like Michael Ignatieff may suggest (1998), due to the effect of too many mass media images of faraway wars and ethnocides? Perhaps, but the growth in grassroots coalitions for change, equity, and health on a worldwide basis suggests that the human faculty for long-distance empathy has not yet been depleted. Is it the concomitant growth in a huge global arms traffic which links small arms and Kalashnikovs to the official state-to-state trade in rockets, tanks, and radar systems in a huge and shady range of deals? Yes, but this tells us only about necessary conditions for global violence and not about sufficient ones.

Or are we in the midst of a vast worldwide Malthusian correction, which works through the idioms of minoritization and ethnicization but is functionally geared to preparing the world for the winners of globalization, minus the inconvenient noise of its losers? Is this a vast form of what we may call econocide, a worldwide tendency (no more perfect in its workings than the market) to arrange the disappearance of the losers in the great drama of globalization? A scary scenario but fortunately lacking in plausible evidence, partly because the world's biggest criminals and tyrants have learned the languages of democracy, dignity, and rights.

So what is it about minorities that seems to attract new forms and scales of violence in many different parts of the world? The first step to an answer is that both minorities and majorities are the products of a distinctly modern world of statistics, censuses, population maps, and other tools of state created mostly since the seventeenth century. Minorities and majorities emerge explicitly in the process of developing ideas of number, representation, and electoral franchise in places affected by the democratic revolutions of the eighteenth century, including satellite spaces in the colonial world.

So, minorities are a recent social and demographic category, and today they activate new worries about rights (human and otherwise), about citizenship, about belonging and autochthony, and about entitlements from the state (or its phantom remnants). And they invite new ways of examining the obligations of states as well as the boundaries of political humanity, falling as they do in the uneasy gray area between citizens proper and humanity in general. It is no surprise that humans viewed as insufficient by others (as for example the disabled, the aged, and the sick) are often the first targets of marginalization or cleansing. That Nazi Germany sought to eliminate all of these categories (iconized by the figure of the Jew) is useful to contemplate.

But minorities do not come preformed. They are produced in the specific circumstances of every nation and every nationalism. They are often the carriers of the unwanted memories of the acts of violence that produced existing states, of forced conscription, or of violent extrusion as new states were formed. And, in addition, as weak claimants on state entitlements or drains on the resources of highly contested national resources, they are also reminders of the failures of various state projects (socialist, developmentalist, and capitalist). They are marks of failure and coercion. They are embarrassments to any state-sponsored image of national purity and state fairness. They are thus scapegoats in the classical sense.

But what is the special status of such scapegoats in the era of globalization? After all, strangers, sick people, nomads, religious dissidents, and similar minor social groups have always been targets of prejudice and xenophobia. Here I suggest a single and simple hypothesis. Given the systemic compromise of national economic sovereignty that is built into the logic of globalization, and given the increasing strain this puts on states to behave as trustees of the interests of a territorially defined and confined "people," minorities are the major site for displacing the anxieties of many states about their own minority or marginality (real or imagined) in a world of a few megastates, of unruly economic flows and compromised sovereignties. Minorities, in a word, are metaphors and reminders of the betrayal of the classical national project. And it is this betrayal – actually rooted in the failure of the nationstate to preserve its promise to be the guarantor of national sovereignty – that underwrites the worldwide impulse to extrude or to eliminate minorities. And this also explains why state military forces are often involved in intrastate ethnocide.

Of course, every case of internal violence against minorities also has its own realist sociology of rising expectations, cruel markets, corrupt state agencies, arrogant interventions from the outside, and deep histories of internal hate and suspicion waiting to be mobilized. But these only account for the characters. We need to look elsewhere for the plot. And the plot – worldwide in its force – is a product of the justified fear that the real world game has escaped the net of state of sovereignty and interstate diplomacy.

And yet, why are minorities targets of this worldwide pattern? Here we may return to the classic anthropological argument by Mary Douglas that "dirt is matter out of place" and that all moral and social taxonomies find abhorrent the items that blur their boundaries (1966). Minorities of the sort that I have described – the infirm, the religiously deviant, the disabled, the mobile, the illegal, and the unwelcome in the space of the nation-state – blur the boundaries between "us" and "them," here and there, in and out, healthy and unhealthy, loyal and disloyal, needed but unwelcome. This last binary is the key to the puzzle. In one way or the other, we need the "minor" groups in our national spaces – if nothing else to clean our latrines and fight our wars. But they are surely also unwelcome because of their anomalous identities and attachments. And in this double quality they embody the core problem of globalization itself for many nation-states: it is both necessary (or at least unavoidable) and it is unwelcome. It is both us (we can own it, control it, and use it, in the optimistic vision) and not us (we can avoid it, reject it, live without it, deny it, and eliminate it, in the pessimistic vision). Thus, from this point of view, the globalization of violence against minorities enacts a deep anxiety about the national project and its own ambiguous relationship to globalization. And globalization, being a force without a face, cannot be the object of ethnocide. But minorities can.

Put more generally, and this is an argument more fully elaborated in chapter 4, minorities are the flash point for a series of uncertainties that mediate between everyday life and its fast-shifting global backdrop. They create uncertainties about the national self and national citizenship because of their mixed status. Their legally ambiguous status puts pressures on constitutions and legal orders. Their movements threaten the policing of borders. Their financial transactions blur the lines between national economies and between legal and criminal transactions. Their languages exacerbate worries about national cultural coherence. Their lifestyles are easy ways to displace widespread tensions in society, especially in urban society. Their politics tend to be multifocal, so they are always sources of anxiety to security states. When they are wealthy, they raise the specter of elite globalization, working as its pariah mediators. And when they are poor, they are convenient symbols of the failure of many forms of development and welfare. Above all, since almost all ideas of nation and peoplehood rely on some idea of ethnic purity or singularity and the suppression of the memories of plurality, ethnic minorities blur the boundaries of national peoplehood. This uncertainty, exacerbated by the inability of many states to secure national economic sovereignty in the era of globalization, can translate into a lack of tolerance of any sort of collective stranger.

It is difficult to know who might emerge as the target minority, the ill-fated stranger. In some cases it seems obvious, in others less so. And that is because minorities are not born but made, historically speaking. In short, it is through specific choices and strategies, often of state elites or political leaders, that particular groups, who have stayed invisible, are rendered visible as minorities against whom campaigns of calumny can be unleashed, leading to explosions of ethnocide. So, rather than saying that minorities produce violence, we could better say that violence, especially at the national level, requires minorities. And this production of minorities requires unearthing some histories and burying others. This process is what accounts for the complex ways in which global issues and clashes gradually "implode" into nations and localities, often in the form of paroxysmal violence in the name of some majority. One classic case is the process by which the Sikhs in India were gradually turned into a problematic minority (Axel 2001). This was not the outcome of any simple form of census politics. It was based on a long twentieth century of regional and national politics and was finally produced in the violence of 1984, the assassination of Indira Gandhi, the state's counterinsurgency campaign against Sikh separatists, and the carnage of the 1984 riots in Delhi and elsewhere. It could be argued that it was in fact the massive unleashing of state and popular violence against Sikhs in 1984 that produced the Sikhs as a cultural and political minority, whose own small terrorist component acquired a general sacrality after these events. So, within a century (and some would say within a decade) a category that was considered a militant auxiliary of the Hindu world turned into its most dangerous internal enemy for at least a decade after 1984.

Consider one last reflection on the links between globalization and violence against minorities. This connection forces one to perform the hardest of analytic exercises, which is to show how forces of great speed, scale, and scope (i.e., the processes of globalization), which are also in many ways very abstract, can be connected to bodily violence of the most intimate sort, framed by the familiarity of everyday relations, the comfort of neighborhood, and the bonds of intimacy. How can friend kill friend, neighbor kill neighbor, even kinsman kill kinsman? These new forms of intimate violence seem especially puzzling in an era of fast technologies, abstract financial instruments, remote forms of power, and large-scale flows of techniques and ideologies.

One way to unravel the horror of the worldwide growth in intimate bodily violence in the context of increased abstraction and circulation of images and technologies is to consider that the relationship is not paradoxical at all. The body, especially the minoritized body, can simultaneously be the mirror and the instrument of those abstractions we fear most. Minorities and their bodies are, after all, the products of high degrees of abstraction in counting, classifying, and surveying populations. So, the body of the historically produced minority combines the seductions of the familiar and the reductions of the abstract in social life, allowing fears of the global to be embodied within it and, when specific situations become overcharged with anxiety, for that body to be annihilated. To be sure, we need to understand a great many specific events and processes in order to get from the vertiginous spin of the global to the intimate heat of local violence. But here is the possibility to consider: that part of the effort to slow down the whirl of the global and its seeming largeness of reach is by holding it still, and making it small, in the body of the violated minor. Such violence, in this perspective, is not about old hatreds and primordial fears. It is an effort to exorcise the new, the emergent, and the uncertain, one name for which is globalization.

REFERENCES

Axel, Brian Keith. 2001. *The Nation's Tortured Body: Violence, Representation, and the Formation of a Sikh "Diaspora."* Durham and London: Duke University Press.

Douglas, Mary. 1966. *Purity and Danger: An Analysis of Concepts of Purity and Taboo.* London: Routledge and Kegan Paul.

Ignatieff, Michael. 1998. *The Warriors Honor: Ethnic War and the Modern Conscience.* New York: Henry Holt.

Jeganathan, Pradeep. 1998. "eelam.com: Place, Nation, and Imagi-Nation in Cyberspace." *Public Culture* 10 (3).

'On Development, Demography and Climate Change: The End of the World as We Know It?'

Population and Environment (2005)

Tim Dyson

Editors' Introduction

Tim Dyson is a Professor in the Development Studies Institute at the London School of Economics and Political Science, where he has taught since 1980. Dyson is one of the best-known demographers working on the global South, though his interests extend well beyond population figures and trends to questions of food supply growth, HIV/AIDS, the demographic effects of continuing warfare against Iraq, and global warming. A social scientist with particular expertise in South Asia, Dyson was elected a Fellow of the British Academy in 2001.

Tim Dyson has always used his demographic expertise to ask important questions about the lives of poorer and more vulnerable people in the developing world. An early paper on the determinants on gender equity in India (with Mick Moore) suggested that kinship structures and purdah impacted more on female autonomy in North than in South India and had significant effects on demographic behaviour (including local dispositions for 'son preference'). The Dyson–Moore hypothesis continues to be debated in the relevant literatures. Dyson also collaborated with Robert Cassen and Leela Visaria to produce a robust account of the prospects for *Twenty-first Century India: Population, Economy, Human Development and the Environment* (2004). More recently, Dyson has begun to address questions of human-induced climate change and global warming. The reading which follows shows Dyson taking the long view, at least with regard to modern development patterns. Dyson suggests that economic development since c.1800 has been based on the burning of fossil fuels, and he sees no reason to suppose this will change any time soon. This leads Dyson to argue that there will be a major rise in atmospheric carbon dioxide through the twenty-first century and with it a significant rise in global temperatures. The effects of this global warming are hard to predict and won't always be negative for all peoples. It is clear, nonetheless, Dyson argues, that human-induced global warming will pose significant threats to significant numbers of the eight or nine billion people who will inhabit the world in fifty or so years' time. Worse, there is little reason to suppose that political or behavioural change will come soon enough to affect 'the broad course of future events [that] is probably now set to run its course'. Dyson has little faith in the technology-led approach to global warming being proposed in the United States, where the Bush administrations have resolutely refused to accept the idea of human-induced global warming, nor in the capacity of the international system to respond in a robust and timely manner to a pressing problem of collective action. We conclude with Dyson because this reading demonstrates how vital collective state-sanctioned intervention is in securing a possible future for all. As things stand, Dyson shows that we can expect state response to climate change to be similar to the disastrous response to HIV/AIDS. Given also that we are collectively entering a phase in which catastrophes may be as likely as normality, questions of development can only become more pressing in the twenty-first century. How ought we to organise markets and empires, or to fight them, in relation to tumultuous natural and cultural processes?

Key references

Tim Dyson (2005) 'On development, demography and climate change: the end of the world as we know it?' *Population and Environment* 27 (2): 117–49.

— (2001) 'The preliminary demography of the 2001 census of India', *Population and Development Review* 27: 341–56.

— (2003) 'HIV/AIDS and urbanization', *Population and Development Review* 29: 427–42.

— (2005) 'Why the world's population will probably be less than 9 billion in 2300', in United Nations Population Division, *World Population to 2300*, New York: United Nations, pp. 145–50.

T. Dyson and M. Moore (1983) 'On kinship structure, female autonomy, and demographic behavior in India', *Population and Development Review* 9 (91): 35–60.

R. Cassen, T. Dyson and L. Visaria (eds) (2004) *Twenty-first Century India: Population, Economy, Human Development and the Environment*, Oxford: Oxford University Press.

INTRODUCTION

Global warming and climate change receive a huge amount of attention. Whether the world is heating up, the implications for the climate, and the possible long run consequences for humanity, are all topics that are never far from the newspaper headlines. It is clear that the issues involved are uncertain, complex, and often the object of controversy. Therefore it might be thought that little can be gained from a general social scientific consideration of the subject – one that starts from a concern with development and demography.

The view taken here, however, is that looking at global warming and climate change in historical perspective, examining the subject in the round (i.e. drawing on material from both the social and the environmental sciences), treating scientific study of it as a form of social activity, comparing human responses to it with those evidenced in relation to broadly analogous issues, and, above all, standing back from the subject – so as not to miss the wood for the trees – can yield fresh insights both about what is happening and about what may happen.

Accordingly, the present paper – which in large part is commentary – is an attempt to provide fresh perspective on global warming and climate change. It adopts an holistic approach, and essentially forwards

five main points. First, that since about 1800 economic development has been based on the burning of fossil fuels, and that this will continue to apply for the foreseeable future. Although there will doubtless be increases in the use of renewable energy sources, and rises in the efficiency with which energy is used, there is no real alternative to the continued use of coal, oil and natural gas for the purpose of economic development. Second, that mainly due to momentum in economic and demographic processes, it is inevitable that there will be a major rise in atmospheric CO_2 during the 21st century. Demographic and CO_2 emissions data will be presented to help substantiate this point. Third, that the available data suggest that the coming rise in global temperatures, which itself will result partly from momentum in climate processes, will be appreciably faster than anything that human populations have experienced during historical times. Moreover, particularly in a system that is being forced, there must be a reasonable chance of the occurrence of an abrupt change in climate. Fourth, that while it is impossible to attach precise probabilities to different scenarios, the chances of an unpleasant climate outcome occurring are at least as great as the chances of a more manageable one. The agricultural, political, economic, demographic, social and other consequences of future climate change could be very considerable. In a more populous world of eight or nine billion people, adverse developments could well occur on several fronts simultaneously, and to cumulative adverse effect. Related to this, it will be argued here that there is a pressing need to improve our ways of thinking about what may happen – because current prognostications tend to be routine, predictable and restricted. Finally, the paper argues that humanity's experience of another difficult 'long' threat – HIV/AIDS – reveals a broadly analogous sequence of human reactions. In short: (i) scientific understanding advances rapidly, but (ii) avoidance, denial, and recrimination characterize the overall societal response, therefore (iii) there is relatively little behavioral change, until (iv) evidence of damage becomes plain. Apropos climate change, however, the opinion expressed here is that major behavioral change to limit world carbon emissions is unlikely to happen in the foreseeable future.

There certainly is uncertainty about what will happen. But the basic data on trends in atmospheric CO_2 and world temperature – presented here – are fairly easy to understand and not in serious dispute. Moreover, despite impressions to the contrary, there is a scientific consensus on the reality of human-induced climate change. It is suggested here that the broad course of future events is probably now set to run its course.

[. . .]

SOCIAL REACTIONS TO THE EVIDENCE ON GLOBAL WARMING

That modern economic growth has raised levels of atmospheric CO_2 – leading to a rise in the Earth's surface temperature and the threat of climate change – is patently unwelcome news. It raises difficult issues about the basis of economic growth. It highlights huge – and morally awkward – disparities in energy use, CO_2 emissions, and living standards between rich and poor. It rears the prospect that some extremely difficult changes in behavior may be required. Indeed, inasmuch as it suggests the need for big cuts in energy consumption, it strikes at the very heart of the modern conception of 'development'.

The view taken here is that the human response to this news has been characterized by a mixture of denial, avoidance and recrimination, and that these reactions are fairly predictable. The social response has been complicated because climate change is commonly seen as a phenomenon which – if indeed it is real – lies far off in the distant future. Most people are preoccupied with the events of their daily lives, they are increasingly distrustful of official sources of information, and they tend to be relatively unconcerned with what may happen over the very long run. Political leaders too have more immediate concerns to occupy their time. They usually avoid difficult issues, being chiefly concerned with the short run – often the period until the next election.

[. . .]

The present view is that the prospects for an enforceable international agreement that will bring about a sustained and significant reduction in annual global CO_2 emissions are very poor. While it may be in the interest of the world as a whole to restrict the burning of fossil fuels, it is in the interest of individual countries to avoid making such changes (i.e. there are elements of a classic isolation paradox). Really major nations, such as the US and China, have considerable capacity to circumvent or ignore international agreements when it suits them. Moreover, the enormous complexities involved – many of them created and informed by matters of interest – will also hinder agreement. Doubtless there will be gains in energy use efficiency, shifts towards less carbon intensive fuels, and greater use of renewable energy sources (e.g. solar, biomass, wind and tidal power). But except for a massive shifts towards nuclear – which has many serious problems attached, and would in any case take decades to bring about – there are limits to what such changes could possibly achieve in terms of CO_2 reduction. Other technological ideas – like the extraction of CO_2 from coal and its sequestration underground (so-called 'carbon capture and storage') or, still more, the development of the so-called 'hydrogen economy' – are remote ideas as large scale and significant solutions to the problem during the foreseeable future (Smil, 2003). Indeed, such notions can themselves be regarded as providing some basis for avoidance inasmuch as they suggest that something is being done. Understandably, poor countries are unlikely to put great effort into constraining their CO_2 emissions – especially in the face of massive discrepancies between themselves and the rich.

In sum, the view taken here is that for the foreseeable future the basic response to global warming will be one of avoidance and, at most, modest change.

That the absolute amount of CO_2 emitted into the atmosphere each year is almost certainly going to *rise* in the coming decades is shown by an examination of basic demographic and emissions data in the next section.

ILLUSTRATIVE CALCULATIONS ON FUTURE CO_2 EMISSIONS

Demographic growth is a useful place to begin when considering future trends in CO_2 emissions. At the start of the 21st century the world's population was about 6.1 billion. The United Nations projects that by 2050 it will be around 9.1 billion (United Nations, 2005). This represents growth of 49% in 50 years. Although this projection is approximate, considerable further demographic growth is inevitable – because of population momentum. Moreover it is worth remarking that the UN has a good record of forecasting the world's total population.

By itself an increase in the world's population of roughly one half (i.e. 49%) will *not* lead to a similar proportional rise in CO_2 emissions from the burning of fossil fuels. The reason is that most of the coming demographic growth will occur in poor countries, which – almost by definition – burn relatively small amounts of coal, oil and natural gas. In this context Table 1 summarizes the situation at the start of the 21st century and provides a way of exploring the future. Column (i) shows the distribution of the world's population in the year 2000. Columns (ii) and (iii) give the corresponding levels of per capita and total CO_2 emissions by region. Notice that in 2000 the world's population of 6.09 billion was releasing about 23.2 billion tons of CO_2 through the combustion of fossil fuels – implying an average annual per capita emissions figure of about 3.8 metric tons. However, the statistics in column (ii) also underscore the enormous variation that exists around this average. Thus in North America (i.e. the United States and Canada) the average level of emissions was about 20.0 tons of CO_2 per person per year, whereas in both sub-Saharan Africa and South-central Asia it was only around 0.9 tons. Column (iii) shows that around the year 2000 the largest absolute regional contribution to total world CO_2 emissions came from North America, followed closely by Europe. Together these two developed regions contained only about 17% of humanity, but at the start of this century they

accounted for around 54% of all CO_2 emissions from fossil fuel burning.

Turning to the future, column (iv) of Table 1 summarizes UN population projections for the year 2050 by region. During the period 2000–2050 the population of sub-Saharan Africa is projected to rise by around 1022 million, and that of South-central Asia (which includes India, Pakistan, and Bangladesh) by 1010 million. Taken together, these two very poor regions are projected to account for about two-thirds of the growth in world population over this time period. Note too that the populations of North America and Oceania are projected to rise by about 123 and 17 million respectively. Only Europe's population is expected to decline in size.

Column (v) of Table 1 shows the total CO_2 emissions that will apply in 2050 if the projected regional populations in column (iv) are combined with the corresponding per capita CO_2 emission figures for 2000 given in column (ii). On this simple and unrealistic assumption (i.e. that of holding per capita emissions in each region constant at the level that prevailed around the year 2000), it can be seen that global CO_2 emissions would rise to about 29.6 billion tons i.e. by 28%. Also, the average level of per capita emissions for the world as a whole would fall from about 3.8 to around 3.3 metric tons per person (i.e. 29,613/9076). The explanation for this fall is that most of the coming demographic growth will occur in poor regions with low per capita emissions – thereby weighting the global per capita emissions figure downwards over time. Precisely the same consideration explains why the projected population increase of 49% leads to a rise in global CO_2 emissions of only 28%. Note from the sub-totals in columns (iii) and (v) that the projected population growth in the developing regions leads to a 42% rise in their total emissions (i.e. from 10.4 to 14.8 billion tons). And for the developed regions too demographic growth produces a 16% rise in total emissions (i.e. from 12.8 to 14.8 billion tons) – despite the projected decline in Europe's population. This helps to underline the fact that in North America, especially, immigration could play a significant role in the growth of future CO_2 emissions.

The rise in annual world CO_2 emissions in the next 50 years may well be appreciably greater than 28%. The huge differentials in current per capita emission levels shown in column (ii) of Table 1 account for this. Although, as comparative newcomers, the developing

Region	Population (millions) 2000 (i)	Per capita CO_2 emissions (metric tons) 2000 (ii)	Total CO_2 emissions (million metric tons) 2000 (iii)	Projected population (millions) 2050 (iv)	Total CO_2 emissions (million metric tons) 2050 (v)
Developing regions					
Sub-Saharan Africa	670	0.9	613.8	1692	1550.1
North Africa/West Asia	335	4.3	1430.8	628	2682.2
Eastern Asia	1479	3.4	5044.6	1587	5412.9
South-central Asia	1485	0.9	1368.2	2495	2298.8
South-eastern Asia	519	1.3	696.1	752	1008.6
Central America and Caribbean	174	2.8	481.2	256	707.9
South America	349	2.2	771.9	527	1165.6
Subtotal	5011	2.1	10,406.6	7937	14,826.2
Developed regions					
Europe	729	8.4	6106.2	653	5469.6
North America	315	20.0	6294.5	438	8752.3
Oceania	31	11.8	365.0	48	565.1
Subtotal	1075	11.9	12,765.7	1139	14,787.0
World	6086	3.8	23,172.2	9076	29,613.2

Table 1 Estimates of regional and global emissions of CO_2 produced by the combustion of fossil fuels for around the year 2000, with illustrative calculations for 2050

Notes: All the figures given above are approximate – especially those relating to CO_2 emissions. The per capita and total emissions statistics shown for 2000 actually pertain to 1999. The regional groupings of countries used are those employed by the World Resources Institute, but with Asia (excluding West Asia) being broken down according to the standard groupings of the United Nations. Here Sudan forms part of sub-Saharan Africa. The regions are designated above as either 'developing' or 'developed' – perhaps the main qualifications being that Japan falls in Eastern Asia, and that Melanesia is part of Oceania. The World Resources Institute provides no regional statistics on CO_2 emissions for sub-Saharan Africa. In 1999, however, South Africa had estimated per capital and total CO_2 emissions of 8.1 tons and 346 million tons respectively. To get the figures shown above for sub-Saharan Africa for the year 2000 it was arbitrarily assumed that per capital emissions for the remainder of the region averaged 0.4 tons (about the levels indicated for Angola and Senegal). Several modest adjustments were required to produce the relatively consistent regional and global picture given above, and therefore some of the figures on CO_2 emissions differ slightly from those published by the World Resources Institute on which they are based. The figures in column (v) are the product of those in (ii) and (iv).
Principal data sources: World Resources Institute (2003: 258–259); United Nations (2005).

regions can expect to benefit from rises in the efficiency with which energy is derived from fossil fuel sources, it is nevertheless virtually inevitable that most of these regions will experience significant rises in their per capita emission levels as they develop economically. Consider, for example, that during 1990–1999 the level of per capita CO_2 emissions rose appreciably in all the developing regions for which data are available. Thus for Asia (excluding West Asia) the increase was about 19.3%; for North Africa/West Asia it was around 19.7%; and for South America it was about 22.5% (World Resources Institute, 2003:258–259). Conservatively, these figures imply a 20% rise in per capita emissions per decade. And, cumulated across five decades, this would translate into an increase in per capita emissions of very roughly 150%. That said, no one knows by how much these per capita emission levels will increase. The degree of uncertainty is substantially greater than that regarding the scale of future demographic growth.

However, the figures in column (v) of Table 1 can be adjusted in a straightforward manner to explore the broad implications of different hypothetical trajectories in future per capita emissions. For example, if during 2000–2050 per capita emissions in the world's more developed regions were to fall by 40% (which many might regard as optimistic) then the total volume of their emissions in 2050 would be about 8.9 billion tons (i.e. 0.6*14,787), and – assuming no change in per capita emissions for the developing regions – then the total volume of world emissions in 2050 would be about 23.7 billion tons (compared to the 23.2 billion that was being emitted around the year 2000). This suggests that a 40% reduction in per capita emissions in the developed regions would be outweighed solely by the effects of demographic growth elsewhere in the world. Alternatively, if per capita emissions were to double (i.e. increase by just 100%) in the developing regions over the same period then their total emissions in 2050 would be around 29.6 billion tons (i.e. 2.0*14,826), and – assuming no alteration in the per capita emission levels of the developed regions – then the total volume of global emissions in 2050 would be about 44.3 billion tons – i.e. a 90% rise compared to the 23.2 billion tons being emitted around the year 2000. This calculation underscores the big influence that increased fossil fuel burning to support economic growth in the developing regions is likely to have on the volume of world CO_2 emissions. Finally, consider the case in which per capita emissions in the developed regions fall by 40% while those in the developing regions double. This combination would produce global CO_2 emissions in 2050 of 38.5 billion tons (i.e. 8.9 + 29.6) – an increase of about 66% compared to the year 2000.

Several conclusions arise from these simple illustrative calculations. First, the period 2000–2050 will see substantial demographic growth – forcing total world CO_2 emissions to rise. Because most of this growth will occur in poor regions, the implied proportional growth in total CO_2 emissions (here 28%) is appreciably less than the population increase (49%). Second, the influence of population growth on future CO_2 emissions will not be confined to the developing world. North America, and to a lesser extent Oceania (which here effectively means Australia/New Zealand), both have very high per capita emission levels and are expected to experience significant demographic growth – much of it due to migration. Consider, for example, that in Table 1: the population of South-central Asia increases by 1010 million in 50 years, which implies the emission of an additional 931 million tons of CO_2; and the population of North America rises by only 123 million, which implies an additional 2458 million tons of CO_2. Third, even should the developed regions make big cuts in their emissions, these will be more than offset by rises elsewhere. Thus the effect of population growth in the developing regions alone would outweigh a 40% reduction in CO_2 emissions in the developed regions. Yet economic development will likely mean that the total emissions of the developing regions will rise by *much* more than is implied just by demographic growth. Finally, as a consequence, it is virtually certain that there will be a significant rise in global CO_2 emissions. This will happen due to population growth, but it will happen much more because of the fueling of economic growth.

Of course, there is great uncertainty about just how big the coming rise in annual world CO_2 emissions will be. The IPCC, for example, has developed many different scenarios for future CO_2 release (both from fossil fuel burning and changes in land use) and explored scenarios for other GHGs and sulfur emissions. This work underscores the critical importance of economic and technological changes for the future evolution of emissions. But the resulting range of variation in emissions between the different scenarios is huge. Thus, towards the extremes, over the period 2000–2050 annual CO_2 emissions from fossil fuel use could increase only slightly or quadruple. Furthermore the IPCC is careful not to assign probabilities to any scenario, nor does it express any preferences with regard to them (IPCC, 2001b).

Particularly in relation to oil – and perhaps natural gas – it seems possible that limits to the available reserves may operate to curb the expansion of their use for energy production in the coming decades (World Energy Council, 2004). However, global reserves of coal are ample, and with China and India investing massively in new coal-fired power stations, growth in world coal use is currently much greater than that for either oil or natural gas (British Petroleum, 2005). It is especially difficult to gauge the extent to which per capita CO_2 emissions from fossil fuel burning will rise with the anticipated future economic expansion of Eastern Asia and South-central Asia. And, to reiterate, there is much uncertainty regarding whether, and to what extent, developed countries will be able to reduce their CO_2 emissions. Even with much greater use of renewable energy sources, and greater use of nuclear, there is little doubt that world use of fossil fuels will rise significantly in the coming decades and that they will continue to dominate in world energy production (e.g. see Bodansky, 2001; International Energy Agency, 2005; Smil, 2003; World Energy Council, 2004). Given the numbers in Table 1, and some simple assumptions, it seems reasonable to hazard that global CO_2 emissions from fossil fuel burning could easily rise by somewhere between a quarter and two-thirds during the first half of the 21st century. But we saw too that a combination of constant per capita CO_2 emissions in the developed regions, and a doubling in the developing regions, would raise annual emissions by 90% – which imply an average annual growth rate in emissions of about 1.3% during 2000–2050. In this context it is worth noting that projections made by the International Energy Agency (2005) suggest that between 2005 and 2030 energy-related CO_2 emissions may rise by 52% – implying an annual growth rate of 1.7%.

In concluding this section the main point is surely very clear: the absolute amount of CO_2 being emitted into the atmosphere each year is almost certainly going to rise appreciably in the coming decades.

[. . .]

THINKING ON THE CONSEQUENCES OF CLIMATE CHANGE

Mainstream thought on the effects of a rise in temperature for the world's climate, and its people, has at one and the same time been valuable, yet restricted. The temperature rises discussed in the previous section may seem small, but their implications could be immense.

So far as the consequences for the climate are concerned, and with reference to its projected range of temperature increase for the year 2100 (i.e. 1.4–5.8 deg/C), the IPCC valuably summarizes the essentials as follows: the land surface temperature rise will probably be greater than the ocean surface tempera-

ture rise; there will probably be more hot days and fewer cold days, but with a reduced diurnal temperature range over most land areas; there will be increases in water vapor in the atmosphere, and rainfall will increase in most locations; in many places there will be more intense rainfall events; in many places there will be an increased risk of drought (e.g. such as those associated with El Niño events); it is likely that there will be increases in the frequency of extreme weather events – like thunderstorms and tornadoes; it is likely that there will be an increase in variability of the rainfall associated with the Asian summer monsoon; glaciers and ice caps will continue to melt; and sea levels will probably continue to rise as the ocean expands due to thermal expansion and the melting of snow and ice – a global mean sea level increase of anywhere between 9 and 88 centimeters over the period 1990–2100 is projected (IPCC, 2001a: 13–16). In relation to all these effects there will be variation by world region, and the effects will generally vary directly with the extent of the coming temperature rise.

The task of gauging what the numerous consequences of these possible changes in climate might be for humanity is probably even greater than that of determining the nature of the likely climate changes themselves. This is partly because of the existence of both regional and socioeconomic variation, and because of the multitude of dimensions both of the environment and of human life. However, key elements of the IPCC's assessment of the implications and consequences of coming changes in climate for human populations include: that natural systems are often limited in the extent to which they can adapt, and that changes in such systems can sometimes be irreversible; that although adverse impacts will probably tend to predominate there will also be beneficial impacts – thus, for example, while the overall effect for world agriculture may be negative, in some locations levels of agricultural production might be raised from some climate changes (e.g. increases in temperature and rainfall); that in most settings – whether between or within countries – the adverse effects of climate change will fall disproportionately upon the poor – for example, '[t]he effects of climate change are expected to be greatest in developing countries in terms of loss of life and relative effects on investment and the economy' (IPCC, 2001c:8), and 'squatter and other informal urban settlements with high population density, poor shelter, little or no access to resources . . . and low adaptive capacity are highly vulnerable [to urban flooding]' (IPCC, 2001c:13); that there will probably be appreciable increases in the geographical areas and human populations that are subject to water stress, to flooding and to food insecurity as a result of climate change; that disaster losses due to extreme weather events are likely to rise substantially; that the adverse impacts of climate change will be greater with more rapid warming; and, lastly, that adaptation is a necessary strategy to complement efforts at climate change mitigation – thus, '[f]or each antici-pated adverse health impact there is a range of social, institutional, technological, and behavioral adaptation options to lessen that impact', and '[a]daptation to climate change presents complex challenges, but also opportunities, to the [insurance and financial services] sector' (IPCC, 2001c:12 and 13).

Given the sheer magnitude of the task, the IPCC's exploration of the likely consequences of the coming change in the world's climate is commendable. How-ever it is open to criticism in several key respects. For example, questions arise about the vocabulary that is used. The single most important theme is usually that of ways of *adapting* to climate change. But 'adaptation', and similar words like 'coping', are not neutral. They presuppose changes to which it will be possible to adjust. Likewise, the analytical perspec-tives that tend to be employed – for example, that there will be 'winners' as well as 'losers' (echoed in some of the preceding extracts), can be criticized in that they presume an element of symmetry – yet it could be that on the basis of some future trajectories of temperature and climate, conditions might deteriorate for almost everyone.

Again, and as one might expect, studies of the consequences of climate change tend to proceed sector by sector – for example, examining the possible implications for agriculture, industry, the ser-vice sector, health, etc. Almost inevitably this means that it is hard to do justice to the manifold possible interactions between different sectors. In fact, in broad terms, the IPCC's assessment of the implica-tions of future climate change starts from a consider-ation of possible ecological changes – for example, relating to water resources, coastal zones, and marine ecosystems – and then proceeds to discuss the implications for the production of goods and services, human settlements, energy, industry, finan-cial services, and health. While this is a reasonable

direction in which to proceed, it is not the only possible one. Thus it is arguably less people-centered than, for example, the recent Millennium Ecosystem Assessment – which more specifically considers eco-systems in terms of the benefits that they provide to people (e.g. in terms of timber, clean air, fibers, food, etc). Moreover, and predictably, the dominant social science perspective in these studies is that of economics. Input from, for example, sociologists or political scientists is negligible in the published IPCC reports. Unfortunately this means that some poten-tially important effects of future climate change receive virtually no consideration at all – for example, as to how people's views of the world might alter (e.g. in terms of religious beliefs) or the ways in which the behavior of nation states in the international arena might change (e.g. towards positions that are even more dominated by instrumentalism and national self-interest than applies now).

A common thread behind the issues raised in the preceding paragraphs is that study of the possible consequences of future climate change tends to shy away from contemplating circumstances that incline in the direction of a rapid and sustained temperature increase or the occurrence of a major 'surprise' (for exceptions see National Research Council (2002) and Stipp (2004). It has been argued here, however, that there is a good chance that such circumstances might arise. This is not the place to consider the possible consequences of more rapidly warming climate scenarios or those involving a major unexpected happening, but a few observations are relevant by way of conclusion.

First, consider that the world's population later in this century will probably be around nine billion. The addition of an extra three billion people will mostly be those who are poor and relatively vulnerable. Second, the continuing process of urbanization will mean that extremely large numbers of people – probably several billion – will be living in low lying, densely populated, coastal areas of the developing world, and their situation is likely to be particularly exposed. Third, probably the most important con-sequence of future climate change for human popula-tions relates to agricultrual production in the world's tropical and semi-tropical regions (IPCC, 2001c). Food production in such regions is an activity that is unlikely to be able to adapt to a rapid rise in tem-perature, and it will almost certainly not be able to cope with any abrupt change in climate. Perhaps no economic generalization is sounder than that small declines in food production can produce big rises in food prices – often with very significant sociopolitical ramifications. Fourth, more thought needs to be given to circumstances in which several adverse changes occur simultaneously and to cumulative adverse effect. This is the matter of how various potential harmful developments might interact. For example, flooding of coastal areas, which might result partly from sea level rise and partly from increased rainfall, could lead to the simultaneous loss of cropland and urban infrastructure, producing food price rises, large scale migration, and possibly significant sociopolitical disruption.

Finally, any really major or abrupt change in the world's climate could well lead to a situation in which virtually everyone loses and *nobody* wins. This could happen, for example, through the likely severe adverse effects on agriculture everywhere. In such circumstances it would be especially naive to believe that only poor countries would be badly affected. Indeed, it is worth considering the notion that the very interdependent complexity and high degree of specialization that characterize the world's most economically advanced countries could well be a potential source of significant vulnerability for some of them. A rapid rise in temperature or sudden change in climate would have consequences that may be almost inconceivable to those of us who have grown up in a generally improving world, one under-pinned by massive increases in the use of fossil fuel energy.

CONCLUSIONS

The view taken in this paper has been that there is a very appreciable chance of major climate change occurring at some point during the present century. To have this view is not to deny that there is scope for technological change to reduce CO_2 emissions per unit of energy; nor is it to deny that changes in human behavior could alter the volume of global CO_2 emissions, and therefore the world's future tem-perature and climate trajectory over the longer run. Furthermore, it is quite possible that future change in the world's climate will be modest, manageable, and perhaps even beneficial for many people. That said, the chance of some sort of large-scale ruinous development happening – such as global warming at

a quite unmanageable rate, or the occurrence of some kind of abrupt climate change – appears to be just as great.

It is all too easy to get distracted by detail. Accord-ingly, as was intimated at the paper's start, the approach adopted here has been to step back, look at the big picture, and try to focus on the essentials. With this in mind, the paper has brought together material on per capita CO_2 emissions and population projections, and it has presented data on levels of atmospheric CO_2 and measures of the world's sur-face temperature. The purpose of doing this is so that, to some extent at least, the reader can make up his or her own mind regarding past and future trends.

It can be predicted with considerable confidence that levels of anthropogenic CO_2 emissions are going to rise significantly. It is virtually certain that the level of CO_2 in the Earth's atmosphere will continue to increase monotonically in the coming decades. It is very hard to envisage that the figure will not exceed 530 ppm by the end of this century. It is likely that the world's surface temperature will continue to rise at a pace that is quite unprecedented during human history. Looking at the more immediate term, it seems very likely that the first decade of the 21st century will supplant the 1990s as the warmest decade since reasonable direct measurement began. There seems to be fair reason to expect that the next secondary peak in the temperature cycle may occur around the year 2010. Indeed, Hansen et al. (2005:4) observe in relation to 2005 that: 'the trend of global temperatures toward global warming is now so steep that in just 7 years the global warming trend has taken temperatures to approximately the level of the abnormally warm year of 1998.' Anyhow, the time series presented in this paper can be updated – since just how these measures change during the coming years should be extremely interesting, and help us to determine just where the world is heading.

Denial and avoidance have also been very signifi-cant themes in the opinion that has been expressed here. Understandably, people don't like to confront difficult issues, nor do they like to change their behavior much. As was intimated at the start, there is a broad parallel here with respect to HIV/AIDS. Within 5 years of its identification, all the main transmission routes of this disease were known, the virus was isolated, tests had been developed, and the first antiretroviral drug was available. It is perhaps worth adding that there was also some optimism among the scientific community about developing a vaccine and 'conquering AIDS' (Curran, 2001). Yet any such optimism was to see things in much too narrow, clinical-medical terms; it was to massively underestimate the human, the *social* side. Denial, avoidance and recrimination were rife with respect to HIV/AIDS – they still are – and, partly as a result, perhaps some 60 million people have either died of the disease or are currently infected. There is much evidence that people only really change their sexual behavior when evidence of damage becomes plain. Similarly, the thrust of the position taken here with respect to global warming and climate change is that people will only really alter their behavior with respect to energy use when they experience serious effects from these phenomena for themselves. That said, the purpose of this piece has been to try to comment objectively on the subject, rather than to try to alter behavior.

That modern economic growth and the demographic transition both began at around the same time in history is hardly coincidental. Population growth, migration, and urbanization all play significant roles in the subject of global warming and climate change. However, the most important part, by far, is that played by fossil energy – coal, oil and natural gas – in fueling economic development. It is important to remember that what still locks so many people in conditions of material poverty is their reliance upon economies that remain overwhelmingly 'organic', i.e. they have no real access to the energy supplied by fossil fuels. If there are major changes to the world's climate in the coming century then the agricultural, economic, political and wider social repercussions could be so great that they impact on the future growth trajectory of the human population. While our children or grandchildren may not face the end of the world, they could well face the end of the world, at least as we have known it.

REFERENCES

British Petroleum. (2005). *BP statistical review of world energy, June 2005*. London: British Petroleum. Available at: http://www.bp.com/statisticalreview. Accessed December 2005.

Bodansky, D. (2001). Global energy problems and prospects. In R. Ragaini (Ed.), *Proceedings of the international seminar on nuclear war and planetary emergencies, 25th session*, Singapore: World Scientific Publishing.

Curran, J. (2001). The eras of AIDS. In R. A. Smith (Ed.), *Encyclopedia of AIDS*, Harmondsworth: Penguin Books.

Hansen, J., Sato, M., Ruedy, R., & Lo, K. (2005). Global temperature. Available at: http://www.columbia.edu/~4jeh1/GlobalTemperatures_03Nov2005.pdf. Accessed December 2005.

International Energy Agency. (2005). *World energy outlook 2005*. Paris: International Energy Agency.

IPCC. (2001a). *Summary for policymakers*. Available at: http://www.ipcc.ch/.

IPCC. (2001b). *IPCC special report on emissions scenarios*. Available at: http://www.ipcc.ch/.

IPCC. (2001c). In J. J. McCarthy, O. F. Canziani, N. A. Leary, D. J. Dokken & K. S. White (Eds.), *Climate change 2001: impacts, adaptation, and vulnerability*, Cambridge: Cambridge University Press.

National Research Council. (2002). *Abrupt climate change: inevitable changes*. Washington, DC: National Academy Press.

Smil, V. (2003). *Energy at the crossroads*. Cambridge, MA: MIT Press.

Stipp, D. (2004). Climate collapse: The Pentagon's weather nightmare. *Fortune Magazine*, 9 February.

United Nations. (2005). *World population prospects: the 2004 revision*. New York: United Nations.

World Energy Council. (2004). *Survey of energy resources, 2004*. London: World Energy Council.

World Resources Institute. (2003). *World resources 2002–2004*. Washington DC: World Resources Institute.

Permissions

Every effort has been made to contact copyright holders for their permission to reprint the readings in this book. The publishers would be grateful to hear from any copyright holder who is not acknowledged and will undertake to rectify any errors or omissions in future editions of the book. The following is the copyright information for the readings herein.

1 THE OBJECT OF DEVELOPMENT

SACHS, MELLINGER and GALLUP Sachs, Jeffrey D., Mellinger, Andrew D. and Gallup, John L., 'The Geography of Poverty and Wealth', *Scientific American* (March 2001), pp. 70–6. Reprinted with permission. Copyright © 2001 by Scientific American, Inc. All rights reserved.

DAVIS Davis, Mike, 'The Origins of the Third World' from *Late Victorian Holocausts* (London: Verso, 2001), pp. 279–301. Reprinted by permission of Verso.

MCCLINTOCK McClintock, Anne, 'The Lay of the Land' from *Imperial Leather: Race, Gender and Sexuality in the Colonial Contest* (London: Routledge, 1995), pp. 46–61. Reprinted with permission.

2 MARKETS, EMPIRE, NATURE AND DIFFERENCE

ARNDT Arndt, Heinz, 'Economic Development: A Semantic History', *Economic Development and Cultural Change* (April 1981), pp. 457–66. Reprinted with permission.

SMITH Smith, Adam, 'Of the Advantages which Europe has derived from the Discovery of America, and from that of a Passage to the East Indies by the Cape of Good Hope' from *An Inquiry into the Nature and Causes of the Wealth of Nations, Book IV* (1776). Public domain.

MARX Marx, Karl, 'The British Rule in India', *New York Daily Tribune* (1853), in *Karl Marx, Frederick Engels Collected Works* (London: Lawrence and Wishart; Moscow: International Publishers, 1979), vol. 12, pp. 125–34. Reprinted with permission.

SPENCER Spencer, Herbert, 'The Organic Analogy Reconsidered' and 'Societal Typologies' from *On Social Evolution: Selected Writings*, edited and with an introduction by J. D. Y. Peel (Chicago: University of Chicago Press, 1972), pp.134–48.

GANDHI Gandhi, M. K. 'Civilisation' and 'What is True Civilisation?' from *Hind Swaraj and Other Writings*, ed. A. J. Parel (Cambridge: Cambridge University Press, 1997), pp. 34–8 and 66–71. Reprinted with permission of the Navajivan Trust.

3 REFORM, REVOLUTION, RESISTANCE

KEYNES Keynes, John Maynard, 'Economic Possibilities for our Grandchildren' from *The Collected Writings of John Maynard Keynes, Vol. IX Essays in Persuasion* (London: Macmillan, Cambridge University Press for the Royal Economic Society, 1985), pp. 321–9. Reproduced with permission of Palgrave Macmillan.

POLANYI Polanyi, Karl, 'Freedom in a Complex Society' from *The Great Transformation: The Political and Economic Origins of Our Time* (Boston: Beacon Press, 2001), pp. 257–68.

FURNIVALL Furnivall, J. S., 'The Background of Colonial Policy and Practice' from *Colonial Policy and Practice: A Comparative Study of Burma and Netherlands India*, pp. 1–10. Copyright © 1948 Cambridge University Press. Reproduced with permission.

GERSCHENKRON Gerschenkron, Alexander, 'The Impasse' from *Bread and Democracy in Germany*, with a new foreword by Charles S. Maier (Ithaca: Cornell University Press, 1989), pp. 81–8. Used by permission of the publisher, Cornell University Press.

FANON Fanon, Frantz, 'This is the Voice of Algeria' from *A Dying Colonialism* (New York: Grove Press, 1965), pp. 71–4, 82–7 and 94–7. Reprinted with permission of the Monthly Review Press.

4 PROMETHEAN VISIONS

ESCOBAR Escobar, Arturo, 'The Problematization of Poverty: The Tale of Three Worlds and Development' from *Encountering Development: The Making and Unmaking of the Third World*, pp. 21–30. © 1995 Princeton University Press. Reprinted by permission of Princeton University Press.

ROSTOW Rostow, W. W., 'Marxism, Communism and the Stages-of-Growth' from *The Stages of Economic Growth: A Non-Communist Manifesto* (Cambridge: Cambridge University Press, 1960), pp. 145–56. © Cambridge University Press 1960, 1971, 1990.

DAVIS Davis, Kingsley, 'Population Policy and the Future' from *The Population of India and Pakistan*, pp. 221–31. © 1951 Princeton University Press, 1979 renewed PUP. Reprinted by permission of Princeton University Press.

LEWIS Lewis, W. A., 'Economic Development with Unlimited Supplies of Labour', *Manchester School* 22 (1954), pp. 139–91, excerpted at first 13 pages. Public domain.

SINGER Singer, H. W., 'The Distribution of Gains between Investing and Borrowing Countries', *American Economic Review* 40 (2) (1950), pp. 473–85, excerpted at 473–80 and 484–5. Reprinted with permission of the American Economic Association.

NOVE Nove, Alec, 'Socialism and the Soviet Experience' from *The Economics of Feasible Socialism* (London: George Allen and Unwin, 1983), pp. 90–7.

MURPHEY Murphey, Rhoads, 'Man and Nature in China', *Modern Asian Studies* 1: 313–33, excerpted at pp. 319–29 (1967). © Cambridge University Press, reproduced with permission.

5 CHALLENGES TO THE MAINSTREAM

BARAN Baran, Paul 'The Steep Ascent' from *The Political Economy of Growth* (Harmondsworth: Penguin, 1973), pp.402–16. Reprinted with permission of the Monthly Review Press.

WOLPE Wolpe, Harold, 'Capitalism and Cheap Labour-Power in South Africa: From Segregation to Apartheid' from *The Articulation of Modes of Production* (London: Routledge and Kegan Paul, 1980), pp. 289, 291–2 and 307–19. Reprinted with permission.

CARSON Carson, Rachel, 'The Obligation to Endure' from *Silent Spring*. Copyright © 1962 by Rachel Carson, renewed 1990 by Roger Christie. Reprinted by permission of Houghton Mifflin Company. All rights reserved.

6 THE HUBRIS OF DEVELOPMENT

7 INSTITUTIONS, GOVERNANCE AND PARTICIPATION

Index

figures and tables are shown in *Italic*